SYNTHESIS

of

PARALLEL

ALGORITHMS

SYNTHESIS

of

PARALLEL

ALGORITHMS

Edited by

John H. Reif

MORGAN KAUFMANN PUBLISHERS
San Mateo, California

Senior Editor: Bruce M. Spatz
Production Manager: Yonie Overton
Production Editor: Carol Leyba
Cover Designer: Studio Silicon
Editorial Assistance: John Hammett, Gary Morris, and Fran Taylor
Composition: SuperScript Typography
Index: Eigentype Compositors
Printer: Edwards Brothers, Inc.

Morgan Kaufmann Publishers, Inc.
Editorial Office:
2929 Campus Drive, Suite 260
San Mateo, CA 94403

© 1993 by Morgan Kaufmann Publishers, Inc.
All rights reserved
Printed in the United States of America

No part of this publication may e reproduced, stored in a retrieval system, or transmitted in any form or by any means—electronic, mechanical, photocopying, recording, or otherwise—without the prior written permission of the publisher.

97 96 95 94 93   5 4 3 2 1

Library of Congress Cataloging-in-Publication Data

Synthesis of parallel algorithms / edited by John H. Reif.
    p.   cm.
  Includes bibliographical references and index.
  ISBN 1-55860-135-x
  1. Parallel programming (Computer science). 2. Computer algorithms.
I. Reif, J. H. (John H.)
QA76.642.S96 1993
005.1'1—dc 20   92-46328
                                                                        CIP

Contents

	Introduction *John H. Reif*	1
PART I	**FUNDAMENTAL PARALLEL GRAPH ALGORITHMS**	**33**
1	Prefix Sums and Their Applications *Guy E. Blelloch*	35
2	Introduction to Parallel Connectivity, List Ranking, and Euler Tour Techniques *Sara Baase*	61
3	List Ranking and Parallel Tree Contraction *Margaret Reid-Miller, Gary L. Miller, and Francesmary Modugno*	115
PART II	**ADVANCED PARALLEL GRAPH ALGORITHMS**	**195**
4	Randomized Parallel Connectivity *Hillel Gazit*	197
5	Advanced Parallel Prefix-sums, List Ranking and Connectivity *Uzi Vishkin*	215
6	Parallel Lowest Common Ancestor Computation *Baruch Schieber*	259
7	Parallel Open Ear Decomposition with Applications to Graph Biconnectivity and Triconnectivity *Vijaya Ramachandran*	275
8	Parallel Algorithms for Chordal Graphs *Philip Klein*	341
PART III	**PARALLEL SORTING AND COMPUTATIONAL GEOMETRY**	**409**
9	Random Sampling Techniques and Parallel Algorithm Design *Sandeep Sen and Sanguthevar Rajasekaran*	411
10	Parallel Merge Sort *Richard Cole*	453

11	Deterministic Parallel Computational Geometry *Mikhail J. Atallah and Michael T. Goodrich*	497
PART IV	**FUNDAMENTAL PARALLEL ALGEBRAIC ALGORITHMS**	**537**
12	Newton Iteration and Integer Division *Stephen R. Tate*	539
13	Parallel Linear Algebra *Joachim von zur Gathen*	573
PART V	**ADVANCED PARALLEL ALGEBRAIC ALGORITHMS**	**619**
14	Parallel Solution of Sparse Linear and Path Systems *Victor Pan*	621
15	Parallel Resultant Computation *Doug Ierardi and Dexter Kozen*	679
PART VI	**EXTENSIONS OF PARALLEL TREE CONTRACTION TO ALGEBRAIC AND LOGICAL PROBLEMS**	**721**
16	Dynamic Parallel Evaluation of Computation DAGs *Erich Kaltofen*	723
17	The Parallel Complexity of Logical Inference *Jeffrey D. Ullman*	759
PART VII	**PARALLEL COMBINATORIAL OPTIMIZATION**	**781**
18	Parallel Graph Matching *Vijay V. Vazirani*	783
19	Parallel Algorithms for Network Flow Problems *Andrew V. Goldberg*	813
PART VIII	**INHERENT LIMITATIONS OF PARALLEL COMPUTATIONS**	**841**
20	The Complexity of Computation on the Parallel Random Access Machine *Faith E. Fich*	843
21	Polynomial Completeness and Parallel Computation *Raymond Greenlaw*	901
PART IX	**ASYNCHRONOUS PARALLEL COMPUTATION**	**955**
22	Asynchronous PRAM Algorithms *Phillip B. Gibbons*	957
INDEX		**999**

Introduction

John H. Reif

Department of Computer Science
Duke University
Durham, NC 27706
reif@cs.duke.edu

1
The Synthesis Approach to Parallel Algorithm Design

One of the great successes of theoretical computer science has been the early discovery of fundamental machine models, which are new paradigms for computing. In the last decade, for example, the computing world has been revolutionized by the use of parallel computers, and a wide variety of parallel machines have been constructed. Many of these machines were inspired by theoretical machine models developed years earlier, including

1. hypercube and butterfly connected parallel machines,
2. parallel machines with prefix computation as a built-in primitive, and
3. parallel machines with hardware support for concurrent random access to shared memory.

Efficient algorithms can also play a crucial role in parallel computing, but only if the computational model fits the existing machine and the problem solved by the algorithm is the problem required by the implementation. In practice, this does not always happen. Certain algorithmic techniques have limited general applicability because they can be applied only to a restricted class of problems and computational models. On the other hand, some algorithmic techniques may be broadly applicable to a wide class of problems and models, even to some that have yet to be discovered.

1.1 Computational Correctness Versus the Diversity of Computational Models

The advantage of a computational model is that it allows us to specify an algorithm precisely and to gauge its computational complexity in terms of well-defined measures, such as time or space. While there are many computational models, including sequential, parallel, and randomized ones, as well as combinations of these, can it be said that there is a single machine model that is correct? We believe not: instead, given models may be applicable in certain contexts. Generally one describes an algorithm in the most compatible machine model (where the algorithm remains efficient) with the least number of assumptions. The wide variety of possible computational models is also an insurance policy, guaranteeing open-ended applicability of algorithms to diverse machines that might be constructed in the future.

Unfortunately, it is not uncommon for researchers to overlook algorithm results that use variants of a given accepted computational model, even if the competing algorithm solves exactly the same problem, with the same or better complexity bounds. This practice has led to a divided research community in algorithms, where individuals pursue research on identical problems, with many elements of similar techniques, but do not accept as valid competitive algorithms that use a slightly different machine model. This clearly is an inefficient use of research effort. It also fails to serve programmers who wish to determine the full range of algorithmic techniques that may be used to solve a particular problem. Assuming that no single computational model can always be the correct one, perhaps a better approach (which we favor) is to investigate interrelations of related techniques in various models.

1.2 The Synthesis Approach

The *derivational and synthesis approach* to algorithm design is to synthesize a family of related algorithms (rather than a single algorithm) from fundamental algorithmic techniques. This approach stresses the development of fundamental techniques that can be utilized for as wide a class of algorithms as possible.

1.3 Structure of the Book

This book resulted from collaborations among many in the parallel algorithm community. The text contains 22 chapters written by 27 researchers. These experts in parallel algorithms worked collaboratively on derivations of parallel algorithms in their areas of expertise. Their efforts were leveraged by a common set of basic techniques used in these derivations. This text draws together the many different principles which have been used to develop the current large collection of parallel algorithms. The specific problems these algorithms solved were carefully chosen to be as fundamental as possible. The text covers parallel algorithms which are theoretically interesting, have practical applications, and are applicable to a wide variety of parallel machine models. Each chapter begins with a careful statement of the fundamental problems, followed by their solutions and the analytic techniques used in deriving the solution. These techniques are related, where possible, to known efficient sequential algorithms. In later sections of each chapter, more sophisticated parallel algorithms are synthesized from the simpler parallel algorithms and techniques discussed earlier. Thus, there is a progression from simple to more complicated (and presumably more efficient) algorithms. This progression reveals the kinds of transformations needed in synthesizing parallel algorithms.

1.4 Impact of the Synthesis Approach

The synthesis approach promises to have a significant impact on the development of parallel algorithms in the future. This text was written for students and researchers engaged in parallel algorithm design. For researchers involved in formal algorithm derivation, the text can serve as a central source, from which the details of the derivation process for some classes of parallel algorithms can be extracted and turned into a tool-set useful for developing future parallel algorithms. This text will also be particularly relevant to software engineers engaged in parallel algorithm implementations on high-performance machines.

1.5 Organization of the Introduction

We first define some asymptotic and graph theoretic terminology used throughout the text. Next we give basic definitions of machine models, beginning with sequential models and then extending these definitions to the corresponding parallel models. As examples, we define some basic computational problems which will arise frequently, and the basic sequential and parallel algorithm techniques utilized to solve these problems.

Following that we describe basic parallel algorithm design techniques, including randomized techniques. Then we describe the organization and interrelation of each of the chapters, stressing the derivation techniques used. To illustrate the relationships between these problems, we give three figures diagramming the various derivations in this paper. (The solid arrows are used to indicate derivations, the dotted arrows are used to indicate alternative derivations.)

1. Synthesis of Parallel Graph Algorithms (Figure 1)
2. Synthesis of Parallel Algorithms for Sorting and Convex Hulls (Figure 2)
3. Synthesis of Parallel Algebraic Algorithms (Figure 3)

Finally, we acknowledge individuals who had impact on the text.

2
Useful Terminology

2.1 Asymptotic Terminology

We will require the following definitions for asymptotic bounds. A function $g(n)$ is said to be $O(f(n))$ ($g(n) = \Omega(f(n))$, respectively) if $g(n)$ is upper

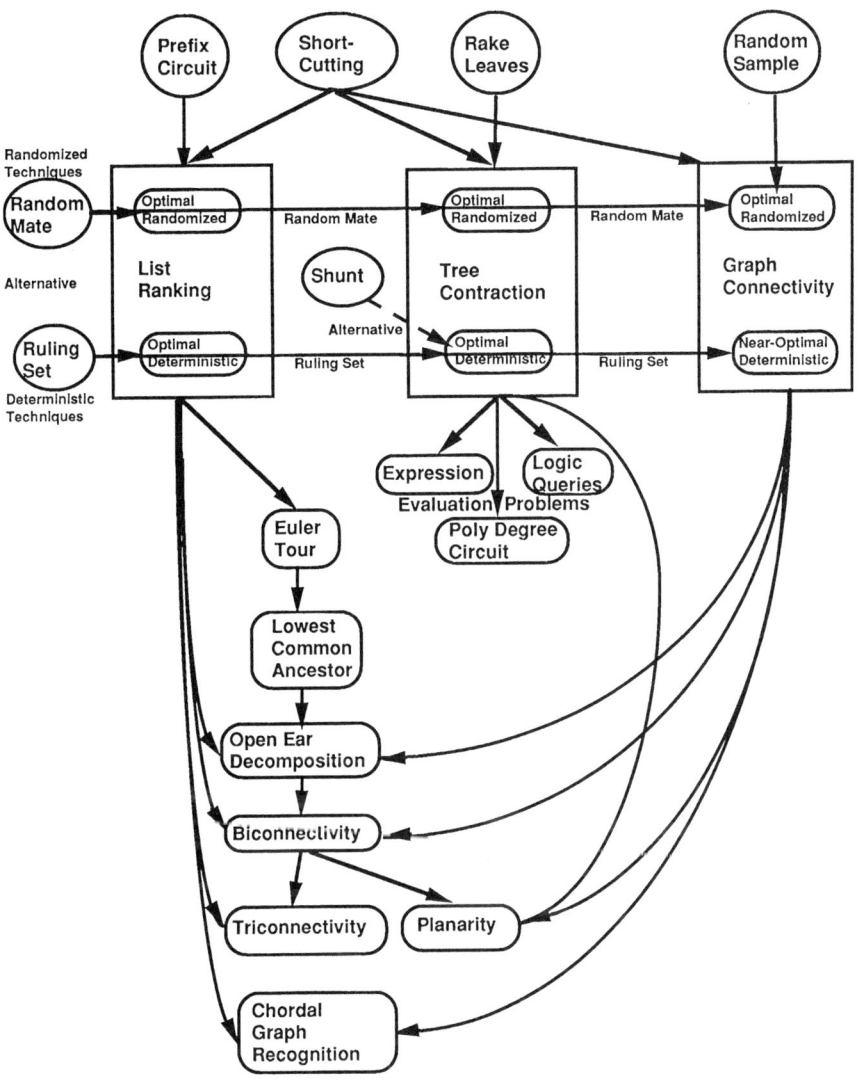

FIGURE 1
Synthesis of Parallel Graph Algorithms

(lower, respectively) bounded by a linear function of $f(n)$. $g(n) = \Theta(f(n))$ if $g(n) = O(f(n))$ and $g(n) = \Omega(f(n))$, and $g(n) = o(f(n))$ if $\lim_{n\to\infty} \frac{g(n)}{f(n)} = 0$.

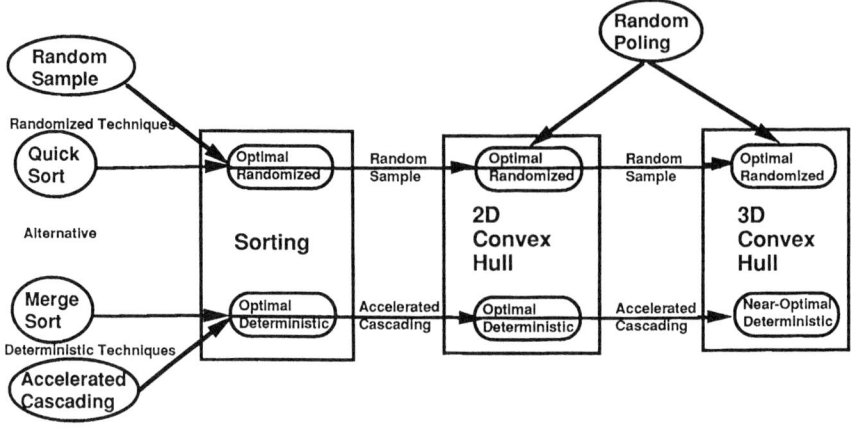

FIGURE 2
Synthesis of Parallel Algorithms for Sorting and Convex Hull

Throughout the text, we will assume $\log n$ to be base 2 unless otherwise indicated. A function $f(n)$ is *polynomial* if $f(n) = O(n^{O(1)})$, it is *linear* if $f(n) = O(n)$, and it is *polylog* if $f(n) = O(log^{O(1)} n)$ and is *logarithmic* if $f(n) = O(\log n)$.

For any real number x, we let $\lceil x \rceil$ denote the smallest integer greater or equal to x and let $\lfloor x \rfloor$ denote the largest integer less than or equal to x.

2.2 Digraph Terminology

A *digraph* is a pair $G = (V, E)$ consisting of a set V of nodes (also known as vertices) and a set E of directed edges between pairs of these nodes. The number of nodes is generally denoted by n, and the number of edges by m. A directed edge departing from node u and entering node v will be denoted (u, v). The edge is a *loop* if $u = v$. A *path* is a sequence of nodes connected by directed edges, and its *length* is the number of edges it traverses. The path is *nontrivial* if it has length > 1. The path is *simple* if it visits no node more than once, except possibly at its ends. A *cycle* is a path beginning and ending at the same node. A *Directed Acyclic Digraph (DAG)* is a digraph with no cycles. The *depth* of the DAG is the length of the longest path. The *fan-in (fan-out)* of the DAG is the maximum number of edges entering (or departing, respectively) any node.

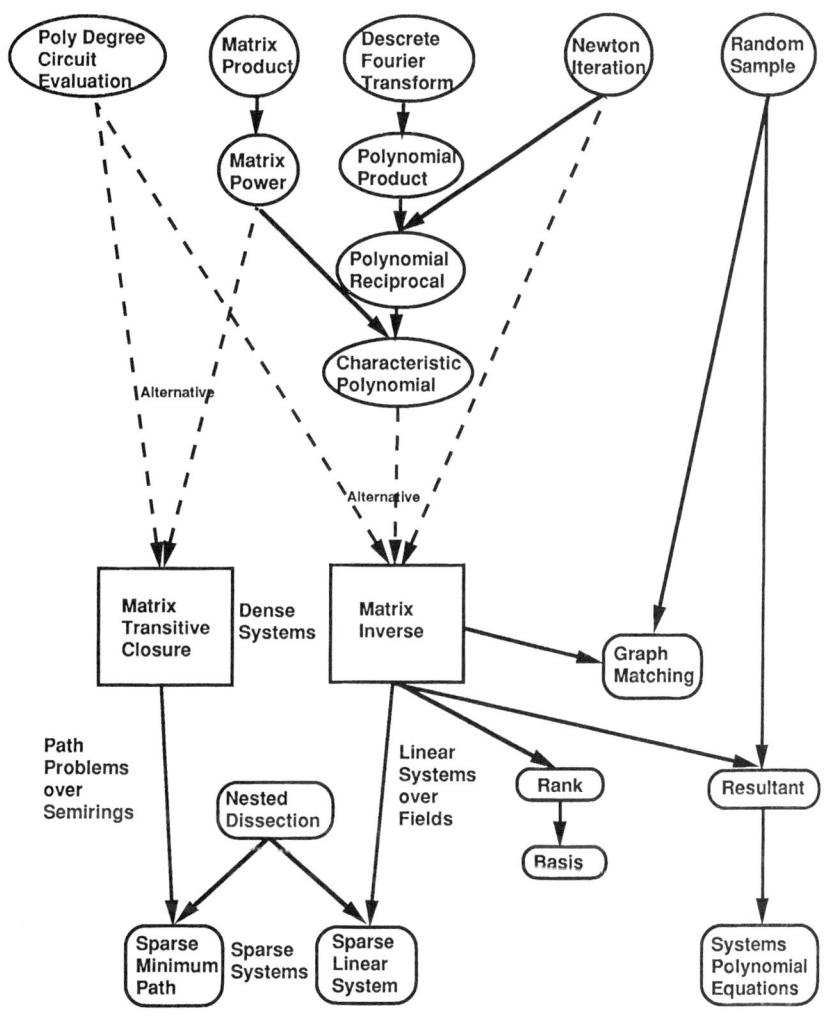

FIGURE 3
Synthesis of Parallel Algebraic Algorithms

2.3 Graph Terminology

A *graph* is a digraph in which the edges are *undirected*; that is, for each edge from node u to node v there is a corresponding edge from v to u. An undirected edge between u and v will be denoted (u,v), and u will be said to be *adjacent* to v. The *degree* of the graph is the maximum number of nodes adjacent to any node. A subset of nodes induces a *node set induced subgraph*, which is a graph containing only the indicated nodes and the original edges between these nodes. A subset of edges induces an *edge set induced subgraph*, which is a graph containing only the indicated edges and their nodes. A graph is *connected* if there is a path between each pair of nodes of the graph. The *connected components* of a graph are the maximum subgraphs which are connected. A graph is *planar* if it can be embedded on the Cartesian plane so that no edges cross. A graph is *bipartite* if the vertex set can be partitioned into two subsets, where no edge has both vertices in the same subset. A *weighted graph* is a graph with a mapping of the edges to real numbers.

2.4 Tree Terminology

A *tree* is a graph with a unique path between each pair of nodes. A *rooted tree* has a distinguished node called the *root*. For each node v in a rooted tree, the *ancestors* of v are those nodes that are on the path from the root to v; the *descendants* of v are those nodes reachable by a path from v passing through no ancestors of v; the *children* of v are the descendants of v reachable by a single edge from v; nodes with no children are called *leaves*. The *depth* of a node v in a rooted tree is the length of the path from the root to v; the *lowest common ancestor* of two nodes is the common ancestor of these nodes which has the largest depth. A tree is *ordered* if the children of each node are linearly ordered.

A *directed tree* is a rooted tree with directed edges. A directed tree is *downward pointing* if there is a unique path from the root to each node of the tree; it is *upward pointing* if there is a unique path from each node to the root. A *(directed) forest* is a collection of (directed) trees. A *spanning tree* of a connected graph is a subgraph which is a tree containing all the nodes.

3
Basic Sequential Computational Models

A *sequential machine* is a machine that executes instructions in sequential order. The *time* of execution of a program on a sequential machine is exactly the same as the *work* performed by the machine, that is, the total number of operations executed. The definition of the *space* used by a program on the machine depends on the model. There is a wide class of sequential machine models in the literature.

3.1 The Straight Line Program and Prefix Computation

Perhaps the most basic machine model is the *straight line program*. The program has a set of variables, with specified subsets of n *input variables* and m *output variables*. The straight line program has a *basis* which is a specified domain of values to which the variables can be assigned, and a specified set of k-ary operations over this domain. Each step of the program consists of assigning the value of a k-ary operation operation (on previously assigned variables) to a non-input variable. The *space* of a straight line program is the total number of non-input variables used, and the *time* is the number of assignments. A straight line program is *Boolean* if the domain and operations are the binary.

Here we will give a simple example of a straight line program for the *prefix-sum* problem, which will arise frequently in the text. The input to the prefix-sum problem is a sequence of n numbers, $X(1), \ldots, X(n)$. For each $i = 1, \ldots, n+1$ the ith output $S(i)$ is the sum of the first $i-1$ input numbers. A straight line program solves the prefix-sum problem in n time by assigning $S(1) := 0$ and for each $i = 2, \ldots, n+1$, assigning the ith output $S(i) := S(i-1) + X(i-1)$. The *partial prefix-sum* problem is defined similarly, except that the $n+1$ output, consisting of the sum of all the inputs, is not included. We will not always distinguish this variant in the chapters, where the context is clear.

A *monoid* is an algebra with an associative binary operation (which we denote by composition) and an identity element, I. The *monoid prefix* problem is a generalization of the prefix-sum problem; the input is a sequence of n elements $X(1), \ldots, X(n)$ of the monoid, the first element of the output is $S(1) = I$, and for each $i = 2, \ldots, n+1$, the ith output $S(i)$ is the composition of the first $i-1$ consecutive inputs. Clearly, the straight line program for the prefix-sum problem extends to monoid prefix problems, with the monoid operation substituted for the addition operation.

3.2 The Random Access Machine and the List-Ranking Problem

The most widely used sequential computation model is the *Random Access Machine (RAM)*. The RAM contains a fixed number of registers and an unlimited number of storage locations referenced by the positive integers. A specified initial segment of the storage locations are *input locations*. The RAM executes a program consisting of a number of possible instructions, including load and store operations to and from memory and/or a register, conditional tests, branching operations, and arithmetic operations on register values. The domain of the arithmetic operations of the RAM will depend on the context. For example, for graph algorithms, the domain is generally assumed to be logarithmic bit integers; for sorting, selection, and computational geometry algorithms, the domain is generally real numbers; and for algebraic algorithms, the arithmetic operations may be over an algebra such as a semiring, ring, or field. This flexibility allows the RAM to be broadly applicable to a wide class of algorithms. Throughout this text, we assume the *unit cost criterion*—each operation has unit-time cost and each non-input memory location used is charged a unit-space cost. Thus the *space* of the RAM is the number of non-input memory locations used and the *time* is the total number of steps.

The RAM model bears a striking similarity to actual sequential machines, which is why it has been very successful as a computational model.

To familiarize the reader with the notation used in describing RAM programs, we give below a simple example of a RAM program for the list-ranking problem, which arises frequently in the text. The input to the list-ranking problem is a linked list of n memory cells, where the last element r of the list satisfies $NEXT(r) = r$ and each other cell v contains a pointer $NEXT(v)$ to the next consecutive element. For each cell v, the output $L(v)$ is the path length from v to r. A RAM program can easily solve the list-ranking problem in linear time. We initialize by constructing backward pointers $PREV$ and using these to reverse chain from r to each other node v, incrementing the path distance and assigning this to $L(v)$. We have written the RAM algorithm using Pidgin PASCAL.

input $NEXT$

1. Initially, **for** each node v **do** $PREV(v) := v$ **od**
2. **for** each node v **do**
   if $NEXT(v) \neq v$ **then**
   do $PREV(NEXT(v)) := v$ **od**
   od
3. $L(r) := 0; v := r$
4. **while** $PREV(v) \neq v$ **do** $L(PREV(v)) := L(v) + 1; v := PREV(v)$
   od

output L

A generalization of the list-ranking problem (known as the *list-prefix* problem) also arises frequently in the text. The input to this generalized list-ranking problem is a linked list of n memory cells, where each cell v contains both a number $X(v)$ and a pointer to the next consecutive element $NEXT(v)$ of the list (again, the last element r of the list has $NEXT(r) = r$). For each cell v, the output $S(v)$ is the sum of the numbers $X(u)$ where u is a node, distinct from v, which is reachable from v by a path of the $NEXT$ pointers. As an example of a very simple algorithm derivation, we can derive a linear time generalized list-ranking from this list-ranking algorithm by replacing the initial assignment of L with $S(r) := X(r)$ and replacing the assignment $L(PREV(v)) := L(v) + 1$ with $S(PREV(v)) := S(v) + X(v)$. Similarly, the algorithm can be extended to monoid prefix problems, where we apply the monoid operation to consecutive elements of the list rather than take sums.

4
Basic Models for Parallel Computation

A *parallel machine* is a machine that can execute instructions in parallel. A wide range of parallel machines have been constructed in the past few years. A parallel machine is *synchronous* if, at each time step, all processors execute an instruction (not necessarily identical) simultaneously, and otherwise is *asynchronous*. A parallel machine is *Single Instruction Multiple Data (SIMD)* if all processors execute the same instruction at the same time, using possibly different inputs for the instruction. A machine is *Multiple Instruction Multiple Data (MIMD)* if processors may execute different instructions simultaneously. Parallel processors may communicate either via shared memory or via fixed connections, and there is a large and diverse set of shared memory primitives and connection networks. What are the correct models for parallel machines? Again, we feel that there is no definite answer to this question. Instead, we need to explore a variety of reasonable parallel models. In the following sections we will define the circuit model and the PRAM model very broadly; the majority of this text will use variants of these models.

4.1 The Circuit Model and Parallel Prefix

Perhaps the most basic parallel machine model is the circuit. A *circuit* is a DAG, with ordered edges and labeled nodes derived from a straight line program as follows. First we replace any variable which is multiply assigned with new distinct variables, so every variable is assigned at most once. The nodes of the circuit are exactly the distinct variables of the resulting straight line program. The input nodes correspond to the input variables. Each non-input node v is labeled with the operation used in its corresponding variable assignment, and has an ordered sequence of edges entering v departing from the sequence of nodes on whose values this assignment depends.

The *time* for parallel evaluation of the circuit is its depth. The *size* of the circuit is the number of its non-input nodes, which correspond to the total computational work done by the circuit. The circuit is *Boolean* (*uniform*, respectively) if the corresponding straight line program has a Boolean basis (is uniform, respectively). The circuit model is a very simple, elegant model for parallel computation, and the circuits can model electrical circuits, e.g., in VLSI.

The *prefix-sum circuit* of fan-in 2 is a useful example of a circuit. Recall that the input to the prefix-sum problem is a sequence of n numbers, $X(1), \ldots, X(n)$ and for each $i = 1, \ldots, n$, the ith output $S(i)$ is the sum of the first $i - 1$ input numbers. Note that the problem requires a circuit of depth 1 and size 2 if the input size is $n \leq 2$. For simplicity, we assume n is a power of 2. A circuit can solve the partial prefix-sum problem by first computing the sum of each consecutive odd and even pair of input elements, and then call the algorithm recursively on the resulting problem which has reduced input size $n/2$. For each $i = 1, \ldots, n/2$, we assign the odd output $S(2i-1) :=$ the ith output of the recursively solved problem, and then assign the next even output $S(2i) := S(2i-1) + X(2i-1)$. Since the input size drops geometrically on each recursive call, the total work is linear and there are $\log n$ recursive levels. The depth is 2 plus the depth of the recursive call, so the total depth is $2 + \log n$. Thus the partial prefix-sum circuit is of logarithmic depth and linear size. The last output of the prefix-sum problem, consisting of the sum of all the inputs, can be gotten by summing the last output of the partial prefix and the last input. Prefix circuits are among the most basic of parallel computations and are frequently used in parallel algorithms. However, it can be laborious to define complex algorithms in the circuit model.

4.2 Fixed Connection Networks

A *fixed connection network* is a collection of processors executing in parallel. The processors are nodes of a fixed digraph, which is called the *communication network*. Communication between processors is via the edges of the network. Generally, the communication network imposes a limit to the communication as edges per step. For example, there may be a requirement that, at most, a single value may be communicated across each edge on each step. Since the fixed connection network may have a cycle, the model is more general than the circuit. There are many useful examples of digraphs which are used for communication networks. A *d-dimensional array (or grid)* is a graph whose node set is $\{1, \ldots, m\}^d$, that is, the set of d-dimension vectors whose elements range over the integers $\{1,\ldots,m\}$. Two nodes, $u = (u_1,\ldots,u_m)$ and $v = (v_1,\ldots,v_m)$ will have an edge between them if $|u_i - v_i| = 1$ for at most one i, and $u_j = v_j$ for all other $j \neq i$. A *hypercube* is a d-dimensional array of n nodes, where $d = \log n$. As another example, the *butterfly* network is a regularly connected network which has degree 4.

This text does not discuss communication networks in great detail. Instead the emphasis of this text will be on the synthesis of parallel algorithms for general parallel models, such as the PRAM. Nevertheless, in many cases these parallel algorithms can be mapped to fixed connection machines.

4.3 The PRAM Model and Parallel List Contraction

The most widely used model for parallel computation is the *Parallel RAM (PRAM)*. The PRAM is a collection of synchronous processors executing in parallel and communicating via shared memory. Each processor is a sequential RAM, and the processors are executing MIMD. The PRAM models allow a variety of methods for resolution of memory access contention. Let *Exclusive Read (ER)* (*Concurrent Read (CR)*, respectively) denote that each memory location can be read by only one processor (by any number of processors, respectively) at a time. Let *Exclusive Write (EW)* (*Concurrent Write (CW)*, respectively) denote that each memory location can be written by only one processor (attempt a write by any number of processors, respectively) at a time. In the case of concurrent write, at any given memory location and at any give step, a single arbitrary processor attempting to write is allowed to succeed in its write request. The CRCW-PRAM allows concurrent write as well as concurrent read whereas the EREW-PRAM requires exclusive write as well as exclusive read.

Here we will give a simple example of a CRCW-PRAM program for the parallel solution of the *list-ranking problem*. Recall that the input to the list-ranking problem is a linked list of n memory cells, each containing a pointer $NEXT(v)$ pointing to the next consecutive element, while the last element r of the list has $NEXT(r) = r$. For each cell v, the output $L(v)$ is the path length from v to r. A PRAM program can solve the list-ranking problem in logarithmic time by repeatedly *shortcutting* (this technique is also known as *pointer jumping* or *list compression* depending on the context) the path from any node v to the root by assigning $NEXT(v) := NEXT(NEXT(v))$. We will associate a processor with each node of the list.

We write the PRAM algorithm using a variant of Pidgin PASCAL, augmented with a primitive operation **in parallel do** to denote parallel execution of multiple processors.

input *NEXT*
for each node v **in parallel do**

1. **if** $v = r$ **then** $L(v) := 0$ **else** $L(v) := 1$
2. **for** $t = 1, \ldots, \log n$ **do**
   if $NEXT(v) \neq NEXT(NEXT(v))$ **then**
   do $L(v) := L(v) + L(NEXT(v));$
   $NEXT(v) := NEXT(NEXT(v))$ **od**
   od

od
output L

Note the distance of the chain of *NEXT* pointers decreases by a factor of 2 each iteration, and thus, at most, $\log n$ iterations of the **for** loop are required. Once the list-ranking problem is solved, then the processors can place the ranked inputs in order in a linear array of length n. The list-prefix problem and its generalizations can be solved in logarithmic time using a linear number of processors by applying the parallel prefix circuit described previously, with this linear array as input.

4.4 Reasonable Machine Models and NC

To specify computations over arbitrary numbers of inputs, a *family* of circuits (straight line programs) is defined, where, for each number n of inputs, there is a distinguished circuit (straight line program, respectively) with n inputs. A straight line program family or circuit family is *uniform* with respect to a given complexity measure if each circuit in the family can be constructed within the specified complexity measure. In the following, we will consider *uniformity* to be constructable within logarithmic sequential space.

A sequential machine model is *reasonable* if it can simulate, in polynomial time, any assignment and Boolean operation and, moreover, each operation of the sequential machine can be simulated in polynomial time by a uniform family of Boolean straight line programs. For example, the RAM operating on logarithmic bit numbers can be easily be shown to be a reasonable sequential model. Generally, a problem is said to be in the class P if there is a polynomial time sequential algorithm for solving the problem. Specifically, we formally define **P** to denote the class of all problems solvable in polynomial time by uniform straight line program families (and thus by any reasonable sequential machine within polynomial time bounds).

A parallel machine model is *reasonable* if it can simulate, in polylog time and polynomial processors, a linear number of Boolean operations and assignments and, moreover, each operation of the parallel machine can be simulated in polylog depth and polynomial processors by a uniform family of Boolean circuits. The PRAM operating on logarithmic bit numbers can be shown to be a reasonable parallel model, by using parallel sorting circuits and arithmetic circuits to simulate basic PRAM operations.

The *work* required by a circuit is the number of its non-input nodes, and the work required by a PRAM is, at most, the product of the number of processors and the time.

Generally, a problem is said to be in the class NC if there is a polylog time, polynomial work algorithm for solving it. Specifically, we define the class **NC** to be the class of all problems solvable simultaneously in polylog time and polynomial work by uniform Boolean circuits (and thus by any reasonable parallel machine within these bounds). For each $i \geq 1$, NC^i denotes the class of all problems solvable by uniform Boolean circuits simultaneously in $O(log^i n)$ time and polynomial work.

5
Design Principles for Fundamental Parallel Algorithms

A parallel algorithm for a problem is *efficient* if the work of the parallel algorithm is within a polylog factor of the work of the best known sequential algorithm for this problem. Ideally we would also like to develop efficient parallel algorithms with at most polylog parallel time bounds. The emphasis of this text is the development of fundamental techniques for such efficient parallel algorithms. We have already presented some fundamental techniques for parallel algorithm design:

1. parallel prefix computation
2. parallel list ranking using list contraction

5.1 Optimal Parallel Algorithms and the Rescheduling Principle

A parallel machine with time bound T and processor bound P satisfies the *rescheduling principle* (also known as the *slowdown principle*) if for any σ, $1 \leq \sigma \leq P$, the time can be slowed down to σT using at most $\lceil P/\sigma \rceil$ processors. Most parallel machines satisfy this slowdown principle, as does the PRAM model. On the other hand, parallel computation models which do not have a uniform cost per access of shared memory may not have this slowdown property. The processors are partitioned into $\lceil P/\sigma \rceil$ groups, each consisting of at most σ processors. Then we let a single processor simulate each distinct group of $\lceil P/\sigma \rceil$ processors.

If a machine has time bound T and processor bound P to solve a problem, then PT can never be less than the work of a single sequential machine to solve the problem. Thus we define a parallel algorithm for a problem to be *(asymptotically) optimal* if the work of the parallel algorithm is within a constant factor of the work of the best known sequential algorithm for this problem.

Surprisingly, the rescheduling principle can often be used to develop optimal parallel algorithms. We illustrate this technique in the case of an optimal PRAM algorithm for the parallel prefix-sum problem with n inputs. Partition the n inputs into P consecutive blocks each of size $\sigma = \lceil \log n \rceil$, except the last block which may be of size $\leq \sigma$. The PRAM algorithm will use $P = \lceil (n+1)/\sigma \rceil$ processors. For $j = 1, \ldots, P$ let the jth processor solve the prefix-sum problem within the jth block of inputs using σ time. Let S^j be the solution of the prefix-sum problem for the jth block. Construct a new problem of input size P, where each consecutive input is the sum of a consecutive block of original inputs. Then use the parallel prefix circuit algorithm to solve this smaller problem in logarithmic time using P processors and let R be the output of this reduced prefix-sum problem. Finally, for each $j = 1, \ldots, P$ and each $i = 1, \ldots, \sigma$ the k the output of the original prefix-sum problem is $S_k = R_j + S_i^j$, where $k = (j-1)\sigma + i \leq n+1$. The total time of this parallel prefix-sum algorithm is logarithmic and the total work is $O(\log n)P = O(n)$, which is asymptotically optimal.

5.2 Randomized Parallel Algorithms

An algorithm is *randomized* if the computation makes use of random numbers. Generally, the random numbers are independent and are drawn from a specified bounded domain. (However, many of the randomized algorithms given in this text do not need truly independent random variables, but instead require only k-way independent random variables, for some k.) We can augment the circuit model to allow for special nodes that give a random value. Similarly, we can augment the RAM and PRAM models to allow a special operation RAND which gives a random value. Note that for a given input, a randomized computation may give differing output values depending on the random choices. In particular, a randomized algorithm for the solution of a given problem may sometimes give an incorrect answer; in this case, the randomized algorithm is termed a *Monte Carlo*; otherwise it is termed a *Las Vegas* algorithm. For a Monte Carlo randomized algorithm to be considered as a solution to a given problem, the success probability must be at least a constant additive factor greater than 1/2.

We will evaluate the performance of randomized algorithms for worst case input. The *expected time* is determined by averaging the time over the random choices. A *high likelihood* bound is one which holds with probability $\geq 1 - 1/n^c$, for some $c \geq 1$. Note that $c \geq 1$ repetitions of a Monte Carlo randomized algorithm holding with success probability $\geq 1 - 1/n$ yields a success probability of $\geq 1 - 1/n^c$.

A randomized parallel machine model is *reasonable* if without randomized choice, it is a reasonable parallel machine model. We let **RNC** denote the class of all problems solvable simultaneously in polylog time and polynomial work by randomized uniform Boolean circuits (and thus by any reasonable randomized parallel machine within these bounds).

The Practical Utility of Randomized Parallel Algorithms

One of the greatest surprises of parallel algorithms is the utility of randomized techniques for parallel algorithms. The randomized techniques are applicable to a wide range of problems. We will describe parallel algorithms for sorting, list ranking, and its generalization to trees, graph connectivity, graph matching, and the solution of certain algebraic problems. We will describe both randomized and purely deterministic techniques for these problems. Since the randomized techniques are generally simpler in concept, we will, in many cases, present randomized techniques first, before describing the more complex deterministic techniques. The randomized techniques generally have smaller constant factor terms than the corresponding purely deterministic techniques. Since the keys to developing efficient and practical parallel algorithm implementations are simplicity and small constant terms, randomized algorithms can be very useful.

Random Sampling

Random sampling is the technique of choosing a random sample from a domain to determine information about the domain. For example, the mean of a random sample of a set of totally ordered keys gives an unbiased estimation for the mean of the entire set of keys, and the variance of the estimated mean decreases quite rapidly with the size of the sample. We use random sampling in this way for our parallel selection algorithms. Also, random sampling can be used to partition the input into nearly equal size parts and is used in this way for the parallel solutions of sorting and computational geometry problems.

Random sampling of input variable values can also be used to determine if a multivariate polynomial over an infinite field is identically 0. This technique is used in the derivation of many parallel algorithms for solving algebraic problems such as resultants, as well as for various combinatorial optimization problems such as parallel matching.

Random Mate

Random mate is a technique of collapsing the nodes of connected components as follows. On each round, each node is labeled randomly to be either *male* or *female*. Then each node that is labeled *male* is collapsed to an arbitrary adjacent node labeled *female*, if there is such a node. It can be shown that after each stage, the expected number of nodes decreases by a constant factor, and thus a logarithmic expected number of stages suffices to collapse the graph to its connected components. Each stage can clearly be executed in constant time using a linear number of processors. Thus the random mating algorithm computes the connected components in expected logarithmic time using a linear number of processors. The extreme simplicity of this connectivity algorithm makes it appealing for implementations. The randomized mate technique can be applied to generate optimal logarithmic time PRAM algorithms for list ranking, tree contraction, and graph connectivity, where these logarithmic time bounds hold with high likelihood.

6
Organization of the Text

Unless mentioned otherwise, we assume the PRAM model when we discuss parallel algorithms in this and the following subsections.

6.1 Fundamental Parallel Graph Algorithms
Parallel Prefix Sums

Part I of the text describes some fundamental parallel graph algorithms. We begin with *Prefix Sums and Their Applications* by Guy Blelloch. There is a wide class of problems that can be solved using prefix primitives. This chapter describes various prefix primitives and prefix-sum circuits including their generalizations to monoid prefix problems and applications.

Introduction to Basic Parallel Graph Techniques

The next chapter, *Introduction to Parallel Connectivity, List Ranking, and Euler Tour Techniques,* by Sara Baase, introduces the fundamental deterministic and randomized algorithms for graph problems.

First, a deterministic logarithmic time algorithm for graph connectivity is derived using a combination of shortcutting and another operation called *hooking.* The algorithm collapses a graph into single nodes representing its connected components in stages, using a linear number of processors. After each stage, the connected components are represented by an upwardly pointing tree whose root points to itself. In each stage, a constant number of pointer jumping operations are applied to contract the length of paths up to the root by a constant factor. Also, the basic operation hooking is used to connect the root of one tree to another tree, if nodes of the two trees are connected by an edge of the input graph. It is shown that a logarithmic number of stages suffices to contract the graph to its connected components using a linear number of processors.

This chapter also defines two basic techniques which, in subsequent chapters, will be of use to derive optimal logarithmic time parallel algorithms for list contraction and related graph problems. One of the inefficiencies of the basic shortcutting algorithm for list-ranking is that on each step, we get two chains, where only one is essential. Various methods have been developed to make list-ranking execute with optimal linear work in logarithmic time. The simplest is random mate, which has been already discussed, and the other is purely deterministic. The deterministic technique makes use of an r-*ruling set*, which is a subset of the nodes of a list, where the subset contains no two adjacent nodes and every node of the subset is within distance r of another element of the subset. An r-ruling set can be found very quickly by an elegant technique called *deterministic coin tossing*, which uses the labels of the adjacent nodes to determine a choice of the r-ruling set. Another problem is that the basic list-ranking algorithm does not work on the EREW mode, since many nodes can point to a given node. These techniques can also be used to make list-ranking execute optimally in the EREW-PRAM. Deterministic ruling set has somewhat greater constants than the random mate technique, which may limit its practical use.

Another very useful technique described in this chapter is the *Euler tour*. An *Euler tour* of a digraph is a path that visits each directed edge exactly once. Any tree has an Euler tour consisting of the face in the planar embedding of the tree. The construction of an Euler tour of a tree is trivial.

Once an Euler tour has been constructed, certain tree problems can be solved by reduction to list-ranking. The applications include, for example, various tree numberings (such as depth first, preorder, and postorder) which are useful for other graph algorithms, such as least common ancestors of trees.

Optimal List Ranking and Parallel Tree Contraction

Chapter 3, *List Ranking and Parallel Tree Contraction*, by Margaret Reid-Miller, Gary L. Miller, and Francesmary Modugno, begins with a description of optimal logarithmic algorithms for list ranking. The optimal algorithms are derived in two ways:

1. from the random mate technique, and
2. from the deterministic ruling set technique.

The *expression tree* is a circuit whose DAG is a tree. The *expression evaluation* problem is to evaluate a given expression tree. While the Euler tour technique is very useful for various problems such as tree numberings, which is essential for many graph algorithms, it cannot solve certain important problems on trees, such as expression evaluation. This chapter also describes *tree contraction*, which is a technique for parallel solution of tree problems including Euler tour, as well as a large class of additional problems, including expression evaluation. In addition to these tree problems, tree contraction has been used extensively to derive many more advanced parallel graph algorithms, such as planarity.

The shortcutting technique may be applied to nodes of degree 2 in a tree, and in this context we will call this operation *list compression*. The *rake* operation applied to a node v of a tree collapses all the leaf children of v into v. *Parallel rake* is the parallel execution of the rake operation simultaneously at all nodes of the tree. Tree contraction uses a combination of parallel list compression and rake operations. It will be shown that a logarithmic number of repetitions of parallel compression and rakes operations suffices to reduce the tree to a single node. This chapter also gives optimal logarithmic time algorithms for parallel tree contraction. The optimal algorithms are derived in three alternative ways:

1. from the random mate technique,
2. using the deterministic ruling set technique, and
3. using an alternative deterministic operation, *shunt*, which combines rake and compression into a single operation.

Shunt is an operation to contract an ordered binary tree to its root. If the tree is not binary, it can be made so by introducing a new node for every child node. Euler-tour can be used to determine the odd-numbered leaves. Shunt is applied only to odd-numbered leaves; it applies rake to these leaves, followed immediately by a compress to their siblings. A logarithmic number of repetitions of parallel shunt suffices to reduce the tree to a single node.

6.2 Advanced Parallel Graph Algorithms

In Part II, we describe some advanced parallel graph algorithms. The first two chapters in this part use some sophisticated and rather complex techniques, which may be best suited for advanced students.

Optimal Randomized Parallel Connectivity

Chapter 4, *Randomized Parallel Connectivity* by Hillel Gazit, derives an optimal logarithmic time randomized algorithm for graph connectivity using the random mate technique as well as random subsampling of the edges.

Optimal Deterministic List Ranking and Parallel Connectivity

This is followed by Chapter 5, *Advanced Parallel Prefix-sums, List Ranking and Connectivity* by Uzi Vishkin. This chapter uses purely deterministic techniques, complementing the previous chapter, which used randomized techniques. Vishkin first gives an independent derivation of optimal logarithmic time deterministic algorithms for list ranking and Euler tours. He then extends these techniques to improve the parallel graph connectivity algorithm given in Chapter 3, decreasing the work to nearly optimal linear while preserving logarithmic time.

The remaining chapters of this part of the text describe some optimal logarithmic time parallel algorithms for a number of additional graph problems. All these algorithms use reductions from the previously described problems, including parallel tree contraction and graph connectivity.

Parallel Lowest Common Ancestor

Recall that the *lowest common ancestor (LCA)* of two nodes in a rooted tree is the common ancestor of these nodes which is the furthest from the root. The LCA computation arises in many graph problems, including the problems of biconnectivity and triconnectivity described below. In Chapter 6, *Parallel Lowest Common Ancestor Computation*, Baruch Schieber gives an optimal logarithmic time algorithm to answer a linear number of parallel LCA queries on a given tree, using a reduction to Euler tour and list ranking.

Parallel Graph Biconnectivity and Triconnectivity

A graph is *biconnected* (*triconnected*) if there are at least two (three, respectively) vertex disjoint paths between each distinct pair of vertices. A *biconnected component* of a graph is a maximal subgraph that is either biconnected or a single edge. Given two adjacent nodes s, t of a graph of n nodes, an *s-t numbering* is an assignment of the nodes to the integers in $\{1, \ldots, n\}$ such that $s = 1, t = n$ and each other node v is numbered so that v is adjacent to at least two nodes u, w such that $u < v < w$. An *open ear decomposition* of a graph is a partitioning of the edges of the graph into a simple cycle p_0, and an ordered sequence of paths $p_1, \ldots, p_k$ where for $i = 1, \ldots k$, the end nodes of path p_i are distinct, the internal path nodes do not appear in any lower numbered path and, moreover, each end node is in some p_j, where $j < i$. A graph is biconnected iff the graph has an s-t numbering iff the graph has an open ear decomposition. In Chapter 7, *Parallel Open Ear Decomposition with Applications to Graph Biconnectivity and Triconnectivity*, Vijaya Ramachandran gives an optimal logarithmic time algorithm for open ear decomposition, using parallel tree contraction to give a reduction from graph connectivity and LCA. The open ear decomposition and Euler tour are used to derive logarithmic time parallel algorithms for biconnectivity and triconnectivity with logarithmic time and linear processors.

The *planarity* problem asks, given a graph, determine if the graph is planar, and if so give a planar embedding. This chapter also cites a logarithmic time, linear processor algorithm for the planarity problem which applies many of the previously mentioned techniques, namely, a reduction to open ear decomposition, LCA, and tree contraction.

Parallel Algorithms for Chordal Graphs

Special classes of graphs arise in the process of Gaussian elimination of a linear system. The *sparsity graph* of an $n \times n$ matrix A of a linear system has for each $i = 1, \ldots, n$ a node i corresponding to the ith variable and an edge (i, j) corresponding to each non-zero coefficient $A_{i,j}$. The Gaussian elimination of the ith variable of the linear system results in a modified matrix and sparsity graph. The node i is deleted and edges are added where needed so that the set of nodes adjacent to i are made into a clique. A *clique* is a subset of nodes where each pair of nodes is adjacent. A sparsity graph is a *perfect elimination* graph if there is an elimination order where no additional edges need be added (i.e., the nodes adjacent to each eliminated node are already a clique). A graph is an *interval* (*chordal*, respectively) graph if the maximal cliques can be totally ordered (placed on a tree, respectively) so that each node of the graph occurs only on a consecutive interval of cliques. It happens that a graph is chordal iff it is a perfect elimination graph. Chapter 8, *Parallel Algorithms for Chordal Graphs* by Philip Klein, gives a logarithmic time linear processor parallel algorithm for recognition of interval graphs and chordal graphs. It describes a divide and conquer reduction to graph connectivity, list ranking, and prefix computation to derive parallel algorithms.

6.3 Parallel Sorting and Applications to Computational Geometry

Part III describes two very different techniques, namely, randomized sampling and deterministic accelerated cascading (to be defined below), for generating parallel algorithms for a wide range of sorting and computational geometry problems. The parallel algorithms based on random sampling are more broadly applicable to machine models ranging from the PRAM to fixed connection machines, and are often somewhat simpler, though randomized. In contrast, the accelerated cascading technique has the advantage of being purely deterministic, but is restricted in application to PRAM algorithms only. This is an excellent example of the advantage of using disparate techniques to solve the same problem: the competing techniques give quite different advantages depending on the computational model.

In the following sorting and selection problems, we assume a key domain with a total ordering. Given a set of n keys, the problem of *sorting* is to list the keys in order. Given the n keys and an integer $i, 1 \leq i \leq n$, the *selection* problem is to find the ith key in the sorted order (without necessarily sorting the keys). The sequential time complexity for sorting and selection is known to be $\Theta(nlogn)$ and $\Theta(n)$, respectively.

Randomized Parallel Sorting

In Chapter 9, *Random Sampling Techniques and Parallel Algorithm Design*, Sanguthevar Rajasekaran and Sandeep Sen give a general description of random sampling techniques and their application on various randomized parallel algorithms for sorting and selection. The parallel selection algorithms choose a random sample of the input and then use this sample to determine the subset of the key to further select from. This process is recursively applied until the subset of keys is constant.

A simple example of a randomized sorting algorithm is the sequential sorting algorithm *quicksort*, which chooses only a single random key to partition the input. The randomized parallel sorting algorithms derived in this chapter choose a random sample of the input, sort this sample, and then use the sorted sample to partition the keys into subsets which are then recursively sorted. The size of the subsample is carefully chosen so as to minimize the parallel time required to solve the subproblems. The **flashsort** parallel sorting algorithm insures that the partitioned subsets are of very nearly equal size by use of *oversampling* by a small factor s; that is, by combining each s consecutive blocks of partitioned keys into a new block. This oversampling technique evens out the slight variations in sizes of the originally partitioned blocks. With high likelihood, oversampling yields blocks of size a factor $1 + o(1)$ of each other. The resulting randomized parallel sorting algorithms execute in logarithmic time on the PRAM using a linear number of processors. This chapter also briefly describes how these randomized sampling techniques can be extended using a more advanced oversampling method to derive an optimal logarithmic time sort known as *flashsort* on regular, bounded degree networks of n processors, such as the butterfly. *Oversampling* combines a small number of consecutive blocks of partitioned keys into a new block to insure that the partitioned subsets are of very nearly equal size; a factor $1 + o(1)$ of each other. This algorithm uses a combination of oversampling, randomized routing techniques and pipelines the logarithmic recursive levels of sorts.

Deterministic Parallel Sorting

In Chapter 10, *Parallel Merge Sort* by Richard Cole, purely deterministic techniques are used to derive parallel selection and sorting algorithms for the EREW-PRAM. Odd-even *mergesort* is a recursive sorting algorithm which, given n keys, subdivides the keys into two subsets, namely the odd and even keys, each of size at most $\lceil n/2 \rceil$, recursively sorts the subsets and then merges the resulting sorted subsequences. Since there are logarithmic recursive levels, and merge takes $O(\log n)$ time, the total time is $O(\log^2 n)$ using n processors. Cole describes an innovative deterministic technique, which is called *accelerated cascading* for reducing the time to do odd-even *mergesort* to logarithmic time using an optimal n processors. A *pipeline* is a sequence of parallel processors, with a linear stream of data passing through the processors, where each data item is processed in sequence by successive processors. Accelerated cascading pipelines the logarithmic recursive levels of merges. It is not known how to extend mergesort, or in fact any deterministic sorting method, to run in logarithmic time on a regular, bounded degree network of linear size.

Parallel Computational Geometry

Computational geometry is the study of computational problems in geometry. Both the randomized (e.g., randomized sampling) and deterministic (e.g., accelerated cascading) techniques can also be used to derive parallel algorithms for computational geometry problems. An example of a computational geometry problem is finding the *d-dimensional(D) convex hull*, which is the minimal convex polyhedron containing n input points in the *Cartesian plane* $\mathbf{R}^d$.

Chapter 11, *Deterministic Parallel Computational Geometry* by Mikhail Atallah and Michael Goodrich, illustrates accelerated cascading techniques similar to those used for *mergesort*, to derive optimal logarithmic time deterministic PRAM algorithms for a 2-D convex hull and various related computational geometry problems in the Cartesian plane. These deterministic algorithms for 2-D convex hull also extend to give efficient (within a logarithmic factor of optimal) nearly logarithmic time PRAM algorithms for 3-D convex hull, and other related problems such as Voronoi diagrams.

Chapter 9 (*Random Sampling Techniques and Parallel Algorithm Design*) uses random sampling techniques similar to the randomized parallel sorting algorithms to derive optimal logarithmic time randomized PRAM algorithms for a 2-D convex hull and related problems. A technique called *polling* is used to insure that the random sample evenly partitions the problems into subproblems of the appropriate size. These optimal randomized parallel algorithms for 2-D convex hull also extend to optimal logarithmic time algorithms for the butterfly network. This chapter also cites optimal randomized logarithmic time PRAM algorithms for 3-D convex hull and Voronoi diagrams which are also derived using the randomized technique of polling.

6.4 Fundamental Parallel Algebraic Algorithms

Part IV describes the synthesis of various basic parallel algebraic algorithms. Both chapters in this part of the text assume the circuit model.

Parallel Algorithms for Polynomial and Integer Arithmetic

Polynomial arithmetic is the basic operations of polynomials, namely, product and reciprocal. Polynomial product can be computed in logarithmic time in a circuit of size $O(n \log n)$ by reduction to circuits for the discrete Fourier transform. *Newton iteration* is a technique of determining the root of a continuous function $f(x)$ from an initial point x_0 by the iteration formula $x_i = x_{i-1} - \frac{f(x_i)}{f'(x_i)}$, where $f'(x) = \frac{df(x)}{dx}$.

The Newton iteration can be generalized in many ways and is a basic technique for the parallel solution of a variety of algebraic problems. Chapter 12 by Stephen Tate, *Newton Iteration and Integer Division*, uses Newton iteration to derive parallel circuits of logarithmic (or near logarithmic) depth for a number of polynomial and integer arithmetic problems, including division and reciprocal.

Parallel Algorithms in Linear Algebra

In Chapter 13, *Parallel Linear Algebra*, Joachim von zur Gathen surveys techniques for deriving **NC** circuits for basic problems in linear algebra and polynomial arithmetic. Let $M(n)$ be the number of processors required to multiply an $n \times n$ dense matrix in $O(\log n)$ time. The current bound for $M(n)$ is approximately $n^{2.376\cdots}$. Let A be an $n \times n$ matrix over a field. A has *rank* r, if there are exactly r linear independent rows of A defining a *basis*. We will assume A is *nonsingular*, that is A has rank n. There are simple **NC**^{2} algorithms using divide and conquer for computing the inverse of a triangular matrix A, which use the fact that the inverse of A can be expressed in terms of products and the inverse of $n/2 \times n/2$ submatrices of A. If A is not triangular, these divide and conquer techniques do not yield **NC** algorithms, so the chapter describes another approach which makes use of the *characteristic polynomial* of a matrix. The characteristic polynomial allows the inverse of the matrix to be expressed in terms of the n powers of the matrix. The inverse is computed **NC**^{2} by computing the powers of the polynomial and computing the coefficients of the characteristic polynomial. The coefficients of the characteristic polynomial can be computed by parallel algorithms for polynomial arithmetic. Thus this derivation of parallel matrix inverse uses reduction to parallel algorithms for polynomial arithmetic, as well as matrix powers. This chapter also gives **NC**^{2} circuits for computation of related problems such as determinants, which reduce to computing the inverse of a matrix.

The problems of rank and basis have no known **NC** algorithms. The **RNC** rank and basis algorithms described in this chapter make an interesting use of random sampling. To test whether an $n \times n$ matrix A is of rank $\geq r$, choose two random matrices B and C, where B is of size $r \times n$ and C is of size $r \times n$ and then test whether BAC are non-singular. If A is of rank $\geq r$, then A is nonsingular with probability $< 1/2$; otherwise A is always singular. This chapter also cites efficient PRAM algorithms for various algebraic problems. Matrix inverse and determinant can be computed efficiently in $O(\log^2 n)$ time using $M(n)$ processors by use of Newton iteration, and rank and basis also have efficient randomized parallel algorithms.

6.5 Advanced Parallel Algebraic Algorithms

Part V of the text describes the use of advanced techniques to derive parallel algorithms for algebraic problems. Many of these techniques are combinatorial, complementing Part VII of this text, which uses, in part, algebraic techniques to solve combinatorial optimization problems such as matching.

Nested Dissection Techniques

A family of graphs closed under the subgraph relation is $s(n)$-*separable* if for each graph of n nodes, there is a subset of $s(n)$ nodes, called the separator, which when deleted disconnects the graph into two subgraphs, each with at most δn nodes, for some $\delta < 1$. For example, planar graphs are $O(\sqrt{n})$-separable, and d-dimensional grid graphs are $O(n^{\frac{d-1}{d}})$-separable. In Chapter 14, *Parallel Solution of Sparse Linear and Path Systems*, Victor Pan gives an efficient parallel algorithm, known as *parallel nested dissection*, for the solution of sparse linear systems, whose sparsity graph is $s(n)$-separable. The parallel nested dissection algorithm uses the separator set to decompose the linear system into a sparse LUD factorization. If $s(n) = n^\epsilon$, for some $\epsilon > 0$, then work is asymptotically bounded by the time to solve an $s(n) \times s(n)$ dense linear system. The number of stages of recursion is logarithmic. The total time of parallel nested dissection is $O(\log^3 n)$ and the number of processors is $M(s(n))/\log n$. This chapter also extends the parallel nested dissection technique to monoid path problems, such as minimal paths in positively weighted graphs, where time reduces to $O(\log n)$ and the number of processors remains the same.

Parallel Algorithms for Resultants

A fundamental problem in algebraic computation is to find the solution of a system of m simultaneous polynomial equations with n variables over an algebraically closed field (for example the complex numbers **C**). The *resultant* is a polynomial in the coefficients of the m polynomials, and the resultant vanishes when a common solution exists for all m of the polynomial equations. Resultants are essential for a number of techniques in finding solutions of these polynomial equation systems. The resultant can be computed through polynomial remainder sequences similar to the remainder sequence in the calculation of greatest common divisors of two polynomials. For example, the resultant of two polynomials is a polynomial defined from a linear recurrence equation involving the coefficients of the two polynomials, which can be solved by a parallel algorithm for solution of linear systems. In Chapter 15, *Parallel Resultant Computation*, Dexter Kozen and Doug Ierardi derive **RNC** algorithms for computing resultants. The key technique is to reduce the resultant computation to basic parallel algorithms in linear algebra, utilizing the special structure of the corresponding polynomial remainder sequences. Random sampling of input variable values is also used in these resultant algorithms to test if certain polynomials are identically 0.

6.6 Extensions of Parallel Tree Contraction to Circuits and Logic Queries

In the two chapters of Part VI, parallel tree contraction is extended to the evaluation of algebraic circuits and to the evaluation of logical queries.

Parallel Evaluation of Algebraic Circuits

Recall that a straight line program of length T can be transformed into a circuit (also called a *computational DAG*) of size and depth T. Note that a circuit over an appropriate domain can be viewed as a multivariate polynomial whose indeterminates are the input variables. The *degree* of the circuit is the maximum total degree of any output variable. In Chapter 16, *Dynamic Parallel Evaluation of Computation DAGs*, Erich Kaltofen gives a **NC** algorithm for the parallel evaluation of a circuit, assuming the circuit has polynomial degree. This parallel circuit evaluation algorithm extends parallel tree contraction techniques to DAGs.

Parallel Evaluation of Logical Queries

In Chapter 17, Jeffrey Ullman uses parallel graph algorithms (in particular, parallel tree contraction) to derive parallel algorithms for logical queries. Prolog is a language for defining logical inference. There is a set of variables which can be assigned *first order terms*, which are expressions. The inference is done by logic queries consisting of a set of rules defining first order rules of term substitution, along with lists of subgoals. A *logic program* is a sequence of logical rules, where each rule has a defined sequence of subgoals. A *logical rule* is of form $p(X_1, \ldots, X_n) \leftarrow S_1, \ldots, S_n$ where the $X_1, \ldots, X_n$ are variables and $S_1, \ldots, S_n$ are subgoals to be satisfied before the rule can be invoked. These subgoals are themselves rules. If a goal is satisfied, then there is a *proof tree* consisting of leaves with base assumptions (*atoms*) and internal nodes consisting of rules where constants are substituted into variables in the logical rule. The proof tree has the *polynomial fringe property* if the number of leaves is polynomial. In his chapter, *Parallel Complexity of Logical Inference*, Ullman gives a **NC** algorithm for the parallel evaluation of the restricted class of logic queries which have the polynomial fringe property. This parallel algorithm uses parallel tree contraction to construct the proof tree quickly in parallel.

6.7 Parallel Algorithms for Combinatorial Optimization

Combinatorial optimization concerns the optimum solution of combinatorial problems arising in the field of operations research. The problems of matching and flow, described below, are among the most fundamental of combinatorial problems, and arise frequently in many practical applications. Part VII of this book derives parallel algorithms for these combinatorial optimization problems.

Parallel Algorithms for Matching

A graph *matching* is a set of edges of a graph where no two edges are adjacent. A graph matching is *perfect* if the matched edges include all vertices. The *graph matching* problem is to find a maximum cardinality matching. The *weighted matching* problem for a weighted graph is to find a maximum cardinality matching of maximum weight.

It is not known if the problem of determining the existence of a given size matching is in **NC**. In Chapter 18, *Parallel Graph Matching*, Vijay Vazirani gives various **RNC** algorithms for graph matching. The parallel algorithms make interesting use of algebraic techniques and random sampling. Given a graph of n nodes $\{1,\ldots,n\}$, we construct an $n \times n$ matrix M where for each $1 \leq i < j \leq n$ we let $M_{i,j}$ = an indeterminant $x_{i,j}$ if (i,j) is an edge of the graph, and $M_{i,j} = 0$ otherwise. We require $x_{i,j} = -x_{j,i}$, so the number of indeterminants is the number of edges. The determinant of matrix M is a multivariate polynomial. It has been proved that the determinant of M is identically 0 iff the graph has a perfect matching. To test if M is identically 0 we can use random sampling: we assign the indeterminants to independently chosen random integer values over $\{1,\ldots,n^{O(1)}\}$. With high likelihood, the resulting integer matrix has determinant 0 iff M has determinant identically 0. The complexity of this randomized parallel algorithm for matching is bounded by the determinant computation, which can be computed in $O(\log^2 n)$ time with $M(n)$ processors by parallel algebraic algorithms described previously. This chapter also derives **RNC** algorithms for constructing graph matchings of regular bipartite graphs and graphs with a polynomial number of matchings, using a similar reduction to parallel algebraic algorithms.

Parallel Algorithms for Network Flow

A *network* is a digraph with distinguished nodes s (the *source*), t (the *sink*), and a labeling of each directed edge with a positive real number called the *capacity*, where each directed edge (i,j) has the same capacity as its *reverse* edge (j, i). A *flow* is a mapping from the directed edges of the network to the real numbers such that

1. the absolute value of the flow at any edge does not exceed the capacity of that edge,
2. the flow of any directed edge (i, j) is the negative of the reverse edge (j, i), and
3. for each node i other than the source and sink, the sum of the flow of all edges entering i is exactly the same as the sum of the flow of all edges departing i.

The *value* of the flow is the sum of the flow on all edges departing the source. The *network flow* problem is to find a maximum flow. There are no known **NC** nor **RNC** algorithms for determining the value of the maximum flow. Andrew Goldberg's chapter, *Parallel Algorithms for Network Flow Problems*, derives parallel algorithms for maximum flow using an extension of an efficient sequential flow algorithm. One of the parallel flow algorithms requires $O(n^2 \log n)$ time and $\sqrt{m}$ processors in a network with n nodes and m edges, and uses a *push-relabel* technique that allows propagation of flow with limited parallelism. The other algorithm uses a combination of push-relabel and *scaling* (a method for adjusting the values of the capacities). If the maximum edge capacity in the network is U, this algorithm requires $O(n^2 \log(U) \log(n))$ time using $\sqrt{m}$ processors.

6.8 Inherent Limitations of Parallel Computation

Part VIII of this text considers the inherent limitations of parallel computation.

Lower Bounds for the PRAM

Chapter 20, *The Parallel Random Access Machine* by Faith Fich, shows that various classes of PRAMs have inherent limitations in the minimum time required to compute various problems. The PRIORITY-PRAM is a strengthening of the CRCW-PRAM which requires a concurrent write to a memory location, such that the highest numbered (rather than arbitrary) processor must succeed in the write attempt. The time to compute parity of n bits on a PRIORITY-PRAM is shown to be at least $\Omega(\log n/\log \log n)$. Therefore, this lower bound also holds for the CRCW-PRAM and EREW-PRAM. Many other lower bounds are also derived for various classes of PRAMs.

P-complete Problems

Some problems in **P** appear not to have **NC** algorithms. A *language recognition problem* is a problem with a single Boolean output. An **NC** reduction is a mapping, computable in **NC**, from a language recognition problem L_1 to a language recognition problem L_2 such that each problem instance in L_1 is mapped by the reduction to a problem instance in L_2 where the new problem instance has the same answer as the solution of the original problem instance in L_1. (This definition of reductions can be easily generalized to functions as well as languages). A problem is **P**-*complete* with respect to **NC** reductions if any problem in **P** has an **NC** reduction to it. The **P**-complete problems appear not to have **NC** algorithms, since if any **P**-complete problem has an **NC** algorithm, then **P** = **NC**. Chapter 21, *Polynomial Completeness and Parallel Computation*, by Raymond Greenlaw, discusses **P**-complete problems and their relation to inherent limitations of parallel computation. The following problems are shown to be **P**-complete:

1. evaluation of a Boolean circuit, even if it is restricted to be planar;
2. maximum flow in networks;
3. linear programming (the *linear programming* problem is to find a solution of a system of m simultaneous linear inequalities of n variables);
4. various graph numberings, including breadth first search, where the numbering algorithm is restricted to a particular order of traversal of the edges in the adjacency lists of the graph (a similarly restricted version of depth first search is also **P**-complete).

6.9 Asynchronous Parallel Computation

Finally, Part IX considers asynchronous parallel algorithms. The standard definition of the PRAM requires the processors to be synchronous. The asynchronous PRAM model is introduced in Chapter 22, *Asynchronous PRAM Algorithms* by Phillip B. Gibbons. This model allows processors to execute asynchronously, and introduces an additional operation called the *synchronization step*. In one variant of this model, the synchronization step is a *barrier*, and processors must temporarily wait until all processors reach the barrier. Memory is partitioned into *local* (also known as *private*) memory blocks, each accessible only by a unique processor, and *global* memory which can be accessed by all processors, but only at the barrier. Between barriers, there cannot be global memory accesses. The execution time is determined by the number of synchronization steps executed. Basic parallel algorithm techniques, including parallel prefix, are derived for the asynchronous PRAM. This chapter also discusses variants of this model, where the barrier restriction is weakened.

7
Acknowledgments

We wish to acknowledge the following individuals who made significant contributions to this text. The research of Robert Paige and Bill Scherlis led us to an understanding of the importance of algorithm derivation. Specifically, Bill inspired this text by a suggestion that the parallel algorithm community should collaborate to produce a single document to bring together the various parallel algorithm derivation techniques. We thank him, as well as Erik Mettala and Ralph Wachter, for their continuous support of this project. The following authors of the text contributed both with respect to their own chapters, and also with respect to reviews of other chapters: Mikhail J. Atallah, Sara Baase, Guy E. Blelloch, Richard Cole, Joachim von zur Gathen, Hillel Gazit, Phillip B. Gibbons, Andrew V. Goldberg, Michael T. Goodrich, Raymond Greenlaw, Doug Ierardi, Erich Kaltofen, Philip Klein, Dexter Kozen, Gary L. Miller, Francesmary Modugno, Victor Pan, Sanguthevar Rajasekaran, Vijaya Ramachandran, Margaret Reid-Miller, Baruch Schieber, Sandeep Sen, Stephen R. Tate, Jeffrey D. Ullman, Vijay V. Vazirani, and Uzi Vishkin. Thanks to Shenfeng Chen, Hongyan Wang, Aki Yoshida and particularly Deganit Armon for improving the presentation of the introduction. We thank Suzanne L. Cecil, Kirk Franklin, and Chris Lane for their assistance in document preparation.

Finally, we dedicate this text to Jane Anderson and Katie Reif.

Part I

Fundamental Parallel Graph Algorithms

1

Prefix Sums and Their Applications

Guy E. Blelloch

School of Computer Science
Carnegie Mellon University
Pittsburgh, PA 15213-3890
blelloch@cs.cmu.edu

1.1 Introduction

Experienced algorithm designers rely heavily on a set of building blocks and on the tools needed to put the blocks together into an algorithm. The understanding of these basic blocks and tools is therefore critical to the understanding of algorithms. Many of the blocks and tools needed for parallel algorithms extend from sequential algorithms, such as dynamic-programming and divide-and-conquer, but others are new.

This chapter introduces one of the simplest and most useful building blocks for parallel algorithms: the *all-prefix-sums* operation. The chapter defines the operation, shows how to implement it on a PRAM and illustrates many applications of the operation. In addition to being a useful building block, the all-prefix-sums operation is a good example of a computation that seems inherently sequential, but for which there is an efficient parallel algorithm. The operation is defined as follows:

DEFINITION 1.1
The all-prefix-sums *operation takes a binary associative operator $\oplus$, and an ordered set of n elements*

$$[a_0, a_1, \ldots, a_{n-1}],$$

and returns the ordered set

$$[a_0, (a_0 \oplus a_1), \ldots, (a_0 \oplus a_1 \oplus \ldots \oplus a_{n-1})].$$

For example, if $\oplus$ is addition, then the all-prefix-sums operation on the ordered set

$$[3 \quad 1 \quad 7 \quad 0 \quad 4 \quad 1 \quad 6 \quad 3],$$

would return

$$[3 \quad 4 \quad 11 \quad 11 \quad 14 \quad 16 \quad 22 \quad 25].$$

The uses of the all-prefix-sums operation are extensive. Here is a list of some of them:

1. To lexically compare strings of characters. For example, to determine that "strategy" should appear before "stratification" in a dictionary (see Exercise 1.2).

2. To add multiprecision numbers. These are numbers that cannot be represented in a single machine word (see Exercise 1.3).
3. To evaluate polynomials (see Exercise 1.6.).
4. To solve recurrences. For example, to solve the recurrences $x_i = a_i x_{i-1} + b_i x_{i-2}$ and $x_i = a_i + b_i/x_{i-1}$ (see Section 1.4).
5. To implement radix sort (see Section 1.3).
6. To implement quicksort (see Section 1.5).
7. To solve tridiagonal linear systems (see Exercise 1.12).
8. To delete marked elements from an array (see Section 1.3).
9. To dynamically allocate processors (see Section 1.6).
10. To perform lexical analysis. For example, to parse a program into tokens.
11. To search for regular expressions. For example, to implement the UNIX grep program.
12. To implement some tree operations. For example, to find the depth of every vertex in a tree (see Chapter 2).
13. To label components in two-dimensional images.

In fact, all-prefix-sums operations using addition, minimum and maximum are so useful in practice that they have been included as primitive instructions in some machines. Researchers have also suggested that a subclass of the all-prefix-sums operation be added to the PRAM model as a "unit time" primitive because of their efficient hardware implementation.

Before describing the implementation we must consider how the definition of the all-prefix-sums operation relates to the PRAM model. The definition states that the operation takes an ordered set, but does not specify how the ordered set is laid out in memory. One way to lay out the elements is in contiguous locations of a vector (a one-dimensional array). Another way is to use a linked-list with pointers from each element to the next. It turns out that both forms of the operation have uses. In the examples listed above, the component labeling and some of the tree operations require the linked-list version, while the other examples can use the vector version.

Sequentially, both versions are easy to compute (see Figure 1.1). The vector version steps down the vector, adding each element into a sum and writing the sum back, while the linked-list version follows the pointers while keeping the running sum and writing it back. The algorithms in Figure 1.1 for both versions are inherently sequential: to calculate a value at any step, the result of the previous step is needed. The algorithms therefore require

```
proc all-prefix-sums(Out, In)          proc all-prefix-sums(Out, In)
    i ← 0                                  i ← 0
    sum ← In[0]                            sum ← In[0].value
    Out[0] ← sum                           Out[0] ← sum
    while (i < length)                     while (In[i].pointer ≠ EOL)
        i ← i + 1                              i ← In[i].pointer
        sum ← sum + In[i]                      sum ← sum + In[i].value
        Out[i] ← sum                           Out[i] ← sum

          Vector Version                            List Version
```

FIGURE 1.1
Sequential algorithms for calculating the all-prefix-sums operation with operator + on a vector and on a linked-list. In the list version, each element of In consists of two fields: a value (.value), and a pointer to the next position in the list (.pointer). EOL means the end-of-list pointer.

$O(n)$ time. To execute the all-prefix-sums operation in parallel, the algorithms must be changed significantly.

The remainder of this chapter is concerned with the vector all-prefix-sums operation. We will henceforth use the term *scan* for this operation.[1]

DEFINITION 1.2
The scan *operation is a vector all-prefix-sums operation.*

Chapters 2, 3 and 5 discuss uses of the linked-list all-prefix-sums operation and derive an optimal deterministic algorithm for the problem on the PRAM.

Sometimes it is useful for each element of the result vector to contain the sum of all the previous elements, but not the element itself. We call such an operation a *prescan*.

DEFINITION 1.3
The prescan *operation takes a binary associative operator* $\oplus$ *with identity* I, *and a vector of* n *elements*

$$[a_0, a_1, \ldots, a_{n-1}],$$

and returns the vector

$$[I, a_0, (a_0 \oplus a_1), \ldots, (a_0 \oplus a_1 \oplus \ldots \oplus a_{n-2})].$$

---
[1]The term *scan* comes from the computer language APL.

A prescan can be generated from a scan by shifting the vector right by one and inserting the identity. Similarly, the scan can be generated from the prescan by shifting left, and inserting at the end the sum of the last element of the prescan and the last element of the original vector.

1.2 Implementation

This section describes an algorithm for calculating the scan operation in parallel. For p processors and a vector of length n on an EREW PRAM, the algorithm has a time complexity of $O(n/p + \lg p)$. The algorithm is simple and well suited for direct implementation in hardware. Chapter 5 shows how the time of the scan operation with certain operators can be reduced to $O(n/p + \lg p / \lg \lg p)$ on a CREW PRAM.

Before describing the scan operation, we consider a simpler problem, that of generating only the final element of the scan. We call this the *reduce* operation.

DEFINITION 1.4
The reduce *operation takes a binary associative operator $\oplus$ with identity i, and an ordered set $[a_0, a_1, \ldots, a_{n-1}]$ of n elements, and returns the value $a_0 \oplus a_1 \oplus \ldots \oplus a_{n-1}$.*

Again we consider only the case where the ordered set is kept in a vector. A balanced binary tree can be used to implement the reduce operation by laying the tree over the values and using $\oplus$ to sum pairs at each vertex (see Figure 1.2a). The correctness of the result relies on $\oplus$ being associative. The operator, however, does not need to be commutative since the order of the operands is maintained. On an EREW PRAM, each level of the tree can be executed in parallel, so the implementation can step from the leaves to the root of the tree (see Figure 1.2b); we call this an up-sweep. Since the tree is of depth $\lceil \lg n \rceil$, and one processor is needed for every pair of elements, the algorithm requires $O(\lg n)$ time and $n/2$ processors.

If we assume a fixed number of processors p, with $n > p$, then each processor can sum an n/p section of the vector to generate a processor sum; the tree technique can then be used to reduce the processor sums (see Figure 1.3). The time taken to generate the processor sums is $\lceil n/p \rceil$, so the total time required on an EREW PRAM is:

$$T_R(n,p) = \lceil n/p \rceil + \lceil \lg p \rceil = O(n/p + \lg p). \tag{1.1}$$

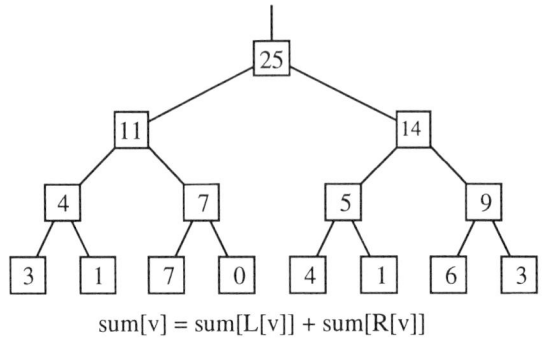

sum[v] = sum[L[v]] + sum[R[v]]

(a) Executing a +-reduce on a tree.

for d **from** 0 **to** $(\lg n) - 1$	
in parallel for i **from** 0 **to** $n - 1$ **by** 2^{d+1}	
$a[i + 2^{d+1} - 1] \leftarrow a[i + 2^d - 1] + a[i + 2^{d+1} - 1]$	

Step	Vector in Memory
0	[3 1 7 0 4 1 6 3]
1	[3 4 7 7 4 5 6 9]
2	[3 4 7 11 4 5 6 14]
3	[3 4 7 11 4 5 6 25]

(b) Executing a +-reduce on a PRAM.

FIGURE 1.2
An example of the reduce operation when $\oplus$ is integer addition. The boxes in (b) show the locations that are modified on each step. The length of the vector is n and must be a power of two. The final result will reside in $a[n-1]$.

When $n/p \geq \lg p$ the complexity is $O(n/p)$. This time is an optimal speedup over the sequential algorithm given in Figure 1.1.

We now return to the scan operation. We actually show how to implement the prescan operation; the scan is then determined by shifting the result and putting the sum at the end. If we look at the tree generated by the reduce operation, it contains many partial sums over regions of the vector. It turns out that these partial sums can be used to generate all the prefix sums. This

> **in parallel for each processor** i
> $\quad$ sum$[i] \leftarrow a[(n/p)i]$
> $\quad$ for j from 1 to n/p
> $\quad\quad$ sum$[i] \leftarrow$ sum$[i]$ + $a[(n/p)i + j]$
> $\quad$ result $\leftarrow$ +-reduce(sum)
>
> $[\underbrace{4\ \ 7\ \ 1}_{\text{processor 0}}\ \ \underbrace{0\ \ 5\ \ 2}_{\text{processor 1}}\ \ \underbrace{6\ \ 4\ \ 8}_{\text{processor 2}}\ \ \underbrace{1\ \ 9\ \ 5}_{\text{processor 3}}]$
>
> Processor Sums $\quad=\quad$ [12 $\quad$ 7 $\quad$ 18 $\quad$ 15]
> Total Sum $\quad=\quad$ 52

FIGURE 1.3
The +-reduce operation with more elements than processors. We assume that n/p is an integer.

requires executing another sweep of the tree with one step per level, but this time starting at the root and going to the leaves (a down-sweep). Initially, the identity element is inserted at the root of the tree. On each step, each vertex at the current level passes to its left child its own value, and it passes to its right child ⊕ applied to the value from the left child from the up-sweep and its own value (see Figure 1.4a).

Let us consider why the down-sweep works. We say that vertex x *precedes* vertex y if x appears before y in the preorder traversal of the tree (depth first, from left to right).

THEOREM 1.1
After a complete down-sweep, each vertex of the tree contains the sum of all the leaf values that precede it.

PROOF
The proof is inductive from the root: we show that if a parent has the correct sum, both children must have the correct sum. The root has no elements preceding it, so its value is correctly the identity element.

Consider Figure 1.5. The left child of any vertex has exactly the same leaves preceding it as the vertex itself (the leaves in region A in the figure). This is because the preorder traversal always visits the left child of

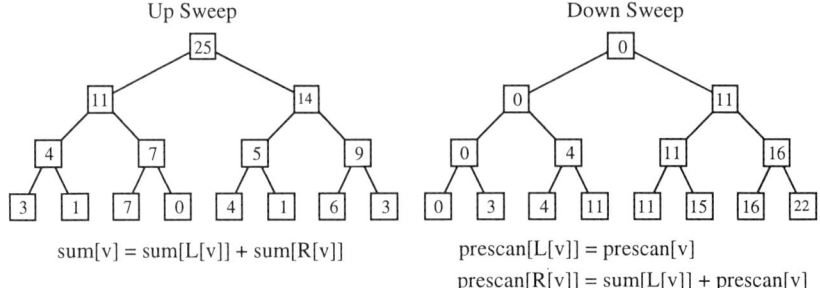

(a) Executing a +-prescan on a tree.

procedure down-sweep(A)
$a[n-1] \leftarrow 0$ % Set the identity
for d **from** $(\lg n) - 1$ **downto** 0
  in parallel for i **from** 0 **to** $n-1$ **by** 2^{d+1}
    $t \leftarrow a[i + 2^d - 1]$ % Save in temporary
    $a[i + 2^d - 1] \leftarrow a[i + 2^{d+1} - 1]$ % Set left child
    $a[i + 2^{d+1} - 1] \leftarrow t + a[i + 2^{d+1} - 1]$ % Set right child

	Step	Vector in Memory							
	0	[3	1	7	0	4	1	6	3]
up	1	[3	4	7	7	4	5	6	9]
	2	[3	4	7	11	4	5	6	14]
	3	[3	4	7	11	4	5	6	25]
clear	4	[3	4	7	11	4	5	6	0]
down	5	[3	4	7	0	4	5	6	11]
	6	[3	0	7	4	4	11	6	16]
	7	[0	3	4	11	11	15	16	22]

(b) Executing a +-prescan on a PRAM.

FIGURE 1.4
A parallel prescan on a tree using integer addition as the associative operator $\oplus$ and 0 as the identity.

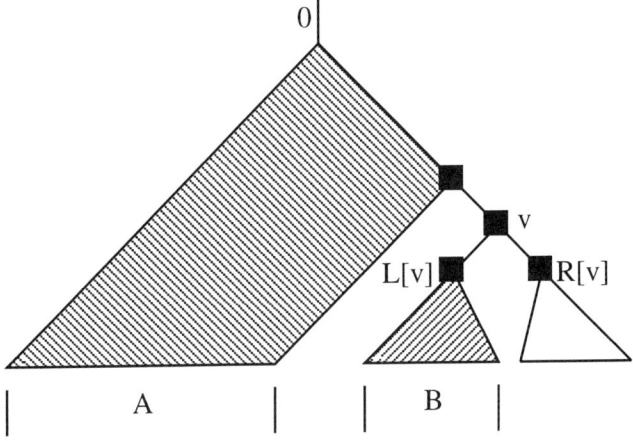

FIGURE 1.5
Illustration for Theorem 1.1.

a vertex immediately after the vertex. By the induction hypothesis, the parent has the correct sum, so it need only copy this sum to the left child.

The right child of any vertex has two sets of leaves preceding it, the leaves preceding the parent (region A), and the leaves at or below the left child (region B). Therefore, by adding the parent's down-sweep value, which is correct by the induction hypothesis, and the left-child's up-sweep value, the right-child will contain the sum of all the leaves preceding it. ∎

Since the leaf values that precede any leaf are the values to the left of it in the scan order, the values at the leaves are the results of a left-to-right prescan. To implement the prescan on an EREW PRAM, the partial sums at each vertex must be kept during the up-sweep so they can be used during the down-sweep. We must therefore be careful not to overwrite them. In fact, this was the motivation for putting the sums on the right during the reduce in Figure 1.2b. Figure 1.4b shows the PRAM code for the down-sweep. Each step can execute in parallel, so the running time is $2\lceil \lg n \rceil$.

If we assume a fixed number of processors p, with $n > p$, we can use a similar method to that in the reduce operation to generate an optimal algorithm. Each processor first sums an n/p section of the vector to generate a processor sum; the tree technique is then used to prescan the processor sums. The results of the prescan of the processor sums are used as an offset

```
         [4   7   1    0   5   2    6   4   8    1   9   5]
         _____/    _____/    _____/    _____/
         processor 0  processor 1  processor 2  processor 3

         Sum        =        [12      7       18      15]
         +-prescan  =        [0       12      19      37]

         [0   4   11    12   12   17    19   25   29    37   38   47]
         _____/     _____/      _____/      _____/
         processor 0    processor 1     processor 2     processor 3
```

FIGURE 1.6
A +-prescan with more elements than processors.

for each processor to prescan within its n/p section (see Figure 1.6). The time complexity of the algorithm is:

$$T_S(n,p) = 2(\lceil n/p \rceil + \lceil \lg p \rceil) = O(n/p + \lg n) \tag{1.2}$$

which is the same order as the reduce operation and is also an optimal speedup over the sequential version when $n/p \geq \lg p$.

This section described how to implement the scan (prescan) operation. The rest of the chapter discusses its applications.

1.3
Line-of-Sight and Radix-Sort

As an example of the use of a scan operation, consider a simple line-of-sight problem. The *line-of-sight* problem is: given a terrain map in the form of a grid of altitudes and an observation point X on the grid, find which points are visible along a ray originating at the observation point (see Figure 1.7).

A point on a ray is visible if and only if no other point between it and the observation point has a greater vertical angle. To find if any previous point has a greater angle, the altitude of each point along the ray is placed in a vector (the *altitude vector*). These altitudes are then converted to angles and placed in the *angle vector* (see Figure 1.7). A prescan using the operator maximum (max-prescan) is then executed on the *angle vector*, which returns to each point the maximum previous angle. To test for visibility each point only needs to

```
procedure line-of-sight(altitude)
  in parallel for each index i
    angle[i] ← arctan(scale × (altitude[i] - altitude[0])/ i)
  max-previous-angle ← max-prescan(angle)
  in parallel for each index i
    if (angle[i] > max-previous-angle[i])
      result[i] ← "visible"
    else
      result[i] ← not "visible"
```

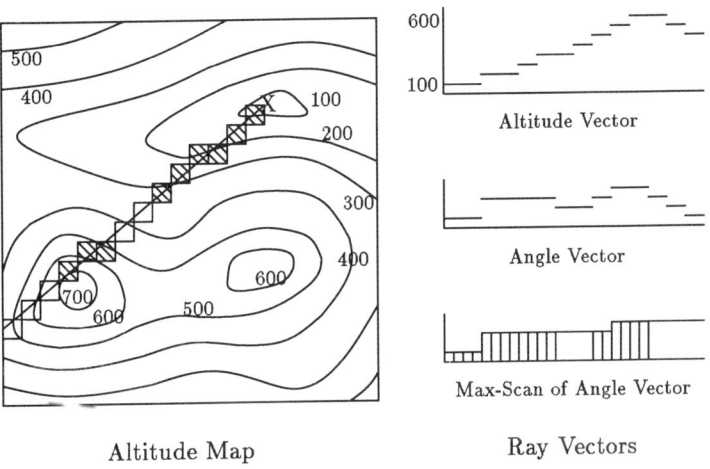

FIGURE 1.7
The line-of-sight algorithm for a single ray. The X marks the observation point. The visible points are shaded. A point on the ray is visible if no previous point has a greater angle.

compare its angle to the result of the max-prescan. This can be generalized to finding all visible points on the grid. For n points on a ray, the complexity of the algorithm is the complexity of the scan, $T_S(n,p) = O(n/p + \lg n)$ on an EREW PRAM.

We now consider another example, a *radix-sort* algorithm. The algorithm loops over the bits of the keys, starting at the lowest bit, executing a split operation on each iteration (assume all keys have the same number of bits). The split operation packs the keys with a 0 in the corresponding bit to

```
       procedure split-radix-sort(A, number-of-bits)
           for i from 0 to (number-of-bits − 1)
              A ← split(A, A⟨i⟩)
```

A	=	[5	7	3	1	4	2	7	2]
A⟨0⟩	=	[1	1	1	1	0	0	1	0]
A ← split(A, A⟨0⟩)	=	[4	2	2	5	7	3	1	7]
A⟨1⟩	=	[0	1	1	0	1	1	0	1]
A ← split(A, A⟨1⟩)	=	[4	5	1	2	2	7	3	7]
A⟨2⟩	=	[1	1	0	0	0	1	0	1]
A ← split(A, A⟨2⟩)	=	[1	2	2	3	4	5	7	7]

FIGURE 1.8
An example of the split radix sort on a vector containing three bit values. The A⟨n⟩ notation signifies extracting the nth bit of each element of the vector A. The split operation packs elements with a 0 flag to the bottom and with a 1 flag to the top.

the bottom of a vector, and packs the keys with a 1 in the bit to the top of the same vector. It maintains the order within both groups. The sort works because each split operation sorts the keys with respect to the current bit (0 down, 1 up) and maintains the sorted order of all the lower bits since we iterate from the bottom bit up. Figure 1.8 shows an example of the sort.

We now consider how the split operation can be implemented using a scan. The basic idea is to determine a new index for each element and then permute the elements to these new indices using an exclusive write. To determine the new indices for elements with a 0 in the bit, we invert the flags and execute a prescan with integer addition. To determine the new indices of elements with a 1 in the bit, we execute a +-scan in reverse order (starting at the top of the vector) and subtract the results from the length of the vector n. Figure 1.9 shows an example of the split operation along with code to implement it.

Since the split operation just requires two scan operations, a few steps of exclusive memory accesses, and a few parallel arithmetic operations, it has the same asymptotic complexity as the scan: $O(n/p + \lg p)$ on an EREW PRAM.[2]

---

[2] On a CREW PRAM we can use the scan described in Chapter 5 to get a time of $O(n/p + \lg p/\lg \lg p)$.

```
procedure split(A, Flags)
    I-down  ← +-prescan(not(Flags))
    I-up    ← n - +-scan(reverse-order(Flags))
    in parallel for each index i
        if (Flags[i])
            Index[i] ← I-up[i]
        else
            Index[i] ← I-down[i]
    result ← permute(A, Index)
```

A	=	[5	7	3	1	4	2	7	2]
Flags	=	[1	1	1	1	0	0	1	0]
I-down	=	[0	0	0	0	0	1	2	2]
I-up	=	[3	4	5	6	6	6	7	7]
Index	=	[3	4	5	6	0	1	7	2]
permute(A, Index)	=	[4	2	2	5	7	3	1	7]

FIGURE 1.9
The split operation packs the elements with a 0 in the corresponding flag position to the bottom of a vector and packs the elements with a 1 to the top of the same vector. The permute writes each element of A to the index specified by the corresponding position in Index.

If we assume that n keys are each $O(\lg n)$ bits long, then the overall algorithm runs in time:
$$O((\frac{n}{p} + \lg p)\lg n) = O(\frac{n}{p}\lg n + \lg n \lg p).$$

1.4
Recurrence Equations

This section shows how various recurrence equations can be solved using the scan operation. A recurrence is a set of equations of the form

$$x_i = f_i(x_{i-1}, x_{i-2}, \cdots, x_{i-m}), \qquad m \leq i < n \qquad (1.3)$$

along with a set of initial values $x_0, \cdots, x_{m-1}$.

The scan operation is the special case of a recurrence of the form

$$x_i = \begin{cases} a_0 & i = 0 \\ x_{i-1} \oplus a_i & 0 < i < n, \end{cases} \quad (1.4)$$

where $\oplus$ is any binary associative operator. This section shows how to reduce a more general class of recurrences to equation (1.4), and therefore how to use the scan algorithm discussed in Section 1.2 to solve these recurrences in parallel.

1.4.1 First-Order Recurrences

We initially consider *first-order* recurrences of the following form

$$x_i = \begin{cases} b_0 & i = 0 \\ (x_{i-1} \otimes a_i) \oplus b_i & 0 < i < n, \end{cases} \quad (1.5)$$

where the a_i's and b_i's are sets of n arbitrary constants (not necessarily scalars) and $\oplus$ and $\otimes$ are arbitrary binary operators that satisfy three restrictions:

1. $\oplus$ is associative (i.e. $(a \oplus b) \oplus c = a \oplus (b \oplus c)$).
2. $\otimes$ is semiassociative (i.e. there exists a binary associative operator $\odot$ such that $(a \otimes b) \otimes c = a \otimes (b \odot c)$).
3. $\otimes$ distributes over $\oplus$ (i.e. $a \otimes (b \oplus c) = (a \otimes b) \oplus (a \otimes c)$).

The operator $\odot$ is called the *companion operator* of $\otimes$. If $\otimes$ is fully associative, then $\odot$ and $\otimes$ are equivalent.

We now show how (1.5) can be reduced to (1.4). Consider the set of pairs

$$c_i = [a_i, b_i] \quad (1.6)$$

and define a new binary operator $\bullet$ as follows:

$$c_i \bullet c_j \equiv [c_{i,a} \odot c_{j,a}, \ (c_{i,b} \otimes c_{j,a}) \oplus c_{j,b}] \quad (1.7)$$

where $c_{i,a}$ and $c_{i,b}$ are the first and second elements of c_i, respectively.

Given the conditions on the operators $\oplus$ and $\otimes$, the operator $\bullet$ is associative as we show below:

$(c_i \bullet c_j) \bullet c_k$
$\quad = \ [c_{i,a} \odot c_{j,a}, \ (c_{i,b} \otimes c_{j,a}) \oplus c_{j,b}] \bullet c_k$
$\quad = \ [(c_{i,a} \odot c_{j,a}) \odot c_{k,a}, \ (((c_{i,b} \otimes c_{j,a}) \oplus c_{j,b}) \otimes c_{k,a}) \oplus c_{k,b}]$

1.4. Recurrence Equations

$$\begin{aligned}
&= [c_{i,a} \odot (c_{j,a} \odot c_{k,a}),\ ((c_{i,b} \otimes c_{j,a}) \otimes c_{k,a}) \oplus ((c_{j,b} \otimes c_{k,a}) \oplus c_{k,b})] \\
&= [c_{i,a} \odot (c_{j,a} \odot c_{k,a}),\ (c_{i,b} \otimes (c_{j,a} \odot c_{k,a})) \oplus ((c_{j,b} \otimes c_{k,a}) \oplus c_{k,b})] \\
&= c_i \bullet [c_{j,a} \odot c_{k,a},\ (c_{j,b} \otimes c_{k,a}) \oplus c_{k,b}] \\
&= c_i \bullet (c_j \bullet c_k)
\end{aligned}$$

We now define the ordered set $s_i = [y_i, x_i]$, where the y_i obey the recurrence

$$y_i = \begin{cases} a_0 & i = 0 \\ y_{i-1} \odot a_i & 0 < i < n, \end{cases} \tag{1.8}$$

and the x_i are from (1.5). Using (1.5), (1.6) and (1.8) we obtain:

$$\begin{aligned}
s_0 &= [y_0, x_0] \\
&= [a_0, b_0] \\
&= c_0 \\
s_i &= [y_i, x_i] \qquad\qquad 0 < i < n \\
&= [y_{i-1} \odot a_i,\ (x_{i-1} \otimes a_i) \oplus b_i] \\
&= [y_{i-1} \odot c_{i,a},\ (x_{i-1} \otimes c_{i,a}) \oplus c_{i,b}] \\
&= [y_{i-1}, x_{i-1}] \bullet c_i \\
&= s_{i-1} \bullet c_i.
\end{aligned}$$

Since $\bullet$ is associative, we have reduced (1.5) to (1.4). The results x_i are just the second values of s_i (the $s_{i,b}$). This allows us to use the scan algorithm of Section 1.2 with operator $\bullet$ to solve any recurrence of the form (1.5) on an EREW PRAM in time:

$$(T_\odot + T_\otimes + T_\oplus) T_S(n, p) = 2(T_\odot + T_\otimes + T_\oplus)(n/p + \lg p) \tag{1.9}$$

where $T_\odot$, $T_\otimes$ and $T_\oplus$ are the times taken by $\odot$, $\otimes$ and $\oplus$ ($\bullet$ makes one call to each). If all that is needed is the final value x_{n-1}, then we can use a reduce instead of scan with the operator $\bullet$, and the running time is:

$$(T_\odot + T_\otimes + T_\oplus) T_R(n, p) = (T_\odot + T_\otimes + T_\oplus)(n/p + \lg p) \tag{1.10}$$

which is asymptotically a factor of 2 faster than (1.9).

Applications of first-order linear recurrences include the simulation of various time-varying linear systems, the backsubstitution phase of tridiagonal linear-systems solvers, and the evaluation of polynomials.

1.4.2 Higher Order Recurrences

We now consider the more general order m recurrences of the form:

$$x_i = \begin{cases} b_i & 0 \leq i < m \\ (x_{i-1} \otimes a_{i,1}) \oplus \cdots \oplus (x_{i-m} \otimes a_{i,m}) \oplus b_i & m \leq i < n \end{cases} \quad (1.11)$$

where $\oplus$ and $\otimes$ are binary operators with the same three restrictions as in (1.5): $\oplus$ is associative, $\otimes$ is semiassociative, and $\otimes$ distributes over $\oplus$.

To convert this equation into the form (1.5), we define the following vector of variables:

$$s_i = [\, x_i \;\; \cdots \;\; x_{i-m+1} \,]. \quad (1.12)$$

Using (1.11) we can write (1.12) as:

$$s_i = [\, x_{i-1} \;\; \cdots \;\; x_{i-m} \,] \otimes_{(v)} \begin{bmatrix} a_{i,1} & 1 & 0 & \cdots & 0 \\ \vdots & 0 & 1 & & \vdots \\ \vdots & \vdots & & \ddots & 0 \\ \vdots & 0 & \cdots & 0 & 1 \\ a_{i,m} & 0 & \cdots & 0 & 0 \end{bmatrix} \oplus_{(v)} [\, b_i \;\; 0 \;\; \cdots \;\; 0 \,]$$

$$= (s_{i-1} \otimes_{(v)} A_i) \oplus_{(v)} B_i \quad (1.13)$$

where $\otimes_{(v)}$ is vector-matrix multiply and $\oplus_{(v)}$ is vector addition. If we use matrix-matrix multiply as the companion operator of $\otimes_{(v)}$, then (1.13) is in the form (1.5). The time taken for solving equations of the form (1.11) on an EREW PRAM is therefore:

$$(T_{m \otimes m}(m) + T_{v \otimes m}(m) + T_{v \oplus v}(m)) T_S(n, p) = O((n/p + \lg p) T_{m \otimes m}(m)) \quad (1.14)$$

where $T_{m \otimes m}(m)$ is the time taken by an $m \otimes m$ matrix multiply. The sequential complexity for solving the equations is $O(nm)$, so the parallel complexity is optimal in n when $n/p \geq \lg p$, but is not optimal in m—the parallel algorithm performs a factor of $O(T_{M \otimes M}(m)/m)$ more work than the sequential algorithm.

Applications of the recurrence (1.11) include solving recurrences of the form $x_i = a_i + b_i / x_{i-1}$ (see Exercise 1.10), and generating the first n Fibonacci numbers $x_0 = x_1 = 1$, $x_i = x_{i-1} + x_{i-2}$ (see Exercise 1.11).

1.5
Segmented Scans

This section shows how the vector operated on by a scan can be broken into segments with flags so that the scan starts again at each segment boundary (see Figure 1.10). Each of these scans takes two vectors of values: a *data* vector and a *flag* vector. The *segmented scan* operations present a convenient way to execute a scan independently over many sets of values. The next section shows how the segmented scans can be used to execute a parallel quicksort, by keeping each recursive call in a separate segment, and using a segmented +-scan to execute a split within each segment.

The segmented scans satisfy the recurrence:

$$x_i = \begin{cases} a_0 & i = 0 \\ \begin{cases} a_i & f_i = 1 \\ (x_{i-1} \oplus a_i) & f_i = 0 \end{cases} & 0 < i < n \end{cases} \quad (1.15)$$

where $\oplus$ is the original associative scan operator. If $\oplus$ has an identity $I_\oplus$, then (1.15) can be written as:

$$x_i = \begin{cases} a_0 & i = 0 \\ (x_{i-1} \times_s f_i) \oplus a_i & 0 < i < n \end{cases} \quad (1.16)$$

where $\times_s$ is defined as:

$$x \times_s f = \begin{cases} I_\oplus & f = 1 \\ x & f = 0. \end{cases} \quad (1.17)$$

This is in the form (1.5) and $\times_s$ is semiassociative with logical **or** as the companion operator (see Exercise 1.9). Since we have reduced (1.15) to the

a	=	[5	1	3	4	3	9	2	6]
f	=	[1	0	1	0	0	0	1	0]
segmented +-scan	=	[5	6	3	7	10	19	2	8]
segmented max-scan	=	[5	5	3	4	4	9	2	6]

FIGURE 1.10
The segmented scan operations restart at the beginning of each segment. The vector f contains flags that mark the beginning of the segments.

form (1.5), we can use the technique described in Section 1.4.1 to execute the segmented scans in time

$$(T_{\text{or}} + T_{\times_s} + T_{\oplus})T_S(n,p) \ . \tag{1.18}$$

This time complexity is only a small constant factor greater than the unsegmented version since **or** and $\times_s$ are trivial operators.

1.5.1 Example: Quicksort

To illustrate the use of segmented scans, we consider a parallel version of quicksort. Similar to the standard sequential version, the parallel version picks one of the keys as a pivot value, splits the keys into three sets—keys less than, equal to and greater than the pivot—and recurses on each set.[3] The parallel algorithm has an expected time complexity of $O(T_S(n,p) \lg n) = O(\frac{n}{p} \lg n + \lg^2 n)$.

The basic intuition of the parallel version is to keep each subset in its own segment and to pick pivot values and split the keys independently within each segment. Figure 1.11 shows pseudocode for the parallel quicksort and gives an example. The steps of the sort are outlined as follows:

1. Check if the keys are sorted and exit the routine if they are.
   Each processor checks to see if the previous processor has a lesser or equal value. We execute a reduce with logical **and** to check if all the elements are in order.
2. Within each segment, pick a pivot and distribute it to the other elements.
   If we pick the first element as a pivot, we can use a segmented scan with the binary operator copy, which returns the first of its two arguments:

   $$a \leftarrow \text{copy}(a, b) \ .$$

   This has the effect of copying the first element of each segment across the segment. The algorithm could also pick a random element within each segment (see Exercise 1.15).
3. Within each segment, compare each element with the pivot and split based on the result of the comparison.
   For the split, we can use a version of the split operation described in Section 1.3, which splits into three sets instead of two, and which is

---
[3] We do not need to recursively sort the keys equal to the pivot, but the algorithm as described below does.

1.5. Segmented Scans

```
procedure quicksort(keys)
   seg-flags[0] ← 1
   while not-sorted(keys)
      pivots     ← seg-copy(keys, seg-flags)
      f          ← pivots <=> keys
      keys       ← seg-split(keys, f, seg-flags)
      seg-flags  ← new-seg-flags(keys, pivots, seg-flags)
```

Key	=	[6.4	9.2	3.4	1.6	8.7	4.1	9.2	3.4]
Seg-Flags	=	[1	0	0	0	0	0	0	0]
Pivots	=	[6.4	6.4	6.4	6.4	6.4	6.4	6.4	6.4]
F	=	[=	>	<	<	>	<	>	<]
Key ← split(Key, F)	=	[3.4	1.6	4.1	3.4	6.4	9.2	8.7	9.2]
Seg-Flags	=	[1	0	0	0	1	1	0	0]
Pivots	=	[3.4	3.4	3.4	3.4	6.4	9.2	9.2	9.2]
F	=	[=	<	>	=	=	=	<	=]
Key ← split(Key, F)	=	[1.6	3.4	3.4	4.1	6.4	8.7	9.2	9.2]
Seg-Flags	=	[1	1	0	1	1	1	1	0]

FIGURE 1.11
An example of parallel quicksort. On each step, within each segment, we distribute the pivot, test whether each element is equal to, less than or greater than the pivot, split into three groups, and generate a new set of segment flags. The operation <=> returns one of three values depending on whether the first argument is less than, equal to or greater than the second.

    segmented. To implement such a segmented split, we can use a segmented version of the +-scan operation to generate indices relative to the beginning of each segment, and we can use a segmented copy-scan to copy the offset of the beginning of each segment across the segment. We then add the offset to the segment indices to generate the location to which we permute each element.

4. Within each segment, insert additional segment flags to separate the split values.
   Knowing the pivot value, each element can determine if it is at the beginning of the segment by looking at the previous element.
5. Return to step 1.

54   Chapter 1. Prefix Sums and Their Applications

Each iteration of this sort requires a constant number of calls to the scans and to the primitives of the PRAM. If we select pivots randomly within each segment, quicksort is expected to complete in $O(\lg n)$ iterations, and therefore has an expected running time of $O(\lg n \cdot T_S(n,p))$.

The technique of recursively breaking segments into subsegments and operating independently within each segment can be used for many other divide-and-conquer algorithms, such as mergesort.

1.6
Allocating Processors

Consider the following problem: given a set of processors, each containing an integer, allocate that integer number of new processors to each initial processor. Such allocation is necessary in the parallel line-drawing routine described in Section 1.6.1. In this line-drawing routine, each processor calculates the number of pixels in the line and dynamically allocates a processor for each pixel. Allocating new elements is also useful for the branching part of many branch-and-bound algorithms. Consider, for example, a brute force chess-playing algorithm that executes a fixed-depth search of possible moves to determine the best next move. We can test or search the moves in parallel by placing each possible move in a separate processor. Since the algorithm dynamically decides how many next moves to generate (depending on the position), we need to dynamically allocate new processing elements.

More formally, given a length l vector A with integer elements a_i, allocation is the task of creating a new vector B of length

$$L = \sum_{i=0}^{l-1} a_i \qquad (1.19)$$

with a_i elements of B assigned to each position i of A. By "assigned to," we mean that there must be some method for distributing a value at position i of a vector to the a_i elements that are assigned to that position. Since there is a one-to-one correspondence between elements of a vector and processors, the original vector requires l processors and the new vector requires L processors. Typically, an algorithm does not operate on the two vectors at the same time, so that we can use the same processors for both.

Allocation can be implemented by assigning a contiguous segment of elements to each position i of A. To allocate segments we execute a +-prescan

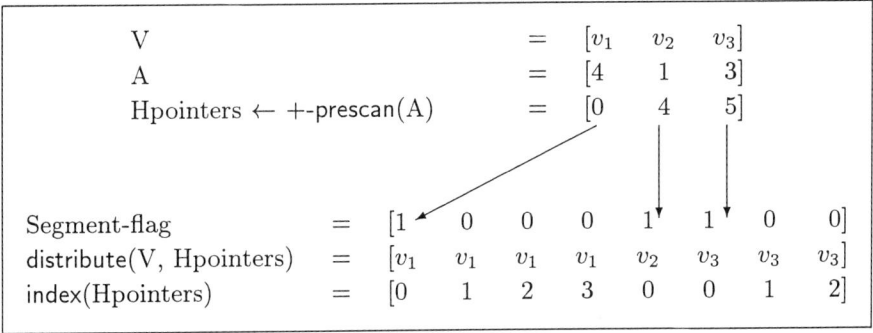

FIGURE 1.12
An example of processor allocation. The vector A specifies how many new elements each position needs. We can allocate a segment to each position by applying a +-prescan to A and using the result as pointers to the beginning of each segment. We can then distribute the values of V to the new elements with a permute to the beginning of the segment and a segmented copy-scan across the segment.

on the vector A that returns a pointer to the start of each segment (see Figure 1.12). We can then generate the appropriate segment flags by writing a flag to the index specified by the pointer. To distribute values from each position i to its segment, we write the values to the beginning of the segments and use a segmented copy-scan operation to copy the values across the segment. Allocation and distribution each require one call to a scan and therefore have complexity $T_S(l, p)$ and $T_S(L, p)$, respectively.

Once a segment has been allocated for each initial element, it is often necessary to generate indices within each segment. We call this the index operation, and it can be implemented with a segmented +-prescan.

1.6.1 Example: Line Drawing

As an example of how allocation is used, consider line drawing. The line-drawing problem is: given a set of pairs of points

$$\langle (x_0, y_0) : (\hat{x}_0, \hat{y}_0) \rangle, \ldots, \langle (x_{n-1}, y_{n-1}) : (\hat{x}_{n-1}, \hat{y}_{n-1}) \rangle ,$$

generate all the locations of pixels that lie between one of the pairs of points. Figure 1.13 illustrates an example. The routine we discuss returns a vector of (x, y) pairs that specify the position of each pixel along every line. If a pixel appears in more than one line, it will appear more than once in the vector.

56 Chapter 1. Prefix Sums and Their Applications

```
procedure line-draw(x, y)
   in parallel for each line i
      % determine the length of the line
      length[i] ← max(|p₂[i].x - p₁[i].x|, |p₂[i].y - p₁[i].y|)

      % determine the x and y increments
      Δ[i].x ← (p₂[i].x - p₁[i].x) / length[i]
      Δ[i].y ← (p₂[i].y - p₁[i].y) / length[i]

   % distribute values and generate index
   p'₁ ← distribute(p₁, lengths)
   Δ' ← distribute(Δ, lengths)
   index ← index(lengths)

   in parallel for each pixel j
      % determine the final position
      result[j].x ← p'₁[j].x + round(index[j] × Δ'[j].x)
      result[j].y ← p'₁[j].y + round(index[j] × Δ'[j].y)
```

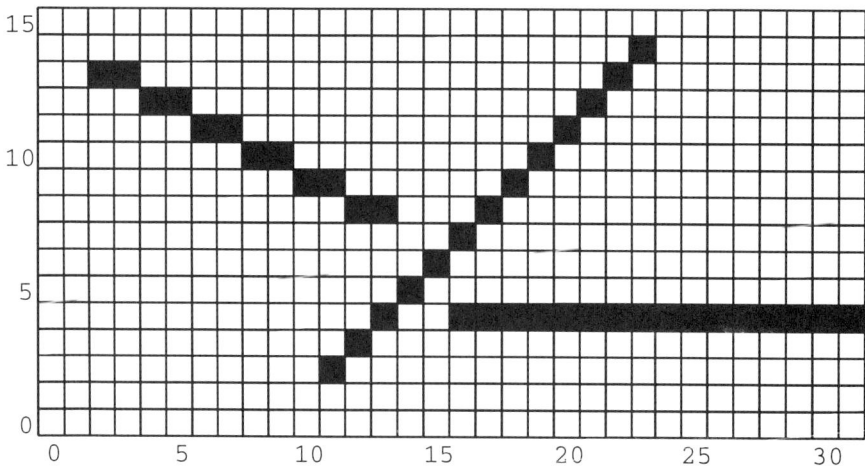

FIGURE 1.13
The pixels generated by a line-drawing routine. In this example the endpoints are $\langle(11,2):(23,14)\rangle$, $\langle(2,13):(13,8)\rangle$, and $\langle(16,4):(31,4)\rangle$. The algorithm allocates 12, 11 and 16 pixels respectively for the three lines.

The routine generates the same set of pixels as generated by the simple digital differential analyzer sequential technique.

The basic idea of the routine is for each line to allocate a processor for each pixel in the line, and then for each allocated pixel to determine, in parallel, its final position in the grid. Figure 1.13 shows the code. To allocate a processor for each pixel, each line must first determine the number of pixels in the line. This number can be calculated by taking the maximum of the x and y differences of the line's endpoints. Each line now allocates a segment of processors for its pixels and distributes one endpoint along with the per-pixel x and y increments across the segment. We now have one processor for each pixel and one segment for each line. We can view the position of a processor in its segment as the position of a pixel in its line. Based on the endpoint, the slope and the position in the line (determined with an `index` operation), each pixel can determine its final (x, y) location in the grid.

This routine has the same complexity as a scan $T_S(m, p)$, where m is the total number of pixels. To actually place the points on a grid, rather than just generating their position, we would need to permute a flag to a position based on the location of the point. In general, this will require the simplest form of concurrent-write (one of the values gets written), since a pixel might appear in more than one line.

1.7
Exercises

1.1 Modify the algorithm in Figure 1.4 to execute a scan instead of a prescan.

1.2 Use the scan operation to compare two strings of length n in $O(n/p + \lg p)$ time on an EREW PRAM.

1.3 Given two vectors of bits that represent nonnegative integers, show how a prescan can be used to add the two numbers (return a vector of bits that represents the sum of the two numbers).

1.4 Trace the steps of the split-radix sort on the vector

[2   11   4   5   9   6   15   3].

1.5 Show that subtraction is semiassociative and find its companion operator.

1.6 Write a recurrence equation of the form (1.5) that evaluates a polynomial

$$y = b_1 x^{n-1} + b_2 x^{n-2} + \cdots + b_{n-1} x + b_n$$

for a given x.

58   Chapter 1. Prefix Sums and Their Applications

1.7 Show that if ⊗ has an inverse, the recurrence of the form (1.5) can be solved with some local operations (not involving communication among processors) and two scan operations (using ⊗ and ⊕ as the operators).

1.8 Prove that vector-matrix multiply is semiassociative.

1.9 Prove that the operator $\times_s$ defined in (1.17) is semiassociative.

1.10 Show how the recurrence $x(i) = a(i) + b(i)/x(i-1)$, where + is numeric addition and / is division, can be converted into the form (1.11) with two terms (m = 2).

1.11 Use a scan to generate the first n Fibonacci numbers.

1.12 Show how to solve a tridiagonal linear-system using the recurrences in Section 1.4. Is the algorithm asymptotically optimal?

1.13 In the language Common Lisp, the % character means that what follows the character up to the end of the line is a comment. Use the scan operation to mark all the comment characters (everything between a % and an end-of-line).

1.14 Trace the steps of the parallel quicksort on the vector

[27   11   51   5   49   36   15   23].

1.15 Describe how quicksort is changed so that it selects a random element within each segment for a pivot.

1.16 Design an algorithm that, given the radius and number of sides on a regular polygon, determines all the pixels that outline the polygon.

Notes and References

The all-prefix-sums operation has been around for centuries as the recurrence $x_i = a_i + x_{i-1}$. A parallel circuit to execute the scan operation was first suggested by Ofman [14] for the addition of binary numbers. A parallel implementation of scans on a perfect shuffle network was later suggested by Stone [16] for polynomial evaluation. The optimal algorithm discussed in Section 1.2 is a slight variation of algorithms suggested by Kogge and Stone [9] and by Stone [18] in the context of recurrence equations.

Ladner and Fischer [10] first showed an efficient general-purpose circuit for implementing the scan operation. Brent and Kung [4], in the context of binary addition, first showed an efficient VLSI layout for a scan circuit. More recent work on implementing scan operations in parallel include the work of Fich [6] and of Lakshmivarahan, Yang and Dhall [11], which give improvements over the circuit of Ladner and Fischer, and of Lubachevsky and Greenberg [12], which demonstrates the implementation of the scan operation on asynchronous machines. Blelloch [1] suggested that certain scan operations be included in the PRAM model as primitives and shows how this affects the complexity of various algorithms. Work on the linked-list-based all-prefix-sums operation is considered and referenced in Chapters 3, 2 and 5.

The line-of-sight and radix-sort algorithms are discussed by Blelloch [1] and Blelloch and Little [3]. The parallel solution of recurrence problems was first discussed by Karp, Miller and Winograd [8], and parallel algorithms to solve them are given by Kogge and Stone [9], Stone [17, 18] and Chen and Kuck [5]. Hyafil and Kung [7] show that the complexity (1.10) is a lower bound.

Schwartz [15] and, independently, Mago [13] first suggested the segmented versions of the scans. Blelloch [2] suggested many uses of these scans including the quicksort algorithm and the line-drawing algorithm presented in Sections 1.5.1 and 1.6.1.

I would like to thank Siddhartha Chatterjee, Jonathan Hardwick and Jay Sipelstein for reading over drafts of this chapter.

Bibliography

[1] Blelloch, G.E., Scans as Primitive Parallel Operations. *IEEE Transactions on Computers*, C-38(11):1526-1538, November 1989.

[2] Blelloch, G.E., *Vector Models for Data-Parallel Computing*. MIT Press, Cambridge, MA, 1990.

[3] Blelloch, G.E., and Little, J.J., Parallel Solutions to Geometric Problems on the Scan Model of Computation. In *Proceedings International Conference on Parallel Processing*, Vol 3: 218-222, August 1988.

[4] Brent, R.P., and Kung, H.T., The Chip Complexity of Binary Arithmetic. In *Proceedings ACM Symposium on Theory of Computing*, pages 190-200, 1980.

[5] Chen, S., and Kuck, D.J., Time and Parallel Processor Bounds for Linear Recurrence Systems. *IEEE Transactions on Computers*, C-24(7), July 1975.

[6] Fich, F.E., New Bounds for Parallel Prefix Circuits. In *Proceedings ACM Symposium on Theory of Computing*, pages 100-109, April 1983.

[7] Hyafil, L., and Kung, H.T., The Complexity of Parallel Evaluation of Linear Recurrences. *Journal of the Association for Computing Machinery*, 24(3):513–521, July 1977.

[8] Karp, R.H., Miller, R.E., and Winograd S., The Organization of Computations for Uniform Recurrence Equations. *Journal of the Association for Computing Machinery*, 14:563–590, 1967.

[9] Kogge, P.M., and Stone, H.S., A Parallel Algorithm for the Efficient Solution of a General Class of Recurrence Equations. *IEEE Transactions on Computers*, C-22(8):786–793, August 1973.

[10] Ladner, R.E., and Fischer, M.J., Parallel Prefix Computation. *Journal of the Association for Computing Machinery*, 27(4):831–838, October 1980.

[11] Lakshmivarahan, S., Yang, C.M., and Dhall, S.K., Optimal Parallel Prefix Circuits with (size + depth) = $2n - n$ and $\lceil \log n \rceil \leq depth \leq \lceil 2 \log n \rceil - 3$. In *Proceedings International Conference on Parallel Processing*, pages 58–65, August 1987.

[12] Lubachevsky, B.D., and Greenberg, A.G., Simple, Efficient Asynchronous Parallel Prefix Algorithms. In *Proceedings International Conference on Parallel Processing*, pages 66–69, August 1987.

[13] Mago, G.A., A network of computers to execute reduction languages. *International Journal of Computer and Information Sciences*, 1979.

[14] Ofman, Y., On the Algorithmic Complexity of Discrete Functions. *Soviet Physics Doklady*, 7(7):589–591, January 1963.

[15] Schwartz, J.T., Ultracomputers. *ACM Transactions on Programming Languages and Systems*, 2(4):484–521, October 1980.

[16] Stone, H.S., Parallel Processing with the Perfect Shuffle. *IEEE Transactions on Computers*, C-20(2):153–161, 1971.

[17] Stone, H.S., An Efficient Parallel Algorithm for the Solution of a Tridiagonal Linear System of Equations. *Journal of the Association for Computing Machinery*, 20(1):27–38, January 1973.

[18] Stone, H.S., Parallel Tridiagonal Equation Solvers. *ACM Transactions on Mathematical Software*, 1(4):289–307, December 1975.

2

Introduction to Parallel Connectivity, List Ranking, and Euler Tour Techniques

Sara Baase

Computer Science Division
Mathematical Sciences Department
San Diego State University
San Diego, CA 92182-0314
baase@cs.sdsu.edu

2.1 Introduction

This chapter presents algorithms for several related problems. The first is the problem of determining the connected components of a graph, one of the most basic graph problems. A parallel algorithm is said to be *optimal* if the product of the number of steps and number of processors is of the same order as the best sequential algorithm. The connectivity algorithm described in this chapter takes $O(\log n)$ time but is not optimal. It introduces the general strategy and some specific techniques (hooking and shortcutting) that are used in the far more complicated optimal algorithms explored in Chapter 5. The same strategy and techniques are used in the randomized connectivity algorithm in Chapter 4.

List ranking is a variation of the prefix sum problem presented in Chapter 1. The shortcutting technique used in connectivity algorithms also appears in the list ranking algorithms. The main list ranking algorithm presented here, while optimal, is not in $O(\log n)$. Techniques introduced here (deterministic coin tossing and 2-ruling sets) are used again in Chapter 5 where an optimal $O(\log n)$ algorithm is described.

List ranking is itself a technique, or subproblem, that appears as part of other algorithms, including Euler tour techniques. Euler tour techniques, introduced in Section 2.4, are used to solve a variety of problems on trees and graphs.

2.2 Finding Connected Components of a Graph

2.2.1 The Problem and Sequential Solutions

Let $G = (V, E)$ be an undirected graph. A *path* in G is a sequence of distinct vertices $v_0, v_1, \ldots, v_k$ in V, for some $k \geq 0$, where for $0 \leq i \leq k-1$, $v_i v_{i+1} \in E$. A graph is *connected* if for every pair of vertices v and w in V, there is a path between v and w. A *connected component* of a graph is a maximal connected subgraph. The graph in Figure 2.1(a) is not connected; it has three connected components.

Let $n = |V|$ and $m = |E|$. We will describe a parallel algorithm to find the connected components of a graph in $O(\log n)$ time using $\max(n, m)$ processors.

This section is adapted with permission from Addison-Wesley Publishing Company from Section 10.4 of *Computer Algorithms: Introduction to Design and Analysis* by Sara Baase, copyright 1988 Addison-Wesley Publishing Company.

2.2. Finding Connected Components of a Graph 63

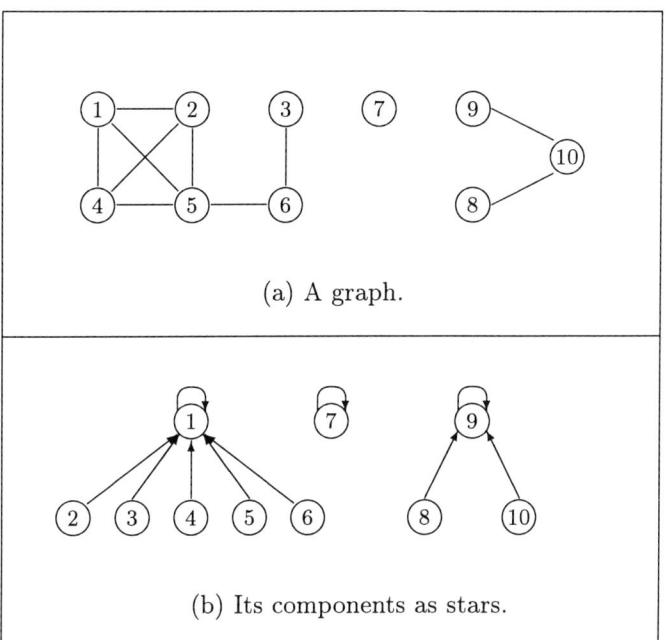

FIGURE 2.1
Connected components turned into stars.

Typical sequential algorithms for finding connected components use depth-first search or breadth-first search. Beginning at an arbitrary vertex, each vertex encountered in the search is marked. The set of vertices encountered in one search is exactly the set of vertices in one connected component. A new search is begun at an unmarked vertex, using a new marker value; this process is repeated until all vertices are marked. Both depth-first search and breadth-first search can be easily implemented to explore every edge and mark every vertex in $\Theta(n+m)$ time.

2.2.2 Strategy and Techniques for Parallel Algorithms

The search techniques used in sequential algorithms for finding connected components are not easy to implement efficiently in parallel. A different approach is used for parallel algorithms.

The general outline and techniques in the algorithm presented in this chapter are common to a series of parallel algorithms, including the one presented in Chapter 5 and the randomized algorithm presented in Chapter 4. Throughout the algorithm, the vertices of the graph are partitioned into rooted trees in which each vertex has a pointer to its parent. (The root points to itself.) Vertices in one tree are all in the same connected component of G. The set of vertices in a tree is called a *supervertex*. Initially, each vertex is in a separate tree. The algorithms repeatedly combine trees whose vertices are in the same connected component, and shorten the trees. Ultimately, each connected component is converted to one tree, or supervertex, of depth at most one, with all the vertices pointing to the root. Such a tree is called a *star*. See Figure 2.1 for an illustration. Once the connected components have been converted to stars, we can determine if two vertices are in the same component in constant time by comparing their pointers (parents).

The algorithm presented here will have write conflicts. In the Arbitrary-Write variant of the CRCW PRAM model, when several processors try to write in the same memory cell at the same time, an arbitrary one of them succeeds. An algorithm for this model must work correctly no matter which processor "wins" the write conflict. The connected component algorithm described here will work on the Arbitrary-Write model.

We begin by explaining the two basic techniques used repeatedly by the algorithm: *shortcutting* and *hooking*. The trees constructed by the algorithm are represented by an array *parent*, such that $parent(v)$ is the parent of vertex v. (The "parent" of a root is the root itself.)

Shortcutting shortens trees. It is also useful in other parallel algorithms (including the list ranking algorithms described later in this chapter). Shortcutting simply changes the parent of each vertex to make the vertex point directly to its grandparent. That is, the shortcutting operation on vertex v is:

$$parent(v) := parent(parent(v)).$$

Shortcutting is applied in parallel to all vertices. To see the speed with which this operation can cut down long paths, consider a simple chain of vertices as in Figure 2.2(a), where $parent(v) = v - 1$ (and $parent(1) = 1$). Figure 2.2(b) shows the parent pointers after the first and second applications of the shortcutting operation. If we start with n vertices in the chain, after $\lceil \log n \rceil$ applications of shortcutting, all vertices have the same parent.

Shortcutting is sometimes called *doubling* or *pointer jumping*.

2.2. Finding Connected Components of a Graph

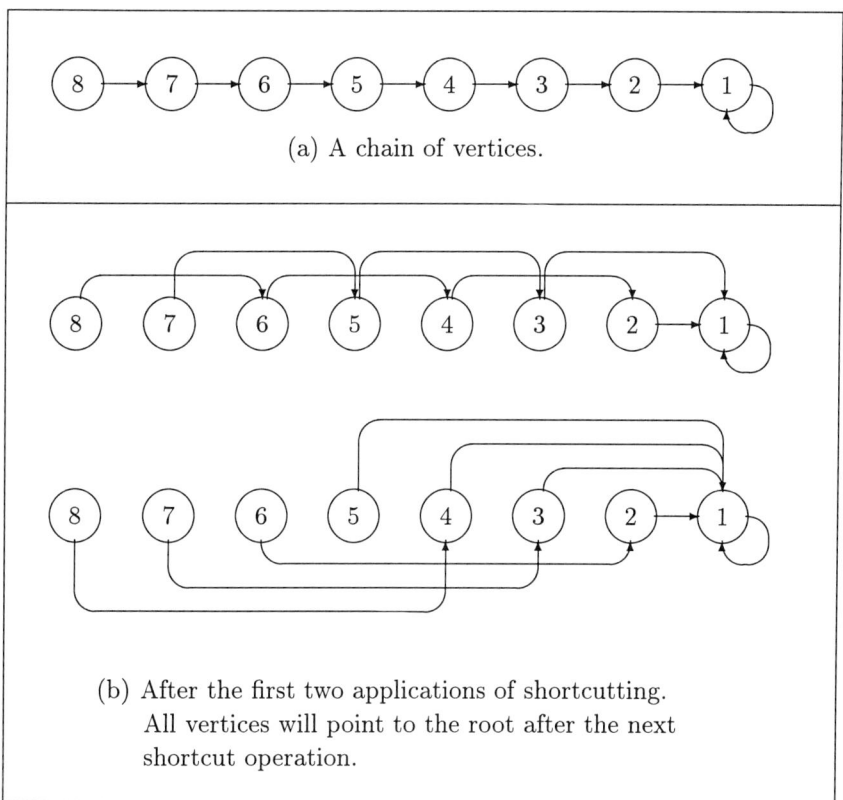

(a) A chain of vertices.

(b) After the first two applications of shortcutting. All vertices will point to the root after the next shortcut operation.

FIGURE 2.2
The effect of shortcutting on a chain of vertices.

Shortcutting never joins two separate trees. We need the hooking operation to connect trees. $Hook(i, j)$ means attach i's root to the parent of j. $Hook(i, j)$ is applied only if i is a root or the child of a root; thus the operation is:

$$parent(parent(i)) := parent(j).$$

There are two versions of hooking in the algorithm. The first is:

Conditional star hooking: If i is in a star, j is adjacent to i (in G), and $parent(i) > parent(j)$, then $Hook(i, j)$.

The requirement that we hook to the smaller of the two parents helps avoid the

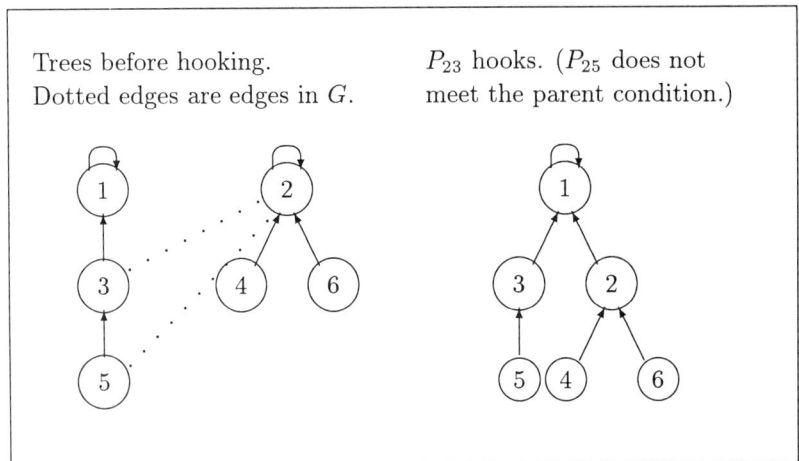

FIGURE 2.3
Illustration of conditional star hooking.

introduction of cycles. Conditional star hooking is illustrated in Figure 2.3. (In Figures 2.3–2.6, P_{ij} denotes the processor that attempts the $Hook(i,j)$ operation.)

The second hooking operation is:

> *Unconditional star hooking*: If i is in a star, j is adjacent to i (in G), and j is not in i's star, then $Hook(i,j)$.

Unconditional star hooking is illustrated in Figure 2.4.

At any one time, for a particular vertex i, there may be several vertices j that satisfy the conditions for hooking, but only one value can be stored as the new parent of i's root. In the parallel algorithm, different processors will be trying the different choices for j, and several processors may try to write in $parent(parent(i))$ at the same time. Only one succeeds in writing, but the algorithm will work properly no matter which one succeeds. By running long enough, the algorithm eventually hooks together all trees that are part of one connected component.

Notice that two trees are hooked only if there is an edge of G whose incident vertices are in the two trees. Thus a supervertex is always a subset of a connected component.

2.2. Finding Connected Components of a Graph

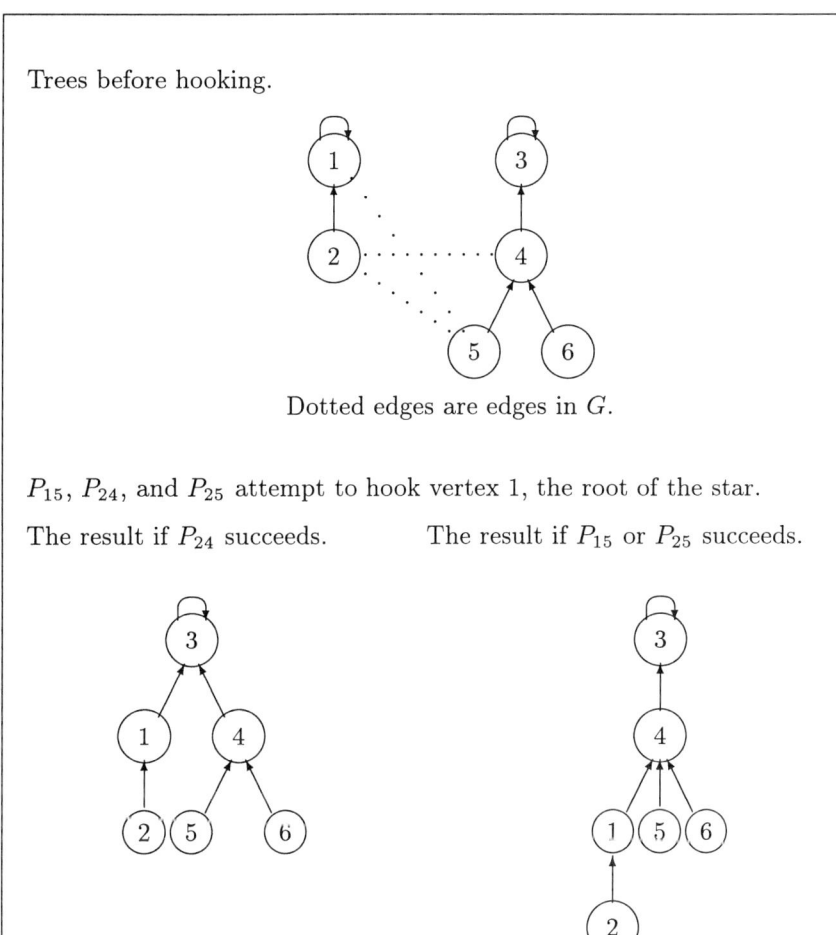

FIGURE 2.4
Illustration of unconditional star hooking.

2.2.3 The Algorithm

We first give a high-level description of the algorithm. Later we will show how the various parts can be implemented on a PRAM. The basic idea of the algorithm, presented next as Algorithm 2.1, almost works, but requires a modification which will be explained afterward.

ALGORITHM 2.1
Finding connected components of a graph.
Input: An undirected graph G.
Output: A forest of directed trees of depth at most one, represented by the array *parent*, indexed by the vertices. Each tree contains the vertices of one connected component.
Comment: An instruction specified for a vertex v is performed in parallel for all vertices. The hooking steps are performed in parallel for all edges (i,j) in G (and only for pairs i and j such that (i,j) is an edge). Each edge, say (x,y), is processed twice, once with x in the role of i, and once with y in the role of i.

(Initialization)
$parent(v) := v$;

repeat

(Conditional star hooking)
if i is in a star **and** $parent(i) > parent(j)$ **then** $Hook(i,j)$
endif

(Unconditional star hooking)
if i is in a star **and** $parent(i) \neq parent(j)$ (j is not in i's star)
then $Hook(i,j)$
endif

(Shortcutting)
if v is not in a star **then** $parent(v) := parent(parent(v))$
endif

until the shortcutting step did not produce any changes.

One of the facts used in the proof that the algorithm works correctly is that conditional and unconditional star hooking do not produce new stars. But unfortunately, on the first pass through the loop, they may do so. Single vertices may form a star when they are hooked together. Then, the unconditional star hooking step might hook two stars to each other in both directions, thus creating a cycle. See Figure 2.5 for an illustration. The problem is eliminated by modifying the algorithm so that on the first pass only singletons

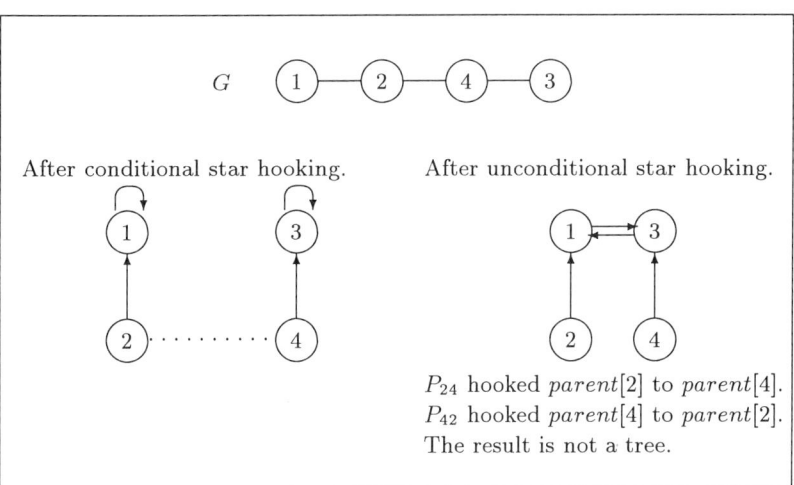

FIGURE 2.5
Introduction of a cycle.

(trees with only one vertex) are hooked. Here is the revision. The loop in Algorithm 2.1 is unchanged.

ALGORITHM 2.2
Finding connected components of a graph (revised initialization).
Comment: Recall that hooking is performed *only* for pairs i and j such that (i, j) is an edge in G.

(Initialization)
$parent(v) := v$;
(Conditional singleton hooking) **if** $i > j$ **then** $Hook(i, j)$;
(Unconditional singleton hooking) **if** i is a singleton **then** $Hook(i, j)$;

Figure 2.6 illustrates the action of the algorithm. The correctness of the algorithm is based on two theorems which in turn are proved in a series of lemmas. The algorithm itself is not very hard to understand if a few examples are worked through, so the reader should examine Figure 2.6 carefully before proceeding. (Note how the modification in Algorithm 2.2 protects the trees rooted at 5 and 7 from the problem illustrated in Figure 2.5.)

70   Chapter 2.   Parallel Connectivity, List Ranking, Euler Tour Techniques

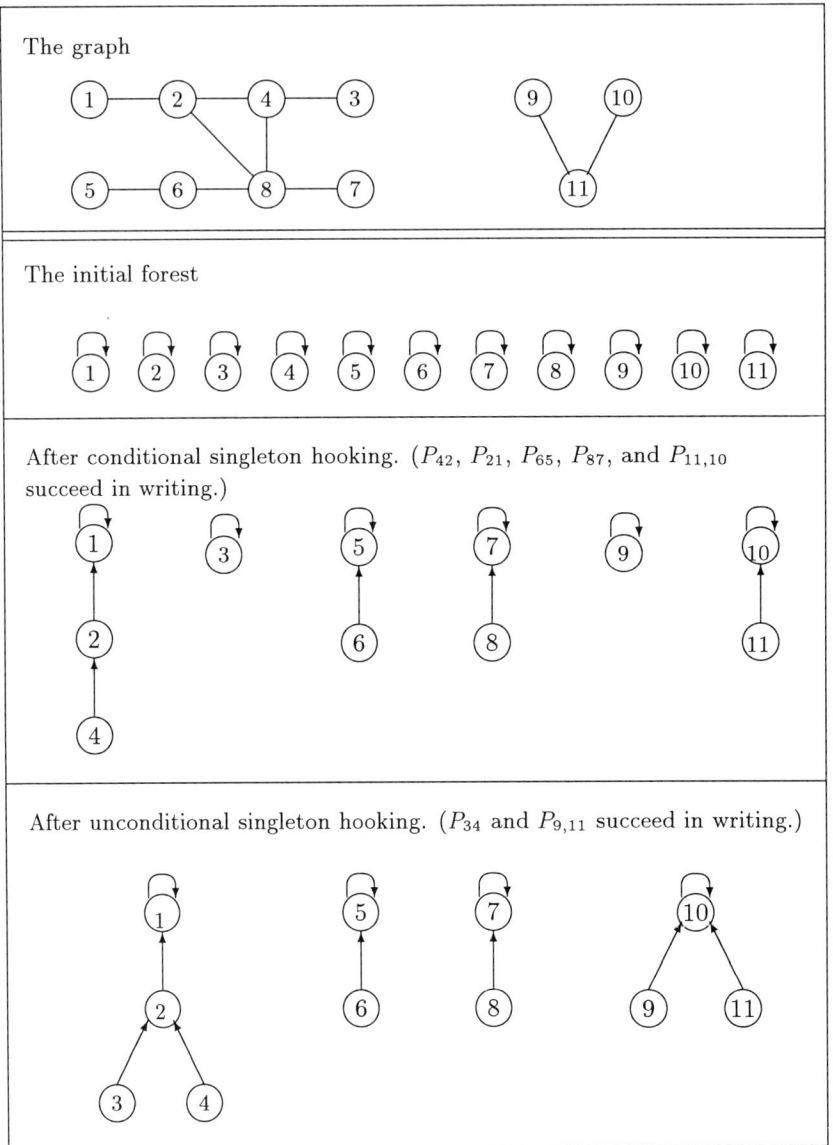

FIGURE 2.6
Illustration of the connected component algorithm.

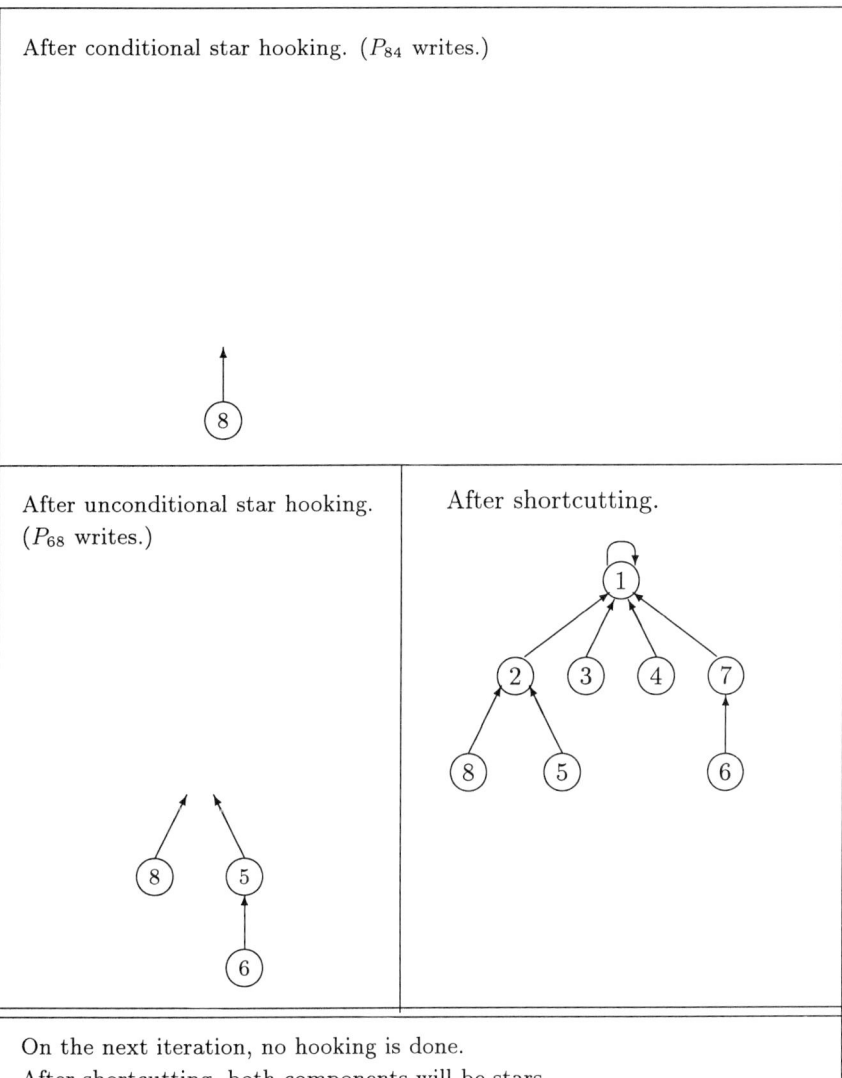

FIGURE 2.6 (continued)
Illustration of the connected component algorithm.

EXERCISE 2.1
In the example in Figure 2.6, when more than one processor tried to write in one memory cell at the same time, we made an arbitrary choice as to which one succeeded. Redo the example making a different valid choice at each step at which there was a write conflict.

THEOREM 2.1
At any time during execution of Algorithm 2.1, the structure defined by the parent pointers is a forest.

THEOREM 2.2
When Algorithm 2.1 terminates, the forest defined by the parent pointers consists only of stars, and the vertices in each star are exactly the vertices of a connected component of G.

The proofs of the theorems use the following lemmas.

LEMMA 2.1
After the initialization, the structure defined by the parent pointers is a forest. All trees have at least two vertices, except for trees consisting of one vertex that is isolated in G (i.e., is a connected component of G).

EXERCISE 2.2
Prove Lemma 2.1.

LEMMA 2.2
Conditional and unconditional star hooking never create new stars.

PROOF
Singletons existing after the initialization will never be hooked to anything else. When the root of a tree with at least two vertices is attached to another tree, the new tree will have depth at least two. ∎

LEMMA 2.3
The unconditional star hooking step never hooks a star onto another star.

PROOF
Suppose it does. Then, at the beginning of the unconditional star hooking step there were two stars, S_1 and S_2, containing vertices i and j, respectively, such that (i,j) is an edge in G. Since conditional star hooking does not create stars (Lemma 2.2), S_1 and S_2 were stars at the beginning of the conditional star hooking step. They satisfy the conditions for conditional star hooking, so the one with the larger root would have been hooked in the conditional star hooking step, and it would no longer be a star. ∎

EXERCISE 2.3

Suppose that in the proof of Lemma 2.3, S_1's root is larger than S_2's root. Would S_1 necessarily have been hooked to S_2? Why?

PROOF OF THEOREM 2.1

The loop starts with trees (Lemma 2.1); we have to show that no step in the loop introduces a cycle. It is clear that shortcutting cannot introduce a cycle. In the hooking steps, if a star is hooked to a non-star, no cycle is introduced because non-stars are not hooked to anything else. Since unconditional star hooking always hooks a star to a non-star (Lemma 2.3), it cannot introduce a cycle. A cycle formed in conditional star hooking must consist entirely of roots of stars. But conditional star hooking hooks a root only to a smaller root, so no cycle can be formed. ∎

LEMMA 2.4

Any star that exists at the end of the unconditional star hooking step must be an entire connected component.

PROOF

By Lemma 2.2, the star was a star at the beginning of the unconditional star hooking step. But if any vertex in the star were adjacent (in G) to a vertex in any other tree, the unconditional star hooking step would have hooked the star to another tree, and it would no longer be a star. ∎

PROOF OF THEOREM 2.2

Since the vertices of G start out in disjoint trees, and two trees are hooked only if they contain vertices i and j that are adjacent, all the vertices in any one tree at any time are in the same connected component of G. The algorithm stops when shortcutting produces no changes. This can happen only when there are no vertices of distance 2 from their roots; i.e., when all vertices are in stars at the end of the star hooking step. By Lemma 2.4, each such star is an entire component. ∎

2.2.4 PRAM Implementation of the Algorithm

Now we consider how the instructions in the algorithm are assigned to processors, and how many PRAM steps are needed to carry out each instruction.

Some processors have two "names." When we perform an operation for each vertex (say, shortcutting), we will use the first n processors, referring to them as P_v ($1 \leq v \leq n$). Because edges are processed in each "direction," it is convenient, for a while, to assume that there are at least $2m$ processors

74   Chapter 2.   Parallel Connectivity, List Ranking, Euler Tour Techniques

(though only m will be needed). When we perform an operation for each edge, we use the first $2m$ processors, referring to them by the name P_{ij}. Since operations on vertices and operations on edges are done in different instructions, each processor will be doing only one thing at a time.

The PRAM algorithm assumes that the input is in the form of a list of edges in the graph G. Each edge appears twice in the list; that is, if $\{i,j\}$ is an edge, the pairs (i,j) and (j,i) are in the input list. Each processor reads a (distinct) memory cell containing an edge, and from then on considers itself to be the processor for that particular (oriented) edge; i.e., it "knows" that its alternative name is P_{ij} if it read (i,j).

The form of the input is not critical to the speed of the algorithm. If the input were provided as an adjacency matrix, we would have n^2 processors read the matrix entries in the first step. Those that read a zero would do no more work for edges. Other variations of the input format are also acceptable.

We present the algorithm again with more implementation detail. The important observation to make here is that each step of the algorithm can be implemented in a constant number of PRAM steps. For simplicity, we denote the operation "write $parent(j)$ in $parent(parent(i))$" by $Hook(i,j)$.

ALGORITHM 2.3
Finding the connected components of a graph.
Input: A list of edges in the graph, each edge listed in both orientations; and n, the number of vertices.
Output: A forest of directed trees of depth at most one, represented by the array *parent*, indexed by the vertices. Each tree contains the vertices of one connected component.
Comment: A Boolean array *star* is used to tell if a vertex is in a star; $star(v)$ is true if and only if v is in a star. The instructions for computing *star* are shown after the main algorithm. The shared Boolean variable *noChange* tells whether or not the shortcutting step has made any changes at each iteration of the loop.

(Initialization)
each edge processor reads an edge;
P_v (for $1 \leq v \leq n$) writes v in $parent(v)$;
(Conditional singleton hooking)
P_{ij} does: **if** $i > j$ **then** $Hook(i,j)$;
(Unconditional singleton hooking)
P_{ij} does: **if** i is a singleton **then** $Hook(i,j)$;

2.2. Finding Connected Components of a Graph

    repeat
        (Conditional star hooking)
        compute $star(v)$ for $1 \leq v \leq n$ (as described later);
        P_{ij} reads $parent(i)$, $parent(j)$, and $star(i)$;
        if $star(i)$ **and** $parent(i) > parent(j)$ **then** $Hook(i,j)$
        endif;
        (Unconditional star hooking)
        compute $star(v)$ for $1 \leq v \leq n$ (as described later);
        P_{ij} reads $parent(i)$, $parent(j)$, and $star(i)$;
        if $star(i)$ **and** $parent(i) \neq parent(j)$ (i.e., j is not in i's star)
        then $Hook(i,j)$
        endif;
        (Shortcutting)
        P_v writes *true* in *noChange*;
        P_v reads $parent(v)$ and $parent(parent(v))$;
        if $parent(parent(v)) \neq parent(v)$ **then**
            P_v writes $parent(parent(v))$ in $parent(v)$;
            P_v writes *false* in *noChange*
        endif
    until *noChange*   (All processors read *noChange* to determine whether to stop.)

Observe that on every iteration of the loop, a processor P_{ij} tests to determine whether it should hook. Sometimes it may try to hook but fail because another processor succeeds in writing in $parent(parent(i))$. P_{ij} actually succeeds in hooking at most one time during the entire course of the algorithm. This observation suggests that it may be possible to speed up the algorithm by organizing the work of the processors in a more efficient way. In any case, as we shall see, this algorithm runs in $O(\log n)$ time.

Computing *star*

A vertex is not in a star if and only if one of the following conditions holds:

1. its parent is not its grandparent,
2. it is the grandparent, but not the parent, of some other vertex, or
3. its parent has a non-trivial grandchild.

Figure 2.7 illustrates all three cases and the computation of *star*. The computation is described in the following algorithm which clearly takes constant time.

ALGORITHM 2.4
Determining if a vertex is in a star.
Comment: These steps are carried out by P_v (for $1 \leq v \leq n$).

>write *true* in $star(v)$;
>read $parent(v)$ and $parent(parent(v))$;
>**if** $parent(v) \neq parent(parent(v))$ **then**
>>write *false* in $star(v)$;
>>write *false* in $star(parent(parent(v)))$
>
>**endif**;
>read $star(parent(v))$;
>**if not** $star(parent(v))$ **then**
>>write *false* in $star(v)$
>
>**endif**

EXERCISE 2.4
Give a method for determining if a vertex is a singleton (in constant time), for the initialization steps.

2.2.5 Analysis

Each of the steps in Algorithm 2.3's main loop can be carried out in constant time on an Arbitrary-Write PRAM, so the number of iterations of the loop determines the order of the running time. All that remains is to show that the number of iterations is in $O(\log n)$.

LEMMA 2.5
Let d be the depth of a non-star tree before a shortcutting step. After shortcutting, its depth is at most $2d/3$.

PROOF
Actually, the depth is cut roughly in half after each shortcutting step. It is easy to show by induction that if d is even, the depth after shortcutting will be exactly $d/2$, and that if d is odd, the depth after shortcutting will be $(d+1)/2$. The worst case is when $d = 3$. The new depth will be 2, which is $2d/3$. ∎

For any connected component C, let $d_C(t)$ be the sum of the depths of all the trees in C at the end of the tth iteration of the loop (for $t \geq 0$).

2.2. Finding Connected Components of a Graph

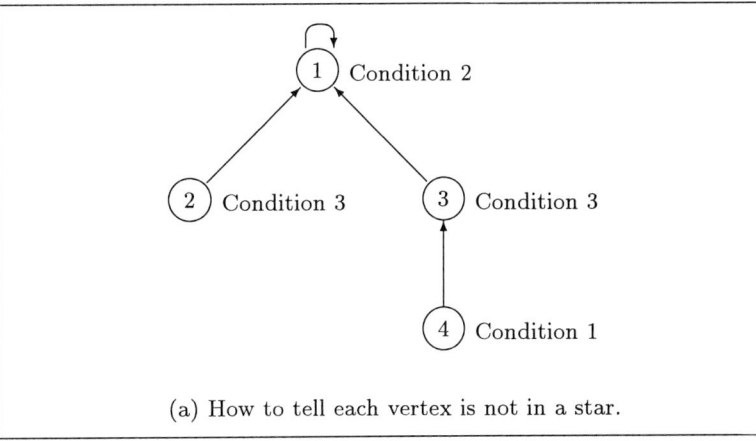

(a) How to tell each vertex is not in a star.

Initial values of *star*.

T	T	T	T
1	2	3	4

If v has a nontrivial grandparent, then $star(v) := \textit{false}$ (condition 1).

T	T	T	F

If v has a nontrivial grandparent,
then $star(v\text{'s grandparent}) := \textit{false}$ (condition 2).

F	T	T	F

In general at this point, if the tree is not a star,
only children of the root may still have $star = \textit{true}$.

If $star(v\text{'s parent})$ is $\textit{false}$, then $star(v) := \textit{false}$ (condition 3).

F	F	F	F

(b) The computation.

FIGURE 2.7
Computation of *star*.

LEMMA 2.6
For any connected component whose vertices do not form a star at the beginning of the tth iteration of the loop (for $t \geq 1$), $d_C(t) \leq 2d_C(t-1)/3$.

PROOF
Consider what happens to the trees of C during the tth iteration. Since a tree is never hooked to a leaf in the loop, the depth of a tree that results from hooking is at most the sum of the depths of the two trees that were hooked. After shortcutting, each tree is at most two-thirds as deep as it was before, so the sum of the depths is at most two-thirds what it was before. ∎

THEOREM 2.3
Algorithm 2.3 runs in $O(\log n)$ time in the worst case on an Arbitrary-Write PRAM with $\max(n, m)$ processors.

PROOF
From Lemma 2.6, for any connected component C, we have

$$d_C(t-1) \geq \frac{3}{2} d_C(t).$$

Iterating this recurrence gives

$$d_C(0) \geq \left(\frac{3}{2}\right)^t d_C(t).$$

Since there are n vertices in G, $d_C(t) < n$ for all C and t, so $d_C(0) < n$. Let T be the number of the first iteration after which the vertices of C are in one star. Then $d_C(T) = 1$. So

$$n > d_C(0) \geq \left(\frac{3}{2}\right)^T d_C(T) = \left(\frac{3}{2}\right)^T,$$

i.e.,

$$n > \left(\frac{3}{2}\right)^T.$$

So

$$T < \log_{3/2} n.$$

Since T is an integer, we conclude that after $\lfloor \log_{3/2} n \rfloor$ iterations, each component is a star. The algorithm does just one more iteration, in which nothing changes, so the total number of iterations, and the running time of the algorithm, is in $O(\log n)$.

With only m processors, we give each processor responsibility for two (oriented) edges. Each hooking step, including the initialization, is performed twice (serially) by each processor, once for each of its edges. This obviously does not change the fact that each step takes constant time. ∎

EXERCISE 2.5
Suppose Algorithm 2.3 has been executed. Write a parallel algorithm to determine how many connected components G has. How much time and how many processors does your algorithm use?

EXERCISE 2.6
Show that when there are write conflicts, the arbitrary choice of which processor succeeds in writing can have an extreme effect on the number of iterations of the loop in Algorithm 2.3. Specifically, find a sequence of connected graphs G_n such that it is possible that all the vertices will be in one star after the initialization steps, but it is also possible that the number of iterations of the loop will be in $O(\log n)$ for these graphs.

EXERCISE 2.7
Let G be the input graph for Algorithm 2.3. Let S be the set of edges (i, j) for which P_{ij} succeeds in performing a $Hook(i, j)$ operation. In other words, $(i, j) \in S$ if and only if P_{ij} is the processor that actually writes in $parent(i)$, not simply one of perhaps several processors that try. Prove that S is a spanning forest for G. (A spanning forest is a set of edges that form one spanning tree for each connected component of G.)

EXERCISE 2.8
Show how to modify Algorithm 2.3 to produce a spanning forest for the input graph G. The output may be in the form of a boolean array SF indexed by the pairs (i, j), where $SF(i, j) = true$ indicates that (i, j) is in the spanning forest.

EXERCISE 2.9
This problem investigates whether or not a small change in Algorithm 2.3 produces an algorithm that will find a minimum spanning forest for a weighted graph.

The initialization step of Algorithm 2.3 is modified so that, before each processor reads an edge from the input list, the edges are sorted in nondecreasing order by weight. We assume that each processor reads the edge from the position in the input list corresponding to its own index. Thus for $k_1 < k_2$, the weight of P_{k_1}'s edge is less than or equal the weight of P_{k_2}'s edge. The revised algorithm will be run on a Priority-Write PRAM. In this model, when more than one processor tries to write in the same shared memory location at the same time, the processor with the lowest index wins.

Assume that the algorithm has been modified as indicated in the previous exercise to produce a spanning forest.

Either prove that the spanning forest produced is always a minimum spanning forest, or show an example where it is not. (In the latter case, try to make what further modifications are needed in the algorithm so that it always produces a minimum spanning forest.)

2.3
List Ranking

2.3.1 The Problem, Some Background, and Applications

Chapter 1 presented an $O(\log n)$ optimal algorithm for the *prefix sums problem* for an associative binary operator $\oplus$. This problem has as input an ordered set of n elements, $a_1, a_2, \ldots, a_n$. The required output is the ordered set of values $a_1, (a_1 \oplus a_2), \ldots, (a_1 \oplus a_2 \oplus \cdots \oplus a_n)$.[1] The algorithm in Chapter 1 works for the case where the input is provided in an array, but not for linked lists. In this chapter we study the list ranking problem (defined below). It is a special case of the problem of computing suffix sums, rather than prefix sums, of elements in a linked list. The ith suffix sum is $(a_i \oplus a_{i+1} \oplus \cdots \oplus a_n)$. The optimal list ranking algorithm presented in this chapter will make heavy use of the array-based prefix sums algorithm of Chapter 1.

DEFINITION 2.1
The rank of an element in a linked list is its distance from the end of the list, or equivalently, the number of elements that follow it in the list.

For a list with n elements, the rank of the first element is $n - 1$, and the rank of the last element is 0. In the *list ranking problem* we are given a linked list and we must find the rank of each element. (It may seem more natural to define rank as position, e.g., where the first element has rank 1, but the rank as defined here is more easily computed, and $position = n - rank$.) The linked list is given in an array of n elements. Each element (except the one that is last in the linked list) contains the array index of its successor in the linked

---
[1] In Chapter 1 the output is slightly different, but the algorithm can be easily modified to provide the output as described here.

2.3. List Ranking

list. The array element for the last item of the linked list contains its own array index. See Figure 2.8 for an illustration. (There is no pointer to the first element in the linked list.)

EXERCISE 2.10
Suppose a linked list with n elements is represented in an array as described above. Show that the index of the predecessor of each element can be found and stored in the shared memory in constant time on an EREW PRAM with n processors.

EXERCISE 2.11
Suppose a linked list with n elements is represented in an array as described above. Show that if each of n processors on a EREW PRAM is assigned to one element, each processor can determine in constant time whether or not its element is the first or last element in the linked list. (Consequently, the indexes of the first and last elements can be stored in the shared memory in constant time.)

The list ranking problem can be solved sequentially in linear time.

EXERCISE 2.12
Give a linear time sequential algorithm for list ranking.

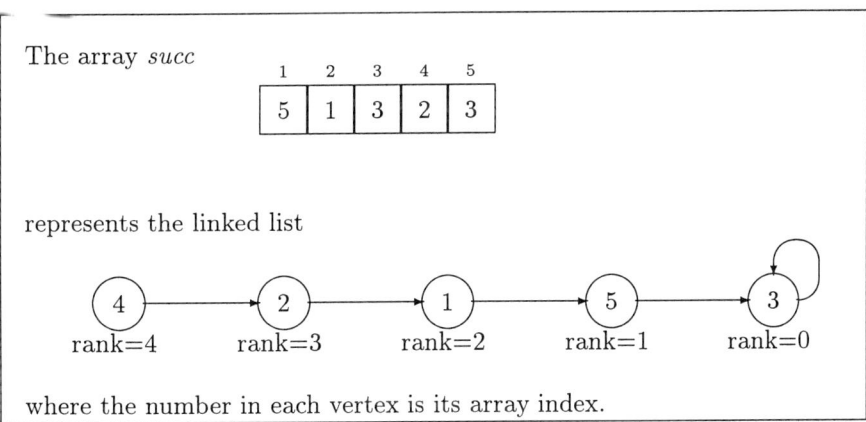

FIGURE 2.8
The linked list in an array.

One immediate application of the list ranking problem is to convert a linked list to an array sorted by position in the list. Once the ranks (and positions) are computed, the elements are simply moved, in one parallel step, into their correct positions. The list ranking problem appears as a subproblem in algorithms for various graph and tree problems. It is used in the Euler tour technique introduced later in this chapter.

The algorithms presented here for list ranking are for EREW PRAMs. It is often easier to describe an algorithm as if concurrent reads were permitted. The following lemma will be used in some places to justify doing so.

LEMMA 2.7
On an EREW PRAM with p processors, all processors can read a value originally in one memory location in $O(\log p)$ steps.

PROOF
Let x be the value to be read. Each time a processor reads x, it writes a copy of x to a new memory location. At each step, the number of copies is doubled, so the number of processors that can read (and write) at the next step doubles. Thus after $\lceil \log p \rceil$ steps, there are enough copies for all p processors. ∎

2.3.2 Easy Parallel Algorithms for List Ranking and Suffix Sums in Linked Lists

The algorithms for list ranking use the technique of "shortcutting," or "pointer jumping" that we used in the connected component algorithm. (See Section 2.2.2.)

The linked list is represented by the array $succ$ where $succ(i)$ is the index of the successor of element i in the linked list. The basic algorithm uses n processors. We initialize the rank to the value 1 for each element other than the last. The rank of the last element is initialized to (and always remains) 0. Then we carry out the shortcutting operation $\lceil \log n \rceil$ times on the $succ$ array, adding the rank of an element's current successor to its own rank each time a pointer is reset. We will show below that at any step $rank(i)$ contains the number of linked list elements from element i's original successor up to (and including) the current $succ(i)$ (for all i except the last linked list element).

ALGORITHM 2.5
Basic list ranking algorithm.
       Each P_i does:

(Initialize rank values)
if $succ(i) = i$ **then** $rank(i) := 0$ **else** $rank(i) := 1$ **endif**;

(Rank computation and shortcutting)
repeat $\lceil \log n \rceil$ times

$$rank(i) := rank(i) + rank(succ(i));$$
$$succ(i) := succ(succ(i))$$

endrepeat

THEOREM 2.4
When the basic list ranking algorithm (Algorithm 2.5) terminates, $rank(i)$ contains the rank of element i, for $1 \leq i \leq n$.

PROOF
Let *last* be the index of the last element of the linked list. As we saw when we considered shortcutting in the connected component algorithm (Section 2.2.2), after $\lceil \log n \rceil$ iterations, $succ(i) = last$ for every index i in the array.

Initially, $rank(last) = 0$ and $succ(last) = last$. Thus $succ(last)$ never changes, and in every iteration of the loop, $rank(last)$ is assigned $rank(last) + rank(last) = 0$.

For $i \neq last$ we show by induction on the number of iterations that after each iteration, $rank(i)$ is the number of elements in the original linked list from the original successor of i up to (and including) the current $succ(i)$. Before any shortcutting, there is one such element, i's successor, and $rank(i) = 1$. Let $succ_j$ denote the $succ$ array after the jth iteration. The original successor of i is $succ_0(i)$. For $j > 0$, if $succ_{j-1}(i) \neq last$, then after the jth iteration $rank(i)$ is the number of elements in the original list from $succ_0(i)$ to $succ_{j-1}(i)$ plus the number of elements from $succ_0(succ_{j-1}(i))$ to $succ_{j-1}(succ_{j-1}(i))$. These two segments are contiguous in the original list, and $succ_{j-1}(succ_{j-1}(i)) = succ_j(i)$, so the conclusion follows. See Figure 2.9 for an illustration. The argument for the case where $succ_{j-1}(i) = last$ is only slightly different, and is left to the reader.

Thus after $\lceil \log n \rceil$ iterations, when $succ(i) = last$ for all i, $rank(i)$ is the rank of element i. ∎

The basic list ranking algorithm has an immediate generalization to the problem of computing suffix sums in a linked list rather than ranks. In the

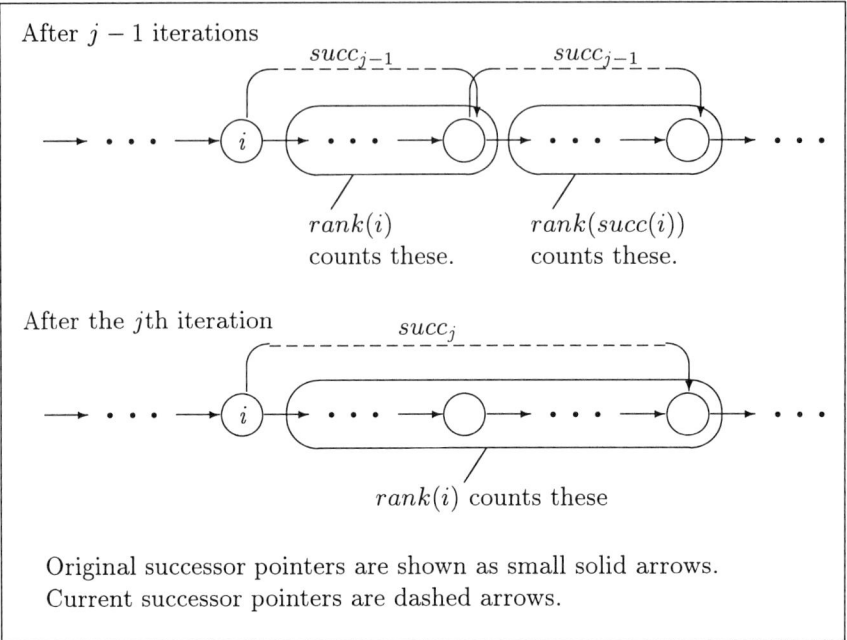

FIGURE 2.9
A step in the basic list ranking algorithm.

suffix sum problem there is a value $value(i)$ associated with element i. The suffix sums (for ordinary addition) are $(a_1+a_2+\cdots+a_n),\ldots,(a_i+a_{i+1}+\cdots+a_n),\ldots,a_n$. If we assign the value 1 to each element except the last, and assign 0 to the last, then the rank of the ith element is the ith suffix sum. The suffix sum algorithm described next is used as a subroutine in the faster list ranking algorithms presented later in this chapter and in Chapter 5. To compute suffix sums in a linked list, the basic list ranking algorithm is modified to initialize the *rank* array with the element values. The list ranking algorithm relies on $rank(last)$ being zero. If $value(last) \neq 0$, a slight "correction" is needed. We can either leave $rank(last) = 0$ in the algorithm, then add $value(last)$ to all the sums at the end, or we can add a dummy element with value 0 to the end of the linked list; we do the former.

2.3. List Ranking 85

ALGORITHM 2.6
Basic suffix sums algorithm for a linked list.
    Each P_i does:

    (Initialize rank values)
    if $succ(i) = i$ **then** $rank(i) := 0$ **else** $rank(i) := value(i)$ **endif**;

    (Rank computation and shortcutting)
    repeat $\lceil \log n \rceil$ times
        $rank(i) := rank(i) + rank(succ(i))$;
        $succ(i) := succ(succ(i))$;
    endrepeat;

    (Correction)
    if $value(last) \neq 0$ **then** $rank(i) := rank(i) + value(last)$ **endif**

The proof of Theorem 2.4 must be modified slightly for suffix sums. We described $rank(i)$ at intermediate steps as the number of elements in the original linked list from the original successor of i up to (and including) the current $succ(i)$. We could have said $rank(i)$ is the number of elements from i up to (and including) the original predecessor of the current $succ(i)$. These numbers are the same; we include exactly one of i or the current $succ(i)$ in the count. The rank of i is the number of elements in the list that *follow* i, so it is natural to use the first description in the proof of Theorem 2.4. However, the suffix sum for i is the sum of the values of elements in the original linked list *from* (including) i to the end of the list. Thus the proof for the modified algorithm should describe $rank(i)$ at an intermediate step as the sum of the values of elements from i up to (and including) the original predecessor of the current $succ(i)$.

THEOREM 2.5
The basic list ranking and suffix sum algorithms (Algorithms 2.5 and 2.6) can each be implemented on an EREW PRAM to run in $O(\log n)$ time using n processors for lists of size n.

PROOF
The algorithms clearly do $O(\log n)$ steps and use n processors, but there are concurrent reads. After shortcutting occurs, several elements will have the same successor and several processors will try to read its rank. In the correction step for the suffix sum algorithm, all processors read $value(last)$.

Observe that if $succ(i) = last$, the operations in the **repeat** loop do not change the values of $rank(i)$ or $succ(i)$, so P_i does not have to do anything in the loop; i.e., the concurrent reads of $rank(last)$ are not needed. To test if $succ(i) = last$, though, P_i must know $last$. By Exercise 2.11 and Lemma 2.7, all the processors can read $last$ with exclusive reads in $O(\log n)$ steps.

Similarly, all processors can read $value(last)$ (in the suffix sum algorithm) with exclusive reads in $O(\log n)$ steps. ∎

Although the basic algorithms are fast, they are not optimal; the total work is in $O(n \log n)$.

2.3.3 A Strategy for More Efficient Algorithms

The array-based prefix sum algorithm of Chapter 1 takes $O(\log n)$ time but does not need n processors at each step. It adds pairs of elements and uses only half as many processors to (recursively) solve the problem for half as many inputs. But generating the prefix sums from the solution to the subproblem requires distinguishing between even- and odd-indexed elements. For the list ranking problem, a processor does not know if its element is at an even position or an odd position in the linked list.

To improve on the basic list ranking algorithm, we will recursively solve the problem on a smaller list, but not one formed from pairs of odd and even neighbors. Here is an overview of the strategy used by several faster algorithms. It uses the suffix sum variant of the basic list ranking algorithm because each element in the subproblem to be solved recursively has an initial value representing a count of some elements omitted from the sublist.

1. Select a subset S of elements from the list L. Let the successor of an element in the sublist L_S be the first element of S that follows it in L. Let the initial value of an element in L_S be the sum of the values of the elements in L from this element up to (and including) the predecessor in L of this element's new successor in L_S. Figure 2.10 shows the result of this step on the first iteration of the algorithm, where the ranks were initially 1 (0 for the last element). Figure 2.11 includes a larger example at a later stage.
2. If L_S is small enough, use the basic suffix sum algorithm on L_S (Algorithm 2.6). Otherwise, recursively solve the problem on L_S. At this point, the values computed for elements of L_S are the correct suffix sums for these elements in L. (See Figure 2.11.)

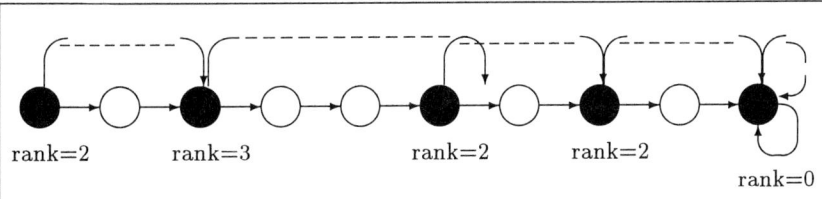

FIGURE 2.10
List contraction.

3. Compute the suffix sums for the elements of $L - L_S$. (This can be done in a straightforward way by traversing each segment of L between elements of S; we will see details later.)

A key to getting a good algorithm is to choose S so that two elements adjacent in L_S are not far apart in L. Then Steps 1 and 3 can be fast. On the other hand, S should not be too dense. We want the size of the list to decrease enough at each iteration (i.e., each recursive invocation of the algorithm) so that the total number of iterations will be small.

There are both randomized and deterministic algorithms that use this strategy. They differ in their methods for selecting the subset S and for constructing the sublist. We will describe one of the algorithms of Cole and Vishkin; it is an optimal deterministic $O(\log n \log \log n)$ algorithm; that is, it uses $n/(\log n \log \log n)$ processors, doing a total of $O(n)$ operations. An optimal $O(\log n)$ algorithm, also due to Cole and Vishkin, is presented in Chapter 5. The algorithm described here introduces some of the techniques used in the faster algorithm as well.

The $O(\log n \log \log n)$-time, $n/(\log n \log \log n)$-processor algorithm is built up from algorithms for subset selection, sublist compaction (both parts of Step 1 above), and the calculations for unselected elements (Step 3). To get the desired time and processor bounds, we want algorithms for the subproblems that run in $O(\log m)$ time on lists of size m and do $O(m)$ operations. We will revert to the basic suffix sum algorithm when the number of elements

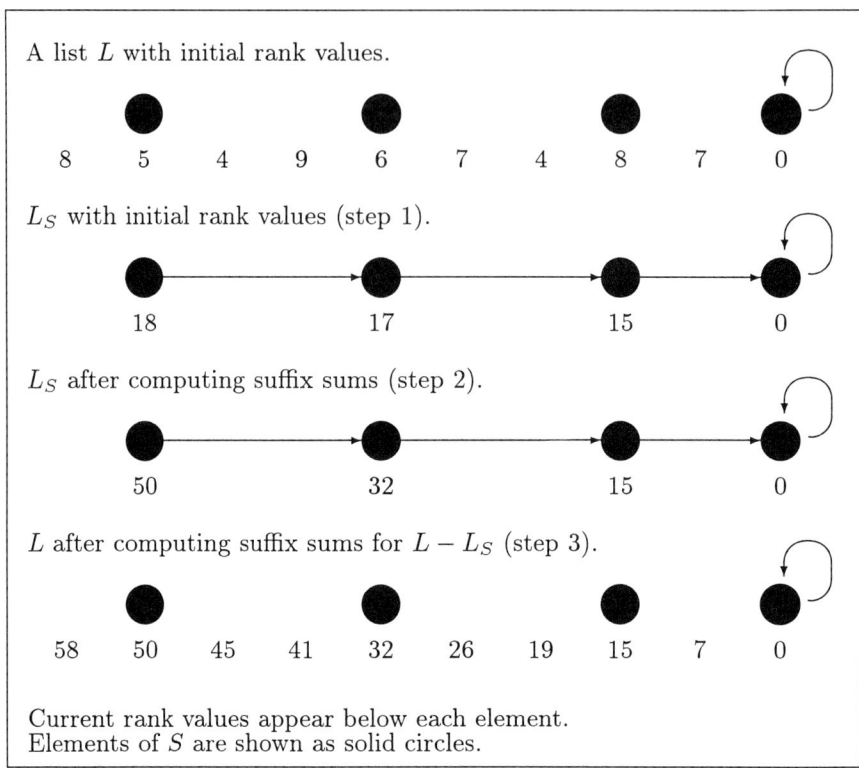

FIGURE 2.11
An example of the recursive algorithm.

in L_S is at most $n/\log n$ so that only $\log \log n$ iterations are needed. (The full analysis, using Brent's scheduling principle to bring in the number of processors, will be described after the complete algorithm.) For the next few sections, we concentrate on solving the subproblems in $O(\log m)$ time and $O(m)$ operations.

2.3.4 Deterministic Coin Tossing

We want to select a subset of the list elements in such a way that we can bound the distance between selected elements (to make Steps 1 and 3 fast) and bound the size of the subset to limit the number of recursive invocations of the algorithm at Step 2. Selecting elements that are at odd (or even) positions in L would satisfy the requirements, but as we observed, a processor cannot tell if its element is in an odd position. In one randomized algorithm each processor randomly selects a value *heads* or *tails*. (This is called random coin tossing, or random mating when the values are described as *male* and *female*.) The subset S includes all elements except those whose "coin" came up *heads* and whose predecessor's coin came up *tails*. No two adjacent elements of L are both eliminated, so the distance between selected elements is at most two. It can be shown that with high probability, the size of S is at most cm for a constant $c < 1$, so we can bound the number of recursive calls. Specifically, if the basic algorithm is used when the list size m is at most $n/\log n$, then the number of recursive list contractions is likely to be in $O(\log \log n)$.

The *deterministic coin tossing* technique, due to Cole and Vishkin, is a clever method for assigning a "tag" (akin to heads or tails) to each element without randomness. It uses variations in the bits of the indexes of neighboring elements to make the assignments. The technique is also used in algorithms for problems other than list ranking: finding connected components and finding a minimum spanning forest for a graph.

ALGORITHM 2.7
Deterministic coin tossing.
Comment: It is convenient here to assume that the array containing the linked list pointers is indexed from 0 to $m - 1$ so that an index can be represented in binary in $\lceil \log m \rceil$ bits. (Usually we assume indexes start at 1.) Let the bit positions be numbered 0 to $\lceil \log m \rceil - 1$ beginning at the rightmost, or least significant, bit. For convenience, assume here that the processors are indexed from 0 to $m - 1$. Processor P_i is assigned to the element with index i.

For $i \neq last$, P_i does:
find the smallest j (rightmost bit position) where bit j of i and bit j of $succ(i)$ differ;
$tag(i) :=$ the ordered pair $(j, \text{bit } j \text{ of } i)$.

P_{last} assigns $tag(last) := (-1, 1)$.

For example, let $i = 25$ $(0 \cdots 011001_2)$ and $succ(i) = 21$ $(0 \cdots 010101_2)$. The first position where i and $succ(i)$ differ is position 2; i has a 0 in bit 2, so $tag(i) = (2, 0)$.

EXERCISE 2.13
Suppose the linked list elements happen to be stored in the array in natural order. What is the tag value for each element?

LEMMA 2.8
For each element $i \neq last$, $tag(i) \neq tag(succ(i))$.

PROOF
If $tag(i) = tag(succ(i))$, both components of these tags are the same. The first component, say j, is a bit position, and the second component is the bit in position j. Thus bit j of i and bit j of $succ(i)$ are the same, but by definition of the tags, j is the first bit position where i and $succ(i)$ differ. Thus $tag(i)$ and $tag(succ(i))$ cannot be equal. ∎

LEMMA 2.9
Deterministic coin tossing (Algorithm 2.7) can be implemented to run in constant time on an EREW PRAM with m processors for a list of size m.

PROOF
Each processor of a PRAM is a RAM. The formal definitions of RAMs allow as basic instructions only arithmetic, test for zero, and a few other very basic operations. Under the uniform cost function, each instruction takes one unit of time (although it acts on $\log n$ bits, where n is the largest integer represented, if we assume a binary representation). The arguments that have been given to show that deterministic coin tossing can be done in constant time assume some instruction beyond the very basic RAM repertoire—an instruction that converts unary to binary, for example, or even an instruction that does the whole problem—i.e., finds the first bit where two integers differ.

We give a calculation that uses boolean instructions and a "find first bit on" instruction (which returns the position number of the first, or least significant, bit that is 1). Boolean instructions are in the instruction set of virtually all general purpose computers and many also include a "find first bit on" instruction.

Here is the calculation performed by P_i ($i \neq last$) to find j, the first position where i and $succ(i)$ differ, and bit, the value of bit j of i.

$$d := i \oplus succ(i); \quad \{ \text{where } \oplus \text{ is exclusive-or. Thus } d \text{ is } \}$$

$$|\leftarrow \ x \ \rightarrow |10\cdots 0$$
$$j \qquad 0 \quad \text{position numbers}$$

$j := FirstBitOn(d);$
b$dd := d$ **and** $i;$ { Thus dd is }

$$| \leftarrow y \rightarrow |b0 \cdots 0$$
$$j0 \quad \text{position numbers}$$

{ where $b = 1$ if and only if bit j of i is 1 }
$k := FirstBitOn(dd);$
if $j = k$ **then** $bit := 1$ **else** $bit := 0$ ∎

2.3.5 Subset Selection: A 2-Ruling Set

We select a subset called a *2-ruling set*. A 2-ruling set is a special case of an *r-ruling set*.

DEFINITION 2.2
An r-ruling set is a subset of the linked list elements that has the following properties:

1. *It is an independent set; i.e., no two neighboring elements are both selected. (Here, and throughout this section, we use the term neighbor to mean the predecessor or successor of an element.)*
2. *The distance from any unselected element to the next selected element is at most r. (This implies that the last element is selected.)*

The first property ensures that the sublist is no more than (roughly) half as large as L.

It is not difficult to construct an $O(\log m)$-ruling set in constant time (beginning with a list of size m). We can select elements i whose tags are local minima, i.e., such that $tag(i) \leq tag(succ(i))$ and $tag(i) \leq tag(pred(i))$, using lexicographical order on the tags (which are ordered pairs). The distance from any element to the next selected element will be at most roughly $4 \log m$. It may seem at first that this is good enough, i.e., that with a $\log m$-ruling set, steps 1 and 3 in the algorithm outline described in Section 2.3.3 can be carried out in $O(\log m)$ time with a total of $O(m)$ operations. We will explain later why this is not true. Construction of a 2-ruling set is more complicated, but it does meet the needs of the algorithm.

EXERCISE 2.14
Give a parallel algorithm to construct an $O(\log \log n)$-ruling set. How much time and how many processors does your algorithm use?

The 2-Ruling Set Algorithm

A 2-ruling set can be constructed by refining and iterating the technique used to construct an $O(\log m)$-ruling set. This approach gives an algorithm that uses $O(\log^* m)$ time with m processors. ($\log^* m$ is the smallest integer k such that $\log^{(k)} m \leq 2$, where $\log^{(k)}$ is the log function iterated k times.) $\log^*$ is a very slow growing function, but the algorithm is not optimal because the total work is in $O(m \log^* m)$. We will describe another approach which gives an optimal $O(\log m)$ algorithm. The first version, presented in Algorithm 2.8, is short and easy to understand, but it uses m processors (and $O(\log m)$ time). Unfortunately, showing how to implement it with only $m/\log m$ processors in $O(\log m)$ time (to obtain the time and operation bounds we want) is not so easy.

ALGORITHM 2.8
Constructing a 2-ruling set.
Comment: Assume the list we are currently working with has size m. The 2-ruling set S will be represented by an array inS where $inS(i) = 1$ if $i \in S$ and $inS(i) = 0$ if $i \notin S$. Assume that predecessor pointers are available in $pred$ (with $pred(first) = first$ for convenience; see Exercise 2.10).

Each P_i does:

Deterministic coin tossing (Algorithm 2.7) to get $tag(i)$;
$inS(i) := 0;$ (S is initially empty.)

    for $tagvalue := (-1, 1)$ **to** $(\lceil \log m \rceil - 1, 1)$ **do**
        if $tag(i) = tagvalue$ **then**
            if $i \notin S$ **and** $pred(i) \notin S$ **and** $succ(i) \notin S$ **then**
                $inS(i) := 1$ **endif**
        endif
    endfor

Note that $last \in S$ because it has the lowest tag value.

LEMMA 2.10
No two neighboring elements are selected.

PROOF
The algorithm tests explicitly to make sure a vertex with a neighbor selected in an earlier iteration of the **for** loop will not be selected. Can two neighbors be selected on the same iteration? No, because all elements considered on one iteration have the same *tag* value (position and bit), and Lemma 2.8 assures us that this cannot happen for neighbors. ∎

LEMMA 2.11
Any unselected element has a selected neighbor.

PROOF
The only reason a vertex is excluded is that it has a selected neighbor. ∎

LEMMA 2.12
For a list of size m, Algorithm 2.8 constructs a 2-ruling set in $O(\log m)$ time with m processors.

PROOF
That the selected elements form a 2-ruling set follows from the previous two lemmas. The $O(\log m)$ time bound is clear from the algorithm. (There are $2\lceil \log m \rceil + 1$ iterations of the loop.) ∎

Optimizing the 2-Ruling Set Algorithm

Implementation of the optimal algorithm for constructing a 2-ruling set is somewhat complicated; the presentation takes several pages. The reader may prefer to skip ahead to Section 2.3.6 now and read the rest of the development of the list ranking algorithm before completing this section.

We now show how to optimize Algorithm 2.8, that is, how to implement it in $O(\log m)$ time with only $m/\log m$ processors, so the total number of operations will be in $O(m)$. For simplicity, we will assume that $\log m$ and $m/\log m$ are integers.

Algorithm 2.8 does not make efficient use of the processors. Each processor P_i is assigned to one element i. It does significant work on only one iteration of the **for** loop: on the iteration where the loop index matches $tag(i)$, it decides whether or not to add i to the set being constructed. On all other iterations, P_i merely checks to see if its "turn" has come. We want to retain the property that all elements processed on the same iteration have the same tag value so that Lemma 2.10 is still true. To reduce the number of processors, we arrange the elements in groups of size at most $m/\log m$, where all elements of one group have the same tag value. Then at each iteration one whole group is processed in parallel by the $m/\log m$ processors. The processors must be able to quickly find the elements they are to process on each iteration, so we store the element indexes in an array *schedule* such that the group of elements to be processed on the kth iteration are stored in the kth segment of size $m/\log m$ in *schedule*. (In other words, group k contains the elements in positions $(k-1)m/\log m + 1$ through $km/\log m$ in *schedule*.)

Each group is limited to $m/\log m$ elements (because we will use $m/\log m$ processors), but we cannot expect each group to be "full," because there may be more or fewer than $m/\log m$ elements with one particular tag value. So we cannot expect all m elements to fit neatly into $\log m$ groups (to be processed in $\log m$ iterations). We spread the elements over $3\log m$ groups with a total of $3m$ slots in *schedule*, some empty. All elements can then be processed in $3\log m$ iterations. Here is an outline of the algorithm:

ALGORITHM 2.9
Optimal algorithm for constructing a 2-ruling set.
Comment: Recall that $tag(i)$ is an ordered pair $(posn, bit)$ (with the last element of the list having the special value (-1,1)). Throughout this algorithm, treat $tag(i)$ as a binary number $2 * posn + bit$. Tag values range from -1 to $2\log m - 1$.

1. Sort the elements by their tag values, assigning to each element i its (unique) position in the sorted list, $sortposn(i)$ (where $1 \leq sortposn(i) \leq m$).
2. "Schedule" the elements by marking all $3m$ slots in *schedule* initially empty, then computing

   $slot(i) := sortposn(i) + (tag(i) + 1)m/\log m;$
   $schedule(slot(i)) := i$

3. Initialize $inS(i) := 0$ for all i. (Easily done in $\log m$ steps with $m/\log m$ processors)
   Each processor P_q (for $1 \leq q \leq m/\log m$) does:
   for $k := 1$ **to** $3\log m$ **do**
   $\quad i := schedule((k-1)m/\log m + q);$
   $\quad$ **if** $i \notin S$ **and** $pred(i) \notin S$ **and** $succ(i) \notin S$
   $\quad$ **then** $inS(i) := 1$ **endif**
   endfor

2.3. List Ranking

It remains for us to show how to implement the sort for step 1, and to prove that no two elements are assigned to the same slot and no group has elements with different tag values. The sort is the most complicated part, so we leave it for last.

LEMMA 2.13
If $i \neq j$, then $slot(i) \neq slot(j)$.

PROOF
Suppose $i \neq j$. Without loss of generality, assume $sortposn(i) < sortposn(j)$. Then $tag(i) \leq tag(j)$, and $tag(j) - tag(i) \geq 0$. Since $sortposn$ values are integers, $sortposn(j) - sortposn(i) \geq 1$. So

$$slot(j) - slot(i) =$$
$$(sortposn(j) - sortposn(i)) + (tag(j) - tag(i))m/\log m \geq 1$$

The slots of i and j are different. ∎

LEMMA 2.14
If elements i and j are in the same group, then $tag(i) = tag(j)$.

PROOF
Suppose $tag(i) \neq tag(j)$. Without loss of generality, assume $tag(i) < tag(j)$. Then $sortposn(i) < sortposn(j)$, and since tag values are integers, $tag(j) - tag(i) > 1$. So

$$slot(j) - slot(i) =$$
$$(sortposn(j) - sortposn(i)) + (tag(j) - tag(i))m/\log m$$
$$\geq m/\log m$$

The slots of i and j are too far apart for i and j to be in the same group. ∎

THEOREM 2.6
If the sorting in step 1 can be done with $m/\log m$ processors in $O(\log m)$ time on an EREW PRAM, then Algorithm 2.9 constructs a 2-ruling set with $m/\log m$ processors in $O(\log m)$ time on an EREW PRAM.

PROOF
That the algorithm constructs a 2-ruling set follows from Lemmas 2.10–2.12 and 2.13–2.14. The time and processor bounds are clear from the algorithm. ∎

Sorting Elements by Their Tag Values

For step 1 in Algorithm 2.9 we need to sort the elements by tag value in $O(\log m)$ time with $m/\log m$ processors.

Since the tag values to be sorted are integers in the range $-1, \ldots, 2\log m - 1$, we can use a bucket sort approach, counting elements that have each tag value. We use the following counters:

$lowertags(i)$ = the number of elements whose tag values are less than element i's tag value ($1 \leq i \leq m$)

$sametag(i)$ = the number of elements with the same tag value as i's, but which appear earlier in the array than i ($1 \leq i \leq m$)

Define $sortposn(i) = lowertags(i) + sametag(i) + 1$. Clearly once $lowertags$ and $sametag$ have been computed, $sortposn$ can be computed in $O(\log m)$ steps using $m/\log m$ processors. Most of the work is in computing the counts. They are computed in three stages, with many applications of the array-based prefix sum algorithm (see Chapter 1) in the second and third stages.

For the first stage, divide the array into segments of $\log m$ elements each. Processor P_q ($1 \leq q \leq m/\log m$) works on the qth segment. P_q will compute the following sums.

$sametag_q(i)$ = the number of elements in the qth segment with the same tag value as element i, but which appear earlier than i (for i in the qth segment)

$count_q(tagvalue)$ = the number of elements in the qth segment with this $tagvalue$ ($-1 \leq tagvalue \leq 2\log m - 1$)

P_q works sequentially through its segment; the counting is straightforward:

(Initialize counters)
for $tagvalue := -1$ **to** $2\log m - 1$ **do**
    $count_q(tagvalue) := 0$
endfor;

(Count)
for $i := (q-1)\log m + 1$ **to** $q \log m$ **do**
    $tagvalue := tag(i)$;
    $sametag_q(i) := count_q(tagvalue)$;
    $count_q(tagvalue) := count_q(tagvalue) + 1$
endfor

Stage 1 clearly runs in $O(\log m)$ time.

Now we have local counts $count_q$ and $sametag_q$. In the second stage, we combine them to get the global totals for $sametag$. Suppose element i is in the qth segment. The formula is:

$$sametag(i) = \sum_{j=1}^{q-1} count_j(tag(i)) + sametag_q(i). \quad (2.1)$$

Thus for each of the $2\log m + 1$ tag values separately, we compute prefix sums for the sequence of local counters $count_q$. That is, we will be applying $2\log m+1$ copies of the parallel array-based prefix sum algorithm, and we want to apply them all in parallel. Recall that prefix sums for n array elements can be computed in $O(\log n)$ time with $n/\log n$ processors (see Chapter 1). Each prefix sum problem has $m/\log m$ keys, so each can be solved in $O(\log m)$ time with $\frac{m/\log m}{\log(m/\log m)} \leq m/(\log m)^2$ processors. Thus all the $O(\log m)$ separate prefix sum problems can be solved in $O(\log m)$ time with $m/\log m$ processors. Finally, to compute $sametag$ using Equation 2.1 (where the summations have already just been computed by the prefix sum algorithms), the processors can be allocated as in stage 1, with P_q computing $sametag$ for the $\log m$ elements in the qth segment of the array. Thus stage 2 takes $O(\log m)$ time.

Note that for each tag value, the last of the prefix sums computed for Equation 2.1 is

$$count(tagvalue) = \sum_{q=1}^{m/\log m} count_q(tagvalue)$$

i.e., the total number of elements with this tag value. This sum is used in the third stage.

In the third stage we compute $lowertags(i)$ for each element i. The relevant formula is:

$$lowertags(i) = \sum_{tagvalue=-1}^{tag(i)-1} count(tagvalue) \quad (2.2)$$

Here we need to compute the prefix sums for the sequence of $2\log m + 1$ tag value counts. This is straightforward; it is done by one processor with one application of the sequential prefix sum algorithm in $O(\log m)$ time. Then, $lowertags(i)$ can be computed by simply reading the correct prefix sum. (The processors can again be allocated as in stage 1, with P_q computing $lowertags$

for the $\log m$ elements in the qth segment of the array.) Thus the computation appears to take $O(\log m)$ time. Note however that when computing *lowertags*, more than one processor may try to read the same prefix sum $\sum_{tagvalue=-1}^{tag(i)-1} count(tagvalue)$ (i.e., for i and i' that have the same tag value) at the same time. The next lemma shows that we can make copies of all the prefix sums for all the processors in $O(\log m)$ time, so that stage 3 can indeed be completed in $O(\log m)$ time with exclusive reads.

LEMMA 2.15
An EREW PRAM with $m/\log m$ processors can make $m/\log m$ copies of each of $\log m$ numbers in the shared memory in $O(\log m)$ time.

PROOF
Divide the processors into groups, one group for each number to be copied. There are $m/(\log m)^2$ processors in each group. Each group reads its number in $O(\log(m/(\log m)^2))$ steps (Lemma 2.7). (Copies written during this stage are ignored.) For the next $\log m$ steps, each processor writes a new copy of the number it read. Since there are $m/(\log m)^2$ processors in each group, a total of $m/\log m$ copies of each number will be written. The total time is in $O(\log(m/(\log m)^2) + \log m) = O(\log m)$. ∎

EXERCISE 2.15
In stage 2 of the sorting process, did more than one processor try to read the same prefix sum at the same time when computing Equation 2.1?

LEMMA 2.16
The m elements of the list L can be sorted by tag values in $O(\log m)$ time using $m/\log m$ processors, for a total of $O(m)$ operations.

PROOF
We verified that each of the three stages of the sort can be done in the stated time and processor bounds. ∎

We have now completed the details for the optimal $O(\log m)$ algorithm to construct a 2-ruling set (Algorithm 2.9).

2.3.6 Sublist Compaction

Suppose the current list has m elements. Suppose also that the subset S on which we will recursively solve the list ranking problem has been determined and is represented by the array inS (where $i \in S$ if $inS(i) = 1$ and $i \notin S$ if $inS(i) = 0$.) We must arrange the elements of L_S into an array of size $|S|$ so that the input for the recursive invocation of the algorithm is in the expected form. This step is called *sublist compaction*.

Part of the task—forming a new, compacted array of selected elements from a given array—is an application of the prefix sum problem. We just apply a prefix sum algorithm to the array inS. Let the array of prefix sums be called *newindex*; for $i \in S$, $newindex(i)$ is the index where element i of the current array belongs in the compacted array.

The rest of the task is to compute new initial successor and rank values for the compacted list. (See Figures 2.10 and 2.11 for examples.)

ALGORITHM 2.10
Sublist compaction.

>(Compute *newindex*.)
>Use a prefix sum algorithm on the array inS to compute $prefixsum(i)$ for all i;
>Each P_i ($1 \leq i \leq m$) does:
>
>**if** $i \in S$ **then**
>>$newindex(i) = prefixsum(i)$;
>>(Initialize successors and ranks for the compacted list.
>>$succ$ and $rank$ are the current successor pointers and ranks;
>>$succ_S$ and $rank_S$ are the arrays for the compacted list.)
>>$rank_S(newindex(i)) := rank(i)$;
>>$succ_S(newindex(i)) := newindex(i)$; (needed only for $i = last$)
>>$j := i$;
>>**repeat** 2 times
>>>$j := succ(j)$;
>>>**if** $j \notin S$ **then**
>>>>$rank_S(newindex(i)) := rank_S(newindex(i)) + rank(j)$;
>>>>$succ_S(newindex(i)) := newindex(succ(j))$
>>>
>>>**endif**
>>
>>**endrepeat**
>
>**endif** (if $i \in S$)

THEOREM 2.7
If L is a list of size m and S is a 2-ruling set, then Algorithm 2.10 constructs L_S in $O(\log m)$ time. The number of operations performed is in $O(m)$.

PROOF

By definition of a 2-ruling set, there can be at most two consecutive elements of L not in S. So the **repeat** loop traverses the elements omitted from L_S and computes the proper rank and successor values. Computing prefix sums can be done in $O(\log m)$ time with a total of $O(m)$ operations, and the rest of the steps in Algorithm 2.10 clearly take constant time. So the number of operations is in $O(m)$. ∎

Why Not Use a $\log n$-Ruling Set?

We saw that an $O(\log m)$-ruling set can be constructed in constant time (by selecting elements whose tags are local minima). The distance from any element to the next selected element would be at most roughly $4\log m$. Before and after the recursive solution for the sublist made up from the ruling set, some computation must be done on the segments of the list between selected elements (as in the **repeat** loop of Algorithm 2.10). With an $O(\log m)$-ruling set, the computation can be done in time $O(\log m)$ with m processors. It appears that since each segment is processed once (serially by one processor), the total number of operations is in $O(m)$. Suppose we try to cut the number of processors to $m/(\log m)$ using Brent's scheduling principle. It seems that the computation can be done in $O(\log m)$ time with $m/(\log m)$ processors. But if each processor is assigned to a set of $\log m$ vertices in advance, it may happen that each vertex assigned to one processor is a selected vertex and is followed by a segment of $\log m$ unselected vertices; the computation would take $\Theta(\log m)^2$ time.

The problem has to do with allocation of the processors to the operations that need to be done. Brent's scheduling principle must be applied with care. It is okay if the actual sequence of operations carried out is not data-dependent. If the work is data-dependent, we must be able to show that the processors can determine quickly enough what they are to do. Sometimes that is not possible.

2.3.7 Finishing Up

For step 3 of the algorithm strategy (Section 2.3.3), we must compute suffix sums for the elements of $L - L_S$ after sums for L_S have been computed in step 2 (by recursive invocation of the algorithm or by the the basic suffix sum algorithm if L_S were small enough). The suffix sums computed for L_S (and stored in $rank_S$) are also the correct suffix sums for the elements of S with respect to the list L because in L_S we initialized $rank_S$ to include the values of elements of $L - L_S$. (Recall the sublist compaction algorithm,

2.3. List Ranking   101

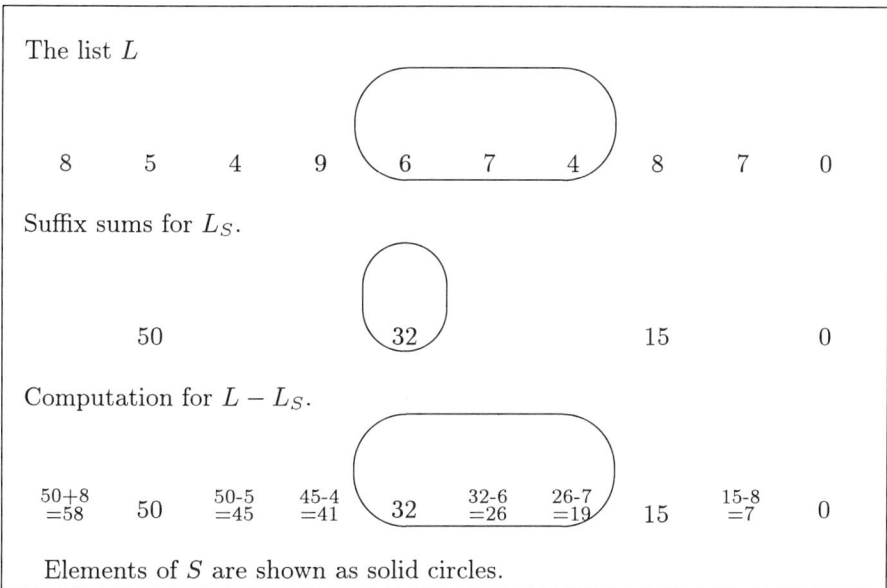

FIGURE 2.12
Computing ranks for $L - L_S$.

Algorithm 2.10.) It is straightforward then to compute the suffix sums for elements of $L - L_S$ (with respect to L). Figure 2.12 shows the computation for the example in Figure 2.11. The algorithm assumes S is a 2-ruling set. It uses *newindex*, the mapping of elements of L into L_S that was computed by the sublist compaction algorithm (Algorithm 2.10).

ALGORITHM 2.11
Finishing rank computation after the recursion.

    Each P_i $(1 \leq i \leq m)$ does:

  if $i \in S$ **then**
    $r := rank_S(newindex(i))$;
    $prev := i; j := succ(i);$ (j is i's successor in L)
    repeat 2 **times**
      if $j \notin S$ **then**

$r := r - rank(prev)$; (subtracting rank value in L)
$rank(j) := r$;
$prev := j; j := succ(j)$
          endif
       endrepeat;
       $rank(i) := rank_S(newindex(i))$
    endif

Comment: If the first element of L was not in S, the steps above will not compute its rank. P_{first} does so as follows.

If $first \notin S$, P_{first} does:

$j := succ(first)$;   (j must be in S)
$rank(first) := rank(first) + rank_S(newindex(j))$

THEOREM 2.8
If L has m elements, S is a 2-ruling set, and $rank_S$ contains the suffix sums for L_S (where ranks were initialized in the sublist compaction algorithm (Algorithm 2.10)), then Algorithm 2.11 computes the suffix sums for L in constant time with m processors, hence $O(m)$ operations.

2.3.8 The $O(\log n \log \log n)$ Algorithm

In this section we put together the pieces in the last four sections to get the $O(\log n \log \log n)$ list ranking algorithm. We are using the strategy of Section 2.3.3.

We are to solve the list ranking problem on an array of n elements. We initially find *last*, the index of the last item of the linked list, and assign $rank(i) := 1$ for $i \neq last$ and $rank(last) := 0$. Also, $m := n$. This initialization is *not* part of the recursive algorithm; it is performed only once. Note that we use the variable name "rank" in the algorithm, although for sublists on which the problem is solved recursively, we are actually computing suffix sums, not simply ranks.

ALGORITHM 2.12
The recursive list ranking (suffix sum) algorithm.

Broadcast m to all processors; determine the indexes *first* and *last*, and broadcast *last* to all processors;

if $m \leq n/\log n$ **then** use the basic suffix sum algorithm (Algorithm 2.6)
else

Establish *pred* pointers;
Do deterministic coin tossing (Algorithm 2.7);
Find a 2-ruling set S (Algorithm 2.9);
Perform sublist compaction with respect to S (Algorithm 2.10);
$m_S := \mathit{prefixsum}(m)$ {from Algorithm 2.10; this is $|S|$};
Recursively apply this algorithm to the compacted list (using m_S, succ_S, and rank_S);
Compute the ranks for the elements of $L-L_S$ (Algorithm 2.11)
endif

THEOREM 2.9
Algorithm 2.12 correctly computes list ranks.

PROOF
The proof is based on the correctness of the overall strategy and the algorithms for the subproblems. The arguments have been made throughout the last several sections. ∎

THEOREM 2.10
The list ranking problem can be solved in $O(\log n \log \log n)$ time with $n/(\log n \log \log n)$ processors on an EREW PRAM.

PROOF
We first determine the number of iterations of the algorithm, then count the work done on each iteration. No two neighboring elements are both included in the 2-ruling set S, so $m_S \leq \lceil m/2 \rceil$. A straightforward calculation (simplified by assuming m is even) shows that after $\lceil \log \log n \rceil$ iterations, $m \leq n/\log n$, so there are at most $\lceil \log \log n \rceil$ iterations.

Table 2.1 summarizes the time and operation bounds already established for the various parts of the algorithm. Ignoring, for the moment, the need for broadcasting data to all processors (to make the algorithm run on a EREW PRAM), the total number of operations at each iteration is in $O(m)$. At each iteration m decreases by half (at least), so the total number of operations for all iterations (before reverting to the basic algorithm) is at most

$$\sum_{i=0}^{\log \log n - 1} n/2^i \in O(n).$$

The time bound for each iteration is $O(\log m)$, so the time for all iterations is in $O(\log n \log \log n)$. The basic suffix sum algorithm is applied

TABLE 2.1
Time and operation bounds for the list ranking algorithm

Subproblem	Time	Operations	Justification
Determine *first*, *last*, and *pred*	$O(1)$	$O(m)$	Exercises 2.10 and 2.11
Broadcast m and *last*			Proof of Theorem 2.10
Deterministic coin tossing	$O(1)$	$O(m)$	Lemma 2.9
Construct 2-ruling set	$O(\log m)$	$O(m)$	Theorem 2.6 and Lemma 2.16
Sublist compaction	$O(\log m)$	$O(m)$	Theorem 2.7
Compute ranks for $L - L_S$	$O(1)$	$O(m)$	Theorem 2.8

to a list of size at most $n/\log n$; this uses time in $O(\log(n/\log n))$ and $O(n/\log n)$ processors (Theorem 2.5). The bound on the number of operations (the product of time and processors) is

$$O\left(\log(n/\log n) \cdot \frac{n}{\log n}\right) = O(n).$$

Thus the time bound, t, for the entire algorithm is $O(\log n \log \log n)$ and the total number of operations, *ops*, is in $O(n)$.

Now suppose we have $p = n/(\log n \log \log n)$ processors. By Brent's scheduling principle, the algorithm can be implemented on p processors in time

$$O(ops/p + t) =$$
$$O\left(\frac{n}{n/(\log n \log \log n)} + \log n \log \log n\right) =$$
$$O(\log n \log \log n).$$

Now, returning to the problem of exclusive reads, we observe that at each iteration of the recursion, the values of a few variables must be read by all p processors. (These include not only those mentioned explicitly in Algorithm 2.12, but also those implied by the implementation of recursion, e.g., a stack pointer.) By Lemma 2.7, this can be done in $O(\log p)$ steps per iteration. For all iterations, the time for broadcasting is in

$$O\left(\log\left(\frac{n}{\log n \log \log n}\right) \log \log n\right) = O(\log n \log \log n). \quad \blacksquare$$

Observe that the subproblems that take time in $O(\log m)$ are construction of the 2-ruling set and sublist compaction, both of which have the prefix sum problem (for arrays, not linked lists) as a subproblem. Some faster algorithms for list ranking use faster array-based prefix sum algorithms.

2.4
Euler Tour Techniques

Many algorithms for tree and graph problems include solutions for basic structural problems (e.g., numbering and counting problems) as subroutines. Examples of such subproblems for trees are:

1. Find the parent of each vertex. (Another way to look at this problem is: Given a tree as an undirected graph with one vertex designated as the root, determine the direction of each edge.)
2. Assign preorder and/or postorder numbers to the vertices.
3. Compute the level of each vertex.
4. Determine the number of descendants of each vertex.

Each of these problems has a linear sequential solution using a tree traversal, or depth-first search. The *Euler tour technique*, devised by Tarjan and Vishkin, is an application of list ranking that can be used to construct efficient parallel algorithms for these problems. When the tree is a depth-first-search tree for a graph or digraph, this technique plays an important part in algorithms for graph problems. Tarjan and Vishkin used it as part of their parallel biconnectivity algorithm.

In this section we introduce the Euler tour technique for trees and show how to apply it to the problems listed above.

2.4.1 The Data Structure

Throughout this section we will assume that the input for each problem is a tree T with vertices $\{1, \ldots, n\}$ and with one vertex *root* designated as the root. We assume that the tree is represented by adjacency lists. That is, for each vertex, there is a list of incident edges. Each edge $\{v, w\}$ appears twice in the structure, on w's list and on v's list. Each edge $\{v, w\}$ in v's edge list has a pointer *next* to the next edge on v's list. (The pointer for the last edge on the list is **nil**.) Another array *adjList* has n entries; $adjList(v)$ is a pointer to the first edge on the list of edges incident with v. There is a pointer in the data structure from each edge, say $\{v, w\}$ in v's list to its copy in w's list. The number of vertices, n, is part of the representation.

For the Euler tour technique we associate with the tree T a directed graph T_D which has n vertices and two directed edges (v, w) and (w, v) in place of each undirected edge $\{v, w\}$ in T. T and T_D are represented by the same data structure; for T_D we simply interpret an edge to be directed from v to w when it appears on v's list, and from w to v when it appears on w's list. The directed edge (w, v) is called the *reversal* of (v, w); the reversal of an edge e is denoted e_R.

The algorithms we describe will use n processors. (We can apply Brent's scheduling principle to reduce the number of processors if desired.) Often we will assign work to processors by assigning them an edge, not a vertex. We assume that each of the first $n-1$ processors is assigned to two specific edges. One way to organize the data structure to make such an assignment simple is to arrange the edges in an array, say E, grouped so that all edges incident with vertex 1 appear first, then all edges incident with vertex 2, and so on. P_i may be assigned to the edges in positions i and $i + n - 1$. This data structure is illustrated in Figure 2.13. (For clarity in the figure, we use names u, v, w and x for the vertices instead of integers.)

With the assignment of processors just described, all edges can be processed in two stages. For algorithms where some work is carried out for the root of the tree, we will assign P_n to do it.

EXERCISE 2.16
Write a parallel algorithm to assign boolean values (i.e., *true* or *false*) to the array *leaf* such that $leaf(v)$ is *true* if and only if v is a leaf in T. Your algorithm should run in constant time.

2.4.2 Euler Tours

Informally, an Euler tour in a graph or directed graph is a path that traverses each edge exactly once and returns to its starting point. Not every graph or directed graph has an Euler tour; however, T_D does.

The algorithm below constructs an Euler tour for T_D in constant time on an EREW PRAM (excluding the time for all processors to read n). Each processor is assigned two edges and is to determine their successors in the tour. Let $e = (u, v)$. The processor for e must have a systematic way to determine which edge from v will follow e without communicating with other processors to coordinate the choice. The trick is to use $e_R = (v, u)$ as the position in the adjacency list for v from which to make the selection. (For

2.4. Euler Tour Techniques   107

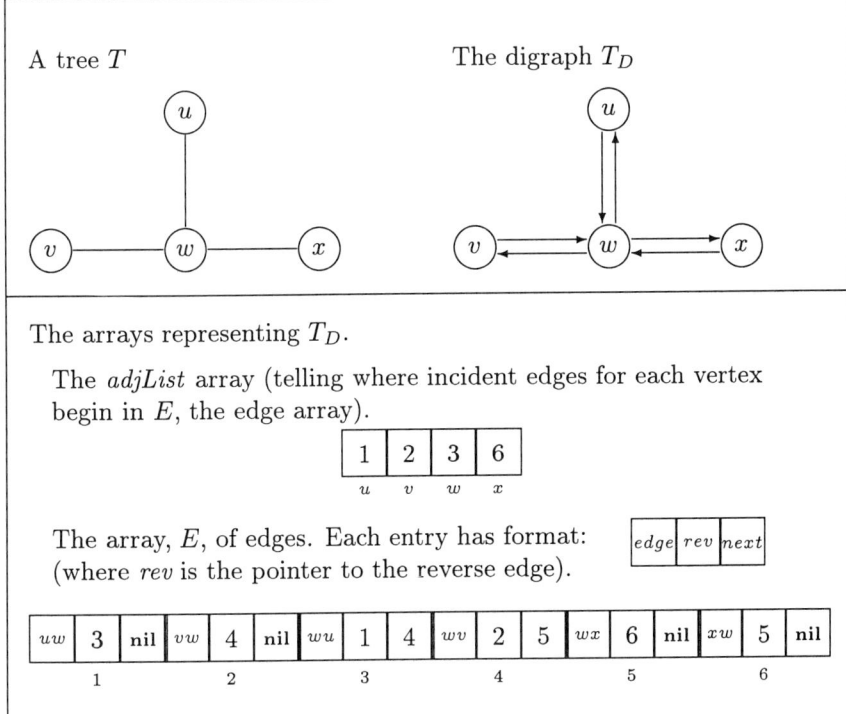

FIGURE 2.13
The data structure for T_D.

simplicity in the algorithms and discussion, we often identify an edge e with its index in the array of edges.)

ALGORITHM 2.13
Construction of a linked Euler tour in a tree.
Input: A tree T in the adjacency list data structure.
Output: An array *etourLink* with $2n - 2$ entries representing an Euler tour of T_D as a linked list. That is, *etourLink*(e) is the index of the edge that follows edge e in the tour.

Each P_i ($1 \leq i \leq n - 1$) does the following steps for its two edges

$e = (v,w)$ in turn:
if $next(e_R) \neq$ **nil then** $etourLink(e) := next(e_R)$
else $etourLink(e) := adjList(w)$
endif

Figure 2.14 shows the Euler tour constructed by Algorithm 2.13 for the digraph T_D in the example in Figure 2.13.

It is clear that Algorithm 2.13 runs in constant time.

EXERCISE 2.17

Show that Algorithm 2.13 does indeed always construct an Euler tour and not a set of separate circuits.

Note that Algorithm 2.13 did not use the root of the tree. The algorithm constructs an Euler tour for an unrooted tree. If we designate an edge incident with *root*, say $(root, v)$ as the first edge of the tour, then the tour constructed traverses the rooted tree in the order of a depth-first search.

We now have an Euler tour represented as a linked list. For some of the applications we need to know the position of each edge in the tour. This is where list ranking is applied. The *etourLinks* are the successor links for the list ranking problem. We observed in Section 2.3.1 that we can establish predecessor pointers in constant time. The predecessor of $(root, v)$ is the last edge in the tour, and we can modify its successor pointer to point to itself, so that the tour has the structure assumed for the list ranking problem.

The *etourLink* array computed by Algorithm 2.13 for the example in Figure 2.13.

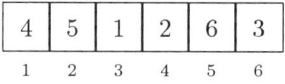

This describes the tour uw, wv, vw, wx, xw, wu.
(The starting point is arbitrary.)

FIGURE 2.14
An Euler tour constructed by Algorithm 2.13.

ALGORITHM 2.14
Construction of an Euler tour with links and position numbers.
Input: A tree T in the adjacency list data structure.
Output: The array *etourLink* from Algorithm 2.13 and an array *posn* where $posn(e)$ is the position of e in the tour.

>Construct a linked Euler tour (Algorithm 2.13);
>P_n reads *root* and assigns $etourLink(e) := e$ for an edge e into the root.
>Apply a list ranking algorithm to the tour to produce the array *rank*;
>(Ranks range from zero for the last edge to $2n - 3$ for the first.)
>Each processor computes $posn(e) := 2n - 2 - rank(e)$ for its two edges

For all the tree problems we discuss in the remainder of this section, the time bounds are determined by the algorithm used for list ranking. The additional work for some algorithms can be done in constant time with n processors; for others, a few additional invocations of a list ranking algorithm are used. The Euler tour technique was developed before the optimal algorithms for list ranking. Thus it originally used the basic list ranking algorithm, giving algorithms for the tree problems that ran in $O(\log n)$ time on n processors. Now, using Algorithm 2.12 gives $O(\log n \log \log n)$ algorithms (with $n/(\log n \log \log n)$ processors) on EREW PRAMs for all the problems. The faster list ranking algorithms of Chapter 5 provide faster optimal algorithms for these applications.

2.4.3 Solving Tree Problems with Euler Tours

For all the algorithms described here, we assume that Algorithm 2.14 has been run to construct an Euler tour of T_D and compute the position of each edge in the tour.

An edge $e = (v, w)$ is a *forward edge* if it is traversed on the Euler tour before e_R, i.e., if $posn(e) < posn(e_R)$. It is a *retreat edge* otherwise. Forward edges are those traversed (from parent to child) in a depth-first search of the tree T; retreat edges correspond to backing up in the depth-first search.

The algorithms are quite straightforward. They all include some simple steps to be performed for each edge. We have seen that such work can be done in parallel for $n - 1$ edges at once, hence in constant time. We will not explicitly mention this in each algorithm.

Finding Parents

In a depth-first search, if (v,w) is a forward edge, i.e., if the tree edge $\{v,w\}$ is traversed first in the direction from v to w, then v is the parent of w in the tree. This observation gives the simple algorithm:

ALGORITHM 2.15
Finding parents.
>   For each edge $e = (v,w)$ **if** $posn(e) < posn(e_R)$ **then**
> $parent(w) := v$
>   **endif**;
>
> P_n does: $parent(root) :=$ **nil**

Algorithms 2.14 and 2.15 together solve the problem of assigning direction to the edges of an undirected, rooted tree.

Numbering and Counting Problems

A key step in the algorithms in the rest of this section is computation of suffix sums for the linked Euler tour using initial weights whose values depend on the particular problem being solved. All the algorithms have the following structure.

>   Initialize weights of the edges;
>   Perform a suffix sum algorithm (e.g., Algorithm 2.12) on
>   the linked Euler tour using the weights as initial values;
>   assume the suffix sums are left in the array *weight*;
>   Calculate the desired result from the values in the weight
>   array.

Assigning Preorder and Postorder Numbers to Vertices

The preorder number of a vertex in a tree is one plus the number of forward edges traversed before encountering the vertex.

ALGORITHM 2.16
Assigning preorder numbers to vertices.
Initialization:
>   **if** e is a forward edge **then** $weight(e) := 1$ **else** $weight(e) := 0$ **endif**;

Compute suffix sums for weights.
Calculate results:

if $e = (v,w)$ is a forward edge **then** $preorder(w) := n - weight(e) + 1$
endif;
P_n does: $preorder(root) := 1$

EXERCISE 2.18
Write a parallel algorithm to assign postorder numbers.

EXERCISE 2.19
Suppose T has k leaves. Write a parallel algorithm to assign the integers 1 through k to the leaves in the order in which they are encountered in the depth-first search represented by the Euler tour. (k is not known to the processors.)

Computing the Number of Descendants of Each Vertex

The number of descendants of a vertex v is the number of (forward) edges in the subtree rooted at v plus one (to count v itself). This is equal to the number of forward edges in the segment of the Euler tour beginning with the forward edge to v and ending with the retreat edge from v. The suffix sum algorithm sums to the end of the tour, so we get the result we need by subtracting the tail segment of the tour.

ALGORITHM 2.17
Computing the number of descendants of each vertex.
Initialization:
    if e is a forward edge **then** $weight(e) := 1$ **else** $weight(e) := 0$ **endif**;
Compute suffix sums for weights.
Calculate results:
    if $e = (v,w)$ is a forward edge **then**
        $descendants(w) := weight((v,w)) - weight((w,v))$
    endif;
P_n does: $descendants(root) := n - 1$

Computing the Level of Each Vertex

The level of a vertex is the number of edges on a path from that vertex up to the root. It is also the difference between the number of retreat edges and forward edges in the tail end of the Euler tour beginning at the vertex of interest.

ALGORITHM 2.18
Computing the level of each vertex.
Initialization:
    if e is a forward edge then $weight(e) := -1$ else $weight(e) := 1$ endif;
Compute suffix sums for weights.
Calculate results:
    if $e = (v, w)$ is a forward edge then $level(w) := weight(e) + 1$ endif;
    P_n does: $level(root) := 0$

Notes and References

The general strategy of the parallel connected component algorithm described in this chapter is due to Hirschberg [8]. It was used in Hirschberg, Chandra & Sarwate [9] to obtain an $O(\log^2 n)$ CREW algorithm using $n\lceil n/\lceil \log n \rceil \rceil$ processors. In Savage & Ja'Ja' [12] a version using $O(m+n \log n)$ processors is given. There is also an algorithm in the same paper that finds connected components in $O(\log n \log d)$ time, where d is the diameter of the graph; it uses $O(n^3/\log n)$ processors.

The $O(\log n)$ algorithm using $\max(n, m)$ processors presented here appeared in a slightly more complicated form in Awerbuch & Shiloach [2] and Shiloach & Vishkin [13]. The simplified version given here is from Awerbuch & Shiloach [3]. A solution for Exercise 2.9 may also be found in that paper.

References for optimal and randomized algorithms for finding connected components appear in the chapters in this book that describe those algorithms.

The basic $O(\log n)$ list ranking algorithm using n processors (Algorithm 2.5) is from Wyllie [16], which also contains a variety of basic parallel algorithms and techniques that have many applications. The deterministic coin tossing technique is introduced in Cole & Vishkin [4], with applications to list ranking, finding connected components of a graph, and finding a minimum spanning forest of a graph. The optimal list ranking algorithm given in this chapter is from Cole and Vishkin [5].

We began the algorithm for deterministic coin tossing with a brief discussion of the basic operations permitted on RAMs and hence PRAMS; for more on this topic, see Cook & Reckhow [6] or Aho, Hopcroft & Ullman [1]. Several methods of constructing 2-ruling sets appear in Cole & Vishkin [5]. The sorting algorithm used in the optimal algorithm for construction of a 2-ruling set uses techniques for small integer sorting like those in Reif [11].

The Euler tour technique was introduced in Tarjan & Vishkin [14] as part of their parallel biconnectivity algorithm. It is also used in Vishkin [15]. The applications described in this chapter appear in those papers. There are early suggestions of the idea of the Euler tour technique in Wyllie [16].

The survey articles Karp & Ramachandran [10] and Eppstein & Galil [7] include overviews of the topics in this chapter and many other topics.

Bibliography

[1] Alfred V. Aho, John E. Hopcroft, and Jeffrey D. Ullman. *The Design and Analysis of Computer Algorithms*. Addison-Wesley, 1974.

[2] Baruch Awerbuch and Yossi Shiloach. New connectivity and msf algorithms for ultracomputer and pram. *Proceedings of the IEEE International Conference on Parallel Processing*, pages 175–179, 1983.

[3] Baruch Awerbuch and Yossi Shiloach. New connectivity and msf algorithms for shuffle-exchange network and pram. *IEEE Transactions on Computers*, C-36(10):1258–1263, October 1987.

[4] Richard Cole and Uzi Vishkin. Deterministic coin tossing and accelerating cascades: Micro and macro techniques for designing parallel algorithms. *Proceedings of the 18th Annual ACM Symposium on Theory of Computing*, pages 206–219, 1986.

[5] Richard Cole and Uzi Vishkin. Deterministic coin tossing with applications to optimal parallel list ranking. *Information and Control*, 70(1):32–53, July 1986.

[6] Stephen A. Cook and Robert A. Reckhow. Time-bounded random access machines. *Journal of Computer and System Sciences*, 7:354–375, 1973.

[7] David Eppstein and Zvi Galil. Parallel algorithmic techniques for combinatorial computation. *Annual Reviews of Computer Science*, pages 233–283, 1988.

[8] Daniel S. Hirschberg. Parallel algorithms for the transitive closure and the connected component problems. *Proceedings of the 8th Annual ACM Symposium on Theory of Computing*, pages 55–57, 1976.

[9] Daniel S. Hirschberg, A. K. Chandra, and D. V. Sarwate. Computing connected components on parallel computers. *Communications of the ACM*, 22:461–464, 1979.

[10] Richard M. Karp and Vijaya Ramachandran. Parallel algorithms for shared-memory machines. In J. van Leeuwen, editor, *Handbook of Theoretical Computer Science, Volume A: Algorithms and Complexity*. MIT Press, 1990.

[11] John H. Reif. An optimal parallel algorithm for integer sorting. *Proceedings of the 26th Annual IEEE Symposium on Foundations of Computer Science*, pages 496–504, 1985.

[12] Carla Savage and Joseph Ja'Ja'. Fast, efficient parallel algorithms for some graph problems. *SIAM Journal on Computing*, 10:682–691, November 1981.

[13] Yossi Shiloach and Uzi Vishkin. An $O(\log n)$ parallel connectivity algorithm. *Journal of Algorithms*, 3:57–67, 1982.

[14] Robert E. Tarjan and Uzi Vishkin. An efficient parallel biconnectivity algorithm. *SIAM Journal on Computing*, 14(4):862–874, November 1985.

[15] Uzi Vishkin. On efficient parallel strong orientation. *Information Processing Letters*, 20:235–240, June 1985.

[16] James C. Wyllie. The complexity of parallel computation. Technical Report 79/387, Cornell University, Ithaca, NY, 1979.

3

List Ranking and Parallel Tree Contraction

Margaret Reid-Miller

School of Computer Science
Carnegie Mellon University
Pittsburgh, PA 15213
mrmiller@cs.cmu.edu

Gary L. Miller

e-mail: glmiller@cs.cmu.edu

Francesmary Modugno

e-mail: fmm@cs.cmu.edu

3.1 Introduction

This chapter discusses parallel algorithms for two problems: list ranking and parallel tree contraction. List ranking is used often as part of a solution to other parallel algorithms, while parallel tree contraction is a technique that has wide application to tree-based problems.

3.2 List Ranking

A common problem in computer science is to find the location of an element in a linked list with respect to the first element of the list, called the **head** of the list. This problem, referred to as list ranking, is a fundamental operation for many applications, such as computing the prefix sum for any associative operation over a linked list; determining the preorder numbering of nodes on trees; and evaluating expressions on trees. On a RAM the problem can be solved by a straightforward sequential algorithm that traverses the linked list in $O(n)$ time.

In this section we discuss various parallel list ranking algorithms. We start by introducing a simple deterministic parallel algorithm due to Wyllie [23]. Wyllie's algorithm is not optimal; the total "work" performed by all the processors is greater than the work performed by a single processor using a sequential algorithm. We then introduce a randomized algorithm by Miller and Reif [17, 18, 19] that optimizes the work, but still needs as many processors as cells in the linked list. This algorithm was improved by Anderson and Miller [4] to use an optimal number of processors and is described next. The final section gives another optimal solution by Anderson and Miller [3], which is deterministic. Chapters 2 and 5 introduce deterministic coin tossing and discuss list ranking further.

3.2.1 The Problem

We consider the slightly modified problem of finding the location, or **rank**, of an element in a linked list with respect to the last element, or **tail** of the list. From solutions to this problem, it is not difficult to obtain solutions to the problem of finding an element's rank with respect to the head of the list.

Formally, the problem is stated as follows:

Given: a linked list of values.

Output: a cell's position with respect to the tail of list.

Assumptions: the linked list is in a continuous block of shared memory, and the tail of the list points to a distinguished value, *nil*.

In order to evaluate the algorithms of this section, we examine their efficiency in terms of the number of processors they require and the total amount of work performed by the processors. An algorithm is *optimal* if it is not possible to reduce simultaneously both the runtime and the number of processors. For the algorithms in this chapter, when we say an algorithm is optimal we mean a stronger notion: that the amount of work required is no more than the amount of work required by an optimal sequential algorithm for the same problem. For list ranking this implies that the work done be $O(n)$. The amount of work done can be measured in two ways: it is equal to the number of processors used times the runtime of the algorithm; or it is equal to the sum of the actual amount of work done by each processor. Very often the two measurements are equal. They differ only when many processors are initially used but become idle as time goes on. In this section we see examples of when the two measurements are and are not equal.

Before beginning, we take a moment to briefly explain some notation. Throughout the entire chapter we describe algorithms in a parallel Algol like code for a PRAM with a host. The meaning of most constructs is clear from context. The "**in Parallel do** *body*" construct is intended to mean that the host broadcasts *body* one instruction at a time to each processor, which then executes each instruction as it arrives. In the "**in Parallel while** *condition* **do** *body*" construction, the *condition* is computed by the host and while it is true, the host broadcasts *body* an instruction at a time to the individual processors. Finally, the special symbol "!" is used to emphasize that the element indexed is local to that processor. Any other index into the array of elements may or may not be local to the processor.

3.2.2 Wyllie's Algorithm for List Ranking

Wyllie presented the simplest algorithm for the list ranking problem. While the algorithm is not optimal, it serves to illustrate a technique that is common to all algorithms for the list ranking problem known as *dereferencing* or *pointer jumping*. Let $v = v_1 \to v_2 \to \ldots \to v_n$ be a link list of cells, not necessarily stored in order. Pointer jumping is the process of reassigning a cell's successor link to point to its successor's successor. That is, if $v_i \to v_{i+1} \to v_{i+2}$, then one application of pointer jumping to v_i will yield $v_i \to v_{i+2}$.

Assume each cell, i, has a pointer, $succ[i]$, that points to its successor in the linked list. The tail of the list points to nil. Each cell i maintains a value $rank[i]$ which is the distance of the cell from its current successor $succ[i]$. Initially, $rank[i]$, is 1. The ranks on the edges keep track of how many pointers were jumped and, in the end, tell us how far from the tail of the list a given cell is. Each processor is assigned one cell.

ALGORITHM 3.1
Wyllie's Algorithm for list ranking
Procedure *Wyllie:*
    In Parallel $rank[!] := 1;$ /* initialize rank */
    In Parallel while $succ[head] \neq nil$ **do**
        if $succ[!] \neq nil$ **do**
            $rank[!] := rank[!] + rank[succ[!]];$ /* update rank */
            $succ[!] := succ[succ[!]];$
        end if
    end in parallel
end *Wyllie*

After each round each linked list is divided into two linked lists of half the length so that after $O(\log n)$ rounds, all the cells point to nil and the $rank[v]$ is the rank of cell v. Figure 3.1 shows two rounds of the algorithm. For illustrative purposes, cells are arranged in their order in the linked list. The ranks of the cells after each round are given on the links. Observe that the algorithm is exclusive read and exclusive write and thus works on an EREW PRAM.

Recall that the sequential algorithm takes time $O(n)$. Since this parallel algorithm uses n processors, each taking $O(\log n)$ time, the total work is $O(n \log n)$, which is not optimal. This is because each processor duplicates the work done by another processor. For example, during Round 1 we produce two chains when only one is needed. Processors for b, d, and f need not have done any work because their final rank can easily be computed once the final rank of their successors are known. Rather than all processors simultaneously computing the final rank of their elements, some processors can wait until the rank of their successor is known and then in unit time compute their own rank.

There are two major issues that arise when trying to devise optimal list ranking algorithms: resolving contention and identifying elements to "jump over." The first issue is that two adjacent list cells can not be jumped over

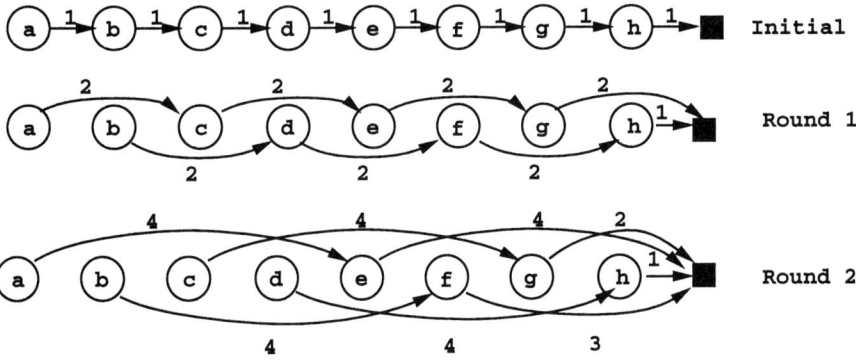

FIGURE 3.1
Two rounds of Wyllie's list ranking algorithm. The black box indicates the value *nil*. The numbers on the links represent the rank of a cell during that round.

at the same time; otherwise two chains would form. To resolve contention Miller and Reif [17, 18, 19] introduced randomization into the *Random_Mate* algorithm, whereas Cole and Vishkin [10] developed deterministic coin tossing based on a cell's address, used by the list ranking algorithms discussed in Chapters 2 and 5. The randomized algorithm is given in the next section. The second issue is addressed by Anderson and Miller [4, 3] in the two following algorithms. In order to splice out elements efficiently, it is necessary to have fewer processors than list cells. A processor needs to identify cells on which to work, once its current cell is spliced out. A problem with naive strategies is that as cells are removed from the linked list, it becomes more difficult to find cells still in the list on which to work. A major feature of the Anderson and Miller algorithms is a rather simple scheduling strategy that allows the processors to keep busy with cells remaining in the linked list.

3.2.3 List Ranking Using Randomization

In this section we present a very simple randomized algorithm for list ranking. It is optimal in the sense that the total work performed is $O(n)$, although it requires one processor for each cell in the linked list and $O(\log n)$ time. The work is optimal because at each round a constant fraction of the processors are freed up to perform other unrelated work. The algorithm uses randomization to resolve contention. Randomization for the list ranking problem has the advantage of being very simple to implement, while always producing the correct answer. However, there is a very small probability that a particular run of the algorithm may take a long time to complete. On average, though, the randomized algorithms are much faster than many deterministic algorithms, because they have much smaller constants.

Our intuition tells us that in order to eliminate excess work, once a processor's cell has been jumped over, the processor should stop jumping cells. That is, at each round half the processors of the previous round ($\lfloor n/2 \rfloor$) should stop jumping cells. Figure 3.2 shows the rounds of this ideal situation. At the final round we have a tree of depth $\log n$, where the numbers on the arcs show the original distance from the cells to which the arcs point. For example, cell a is distance 8 from nil, and cell c is distance 2 from cell e. Following this splicing out phase is a reconstruction phase, in which the linked list is reconstructed and the distance of each cell from the end of the list is computed to obtain its rank. The reconstruction phase is simpler than the first phase since the reconstruction is performed by undoing the work performed in the first phase. Thus, the rank is computed for each cell in reverse order of their removal. For example, first the rank is found for cell e, then cells c and g, and finally for cells b, d, f, and h.

How does a processor know whether it should jump a cell? One approach is to use randomization to determine whether a processor should jump a

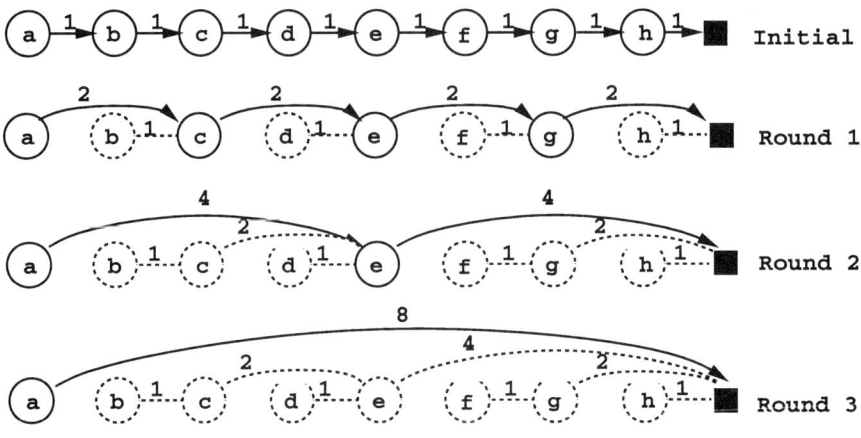

FIGURE 3.2
List ranking with minimal parallel work. The dotted cells indicate that corresponding processors were inactive during the pointer jumping.

cell or not. Once a processor's cell is jumped over, it stops working on this ranking problem and is free to work on other problems until it is needed for a reconstruction phase.

On each round, a processor determines whether to jump a cell by flipping a coin and assigning a sex, either male or female. If the sex of a processor's cell is female and the next cell in the linked list is male, then the processor can dereference its cell. Otherwise, the processor waits for the next round. We refer to this process as *random mating*. It is clear that no two adjacent cells will be jumped over in the same round. Once a cell is jumped over its processor becomes inactive until the reconstruction phase. As in Wyllie's algorithm, during the splicing out phase we maintain the cell's distance in the original linked list to its new successor by adding to its distance the distance of its old successor. At the end of the pointer jumping phase, cells either point to *nil* and have their final value for *rank* or are inactive and their rank is their distance in the original linked list to their current successor.

After the pointer jumping phase is the reconstruction phase, in which processors are reactivated in the reverse order in which they became inactive. The distance of a cell from the tail of the list is equal to the distance to its successor cell plus the distance of its successor cell to the tail of the list. Because we reactivate the processors in the reverse order in which they became inactive, when a processor is reactivated, its successor processor has already been reactivated and has its final rank. Thus, a cell's final rank is simply the rank of itself plus the rank of its successor. In order to determine when to reactivate a cell we use a time stamp, indicating which round a cell was deactivated.

Each processor runs the same program and all the processors start each iteration of the while loops at the same time. Algorithm 3.2 gives the parallel Algol code for the *Random_Mate* algorithm.

ALGORITHM 3.2
A randomized list ranking algorithm
Procedure *Random_Mate:*
    In Parallel $rank[!] := 1$; $active[!] := true$;   /* Initialize */
    host set t := 1;

    In Parallel while $succ[head] \neq nil$ **do**   /* Pointer jumping phase */
        if $active[!] = true$ **and** $succ[!] \neq nil$ **then do**
            $sex[!] := Random\{M, F\}$;
            if $sex[!] := F$ **and** $sex[succ[!]] = M$ **then do**

```
                    time[succ[!]] := t;
                    active[succ[!]] := false;
                    rank[!] := rank[!] + rank[succ[!]];
                    succ[!] := succ[succ[!]];
                end then
                t := t + 1;
            end then

    In Parallel while t > 0 do                  /* Reconstruction phase */
        if time[!] = t and succ[!] ≠ nil then
            rank[!] := rank[!] + rank[succ[!]];
        t := t - 1;
    end in parallel
end Random_Mate
```

Analysis

Using this algorithm, we guarantee that no two adjacent cells in the linked list are jumped over in the same round. The probability a cell is jumped over is $\frac{1}{4}$ because, with probability $\frac{1}{2}$, $sex(i) = F$ and, with probability $\frac{1}{2}$, $sex[succ[i]] = M$. How many rounds of random mate are needed to jump over all cells?

THEOREM 3.1
Simple Random Mate computes the rank of each element of a linked list of length n in $O(\log n)$ time using n processors on a EREW PRAM.

PROOF
Note that the cells are not statistically independent of each other. If a cell i is jumped over then, with probability one, the cell $succ[i]$ is *not* jumped over in the same round. If cell i is not jumped over, then the cell $succ[i]$ is jumped over in the same round with probability $1/3$. This is because the probability that $succ[i]$ is jumped over given that i is not jumped over is equal to the probability that both the $succ[i]$ is jumped over and i is not jumped over $(1/4)$, divided by the probability that i is not jumped over $(3/4)$. However, the probability that a cell is not jumped over in one round is independent of whether it is jumped over in the next round. Let P_i be the probability that the ith cell is still not jumped over after k rounds. Then

$$P_i = (3/4)^k, i = 2, \ldots, n$$

If we choose k so that $P_i = 1/n^c, c \geq 2$, then

$$k = c\lceil \log_{\frac{4}{3}} n \rceil$$

The probability that at least one cell has not been jumped over after k rounds is the disjunction of the probabilities that each cell is not jumped over after k rounds. This disjunction is bounded from above by the sum of the probabilities of each cell not having been jumped over after k rounds. That is,

$$\begin{aligned} \text{Prob (number of cells not jumped over after } k \text{ rounds} > 0) \\ = \quad & P_1 \vee P_2 \vee \ldots \vee P_n \\ \leq \quad & \sum_{i=1}^{n} P_i \\ = \quad & 1/n^{(c-1)} \end{aligned}$$

∎

Thus, for large n, the probability that the algorithm runs for more than $c \log n$ rounds is small. The amount of work done is $n + \frac{3}{4}n + (\frac{3}{4})^2 n + \ldots = O(n)$, which is optimal, although the product of the runtime and processor count is not.

3.2.4 A Simple, Optimal Randomized Algorithm for List Ranking

The problem with the algorithm of the previous section is that it is not optimal in the sense that the number of processors that are actively working on the list ranking problem decreases geometrically each round. Once a processor's cell is jumped over it is free to do other work, until it is needed during the reconstruction phase. However, scheduling these freed processors with other work introduces overhead and in some Single Instruction Multiple Data (SIMD) architectures these processors must remain idle.

In this section we introduce an approach to keep a fixed set of processors busy most of the time. If we assume that each round takes $O(1)$ time, then in order to obtain an algorithm that takes $O(\log n)$ time and $n/\log n$ processors we must remove $O(n/\log n)$ cells per round.

One approach would be to simulate Random Mate using $n/\log n$ processors, by letting each processor do the work of $\log n$ virtual processors. We assign each processor $\log n$ adjacent memory cells. Note that adjacent cells in memory need not be adjacent cells in the linked list. Each processor assigns the cells in its set a sex, and jumps over all males pointed to by a female.

However, a processor may be unlucky and have all its $\log n$ cells assigned female, each of which points to a male. Thus, the unlucky processor jumps over other processors' cells, but none of its own cells are eliminated. If this happens repeatedly, we get an $O((\log n)^2)$ algorithm. However, it is not unrealistic to assume that prefix sum is a unit time operation [5]. In this model it is then possible to rebalance the work among the processors using prefix sum operations, while maintaining a $O(\log n)$ running time.

Another approach is to think of a processor's $\log n$ cells as a queue. A processor looks only at the top of the queue in each round. During each round of the list ranking algorithm, each processor attempts to jump over the successor of the top of its queue. When the top of a queue is jumped over, the processor moves to the next element, etc. However, we need to prevent two processors from jumping over adjacent cells in the linked list to avoid contention. Contention occurs when both i and $succ[i]$ are at the top of queues. When several tops of queues are adjacent cells in the linked list, we call the set of adjacent cells a **chain**. Cells of a chain cannot all be spliced out in the same round. Again, contention can be avoided if, during each round, each processor chooses independently and randomly a sex for the top of its queue. If during the round, two adjacent cells are at the tops of queues, then $succ[i]$ is jumped over only if i is female and $succ[i]$ is male. Since only cells that are at the top of a queue are attempting to jump cells, we can assume that all cells not at the top of a queue have a male sex. A given round is shown in Figure 3.3. A solid arrow head indicates the top of the queue.

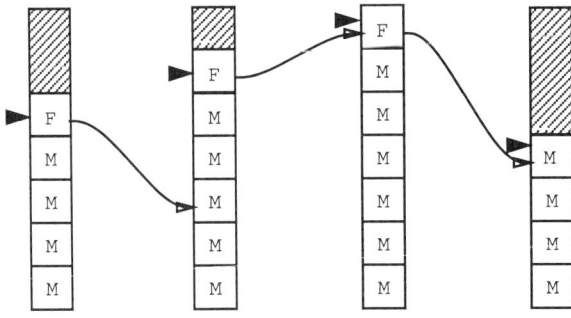

FIGURE 3.3
Random Mate using queues. Solid arrow heads indicate the top of the queues.

Unfortunately, this algorithm can also become imbalanced. If a large number of tops of queues all point to queues of only a few processors, then these few processors have most of their cells jumped over and have little work to do, while the remaining processors have almost all the work to do. If these same tops of queues again point to queues of another small number of processors, then a few more processors have little work remaining to do. Eventually a small number of processors would have all the remaining work to do and a load balancing step would be required in order to keep the running time down. Figure 3.4 illustrates how cells might be eliminated in this fashion, leading to imbalanced queues. The problem here is that some processors may be lucky and have many of their cells eliminated, either by themselves or by the work of other processors. While other processors are unlucky and work on eliminating other processors' cells and only have a few of their own eliminated.

To avoid this imbalance we let each processor **splice** out the cells assigned to *itself*. We distinguish here between *splicing* out one's *own* cell and *jumping over* a successor cell. A processor can splice its own cell if the linked list is doubly linked, i.e., each cell, i, has a pointer to both its predecessor, $pred[i]$, and its successor, $succ[i]$, and the successor of the tail of the linked list and the predecessor of the head of the list point to nil. If we are splicing out cell i, we set $pred[succ[i]] := pred[i]$ and $succ[pred[i]] := succ[i]$. As with

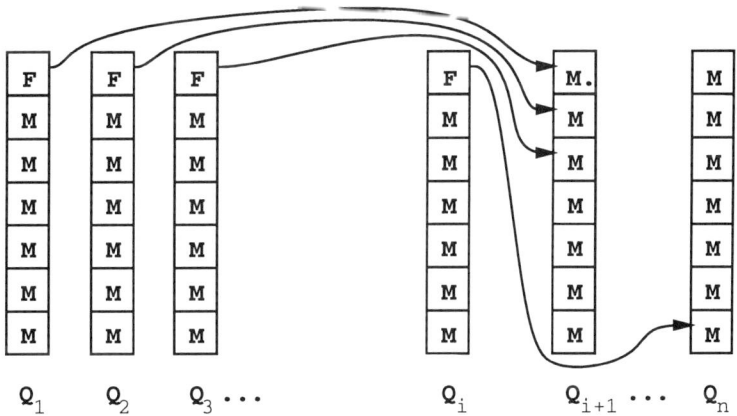

FIGURE 3.4
Imbalance caused by jumping over successor cells.

the previous algorithms the head of the list is never spliced out, and when the tail of the list is spliced out the successor of the new tail points to *nil*.

EXERCISE 3.1

Show how, if the cells are singly linked, they can be doubly linked in $\log n$ steps using $n/\log n$ processors on an EREW PRAM.

Again, each processor is assigned $\log n$ consecutive cells in memory and treated as a queue, and the algorithm has two phases. During the first phase, all cells are spliced out of the linked list. When all the processors have completed the first phase, the second phase starts. The cells are put back into the linked list in the reverse order of their removal, and the distance of the cells from the tail of the list is computed. Code for the algorithm is given in Algorithm 3.3.

The linked list is represented as a two-dimensional array, indexed by the processor ID and position in the queue. We represent this index pair using the infix operator at (@). For example the index pair 7@23 means the 23rd element of processor 7's queue. If i is an index pair, we use $i.0$ to indicate the first element of the pair, namely the processor ID, and $i.1$ to indicate the second element of the index pair, namely the position of the processor's queue. Again we use ! to represent a processor's ID. The variable *top* is an index pair, where *top*[!] is the top of the queue for the current processor. We process the queue by increasing positions. That is, *top*[!] is initialized to !@1 and proceeds to !@2, and so on.

A cell is spliced out only if it is a male pointed to by a female. Since only males are spliced out, all cells not at the top can be assumed to be female. During the splice out phase, each cell maintains a record of its original distance, *rank*[i], from the cell to which it points, so that we may reverse the first phase in order to compute the cell's final rank. Initially, *rank*[i] := 1 for all cells. If during a round i is spliced out, then we add *rank*[i] to *rank*[*pred*[i]]. During the reconstruction phase the rank of a cell is computed in reverse order of removal. When i is added back to the linked list the rank is *rank*[i] := *rank*[i] + *rank*[*succ*[i]].

ALGORITHM 3.3

An optimal randomized list ranking algorithm

Procedure *splice_out*(i)
    rank[*pred*[i]] := *rank*[*pred*[i]] + *rank*[i];
    succ[*pred*[i]] := *succ*[i];
    if *succ*[i] $\neq$ *nil* **then** *pred*[*succ*[i]] := *pred*[i];
end *splice_out*

3.2. List Ranking

Procedure *Optimal_Random_Mate*
    In Parallel do   /* Initialize */
        for $i =!@1$ **to** $!@\lceil \log n \rceil$ **do**
            $rank[i] := 1;\ sex[i] := F;$
        end for
        $top[!] :=!@1;$
    end in parallel
    host set $sex[nil] := M;\ t := 1;$

    In Parallel while $succ[head] \neq nil$ **do**   /* Pointer jumping phase */
        if $top[!].1 \leq \lceil \log n \rceil$ **then do**
            $sex[top[!]] := Random\{M, F\};$
            if $(sex[pred[top[!]]] = F$ **and** $sex[top[!]] := M)$ **then do**
                $splice\_out(top[!]);$
                $top[!].1 := top[!].1 + 1;\ splicetime[top[!]] := t;$
            end then
        end then
        $t := t + 1;$
    end in parallel

    In Parallel while $t \geq 0$ **do**   /* Reconstruction phase */
        $top[!].1 := \lceil \log n \rceil;$
        if $(splicetime[top[!]] = t$ **and** $succ[top[!]] \neq nil)$ **then do**
            $rank[top[!]] := rank[top[!]] + rank[succ[top[!]]];$
            $top[!].1 := top[!].1 - 1;$
        end then
        $t := t - 1;$
    end in parallel
end *Optimal_Random_Mate*

Analysis

THEOREM 3.2
Optimal_Random_Mate computes the rank of each element of a linked list of length n in $O(\log n)$ time using $n/\log n$ processors on an EREW PRAM.

128 Chapter 3. List Ranking and Parallel Tree Contraction

PROOF

The probability that the top of a queue is removed during any given round is at least $\frac{1}{4}$. To make the analysis easy, we assume that if a cell is not adjacent to another cell at the top of a queue, it only has probability of $\frac{1}{4}$ of being removed. The number of items that are removed from a queue after t rounds can be viewed as a binomial random variable $S_t^p, p = \frac{1}{4}$ (i.e., sum of t independent Bernoulli trials with success probability p). The *expected* time for a queue to become empty is at most $4\log n$. Chernoff [8] shows that S_t^p is substantially less than its expected value, pt, with small probability:

$$\operatorname{Prob}[S_t^p < (1 - \beta)pt] < e^{-\beta^2 pt/2}, \text{ for } 0 \leq \beta \leq 1$$

If we take $p = \frac{1}{4}$, $t = 16\log n$, and $\beta = 3/4$, we have:

$$\operatorname{Prob}[S_t^p < \log n] < e^{-\frac{9}{8}\log n} < 1/n$$

Thus, the probability that a particular queue is not empty after $16\log n$ rounds is less than $1/n$. Since there are $n/\log n$ queues, the probability that there is a nonempty queue at time $16\log n$ is less that $1/\log n$. It follows that the expected runtime is $O(\log n)$. ∎

It is possible to improve the runtime of this algorithm by a constant factor by splicing out elements from the linked list until there are $n/\log n$ elements left. This takes just a little over $4\log n$ rounds. Then using the $n/\log n$ processors, the elements can be spliced out using Wyllie's algorithm in time $O(\log n)$. Finally, the linked list is reconstructed in the reverse order the cells were spliced out in the first phase. Reducing the problem to a smaller linked list that can use Wyllie's algorithm is a common approach to optimal list ranking algorithms. That is, the steps are:

1. Reduce the problem to $O(n/\log n)$ elements.
2. Solve the list ranking problem on the reduced linked list with Wyllie's algorithm.
3. Fill in the ranks for the remaining elements.

3.2.5 An Optimal Deterministic List Ranking Algorithm

In the previous algorithm, when the top cells of two processor queues are adjacent in the linked list, we avoided attempting to splice out both top cells simultaneously by randomly tossing a male/female coin. In this section we use a variant of the deterministic coin tossing technique devised by Cole and Vishkin for breaking symmetry in parallel algorithms. Unlike the optimal CRCW PRAM list ranking algorithms presented in Chapters 2 and 5, which use 2-ruling sets, the algorithm presented here only needs to find $\log \log n$-ruling sets to get an optimal $O(\log n)$ time $n/\log n$ processor EREW PRAM algorithm. Finding $\log \log n$-ruling sets is substantially simpler than finding 2-ruling sets, which requires a complicated sorting step, and hence much larger constants. In addition, the scheduling step of the algorithm in this section is simple and has the advantage that cells are reallocated to processors only once. With local memory architectures reallocation can add sizable overhead to an algorithm. In this section we describe the basic deterministic algorithm. In the next section we show how to find $\log \log n$ ruling sets.

The basic idea of the deterministic algorithm is the same as with the *Optimal_Random_Mate* algorithm of the previous section. Here we also assume that we have $n/\log n$ processors, each originally assigned $\log n$ continuous blocks of memory. The main difference between the two algorithms is that here we find a k-ruling set (defined below) to resolve contention instead of randomization.

The algorithm consists of three phases: First, a deterministic list ranking algorithm splices out items from the linked list until $n/\log n$ items remain. At this point in the algorithm the linked list has been reduced to an $n/\log n$ cell *weighted* linked list. Second, we use a prefix sum algorithm to assign one pointer to each of the $n/\log n$ processors and then apply Wyllie's algorithm with one processor per list cell. Finally the ranks of the cells are computed by adding them back into the linked list in the reverse of the order that they were removed.

Ruling Set and Graph Colorings

In this subsection we show how to find colors and ruling sets for linked lists. They need not be doubly linked. We begin by defining the notion of a k-ruling set, which is also defined in Chapter 2. We treat a linked list L as a directed graph. If $V_1, \ldots, V_n$ are the cells of L then we represent L as the directed graph $G = (\{V_1, \ldots, V_n\}, E)$, where E is the set of edges $\{e_{ij}\}$ such that V_i is linked to V_j in L.

DEFINITION 3.1
Let $G = (V, E)$ be the directed graph representation of a linked list L with

vertices $V_1, \ldots, V_n$. We define a subset S of V to be a **k-ruling set** if:

1. No two vertices in S are adjacent.
2. For each vertex V_i in V there is a directed path from some vertex in S to V_i with length at most k.

This definition implies that there are at most k vertices between any two vertices in a k-ruling set. The **kingdom** of a vertex V_i in S is the list of all vertices in V that lie between V_i and the next vertex in S. The vertices in S are called the **rulers** and the vertices between rulers are called the **subjects**.

$$\underbrace{\overset{ruler}{V_1} \to \overset{sbjct}{V_2} \to \overset{sbjct}{V_3} \to \ldots \to}_{kingdom} \overset{ruler}{V_i} \to \ldots \to \overset{ruler}{V_j} \to \ldots$$

In the notation of the figure above, the vertices in the k-ruling set are marked with *ruler*. Note that a kingdom has at least one subject (not including the vertex in the ruling set) and at most k subjects.

The notion of k-ruling set is similar to the concept of a graph coloring. Let $G = (V, E)$ be an undirected graph. A map $C : V \to \{0, \ldots, k-1\}$ is a **k-coloring** of G if $(x, y) \in E$ implies that $C[x] \neq C[y]$. We first observe that given a k-coloring of the vertices, a $2k$-ruling set can be obtained in unit time using n processors. Then we show how to find a coloring for a linked list when k is small.

Let C be a k coloring of a linked list L. A *2k-ruling set* is defined as follows:

1. The head of the linked list, *head*, is a ruler.
2. Element x is a ruler if $pred[x] \neq head$ and $C[pred[x]] > C[x]$ and $C[succ[x]] > C[x]$.

Observe that the head of the linked list can have the largest possible kingdom. This happens when the color of the head is greater than zero and its subjects have colors $0, 1, \ldots, k-1, k-2, \ldots, 1$, respectively. Note that we get a slightly simpler construction if we assume that the head has color zero. We have shown that the problem of constructing a $2k$-ruling set can be reduced to finding k-colorings.

EXERCISE 3.2

Show how to construct a k-ruling set from a k-coloring in constant time using n EREW processors.

3.2. List Ranking

EXERCISE 3.3

Show how to define a notion of ruling sets for an arbitrary graph. Can you get small kingdoms from a coloring using a small number of colors?

Small Colorings

We say that a coloring C is an m-bit coloring if each color is written as a length m binary string of zeros and ones. Let C be an m-bit coloring of our linked list L. For example, we could use the processor ID as the coloring, where $m = \log P$ and P is the number of processors. We assume that *head* has color zero, i.e., a bit string of m zeros. Since C is a coloring there must be some bit of $C[x]$ that is not equal to the corresponding bit of $C[pred[x]]$. Let $B'[x]$ be the index of this differing bit and, for simplicity, let $B'[x]$ be the smallest such index. Thus, $B'[x]$ is a number between 0 and $m-1$ for $x \neq head$. We write $B'[x]$ in binary. Let $B[x] = a \cdot B'[x]$, where a is the $B'[x]^{th}$ bit of $C[x]$ and $\cdot$ is concatenation. $B[x]$ simply describes the first bit of $C[x]$ that differs from the corresponding bit of $C[pred[x]]$. A $\lceil \log m \rceil + 1$-bit coloring C' is obtained as follows:

$$C'[x] = \begin{cases} B[x] & \text{if } x \neq head \\ 0 & \text{if } x = head \end{cases} \quad (3.1)$$

EXERCISE 3.4

Show how to obtain a $\log \log n$-bit coloring in constant time.

EXERCISE 3.5

Show that equation 3.1 also correctly colors rooted trees where *head* is now the root.

EXERCISE 3.6

Show how to modify equation 3.1 to properly color bounded degree graphs.

Deterministic List Ranking

As before, each of $n/\log n$ processors is assigned $\log n$ cells to remove. When a cell is at the top of a queue and neither its successor nor its predecessor are at the top of a queue, we call the cell an **isolated cell**. There is no contention with isolated cells. The difficulty comes when the cell is part of a chain of cells at the top of several queues. Before, we broke the symmetry by flipping a male/female coin and spliced out a cell when it was a male to which a female pointed. Otherwise the processor remained idle until the next round. Here we break symmetry by using deterministic coin tossing to obtain

$\log \log n$ ruling sets among the cells. Since a chain can have at most $n \log n$ cells and we have $n \log n$ processors, we can use the results of the previous section to obtain $\log \log n$ ruling sets in constant time.

Ruling sets divide a chain of cells into sublists where the head is the *ruler* and the remaining cells are the *subjects*. The key idea is that the rulers are assigned the task of splicing out *all* of their subjects. This means that each processor that finds a subject at the top of its queue can assume that its subject will be spliced out by another processor. Therefore, it can skip over its subject by adjusting the top of the queue to the next cell in the queue. In the next round, this processor can then continue working using its new top. In the mean time, each ruler splices out one subject per round, for up to $\log \log n$ rounds.

A processor stops working when either its queue becomes empty or it has executed $5 \log n$ rounds. As we show later, $5 \log n$ rounds is sufficient to reduce the total number of cells remaining to at most $n / \log n$ cells. Once all processors have stopped, then the remaining cells are allocated to the $n / \log n$ processors, one cell each, and the processors proceed with Wyllie's algorithm. Finally, there is the reconstruction phase.

We give a more formal description of splice out phase of the algorithm by giving the code that each processor executes in Algorithm 3.4. As before, the queue for the processor is represented by an array, with *top* pointing to the cell at the top of the queue. The linked list is assumed to be doubly linked. The subprocedure *splice_out* is the same as in the *Optimal_Random_Mate* algorithm.

ALGORITHM 3.4
The splice out phase of an optimal deterministic list ranking algorithm
Procedure *increment_top*
    $top[!].1 := top[!].1 + 1;$
    if $top[!].1 < \log n$ **then if** $pred[top[!]] = nil$ **then**
        $top[!].1 := top[!].1 + 1;$ /* Don't splice out head of list */
    if $top[!].1 < \log n$ **then** $status[top[!]] := A;$
end *increment_top*

Procedure *Deterministic_List_Ranking*
    In Parallel do   /* Initialize */
        for $i = 1$ **to** $\lceil \log n \rceil$ **do**
            $rank[!@i] := 1; \ status[!@i] := I;$
        end for

$top[!] := !@0;$ *increment_top*;
end in parallel
host set $status[nil] := I$;

In Parallel for $t = 1$ **to** $6 \log n$ **do**
    if $top[!].1 < \log n$ **then do**   /* Subdivide chain */
        if $status[top[!]] = A$ **and**
            $(status[pred[top[!]]] = A$ **or** $status[succ[top[!]]] = A)$ **then**
            $status[top[!]] := Find\_Ruling\_Sets(top[!])$;

        if $status[top[!]] = A$ **then do**
            $splice\_out(top[!])$;   /* Splice out isolated cell */
            increment_top;
        end then

        if $status[top[!]] = S$ **then**
            increment_top;   /* Advance top past subjects */

        if $status[top[!]] = R$ **then do**
            $splice\_out(next[top[!]])$;   /* Splice out a subject */
            if $next[!] = nil$ **then**   /* No more subjects */
                $status[top[!]] := A$;
        end then
    end then
end Parallel for
end *Deterministic_List_Ranking*

A cell can have the status $\{A, I, R, S\}$ denoting active, inactive, ruler, or subject, respectively. Initially, the top of each queue has status *active* and all other cells have status *inactive*. At the beginning of each round the top of a queue can either be active or a ruler.

If the top of a queue is active, the first step is to determine whether it is part of a chain. If it is, it calls the subprocedure *Find_Ruling_Sets*, which subdivides chains into short linked lists, linked by *next*, such that the heads of the linked lists are rulers and the remaining cells are subjects. At this point the top of a queue is either active, a subject, or a ruler. Active cells do not have adjacent cells that are active and, therefore, can splice themselves out and advance the top of the queue to the next cell in the queue. Subjects advance the queue top to the next cell in the queue since they will be spliced out by their ruler.

134 Chapter 3. List Ranking and Parallel Tree Contraction

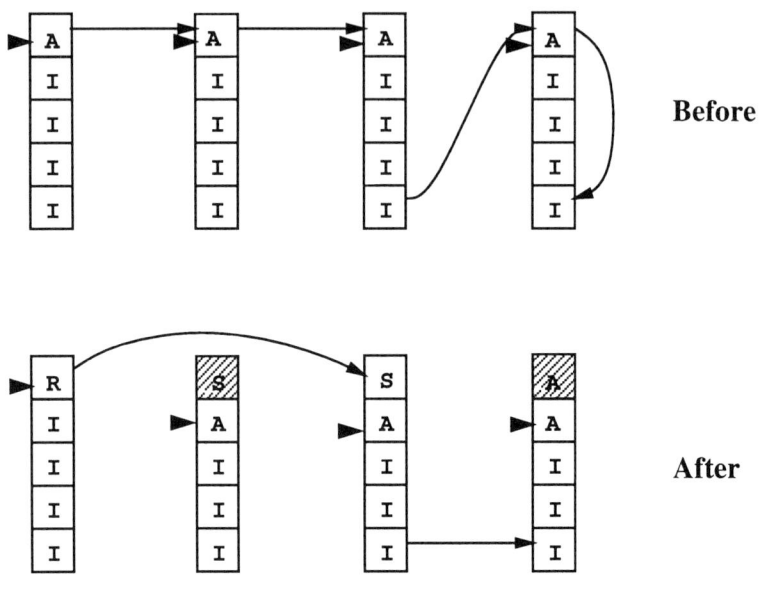

FIGURE 3.5
Status of cells after one application of *Deterministic_List_Ranking*.

Finally rulers splice out one of their subjects. Once a ruler has spliced out all of its subjects it puts itself back into the active status and in the next round checks whether it is part of a new chain. Figure 3.5 shows an example of the status of the elements of a linked list after one application of *Deterministic_List_Ranking*.

EXERCISE 3.7
Can the three cases for **if** *status* ... be run in parallel? If not, show when parallel execution fails.

The correctness of the algorithm follows from the facts that adjacent cells are never removed at the same time, and every cell is eventually looked at by a processor. In the next section we discuss the scheduling that ensures that the running time of the algorithm is $O(\log n)$.

Scheduling

In order to keep all the processors busy we need to be sure that the queues become empty at approximately the same time. Queues can become imbalanced because the length of a queue is not reduced while its top is a ruler removing subjects. The solution to this problem has two parts: we insure that chains do not become too long by breaking them into $\log\log n$ ruling sets, and we insure that rulers are assigned to shorter queues as opposed to longer queues.

In the section on small colorings we showed how to find $\log\log n$ ruling sets. In this section we consider the second problem, that a long queue could be assigned a ruler. A processor needs to splice out all of its subjects before it can continue working on its own queue, which can lead to an imbalance in queue lengths. We would like the processor with the smallest queue to become a ruler. Therefore, we modify *Find_Ruling_Sets* to subdivide chain into ones that have either monotone increasing or monotone decreasing queue lengths and then find $\log\log n$ ruling sets within the monotone chains. If a ruling set has monotone decreasing queue lengths, we reverse the pointers in constant time so that the chain is traversed backwards instead of forwards to splice out elements. This allows rulers to have minimum height queues and remove cells working uphill. Algorithm 3.5 shows the code for this modified version of *Find_Ruling_Sets* and its subprocedure *Find_Increasing_Chains*. Recall that *Dererministic_List_Ranking* calls *Find_Ruling_Sets* with $top[!]$ as an argument, which is an index pair for two-dimensional arrays. The code for *Find_Ruling_Sets* uses the parameter i for this index pair.

ALGORITHM 3.5
Algorithm to find chains with monotone increasing queue lengths
Procedure *Find_Increasing_Chains(i)*
    In Parallel do

$$inchain[i] := true;\ next[i] := nil;\ covered[i] := false;$$
if $succ[i] \neq nil$ **then if** $inchain[succ[i]]$ **then do**
    if $top[i.0].1 \leq top[succ[i].0].1$ **then do**
        $next[i] := succ[i];$ /* Increasing or equal queue lengths */
        $covered[i] := covered[next[i]] := true;$
    end then
    else   /* Decreasing queue lengths */
        if not $(covered[i]$ **or** $covered[pred[i]])$ **then**
            $next[i] := pred[i];$ /* Reverse pointers */
end then
    end in parallel
end *Find_Increasing_Chains*

Procedure *Find_Ruling_Sets(i)*
    In Parallel do
        Find_Increasing_Chains(i);
        Find_Coloring(i);
        /* Subdivide chains at local minima */
        if $next[i] \neq nil$ **then if** $next[next[i]] \neq nil$ **then**
            if $(color[i] > color[next[i]]$ **and**
                $color[next[i]] < color[next[next[i]]])$ **then**
                $next[i] := nil$;
        $status[i] := R$;   /* Determine status */
        if $next[i] \neq nil$ **then** $status[next[i]] := S$;
        if $next[i] = nil$ **and** $status[i] = R$ **then** $status[i]] = A$;
    end in parallel
end *Find_Ruling_Sets*

Find_Increasing_Chains subdivides each chain into subchains that are monotone with respect to queue lengths; each subchain is linked through the pointers *next* and the tails point to *nil*. For those parts of the chain that are monotone increasing or strictly equal in queue length, *next* is the same as the *succ* pointer. For parts that are strictly decreasing, *next* is the same as the *pred* pointers. In this way, as one traverses these subchains along the *next* pointers the queue lengths increase monotonically.

We use the boolean variable *inchain* to mark which cells are part of a chain and which are singletons or inactive. Those cells that do not have *inchain* set to *true* are not in a chain. We initialize all *next* pointers to *nil*. The boolean *covered* indicates whether we have determined to which subchain the cell belongs and we initialize it to *false*. First we find subchains from the parts of the chain that have monotone increasing queue lengths. Recall that $top[i.0].1$ is the length of the queue $i.0$. The *next* pointers are set in the forward direction using the *succ* pointers. The local maxima are the tails of these subchains. We mark cells in these subchains, including the tails, as *covered*. The remaining uncovered cells have strictly increasing queue lengths. If a pair of cells is uncovered, the second cell in the pair sets its *next* pointer to the *pred* so that the pointers go in the reverse direction. Notice that the first or last cell in the original chain may become a singleton chain.

Find_Ruling_Sets takes these monotone chains and subdivides them into $2 \log \log n$ rulings sets. First it finds a $\log \log n$ coloring as shown in

3.2. List Ranking

Section 3.2.5. Then it breaks the chains at local minimum colorings, so that chains are at most $2 \log \log n$ long. Finally it set the heads of the subchains as rulers and the remaining cells as subjects. Any ruler that does not have a subject is isolated from active cells so it too is set to active.

The effect of these modifications is to have the following two constraints on rulers and their chain of subjects: a ruler never has more that $2 \log \log n$ subjects, and the height of a ruler is no greater that the height of its subjects. The performance analysis given in the next section shows that these conditions are sufficient to make the list ranking algorithm an $O(\log n)$ algorithm.

There is one final detail. Because we have reversed the direction that some rulers splice out their subjects, we can have two rulers that splice out adjacent subjects. Therefore, we modify *Deterministic_List_Ranking* so that when rulers splice out their subjects they use the subprocedure *splice_Next* instead of *splice_Out*. The *splice_Next* algorithm, shown in Algorithm 3.6, is written from the point of view of the ruler, who is actually doing the work and lets rulers who have chains running forward to splice first, and then rulers with backward chains to splice next.

ALGORITHM 3.6
A ruler splicing out its next subject
Procedure *splice_Next(i)*;
    In Parallel do
        if $next[i] = succ[i]$ **then do**   /* Forward chains */
            $rank[i] := rank[i] + rank[succ[i]]$;
            $succ[i] := succ[succ[i]]$;
            if $succ[i] \neq nil$ **then** $pred[succ[i]] := i$;
        end then
        else do   /* Backward chains */
            $rank[pred[pred[i]]] := rank[pred[pred[i]]] + rank[pred[i]]$;
            $pred[i] := pred[pred[i]]$;
            $succ[pred[i]] := i$;
        end else
        $next[i] := next[next[i]]$;
    end in parallel
end *splice_Next*

Analysis

In order to analyze the running time of these algorithms, we provide an amortization scheme that assigns weights to the items for which a processor is responsible. Then we show that after a given round the weight of the entire system is reduced by a constant factor. In order to determine the efficiency of this algorithm, we use the following accounting scheme. We think of a queue as being stacked vertically, and the height of an element as its distance from the bottom of the queue. We refer to the height of the queue as the number of elements in the queue remaining to be processed. For a given processor queue, each item is assigned a weight relative to its position in the queue. The i-th item from the top of the queue is assigned weight $(1-\alpha)^i$, where $\alpha = 1/\log\log n$.

The weight change to the system is as follows:

1. the removal of an isolated item removes the item's entire weight.
2. the removal of a subject removes half its weight.
3. the identification of a subject removes half its weight.
4. the identification of a ruler removes nothing.

We show that each round reduces the total weight by a factor of at least $1-\frac{\alpha}{4}$. This rate of reduction allows us to bound the number of rounds required to reduce the number of items remaining in the linked list to $n/\log n$.

LEMMA 3.1
A single round of the algorithm reduces the weight by a factor of at least $1-\frac{\alpha}{4}$.

PROOF
To facilitate the argument, we use the following bookkeeping trick: the weight of a processor's queue is the sum of the weights of the remaining items on the queue, plus one half the initial weights of any subjects for which it is responsible. For example, in Figure 3.6, queue Q_i has weight $(1-\alpha)^2 + (1-\alpha)^3 + (1-\alpha)^4 + \frac{1}{2}(1-\alpha)^2 + \frac{1}{2}(1-\alpha)^1$, with the first three terms coming from the items on queue Q_i and the fourth and fifth terms coming from the subjects on queues Q_j and Q_k, respectively.

The following facts are useful in the analysis that follows:

1. the number of queues is $n/\log n$.
2. initially, the weight of each queue is:

$$\sum_{i=0}^{\log n}(1-\alpha)^i \leq \sum_{i=0}^{\infty}(1-\alpha)^i = 1/(1-(1-\alpha)) = 1/\alpha = \log\log n.$$

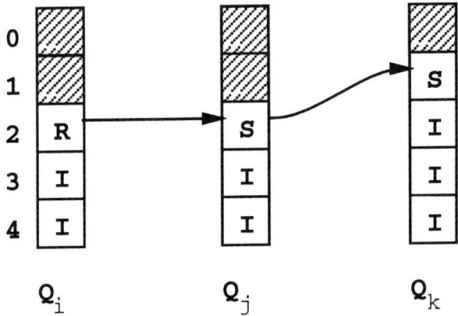

FIGURE 3.6
The weight of Q_i is equal to the weight of the remaining items on the queue plus half the weights of the subjects of the chain.

3. initially, the total weight of the system is:
$$\leq \frac{n \log \log n}{\log n}.$$

In each round of the algorithm a processor either 1) identifies the top of the queue as an isolated cell, which it splices out, 2) identifies the elements of a chain as either subjects or rulers, or 3) finds the top of its queue is a ruler and splices out one subject. Thus, we divide the total weight into three corresponding categories: 1) isolated active cells and the inactive cells below them, 2) active cells in chains and the inactive cells below them, and 3) rulers and inactive cells below them and subjects, as shown in Figure 3.7. Every cell remaining in the system can be placed in exactly one of these categories. We examine the rate at which the weight is reduced for each case during one round of the algorithm and show that each case reduces the weight of its category by at least a factor of $1 - \frac{\alpha}{4}$, and hence the total weight by at least a factor of $1 - \frac{\alpha}{4}$.

Case 1. Removal of an isolated cell (Figure 3.8)
For each isolated cell, assume the weight of the cell is $(1 - \alpha)^i$. Initially, the weight of the queue is
$$Before = \sum_{j=i}^{\log n} (1 - \alpha)^j.$$

140   Chapter 3.   List Ranking and Parallel Tree Contraction

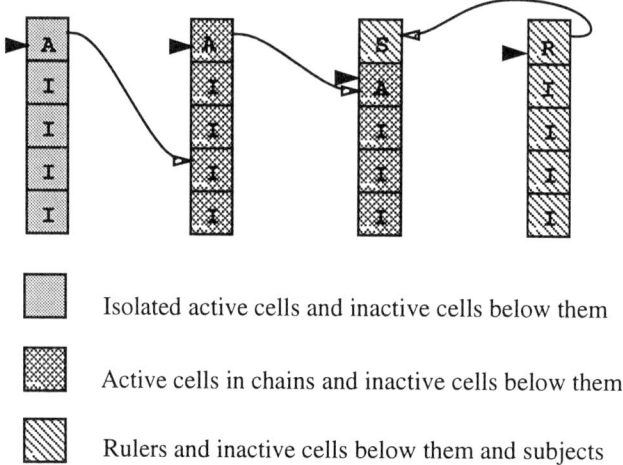

FIGURE 3.7
Categories to which cells belong.

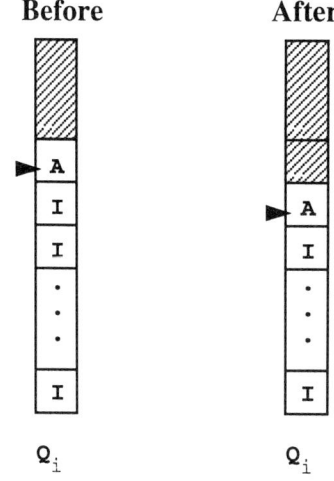

FIGURE 3.8
Removal of an isolated cell.

After the cell is removed the weight of the queue becomes

$$After = \sum_{j=i+1}^{\log n}(1-\alpha)^j.$$

$$\begin{aligned}\frac{After}{Before} &= \frac{Before - (1-\alpha)^i}{Before} \\ &= 1 - \frac{(1-\alpha)^i}{Before} \\ &\leq 1 - \frac{(1-\alpha)^i}{(1-\alpha)^i \sum_{j=0}^{\infty}(1-\alpha)^j} \\ &= 1 - \alpha\end{aligned}$$

Case 2. Identification of a Subject (Figure 3.9)
We account for the weight of all of the queues associated with the chain. The queues that have cells that become subjects lose one cell each, and the queue that has the ruler picks up the subject cells. Cells that are identified as subjects lose half of their weight. When identifying subjects, the weight of the ruler's queue will reduce the least when the ruler has only one subject and both the ruler and the subject are at the same height, i. Initially, the weight of the two queues is

$$Before = 2\sum_{j=i}^{\log n}(1-\alpha)^j.$$

After the cell is identified as a subject and half of its weight is removed from system the weight of the two queues becomes

$$After = Before - \frac{1}{2}(1-\alpha)^i.$$

Therefore,

$$\begin{aligned}\frac{After}{Before} &= 1 - \frac{\frac{1}{2}(1-\alpha)^i}{Before} \\ &\leq 1 - \frac{\frac{1}{2}(1-\alpha)^i}{2(1-\alpha)^i \frac{1}{\alpha}} \\ &= 1 - \frac{\alpha}{4}\end{aligned}$$

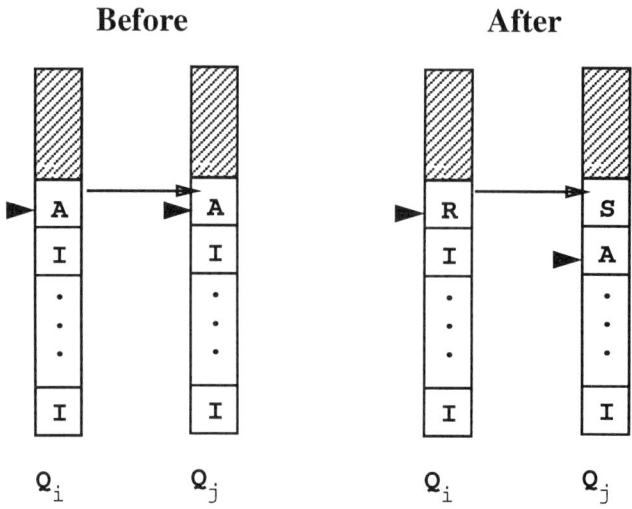

FIGURE 3.9
Identification of a subject.

Case 3. Removal of a Subject (Figure 3.10) When a subject is removed only the weight of the ruler's queue changes. Recall that in our chains, the ruler has the shortest queue. For the purpose of the analysis, we can think of a ruler as splicing out the subject with the greatest height first. In this way, the weight changes the least when the ruler has the maximum number of subjects, $\log \log n$, and each subject is at the same height as the ruler. Initially, the weight of the ruler is the weight of the ruler's queue plus the weight of the $\log \log n$ subjects:

$$Before = \sum_{j=i}^{\log n}(1-\alpha)^j + \frac{1}{2}\log \log n(1-\alpha)^i.$$

Afterward, half of the initial weight of a subject is removed from system, and the weight of the two queues becomes

$$After = Before - \frac{1}{2}(1-\alpha)^i.$$

3.2. List Ranking 143

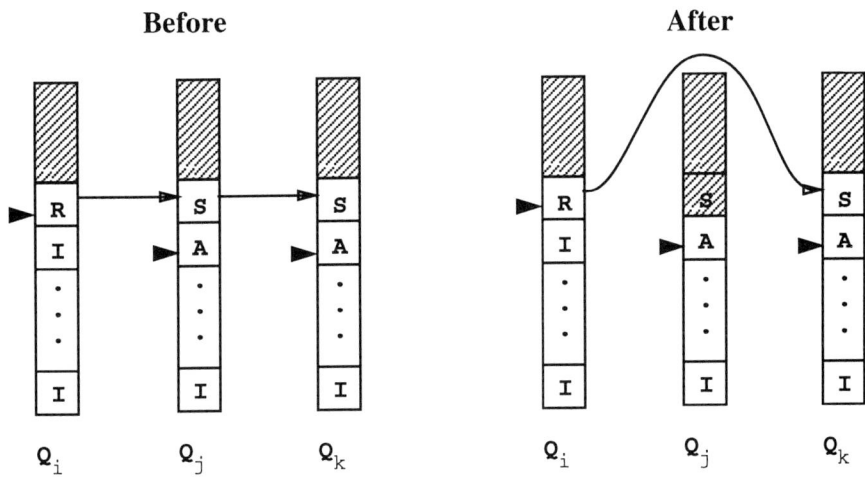

FIGURE 3.10
Removal of a subject.

Therefore,

$$\begin{aligned}
\frac{After}{Before} &= 1 - \frac{\frac{1}{2}(1-\alpha)^i}{\sum_{j=i}^{\log n}(1-\alpha)^j + \frac{1}{2}\log\log n(1-\alpha)^i} \\
&\leq 1 - \frac{\frac{1}{2}(1-\alpha)^i}{(1-\alpha)^i \frac{1}{\alpha} + \frac{1}{2\alpha}(1-\alpha)^i} \\
&= 1 - \frac{\alpha}{3}.
\end{aligned}$$

Hence, as long as the queues are not empty every queue is reduced by a factor of at least $(1 - \frac{\alpha}{4})$. ∎

THEOREM 3.3
The number of list cells remaining after $6\log n$ applications of Deterministic_List_Ranking is at most $n/\log n$.

PROOF
First we show

$$\alpha > \left(1 - \frac{\alpha}{4}\right)^{\log n}. \tag{3.2}$$

Chapter 3. List Ranking and Parallel Tree Contraction

For sufficiently large n

$$\begin{aligned} \log n & > 4(\log \log n)^2 \\ & > 4 \log \log n \cdot \log \log \log n \\ & = \frac{4}{\alpha} \log \log \log n. \end{aligned}$$

Thus,

$$\log \log \log n < \frac{\alpha}{4} \log n.$$

Exponentiating both sides:

$$\log \log n < e^{\frac{\alpha}{4} \log n}. \tag{3.3}$$

Taking reciprocals and using the facts that $\lim_{\alpha \to 0} (1 - \alpha/4)^{1/\alpha} = e^{-1/4}$ and $(1 - \alpha/4)^{1/\alpha}$ is monotone decreasing:

$$\begin{aligned} \alpha & > e^{-\frac{1}{4} \alpha \log n} \\ & > \left[\left(1 - \frac{\alpha}{4}\right)^{1/\alpha} \right]^{\alpha \log n} \\ & = \left(1 - \frac{\alpha}{4}\right)^{\log n}. \end{aligned}$$

Recall by the facts given at the previous lemma that the total weight of the system is less than $\frac{n}{\log n} \frac{1}{\alpha}$. Since the total weight is reduced by a factor of at least $\left(1 - \frac{\alpha}{4}\right)$, the weight after $6 \log n$ rounds is at most

$$\begin{aligned} \frac{n}{\log n} \frac{1}{\alpha} \left(1 - \frac{\alpha}{4}\right)^{6 \log n} & < \frac{n}{\log n} \left(1 - \frac{\alpha}{4}\right)^{5 \log n} \\ & < \frac{n}{\log n} \left(1 - \alpha - \frac{\alpha}{4} + \frac{5\alpha^2}{8} - \frac{5\alpha^3}{32} \cdots \right)^{\log n} \\ & < \frac{n}{\log n} (1 - \alpha)^{\log n}. \end{aligned}$$

Since there are $n/\log n$ queues and the smallest weight of an item is $(1-\alpha)^{\log n}$, there are at most $n/\log n$ cells remaining after $6 \log n$ rounds. ∎

We have shown that by using $n/\log n$ processors we can reduce a linked list of size n to one of size $n/\log n$ in $O(\log n)$ time. Now we can apply Wyllie's algorithm to reduce the linked list to a single cell. Therefore, this algorithm is optimal up to a constant factor.

EXERCISE 3.8
Show that if a chain is *strictly* increasing in queue size, then it is not necessary to find $\log \log n$ ruling sets. Write an algorithm taking advantage of this observation. Note: It is still necessary to find $\log \log n$ ruling sets for chains that have constant size queues.

EXERCISE 3.9
By adjusting how much identifying a subject reduces a queue's weight and refining the analysis in Theorem 3.3, show that only $4 \log n$ rounds are necessary to reduce the number of list cells to at most $n/\log n$.

Conclusion

In this section we introduced a number of list ranking algorithms. The simplest and probably the most practical is an algorithm due to Wyllie. It is not optimal, with respect to the work done, but has very small constants. The next algorithm, Random Mate, introduced randomization to resolve contention. This algorithm is also very simple and, although optimal with respect to the work done, still requires n processors. The Optimal Random Mate algorithm reduces the number of processors needed to $n/\log n$ by providing a simple scheduling scheme in addition to randomization. Finally, we gave an optimal deterministic list ranking algorithm that is quite simple relative to other deterministic algorithms in the literature. The randomization algorithms tend to have much smaller constants than deterministic algorithms. However, on some architectures getting the random bits to the processors fast enough may limit their practicality.

TABLE 3.1
Time and processor count for the list ranking algorithms discussed in this section.

Problem	Time	Processors	Work
Wyllie's algorithm	$O(\log n)$	n	$O(n \log n)$
random mate	$O(\log n)$	n	$O(n)$
optimal random mate	$O(\log n)$	$n/\log n$	$O(n)$
optimal deterministic algorithm	$O(\log n)$	$n/\log n$	$O(n)$

3.3 Parallel Tree Contraction

3.3.1 Top Down versus Bottom Up Tree Algorithms

Trees play a fundamental role in many computations, both for sequential as well as parallel problems. For sequential algorithms the classic paradigms for trees are depth-first-search and breadth-first-search. However, for processor efficient parallel algorithms, depth-first-search and breadth-first-search have not appeared to be successful in general. Typically, parallel algorithms involving trees use either divide-and-conquer or parallel tree contraction. Divide-and-conquer is a "top-down" approach, where first one finds a vertex that separates the tree into two subtrees roughly the same size, and recursively solves the two subproblems. A now classic example is Brent's work of parallel evaluation of arithmetic expressions [7]. In this case one subtree is a proper expression tree, which can fully determine its value, while the other subtree has a "scar," where it has to wait for the value of the first subtree before it can finish computing its own value. The main problem with the divide-and-conquer approach is finding the separators that separate the tree into components with a size not more than 2/3 of the original size. If we require that we find the separators on-line, that is, that no preprocessing is allowed, then finding the separators seems to add a factor of $\log n$ to the running time of most algorithms. The second approach is to use parallel tree contraction, which is a technique for constructing parallel algorithms on trees working from the bottom up. That is, all modifications to the tree are done locally. This "bottom-up" approach, which is called CONTRACT, has two major advantages over the top-down approach: (1) the control structure is straightforward and easier to implement, facilitating new algorithms using fewer processors and less time; (2) problems for which it was too difficult or too complicated to find polylog parallel algorithms are now easy. It has already been applied to finding small separators for planar graphs in parallel [16] as well as numerous other applications [19].

We start by introducing the basic parallel tree contraction paradigm and show that it takes $O(\log n)$ contractions to reduce a tree to its root. Processors assigned to each vertex in the tree work on the tree from the bottom up. During each contraction, they process the leaves in parallel and then remove them from the tree, creating new leaves that are processed at the next round. Removing leaves is called the RAKE operation. Clearly, removing leaves is not sufficient for a fast algorithm; a tree that is a simple list would take a linear number of rounds to reduce the tree to a point. Thus, a second operation is introduced, called COMPRESS, which reduces a chain of vertices, each with a single child, to a chain of half the length. Like list ranking, COMPRESS uses pointer jumping. Since RAKE and COMPRESS work on different parts of the tree, they can be run simultaneously. As the algorithm is running, the RAKE operation tends to produce chains that the COMPRESS operation then reduces. Thus, enough processors are kept busy to make the algorithm run quickly. The advantage of this approach to tree-based parallel algorithm design is that the processing of leaves can be designed separately from the processing of the internal nodes with single children, simplifying the algorithms.

Initially we restrict ourselves to trees with bounded degree and consider the changes necessary for unbounded degree at the end. First we describe the abstract parallel tree contraction paradigm and then we give the basic form of the implementation on a CRCW PRAM. As an example of its use, we implement arithmetic expressions evaluation. However, the algorithm is not optimal, in that it uses $O(\log n)$ time and n processors, and uses concurrent reads and writes. As with list ranking, we can reduce the number of processors to $n/\log n$ by using randomization. But unlike list ranking, it seems that a complicated load balancing step is also required. Therefore, we do not show it here.

When the tree is restricted to being binary there is a simple optimal EREW PRAM algorithm, which we present next. Each RAKE operation is immediately followed by a COMPRESS operation so that chains never form. The trick is to work on alternate leaves so that processors do not interfere with each other. Determining which leaves to work on is made easy using prefix-sums.

Next we present an optimal EREW PRAM algorithm for any bounded-degree tree structure. It consists of two stages. The first stage subdivides the tree into subtrees and assigns each subtree to a processor. A processor then reduces its subtree to a single vertex. In this way P processors reduce a tree of size n to one of size P in $O(n/P)$ time. The second stage contracts the tree of size P to its root in $O(\log P)$ time. Since the size of the tree has been reduced by the first stage, a processor can be assigned a single vertex; thus the second stage of the algorithm in and of itself need not be optimal. The trick is to ensure that no concurrent reads or writes take place.

Finally, we consider the modifications required by the basic and optimal tree contractions algorithms when the tree has unbounded degree. For the nonoptimal contraction algorithms the concern is that the RAKE operation may not be constant time, but depends on the number of children of a node, which is unbounded. For the optimal contraction algorithm, the assignment of subtrees in the first stage needs to be modified so that no processor gets an unbounded number.

In order to simplify the code, we stop using the ! notation in this section and use the index v to refer to a vertex of the tree. We leave it to the implementor to modify the code to handle the indexing by each processor.

3.3.2 The RAKE and COMPRESS Operations

In this section we introduce two abstract parallel tree contraction operations, RAKE and COMPRESS, and note how they reduce the size of a tree. We then show that if both operations are applied $O(\log n)$ times to a tree, the tree reduces to a point. In the next section we give a suboptimal deterministic CRCW PRAM implementation using $O(\log n)$ time and $O(n)$ processors.

Let $T = (V, E)$ be a rooted tree with n vertices and root r. In order to describe the RAKE and COMPRESS operations, we first introduce the definition of a chain.

DEFINITION 3.2
Let $v_1, \ldots v_k$ be vertices of a rooted tree. Then $v_1, \ldots, v_k$ is **chain** *of length k if:*

- *v_{i+1}, is only child of v_i $1 \le i < k$, and*
- *v_k has only 1 child and it is not a leaf.*

A chain is **maximal** *if it is not possible to add more vertices to the chain.*

We now define RAKE and COMPRESS and introduce a new operation, CONTRACT.

RAKE: Let RAKE be the operation that removes all leaves from T. An example of a single RAKE operation is shown in Figure 3.11. It is easy to see that if the tree is highly imbalanced, for example a simple linked list, RAKE would need to be applied a linear number of times in order to reduce T to a single vertex. We can circumvent this problem by adding one more operation.

COMPRESS: In one parallel step, we *compress* all chains by identifying v_i with v_{i+1} for i odd and $1 \le i < k$, whenever $v_1 \ldots v_k$ is a chain. Thus, the chain $v_1, \ldots, v_k$ is replaced with a chain $v'_1, \ldots, v'_{\lceil k/2 \rceil}$. Let COMPRESS be the operation on T which "compresses" all maximal chains of T in one step. Observe that a maximal chain of length one is not affected by COMPRESS. An example of the COMPRESS operation is shown in Figure 3.12.

3.3. Parallel Tree Contraction   **149**

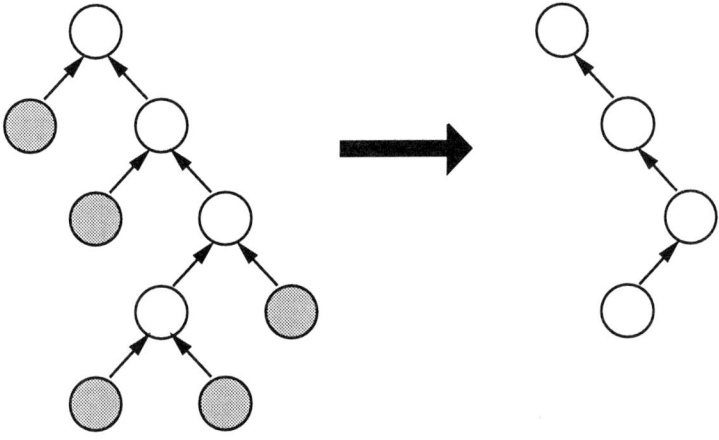

FIGURE 3.11
Result of a single RAKE operation. The shaded nodes of the tree on the left are the nodes that are deleted in order to obtain the tree on the right.

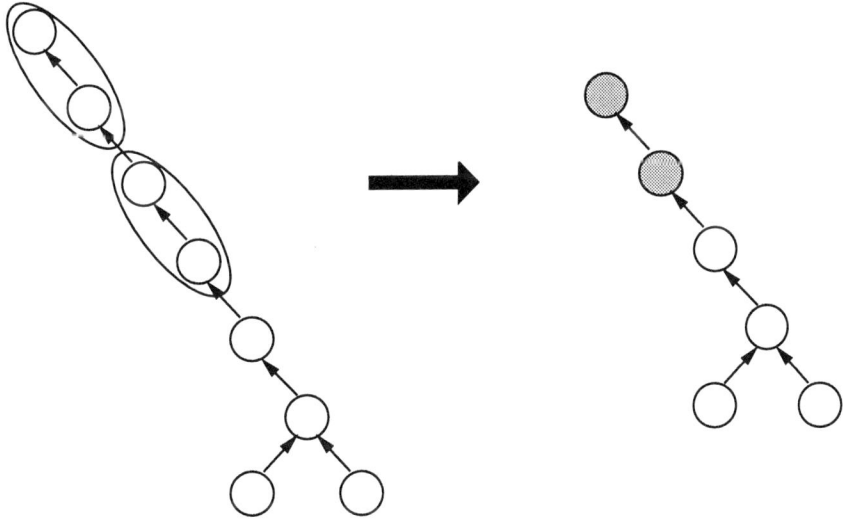

FIGURE 3.12
Result of a single COMPRESS operation. The shaded nodes represent the nodes that replace the pairs of nodes in the tree on the left.

CONTRACT: Let CONTRACT be the simultaneous application of RAKE and COMPRESS to the entire tree. We next show that the CONTRACT operation needs only be executed $O(\log n)$ times to reduce T to its root. In particular, we show:

$$|\text{CONTRACT }(T)| \leq \frac{4}{5}|T|$$

THEOREM 3.4
After $\lceil \log_{5/4} n \rceil$ applications of CONTRACT to a tree with n vertices, the tree is reduced to its root.

PROOF
We partition the vertices of T into two sets Ra and Com such that $|Ra|$ decreases by a factor of 4/5 after an execution of RAKE and $|Com|$ decreases by a factor of 1/2 after COMPRESS.

Let

V_0 be the set of leaf nodes of T.
V_1 be the set of nodes with 1 child.
V_2 be the set of nodes with 2 or more children.

Next, we subdivide V_1 into:

$C_0 = \{v \in V_1 \mid v's \text{ child is in } V_0\}$.
$C_1 = \{v \in V_1 \mid v's \text{ child is in } V_1\}$.
$C_2 = \{v \in V_1 \mid v's \text{ child is in } V_2\}$.

Finally, we consider a subset of C_1:

$GC_0 = \{v \in C_1 \mid v's \text{ grandchild is in } V_0\}$.

All vertices in V_1 except those of C_0 belong to a chain, by definition of a chain. In order for Com to decrease by a constant factor after each COMPRESS, we want to exclude chains of length one. Notice, for example, in Figure 3.13 all the chains are of length one and COMPRESS does not remove any vertex. These chains consists of a vertex in either C_2 or in GC_0. Therefore, we exclude C_2 and GC_0 from Com. We put all the remaining vertices in Ra.

Thus, let

$Com = V_1 - C_0 - C_2 - GC_0 = C_1 - GC_0$
$Ra = V - Com = V_0 \cup V_2 \cup C_2 \cup C_0 \cup GC_0$

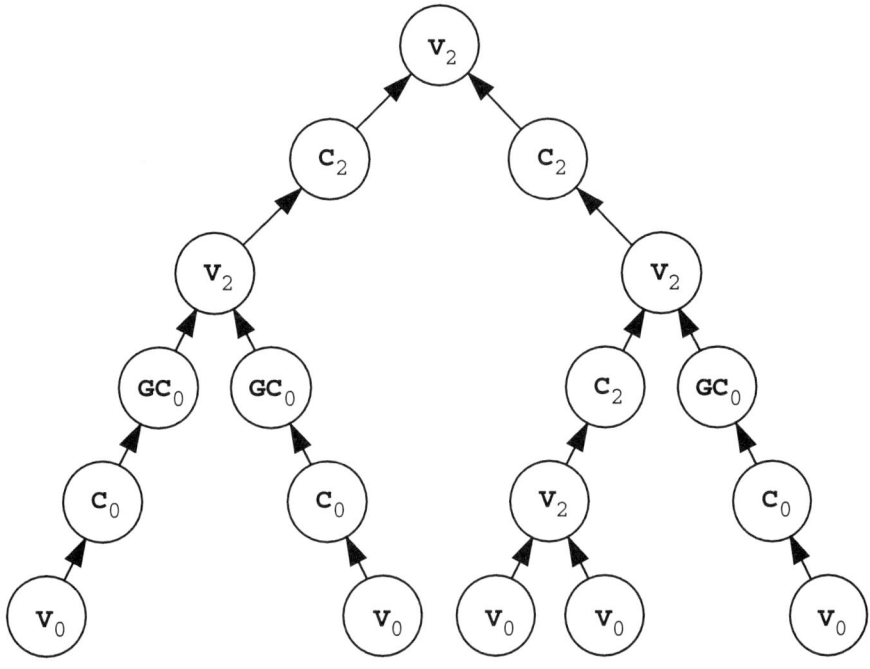

FIGURE 3.13
Vertex classes of a tree.

In this way, every vertex of Com belongs to some maximal chain. Because we have excluded C_2 and GC_0 from Com, if $v_1, \ldots, v_k$ are the vertices of a maximal chain then $v_1, \ldots, v_{k-1}$ are the only elements in the chain belonging to Com. Thus, the number of elements in Com decreases by at least a factor of $1/2$ after COMPRESS.

A RAKE operation removes all vertices in V_0. To see that the size of Ra decreases by a factor of $1/5$ after each RAKE we show that $|Ra| \leq 5|V_0|$. But this inequality follows by observing the following inequalities: $|C_0| \leq |V_0|$, $|GC_0| \leq |V_0|$, and $|C_2| \leq |V_2| < |V_0|$. ∎

3.3.3 The Basic Tree Contraction Algorithm

In this section we describe in more detail a CRCW PRAM implementation of CONTRACT. This basic algorithm requires $O(n)$ processors to achieve $O(\log n)$ time. For now we assume that the trees are of bounded degree. The analysis of parallel tree contraction on trees of unbounded degree is in Section 3.3.6. First we describe a particular application, namely expression evaluation in more detail. Next we show that by reversing the contraction algorithm we can expand the tree back to its original structure and in the process compute results for every vertex, not just the root. Thus, we use the same resource bounds to compute over all subtrees as to compute over the tree as a whole. This result is a natural generalization of parallel prefix evaluation [15, 11, 22]. Finally, we give several applications and their implementations.

There are many useful applications of parallel tree contraction and expansion. For each given application, we associate a certain procedure with each RAKE and COMPRESS operation, which we assume can be computed in parallel in constant time. We denote applying these procedures on a vertex v by $rake(v)$ and $compress(v)$. Typically the vertices of the tree T contain variables storing information relevant to the given application. The *rake* and *compress* procedures modify these variables while the overall CONTRACT procedure modifies the tree structure itself.

Let T be a rooted tree with vertex set $V =$, $|V| = n$, and root $r \in V$. We view each vertex that is not a leaf as a function to be computed. The children of the vertex supply all the information needed to compute its function. Initially, only children that are leaves can supply the necessary information to their parents. We first need to consider how a vertex determines whether it is a leaf vertex, which can evaluate its function, or a vertex with only one missing argument and potentially part of a chain. The approach is to have those vertices that have computed their values tell their parents they are done. In particular, these vertices mark a space reserved for them by their parents. The parents need only check whether all or all except one child have marked their spaces in order to determine if they are now leaves or part of a chain.

In order to reserve space at the parent for each child, we need to assume that the tree is ordered so that for each vertex v the children of v are ordered $v_1, \ldots, v_k$ and each child knows its index. That is, let $index[v_i]$ be the index of v_i in this ordering of children, i.e., $index[v_i] = i$. For each vertex v we set aside k locations $label[v, i], i = 1, \ldots, k$ in shared memory that the children can mark. Initially each $label[v, i]$ is empty or *unmarked*. Let $Arg(v)$ compute the number of unmarked labels for v. Thus, initially $Arg(v) = k$, the number of children of v. When the function at v_i can be computed, indicated by $Arg(v_i) = 0$, we apply the *rake* procedure and mark $label[v, i]$, which we denote by $mark(label[v, i])$. Let $P[v]$ be the vertex that is the sole parent of v. When a vertex v and its parent $P[v]$ have only one unevaluated argument, then v and $P[v]$ are members of a chain. The *compress* procedure is applied to v, and $P[v]$ is jumped over, i.e., $P[v] = P[P[v]]$. Algorithm 3.7 shows the *Basic_Contract* procedure.

ALGORITHM 3.7
The Basic Contract Phase
Procedure *Basic_Contract*
   In Parallel *Initialize(v)*
   In Parallel while $Arg\ (root) > 0$ **do**
      if $P[v] \neq nil$ **then do**
         Parallel Case $Arg\ (v)$ **equals**
         0) *rake* (v);   /* RAKE */
             mark $(label[P[v], index[v]])$;
             $P[v] := nil$;
         1) **If** $Arg\ (P[v]) = 1$ **then**   /* COMPRESS */
             compress (v);
             $P[v] := P[P[v]]$;
         end case
      end then
   end in parallel
   host set *rake* $(root)$;
end *Basic_Contract*

The algorithm is equivalent to one application of CONTRACT if one notes that case 0 is RAKE and case 1 is COMPRESS.

THEOREM 3.5
After $O(\log_{4/3} n)$ applications to a tree with n vertices, Basic_Contract reduces the tree to its root. If RAKE *and* COMPRESS *take $O(1)$ time, then the time to reduce a tree to its root is $O(\log n)$.*

PROOF
Observe that after *Basic_Contract* every maximal chain decomposes into two chains, one *essential chain* corresponding to COMPRESS and an unnecessary chain that is out of phase. The head of this second chain is unevaluated. For the purpose of analysis we can discard the second chain, since it will never be evaluated.

Note that *Basic_Contract* is slightly faster than CONTRACT, since it does not test if the only child of a vertex is a leaf or not. Thus, some pointer jumping occurs in *Basic_Contract* that does not occur in CONTRACT. That is, chains in *Basic_Contract* contain all V_1 vertices, including C_0

vertices. Therefore, Com for $Basic\_Contract$ can also contain GC_0 vertices and still reduce the size of Com by at least $1/2$ after every phase. Since Ra does not contain GC_0 vertices, the number of vertices in Ra reduces by at least a factor of $1/4$ after every phase of $Basic\_Contract$. Thus, after $O(\log_{4/3} n)$ applications of $Basic\_Contract$ to a tree with n vertices, $Basic\_Contract$ reduces the tree to its root. ∎

Expression Tree Evaluation

More intuition can be gained by seeing $Basic\_Contract$ applied to expression evaluation over $\{+, \times\}$. Let T be a binary expression tree in which the internal nodes hold the operators and the leaves hold the operands. The *value of a leaf* is the constant assigned to it. The *value of an internal node* is defined recursively as the operation at that node applied to the value of its children.

For reasons that will become clear later, we modify the definition of an expression tree so that associated with each edge $(v, P[v])$ is a function, f_v, in one variable. For expression trees over $\{+, \times\}$, these functions would be linear forms, $aX + b$ where X is an indeterminant and a and b would be constants. The *value of leaf* remains the value initially assigned to it. If v is an internal node with left and right children L and R and operation $\odot_v$ then the *value of an internal node* is

$$val(v) = f_L(val(L)) \odot_v f_R(val(R)),$$

where f_L and f_R are the unary functions for edges (L, v) and (R, v), respectively. Initially, the function on every edge is simply the identity function, so that every vertex of this modified expression tree has the same value as the original expression tree. We can think of $f_L(val(L))$ as the *contribution* of L to its parent's value. Let *rake* correspond to computing a vertex's value and *mark* to computing its contribution to its parent.

Next, let us consider the COMPRESS operation. After every application of COMPRESS we want an expression tree in which the value of every vertex in the current tree is the same as the value of the same vertex in the original tree. Suppose $Arg(v) = Arg(L) = 1$. In particular, suppose the value of R has been computed but L is still missing one argument. Figure 3.14 depicts this situation. During the COMPRESS operation L pointer jumps over v and points to $P[v]$, the parent of v. However, we want to be sure that $P[v]$ computes the same value after the COMPRESS operation as before. Let C_v represent the contribution of v to the value of $P[v]$ and C_R represent the contribution of R to

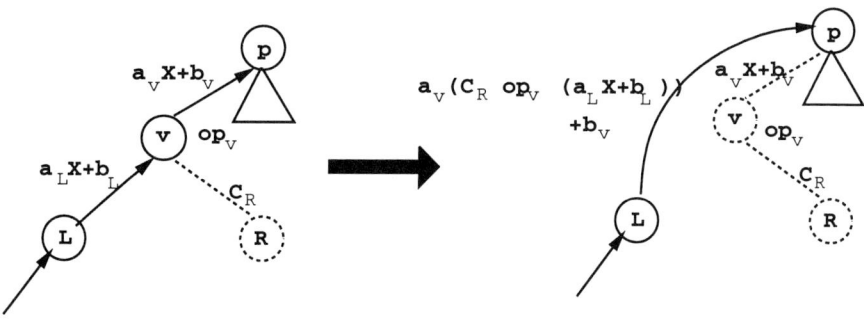

FIGURE 3.14
COMPRESS on an expression tree. The dotted nodes represent nodes that have been removed.

the value of v. Since $val(R)$ is known, C_R is also known and is a constant. Let

$$val(v) = f_L(X) \odot_v C_R = f_\odot(X). \tag{3.4}$$

Thus, $val(v)$ is a linear form in X, which we denote by $f_\odot(X)$.

But then C_v is the composition of the two linear forms f_v and $f_\odot$. This composition is also a linear form. Let f'_L denote this linear form. That is,

$$C_v = f_v(val(v)) = f_v(f_\odot(X)) = f'_L(X)$$

Therefore, after pointer jumping, the correct function on the edge from v_L to $P[v]$ is f'_L. Observe that the contribution of L to $P[v]$ after the COMPRESS operation is the same as the contribution of v to $P[v]$ before the COMPRESS operation. This observation motivates the use of the modified expression tree. Figure 3.14 depicts the case in which the function of an edge depends only on the value of a node and its children. In general, however, the function value may be a composition of the operations of all the nodes that have been jumped over on the path between a node and its new parent.

Given this information, we now consider an implementation for expression evaluation over $\{+, \times\}$. Algorithm 3.8 gives the implementation, and Figure 3.15 shows the initialization and two applications of this implementation on an expression tree. Initially, the operands of the expression tree at the leaves are stored in *val* and the operators at internal vertices are stored in *op*. Every vertex keeps a pointer to its parent, an index *side* indicating whether it is a left or right child of its parent, and an index *sib* of its sibling.

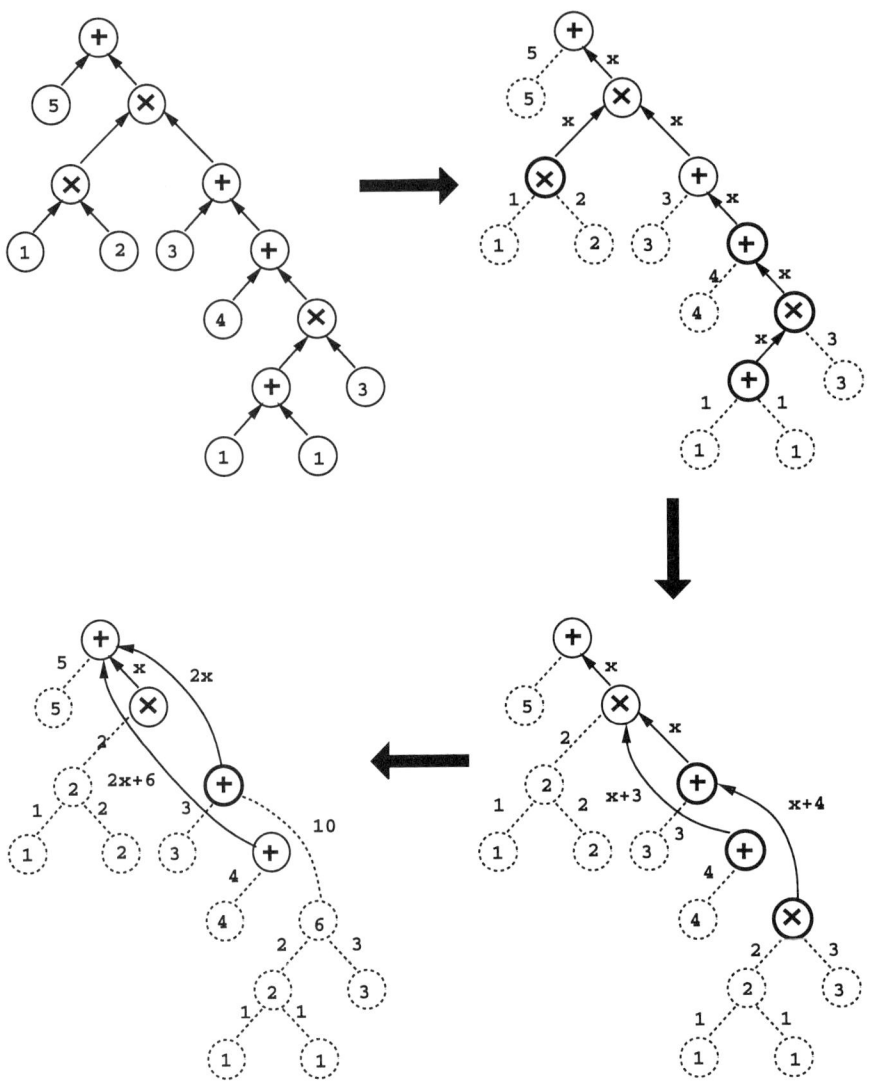

FIGURE 3.15
Initialization and two applications of *Expression_Contraction*. Vertices with heavy lines are working to generate the next version of the tree. Dotted vertices have been raked. The solid arcs show the linear forms for the child vertices. The dotted arcs show the labels sent to the parent vertices.

Since the functions on an edge $(v, P[v])$ is a linear form $aX + b$, we associate with each vertex v a pair of numbers (a, b) that represent the linear function $a[v]X + b[v]$. The linear form at the root is irrelevant. Initially the linear function is simply X. We also set aside storage for the value of the vertex, $val[v]$, and the contributions of its left and right children, $label[v, L]$ and $label[v, R]$. The children supply the values for $label$. We use the function $eval$ to evaluate the value of a vertex, given its operator and the contributions of its left and right children. We use the function $simplify$ to find the new linear form for v, given (a, b) of itself and its parent, the operator of its parent, and the contribution of its sibling. The contribution of the sibling of v is stored in $label[P[v], sib[v]]$. That is, $simplify$ finds the new linear form $f'_v(X)$.

ALGORITHM 3.8
Expression Evaluation Contraction Phase
Procedure *Expression_Contraction*:
   In Parallel do   /* Initialize */
      if $Arg(v) = 0$ **then do**   /* Leaves */
         $label[P[v], side[v]] := val[v]$;
         $P[v] := nil$;
      end then
      else $(a, b)[v] := (1, 0)$;   /* Internal vertices */
   end in parallel

   In Parallel while $Arg\ (root) > 0$ **do**
      if $P[v] \neq nil$ **then do**
         Parallel Case $Arg\ (v)$ **equals**
         0) $val[v] := eval\ (op[v], label[v, L], label[v, R])$ /* Rake */
            $label[P[v], side[v]] := a[v] * val[v] + b[v]$;   /* Mark */
            $P[v] := nil$;
         1) **if** $Arg(P[v]) = 1$ **then**   /* Compress */
            $(a, b)[v] := simplify\ ((a, b)[v], (a, b)[P[v]],$
                             $op[P[v]], label[P[v], sib[v]])$;
            $P[v] := P[P[v]]$;
         end case
      end then
   end in parallel

   host set $val[root] := eval\ (op[root], label[root, L], label[root, R])$;
end *Expression_Contraction*

We used the more conservative definition of a chain in CONTRACT since, for some applications, a vertex with a leaf as a child can use the time at this stage to incorporate the value of the child in its own value rather than pointer jumping. That is, in some implementations, it may be preferable to have the RAKE operation not only compute the value of a leaf vertex, but also do some computation at the parent given that the value of a child is now known. For example, in expression tree evaluation, evaluating a vertex is a simple, fast process. Once the value of the vertex is known the parent can partially evaluate its vertex, i.e., to find $f_\odot(X)$ in equation 3.4. Thus, the *simplify* function is divided into two parts: one part is the partial evaluation of a vertex done by the processors performing the *rake* procedure, and the other part that composes two linear forms and returns a linear form, done by the processors performing the *contract* procedure. In this way, the *rake* and *compress* procedures may be better balanced in terms of the time it takes to complete the two subprocedures.

All Subexpression Evaluation

Note that many vertices are not evaluated. That is, for many vertices v the value $Arg(v)$ is never set to 0 during any stage of *Basic_Contract*. We define a new procedure *Basic_Expand* that allows the evaluation of all vertices, i.e., each vertex eventually has all its arguments after completion of the procedure. We modify *Basic_Contract* so that each vertex keeps a push-down store $parentStore_v$ of all the previous values of $P[v]$ and add a line before the start of the **Parallel Case** statement of *Basic_Contract*:

Push $P[v]$ **onto** $parentStore[v]$;

We also include a counter t, which counts the number of iterations required to contract the tree to its root.

We now apply *Basic_Contract*, which computes the value of the root r, followed by *Basic_Expand*, given in Algorithm 3.9, which computes the value of all vertices. At each iteration, we reintroduce vertices that were either raked or were jumped over in the corresponding iteration of *Basic_Contract*. We expand the tree for t iterations to reconstruct the whole tree.

ALGORITHM 3.9
The Basic Expansion Phase
Procedure *Basic_Expand*:
    In Parallel $done[v] := false$;   /* Initialize */
    Host set $done[root] := true$;

In Parallel for $i := 1$ **to** t **do**
    if not $empty(parentStore[v])$ **then do**   /* Get new parent */
        $oldP[v] := P[v]$;
        $P[v] := Pop(parentStore[v])$;

        if $P[v] \neq oldP[v]$ **then**
            if $oldP[v] = nil$ **then do**   /* Leaf reintroduced */
                $unrake\ (v)$;
                $done[v] := true$;
            end then
            else if $done[v]$ **then**   /* Child of spliced node */
                $mark(label[P[v], index[v])$;

        if not ($done[v]$ **and**
            $Arg(v) = 0$) **then do** /* Spliced node reintroduced */
        $uncompress\ (v)$;
        $done[v] := true$;
    end then

    end then
  end in parallel
end *Basic_Expand*

THEOREM 3.6
At the completion of Basic_Expand all vertices have their arguments.

PROOF
As in the proof of Theorem 3.5 we can discard those chains where the leaves are unevaluated and consider only essential chains. The proof is by induction on the trees with only essential chains, starting from the trivial tree consisting of a singleton vertex r and finishing with the original tree T. Let $\{r\} = T_0, \ldots, T_t = T$. The structure of these trees correspond one to one with the trees defined during the tree contraction, but in reverse order. Assume that at end of the ith application of *Basic_Expand* tree T_i has all its vertices evaluated. The vertices added to T_{i+1} are either leaves that were evaluated by a *rake*, or vertices that were jumped over by *compress* during the corresponding contraction phase. Thus, every new vertex in T_{i+1} is either a leaf, in which case we know its value, or is missing one argument, the value of a vertex in T_i, which is also known. In the latter case the value of the reintroduced vertex can then be computed. ∎

THEOREM 3.7
At most $\lceil \log_{5/4} n \rceil$ applications of basic tree contraction and $\lceil \log_{5/4} n \rceil$ applications of basic tree expansion are needed to evaluate all the vertices.

In the implementation, we need to be able to distinguish between vertices on essential chains that can compute their value, and vertices not on essential chains. We use the boolean value *done*, which indicates the vertices in T_i when we are currently generating T_{i+1}. Initially only the root is in T_0. By popping $P[v]$ from the stack at each round, we get the structure of the tree during the corresponding contraction phase. We use the difference in the structure of the new tree from the old tree to note vertices reintroduced. Whenever the parent of a vertex changes either the vertex was a leaf that was raked or it spliced out its parent in the corresponding contraction phase. In the former case, leaves introduced are vertices that previously had been removed from the tree and now are connected to their parents. These leaves are "unraked" and set as done. In the latter case, those vertices that are already done are on an essential chain and can mark their new parents, which are the nodes being reintroduced this round. These parents are nodes that are not done and have zero missing arguments. Because these parents have just been marked, they can evaluate themselves using *uncompress* and set themselves done.

Consider expression evaluation over $\{+, \times\}$ as an example. When a vertex is raked it knows its final value, so *unrake* does nothing. In order to mark a parent, we need to send the contribution of the vertex to its parent. To compute the contribution, we need to know the linear form of the vertex prior to it splicing out its parent during the corresponding contraction phase. We therefore need to modify *compress* in *Expression_Contract* to save its linear form on a stack. We call this stack *envirStore*. We get the old value of the linear form by popping the stack and then computing the contribution to its parent. Finally, vertices that now have all their arguments can compute their own values, using the contributions of their children. Algorithm 3.10 shows the code for the expansion phase of expression evaluation.

ALGORITHM 3.10
The expansion phase for expression evaluation
Procedure *Expression_Expand*:
    In Parallel $done[v] := false$;
    Host set $done[root] := true$;

    In Parallel for $i := 1$ **to** t **do**
        if not $empty(parentStore[v1])$ **then do**   /* Get new parent */

$oldP[v] := P[v];$
$P[v] := Pop(parentStore[v]);$

if $P[v] \neq oldP[v]$ **then do**
    if $oldP[v] = nil$ **then**
        $done[v] := true;$ /* leaf reintroduced */
        else if $done[v]$ **then do**   /* child of node reintroduced */
            $(a, b)[v] := Pop(envirStore[v]);$
            $label[P[v], side[v]] := a[v] * val[v] + b[v];$ /* mark */
        end then
    end then

    if not $(done[v]$ **and**
        $Arg\ (v) = 0)$ **then do** /* spliced node reintroduced */
        $val[v] := eval(op[v], label[v, L], label[v, R]);$ /* uncompress */
        $done[v] := true;$
    end then

    end then
  end in parallel
end *Expression_Expand*

EXERCISE 3.10
Show the expansion phase for the expression tree in Figure 3.15

Applications of Tree Contraction

Expression Evaluation   Let T be a tree with vertex set V and root r. We assume each leaf is initially assigned a constant from the domain $\mathcal{D}$, and each internal vertex v, with children $u_1, \ldots, u_k$, has an operator on $\mathcal{D}$ of the form $\odot(u_1, \ldots, u_k)$. To apply parallel tree contraction to an expression problem seems to require finding a general form for implementing and storing the composition of unary functions. A function of the form $f : \mathcal{D} \to \mathcal{D}$ is a **unary function** over the domain $\mathcal{D}$. The following two closure properties of unary function classes are important to using parallel tree contraction [20].

DEFINITION 3.3
(**Composition**) *A unary function class* $\mathcal{F}$ *is* **closed under composition** *if, for all* $f_1, f_2 \in \mathcal{F}$, $f_2 \circ f_1 \in \mathcal{F}$.

DEFINITION 3.4
(**Projection**) *A unary function class $\mathcal{F}$ is* **closed under projection** *if for all operators $\odot$, for all $a_1, \ldots, a_k \in \mathcal{D}$, and for all $i, 1 \leq i \leq k$:*

$$\odot(a_1, \ldots, a_{i-1}, x, a_{i+1}, \ldots, a_k) \in \mathcal{F}.$$

Consider, for example, arithmetic expression trees over the reals with operators $\odot \in \{+, -, \times, \div\}$. The operations $\{+, -, \times, \div\}$ have their usual interpretations e.g., $a/b + c/d = (ad + bc)/bd$. We assume that the number of arguments at a vertex is at most 2. If not, we assume that in $O(\log n)$ time we can convert it into such a tree. In order to perform *compress* we need a representation for unary functions that is closed under projection and is closed under composition. Consider $\mathcal{F}$, the ratio of a pair of linear functions of the form $(ax + b)/(cx + d)$. The values stored or manipulated are sums, products, and differences of the initial leaf values $val(v)$. The function is the ratio of these elements. $\mathcal{F}$ is closed under composition, because for all $f(x) = (ax + b)/(cx + d)$ and $f'(x) = (a'x + b')/(c'x + d')$

$$f' \circ f = \frac{a'(ax+b)/(cx+d) + b'}{c'(ax+b)/(cx+d) + d'} = \frac{a''x + b''}{c''x + d''}$$

$\mathcal{F}$ is closed under projection, because $+(a, x) = +(x, a) = x + a$, $-(a, x) = -x + a$, $-(x, a) = x - a$, $\times(a, x) = \times(x, a) = ax$, $\div(a, x) = a/x$, and $\div(x, a) = x/a$.

EXERCISE 3.11
Find the maximum independent set of a tree. (*Hint:* Show that a greedy algorithm is sufficient.) Show how this problem is equivalent to evaluating an expression tree.

EXERCISE 3.12
Find the minimum number of registers needed to evaluate an expression tree. (*Hint:* Find the equivalent expression tree and find a function class closed under projection and composition.)

EXERCISE 3.13
Compute the *height* of each vertex in a tree, where height(v) is the length of the longest path from v to a leaf of the tree.

Ancestors' Maximum Value Given a tree with values at each vertex, for each vertex find the maximum value of the ancestors of that vertex.

This problem is slightly different form the ones we considered before because the information needed to compute the maximums comes from the ancestors, not the descendants, of the vertex. That is, the information needs to flow down the tree, not up. Thus, *rake* cannot contribute to the solution. But *unrake* does, by using the information filtered down to its parent.

Solution: We use parallel tree contraction with the following operations:
Initialize: $max[v] := value[P[v]]$
Rake: /* null operation */
Compress: $max[v] := maximum(max[v], max[P[v]])$
Unrake: $max[v] := maximum(max[v], max[P[v]])$
Uncompress: $max[v] := maximum(max[v], max[P[v]])$

During the contract phase max is the maximum of the chain of values between a vertex and its current parent. Thus, the maximum value of the ancestors of a vertex v is the maximum of $max[v]$ and the maximum value of the ancestors of $P[v]$. During the expansion, for a vertex that is done, max is the maximum of all the ancestors of the vertex. Note that when a vertex is unraked its parent is done. When a vertex in a chain is reintroduced, not only was its child in an essential chain, so was its parent. Hence, it, too, is done.

EXERCISE 3.14
Given a tree with connections between pairs of vertices, for each connection give the vertex that is the lowest common ancestor of the vertices at the ends of the connection.

EXERCISE 3.15
Given T and B, the tree and back edges produced by a depth-first search of a connected, undirected graph $G = (V, E)$, find the low point number, Low, for the each vertex v in V. That is, assume that the vertices are labeled by their depth-first numbers. Then

$$Low[v] = \min \left(\{v\} \cup \left\{ w \mid \begin{array}{l} \text{there exists a back edge } (x, w) \in \\ B \text{ such that } x \text{ is a descendant of} \\ v, \text{ and } w \text{ an ancestor of } v \text{ in the} \\ \text{depth first spanning forest } (V, T) \end{array} \right\} \right)$$

For a depth-first search sequential algorithm, see [2].

3.3.4 Optimal Parallel Tree Contraction for Binary Trees

In this section we show a new operation, called SHUNT, to contract an ordered binary tree to its root [1, 13]. If the tree is not binary it sometimes can be interpreted as a binary tree by introducing a new vertex for every child vertex (see [1]). Otherwise one of the more general algorithms must be used. Parallel tree contraction using shunting can reduce a tree to its root in $O(\log n)$ time using $n/\log n$ processors on an EREW PRAM. Note that the *Basic_Contract* algorithm is not optimal because the time and processor product is greater than the time for the optimal sequential algorithm. One of the inefficiencies is that by using Wyllie's pointer jumping for chains we get two chains where only one is essential.

Another problem with *Basic_Contract* is that it does not work on the EREW model. The problem arises at the parent of a chain (a vertex in V_2 with a child in V_1 is called the **parent** of a chain). If v is a parent of a chain, then by using the pointer-jumping algorithm of Wyllie, we encounter the problem that over time many vertices in the chain may eventually point to v. Now, if v later becomes a vertex in V_1, then all these vertices will want to jump over v and, therefore, must read $P[v]$, which requires a concurrent read.

We can avoid using Wyllie's algorithm if we prevent chains from forming. A chain is produced when the tree contains a binary subtree where each internal vertex has one child that is a leaf and one that is not. After a RAKE operation, the subtree becomes a chain. If we apply RAKE to a leaf followed immediately by a COMPRESS on its sibling we prevent chains from forming. We call this operation pair **shunt**. That is, to apply SHUNT to a leaf vertex, v, we delete v and $P[v]$ and set $P[v']$ to $P[P[v]]$, where v' is sibling of v. Figure 3.16 shows an application of SHUNT to a leaf. Note that shunt applied to a leaf v does not produce any new leaves. The only effect on leaves is that

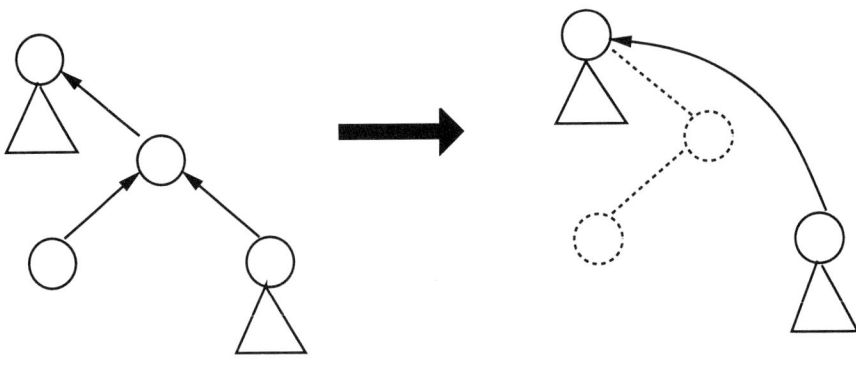

FIGURE 3.16
SHUNT applied to a left leaf. The dotted vertices have been removed from the tree.

it removes v, so that each application of shunt to a leaf reduces the number of leaves by one. Also note that shunt is not defined for a child of the root vertex, because we can not apply COMPRESS to the root.

However, we can not apply shunt to all leaves simultaneously. To prevent concurrent reads we cannot apply SHUNT to two leaves with the same parent. Otherwise the two leaves would attempt to reconnect their siblings to their grandparent by jumping over their common parent at the same time. But we also have to be careful not to apply SHUNT to leaves with different parents that are also consecutive in a left-to-right ordering of the leaves. If we do apply SHUNT to two such leaves we end up with two disconnected subtrees, as shown in Figure 3.17. Therefore, we apply SHUNT to the odd-numbered leaves only. However, we still can get disconnected subtrees. Figure 3.18 shows such a situation.

EXERCISE 3.16
Number the leaves of an ordered binary tree in $O(\log n)$ time using $n/\log n$ processors on an EREW PRAM. (*Hint*: Use the Euler Tour of the tree.)

To prevent the tree from becoming disconnected, we apply SHUNT first to all the left children that are odd numbered and then to all the right children that are odd numbered. We exclude the children of the root since SHUNT is un-

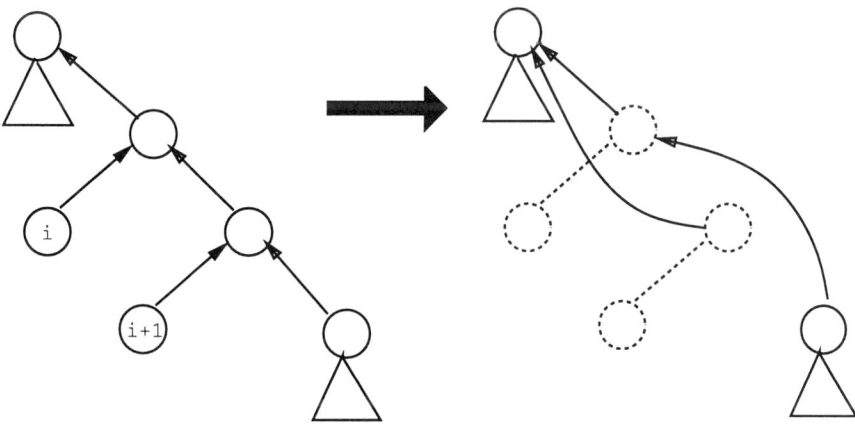

FIGURE 3.17
SHUNT applied to two consecutive leaves. The vertex labels give the leaf numbering.

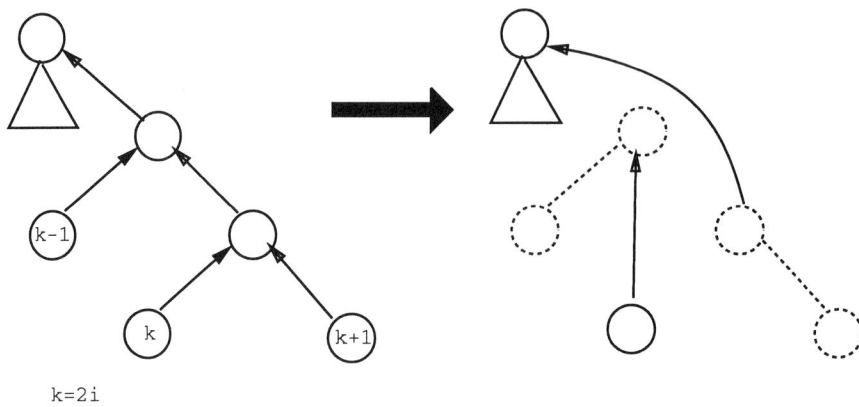

FIGURE 3.18
SHUNT applied to odd-numbered leaves. The vertex labels give the leaf numbering.

defined for these vertices. Notice that the relative order of the leaves remaining stays unchanged. Therefore, we can renumber the leaves by simply dividing the leaf numbers by 2. We repeat the algorithm until all leaves are removed except for the children of the root at which point we evaluate the root. The basic parallel tree contraction algorithm using SHUNT, called *Shunt_Contract*, is shown in Algorithm 3.11. We use *number* as the index of a leaf vertex in a left-to-right numbering of leaves, starting at 0. By starting the numbering at 0, SHUNT is never applied to the leftmost leaf, and this leaf eventually becomes a child of the root. Nonleaf vertices have an index of 0. The boolean *side* indicates whether a vertex is the left or right child of its parent. The procedure $shunt(v)$ applies *rake* to v and *compress* to $sib[v]$ and sets $P[sib[v]]$ to $P[P[v]]$, where sib is the sibling of v. Initially only leaves are active. The result of a *shunt* operation on a leaf of an $\{+, \times\}$ expression tree is shown in Figure 3.19.

ALGORITHM 3.11
The basic parallel tree contraction algorithm using shunting
Procedure $shunt(v)$
    $rake\ (v);\ active[v] := false;$
    $active[P[v]] := false;$
    $compress(sib[v]);$
    $P[sib[v]] := P[P[v]];$
end $shunt$

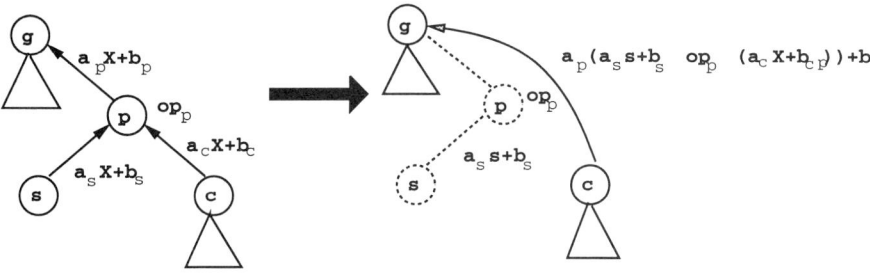

FIGURE 3.19
The SHUNT operation applied to a leaf of a $\{+, \times\}$ expression tree.

Procedure *Shunt_Contract*
   In parallel do   /* Initialize */
      if *isLeaf*[v] **then**
         active[v] := *true*;
      else
         active[v] := *false*;
      number[v] := *numberLeaves*(v);
   end in parallel

   In parallel for $i = 1$ **to** $\lceil \log n \rceil$ **do**   /* Contraction */
      if $v \neq root$ **and** *active*[v] **then**
         if *isOdd* (*number*[v]) **and** $P[v] \neq root$ **then do**
            if *side*[v] = *left* **then** *shunt* (v);
            if *side*[v] = *right* **then** *shunt* (v);
         end then
      else
         number[v] := *number*[v]/2;
   end in parallel
end *Shunt_Contract*

Figure 3.20 shows the contraction of an expression tree to its root, using *Shunt_Contract*. The linear form saved at a vertex is shown at the arc to the parent. The linear forms in parentheses will be explained in the next section.

LEMMA 3.2
Shunt_Contract runs correctly on an EREW PRAM.

168 Chapter 3. List Ranking and Parallel Tree Contraction

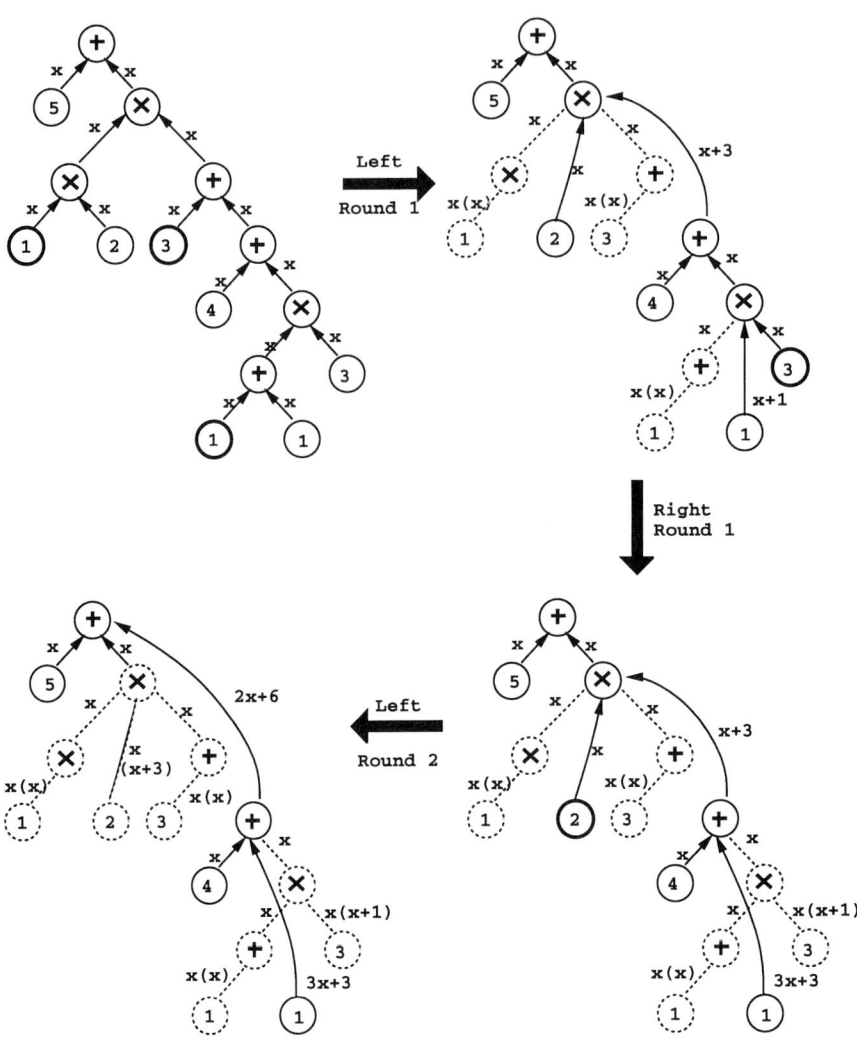

FIGURE 3.20
Parallel tree contraction using *Shunt_Contract*. At each round SHUNT is applied to the highlighted vertices. The expressions in parenthesis are saved for use during the expansion phase.

PROOF

Let v_1 and v_2 be nonconsecutive left (right) leaves. Then $P[v_1]$ and $P[v_2]$ are not identical since v_1 and v_2 are both left (right) leaves, nor are $P[v_1]$ and $P[v_2]$ a parent of the other, since v_1 and v_2 are not consecutive leaves. Therefore, v_1 and v_2 can read and write all the information associated with themselves, their parents, and their siblings without conflict. ∎

THEOREM 3.8

After $O(\log n)$ applications of Shunt_Contract on a EREW PRAM with $n/\log n$ processors a tree with n vertices is reduced to its root. If the RAKE *and* COMPRESS *operations run in $O(1)$ time then the overall running time is $O(\log n)$.*

PROOF

Assign each processor $\log((n+1)/2)$ consecutive leaves. During the ith application of *Shunt_Contract* each processor eliminates $1/2$ of its $\log((n+1)/2^i)$ vertices. Hence, after $\log n$ applications, each processor has eliminated all of its leaves. If RAKE and COMPRESS run in $O(1)$ time then the overall running time is $O(\log n)$ time. ∎

Note that the *rake* and *compress* operations cannot be performed in parallel as they could in *Basic_Contract* because both operations can simultaneously be operating on the same node. Also note that each round of *Shunt_Contract* works on left leaves followed by right leaves. Therefore one round of *Shunt_Contract* takes at least twice as long as one round of *Basic_Contract*.

All Subexpression Evaluation Using SHUNT

As before, we can easily modify *Shunt_Contract* to enable us to evaluate all subexpressions of the tree by applying an expansion phase following the contraction phase. Note that when we applied SHUNT to a leaf v, $P[v]$ never receives its argument, $sib[v]$, since $P[v]$ is deleted from the tree. However, if we expand the tree by running the contraction phase in reverse, when v and $P[v]$ are reintroduced the value of $sib[v]$ has already been computed. As before, we can prove this fact by using induction on the list of trees formed during the expansion phase and noting that after every expansion all subexpressions of the current tree have been computed. Thus, during the expansion phase, we reintroduce the vertices eliminated during the corresponding application of the contraction phase. The state of the vertices is restored so that the value of the reintroduced internal nodes can be computed.

Recall that for arithmetic subexpressions, *Compress* changes the linear form saved at the vertex. Therefore in *Basic_Contract* we saved the current state of a vertex on a stack each time we applied *compress*. Then, at the corresponding point during the expansion phase, that state is popped off the stack. Similarly for *Shunt_Contract*, we need to save the current state of $sib[v]$ before we apply the *compress* procedure. As it turns out, when we use *shunt*, we can save the current state of $sib[v]$ at vertex v before we apply *compress* to $sib[v]$. We can eliminate the stack because every time we apply *compress* to a vertex, we know we have just applied *rake* to its sibling and the sibling is different every time. Thus, during the expansion phase when v and $P[v]$ are reintroduced, v has the correct state information for $sib[v]$. Recall that $sib[v]$ has already been computed so we can compute its contribution to $P[v]$, given this state information. Then we can apply *unrake* to v and *uncompress* to $P[v]$. The saved linear forms of the siblings appear in parentheses in Figure 3.20.

EXERCISE 3.17
Show the expansion phase for the example in Figure 3.20.

3.3.5 Optimal EREW PRAM Parallel Tree Contraction Algorithm

In this section we exhibit an optimal deterministic EREW PRAM parallel tree contraction algorithm using $O(n/P)$ time and P ($P \leq n/\log n$) processors [12]. The algorithm has two stages. The first stage uses a new reduction technique called M-CONTRACTION. The basic idea is to dynamically divide the tree into subtrees, each of which has at most one unknown leaf, and then assign the subtrees to P processors, which *partially* evaluate the subtrees and succinctly compress the information to single vertices. In this way a tree of size n is reduced to one of size P in $O(n/P)$ time, using P processors on an EREW PRAM. The second stage uses a technique called ISOLATION to contract a tree of size P to its root in $O(\log P)$ time, using P processors on an EREW PRAM. Isolation eliminates the concurrent reads needed by the Wyllie approach used in the *Basic_Contract* procedure.

This section consists of three subsections. The first subsection contains the basic graph theoretic results and definitions that we need in the following subsection. In that subsection we show how to reduce the problem of size n to one of size P, where P is the number of processors. In the last subsection we show the isolation technique used to implement parallel tree contraction on a deterministic EREW PRAM in $O(\log n)$ time using n processors. We then discuss some implementation techniques and the expansion phase for all subexpression evaluation.

Basic Graph Theoretic Results

In this section we give some graph theoretic results that can be used to find a set of vertices that subdivide a tree into independent subtrees of approximately equal size. From these subtrees we can define the m-contraction of a tree to reduce the size of the tree.

First we consider the decomposition of a tree T into subtrees by finding vertices that partition the edges of T in a natural way. The vertices we consider are called m-critical vertices. Subtrees (subgraphs) are then formed out of each partition of edges by reintroducing the vertices at the end points of the edges. These subgraphs are known as **bridges**. We give the formal definitions needed to define m-critical vertices and bridges next.

Let $T = (V, E)$ be a directed graph in which every vertex, except the root, points to its unique parent. The **weight** of a vertex v in T is the number of vertices in the subtree rooted at v, denoted by $W(v)$. If n equals the number of vertices in T, the weight of the root r is n.

Let m be any integer such that $1 < m \leq n$. In the next subsection we let $m = 2n/P$, where P is the number of processors. A vertex, v, is m-**critical** if

1. v is not a leaf, and

2. $\lceil \frac{W(v)}{m} \rceil > \lceil \frac{W(v')}{m} \rceil$ for all $v' \in$ children (v).

LEMMA 3.3
If v_1 and v_2 are m-critical, then their least common ancestor is m-critical.

PROOF
If either v_1 or v_2 is an ancestor of the other, then the lemma is trivially true. We therefore consider the case when neither is an ancestor of the other. Let v be an m-critical vertex and let w be a child of v. Since $\lceil \frac{W(v)}{m} \rceil > \lceil \frac{W(w)}{m} \rceil \geq 1$, the weight of v must be greater than m. Therefore, if v_1 and v_2 are m-critical then both their weights are greater than m. Then u, the least common ancestor of v_1 and v_2, must have weight greater than $2m$ since it has two descendants with weight greater than m. Each child of u cannot have weight greater than $W(u) - m$

because at least one of v_1 and v_2 is not among the child's descendants, implying that u is m-critical. ∎

Let $G = (V, E)$ be a graph and let $C \subset V$. Two edges, e and e' of G, are **C-equivalent** if there exists a path from e to e' that avoids the vertices C. Also, let $E' \subset E$. E' induces a subgraph, $G' = (E', V')$, with $V' = \{v \in V \mid v$ is an endpoint in $E'\}$. That is, the endpoints of E' are included in G'. The graphs induced by the equivalence classes of the C-equivalent edges are called the **bridges** of C. A bridge is **trivial** if it consists of a single edge. The **attachments** of a bridge B are those vertices of B that are also in C. An example of the C-equivalent classes and their induced bridges of a graph are shown in Figures 3.21 and 3.22.

The m-**bridges** of a tree T are bridges of C, where C is the set of m-critical vertices of T. Note that the attachments of an m-bridge B are either

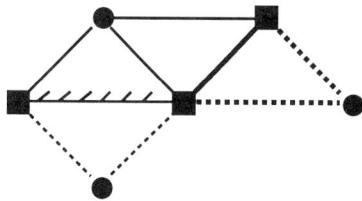

FIGURE 3.21
Square vertices represent members of C. Each line type represents another C-equivalent class.

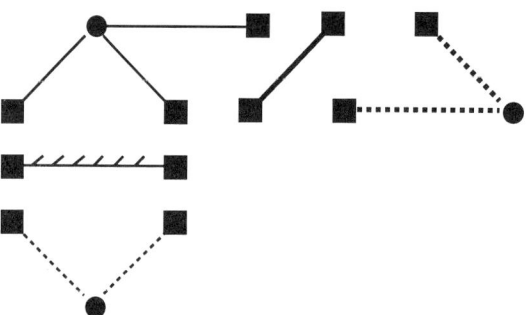

FIGURE 3.22
The bridges of C. Square vertices are the attachments.

3.3. Parallel Tree Contraction   173

the root of B and/or one of its leaves. In Figure 3.23 we give a tree and its decomposition into its 5-bridges. The vertices represented by boxes are the 5-critical vertices, and the numbers next to these vertices are their weights.

LEMMA 3.4
If B is an m-bridge of a tree T, then B can have at most one leaf attachment.

PROOF
The proof is by contradiction. We assume that B is an m-bridge of a tree T, and v_1 and v_2 are two leaves of B that are also m-critical. We prove that this is impossible. Let w be the lowest common ancestor of v_1 and v_2 in T. Since B is connected, w must be a vertex of B and there must be a path from v_1 to w and from w to v_2. Therefore, by definition of an m-bridge w cannot be m-critical. On the other hand, w is m-critical by the above lemma.   ∎

From Lemma 3.4 one can see that there are three types of m-bridges: (1) a **leaf bridge** which is attached by its root; (2) an **edge bridge** which is attached by its root and one leaf; and (3) a **top bridge**, containing the root of T, which exists only when the root is not m-critical. Except for the top bridge, the root of each m-bridge has a unique child. The edge from this child to its root is called the **leading edge** of the bridge.

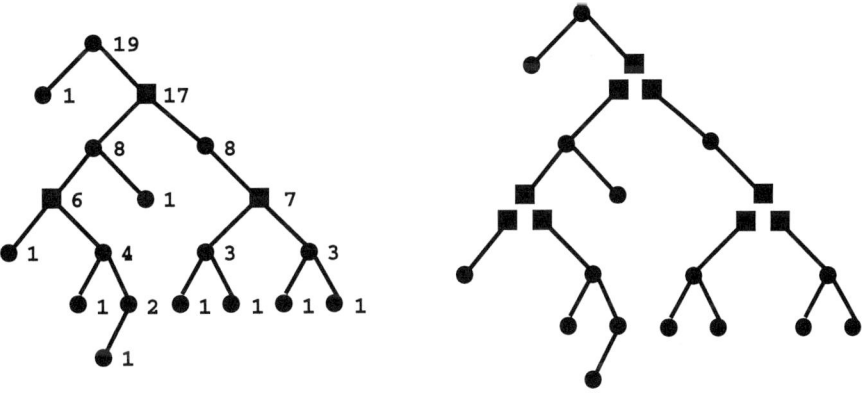

FIGURE 3.23
The decomposition of a tree into its 5-bridges.

LEMMA 3.5
The number of vertices of an m-bridge is at most $m + 1$.

PROOF
Consider the three types of m-bridges: leaf, edge, and top. Suppose B is a leaf bridge with root r'. Then r' is the only m-critical vertex in B and has weight $\geq m$. Because $m > 1$, r' must have a child, w, and this child is unique in B. We claim that w has weight $< m$. Suppose it has weight $\geq m$. Then we claim there is at least one m-critical vertex in the subtree rooted at w. Because all paths from w to a leaf vertex have strictly decreasing weight and the leaf has weight 1, there exists some vertex with weight $\geq m$ and all of its children have weight $< m$. By definition this vertex is m-critical, which contradicts that r' is the only m-critical vertex in B. Thus, w must have weight $< m$ and the number of vertices of B is at most m. If B is an edge bridge with m-critical root r' and m-critical leaf u, then r' will have a unique child w in B. Since u is not a leaf of T and all the vertices in the subtree rooted in u are not in B, the number of vertices in B is $W(w) - W(u) + 2$ (we add 2 to include vertices r' and u). Since w is not m-critical, $W(w) - W(u) < m$ by arguments similar to the above. Thus, the number of vertices in B is $\leq m + 1$. The case for a top bridge follows by similar arguments. ∎

To devise a parallel algorithm, it would be convenient to have few m-bridges [i.e., $O(n/m)$]. However, that is not always the case. For example, consider an unbounded degree tree of height 1, where $m < n$ and every edge is an m-bridge. The following lemma shows, however, that the number of m-critical vertices is not large.

LEMMA 3.6
The number of m-critical vertices in a tree of size n is at most $2n/m - 1$ for $n \geq m$.

PROOF
Let n_k be the number of vertices in a minimum size tree with k m-critical vertices. The lemma is equivalent to the statement:

$$n_k \geq (\frac{k+1}{2})m, \quad \text{for} \quad k \geq 1. \tag{3.5}$$

We prove inequality 3.5 by induction on k. If v is m-critical, then its weight must be at least m. This proves 3.5 for $k = 1$. Suppose that 3.5 is true for $k \geq 1$ and all smaller values of k. We prove 3.5 for $k + 1$. Suppose that T is a minimum size tree with $k + 1$ m-critical vertices.

The root r of T must be m-critical for it to be of minimal size, because we can discard all of the tree above the first m-critical vertex (the root bridge) without affecting the number of critical vertices. Assuming r is m-critical, there are two possible cases for the children of root r: (1) r has two or more children, $u_1, \ldots, u_t$, and each of their subtrees contains an m-critical vertex; or (2) r has exactly one child u whose subtree contains an m-critical vertex.

We first consider Case 1. Let n_i be the number of vertices, and k_i the number of m-critical vertices in the subtree of u_i, for $1 \leq i \leq t$. Since T is of minimum size, $u_1, \ldots, u_t$ must be the only children of r. Since T has $k+1$ m-critical vertices, one of which is the root, $k = \sum_{i=1}^{t} k_i$ and $\sum_{i=1}^{t} n_i \leq n_{k+1}$. Using these two inequalities and the inductive hypothesis we get the following chain of inequalities:

$$n_{k+1} \geq \sum_{i=1}^{t} n_i \geq \sum_{i=1}^{t} \left(\frac{k_i+1}{2}\right) m \geq \left(\frac{(\sum_{i=1}^{t} k_i)+t}{2}\right) m,$$

$$= \left(\frac{k+t}{2}\right) m \geq \left(\frac{k+2}{2}\right) m = \left(\frac{(k+1)+1}{2}\right) m.$$

This proves Case 1.

In Case 2 the subtree rooted at u contains a unique maximal vertex w which is m-critical, and the subtree of w contains k m-critical vertices. Thus, the induction hypothesis shows that $W(u) \geq (\frac{k+1}{2})m$. We consider two subcases of Case 2, when k is odd and even. If k is odd then $(\frac{k+1}{2})m$ is an integral multiple of m. In order for $W(r)$ to be an integral multiple of m greater than $W(u)$,

$$W(r) \geq (\frac{k+1}{2})m + m \geq (\frac{k+2}{2})m.$$

If k is even then $\frac{k}{2}m$ is an integral multiple of m. Again in order for $W(r)$ to be an integral multiple of m greater than $W(u)$,

$$W(r) \geq (\frac{k+1}{2})m + \frac{m}{2} = (\frac{k+2}{2})m \qquad \blacksquare$$

The m-**contraction** of a tree T with root r is a tree $T_m = (V', E')$, such that the vertices V' are the m-critical vertices of T union r. Two vertices v_1 and v_2 in V' are connected by an edge in T_m if there is an m-bridge in T which contains both v_1 and v_2. Note that every edge in T_m corresponds to a unique m-bridge in T which is either an edge bridge or the top bridge. Thus

176   Chapter 3.   List Ranking and Parallel Tree Contraction

by Lemma 3.6, T_m is a tree with at most $2n/m$ vertices. In the next section we show how to reduce a tree to its m-contraction, where $m = 2n/P$, in $O(m + \log n)$ time on a EREW PRAM.

Reduction from Size n to Size n/m

In this section we show how to contract a tree of size n to one of size $2n/m$ in $O(m)$ time using n/m processors, for $m \geq \log n$. If we set $m = \lceil 2n/P \rceil$, then this gives us a reduction of a problem of size n to one of size P. In the next section we show how to contract a tree of size P to a point.

From the previous section, we learned that there are at most $2n/m - 1$, m-critical vertices, but possibly many more m-bridges. Since we have no bound on the number of m-bridges in a tree, we cannot simply assign an m-bridge to each processor. However, since we are assuming that the tree is of bounded degree d, there can be at most d m-bridges common to and below an m-critical vertex of T. Therefore, to perform the reduction, we need only find the m-critical vertices and efficiently assign them to processors. A processor is assigned to each m-critical vertex and computes the value (function) of the (at most d) m-bridges below it. A processor is also assigned to the top bridge if the root is not m-critical, and computes the function for the top bridge. Since each m-bridge has at most $m + 1$ vertices, a processor can sequentially compute the value or function of its $O(1)$ m-bridges in $O(m)$ time. The sketch of the algorithm is in Algorithm 3.12.

ALGORITHM 3.12
Sketch of m-Contract
Procedure *m-Contract* (T)

1. $m := \lceil 2n/P \rceil$
2. Compute $W[v]$ and $\lceil W[v]/m \rceil$ for all vertices v in T
3. Determine the m-critical vertices in T
4. Assign a processor to each m-critical vertex and one to the root
5. Each processor computes the value of the leaf bridges or the unary function of the edge or top bridges below the m-critical vertex or root assigned to it
6. Return the m-contraction of T

The following example illustrates this procedure. Consider again an expression tree. The evaluation of a leaf bridge is the value of all the vertices of the subtree. An edge or top bridge can be considered as a unary function, with the leaf attachment as the indeterminate. That is, the evaluation of the edge or top bridge, with leaf attachment vertex l_s and root r_s, is a unary function f_s such that $value[r_s] = f_s(value[l_s])$. Figure 3.24 shows a expression tree over $\{+, *, -\}$, its division into 5-bridges, and its 5-contraction.

Assume that a tree is given as a set of pointers from each child to its parent and that the tree is ordered. That is, the children of a vertex are ordered from left to right and each child knows its position (index) in that ordering. Furthermore, assume that each parent has a consecutive block of memory cells, one for each child, so that each child can write its value, when known, into its location using its index. This last assumption permits us to compute the maximum value of each set of siblings needed to determine the m-critical vertices.

The Bounded Degree Case In this subsection we consider each step in more detail for trees of bounded degree. In order to implement many of the steps in the m-$Contract$ procedure we use the Euler tour of a tree, which is a list of both forward and backward edges of the tree in depth first order. Therefore, we add finding the Euler tour of the tree T to Step 1.1. For a more detailed discussion on Euler tours and their construction and use, see Chapter 2.

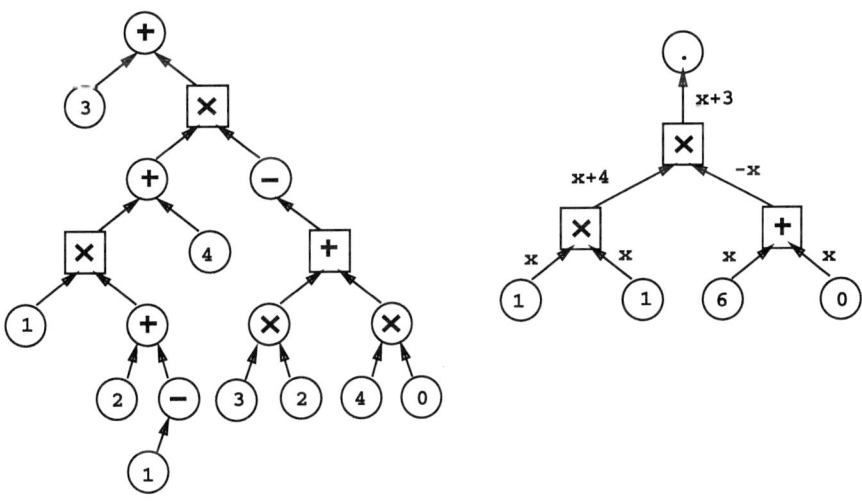

FIGURE 3.24
The 5-contraction of an expression tree.

Step 1

Compute m and find the Euler tour of T. Next we use list ranking to order the edges of T, which we then use to map the ith edge of the tree to the ith element of an array. By ordering the edges in an array, rather than in a linked list, we can perform the All Prefix Sum operations on the array without traversing the linked list each time.

Implementation note: If we are working on a distributed memory parallel computer, we need to send to each processor the information for the edges corresponding to a consecutive block of the Euler tour array. Karlin and Upfal [14] show that once the Euler tour numbering is known the information corresponding to an edge can be moved to its correct location in $O(\log n)$ time using a randomized algorithm. Ranade [21] gives an algorithm for moving the data on a Butterfly network.

Implementation note: Although the running time of All Prefix Sum calculations on a linked list is the same as on an array for a PRAM, on a fixed connection machine, the array representation can result in a $\log n$ improvement in running time. In particular, it can be shown that the All Prefix Sums can be computed in $6 \log n$ time on a binary N-cube parallel computer, where $N = n/\log n$, whereas list ranking on the same size N-cube computer takes $O(\log^2 n)$ time. Therefore, the *m-Contract* algorithm does a single list ranking, converts the linked list to an array, and then performs All Prefix Sum operations on the array.

Step 2

We need to compute the weight of each vertex, that is, the number of vertices in the subtree root at that vertex. Similar to the number of descendants computed in the chapter on Euler tours, the weights can be computed by numbering every forward edge with 1 and backward edge with 0, finding the All Prefix Sums using addition, and computing the weight of a vertex as one plus the difference between the prefix sum of backward edge leaving the vertex and the prefix sum of forward edge entering the vertex.

Step 3

We need to revert back to the original representation of the tree in order to determine which vertices are m-critical. This is easy to do if we save the location of the vertex in the original representation with the forward edge of the Euler tour representation. A processor is responsible for

$\lceil n/P \rceil$ vertices. Each vertex writes its weight to the memory locations reserved by its parent, denoted by wt. That is, vertex v writes $weight[v]$ to $wt[P[v], index[v]]$. In this way, each vertex has the weights of all its children in a subarray. We can then perform a Segmented All Prefix Sum operation using the max operator so that the All Prefix Sum starts afresh each time it reaches a new subarray of the wt array. Let $maxwt$ be the result of the Segmented All Prefix Sum. A vertex is m-critical if $\lceil weight[v]/m \rceil > \lceil maxwt[v]/m \rceil$.

Step 4

We enumerate the m-critical vertices, by assigning a 1 to m-critical vertices and 0 to all others, and then compute the All Prefix Sum over the addition operator. Processor i is responsible for the ith m-critical vertex and Processor 0 is responsible for the root if the root is not m-critical.

Step 5

Each processor evaluates the m-bridges below the m-critical vertex assigned to it. The number of m-bridges below an m-critical vertex is equal to the number of children of that vertex and is bounded by d, the degree of the tree. If there is a depth-first-search sequential algorithm to evaluate an m-bridge, then a processor can evaluate the m-bridges by simply scanning the Euler tour of the tree starting at the forward edge of the first child of the m-critical vertex. Traversing the Euler tour is equivalent to a depth first traversal. Note that all the edges of a leaf bridge are in consecutive edges of the Euler tour. Thus, if the m-bridge is a leaf bridge, then when the processor returns back to its m-critical vertex it will have completely evaluated the subtree. The next forward edge on the Euler tour is the first edge of the next m-bridge for which the processor is responsible. On the other hand, the edges of an edge or top bridge are in two separate consecutive edges of the Euler tour, the intervening edges being part of the subtree routed at the leaf attachment. Therefore, if the m-branch is an edge or top bridge, the processor will reach an m-critical vertex that is not its own. This vertex represents the indeterminate of the edge or top bridge function. To continue its tour of the m-bridge the processor needs to jump to the corresponding backwards edge. Again, when it reaches its own m-critical vertex it has completed the evaluation of the m-bridge and is ready to proceed with the next m-bridge, if there is one. There is another m-bridge if the next edge is a forward edge. Otherwise, there are no more m-bridges.

Since each m-bridge has at most $m+1$ vertices, if the sequential running time of evaluating an tree of m vertices is $T(m)$, then a processor can evaluate all its m-bridges in $O(T(m))$. For example, the sequential evaluation of expression trees over the operators $\{+, \times\}$ is linear. Therefore, evaluating the m-bridges takes $O(m)$ time.

Implementation note: If the m-*Contract* is implemented on a distributed memory parallel machine then each processor maintains the data corresponding to n/P vertices and $2n/P = m$ consecutive edges of the Euler tour. The m-bridges for which a processor is responsible are located in at most $d+1$ separate portions of the Euler tour. Each separate portion can contain at most $2m$ edges and, therefore, can only be located in the memory of at most 3 processors. Thus, a processor needs to access the memory of at most $3d + 3$ processors.

Step 6
We return the Euler tour of m-contraction of the tree. The Euler tour is easily computed using the All Prefix Sums of the m-critical vertices.

The Unbounded Degree Case In this subsection we show how to compute the m-contraction of an unbounded degree tree. In the unbounded case the number of m-bridges immediately below an m-critical vertex may be large and in particular the number of leaf m-bridges may be much larger than the number of processors. Therefore, we cannot load balance by simply assigning a processor to all the m-bridges immediately below an m-vertex. On the other hand, the total number of bridges that are either a top bridge or an edge bridge is bounded by the number of m-critical vertices which, in turn, is bounded by $2n/m$. To handle the unbounded degree case, we change Step 5 of m-*Contract* (Algorithm 3.12) into 3 substeps, as shown in Algorithm 3.13.

Each processor is assigned one edge or top bridge and some number of leaf bridges, depending on their size. To evenly divide the leaf bridges among the processors we compute the All Prefix Sums of their weights. That is, we assign the leading edge of a leaf bridge a value equal to the weight of the child vertex of the leading edge. To all other edges, assign a value of zero. Let $S[e]$ be the sum up to edge e. We would like to assign each processor an equal number of vertices. We approximate this by assigning processor i all leaf bridges with leading edge e such that $\lceil (i-1) \cdot n/P \rceil < S[e] \leq \lceil i \cdot n/P \rceil$.

A processor need only know the first leaf bridge in its interval for which it is responsible. That is, let $proc[e]$ be the processor number responsible for edge e. Then $proc[e] = \lceil S[e]P/n \rceil$ and processor i is responsible for all leaf bridges with edges e such that $proc[e] = i$. The first leaf bridge in the interval has leading edge e' such that $proc[e'] > proc[e' - 1]$.

In the unbounded degree case, a processor may be required to evaluate many small leaf bridges, since there may be a large number of them.

ALGORITHM 3.13
Step 5 of m-Contract for unbounded degree trees
  5. Assign m-bridges to processors:

   a) Assign to each leading edge of a leaf bridge a value equal to the weight of its bridge. To all other edges, assign a value of zero. Compute All-Prefix-Sums of value; let $S(e)$ be the sum up to e and $proc[e] = \lceil S[e]P/n \rceil$.

   b) **if** $proc[e] > proc[e - 1]$ **then** $firstLeafBridge[proc[e]] = e$

   c) Using the All-Prefix-Sums procedure, assign a new processor to each edge or root m-bridge.

Isolation and EREW Parallel Tree Contraction

In the previous section we showed that if we find an EREW parallel tree contraction algorithm, which takes $O(\log n)$ time and uses n processors, then we get an $O(\log n)$ time, $n/\log n$ processor EREW PRAM algorithm for parallel tree contraction, by first applying *m-Contract*. Thus, we may restrict our attention to $O(n)$ processor algorithms. In this section we present a technique called ISOLATION and use it to implement parallel tree contraction on an EREW PRAM without increasing the time and processor count of *Basic_Contract*.

Recall that the *Basic_Contract* algorithm seems to require concurrent reads because the pointer jumping technique of Wyllie causes tails of essential and nonessential chains to have the same parent. A concurrent read occurs when this parent eventually has only one unevaluated argument and extends these chains; all the vertices that were the tails of the chains want to jump over the parent. We call a chain an **isolated chain** if no chain can join it and it cannot join another chain in any round of contraction until the chain is compressed to a single vertex. One way to avoid concurrent reads is to compress an isolated chain to a single vertex before allowing it to join another chain. When a vertex has only a single unevaluated argument and it is not part of an isolated chain, we say the vertex is **free**, because it is free to form a new isolated chain. Once an isolated chain is compressed to a single vertex, that vertex is made free. Algorithm 3.14 displays a high level description of *Isolate_Contract*, a deterministic algorithm for parallel tree contraction. We use the subprocedure *isolate* to prevent vertices not in the chain to become part of the chain and the variable *inChain* to indicate whether a vertex is part of an isolated chain or not. We use the subprocedure *isSingleton* to test whether a chain has been reduced to a single vertex.

ALGORITHM 3.14
ISOLATE *and* CONTRACT *for Deterministic Parallel Tree Contraction*
Procedure *Isolate_Contract*
    In Parallel $inChain[v] := false$;   /* Initialize */
    end in parallel

    In Parallel while $Arg(v) = 0$ **do**
        if $P[v] \neq nil$ **then do**

            Parallel Case $Arg(v)$ **equals**
            0) $rake(v)$;   /* RAKE */
                $mark(label[P[v], index[v]])$;
                $P[v] := nil$;

            1) **if not** $inChain[v]$ **then**   /* ISOLATE */
                $isolate(v)$;
                $inChain[v] := true$;
              else   /* LOCAL COMPRESS */
                $compress(v)$;
                if $isSingleton(v)$ **then** $inChain[v] := false$;
            end case

        end then
    end in parallel
end *Isolate_Contract*

The difference between the contraction phase used in this algorithm and the contraction phase of *Basic_Contract* is that the COMPRESS is replaced by two operations: ISOLATE and LOCAL COMPRESS. ISOLATE marks each chain so that it cannot become part of another chain. Each LOCAL COMPRESS applies one conventional COMPRESS operation to an isolated chain during each contraction phase.

Implementation Techniques We present one method of implementing the generic contraction phase on an EREW model in $O(\log n)$ time, using n processors. Recall that when we compress a chain using Wyllie's algorithm, two chains are created, one essential and one not useful. We modify Wyllie's pointer-jumping algorithm so that processors that pointer jump over nonessential chains eventually stop before a concurrent read takes place. Any chain found in one round is isolated so that it cannot join any other chain found in succeeding rounds. By isolating any chain of length two or more and compressing it until it becomes a single vertex, we can ensure that all jumping over nonessential chains stops before this single vertex is free to become part of another chain.

One way to isolate a chain is to mark the vertices of the chain as being either the *head*, a *middle* vertex, or the *tail* of the chain. The head of the chain is the first vertex in the chain and its child is either a leaf or has more than one child. The tail of the chain is the last vertex in the chain and its parent has more than one child. A middle vertex lies somewhere between the head and the tail of the chain.

An essential chain contains the head of the original chain which will eventually be evaluated during the contraction phase. After several rounds of pointer jumping, all the vertices of the chain will point to v, the parent of the chain. To avoid having all vertices of the chain trying to find the parent of v, we only allow the head of the chain to jump over v. That is, only the head of the chain can be free to form part of a new chain. All other vertices of the chain stop pointer jumping.

Figure 3.25 shows the isolation and compression of a chain and the tagging of vertices as described below. To isolate a chain we **tag** the vertices of the chain with one of three possibilities: R, M, or T, depending on whether it is the *tail*, *middle*, or *head* of a chain. If a vertex is not part of an isolated chain it is tagged with $\emptyset$. All vertices are initially tagged with $\emptyset$. During the COMPRESS phase, the tail of a chain does no pointer jumping, because it is isolated from any newly isolated chains in front of it. Whenever a middle vertex of a chain jumps over the tail, that middle vertex becomes a tail of a new chain. That is, every round of pointer jumping creates one new chain with a new tail. Eventually, every middle vertex becomes a tail of some nonessential chain and stops pointer jumping. At the same time or at the next round, the head of the chain jumps over the tail; the essential chain is a single vertex,

184 Chapter 3. List Ranking and Parallel Tree Contraction

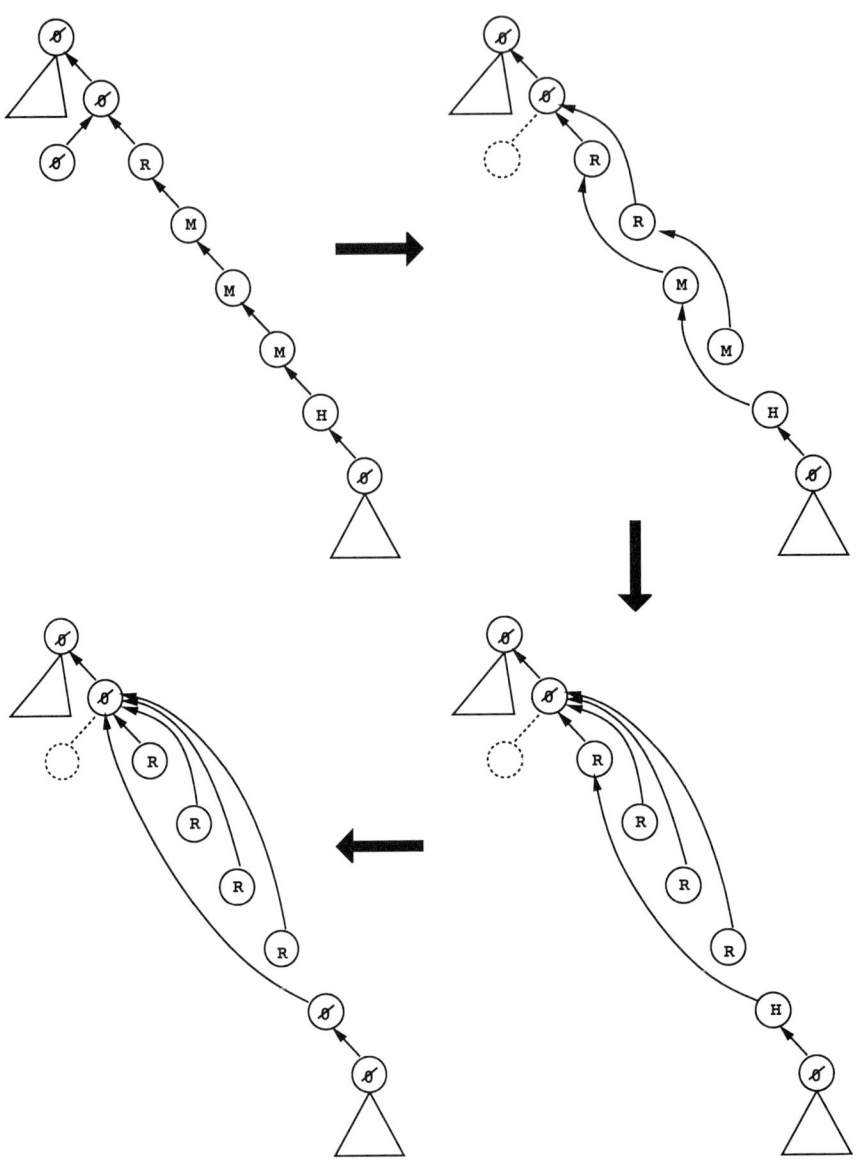

FIGURE 3.25
Three applications of *Isolate_Contract*. Root, middle and head nodes of an isolated chain are labeled R, M and T respectively. Nodes not part of an isolated chain are tagged with $\emptyset$.

the head of the chain. It is at this point the head becomes "free" to join a new isolated chain. A vertex v is *free* if $Arg(v) = 1$ and $Tag = \emptyset$; otherwise v is *not free*. Thus, the tag of the head is set to $\emptyset$ to indicate that it is now not part of an isolated chain.

When a vertex v is free to become part of a new isolated chain, we need to determine its new tag. A vertex is the tail of a chain if its parent is not free and its child is free; it is the head of a chain if its parent is free and its child is not free; and it is a middle of a chain if both its parent and child are free. In this way, isolated chains contain at least two vertices, a tail and a head. To determine whether a child is free, each child vertex maintains a Boolean variable, *free*, that its parent can read to determine whether it is free or not. No concurrent reads can take place because each vertex has only one parent. But, because a vertex can be the parent of several chains, we must be sure that the tails of new isolated chains do not attempt a concurrent read. Therefore, each parent vertex v also maintains an array *pFree* that indicates whether v is free or not multiple times and is indexed by the index of its children (i.e., each child reads one array entry exclusively). Initially all the elements of the array indicate that a vertex is not free. When a parent vertex becomes free it has only one unevaluated child remaining. We assume that the index of this child is set into the variable *child* by the procedure *Arg*. Thus, the free parent need only update the array entry indexed *child*, since there are no other children to read the other entries. Algorithm 3.15 shows an implementation of the *Isolate_Contract* procedure.

ALGORITHM 3.15
An implementation of Isolate_Contract
Procedure *Isolate_Contract_by_Tagging*
   In Parallel $Tag[v] := \emptyset$;

    In Parallel while $Arg(v) = 0$ **do**
       if $P[v] \neq nil$ **then do**

          Parallel Case $Arg(v)$ **equals**
          0)   *rake* (v);   /* RAKE */
              mark $(label[v])$;
              $P[v] := nil$;

          1) **Parallel Case** $Tag[v]$ **equals**
              $\emptyset$) $free[v] = true$;   /* ISOLATE */
                    $pFree[v, child[v]] = true$;

                    if $pFree[P[v], index[v]]$ then
                        if $free[child[v]]$ then
                            $Tag[v] := M$;
                        else
                            $Tag[v] := H$;
                    else if $free[child[v]]$ then
                        $Tag[v] := R$;

        M) if $Tag[P[v]] = R$ then $Tag[v] := R$;
           compress (v); $P[v] := P[P[v]]$; /*LOCAL COMPRESS*/

        H) if $Tag[P[v]] = R$ then $Tag[v] := \emptyset$;   /* isSingleton */
           compress (v); $P[v] := P[P[v]]$; /*LOCAL COMPRESS*/
        end case
    end case

    end then
  end in parallel
end *Isolate_Contract_by_Tagging*

LEMMA 3.7
After each application of Isolate_Contract $|Com|$ decreases by a factor of at least 1/4.

PROOF
By the way the ISOLATE operation is implemented, each isolated chain has a length of at least 2. Moreover, every pair of free vertices is part of an isolated chain, i.e., no two consecutive vertices are singleton vertices. Thus, at every phase of the algorithm, the whole length of a chain consists of a sequence of isolated chains interspersed with singleton vertices. In the worst case, a complete chain consists of an alternating sequence of a singleton and an isolated chain of length three, possibly followed by a singleton. Recall that Com does not include the head of a chain so that the final singleton can be ignored. In such a case, after a single COMPRESS only one vertex in each isolated chain is eliminated and all the other vertices in the chain remain; that is, only one in four vertices is eliminated. ∎

Since step RAKE removes 1/5 of Ra and steps ISOLATE and LOCAL COMPRESS remove 1/4 of Com, together they must remove 1/5 of the vertices. This gives the following theorem.

THEOREM 3.9
A tree of n vertices is reduced to its root after one applies Isolate_Contract $\lceil \log_{5/4} n \rceil$ times.

THEOREM 3.10
Tree contraction can be performed, deterministically, in $O(n/P)$ time using P processors on an EREW PRAM for all $P \leq n/\log n$.

The Expansion Phase If we use procedure *Isolate_Contract* to evaluate an arithmetic expression it will not return the value of all subexpressions. To compute the value of all subexpressions we run the contraction phase "backwards" in a parallel tree expansion phase. Because *Basic_Expand* in Section 3.3.3 only expands along essential chains, *Basic_Expand* uses only exclusive reads and writes. Therefore, we can follow *Isolate_Contract* with *Basic_Expand* to obtain the value of all subexpressions.

If we used *Isolate_Contract* in conjunction with *m-Contract* we need to compute the value of all subexpressions for edge and top bridges. Again, processors can simply re-evaluate the bridges for which they are responsible, as they would leaf bridges, using the now known values of their leaf attachments.

3.3.6 Parallel Tree Contraction for Trees of Unbounded Degree

The discussion on *Basic_Contract* in Section 3.3.3 assumed that the tree was of bounded degree. When the tree has unbounded degree, two problems can arise with *Basic_Contract*. One is that we need a way to test whether $Arg(v)$ equals 0 or 1 in constant time. The other is that the time to perform a RAKE may depend on the number of children of a vertex, and hence is not constant time. For this latter problem we show that by simply running RAKE and COMPRESS asynchronously we continue to get $O(\log n)$ running time.

If the number of arguments per vertex is bounded we can test whether $Arg(v)$ equals 0 or 1 in constant time by counting the number of unmarked labels for v using the processor assigned to it. But when the number of arguments is unbounded we need to use the processors of the children vertices to compute $Arg(v)$. We start by setting aside a memory location $argIndex[v] = null$. Each processor that does not know the value of its vertex writes its index into the memory location of its parent. That is, all children that do not know their values do a concurrent write to their parent $argIndex[v]$ as shown in Algorithm 3.16. Assume that one of these children succeeds in writing its index. If $argIndex[v] = null$ then all the children know their value and $Arg(v) = 0$. To test whether $Arg(v) = 1$, each processor that doesn't know its value reads *argIndex* of its parent and if the value is not the same as its own index it rewrites its index to *argIndex* of its parent. If the value of *argIndex* does not change then only one child attempted to write to it initially, implying that $Arg(v) = 1$. Otherwise, $Arg(v) \geq 2$.

ALGORITHM 3.16
Determining the number of unknown arguments of a vertex of a tree with unbounded degree on a CRCW PRAM

> **Function** $Arg(v)$
> $argIndex[v] := null$;
>   **if** $val[v] = null$ **then**   /* concurrent write to parent */
> $argIndex[P[v]] := index[v]$;
>   **if** $argIndex[v] = null$ **then**   /* all args known */
>   **return** 0;
> $oldArg[v] := argIndex[v]$;
>   **if** $val[v] = null$ **then**
>   **if** $argIndex[P[v]] \neq index[v]$ **then**   /* concurrent read */
> $argIndex[P[v]] := index[v]$;   /* concurrent rewrite */
>   **if** $oldArg[v] = argIndex[v]$ **then**   /* no rewrites*/
>   **return** 1;
>   **else**   /* 1 or more rewrites */
>   **return** 2;
> **end** $Arg(v)$

Up until now we have assumed that the *rake* operation could be performed in constant time. For many applications this is not the case. Miller and Reif [19] show that finding canonical labels for trees has a *rake* operation that is considerably more complicated than just deletion. It requires sorting the labels assigned to the children of a vertex, which requires $O(\log n)$ time. Thus, the parallel time of raking the leaves of a vertex with k children is $O(\log k)$. If we require one application of CONTRACT to finish completely before we start the next application of CONTRACT, then total cost to reduce a tree to its root is the cost for *rake* times the logarithm of the size of the tree. Thus, the naive analysis for canonical labels would be that it runs for $O(\log^2 n)$ time. We improve the running time by a factor of $\log n$ below.

We modify parallel tree contraction so that for those parts of the tree where CONTRACT has already finished we implement a new round of CONTRACT, i.e., each processor executes CONTRACT asynchronously. We shall assume that the time used to remove the leaves of a given vertex is only a function of the number of leaves at that vertex. We should point out that the synchronous and asynchronous versions of CONTRACT may return very different answers. For example, when computing canonical forms for trees by sorting leaves, both the synchronous and asynchronous algorithms are correct. However, the two algorithms produce different sets of canonical labels. In addition, the asynchronous version is faster.

Asynchronous_Contract can be described graph theoretically by viewing it as operating on trees with special leaves which we call *phantom leaves*. The algorithm runs in stages. Initially the tree T has no phantom leaves. We apply the procedure CONTRACT to T to obtain the tree T'. If a given vertex $v \in T$ has $k \geq 2$ children that are nonphantom leaves then we replace them with a new phantom leaf $w \in T'$. Furthermore, if the time required for *Asynchronous_Contract* to process these k children of v is t then the phantom vertex w persists for t stages, at which time it simply disappears. Every time a new block of children of v become leaves, a new phantom child replaces them. In this way a vertex may have several phantom children. Until all leaves, including phantom leaves, of a vertex are removed, the vertex is not a leaf. The time to execute *Asynchronous_Contract* is the number of stages it takes to reduce the tree to its root and for all phantom leaves to disappear.

THEOREM 3.11

If the cost to rake a vertex with k children is bounded by $O(\log k)$ then Asynchronous_Contract requires only $O(\log n)$ time.

PROOF

Suppose the time to *rake* k children of a vertex is bounded by $c \log k$ for $k \geq 2$ and *rake* for a single child can be performed in unit time. We analyze the time used by *Asynchronous_Contract* using an amortization argument. We assign weights to the vertices of the tree such that, at any stage of the algorithm, the weight of the tree reflects the progress made so far. We show that the weight for the tree as a whole decreases by a constant proportion at each stage.

The problem is that *Asynchronous_Contract* can go through several stages removing no leaves, and then remove many leaves. We want to be able to say that at every stage a constant proportion of the work is done. By introducing phantom leaves for accounting purposes, we can take some credit at each stage for the work performed by *rake*, while making sure that the overall credit for raking k leaves remains k. Therefore, we introduce the notion of a weighted tree. A **weighted tree** is a tree with weights assigned to the vertices. The weight of a tree is the sum of the weights of the vertices in the tree. In this application all vertices have weight 1, except phantom leaves which may have arbitrary real weights greater than or equal to 1. Initially, the weight of the tree is the size of the tree.

First we describe in more detail how weights are assigned to phantom leaves. Suppose the time required to rake the k nonphantom leaves of a vertex v is $f(k)$. There is a subtlety here; if the time to rake k leaves of a vertex varies from vertex to vertex, the way the tree contracts may vary dramatically. Our analysis only depends on an upper estimate for the time to rake the children of a vertex. We define β to be a function of k, such that $\beta^{f(k)-1}(k) = 1$ for $f(k) > 0$. Hence $\beta(k) < 1$ for all $k \geq 2$. The constant $\beta(k)$ is the rate at which the phantom leaf decays. We replace the k leaves of vertex v with a phantom leaf w, which we give weight k. After each successive stage we decrease the weight on w by a factor of $\beta(k)$ until the weight equals 1. In the next stage we simply delete the phantom leaf w. Thus, the phantom leaf w exists for $f(k)$ stages, at which time it is deleted. Note that the weight of a phantom vertex is always at least 1.

As in the proof of Theorem 3.4 we partition the vertices of T into two sets, Ra and Com. We claim that the weight of Com decreases by a factor of $1/2$ at each stage while the weight of Ra decreases by a factor of at least $(4+\beta)/5$ at each stage, where $\beta = \max\{\beta(k) | 1 \leq k \leq n\}$. Note that different phantom leaves decay at different rates. We have picked β to be the slowest such rate. The fact that Com decreases by $1/2$ follows by noting that the vertices in Com are processed the same way as in CONTRACT and their weights are all 1. Next we consider the case of Ra. Recall that $Ra = V_0 \cup V_2 \cup C_0 \cup C_2 \cup GC_2$, where V_0 is the set of leaves and phantom leaves. Since the weight on any vertex in V_0 is at least 1 and the weight of any vertex not in V_0 is 1, we see that that weight of V_0 is at least $1/5$ of the weight of Ra. On the other hand the weight of V_0 decreases by at least β at each stage. Thus, the weight of Ra decreases by at least a factor of $4/5 + \beta/5$ at each stage.

This shows that the number of stages is bounded by $\log n$ base $5/(4+\beta)$. For a particular case of interest, when $f(k) \leq c \log k$ for some constant c and $k \geq 2$, we see that β is bounded away from 1 for all n. This proves the Theorem. ∎

3.3.7 Conclusions

In this section we have shown the paradigm of parallel tree contraction, which can be used to perform computations over trees efficiently and is quite general. This paradigm often supplies computationally superior parallel algorithms than algorithms that use divide-and-conquer. In addition, the algorithms are usually easier to devise and understand.

We introduced several concepts that were important in the design of parallel tree algorithms. First we introduced RAKE and COMPRESS which allowed us to separate the processing of leaf vertices from internal vertices. For many applications, finding the *rake* subprocedure is obvious, especially when the natural flow of information is from the bottom up. COMPRESS requires finding a way to combine the information for a pair of vertices in a chain succinctly. For example, for expression trees defining the *compress* subprocedure simply requires finding an efficient representation for the unary function that represents an expression tree with one unknown leaf, and then applying function projection and composition.

When an application either requires results for all subtrees or the natural flow of information is from the top down, an additional expansion phase is required. This expansion phase is simply the contraction phase run in reverse using *unrake* and *uncompress* subprocedures. Often they are similar to their counterparts. However, when information flows from the bottom up and the top down, *unrake* can be quite different from *rake*.

The next major concept we introduced was SHUNT for binary trees. It is possible to use shunt because there is an efficient algorithm to number leaves. SHUNT was applied only to odd-numbered leaves, and combined RAKE and COMPRESS into a single operation. However, in designing a *shunt* subprocedure, it is usually easier to think of it as a rake followed by a compress on the sibling vertex.

The next major concept was the notion of dividing a tree into blocks of subtrees on which processors could work. The processors reduce the subtrees to single edges forming a smaller tree. The number of vertices in this tree equals the number of processors so that, at this point, a nonoptimal algorithm can be used to reduce the tree to the root.

To obtain an EREW algorithm we introduced the concept of isolating a chain, so that new vertices cannot join the chain, and reducing the chain to a point before allowing it to become part of a new chain. By isolating chains we prevented concurrent reads and writes.

The basic tree contraction algorithm is not optimal and uses concurrent reads and writes. It runs in $O(\log n)$ time using n processors, as long as *rake* and *compress* take $O(1)$ time. If *rake* takes as much as $O(\log n)$ time and we can rake and compress different parts of the tree asynchronously, then we still only use $O(\log n)$ time overall. The remaining algorithms are optimal and use only exclusive reads and writes. *Shunt* has somewhat greater constants and can only be used on binary trees, which limits its applicability, whereas the constants for the m-contract/isolation algorithms are greater than the other tree contraction algorithms, making them less practical.

TABLE 3.2
Time and processor count for the parallel tree contraction algorithms discussed in this section.

Problem	Time	Processors	Work
basic tree contraction	$O(\log n)$	$O(n)$	$O(n \log n)$
tree contraction for binary trees	$O(\log n)$	$n/\log n$	$O(n)$
deterministic tree contraction	$O(n/P)$	$P, P \leq n/\log n$	$O(n)$
tree contraction, unbounded degree	$O(\log n)$	$n/\log n$	$O(n)$

Bibliography

[1] K. Abahamson, N. Dadoun, D. K. Kirkpatrick, and T. Przytycka. A simple parallel tree contraction algorithm (preliminary version). In *Proceedings of the 25th Annual Allerton Conference on Communication, Control, and Cumputing*, pages 624–633, Monticello, Illinois, Sept/Oct 1987.

[2] A. Aho, J. Hopcroft, and J. Ullman. *The Design and Analysis of Computer Algorithms*. Addison-Wesley, 1974.

[3] Richard J. Anderson and Gary L. Miller. Deterministic parallel list ranking. In J. H. Reif, editor, *VLSI Algorithms and Architectures: 3rd Aegean Workshop on Computing, AWOC 88*, pages 81–90, N.Y., June/July 1988. Springer-Verlag. Lecture Notes in Computer Science, Vol. 319.

[4] Richard J. Anderson and Gary L. Miller. A simple randomized parallel algorithm for list-ranking. *Information Processing Letters*, 33(5):269–273, January 1990.

[5] Guy E. Blelloch. *Vector Models for Data-Parallel Computing*. MIT Press, 1990.

[6] I. Bar-On and U. Vishkin. Optimal parallel generation of a computation tree form. *ACM Transactions on Programming Languages and Systems*, 7(2):348–357, April 1985.

[7] R. P. Brent. The parallel evaluation of general arithmetic expressions. *Journal Assoc. Computing Machinery*, 21(2):201–208, April 1974.

[8] H. Chernoff. A measure of asymptotic efficiency for tests of a hypothesis based on the sum of observations. *Annals of Math. Statistics*, 23, 1952.

[9] R. Cole and U. Vishkin. Approximate and exact parallel scheduling with applications to list, tree, and graph problems. In *27th Annual Symposium on Foundations of Computer Science*, pages 478–491, Toronto, Oct 1986. IEEE.

[10] Richard Cole and Uzi Vishkin. Deterministic coin tossing with applications to optimal list ranking. *Information and Control*, 70(1):32–53, 1986.

[11] Faith E. Fich. New bounds for parallel prefix circuits. In *Proceedings of the 15th Annual ACM Symposium on Theory of Computing*, pages 100–109, Boston, MA, April 1983. ACM.

[12] H. Gazit, G. L. Miller, and S-H Teng. Optimal tree contraction in an EREW model. In S. K. Tewksbury, B. W. Dickinson, and S. C. Schwartz, editors, *Concurrent Computations: Algorithms, Architecture and Technology*, pages 139–156, New York, 1988. Plenum Press. Princeton Workshop on Algorithms, Architecture and Technology Issues for Models of Concurrent Computation.

[13] S. R. Kosaraju and A. L. Delcher. Optimal parallel evaluation of tree-structured computation by ranking (extended abstract). In J. H. Reif, editor, *VLSI Algorithms and Architectures: 3rd Aegean Workshop on Computing, AWOC 88*, pages 101–110. Springer-Verlag, N.Y., June/July 1988. Lecture Notes in Computer Science, Vol. 319.

[14] Anna Karlin and Eli Upfal. Parallel hashing–an efficient implementation of shared memory. In *Proceedings of the 18th Annual ACM Symposium on Theory of Computing*, pages 160–168, Berkeley, May 1986. ACM.

[15] Richard E. Ladner and Michael J. Fisher. Parallel prefix computation. *Journal Assoc. Computing Machinery*, 27(4):831–838, October 1980.

[16] Gary L. Miller. Finding small simple cycle separators for 2-connected planar graphs. *Journal of Computer and System Sciences*, 32(3):265–279, June 1986. Invited publication.

[17] Gary L. Miller and John H. Reif. Parallel tree contraction and its application. In *26th Symposium on Foundations of Computer Science*, pages 478–489, Portland, Oregon, October 1985. IEEE.

[18] Gary L. Miller and John H. Reif. Parallel tree contraction part 1: Fundamentals. In *Advances in Computing Research*, Vol. 5, pp. 47–72, 1989.

[19] Gary L. Miller and John H. Reif. Parallel tree contraction part 2: Further applications. *SIAM J. Comput.* 20(6):1128–1147, December 1991.

[20] Gary L. Miller and Shang-Hua Teng. Systematic methods for tree based parallel algorithm development. In *Second International Conference on Supercomputing*, pages 392–403, Santa Clara, May 1987.

[21] A. Ranade. How to emulate shared memory. In *28th Annual Symposium on Foundations of Computer Science*, pages 185–194, Los Angeles, Oct 1987. IEEE.

[22] U. Vishkin. Randomized speed-ups in parallel computation. In *Proc. of the 16th Annual ACM Symp. on Theory of Computing*, pages 230–239, Washington D.C., April 1984. ACM.

[23] J. C. Wyllie. The complexity of parallel computation. Technical Report TR 79-387, Department of Computer Science, Cornell University, Ithaca, New York, 1979.

Part II

Advanced Parallel Graph Algorithms

4

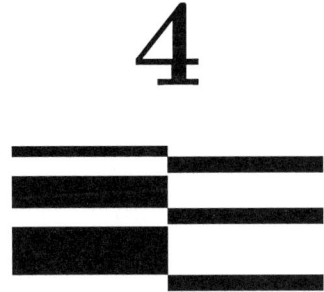

Randomized Parallel Connectivity

Hillel Gazit

Advanced Computing Support Center
Los Angeles, CA
hillel@acsc.com

4.1
Introduction

In this chapter we describe parallel randomized connectivity algorithms, all of which use the *CRCW PRAM* model.

Parallel connectivity algorithms use randomization to achieve either of the two following goals:

1. Break symmetry.
2. Distinguish between high degree vertices and low degree vertices in constant time.

Two parallel randomized algorithms for finding the connected components of an undirected graph are presented. Both algorithms take $O(\log(n))$ time in *CRCW PRAM* model. In this parallel model all the processors can read and write into a common memory. In case of concurrent writes into the same memory location, it is assumed that an arbitrary processor succeeds.

The first algorithm is very simple but requires $m+n$ processors. The second algorithm requires $P = O(\frac{m+n}{\log(n)})$ processors, where m is the number of edges and n is the number of vertices. This algorithm is *optimal* in the sense that the product $P \cdot T$ is a linear function of the input size. Both algorithms are *optimal* in space.

Reif [10] was the first to present a simple connected components algorithm that uses randomization to break symmetry between vertices. Later Gazit [5, 6] presented a more complicated connected components algorithm that joins high degree vertices in some stages and low degree vertices in other stages. The second algorithm is the first optimal connected components algorithm.

4.2
Preliminaries

4.2.1 Definitions

An undirected graph $G = (V, E)$ consists of a set of *vertices* V of size n and a set of *edges* E of size m. Each edge is an unordered pair (v, w) of distinct vertices v and w. In this chapter it is assumed that there may be more than one edge between two vertices. An edge (v, w) *hits* vertices v and w. The *degree* of a vertex v is the number of edges that hit v. A *path* joining v_1 and v_k in G is a sequence of vertices $v_1, v_2, ..., v_k$ such that $(v_i, v_{i+1}) \in E$.

A set of vertices $U \subseteq V$ is a *connected subgraph* if for every pair of vertices $v, u \in U$, there is a path in U joining v and u. A *connected component* is a maximal connected subgraph.

The following notations are used:

1. P = Number of processors.
2. n = Number of vertices.
3. m = Number of edges.

Assuming there are at least $\frac{n+m}{\log(n)}$ processors, the following definitions are given:

- An *easy graph* is a graph such that $n + m \leq P$.

- A *dense graph* is a graph such that $n \leq P$.

- A *sparse graph* is a graph such that $n > P$.

A *supervertex* is defined as a tree rooted at some vertex. Every vertex has a unique pointer to its parent. There are no cycles, but there is one self-loop for every supervertex tree, since its root is a vertex that points onto itself. There is a directed path from every vertex in a tree to the root of that tree. The *root* of each vertex v is defined as the root of the supervertex tree to which v belongs, denoted by $root(v)$. If a tree is of height 1 or 0, the supervertex is a *star*.

We will later show that a supervertex is the natural data structure for solving the connected components problem. The majority of recent connectivity algorithms ([11, 2, 5]) use supervertices as their basic data structure.

Supervertices are viewed as vertex disjoint partial connected components. In any stage of the algorithm the graph is a forest of supervertices. An edge (v, u) is a *live* edge if u and v belong to different supervertices. There can be more than one edge between two supervertices. A supervertex is *live* if there is a live edge incident on any of its vertices. The *degree* of a supervertex is the number of live edges that hit its vertices.

4.3
A Simple Connected Components Algorithm

A simple and easy to analyze connected components algorithm is presented in this section. This is the *Random-Mate* algorithm. In this algorithm

procedure *random-mate* (V, E)

for all $v \in V$ **in parallel do** $parent(v) := v$ **od**
while there is a live edge in the graph **do**
    for every vertex v **in parallel do**
        v chooses in random a *sex* in $\{M, F\}$
    od
    for every live edge (u, v) **in parallel do**
        if $sex(parent(u)) = M$ **and** $sex(parent(v)) = F$ **then**
            $parent(parent(u)) := parent(v)$ **fi**
        if $sex(parent(v)) = M$ **and** $sex(parent(u)) = F$ **then**
            $parent(parent(v)) := parent(u)$ **fi**
    for every vertex v **in parallel do**
        $parent(v) := parent(parent(v))$
    od
od

end *random-mate*

FIGURE 4.1
Connected Components: Simple Case

each supervertex chooses a sex at random. Every male vertex then mates with any live female neighbor. No cycles are possible, all chains are of length 2 and the subgraph size is reduced by a constant factor.

This algorithm was given by Reif [10], and is contained in Figure 4.1.

The basic data structures of the algorithm are supervertices that are maintained as trees of height one or zero (stars). All vertices having a live edge can be notified about it in constant time using $O(m)$ processors. If an M (male) supervertex v has an edge to an F (female) supervertex u, then v becomes the child of u. It is easy to see that this method cannot create long chains because an F supervertex does not get a new parent, thus all chains are of length 1.

LEMMA 4.1
After each iteration all the supervertices are stars.

PROOF
Induction on the number of iterations.

The induction basis is trivial because initially every vertex is a supervertex with height 0.

Assume that the Lemma is correct after the kth iteration and prove for the $(k+1)^{th}$ iteration. After the kth iteration every vertex v has $parent(v) = root(v)$ because by the induction hypothesis every supervertex is a star. In the $(k+1)^{th}$ iteration we may assign v a new parent which is the root of some other supervertex. By doing so we create a tree with maximum height 2, which we reduce to 1 by setting $parent(v) := parent(parent(v))$. ∎

LEMMA 4.2
Every iteration takes constant time using $m + n$ processors.

PROOF
Because all the supervertices are stars we can check if an edge (u,v) is live by checking if $parent(v) = parent(u)$. The rest of the operations are straightforward in the $CRCW$ model. ∎

LEMMA 4.3
The probability that a live supervertex be joined to some other supervertex, in some iteration i, is at least $\frac{1}{4}$.

PROOF
A live supervertex has at least one live edge. The supervertex will get a new root if and only if its sex is M and it has a live edge to an F supervertex. The probability of this event is the probability that the sex of this supervertex is M multiplied by the probability that the other vertex will be F, that is, $\frac{1}{2} \cdot \frac{1}{2} = \frac{1}{4}$. ∎

LEMMA 4.4
The probability that a vertex will be a live root after $5 \cdot \log(n)$ iterations is at most $\frac{1}{n^2}$.

PROOF
By Lemma 4.3 a supervertex that is live after iteration i has a probability of at most $\frac{3}{4}$ to remain live after iteration $i+1$. Therefore the probability to be live after $5 \cdot \log(n)$ iterations is bounded by $\left(\frac{3}{4}\right)^{5 \cdot \log(n)} < \frac{1}{n^2}$. ∎

LEMMA 4.5
The expected number of live supervertices after $5 \cdot \log(n)$ iterations is at most $\frac{1}{n}$.

PROOF
We compute the expected number of live roots by summing up the prob-

ability of each vertex to be a live root. By Lemma 4.4 this is equal to $\sum_{i=1}^{n} \frac{1}{n^2} = \frac{1}{n}$. ∎

THEOREM 4.1
With probability $\frac{1}{n}$ the algorithm will be finished after $5 \cdot \log(n)$ iterations.

PROOF
The number of live supervertices is a non-negative integer. We define the probability of having k live supervertices after $5 \cdot \log(n)$ iterations as p_k. By Lemma 4.5 and by definition of expectation

$$\frac{1}{n} \geq \sum_{k=0}^{n} k \cdot p_k = \sum_{k=1}^{n} k \cdot p_k \implies \sum_{k=1}^{n} p_k \leq \frac{1}{n}.$$

The Theorem is correct because the algorithm stops when the number of live supervertices is zero. ∎

4.4 Easy Case

In this section we describe a deterministic algorithm for finding connected components when there is a processor for every edge and a processor for every vertex. This algorithm is a variant of Shiloach and Vishkin's algorithm [11] and is used as a subroutine by the randomized algorithm. An outline is presented in Figure 4.2. Some results from [11] that are needed later in this chapter are presented here. The detailed proofs can be found in the original paper.

LEMMA 4.6
Checking if an edge is live before using it, can be done in constant time using one processor.

PROOF
If an edge is used in step 2, then it should have connected two roots and therefore it is easy to check whether or not it is live. The trees that are joined in step 3 were not changed in step 1 and are therefore stars; checking if an edge connected to a vertex in a star is live can be done in constant time. ∎

LEMMA 4.7
During the execution of the algorithm, the partial connected components are always kept as supervertices.

procedure *easy-case* (V, E)

for all $v \in V$ **in parallel do** $parent(v) := v$ **od**
while there is a live edge in the graph or some tree is not a star **do**
/* Step 1 */
    for all $v \in V$ **in parallel do** $parent(v) := parent(parent(v))$ **od**
    for every live edge (u, v) using concurrent write **in parallel do**
/* Step 2 */
        if $parent(parent(v)) = parent(v)$ **and** $parent(parent(u)) = parent(u)$
        then if $parent(u) > parent(v)$ **then** $parent(parent(u)) := parent(v)$
            else $parent(parent(v)) := parent(u)$
            fi
        fi
/* Step 3 */
        if $parent(u) = parent(parent(u))$ **and**
            $parent(u)$ did not get new links in the first two operations
        then $parent(parent(u)) := parent(v)$
        fi
        if $parent(v) = parent(parent(v))$ **and**
            $parent(v)$ did not get new links in the first two operations
        then $parent(parent(v)) := parent(u)$
        fi
    od
/* Step 4 */
    for all $v \in V$ **in parallel do** $parent(v) := parent(parent(v))$ **od**
od

end *easy-case*

FIGURE 4.2
Finding Connected Components: Easy Case

PROOF
The proof is similar to the proof of Theorem 2.1 in Baase's chapter. ∎

LEMMA 4.8
If a tree T has not been changed during an entire iteration it remains unchanged until the end. This tree is a star and its vertices are a connected component.

PROOF
Every unchanged tree must be a star, otherwise it would have some descendant v in depth 2, and then the operation $parent(v) := parent(parent(v))$ would change the tree. By Lemma 4.7 this star represents a partial connected component. It is maximal, otherwise it would have a live edge e that would be used in the third step, thus changing it. ∎

DEFINITION 4.1
The height of the forest is the **sum** *of the heights of all the trees that are either live or have height greater than one.*

LEMMA 4.9
After the first iteration the height of the forest is at most $\frac{2 \cdot n}{3}$.

PROOF
After the first three steps every vertex with a live edge belongs to some tree. The total height of all the stars is bounded above by the number of vertices in the stars over two (this is the case when all stars are of size 2). The total height of the non-star trees is bounded by the number of vertices in these trees. After the fourth step each of those trees has at most $\frac{2}{3}$ of its original height. ∎

LEMMA 4.10
The height of the forest is reduced by a factor of at least $\frac{1}{3}$ in each iteration (after the first iteration).

PROOF
The proof is similar to the proof of Lemma 2.5 in Baase's chapter. ∎

COROLLARY 4.1
The number of the live trees after iteration i is bounded by $n \cdot \left(\frac{2}{3}\right)^i$.

PROOF
The proof follows immediately from Lemma 4.10. The upper bound may happen only if all the trees are stars. ∎

THEOREM 4.2
(Main Theorem of [11])

1. The algorithm terminates after $\lceil \log_{\frac{3}{2}}(n) \rceil$ iterations.
2. $parent(u) = parent(v)$ if and only if u and v are in the same connected component.

PROOF
The proof is similar to the proof of Theorem 2.3 in Baase's chapter. ∎

4.5 Dense-to-Easy Reduction

4.5.1 Informal Description

A dense graph is a graph for which there is a processor for every vertex and a processor for every $\log(n)$ edges. The edges processors and the vertices processors are not necessarily distinct. The method presented in this section repeatedly chooses a random sample of edges. A vertex with a high degree has a better probability to have at least one of its edges chosen than a vertex with a low degree. It is possible that a vertex has a live edge that was not chosen in some iteration, but as the number of its live edges increases, so does the probability that at least one of them will be chosen. Therefore, the probability that a vertex having many live edges (a dense vertex) is joined, is greater than that of a vertex with few live edges (a sparse vertex). The goal is to join dense vertices among themselves and leave out the sparse ones.

After the dense-to-easy algorithm is completed, all the remaining live edges are connected to sparse vertices. It will be shown later that the expected number of live edges in the graph after the algorithm has completed is $O(n)$, and therefore the *easy-case* algorithm can be used on the remaining graph.

The dense vertices can be "forced" to mate among themselves and not with the sparse vertices. As noted above, when selecting an edge and a vertex, the probability of the edge hitting this vertex is higher if the vertex has many edges. If one randomly picks a large enough sample of edges and some vertex v, and none of these edges hit the vertex, then there is a high probability that v is sparse. For each edge in the sample an attempt is made to join its end-vertices; another (smaller) sample is chosen, and the process is repeated. Note that several edges in the sample may hit the same vertex, but only one of them is used.

The size of the sample we choose decreases geometrically (so as to reduce the time complexity of successive iterations). This means that the probability for a given supervertex to have no incident edge selected (and therefore to be classified as sparse) increases. However, we will later show that the expected number of live edges that hit a supervertex classified as sparse is inversely proportional to the sample size.

The number of dense supervertices also decreases geometrically. By choosing the samples so that sample size decreases more slowly than the number of supervertices we achieve two desirable results:

- For each iteration, consider the set of all supervertices that were classified as sparse during this iteration and the expected total number of live edges that hit this set; we will show that this number decreases geometrically.

- The time complexity of each step is proportional to the sample size. The sample size decreases geometrically but not as fast as the number of dense super vertices. Therefore the time complexity of each step decreases geometrically. The first element in a geometrically decreasing sequence dominates the sum, and this element is $\frac{m}{P} = O(\log(n))$; therefore the time complexity is optimal.

The algorithm presented in Figure 4.3 reduces a dense graph to the easy graph case. After execution of the algorithm, every vertex in the graph points to the root of the partial connected component to which it belongs. All live edges connect roots; and the expected number of live edges is $O(n)$. Therefore, the easy case algorithm can be applied. The reduction algorithm always halts, but the number of live edges that remain in the graph at the end may be greater than expected. The time complexity is $O(\log(n))$. The probability to fail is less than $\lceil \log_{\frac{3}{2}}(n) \rceil \cdot \left(\frac{2}{e}\right)^{\frac{n}{\log^3(n)}}$.

4.5.2 Data Structure

In addition, for the supervertex tree structure that was described before every vertex v has a single bit flag called ext_v. The flag of vertex v is 1 if v is assumed to be a dense vertex.

4.5.3 Analysis of the Dense-to-Easy Reduction

DEFINITIONS:

1. An *Extrovert root* r is a root with $ext_r = 1$. At the beginning of the *dense-to-easy* procedure, every root is an *Extrovert* root.
2. An *Extrovert supervertex* v is a supervertex with an *Extrovert* root.
3. An *Extrovert vertex* v is a vertex that belongs to a supervertex whose root is an *Extrovert* root.

procedure *dense-to-easy* $(G(V, E))$

for all $v \in V$ **in parallel do** $ext_v := 1$ **od**
for all $v \in V$ **in parallel do** $parent(v) := v$ **od**
$i := -1$
while there is a root r such that $ext_r = 1$ **do**
    $i := i + 1$
    for all $v \in V$ **in parallel do** $parent(v) := parent(parent(v))$ **od**
    Pick a sample of edges of size $\max(P, \lceil m \cdot \alpha^i \rceil)$
    for every live edge (u, v) in the sample using concurrent write **in parallel do**
        if $ext_{parent(u)} = ext_{parent(v)} = 1$ **then**
            if $parent(parent(v)) = parent(v)$ **and**
                $parent(parent(u)) = parent(u)$ **and**
            then if $parent(u) > parent(v)$ **then** $parent(parent(u)) := parent(v)$
                else $parent(v) := parent(u)$
            fi
            if $parent(u) = parent(parent(u))$ **and**
                $parent(u)$ did not get new links in the previous two operations
            then $parent(parent(u)) := parent(v)$
            fi
            if $parent(v) = parent(parent(v))$ **and**
                $parent(v)$ did not get new links in the previous two operations
            then $parent(parent(v)) := parent(u)$
            fi
        fi
    od
    for all $v \in V$ **in parallel do** $parent(v) := parent(parent(v))$ **od**
    for every root r **in parallel do**
        if the tree rooted in r was not changed during the iteration
        then $ext_r := 0$ **fi**
    od
od
for every edge (u, v) in the graph **in parallel do**
    if (u, v) is a live edge **then** replace it by an edge $(root(u), root(v))$ **fi**
od

end *dense-to-easy*

FIGURE 4.3
Reducing Dense Graph to Easy Graph

4. An *Introvert vertex* v is a vertex that belongs to a star and $ext_{parent(v)} = 0$. At the end of the *dense-to-easy* procedure all the vertices are *Introvert*.
5. An *Extrovert Edge* is a live edge which connects two *Extrovert* vertices.
6. An *Extrovert Edge* becomes an *Introvert edge* if one of its end supervertices becomes *Introvert*.
7. n_i is a random variable representing the number of *Extrovert* supervertices after iteration i.
8. $\alpha = \sqrt{\frac{2}{3}}$ a constant, explained later.

LEMMA 4.11
The **while** loop of the dense-to-easy algorithm will be finished after at most $\lceil \log_{\frac{3}{2}}(n) \rceil$ iterations.

PROOF
The proof is the same as for Theorem 4.2. ∎

Let S be a subset of the *Extrovert* stars. Define $deg(S)$ as the number of *Extrovert* edges that hit supervertices in S; that is

$$deg(S) = |\{(u,v)|(u,v) \text{ is an extrovert edge } \wedge (u \in S \vee v \in S)\}|.$$

LEMMA 4.12
Given a subset of stars S, such that $deg(S) = x$, the probability that none of these x edges is chosen in iteration i (and so all the vertices of S and all the x Extrovert edges will become Introvert) is less than or equal to $e^{-x \cdot \frac{y}{m}}$, where $y = \max(P, \lceil m \cdot \alpha^i \rceil)$ is the size of the sample of edges chosen in procedure dense-to-easy.

PROOF
The probability of a particular edge being chosen by one processor is $\frac{1}{m}$. The probability of an edge not being chosen for the sample is bounded by $(1 - \frac{1}{m})^y$, where y is the sample size. The probability that none of the x edges is chosen is at most:

$$\left[\left(1 - \frac{1}{m}\right)^y\right]^x = \left(1 - \frac{1}{m}\right)^{x \cdot m \cdot \frac{y}{m}} \leq (e^{-1})^{x \cdot \frac{y}{m}} = e^{-x \cdot \frac{y}{m}}.$$ ∎

LEMMA 4.13
For a given x, the probability that at least x edges willbecome Introvert in

iteration i is bounded by $e^{-x \cdot \frac{y}{m}} \cdot 2^{n_i}$, where n_i is the number of Extrovert supervertices after iteration i and $y = \max(P, \lceil m \cdot \alpha^i \rceil)$.

PROOF

A subset of edges will become *Introvert* only if every one of them hits at least one supervertex that has become *Introvert* during this iteration. The number of such subsets is bounded by 2^{n_i}. The probability of such a subset to become *Introvert* is given in Lemma 4.12. Only subsets S with at least x Extrovert edges, of which none were chosen, are considered. Therefore, an upper limit to the probability is $e^{-x \cdot \frac{y}{m}} \cdot 2^{n_i}$. ∎

LEMMA 4.14

The number of Extrovert supervertices, after every iteration i, is at most $n \cdot \left(\frac{2}{3}\right)^i$.

PROOF

The proof is similar to the proof of Corollary 4.1. ∎

LEMMA 4.15

For every i, the probability that more than $n \cdot \alpha^i$ Extrovert edges become Introvert during iteration i is bounded by $\left(\frac{2}{e}\right)^{n \cdot \alpha^{2i}}$.

PROOF

Set $x = n \cdot \alpha^i$, $z = m \cdot \alpha^i$. By Lemma 4.14 $n_i \leq n \cdot \alpha^{2 \cdot i}$. By Lemma 4.13 the probability with $y \geq z$ is $e^{-n \cdot \alpha^i \cdot \alpha^i} \cdot 2^{n \cdot \alpha^{2 \cdot i}} = \left(\frac{2}{e}\right)^{n \cdot \alpha^{2 \cdot i}}$. ∎

Define $\gamma_n = \lceil 3 \frac{\log(\log(n))}{\log(3/2)} \rceil$. Note that α^{γ_n} is about $\log^{-3}(n)$.

LEMMA 4.16

The probability that more than $\frac{n}{\log^2(n)}$ Extrovert edges will become Introvert during iteration $i \geq \gamma_n$ is bounded by $\left(\frac{2}{e}\right)^{\frac{n}{\log^3(n)}}$.

PROOF

Set $x = \frac{n}{\log^2(n)}$, $y = P$, and $n_i = n \cdot \alpha^{2 \cdot \gamma_n} = \frac{n}{\log^3(n)}$ (by Lemma 4.14). Then, $x \cdot \frac{y}{m} \geq \frac{n}{\log^3(n)}$; and by substitution in the formula of Lemma 4.13, the result is obtained. ∎

LEMMA 4.17

With probability greater than or equal to $1 - \lceil (\log_{\frac{3}{2}}(n)) \rceil \cdot \left(\frac{2}{e}\right)^{\frac{n}{\log^3(n)}}$, the number of edges that become Introvert is less than $6.5 \cdot n$.

PROOF

Assume that the results expected in Lemmas 4.15, 4.16 are obtained. (The probability for this will be computed later).

Chapter 4. Randomized Parallel Connectivity

Lemma 4.15 is used to compute the number of edges that become *Introvert* in the first iterations, and Lemma 4.16 is used to compute this number for the last iterations.

The sum is bounded by:

$$\left(\sum_{i=0}^{\gamma_n} n \cdot \alpha^i\right) + \sum_{\gamma_n+1}^{\lceil \log_{\frac{3}{2}}(n) \rceil} \frac{n}{\log^2(n)}.$$

For a large enough n, the sum can be bounded by

$$\left(\sum_{i=0}^{\infty} n \cdot \alpha^i\right) + \frac{n}{\log^2(n)} \cdot \log^2(n) =$$

$$= n + n \cdot \sum_{i=0}^{\infty} \alpha^i = n + \frac{n}{1-\sqrt{2/3}} < 6.5 \cdot n.$$

We now compute an upper limit to the probability of failing.
Failure can happen if:

1. The number of edges that become *Introvert* in any of the first γ_n iterations is more than $n \cdot \alpha^i$. By Lemma 4.15 the probability is bounded by $\left(\frac{2}{e}\right)^{\frac{n}{\log^3(n)}}$.
2. The number of edges that become *Introvert* after the first γ_n iterations is more than $n \cdot \alpha^i$. By Lemma 4.16, the probability is bounded by $\left(\frac{2}{e}\right)^{\frac{n}{\log^3(n)}}$.

An upper bound to the number of iterations ($\lceil \log_{\frac{3}{2}}(n) \rceil$) is given by Lemma 4.11. Let us assume that failure in any iteration causes the algorithm to fail. This assumption may be false, but it gives us an upper bound on the probability to fail. By adding up the probabilities to fail in all the iterations we get an upper bound for the probability of the algorithm to fail. Multiplying the upper bound for the number of iterations by the upper bound for the probability to fail in each iteration yields the claimed probability. ∎

THEOREM 4.3
The time complexity of the dense-to-easy algorithm is $O(\log(n))$.

PROOF
The time complexity of iteration i is $O(1 + \frac{m}{P} \cdot \alpha^i)$. By Lemma 4.11 the number of iterations is $O(\log(n))$. Therefore, the complexity of the **for** loop is

$$O(\sum_{i=0}^{O(\log(n))} 1 + \frac{m}{P} \cdot \alpha^i) = O(\log(n)).$$

∎

COROLLARY 4.2
With probability greater than or equal to $1 - \lceil \log_{\frac{3}{2}}(n) \rceil \cdot (\frac{2}{e})^{\frac{n}{\log^3(n)}}$, *the running time of the connectivity algorithm for a dense graph is* $O(\log(n))$.

PROOF
The time complexity of the dense-to-easy algorithm is $O(\log(n))$ (by Theorem 4.3). The probability to have more than $O(n)$ edges was bounded above by Lemma 4.17. In a dense graph $n \leq O(P)$, therefore by Theorem 4.2 the easy-case algorithm will run in $O(\log(n))$ time.

∎

4.6
Sparse-to-Dense Reduction

In a sparse graph $P < n$ and therefore we cannot assign a processor for every vertex. Since the dense graph reduction procedure attempts to find a sparse subgraph, it is certainly not applicable to the sparse case. An algorithm for the sparse case is described in [5, 6], but due to its length and complexity is not presented here and only the general idea is explained.

There are three procedures in the algorithm.

1. Deterministic Mate. Similar to Reif's random mate which we described earlier this procedure creates trees of depth one, but in a deterministic way. Reif's algorithm is randomized, and has a probability of $\frac{1}{2}$ to create very few new links between supervertices. In order to reduce the probability to fail we would like to have a deterministic algorithm that joins the supervertices and always succeeds.
2. Partition. The second procedure partitions a graph with n_i vertices into two subgraphs, *Extrovert* and *Introvert*. The method is similar to the one used in the dense-to-easy algorithm. We sample the

edges and use the live ones to join the *Extrovert* vertices. There is one important difference, the procedure is run only $2 \cdot \log(\log(n))$ iterations, resultingwith high probability, the following two subgraphs: An *Extrovert* subgraph with at most $\frac{n_i}{\log(n)}$ vertices and an *Introvert* subgraph with at most $\alpha^2 \cdot n_i$ vertices and $O(n_i)$ edges. We cannot reduce the height of the trees by performing $parent(v) := parent(parent(v))$ because we do not have a processor for every vertex, therefore we would like to keep the trees with a low depth. That is done by joining the vertices using the deterministic mating procedure that creates trees with height 1.

3. This is the main sparse graphs connectivity procedure. It calls the partition procedure $2 \cdot \log(\log(n))$ times, each time sending the *Introvert* subgraph resulting from the previous call as input to be partitioned again.

The complexity of each iteration is bounded by the number of edges in the sparse subgraph over the number of processors. However the number of edges in the sparse subgraph decreases geometrically. Therefore, by rule of geometric series the time complexity of the sparse-to-dense algorithm is equal to the complexity of the first time that we partition the graph.

In each application of the partition procedure the number of vertices in the *Introvert* subgraph is reduced by a factor of α^2. Thus, after $2 \cdot \log(\log(n))$ such applications the number of vertices in *Introvert* is bounded by $\frac{n}{\log(n)}$.

Also, in each application of partition we separate a small *Extrovert* graph from the sparse graph. The sizes of the *Extrovert* graphs also decrease geometrically because as said before, the sizes of the sparse graphs decrease geometrically and the size of the *Extrovert* graph is bounded by the size of the sparse graph over $\log(n)$. After the ith application of the partition algorithms, at most $O(\frac{n_i}{\log(n)})$ vertices are added to the *Extrovert* subgraph, where n_i is the number of sparse vertices in iteration i. Since n_i decreases geometrically, after $2 \cdot \log(\log(n))$ repeated applications of partition, the number of dense vertices is at most $\sum_{i=0}^{\infty} \alpha^{2 \cdot i} \cdot \frac{n}{\log(n)} = O(\frac{n}{\log(n)})$.

The number of vertices in the resulting graph is the sum of the size of the last *Introvert* subgraph and the sizes of all the *Extrovert* subgraphs. This sum is bounded by $O(\frac{n}{\log(n)})$, and therefore qualifies the graph for the *dense-to-easy* algorithm.

4.7
Exercises

4.1 We define $D(u,v)$ to be the distance between u and v if there is a path between them, and 0 otherwise. The diameter of a graph d is defined as $d = \max_{u,v \in V} D(u,v)$.

   a) What is the expected running time of the random-mate algorithm on a graph with a constant diameter?

   b) Find a deterministic algorithm that uses $n + m$ processors and find the connected components of a graph with a constant diameter in a constant time. (*Hint:* Use the fact the minimum can be computed in a constant time using a linear number of processors.)

4.2 a) Find an algorithm that reduces the case where $P = n \cdot \frac{\log(\log(n))}{\log(n)} + \frac{m}{\log(n)}$ to the dense graph case in $O(\log(n))$ time.

   b) Find an algorithm that computes connected components to all graphs using $P = \frac{n+m}{\log(n)}$ in $O(log(n) \cdot \log^*(n))$ time.

   c) Find an algorithm that computes connectivity to all graphs using $P = \frac{n+m}{\log(n)} \cdot \log^*(n)$ in $O(log(n))$ time.

Bibliography

[1] D.S. Hirschberg, A.K. Chandra, and D.V. Sarwate. Computing connected components on parallel computers. *Communication of the ACM*, 22(8):461–464, August 1979.

[2] R. Cole and U. Vishkin. Approximate and exact parallel scheduling with applications to list tree, and graph problems. In *27st Annual Symposium on Foundation of Computer Science*, pages 478–491. IEEE, Oct 1986.

[3] R. Cole and U. Vishkin. Deterministic coin tossing and accelerating cascades: micro and macro techniques for designing parallel algorithms. In *Proc. of the Eighteenth Annual ACM Symp. On Theory of Computing*, pages 206–219. ACM, 1986.

[4] R. Cole and U. Vishkin. Faster optimal parallel prefix sums and list ranking. Submitted to *Information and Computation*, 1987.

[5] H. Gazit. An optimal randomized parallel algorithm for finding connected components in a graph. In *27st Annual Symposium on Foundation of Computer Science*, pages 492–501. IEEE, Oct 1986.

[6] H. Gazit. An optimal randomized parallel algorithm for finding connected components in a graph (extended version). *Accepted to SIAM Journal of Computing*, 1989.

[7] T. Hagerup. Optimal parallel algorithms on planar graphs. In *Aegean Workshop on Computing*, pages 24–32, 1988.

[8] Arkady Kanevsky and Vijya Ramachandran. Improved algorithms for graph four-connectivity. In *28th Symposium on Foundations of Computer Science*, pages 252–259, Los Angeles, October 1987. IEEE.

[9] Gary L. Miller and Vijya Ramachandran. A new graph triconnectivity algorithm and its parallelization. In *19st Annual ACM Symposium on Theory Of Computing*, pages 335–344, New York, May 1987. ACM.

[10] J. H. Reif. Optimal parallel algorithms for graph connectivity. Center for Computing Research TR-08-84, Harvard University, 1984.

[11] Y. Shiloach and U. Vishkin. An $O(\log n)$ parallel connectivity algorithm. *J. of Algorithms*, 3(1):57–67, 1983.

[12] R.E. Tarjan and U. Vishkin. An efficient parallel biconnectivity algorithm. *SIAM J. Comput.*, 14(4):862–874, 1985.

[13] J. C. Wyllie. The complexity of parallel computation. Department of Computer Science TR-79-387, Cornell University, 1979.

5

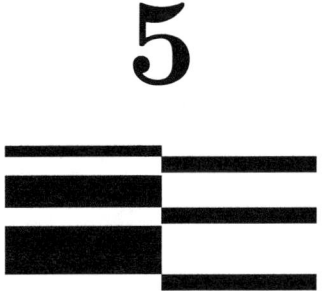

Advanced Parallel Prefix-sums, List Ranking and Connectivity

Uzi Vishkin

UMIACS
University of Maryland
College Park, MD 20742-3251
vishkin@umiacs.umd.edu
and
Department of Computer Science
Tel Aviv University
Tel Aviv, Israel

5.1 Introduction

The model of parallel computation used in this chapter is the arbitrary concurrent-read concurrent-write (CRCW) parallel random access machine (PRAM).

In Section 5.2, a parallel algorithm for the prefix sums problem is presented. The algorithm runs in time $O(\log n/\log \log n)$ using $n \log \log n/\log n$ processors. This algorithm leads to a parallel list ranking algorithm which runs in $O(\log n)$ time using $n/\log n$ processors, described in Section 5.3. The first link in a chain of advanced parallel connectivity algorithms is given in Section 5.4. For an undirected graph with n vertices and $m \geq n \log n \log^{(3)} n$ edges, it runs in $O(\log n \log^{(3)} n)$ time, using $m/(\log n \log^{(3)} n)$ processors. The bibliography section gives reference to the final result.

5.2 Fast Parallel Prefix Sums

ALGORITHM 5.1
Input. An array of n numbers $A(1), A(2), \ldots, A(n)$. *The prefix sums problem.*
Compute

$$\sum_{j=1}^{i} A(j) \quad for\ all\ 1 \leq i \leq n$$

We will assume familiarity with Brent's theorem for assignment of processors to jobs and with the Ladner-Fischer standard logarithmic time optimal parallel algorithm for the prefix sums problem. For simplicity, we will assume throughout this chapter that $\log n$, $\log \log n$, $\log n/\log \log n$ and n over each of these numbers are all integers.

We present an algorithm to compute the prefix sums of n numbers, each of $\log n$ bits, using $O(n)$ operations and $O(\log n/\log^{(2)} n)$ time. Later, we mention how to achieve the same complexity result for the prefix sums problem, where each number is of $O(\log n)$ bits.

ALGORITHM 5.2
The new prefix sums algorithm
Input: n numbers, each of $\log n$ bits.

5.2. Fast Parallel Prefix Sums

The prefix sums algorithm computes the n prefix sums of these numbers in time $O(\log n/\log^{(2)} n)$, using $O(n)$ operations, that is, using $n\log^{(2)} n/\log n$ processors, on the CRCW PRAM. The algorithm has five steps.

Step 1 Partition the n numbers into groups of $\log n/\log^{(2)} n$ numbers each. For each group add its numbers in a serial fashion. This is done in parallel for all groups. The output of step 1 consists of $n\log^{(2)} n/\log n$ numbers each with at most $\log^{(2)} n \log n$ bits.

Step 2 Add together sets of $x = 4\log^{2/3} n\, 2^{\log^{2/3} n}$ numbers output by Step 1, in time $O(\log x) = O(\log^{2/3} n)$. (Use the standard parallel prefix sums algorithm.)

Let y be $n\log^{(2)} n/(4\log^{2/3} n\, 2^{\log^{2/3} n} \log n)$. The output of step 2 consists of y numbers each with at most $2\log n$ bits. For large enough n, $2\log^{1/3} y \geq \log^{1/3} n$ (see Exercise 5.1 below). For simplicity, we assume that this inequality holds and therefore each of these y numbers consists of at most $4\log^{2/3} n\, \log^{1/3} y$ bits.

Step 3 Divide each number, output by Step 2, into (at most) $2\log^{2/3} n$ pieces, each of $2\log^{1/3} y$ bits, namely the trailing $2\log^{1/3} y$ bits, the next least significant $2\log^{1/3} y$ bits, and so on. In parallel for $1 \leq i \leq 2\log^{2/3} n$, compute the prefix sums for the set of ith pieces using the "Main routine", which is presented below. There are $2\log^{2/3} n$ prefix sum subproblems, each of size y.

Step 4 For each of the y numbers, combine (serially) the $2\log^{2/3} n$ prefix sums computed for that number (one per piece). This uses $O(\log^{2/3} n)$ time and y processors.

Step 5 "Backtrack" through Steps 2 and 1 to compute the prefix sums for each number input to Step 1. Similar backtracking is done in the standard parallel prefix sums algorithm which is described elsewhere in this book. The complexity of the backtracking is dominated by the complexity of Steps 1 and 2.

218   Chapter 5.   Advanced Parallel Prefix-sums, List Ranking and Connectivity

EXERCISE 5.1
Show that for large enough n, $2\log^{1/3} y \geq \log^{1/3} n$. What is the smallest value of n for which this inequality holds?

We saw that the prefix sums algorithm employs a routine called the *Main* routine. The Main routine employs a recursive routine called the *Auxiliary* routine, which employs the *Basic* routine. We describe the Basic routine first, then the Main routine, and finally the Auxiliary routine.

THE BASIC ROUTINE

Input: m numbers, each of m bits, and $m \cdot 2^{m^2}$ processors.

The Basic routine finds all m prefix sums in $O(1)$ time. Prior to an application of the Basic routine we must apply a routine called the *Precomputation* routine. The Precomputation routine initializes a table, called the *Precomputation table*, which is then used in the Basic routine.

REMARK 5.1
In all applications of the Basic routine, m will have the same value and therefore the Precomputation routine has to be performed only once throughout the prefix sums algorithm. It might be useful to know at this stage that $m \leq \log^{1/3} n$. We also note that the assumption used for our PRAM model is that a single processor can handle a word of $O(\log n)$ bits in $O(1)$ time.

We first describe the Precomputation routine.

THE PRECOMPUTATION ROUTINE

There are 2^{m^2} different possible inputs for the Basic routine. Due to the nature of the Basic routine, we call each possible input (which is a string of m^2 bits) a *hypothetical input*. The domain of the Precomputation table has an entry for each hypothetical input. The range of each entry comprises all the m prefix sums of the entry. The Precomputation routine simply applies the standard parallel prefix sums algorithm in parallel to each entry of the Precomputation table. It should be clear that this requires $O(m)$ operations and $O(\log m)$ time per entry, or a total of $O(m \cdot 2^{m^2})$ operations and $O(\log m)$ time.

We return to the Basic routine. We provide a set of m processors for each of the 2^{m^2} hypothetical inputs. The task, for each set of processors, is to discover if the actual input is identical to its hypothetical input. If they are not identical, the set of processors has completed its task. If they are identical, then the ith processor from the set simply looks up the ith prefix sum and outputs it in $O(1)$ time. We determine if the actual and hypothetical inputs are equal, as follows. Each processor in the set is responsible for one of the m numbers of the hypothetical input: this number in the actual input is checked against its counterpart in the hypothetical input. If they are unequal the processor writes a "fail" message to a memory location for its set of processors. Thus, only if there is no fail message in this memory location is the actual input identical to the hypothetical input for this set of processors. There will be exactly one set of processors for which the actual and hypothetical inputs are identical; this is the set of processors that proceed to output the prefix sums. We have left to the reader the trivial details of how to assign the processors to their jobs. So the Basic routine requires a total of $O(m 2^{m^2})$ operations and $O(1)$ time. The Basic routine will be employed many successive times; this (more than) balances the $O(\log m)$ running time of the Precomputation routine.

THE MAIN ROUTINE

Input: y numbers, each of at most $2 \log^{1/3} y$ bits and $y \cdot 2^{\log^{2/3} y}$ processors.

REMARK 5.2
As we saw in the prefix sums algorithm,
$y = n \log^{(2)} n / (4 \log^{2/3} n \, 2^{\log^{2/3} n} \log n)$.

The Main routine finds the prefix sums of these y numbers in two steps.

Step 1 Set m to be $\log^{1/3} y$ and compute the Precomputation table as in the Precomputation routine.

The Precomputation routine is needed for a subsequent use of the Basic routine.

Step 2 Call the "Auxiliary routine", which is a recursive routine described below, to find the prefix sums of the y numbers.

THE AUXILIARY ROUTINE

Input. z numbers, each of at most $2 \log^{1/3} y$ bits, where $z \geq \log^{1/3} y$. Consider dividing the z numbers into groups of $\log^{1/3} y$ numbers each. Another input requirement is that we can provide each group with $\log^{1/3} y \cdot 2^{\log^{2/3} y}$ processors.

The Auxiliary routine recursively finds the prefix sums of these z numbers. It will be convenient to add bits so that each of the z input numbers has exactly $2\log^{1/3} y$ bits. The Auxiliary routine has five steps.

Step 1 Divide each of the numbers into two numbers of $\log^{1/3} y$ bits each, as follows. The *trailing* number consists of the trailing $\log^{1/3} y$ bits and the *leading* number consists of the leading $\log^{1/3} y$ bits.

We then compute the prefix sums separately for the leading numbers and for the trailing numbers. We describe how to compute the prefix sums for the trailing numbers. The computation for the leading numbers is identical.

Step 2 Divide the z numbers into $z/\log^{1/3} y$ groups of $\log^{1/3} y$ numbers, each number of $\log^{1/3} y$ bits. We provide each group with $\log^{1/3} y \cdot 2^{\log^{2/3} y}$ processors. The (local) prefix sums with respect to each group are computed in $O(1)$ time, using the Basic routine.

Step 3 Call the Auxiliary routine recursively to compute the (global) prefix sums with respect to the $\frac{z}{\log^{1/3} y}$ sums of the groups.

Step 4 For each of the z trailing numbers, compute its prefix sum by adding its local prefix sum to the global prefix sum of the preceding group.

Steps 2, 3, and 4 are performed with respect to the leading numbers, as well, and we have their prefix sums also.

Step 5 For each of the z input numbers, we combine the prefix sums of its trailing and leading numbers, in a further $O(1)$ time.

We evaluate the complexity of the Main and Auxiliary routines together. First, we make the following two observations: (1) The ith recursive level of the Auxiliary routine issues at most 2^i calls to the $i + 1$st recursive level. (For instance, the single call at the first level issues one call (for the second recursive level) in the prefix sums computation of the trailing bits and another in the prefix sums computation of the leading bits). (2) Consider the calls to the Auxiliary routine at the ith recursive level. There are $\frac{y}{\log^{(i-1)/3} y}$ input numbers in each such call. This leads to the following two conclusions: (a) The depth of the recursion is $O(\log y/\log^{(2)} y)$. Since each recursive level needs $O(1)$ time, the Main routine runs in $O(\log y/\log^{(2)} y)$ time. (b) The number of processors used for the first recursive level is $y2^{\log^{2/3} y}$. We note that the number of input elements for each successive recursive call decreases by a factor of $\log^{1/3} y$, and the number of calls increases only by a factor of (at most) two. Therefore, the number of processors for the first recursive level suffices for implementing each of the deeper levels in $O(1)$ time. We conclude,

THEOREM 5.1
The Main routine uses $O(\log y/\log^{(2)} y)$ time and $y2^{\log^{2/3} y}$ processors.

The theorem implies that Step 3 of the new prefix sums algorithm requires $O(\log n/\log^{(2)} n)$ time and $O(n)$ operations. We conclude that overall, the five steps of the new prefix sums algorithm require $O(\log n/\log^{(2)} n)$ time and $O(n)$ operations.

REMARK 5.3
It is easy to extend the prefix sums algorithm to numbers consisting of $c \log n$ bits each, for any fixed integer $c > 0$. Divide each number into c pieces, each of $\log n$ bits, namely the trailing $\log n$ bits, the next least significant $\log n$ bits, and so on. For $1 \leq i \leq c$, compute the prefix sums for the set of the ith pieces using the above parallel algorithm. For each number combine the c prefix sums computed for this number.

We conclude,

THEOREM 5.2
The prefix sums of n numbers, of $O(\log n)$ bits each, can be computed in $O(\log n/\log^{(2)} n)$ time using $n \log^{(2)} n/\log n$ processors.

5.3
2-Ruling Set and List Ranking

Section 5.3.1 revisits the deterministic coin tossing technique which is presented in Chapter 2 (Algorithm 2.7). Section 5.3.2 gives a new 2-ruling set algorithm, where the deterministic coin tossing technique is used. Finally, Section 5.3.3 applies the 2-ruling set algorithm to provide a list ranking algorithm.

Input. A connected directed graph $G(V, E)$. The in-degree of each vertex is exactly one. The out-degree of each vertex is exactly one. Note that the graph forms a directed circuit. The vertices are given in an array of size n.

Each vertex has a pointer to the next vertex (representing its outgoing edge). We define a subset U of V to be an $r - ruling\ set$ of G if:

1. No two vertices of U are adjacent.
2. For each vertex v in V there is a directed path from v to some vertex in U whose edge length is at most r.

The r-ruling set problem. Find an r-ruling set of V.

5.3.1 The Deterministic Coin Tossing Technique Revisited

Recall that we needed the following assumptions about the input representation. The vertices are given in an array of length n. The entries of the array are numbered from 0 to $n - 1$. The numbers are represented as binary strings of length $\lceil \log n \rceil$. We refer to each binary symbol (bit) of this representation by a number between 0 and $\lceil \log n \rceil - 1$. The rightmost (least significant) bit is called bit number 0 and the leftmost bit is called bit number $\lceil \log n \rceil - 1$. Each vertex v has a pointer $succ[v]$ to its successor in the ring (representing its outgoing edge) and a pointer $pred[v]$ to its predecessor in the ring. Application of the deterministic coin tossing technique resulted in assigning new numbers, denoted tag, to each vertex. The tag values had the following properties:

1. For each element i, $tag[i] \neq tag[succ[i]]$.
2. The range of tag values is $(-1, 1), (0, 0), (0, 1), (1, 0), (1, 1) \ldots (\lceil \log n \rceil - 1, 0), (\lceil \log n \rceil - 1, 1)$. This range can be shifted to $0, 1, \ldots, 2\lceil \log n \rceil$. Each tag value can then be represented as a binary string of at most $\lceil \log \log n \rceil + 2$ bits.
3. Each tag value can be computed in $O(1)$ time using a single processor.

These three properties allow a recursive application of the deterministic coin tossing technique relative to the tag values yielding $newtag$ values. The $newtag$ values have the following properties.

1. For each element i, $newtag[i] \neq newtag[succ[i]]$.
2. The range of $newtag$ values is $(-1, 1), (0, 0), (0, 1), (1, 0), (1, 1) \ldots (\lceil \log \log n \rceil + 1, 0), (\lceil \log \log n \rceil + 1, 1)$.

LEMMA 5.1
Computation of the newtag values takes $O(1)$ time and $O(n)$ operations.

5.3.2 2-Ruling Set Algorithm

Now we describe an $O(\log m / \log \log m)$ time algorithm for the 2-ruling set problem whose total number of operations is $O(m)$. For simplicity we will also assume that $2 \log \log m + 3 \leq \log m / \log \log m$ (otherwise, m must be smaller than some constant).

EXERCISE 5.2
Find where this assumption is used. Try to achieve the same complexity result without this assumption.

We arrange the elements in groups of size $m \log \log m / \log m$, where all elements of one group have the same newtag value. Then at each iteration one whole group is processed in parallel by the $m \log \log m / \log m$ processors. The processors must be able to quickly find the elements they are to process on each iteration, so we store the element indexes in an array *schedule* such that the group of elements to be processed on the kth iteration are stored in the kth segment of size $m \log \log m / \log m$ in *schedule*. (In other words, group k contains the elements in positions $(k-1) m \log \log m / \log m + 1$ through $k m \log \log m / \log m$ in *schedule*.)

Each group is limited to $m \log \log m / \log m$ elements (because we use this number of processors), but we cannot expect each group to be "full". We spread the elements over $2 \log m / \log \log m$ groups with a total of $2m$ slots in *schedule*, some empty. All elements can then be processed in $2 \log m / \log \log m$ iterations. Following is an outline of the algorithm.

ALGORITHM 5.3
Optimal algorithm for constructing a 2-ruling set
Recall that $newtag[i]$ is an ordered pair $[posn, bit]$. Throughout this algorithm, treat $newtag[i]$ as a binary number $2 * posn + bit + 1$. $newtag$ values range from zero to $2 \log \log m + 2$.

1. Sort the elements by their newtag values, assigning to each element i its (unique) position in the sorted list, $sortposn[i]$ (where $1 \leq sortposn[i] \leq m$).
2. "Schedule" the elements by marking all $2m$ slots in *schedule* initially empty, then computing
    $slot[i] := sortposn[i] + newtag[i] m \log \log m / \log m;$
    $schedule[slot[i]] := i$
3. Construct the 2-ruling set S:

Initialize $inS[i] := 0$ for all i. { Easily done in $\log m / \log \log m$
   steps with $m \log \log m / \log m$ processors }
Each processor P_q (for $1 \leq q \leq m \log \log m / \log m$) does:
for $k := 1$ **to** $2 \log m / \log \log m$ **do**
   $i := schedule[(k-1)m \log \log m / \log m + q]$;
   if $i \notin S$ **and** $pred[i] \notin S$ **and** $succ[i] \notin S$ **then**
      $inS[i] := 1$ **endif**
endfor

Proving that no two elements are assigned to the same slot and no group has elements with different newtag values is similar to the 2-ruling set algorithm of Chapter 2. We conclude:

THEOREM 5.3
If the sorting in step 1 can be done with $m \log \log m / \log m$ processors in $O(\log m / \log \log m)$ time on the CRCW PRAM, then the above algorithm constructs a 2-ruling set with $m \log \log m / \log m$ processors in $O(\log m / \log \log m)$ time on the CRCW PRAM.

We show how to implement the sort for step 1.

Sorting Elements by Their Newtag Values

For the above step 1, it remains to sort the elements by newtag value in $O(\log m / \log \log m)$ time with $m \log \log m / \log m$ processors.

Since the newtag values to be sorted are integers in the range $0, \ldots, \log m / \log \log m$ (hint: see Exercise 5.2), we can use a bucket sort approach, counting elements that have the same newtag value. We use the following counters:

$lowernewtags[i]$ = the number of elements whose newtag values are less than element i's newtag value ($1 \leq i \leq m$)
$samenewtag[i]$ = the number of elements with the same newtag value as i's, but which appear earlier in the array than i ($1 \leq i \leq m$)

Define $sortposn[i] = lowernewtags[i] + samenewtag[i] + 1$. Once $lowernewtag$ and $samenewtag$ values have been computed, $sortposn$ can be computed within our desired bounds for time and number of operations. The counts are computed in three stages, with applications of the CRCW prefix sum algorithm, given above, in the second and third stages.

5.3. 2-Ruling Set and List Ranking

For the first stage, divide the array into segments of $\log m / \log \log m$ elements each. Processor P_q ($1 \leq q \leq m \log \log m / \log m$) works on the qth segment. P_q will compute the following sums.

$samenewtag_q[i]$ = the number of elements in the qth segment with the same newtag value as element i, but which appear earlier than i (for i in the qth segment)

$count_q[newtagvalue]$ = the number of elements in the qth segment with this $newtagvalue$ ($0 \leq newtagvalue \leq \log m / \log \log m$)

Stage 1 takes $O(\log m / \log \log m)$ time.

Now we have local counts $count_q$ and $samenewtag_q$. In the second stage, we combine them to get the global totals for $samenewtag$. Suppose element i is in the qth segment. The formula is:

$$samenewtag[i] = \sum_{j=1}^{q-1} count_j[newtag[i]] + samenewtag_q[i]. \qquad (5.1)$$

Thus for each of the $\log m / \log \log m$ newtag values separately, we compute prefix sums for the sequence of local counters $count_q$. That is, we will be applying $\log m / \log \log m$ copies of the parallel prefix sum algorithm, and we want to apply them all in parallel. In a previous section we gave a CRCW PRAM algorithm for the prefix sums problem that needs $O(n)$ operations and takes $O(\log n / \log \log n)$ time. The number of operations for each of our prefix sums problem is linear in its size, that is $O(m \log \log m / \log m)$. The time is bounded by $O(\log m / \log \log m)$. Since the total size of the prefix sums problems is m, we get a total of $O(m)$ operations and $O(\log m / \log \log m)$ time for stage 2. No new ideas are needed in order to apply Brent's scheduling principle in order to achieve the same time bound with $m \log \log m / \log m$ processors.

For each newtag value, the above prefixs sums also yield

$$count[newtagvalue] = \sum_{q=1}^{m/\log m} count_q[newtagvalue]$$

i.e., the total number of elements with this newtag value. This sum is used in the third stage.

In the third stage we compute $lowernewtags[i]$ for each element i. The relevant formula is:

$$lowernewtags[i] = \sum_{newtagvalue=0}^{newtag[i]-1} count[newtagvalue] \qquad (5.2)$$

In words, we compute the prefix sums for the sequence of $\log m / \log \log m$ newtag value counts. This is done with one application of the sequential prefix sums algorithm in $O(\log m / \log \log m)$ time, by one processor. Then, *lowernewtags*[i] can be computed by simply reading the correct prefix sum. The processors can again be allocated as in stage 1, with P_q computing *lowernewtags* for the $\log m / \log \log m$ elements in the qth segment of the array.

LEMMA 5.2
Sorting by the newtag values takes $O(\log m / \log \log m)$ time using $m \log \log m / \log m$ processors on a CRCW PRAM.

PROOF
We verified that each of the three stages of the sort can be done in the stated time and processor bounds. ∎

We have now completed the details for the optimal $O(\log m / \log \log m)$ algorithm to construct a 2-ruling set.

5.3.3 The Logarithmic Time List Ranking Algorithm

We use the above $O(\log n / \log \log n)$ time, optimal parallel algorithms for the prefix sums problem and the 2-ruling set problem. Given these algorithms, the changes, with respect to the $O(\log n \log \log n)$ list ranking algorithm of Chapter 2 are minimal.

The input for the list ranking problem is an array of n elements.

ALGORITHM 5.4
The Recursive List Ranking (Suffix Sum) Algorithm
    The input for the recursive algorithm is a linked list L of length m
        stored in an array of m entries
if $m \leq n / \log n$
then use the basic suffix sum algorithm of chapter 2
else
    Find a 2-ruling set S (that defines a sublist L_S);
    Perform sublist compaction on L to get L_S using a prefix sums
        algorithm; $m_{new} := |S|$;
    Recursively apply this algorithm to the compacted list L_S;
    Compute the ranks for the elements of $L - L_S$;
endif

The algorithm correctly computes list ranks.

THEOREM 5.4

The list ranking problem can be solved in $O(\log n)$ time with $n/(\log n)$ processors on a CRCW PRAM.

PROOF

We first determine the number of iterations of the algorithm, then count the work done on each iteration. No two neighboring elements are both included in the 2-ruling set S, implying $m_{new} \leq m/2$. Thus, after $\lceil \log \log n \rceil$ iterations, $m \leq n/\log n$, so there are at most $\lceil \log \log n \rceil$ iterations.

Table 5.1 summarizes the time and operation bounds already established for the various parts of the algorithm. The total number of operations at each iteration is $O(m)$. At each iteration m decreases by half (at least), so the total number of operations for all iterations (before reverting to the basic algorithm) is at most

$$\sum_{i=0}^{\log \log n - 1} n/2^i = O(n).$$

The time bound for each iteration is $O(\log n / \log \log n)$, so the time for all iterations is $O(\log n)$. The basic suffix sum algorithm is applied to a list of size at most $n/\log n$; this uses time in $O(\log(n/\log n))$ and $O(n)$ operations (see Chapter 2). Thus, the time bound for the entire algorithm is $O(\log n)$ and the total number of operations is $O(n)$. By easy application of Brent's scheduling principle, the algorithm can be implemented on p processors in time

$$O(n/p + \log n). \qquad \blacksquare$$

TABLE 5.1
Time and operation bounds for the list ranking algorithm

Subproblem	Time	Operations	Justification
Deterministic coin tossing	$O(1)$	$O(m)$	Section 5.3.1
Construct 2-ruling set	$O(\log m / \log \log m)$	$O(m)$	Section 5.3.2
Sublist compaction	$O(\log m / \log \log m)$	$O(m)$	Section 5.2 + Chapter 2
Compute ranks for $L - L_S$	$O(1)$	$O(m)$	See Chapter 2

5.4 Graph Connectivity

5.4.1 Basic Techniques and Previous Work

We start with a brief review of the $O(\log^2 |V|)$ time parallel connectivity algorithm of Hirschberg, Chandra and Sarwate, and the $O(\log |V|)$ time algorithm of Shiloach and Vishkin. Chapter 2 relates to these algorithms. The purpose of the description below is to extract features of these algorithms which are relevant for the presentation of the present chapter. One purpose of this review is to enable a discussion of the obstacles to deriving optimal speed-up implementations from them.

Our problem is to compute the connected components of a graph $G = (V, E)$ which is given as follows.

Input form. Let $V = \{1, \ldots, n\}$ and $|E| = m$. We assume that the edges are given in a vector of length $2m$. The vector contains first all the edges incident on vertex 1, then all the edges incident on vertex 2, and so on. Each edge appears twice in this vector. We also need the two copies of each edge to be linked; we discuss how to achieve this linking, if it is not provided as part of the input, at the end of Section 5.4.2.

DEFINITIONS

A rooted tree is a directed graph satisfying:

> *(a) The undirected graph which is obtained by removing directions from the edges is a tree.*

> *(b) It has a vertex r called the root such that for each vertex v there exists a directed path from v to r.*

A rooted star is a rooted tree in which the path from each vertex to the root comprises (at most) one edge.

The following is common to both the HCS and SV connectivity algorithms and to the new connectivity algorithm presented here. At each step during the algorithms each vertex v has a pointer $D(v)$ through which it points to another vertex or to no vertex. One can regard the directed edge $(v, D(v))$ as a directed edge in an auxiliary graph, called the *pointer graph*. Initially, for each vertex v, $D(v)$ points to no vertex and therefore the initial pointer graph consists of only the vertices, but has no edges. The pointer graph keeps changing during the course of the algorithms. However, at each step of each of these three algorithms the pointer graph consists of rooted trees. It will be convenient to refer to a set of vertices comprising a tree as a *supervertex*.

Sometimes, we identify a supervertex with the root of its tree. No confusion will arise. As the algorithms proceed, the number of trees (supervertices) decreases. This is achieved by (possibly simultaneous) *hooking* operations. In each hooking a root r of a tree is 'hooked' onto a vertex v of another tree (that is, $D(r) := v$). A careful look at each of these connectivity algorithms reveals that:

1. Each such hooking is performed only after the algorithm "identified" an edge connecting a vertex in the supervertex of r with a vertex in the supervertex of v. Let us call such a connecting edge the *causing edge* of its hooking. We illustrate this notion of causing edge in the description of the HCS algorithm given below.
2. (*The spanning forest property*) For each supervertex, consider the collection of the causing edges that connect pairs of its vertices. This collection forms a spanning tree of the vertices comprising the supervertex. Thus, the collection of the causing edges, throughout each of these connectivity algorithms, forms a spanning forest of the input graph.

   The trees are also subject to a *shortcut* operation. That is, for every vertex v of the tree,

   if $D(D(v))$ is some vertex (as opposed to no vertex)

   then $D(v) := D(D(v))$.

The shortcut operation (approximately) halves the height of a tree. Shortcuts do not introduce cycles into the pointer graph, as can be readily verified. Simultaneous hookings are performed in each of these algorithms in such a way that no cycles are introduced into the pointer graph.

The algorithms also use the following graph. Each edge (u, v) in the input graph induces an edge connecting the supervertex containing u with the supervertex containing v. The graph whose vertices are the supervertices and whose edges are these induced edges is called the *supervertex graph*.

At the end of each of these algorithms the vertices of each connected component form a rooted star (which is, in particular, a single supervertex) in the pointer graph. As a result, a single processor can answer a query of the form "do vertices v and w belong to the same connected component?" in constant time.

The HCS parallel connectivity algorithm works in $O(\log n)$ iterations. Upon starting an iteration each supervertex is represented by a rooted star in the pointer graph. Each root hooks itself onto a minimal root which is adjacent to it in the supervertex graph. In case two roots are hooked on one another, we cancel the hooking of the smaller (numbered) root. As a result several rooted stars form a rooted tree; the root of one of these rooted stars becomes the root of the new tree. An iteration finishes with $O(\log n)$ shortcuts. It remains to identify the causing edges in this algorithm. In the HCS algorithm a root hooks itself onto a minimal adjacent root. There might be more than one edge connecting the two supervertices, but the algorithm selects precisely one of these edges. The selected edge, which then induces the hooking in the course of the computation, is the causing edge of the hooking.

The SV parallel algorithm also works in $O(\log n)$ iterations. Unlike the HCS algorithm: (i) An iteration of SV takes constant time, and (ii) The pointer graph at the beginning of an iteration is a collection of rooted trees (which are not necessarily stars). In principle, an iteration comprises the following steps.

1. Each rooted star is hooked onto a smaller vertex that is adjacent (in G) to some vertex of its supervertex (if there is any such smaller vertex).
2. Consider the rooted stars that did not hook and were not hooked upon in step (1); each such rooted star is hooked onto a vertex that is adjacent (in G) to some vertex of its supervertex.
3. Shortcuts.

The algorithm employs a processor for each vertex and each edge of the graph. This amounts to $n+m$ processors. It has been proven (see Chapter 2) that the total height of "still active" trees decreases by a factor of at least $1/3$ per iteration, implying that only $O(\log n)$ iterations are needed and, therefore, the algorithm runs in $O(\log n)$ time. The algorithm does not achieve optimal speed up.

5.4.2 The New Algorithm

The algorithm presented here is the first in a chain of connectivity algorithms. The ultimate algorithm runs in time $O(\log n)$. It achieves optimal speed up for graphs with $m \geq n \log^* n$, and almost optimal speed up in general. The ultimate algorithm has two parts.

The second (and more basic) part is an algorithm for relatively dense graphs ($m \geq n \log n \log^{(3)} n$). (Implicitly, we are assuming $n \geq 16$, to ensure

$\log^{(3)} n \geq 1$.) This second part can be viewed as a two level improvement in efficiency relative to the SV algorithm. The first level achieves optimal speed up with a parallel time of $O(\log n \, \log^{(3)} n)$; the second level achieves optimal speed up with a time of $O(\log n)$. The second level is asymptotically better; however, it involves the use of expander graphs, and consequently the "big oh" notation hides considerably larger constants. In this chapter, only the first level of the second part is described.

Our algorithm uses a number of parameters that strictly speaking are not integers, but which nonetheless we treat as integers (for instance, $\log n$). To justify this we view all parameters as being rounded up to the nearest integer power of 2. In more complex expressions, namely products and ratios (such as $\log n / \log^{(2)} n$), we round each of the basic terms separately ($\log n$ and $\log^{(2)} n$ here) and then compute the expression with these rounded terms; in our expressions the result is always a non-negative integer power of 2. Also, it is convenient to interpret $\log 1$ to be 1.

High-Level Description of the Connectivity Algorithm

We state the opportunity and main difficulty in obtaining an optimal speed up algorithm from the SV algorithm. In the SV algorithm, a processor is standing by each edge. The opportunity is that for each processor there is at most one step during the whole algorithm during which its edge is used for hooking; the difficulty is that we do not know in advance when this step comes.

The connectivity algorithm applies the SV connectivity algorithm. The key to obtaining an optimal algorithm is the following strategy: The SV algorithm is applied only to selected subsets of the edges of the graph, the subsets changing as the algorithm proceeds. More specifically, the $O(\log n)$ iterations of the SV algorithm are now divided into $O(\log \log n)$ phases. At the start of each phase new edges subsets are selected. Roughly speaking, each phase includes about as many iterations of the SV algorithm as all the preceding phases put together. Below, we characterize each phase by an input/output relation. The parameter used for this characterization is denoted by the somewhat awkward notation $d_{squared}$. The remark below justifies this awkwardness.

INPUT OF A PHASE

For each supervertex, its constituent vertices have $\geq d_{squared}$ incident edges from G, where $d_{squared}$ is an integer, $d_{squared} \geq 2$, and an edge with both endpoints in the supervertex is counted twice.

OUTPUT OF A PHASE
For each supervertex, its constituent vertices have $\geq d_{squared}^{1.5}$ incident edges from G.

REMARK 5.4
The remark on rounding applies to these input and output parameters $d_{squared}$ and $d_{squared}^{1.5}$ in the following way. Inductively, $d_{squared}$ was realized as a power of 2. A key parameter is $d = (d_{squared})^{1/2}$. It is realized simply as the next largest power of 2 of the real number $(d_{squared})^{1/2}$. $d_{squared}^{1.5}$ is realized as $d \cdot d_{squared}$, which is a power of two. $d_{squared}^{1.5}$ becomes $d_{squared}$ of the next phase. Finally, note the fact $d_{squared} \leq d^2 \leq 2 d_{squared}$. This fact will be used later in the text.

Before outlining the implementation of a phase, we need a few definitions for classifying the edges of G with respect to the supervertex graph. Given a supervertex graph, an edge of G is *redundant* (with respect to the supervertex graph) if both its endpoints lie in the same supervertex. An edge is an *outedge* if it is not redundant. If several outedges connect the same pair of supervertices, one of these outedges is chosen to be the *actual* outedge; the other outedges are called *duplicate* outedges. (The rule for choosing the actual outedge is specified below, in the detailed description of a phase.) The *degree* of a supervertex v is defined to be the number of actual outedges incident on v. Also, in each phase, we classify the supervertices according to whether they already satisfy the output condition; thus we define a supervertex to be *large* if it is known to have at least $d \cdot d_{squared}$ incident edges, and to be *growing* otherwise. A phase comprises the following three steps.

Step 1 Select an edge set such that

(i) For each growing supervertex of degree $\leq d$, all the actual outedges are selected.

(ii) For each growing supervertex of degree $\geq d$, exactly d actual outedges are selected.

Step 2 Run the SV connectivity algorithm for $\lfloor \log_{3/2} d \rfloor + 1$ iterations on the graph induced by the edges selected in Step 1. (This is the number of iterations needed to guarantee that the supervertices being created all satisfy the output condition for the phase, as we show later in the detailed analysis of Step 2.)

Step 3 This step is applied only to those new supervertices (rooted trees from Step 2) that will be growing in the next phase. This includes each new supervertex with fewer than $(d \cdot d_{squared})^{1.5}$ incident edges. For each such supervertex, we form an adjacency list of its incident edges (this is needed for Step 1 of the next phase). We call this *contracting* the supervertex (for one can view this step as contracting the tree for a new growing supervertex to a single node).

Later we give a detailed description and analysis of each step.

For the purposes of the analysis, it is useful to guarantee that each vertex has degree an integer multiple of $\log n \log^{(3)} n$. To achieve this, we add up to $\log n \log^{(3)} n$ self loops per vertex; each instance of a self loop (v, v) is recorded exactly once on v's adjacency list (rather than twice, once for each endpoint). We add at most m edges (recall that, by assumption, $m \geq n \log n \log^{(3)} n$). In fact, the self loops do not have to be added; it suffices to pretend that they are present for the purposes of the analysis. Also, each self loop is defined to contribute one incident edge to the supervertex to which it belongs.

Observations

a) There are at most $3m$ incident edges in the graph (where each instance of each edge is counted, i.e., an edge is counted once in each adjacency list in which it occurs).

b) It is safe to set $d_{squared} = \log n \log^{(3)} n$ initially, since each vertex has degree $\geq \log n \log^{(3)} n$. After $O(\log \log n)$ phases there will be only one supervertex comprising all the vertices in each connected component of the graph. (Actually, we could choose a smaller initial value for $d_{squared}$. The present implementation of Step 3.1.1.2 requires $d_{squared} \geq \log n / \log^{(2)} n$ (initially) to achieve an optimal algorithm; with an alternative implementation, we could reduce the initial value of $d_{squared}$ somewhat. However, as we are trying to make $d_{squared}$ grow, we might as well initialize it with the largest possible value.)

c) In Step 1, $O(m/d)$ edges are selected. (Since the input to the phase has $\leq 3m/d_{squared} \leq 6m/d^2$ supervertices).

d) Consider the components of the graph comprising the supervertices and the edge set selected in Step 1, above. Each growing supervertex of degree less than d has all its incident actual outedges selected. Thus, each component either includes a supervertex with at least d actual outedges, or it includes a large supervertex, or one supervertex comprises the whole component. As we show (in the detailed description of Step 2) each supervertex formed in Step 2 either comprises $\geq d$ input supervertices (and so has $\geq d \cdot d_{squared}$ incident edges), or it contains a large input supervertex (and so has $\geq d \cdot d_{squared}$ incident edges), or it comprises all the vertices of a component.

e) Contracting all supervertices with fewer than $(d \cdot d_{squared})^{1.5}$ incident edges guarantees that any supervertex that is not contracted in step 3 is *large* for the next phase. If a supervertex has sufficiently many incident edges it may be large for several consecutive phases.

High Level Analysis

Step 1 Its analysis is deferred to the detailed description of this step.

Step 2 Per phase, it processes $O(m/d)$ edges for $O(\log d)$ steps. Thus, per phase, it performs $O((m/d)\log d)$ operations in $O(\log d)$ time. Over the whole algorithm this is $O(m \log^{(2)} n / (\log n \log^{(3)} n)^{1/2}) = O(m)$ operations and $O(\log n)$ time.

Step 3 For those new supervertices that will be large in the next phase, not only is it unnecessary to contract them, it is too expensive (in time and operations) to do so. In the current phase we allocate $\Theta(\log d)$ time and $\Theta(m \log d / \log n)$ operations for performing contractions. We call this resource allocation the *increasing budget* for time and operations. The increasing budget may not be big enough to complete the contraction of a large supervertex in the current phase. However, as later phases have larger increasing budgets, a later phase, in which the supervertex is no longer large, will be able to do the contraction.

There are other parts of Step 3 whose cost is not covered by the increasing budget. These parts will be dealt with when the detailed description of Step 3 is given.

We have already introduced one budget used in the analysis. Altogether, the analysis uses four kinds of budgets (for time and number of operations):

1. An *increasing* budget (for each phase). The increasing budget for each of the two items (namely, time and number of operations) increases from phase to phase; specifically, it is $O(\log d)$ time and $O(m \log d / \log n)$ operations.

2. A *fixed* budget (for each phase). The fixed budget is the same for each phase: $O(\log n \log^{(3)} n / \log^{(2)} n)$ time and $O(m/\log^{(2)} n)$ operations.
3. A *general-pool* budget (for the whole algorithm): $O(\log n \log^{(3)} n)$ time and $O(m)$ operations.
4. *Miscellaneous* budgets (for each phase). These are specified as they are needed (one miscellaneous budget was used in the analysis of Step 2 above).

The *total* budget for the whole algorithm is the sum of these budgets over all the phases.

The fixed and general pool budgets are used in the analysis of Step 1, the edge selection. We note that the general-pool budget is independent of the budget for any individual phase; in other words, the edge selection procedure has an interesting amortized complexity analysis.

Actually we do not need to use the budgets in our analysis below. The reason for specifying the budgets, and referring to them in the analysis, is to emphasize the different (time, operation) combinations used in the various substeps of the algorithm.

Next, we describe in detail and analyze each step in turn.

Analysis of Step 1, the Edge Selection

Here, we give a procedure for edge selection that performs $O(m)$ operations in time $O(\log n \log^{(3)} n)$ over the course of the whole algorithm.

In this step the processors are reallocated every $\Theta(\log n / \log^{(2)} n)$ time units, using a parallel prefix sum algorithm. We do not know how to reallocate the processors faster; consequently, we will process the edges in units of size $\log n / \log^{(2)} n$, called *blocks*. Thus, at the start of the algorithm, the edges incident on each vertex are divided into blocks of $\log n / \log^{(2)} n$ edges each (for recall that by the addition of self loops we ensured that each vertex has degree a multiple of $\log n \log^{(3)} n$). In general, the edges incident on each supervertex are divided into blocks of $\log n / \log^{(2)} n$ edges, but with possibly some incomplete blocks (i.e., blocks having too few edges). However, there will always be at most $3m/\log n$ incomplete blocks (this is certainly true initially, and this invariant will be maintained in Step 3). The division into blocks may change from phase to phase. More precisely, new blocks are created by combining and repartitioning blocks that belong to the same growing supervertex. The total number of blocks at hand is always bounded by $3m \log^{(2)} n / \log n$, as we will show in the description of Step 3.

236 Chapter 5. Advanced Parallel Prefix-sums, List Ranking and Connectivity

Data Structures

$T(v)$ is the name of the supervertex currently containing vertex v of G (T is the *vertex table*).

$ACTUAL(u, v)$ is an $n \times n$ array; it records if an edge connecting supervertex u and supervertex v has been found in the current phase. Specifically, $ACTUAL(u, v)$ comprises a pointer to an edge plus a timestamp; the timestamp is a phase number. (We explain at the end of this section how to implement ACTUAL in space $O(min[n^2, mn^\epsilon])$ for any fixed $\epsilon > 0$; evaluating $ACTUAL(u,v)$ takes time $O(1/\epsilon)$).

The block headers (i.e., names) are stored in a single array, with the blocks for each growing supervertex occupying a contiguous portion of the array. The edges in each block are stored in a linked list. The growing supervertex headers are stored in an array. Each supervertex header records the span occupied by its block headers. This array also contains headers for each growing supervertex from a previous phase that is part of a large supervertex; more precisely, each such growing supervertex remains in this array until it becomes part of a new larger growing supervertex. We call such a growing supervertex an *out-of-date growing supervertex*.

We say growing supervertex v is *active* for the current iteration of Step 1.1 (below) of the edge selection procedure unless either we have found d actual outedges incident on v, or we have checked all the edges incident on v. A block is active if its supervertex is active. Let b_A (resp. b'_A) be the number of active blocks at the start (resp. end) of the current iteration of Step 1.1.

In the procedure below, the active blocks (or rather, their headers) are kept in an array with blocks belonging to the same supervertex being contiguous. This procedure assumes that $p = \frac{3m}{\log n \log^{(3)} n}$ processors are available. (This choice for p arises because we are aiming for an optimal algorithm, and because even if we have $\Theta(m/\log n)$ processors available, the procedure below will still require $\Theta(\log n \log^{(3)} n)$ time, as we justify later. The choice of the constant 3 is unimportant; it simplifies some constants in the analysis.)

Step 1.1 while $b_A > 0$ do begin

Step 1.1.1 Process every $\lceil b_A/p \rceil$th block in the array of active blocks.

Processing a block using a single processor: For each edge $e = (i, j)$ in the block, determine whether it is redundant, duplicate, or actual. In the first two cases eliminate the edge from further consideration; in the last case add the edge to the set of selected edges. Proceed as follows.

a) Let $u = T(i), v = T(j)$. If $u = v$, the edge is redundant. If not, continue with step (b).
b) Check $ACTUAL(u,v)$. If an edge is recorded with the current timestamp then e is a duplicate outedge. Otherwise continue with step (c).
c) Edge e attempts to write its address and the current phase number to $ACTUAL(u,v)$. If the write is successful e is selected (e is an actual outedge); otherwise e is a duplicate outedge.

Step 1.1.2 For each growing supervertex, determine if it is still active. (This step is easily performed using a parallel summation algorithm with respect to the block headers.) Then compress the (headers of the) still active blocks to the start of the active block array. (Detecting the active blocks is facilitated by marking the first and last block for each growing supervertex; it then merely requires a parallel prefix sum computation to separate the active blocks.)

end

Step 1.2 For each growing supervertex, among the edges selected in Step 1.1, we will select d (or all, if $< d$ are available). The selected edges will be placed in an array of size $O(m/d)$. First, a parallel prefix sum computation with respect to the block headers determines how many edges each block provides, and for each block provides a range of serial numbers for the edges contributed by that block. Then the blocks that might contribute edges are processed in the same order as in Step 1.1 so as to place the selected edges in the array of size $O(m/d)$.

Analysis. We note that each iteration of Step 1.1 takes time $O(\log n/\log^{(2)} n)$ and performs $O(m/(\log^{(2)} n \log^{(3)} n))$ operations. We show that over the whole algorithm, these iterations perform $O(m)$ operations in time $O(\log n \log^{(3)} n)$.

We first present a special case of the analysis; it provides the intuition and an overview for the general case. So, for now, suppose that all the blocks are complete and all the supervertices have an integer multiple of $\lceil b_A/p \rceil$ blocks. Then we can partition the iterations into two cases.

Case 1. $b'_A \leq 1/2\, b_A$ – a *reducing* iteration. (Recall b_A and b'_A are the number of blocks active, respectively, at the start and at the end of the iteration). Below, we show that, per phase, there are $O(\log^{(3)} n)$ reducing iterations (Claim 5.1). Thus, per phase, the reducing iterations use $O(\log n \, \log^{(3)} n / \log^{(2)} n)$ time, and perform $O(m/\log^{(2)} n)$ operations (to obtain this multiply $O(\log^{(3)} n)$ iterations by $O(m/(\log^{(2)} n \, \log^{(3)} n))$ operations). This is charged to the fixed budget. Over the whole algorithm this is $O(\log n \, \log^{(3)} n)$ time and $O(m)$ operations.

Case 2. $b'_A > 1/2 \, b_A$. Then at least half the blocks processed belong to still active supervertices; we call these the *productive* blocks. Each edge in a productive block is either eliminated or found to be an actual outedge. We partition non-reducing iterations into eliminating and selecting iterations.

Case 2a. At least half the edges processed, among the productive blocks, were eliminated—an *eliminating* iteration. The number of edges eliminated in an eliminating iteration is at least $3m/(4\log^{(2)} n \, \log^{(3)} n)$ (since for at least half the blocks, namely the productive blocks, at least half the edges were eliminated). But at most $3m$ edges can be eliminated (each edge in each adjacency list in which it is present). Thus there are at most $4\log^{(2)} n \, \log^{(3)} n$ eliminating iterations over the course of the algorithm. The cost of these iterations ($O(\log n \, \log^{(3)} n)$ time and $O(m)$ operations) is charged to the general pool budget.

Case 2b. More than half the edges processed, among the productive blocks, were actual outedges (i.e., not case 2a)—a *selecting* iteration. The operations of a selecting iteration are charged to the actual edges found which belong to supervertices that remain active; we call these edges *charged edges*. Clearly, there are at least $3m/(4\log^{(2)} n \, \log^{(3)} n)$ charged edges per selecting iteration (the actual edges provided by the productive blocks). We show that there are fewer than $6m/d$ charged edges over the course of the phase (Claim 5.2). Thus there are at most $\lfloor 8 \log^{(2)} n \, \log^{(3)} n / d \rfloor = O(1)$ selecting iterations per phase (since $d \geq (\log n \, \log^{(3)} n)^{1/2}$; in fact, for large enough n there are 0 selecting iterations per phase). Per phase, these iterations cost $O(\log n / \log^{(2)} n)$ time and $O(m/(\log^{(2)} n \, \log^{(3)} n))$ operations. This is charged to the fixed budget. Over the whole algorithm this is $O(\log n)$ time and $O(m/\log^{(3)} n)$ or $O(m)$ operations.

But rounding makes everything messier. The problems are caused by two sorts of blocks, which we call *non-productive* blocks. First, incomplete blocks, of which there are at most $3m/\log n$, as asserted at the start of the description of Step 1. Second, blocks which were processed in excess disproportion to the supervertex size, for supervertices that become inactive; that is, if v has $k\lceil b_A/p\rceil + l$ blocks, where $l < \lceil b_A/p\rceil$, $k+1$ of v's blocks are processed, and v becomes inactive, then with one of v's processed blocks we are unable to associate a further $\lceil b_A/p\rceil - 1$ of v's blocks that became inactive; call this block the *excess* block (we arbitrarily choose the excess block to be a specific block among v's blocks, say the block of smallest index in the active block array, processed in this iteration). There is at most one excess block per supervertex over the course of a phase; that is, at most $3m/d_{squared}$ excess blocks, per phase. In order to incorporate non-productive blocks into the analysis we change Case 2 and introduce Case 3.

New Case 2. At least a quarter of the blocks processed are productive.

New Case 2a. At least half the edges processed, among the productive blocks, were eliminated—an *eliminating* iteration. Clearly, we at most double the number of eliminating iterations as compared to the previous analysis (since each eliminating iteration need only eliminate half as many edges as before). So the cost of the eliminating iterations is still $O(\log n \log^{(3)} n)$ time and $O(m)$ operations.

New Case 2b. More than half the edges processed, among the productive blocks, were actual outedges (i.e., not New Case 2a)—a *selecting* iteration. Again, we at most double the number of selecting iterations per phase. So, over the whole algorithm, the cost of the selecting iterations is still $O(\log n)$ time and $O(m/\log^{(3)} n)$, which is $O(m)$ operations.

Case 3. Otherwise; that is, $b'_A > 1/2\, b_A$ and fewer than a quarter of the blocks processed are productive—a *removing* iteration. We show that at least a quarter of the processed blocks are non-productive (Claim 5.3). As argued above, in the current phase there can be at most $3m/\log n + 3m/d_{squared} \leq 6m/\log n$ non-productive blocks. Since $3m/(\log n \log^{(3)} n)$ blocks are processed in each iteration, and as the iteration is removing only if at least a quarter of the processed blocks are non-productive, we conclude that there are at most $8\log^{(3)} n$ removing iterations per phase. Thus, the removing iterations perform $O(m/\log^{(2)} n)$ operations in time $O(\log n \log^{(3)} n/\log^{(2)} n)$ per phase. This is charged to the fixed budget. Over the whole algorithm, this is $O(m)$ operations and $O(\log n \log^{(3)} n)$ time.

So, over the whole algorithm, all four types of iterations perform $O(m)$ operations in $O(\log n \, \log^{(3)} n)$ time.

Reducing Iterations

CLAIM 5.1
There are only $O(\log^{(3)} n)$ reducing iterations in a phase.

PROOF OF CLAIM 5.1
After $\log^{(3)} n + \log^{(4)} n$ reducing iterations the number of active blocks decreases by a factor of at least $\log^{(2)} n \, \log^{(3)} n$. Thus, in the next iteration a processor will be assigned to each (presently) active block and following this iteration there will be no active blocks left. ∎

Selecting Iterations

CLAIM 5.2
There are fewer than $6m/d$ charged edges over the course of the phase.

PROOF OF CLAIM 5.2
Consider a charged edge e and the iteration t in which it was charged. Let e be in a block of supervertex v. At the end of iteration t the supervertex v remained active. Thus fewer than d actual edges incident on v had been discovered at the end of iteration t. We conclude that the last charged edge, in this phase, for supervertex v is at most the $d-1$st charged edge for supervertex v. As there are at most $6m/d^2$ supervertices in this phase the claim follows. ∎

Removing Iterations

CLAIM 5.3
In a removing iteration at least a quarter of the processed blocks are non-productive.

PROOF OF CLAIM 5.3
Recall that in a removing iteration $b'_A > 1/2 \, b_A$ and fewer than a quarter of the blocks processed are productive. Consider a processed block B; let B belong to supervertex v. There are four possibilities.

Case A. B is productive. Since the iteration is removing, fewer than a quarter of the blocks fit this case.

Case B. B becomes inactive, but B is not the excess block for v. Since the iteration is not reducing, fewer than half the blocks fit this case (for recall we can associate a further $\lceil b_A/p \rceil - 1$ of v's blocks with B, and all these associated blocks become inactive).

Case C. B is the excess block for v, if any.

Case D. B is incomplete.

Cases C and D include only non-productive blocks. Clearly, in a removing iteration, Cases C and D must include at least a quarter of the processed blocks. ∎

It is clear that the complexity of the Step 1.2 is dominated by that of Step 1.1.

COMMENT
With more processors, but still using $O(m/\log n)$ processors, it is possible to perform all the eliminating, selecting and removing iterations in $O(\log n)$ time; however, this does not apply to the reducing iterations. In each phase, there may be $\Theta(\log^{(3)} n)$ reducing iterations, even using $m/\log n$ processors. As mentioned above, this guides the choice of $p = \Theta(m/(\log n \log^{(3)} n))$. Thus, varying the parameters will not speed up Step 1 by more than a constant factor, at least so long as we aim for an optimal algorithm.

Also, this is why the algorithm requires $m \geq n \log n \log^{(3)} n$. In the first phase, in each iteration of Step 1.1, at least $\Theta(n \log n / \log^{(2)} n)$ operations are performed (to process one block per vertex); so in the first phase, the reducing iterations may perform $\Theta(n \log n \log^{(3)} n / \log^{(2)} n)$ operations. With a uniform implementation of the edge selection in each phase this implies that $\Theta(n \log n \log^{(3)} n)$ operations may be performed. For an optimal algorithm, we therefore require $m \geq n \log n \log^{(3)} n$.

Implementation of the Array ACTUAL

We show how to implement the array $ACTUAL$ in $O(mn^\epsilon)$ space, for any fixed $\epsilon > 0$. Since an alternative implementation in $O(n^2)$ space is obvious, we conclude that array $ACTUAL$ can be implemented in $O(min(n^2, mn^\epsilon))$ space. Our description below ignores the timestamps attached to each edge.

We use two arrays A and B. Array A is of size $n \times n^\epsilon$ and array B is of size $m \times n^\epsilon$. Suppose the i-th edge of the input graph connects supervertices u and v, where $1 \leq i \leq m$. We show how to store this edge in these arrays using a single processor. This parallel procedure will take at most $1/\epsilon$ steps. Each of u and v is a number between 1 and n. Suppose, without loss of generality, that (the number) u is smaller than (the number) v. Let $v_1 v_2 \ldots v_{1/\epsilon}$ be the representation of v with respect to base n^ϵ. (If $1/\epsilon$ is not an integer take $\lceil 1/\epsilon \rceil$ instead.)

Step 1 Write i in location $A(u, v_1)$. Observe that concurrent writes may occur. Suppose that edge j_1 was actually written into $A(u, v_1)$. If $i = j_1$ then we have finished storing edge i. Otherwise,

Step $l, 1 < l \leq 1/\epsilon$. Write i in location $B(j_{l-1}, v_l)$. Again, concurrent writes may occur. (The following is not required in Step $1/\epsilon$). Suppose that edge j_l was actually written into $B(j_{l-1}, v_l)$. If $i = j_l$ then we have finished storing edge i. Otherwise, proceed to Step $l + 1$.

It is easy to verify that all the edges being inserted will be stored by the end of Step $1/\epsilon$. Also, given two vertices u and v, it takes at most $1/\epsilon$ steps to find out whether there exists an edge connecting u and v, using the information in arrays A and B.

The array ACTUAL is also used to link the two copies of each edge at the start of the algorithm. We proceed in two stages. First, for each edge (u, v) with $u < v$ we insert the edge into the array ACTUAL. Next, for each edge (u, v) with $u > v$ we search for the edge (v, u), which is already in the array ACTUAL. The two copies of the edge are then readily linked. It is clear this requires $O(1/\epsilon)$ time and $O(1/\epsilon \cdot m)$ operations.

Analysis of Step 2

As stated above, in the dth phase we apply the SV connectivity algorithm for $O(\log d)$ time to the graph specified by the supervertices and the edges selected in Step 1. Actually, we need to make one minor change to the algorithm: only growing supervertices participate in an active way in the algorithm; the large supervertices, or rather their constituent out-of-date growing supervertices, simply provide nodes for hooking onto. This implies that when the pointer of a growing supervertex v points to an out-of-date growing supervertex, then v no longer performs pointer jumping; i.e., henceforth v participates only in a passive way in the SV algorithm. Otherwise the growing supervertices perform the algorithm in the standard way. Note that the array of growing and out-of-date growing supervertices provides the vertex set for the SV algorithm here.

The SV connectivity algorithm requires that the edges be accessible via an array whose size is twice the number of the edges and that each edge appears there once in each of its two directions. It is easy to achieve this. In Step 1.2, an edge is placed in the selected edge array in both of its directions. (Even if an edge was selected by both its endpoints, no damage will be caused by the redundant edges).

The SV algorithm maintains the following invariant: Let T be a tree constructed by the algorithm in t steps, and let l be the length (in edges) of the longest path from a leaf to the root in tree T; then either $|T| \geq (3/2)^{t-1} \cdot l$, for $t \geq 1$, or T is a spanning tree of a connected component. Thus after $\lfloor \log_{3/2} d \rfloor + 1$ steps, any tree built by the SV algorithm either will contain at least d vertices, or it will comprise a connected component.

We deduce that after $\lfloor \log_{3/2} d \rfloor + 1 = O(\log d)$ iterations, any tree built by the algorithm, comprising only growing supervertices, includes either at least d old supervertices or all the old supervertices of a component, as claimed in observation (d) above (the only other possibility is that the tree contains a large supervertex).

The processor allocation for Step 2 is straightforward: simply assign one processor to each edge in the array computed in Step 1.2. Step 2 performs $O(m \log d / d)$ operations in $O(\log d)$ time using the formulation of [22], Step 2 uses $md/d_{squared}$ (which is $O(m/d)$) processors and $O(\log d)$ time. This is charged to a miscellaneous budget. Over the whole algorithm this is $O(m)$ operations and $O(\log n)$ time.

Analysis of Step 3

For each new supervertex (a tree constructed in Step 2) that will be growing in the next phase, Step 3 forms an adjacency list of its edges, divided into blocks. Basically, first, we form an Euler circuit of each tree constructed in Step 2; second, for each tree, we combine the adjacency lists of the nodes in the tree with the help of the Euler circuits; third, for each block of an adjacency list L such that L is not too long, we recognize that the block belongs to L. This recognition is performed by applying Wyllie's list ranking algorithm [27] to the adjacency lists. In using Wyllie's algorithm, we perform pointer jumping for each unit in the lists being processed; over the course of the algorithm, there will be $\Theta(\log n)$ pointer jumping steps per unit processed. In order to obtain an optimal algorithm we need to process the adjacency lists in units of size $\Omega(\log n)$ edges.

This motivates us to create *clusters*. For each supervertex, its edges are partitioned into clusters of at most $\log n$ edges each; each cluster comprises up to $\log^{(2)} n$ blocks. Initially, for each vertex, each cluster contains exactly $\log^{(2)} n$ blocks (recall that each vertex initially has a degree that is an integer multiple of $\log n \log^{(3)} n$). Each cluster can only lose edges (from the edge elimination); a cluster never gains edges. However, the edges within a cluster may be repartitioned among its blocks; thus a block may lose and gain edges, but only from within its cluster. There will be at most one incomplete block per cluster. There are at most $3m/\log n$ clusters (since there are at most $3m$ edges initially); hence there are at most $3m/\log n$ incomplete blocks and at most $3m \log^{(2)} n/\log n$ blocks. We keep an array of cluster headers, with clusters belonging to the same growing supervertex being contiguous. Each cluster header records the span of the headers of its blocks in the block header array (it will always be the case that a cluster's blocks are contiguous in the block header array). Also, for each supervertex, we keep its clusters in a circular linked list.

The input for Step 3 comprises:
a) For each processor, the ordered list of the blocks it processed in Step 1.
b) The array of the edges used in Step 2.
c) The array of clusters. This array also includes dummy clusters, whose role will become clear later. There are $O(m/\log n)$ dummy clusters (Claim 5.4).
d) The array of growing supervertices.

Next we outline the procedure for Step 3.

Step 3.1 For each supervertex (tree) created in Step 2, form a single linked list of its clusters. This will require the introduction of dummy clusters, and will be based on an Euler Tour of each tree.

Step 3.2 Separate the new growing and new large supervertices. That is, for each cluster recognize the type of its new supervertex, and if it is growing, in addition, find out the name of the new growing supervertex. This step is performed by applying Wyllie's list ranking algorithm to the lists formed in Step 3.1.

Step 3.3

Step 3.3.1 Remove those edges eliminated by the edge selection in Step 1 and form new blocks so that there is at most one incomplete block in each cluster.

Step 3.3.2 Rearrange the arrays of block headers and cluster headers so that for each growing supervertex, its associated blocks (resp. clusters) are contiguous. Also, update the array of growing supervertices.

Step 3.4 Update the vertex table.

Below we alternately use several patterns for the assignments of processors to jobs. One pattern is called *the pattern of Step 1*: each processor processes one block at a time, in the same order as in Step 1 (this is easily obtained from item (a) of the input to Step 3). The other patterns will be described as they are used.

Step 3.1

Step 3.1.1 In order to carry out Step 3.1 we need to compute two numberings for each edge that became a causing edge in this phase (recall the causing edges are the edges that induce the hookings in Step 2). First, we number all the causing edges with a single serial numbering. Second, for each cluster, we number its causing edges with a serial numbering. Observe that while a growing supervertex v seeks to hook itself onto only one other supervertex in Step 2, several of its incident edges may become causing edges that hook the supervertex at their other endpoint onto v.

In order to simplify the exposition we neglected to add to Step 2 an instruction for marking both copies of a causing edge (immediately after it is used for hooking). So we assume that each causing edge is so marked. Whenever, in Step 3, we refer to a causing edge, we mean a copy of a causing edge introduced in Step 2 of the present phase.

The numberings are computed as follows. In Step 3.1.1.1 we assign a processor to each causing edge. In Step 3.1.1.2 we assign serial numbers to the causing edges in each block. In Step 3.1.1.3 we assign the two desired serial numbers to each causing edge.

Step 3.1.1.1 By means of a prefix sum algorithm with respect to the array of edges used in Step 2 we assign one processor to each of the two copies of each causing edge; this takes $O(\log n/\log^{(2)} n)$ time and $O(m/d)$ operations, per phase. There are at most $12m/d^2$ copies of causing edges, since there are at most $6m/d^2$ growing supervertices, each providing at most one causing edge and each causing edge having two copies.

Step 3.1.1.2 For each block, we compute both the number of its causing edges and the serial number of each of them. For each copy of a causing edge, its processor traverses the list of edges in the block of this copy, till it comes to the front of the list (which has a pointer to the block header). Since there may be several processors traversing a block only the processor of the first causing edge in the block (using the original order of the block) remains to take care of the block. It first assigns serial numbers to each causing edge in the block and later writes their total number to a variable associated with the block header; this takes $O(\log n/\log^{(2)} n)$ time and $O(m \log n/(d^2 \log^{(2)} n))$ operations per phase.

Step 3.1.1.3 For each cluster, we visit its block headers in turn (i.e., serially) and assign to each block a range for the serial numbers of its causing edges relative to the other causing edges of the cluster. This takes $O(\log^{(2)} n)$ time and $O(m \log^{(2)} n/\log n)$ operations per phase.

Then, by means of a prefix sum computation with respect to the block headers, we assign to each block a range for the serial numbers of its causing edges relative to the whole set of causing edges. This takes $O(\log n/\log^{(2)} n)$ time and $O(m \log^{(2)} n/\log n)$ operations per phase.

Now, using the processor assignment pattern computed in Step 3.1.1.1 (recall that it provides a separate processor to each causing edge) we assign the two serial numbers to each causing edge. This takes $O(1)$ time and $O(m/d^2)$ operations per phase.

Step 3.1.2 Our goal here is to provide, for each new supervertex S, a (circular) linked list that goes through all the clusters in S. Recall that at the beginning of a phase the clusters of each (old) supervertex were arranged in a circular linked list. Let C be a cluster. Suppose C has h causing edges. In this circular list of clusters, we replace cluster C by a list comprising cluster C plus h dummy clusters; each of the dummy clusters represents one of the h causing edges introduced in the present phase. Each old supervertex now has a circular list comprising all the clusters it had previously plus some additional dummy clusters. (Note that to order the dummy clusters associated with cluster C we need the second of the serial numbers computed in Step 3.1.1.3).

CLAIM 5.4
The number of dummy clusters created during the whole algorithm is bounded by $2(n-1)$ which is $O(m/\log n)$.

PROOF OF CLAIM 5.4
Recall that the input vertices and causing edges form a spanning forest; thus there are at most $n-1$ causing edges. ∎

Next, we show how to form a single circular list for each new supervertex. This new list will include all the clusters from the old supervertices that form the new supervertex. To achieve this we apply an idea of [3] for "stitching" the circular lists at the causing edges. We note that a causing edge has a copy (which is a dummy cluster) in two of these circular lists. Each copy has a successor in its own list. In parallel, we *make the successor of each copy of each causing edge the successor of the other copy of the same causing edge.* This gives a single circular list for each new supervertex (see Exercise 5.3 below).

EXERCISE 5.3

Use an argument as in [3] to show that we indeed get a single circular list for each new supervertex.

The number of operations required is proportional to the number of causing edges, which is $O(m/d^2)$. The time is $O(1)$.

It is convenient to place the dummy clusters introduced in this phase into the array of clusters. To do this we need to assign a serial number to each such dummy cluster; we simply add the first serial number associated with the corresponding causing edge to the current size of the cluster array. Per phase this takes $O(1)$ time and $O(m/d^2)$ operations.

Henceforth, when we refer to dummy clusters, we intend all the dummy clusters that are present, and not just those created in the current phase (In Step 3.2 we will see why dummy clusters from previous phases might be present).

Step 3.2 Our goal is to separate new growing from new large supervertices. It will be helpful to attach a distinct identifier to each cluster. Thus, each dummy cluster, on creation, is given as identifier the pair $(|V| + u, v)$, where (u, v) is the causing edge that caused the dummy cluster to be created. Likewise, each actual (non-dummy) cluster is given the identifier (u, v), where (u, v) is the first edge on the edge list of its first block.

It is helpful to note that if a supervertex has γ actual clusters it can have at most 2γ dummy clusters. For the causing edges in the supervertex form a spanning tree of its constituent vertices and so there is one fewer causing edge than constituent vertices. It remains to note that each vertex contributed at least one actual cluster, while each causing edge produced just two dummy clusters.

A supervertex will be large in the next phase if it has at least $(d \cdot d_{squared})^{1.5}$ incident edges. Let $\gamma_d = (d \cdot d_{squared})^{1.5}/\log n$. Thus, for each new growing supervertex, the circular list of clusters, actual and dummy, comprises fewer than $3\gamma_d$ nodes (at most γ_d actual clusters and fewer than $2\gamma_d$ dummy clusters). It is convenient to redefine a large (resp. growing) new supervertex to be one with at least (resp. fewer than) $4\gamma_d$ nodes in its circular list of clusters. This implies that a new large supervertex has at least $4/3 (d \cdot d_{squared})^{1.5}$ incident edges (since at least $1/3$ of the nodes are actual clusters and each actual cluster contains $\log n$ edges), and a new growing supervertex has fewer than $4(d \cdot d_{squared})^{1.5}$ incident edges. For each of our lists we determine whether it represents a large or growing new supervertex by the following computation.

a) We iterate at each cluster, in parallel, the following basic (doubling) operation $\beta_d = \log(2\gamma_d)$ times.

  1. The new identifier of the cluster is the (lexicographic) minimum between its own identifier and the identifier of its successor.
  2. Doubling. (i.e., $D(C) := D(D(C))$, where initially $D(C)$ is the successor of cluster C in the circular list of clusters.)

Following this computation each cluster holds the minimum identifier among its own (original) identifier and the (original) identifier of its $2^{\beta_d} - 1$ original successors.

b) We check at each cluster whether its minimum identifier and the minimum identifier of its (present) successor are equal.

We observe that the cyclicity of each list implies that if the minimum identifiers of some cluster and its present successor are equal then: (i) This minimum identifier must be the minimum identifier of their list. (ii) The list contains at most $2^{\beta_d+1} - 1$ clusters and therefore the supervertex is growing.

c) We apply β_d+1 parallel doublings in order to "broadcast" this minimum. In each list in which no minimum identifier was found, nothing is being broadcast and each node in such a list can conclude that it is part of a new large supervertex.

d) Using β_d+1 parallel doublings, for the lists of growing supervertices, we shortcut over, and thereby discard, the dummy clusters.

e) Next we give serial numbers, starting from one, to the actual clusters of each growing supervertex, as follows. Using $\beta_d + 1$ parallel doublings, for the lists of growing supervertices, we rank each actual cluster with respect to the cluster whose identifier is minimum in its list.

The processor allocation for Step 3.2 is to provide one processor to each cluster and dummy cluster ($O(m/\log n)$ processors). Thus, per phase, Step 3.2 takes $O(\log d)$ time and $O(m \log d/\log n)$ operations.

Step 3.3

Step 3.3.1 The goal is to remove those edges eliminated by the edge selection in Step 1 and to form new blocks so that there is at most one incomplete block per cluster.

Step 3.3.1.1 For each block processed in Step 1 we remove those edges eliminated in Step 1, forming a list of the remaining edges. Also, for each block, we record the number of edges still present.

The processor allocation for this Step is given by the pattern of Step 1; its complexity is dominated by that of Step 1.

Step 3.3.1.2 For each cluster we form a list of its blocks. We partition each such list into two sublists. The first sublist comprises the non-empty blocks that were processed in Step 1 and incomplete blocks (there is at most one incomplete block for each cluster). The second sublist contains the complete blocks that were not processed in Step 1. Empty blocks are discarded. This separation is performed in parallel for each cluster by a sequential scan of the block headers in the cluster. Next, for each cluster, for each first list, we compute the prefix sums of the block sizes. This step uses $O(\log^{(2)} n)$ time and $O(m \log^{(2)} n/\log n)$ operations, per phase.

Step 3.3.1.3 For each sublist of the first type, we form new blocks, as follows. Using the prefix sums, we determine the new block boundaries. (This requires scanning the edges in the blocks containing such boundaries.) Then we append each list of edges to the end of the list for the preceding block. For each cluster, the edges are now divided into complete blocks plus at most one incomplete block. This takes $O(\log n/\log^{(2)} n)$ time and, per block processed, $O(\log n/\log^{(2)} n)$ operations.

The processor allocation for this step follows the pattern of Step 1, except that in addition we need to process each old incomplete block, of which there is at most one per cluster; but this just requires an additional allocation of one processor per cluster, which is straightforward. Thus the complexity of this step is dominated by that of Step 1.

Step 3.3.1.4 For each new block, it remains to place its header in the portion of the block array belonging to its cluster. For each new block B, we choose an old block B' being removed from the same cluster; B's header will take the place occupied by B''s header. B' can be chosen according to the following rule: it is the first old block which overlaps with B. Next, for each cluster, we compress its block headers to a contiguous portion of the block array, by means of a prefix sum computation with respect to the array of block headers. We also record, for each cluster, the span occupied by its present block headers. This takes $O(\log n/\log^{(2)} n)$ time and $O(m \log^{(2)} n/\log n)$ operations per phase.

Step 3.3.2 Each new supervertex is presently represented by a circular list of clusters; the lists for large supervertices include dummy clusters, while the lists for growing supervertices do not (Step 3.2). During this step, the array of cluster headers is reordered so that clusters belonging to the same new growing supervertex are in contiguous locations. The array of blocks headers is reordered correspondingly.

We compute the new location for each cluster header as follows. From part (e) of Step 3.2, we have already determined, for each new growing supervertex, the number of clusters it contains. This number is placed in the 'first' cluster for the supervertex; every other cluster in a new growing supervertex is assigned the number zero. The clusters in new large supervertices are all assigned the number one. By performing a prefix sum computation over these numbers with respect to the array of clusters, we obtain the new location of each first cluster of a new growing supervertex and of each cluster of a new large supervertex. Finally, each cluster in a new growing supervertex computes its location relative to the first cluster in its new growing supervertex. This step uses $O(\log n/\log^{(2)} n)$ time and $O(m/\log n)$ operations per phase.

Next, we rearrange the block headers so as to match the new cluster ordering. To do this, each cluster determines how many blocks it has; then by means of a prefix sum computation with respect to the cluster headers it determines the new locations of its block headers. Then, each block header is given its new location by its cluster. Finally, the block headers are relocated simultaneously. Per phase, this takes $O(\log n/\log^{(2)} n)$ time and $O(m \log^{(2)} n/\log n)$ operations.

The array of growing supervertices is readily updated with the help of a parallel prefix sum computation with respect to the array of clusters; it sums the number of "first" clusters, including the "first" clusters for out-of-date growing supervertices. Per phase, this takes $O(\log n/\log^{(2)} n)$ time and $O(m/\log n)$ operations.

Step 3.4 Initially, we associate with each vertex v its "first" cluster C_v (chosen to be the minimum among its clusters when ordered by their identifiers). C_v will be responsible for updating $T(v)$. A growing supervertex is given the vertex name associated with its "first" cluster.

To update the vertex table T, if C_v is part of a growing supervertex S, it performs the following operation. Let w be the vertex name associated with S (in Step 3.2(c) each cluster belonging to S learned this name); C_v performs the assignment $T(v) := w$.

The processor assignment for this step is to provide one processor to each cluster; then this step takes $O(1)$ time and $O(m/\log n)$ operations per phase.

Analysis We have shown that, per phase, the complexity of Step 3 is dominated by the sum of three components:

(i) the complexity of Step 1,

and complexities of:

(ii) $O(\log n/\log^{(2)} n)$ time and $O(m/\log^{(2)} n)$ operations,

(iii) $O(\log d)$ time and $O(m \log d/\log n)$ operations.

(To verify this, it suffices to recall that $d \geq (\log n \, \log^{(3)} n)^{1/2}$.)

Summing over all the phases, we obtain that over the whole algorithm, the complexity of Step 3 is bounded by the complexity of Step 1, plus a complexity of $O(m)$ operations and $O(\log n)$ time. Hence Step 3 performs $O(m)$ operations in $O(\log n \, \log^{(3)} n)$ time.

We conclude that,

THEOREM 5.5
For $m \geq \log n \, \log^{(3)} n$ there is an optimal connectivity algorithm in the CRCW PRAM model that performs $O(m + n)$ operations in time $O(\log n \, \log^{(3)} n)$ on $(m + n)/(\log n \, \log^{(3)} n)$ processors.

Notes and References

The standard logarithmic time optimal parallel prefix-sums algorithm, due to Stone [24] and Ladner and Fischer [19], is given earlier in this book. Wyllie's [27] standard parallel list ranking algorithm is reviewed in Chapter 2. Standard parallel connectivity algorithms, due to Hirschberg, Chandra and Sarwate [17], and Shiloach and Vishkin [22] are reviewed briefly in this chapter. For more on them, see chapter 2. The material on prefix-sums and list ranking is from Cole and Vishkin [9] and on connectivity is from Cole and Vishkin [10].

Parallel algorithms for the prefix-sums problem are apparently the most heavily used routines in parallel computation. The prefix sums algorithm improved a similar result due to Reif 1985. The result for the prefix sums problem is tight. That is, any algorithm using a polynomial number of processors needs $\Omega(\log n/\log \log n)$ time. This follows from the lower bound of Hastad [14] for circuits together with the general simulation result of Stockmeyer and Vishkin [23] between PRAMs and circuits. A direct proof was given recently in Beame and Hastad [4].

The first logarithmic time optimal parallel algorithm for list ranking was given in Cole and Vishkin [8] (on an EREW PRAM). It is based on a general method for assigning processors to jobs (called approximate task scheduling) which uses expander graphs. The algorithm presented here is considerably simpler and its time bounds have small constants. This result invalidated Wyllie's conjecture that such a result is impossible. A very interesting algorithm with the same performance is due to Anderson and Miller [2]. All deterministic polylogarithmic time optimal parallel algorithms for list ranking are based on the deterministic coin tossing technique, of Cole and Vishkin [6], that was presented in chapter 2. An optimal parallel algorithm that runs in $O(n^\epsilon)$ using $n^{1-\epsilon}$ processors for fixed $1 \geq \epsilon > 0$, was given by Kruskal, Rudolph and Snir [18]. Before the deterministic coin tossing technique was known, Vishkin [26] suggested using randomization in order to achieve optimal speed-up parallel algorithms for list ranking. One of these randomized algorithms is very simple, is optimal and runs in $O(\log n \log \log n)$ time. A step of this simple algorithm consists of applying the standard logarithmic time optimal parallel prefix sums algorithm $O(\log \log n)$ times. Applying the parallel prefix-sums algorithm of this chapter is the only modification needed in order to enhance this algorithm into logarithmic time and optimal speed-up. Miller and Reif [20], provide another logarithmic time randomized algorithm for list ranking. Yet another (EREW) simple randomized logarithmic time optimal parallel algorithm is due to Anderson and Miller [1].

The presently best asymptotic parallel result for connectivity is logarithmic time and optimal speed up for $m \geq n \log^* n$. More specifically, the result for the CRCW PRAM is $T = O(\log n)$ time using $O((n+m)\alpha(m,n)/T)$ processors, where $\alpha(m,n)$ is the inverse Ackerman function. The algorithm of Cole and Vishkin [10], that achieves this result, requires space $O(min[n^2, mn^\epsilon])$, where ϵ can be any constant satisfying $0 < \epsilon < 1$. The algorithm is rather involved. Only its first part is described in this chapter.

We mention a few parallel connectivity algorithms in addition to the ones that were described in Chapter 2. Kruskal, Rudolph and Snir [18] gave an efficient algorithm for relatively slow times (non poly-log) and non-sparse graphs. Gazit [12] gave a randomized connectivity algorithm which runs, with high probability, in logarithmic time using an optimal number of processors.

The only deterministic poly-log time optimal parallel algorithms are based on either the assumption that the graph is given by its adjacency matrix or the assumption that the graph is planar. For the first kind of graphs the fastest serial connectivity algorithms use $O(n^2)$ time and the parallel algorithms of Chin, Lam and Chen [5] and Vishkin [25] achieve $O(\log^2 n)$ time using $\frac{n^2}{\log^2 n}$ processors. These algorithms, which are variants of the algorithm of Hirschberg, Chandra and Sarwate, operate by reducing the size of the adjacency matrix to represent only the shrunken graph at hand. For planar graphs, the main observation needed is that the number of edges can be at most linear in the number of vertices, as follows from Euler's theorem (for instance, see Even [11]). The parallel algorithms of Hagerup, Chrobak and Diks [16] and Hagerup [15] run in $O(\log n \log^* n)$ and logarithmic time, respectively, using an optimal number of processors on planar graphs.

Quite a few parallel algorithms are based on reductions into the problems considered in this chapter. The fundamental role of these problems is illustrated in Figure 5.1, in which there is a link going down from each (relatively) involved problem to simpler problems; routines for the simpler problems were used in algorithms for the more involved problems.

We avoid listing algorithms that apply prefix-sums computation, since the list is too long.

Algorithmic techniques and problems whose algorithms are based on deterministic coin tossing and list ranking:

1. Euler tour technique on trees.
2. Tree contractions.

254 Chapter 5. Advanced Parallel Prefix-sums, List Ranking and Connectivity

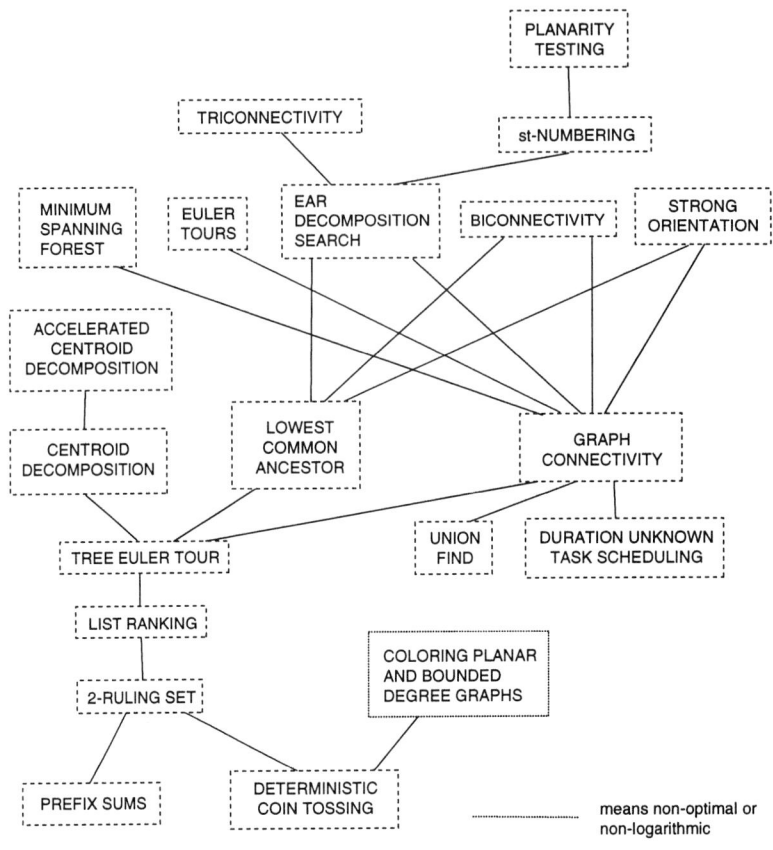

FIGURE 5.1
List, tree and graph algorithms.

   3. Problems on sparse graphs (such as in Goldberg, Plotkin and Shannon [13]).
   4. Graph connectivity.

We mention a few problems whose parallel algorithms are based on parallel connectivity:

   1. Finding biconnected components of undirected graphs.
   2. Finding triconnected components of undirected graphs.
   3. Testing planarity of graphs.

4. Strong orientation: Orienting the edges of a connected bridgeless undirected graph so that the resulting directed graph is strongly connected.
5. Ear decomposition and finding st-numbering of biconnected graphs.
6. Finding Euler tours in directed and undirected graphs (or determining that they do not exist).

Acknowledgement

Helpful comments by Richard Cole and two anonymous referees are gratefully acknowledged.

Bibliography

[1] R.J. Anderson and G.L. Miller, "Optimal parallel algorithms for list ranking", extended abstract, 1986.

[2] R.J. Anderson and G.L. Miller, "Deterministic parallel list ranking", Proc. AWOC 1988, 81–90.

[3] M.J. Atallah and U. Vishkin, "Finding Euler tours in parallel", *J. Computer and Systems Sciences* 29 (1984), 330–337.

[4] P. Beame and J. Hastad, "Optimal bounds for decision problems on the CRCW PRAM", *Proc. 19th ACM Symp. on Theory of Computing*, 1987, 83–93.

[5] F.Y. Chin, J. Lam and I. Chen, "Efficient parallel algorithms for some graph problems", *Comm. ACM* 25 (1982), 659–665.

[6] R. Cole and U. Vishkin, "Deterministic coin tossing with applications to optimal parallel list ranking", *Information and Control* 70 (1986), 32–53.

[7] R. Cole and U. Vishkin, "Approximate and exact parallel scheduling with applications to list, tree and graph problems", *Proc. Twenty Seventh Annual Symp. on Foundations of Computer Science*, 1986, 478–491.

[8] R. Cole and U. Vishkin, "Approximate parallel scheduling. Part I: The basic technique with applications to optimal parallel list ranking in logarithmic time", *SIAM J. on Computing* 17 (1988), 128–142.

[9] R. Cole and U. Vishkin, "Faster optimal parallel prefix sums and list ranking", *Information and Computation* 81 (1989), 334–352.

[10] R. Cole and U. Vishkin, "Approximate parallel scheduling. Part II: Applications to optimal parallel graph algorithms in logarithmic time", *Information and Computation*, 91, 1 (1991), 1–47.

[11] S. Even, *Graph Algorithms*, Computer Science Press, 1979.

[12] H. Gazit, "An optimal randomized parallel algorithm for finding connected components in a graph", *Proc. Twenty Seventh Annual Symp. on Foundations of Computer Science*, 1986, 492–501.

[13] A.V. Goldberg, S.A. Plotkin and G.E. Shannon, "Parallel symmetry-breaking in sparse graphs", *Proc. 19th ACM Symp. on Theory of Computing*, 1987, 315–324.

[14] J. Hastad, "Almost optimal lower bounds for small depth circuits", *Proc. 18th ACM Symp. on Theory of Computing*, 1986, 6–20.

[15] T. Hagerup, "Optimal parallel algorithms on planar graphs", Proc. 3rd AWOC, Lecture Notes in Computer Science, Springer-Verlag, 1988, 24–32.

[16] T. Hagerup, M. Chrobak, K. Diks, "Optimal parallel 5-coloring of planar graphs", Proc. 14th *ICALP*, 1987, 304–313.

[17] D.S. Hirschberg, A.K. Chandra and D.V. Sarwate, "Computing connected components on parallel computers", *Comm. ACM* 22 (1979), 461–464.

[18] C.P. Kruskal, L. Rudolph and M. Snir, "Efficient parallel algorithms for graph problems", *Algorithmica*, in press. The conference version is in *Proc. Int. Conf. on Parallel Processing*, 1985, 180–195.

[19] R. Ladner and M. Fischer, "Parallel prefix computation", *JACM*, 27 (1980), 831–838.

[20] G.L. Miller and J.H. Reif, "Parallel tree contraction and its application", Proc. 26th Symp. on Foundations of Computer Science, 1985, 478–489.

[21] S. Rajasekaran and J.H. Reif, "Optimal and sublogarithmic time randomized parallel integer sorting", *SIAM J. Comput.* 18 (1989), 594–607.

[22] Y. Shiloach and U. Vishkin, "An $O(\log n)$ parallel connectivity algorithm", *J. Algorithms* 3 (1982), 57–67.

[23] L.J. Stockmeyer and U. Vishkin, "Simulation of parallel random access machines by circuits", *SIAM J. Comput.* 13 (1984), 409–422.

[24] H. S. Stone, "Parallel tridiagonal equation solvers", *Transactions on Mathematical Software*, 1 (1975), 289–307.

[25] U. Vishkin, "An optimal parallel connectivity algorithm", *Discrete Applied Math.* 9 (1984), 197–207.

[26] U. Vishkin, "Randomized speed-ups in parallel computation", *Proc. 16th ACM Symp. on Theory of Computing*, 1984, 230–239. For a journal version of this paper see "Randomized speed-ups for list ranking", *J. Parallel and Distributed Computing*, 4,3 (1987), 319–333.

[27] J.C. Wyllie, "The complexity of parallel computation", TR 79-387, Department of Computer Science, Cornell University, Ithaca, New York, 1979.

6

Parallel Lowest Common Ancestor Computation

Baruch Schieber

IBM - Research Division
T.J. Watson Research Center
Yorktown Heights, NY 10598
sbar@watson.ibm.com

6.1
Introduction

Suppose that we are given a rooted tree $T(V, E)$ for preprocessing. After the preprocessing we have to answer on-line queries of the form, "Which vertex is the Lowest Common Ancestor (LCA) of x and y?" for any pair of vertices x, y in T. (We denote such a query $LCA(x, y)$.) We present an optimal parallel preprocessing algorithm that runs in $O(\log n)$ time using $n/\log n$ processors on an EREW PRAM, where n is the number of vertices in T. Given this preprocessing, we show how to process each LCA query in constant time using a single processor. Parallelizing the query processing is straightforward provided read conflicts are allowed: k queries can be processed in $O(1)$ time using k processors on a CREW PRAM.

The algorithm presented here was originally given in [7]. Less efficient parallel algorithms for the same problem are given in [8, 10].

The parallel LCA algorithm can be applied to improve the complexity of several other parallel algorithms. We give here only two examples.

1. *Open ear decomposition.* Given an undirected biconnected graph, compute its open ear decomposition. (See Chapter 7 and [6].)
2. *The strong orientation problem.* Given an undirected bridgeless graph, orient its edges so that the resulting directed graph is strongly connected. (See [10].)

In both algorithms a spanning tree is computed, and then the LCA has to be computed for $O(m)$ pairs of vertices, where m is the number of edges in the input graph. Using previously known algorithms this required $O(m \log n)$ operations when implemented to run in logarithmic time. Using the algorithm presented here this can be done optimally in $O(\log n)$ time and $(m + n)/\log n$ processors. As a result, the complexity bounds of both algorithms is the same as the complexity bounds of the parallel algorithm for computing connected components of an undirected graph.

One of the advantages of the algorithm is its simplicity. Its (linear time and space) sequential version is much simpler than the original sequential algorithm given to this problem in [5].

The LCA algorithm for arbitrary rooted trees is based on the algorithms for two special cases: chains and complete binary trees.

Suppose that the input tree is a chain. In this case it is easy to see that for any pair of vertices, the vertex that is closer to the root is the ancestor of the other vertex, and hence it is also the lowest common ancestor. Thus, after precomputing the *level* of each vertex (i.e., its distance from the root), we can answer any LCA query in constant time. The precomputation of the levels can be done in $O(\log n)$ time using $O(n/\log n)$ processors on an EREW PRAM using the Euler tour technique ([9, 10] and Chapter 2) together with the optimal logarithmic time parallel list ranking algorithm ([4, 2] and Chapters 2 and 3). (See Exercise 6.1.)

In the next section we show how the precomputation of the inorder numbering of a complete binary tree can be used for answering LCA queries in this tree in constant time. Then, we show how to integrate the algorithms for chains and complete binary trees to get an algorithm for arbitrary rooted trees.

6.2
The Algorithm for Complete Binary Trees

A complete binary tree is a rooted tree in which each internal vertex has exactly two children and all the leaves are in the same level. See Fig. 6.1 for an example of a complete binary tree.

Consider a complete binary tree in which the leaves are at level l. It is easy to see that the number of vertices at level $0 \leq d \leq l$ of the tree is 2^d, and that the total number of vertices is $\sum_{d=0}^{l} 2^d = 2^{l+1} - 1$. (In Fig. 6.1 the level of the leaves is four and the total number of vertices is $2^5 - 1$.)

The *inorder traversal* of a rooted binary tree is defined recursively as follows (cf. [1], pp. 54-55).

1. Visit in inorder the left subtree of the root r (if it exists).
2. Visit the root r.
3. Visit in inorder the right subtree of the root r (if it exists).

The *inorder number* of a vertex in the tree is its number in the inorder traversal (starting from the number one).

The inorder numbering of the complete binary tree with 31 vertices is given in Fig. 6.1.

The inorder numbering of a complete binary tree has several properties. To state these properties it is convenient to consider the binary representation of the inorder numbers. (The binary representations are also given in Fig. 6.1.) Let l be the level of the leaves of the tree. We index the bits in the binary representation from the least significant bit whose index is 0 to the most significant bit whose index is l.

262 Chapter 6. Parallel Lowest Common Ancestor Computation

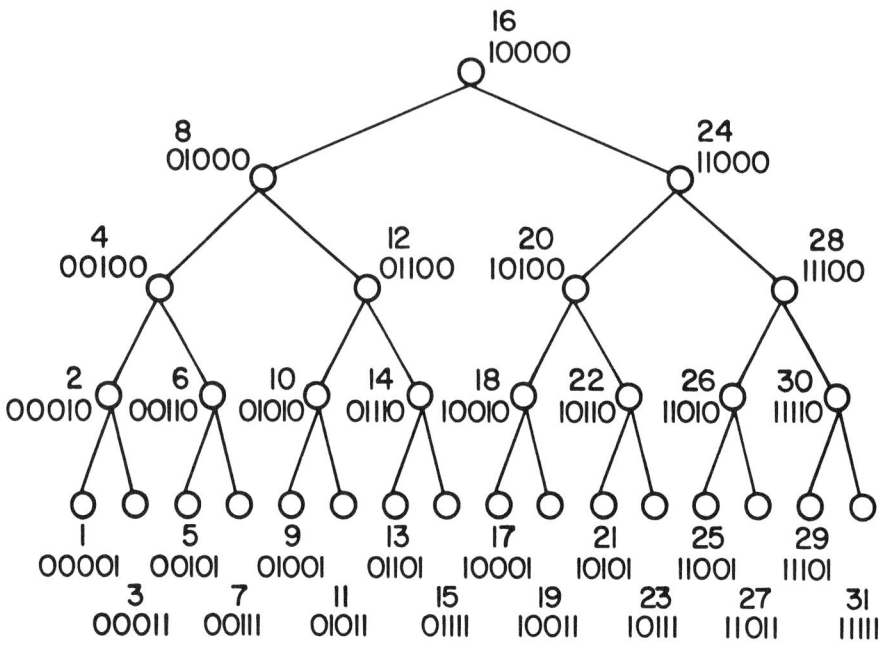

FIGURE 6.1
A complete binary tree with 31 vertices and its inorder numbering.

Consider a vertex with inorder number b. (From now on we are going to identify each vertex in the complete binary tree with its inorder number.) Let i be the index of rightmost "1" bit in b. That is, b consists of $l - i$ leftmost bits, followed by a single "1" and i "0"s. The following properties can be easily verified by induction.

Property 1:   The vertex b is in level $l - i$.

Property 2:   Let b_L and b_R be the left and right children of b, respectively. (i) b_L consists of the $l-i$ leftmost bits of b followed by a single "0", a single "1" and $i - 1$ "0"s, and (ii) b_R consists of the $l - i$ leftmost bits of b followed by two "1"s and $i - 1$ "0"s.

The above properties readily imply the following Lemma.

LEMMA 6.1
The vertex b is an ancestor of a vertex c if and only if (1) the $l - i$ leftmost bits of c are the same as the $l - i$ leftmost bits of b, and (2) the

6.2. The Algorithm for Complete Binary Trees

index of the rightmost "1" bit in c is at most i. (A vertex is considered both an ancestor and a descendant of itself.)

From Lemma 6.1 it follows that b is a common ancestor of two given vertices x and y if and only if (1) the $l - i$ leftmost bits of both x and y are the same as the $l - i$ leftmost bits of b, and (2) the index of the rightmost "1" bit in both x and y is at most i. Thus, in order for b to be the *lowest* common ancestor of x and y, b has to be the vertex with the *minimum* index i (that is, with the *maximum* level) satisfying both conditions.

We show how to compute $b = LCA(x, y)$, where x and y are any two (inorder numbers of) vertices in the complete binary tree. Notice that to compute b it is sufficient to compute i, the index of the rightmost "1" bit in b. This is because b consists of the $l - i$ leftmost bits of x (or y), followed by a single "1" and i "0"s. The computation of i, followed by the discussion of its implementation is given below.

Step 1 Check whether x is either an ancestor or a descendant of y. For this we just have to check whether the conditions of Lemma 6.1 are satisfied. Specifically, we compute i_x and i_y, the indices of the rightmost "1" bit in x and y, respectively. Let $i' = \max\{i_x, i_y\}$. If the leftmost $l - i'$ bits in x and y are the same, then x is either an ancestor or a descendant of y. If $i_x \geq i_y$ then x is an ancestor of y, and hence $LCA(x, y) = x$, otherwise, i.e., if $i_x < i_y$, x is a descendant of y, and hence $LCA(x, y) = y$.

Suppose that x is neither an ancestor nor a descendant of y. That is, the leftmost $l - i'$ bits in x and y are different.

Step 2 Compute the index i of the leftmost bit in which x and y differ. $LCA(x, y)$ consists of the $l - i$ leftmost bits of x (or y), followed by a single "1" and i "0"s.

The implementation of the above computation involves several bitwise operations. To compute i_x, we first compute the number that consists of $l - i_x$ leftmost "0"s, a single "1" and i_x "0"s. It can be verified that this number is given by $x - (x \wedge (x - 1))$, where " $\wedge$ " denotes the bitwise AND operation. It follows that i_x is the (base two) logarithm of this number. The index i_y is computed similarly. To compute the index of the leftmost bit in which x

and y differ (Step 2), we first compute $x \oplus y$, where " $\oplus$ " denotes the bitwise exclusive OR operation. The bitwise exclusive OR of x and y assigns "1" to each bit in which x and y differ. The floor of the (base two) logarithm of $x \oplus y$ gives the index of the leftmost bit of difference (starting from the least significant bit whose index is zero).

6.3
The Algorithm for Arbitrary Trees

We show how to integrate the LCA algorithms for chains and for complete binary trees to get an LCA algorithm for arbitrary trees.

First, we give an overview of the algorithm. In the preprocessing stage we decompose the input tree $T(V, E)$ into vertex disjoint chains. Given a query $LCA(x, y)$, if x and y belong to the same chain, then, as shown above, we can answer the query using the precomputed levels of the vertices.

In the rest of the discussion we assume that x and y do not belong to the same chain. Let C_x and C_y be the chains in the decomposition of T that contain x and y, respectively.

To deal with this case, we compute in the preprocessing stage an injective mapping from the set of chains to the set of vertices of a complete binary tree with $2^{\lfloor \log_2 n \rfloor + 1} - 1$ vertices. (Notice that the complete binary tree with $2^{\lfloor \log_2 n \rfloor + 1} - 1$ vertices is the smallest such tree that has at least n vertices.) Denote this complete binary tree by B.

The mapping has the following *Ascendance-Preservation* property. Let v and u be any two vertices in T, and let C_v and C_u be the chains that contain v and u, respectively. (Notice that C_v may be the same as C_u.)

The Ascendance-Preservation Property: If v is an *ancestor* of u, then C_v is mapped to an *ancestor* of the vertex to which C_u is mapped in B.

Given a query $LCA(x, y)$, we use the LCA algorithm for complete binary trees to compute the LCA (in B) of the two vertices to which C_x and C_y are mapped. Then, using some additional information, as explained later, we can identify the chain in T that contains $LCA(x, y)$, and finally, we can find $LCA(x, y)$ itself.

We turn now to describe the algorithm in more detail: first, the preprocessing stage, and then the query processing.

6.3.1 The Preprocessing Stage

The outcome of the preprocessing stage consists of labels that are assigned to the vertices of T and a look-up table, called HEAD. The label of each vertex $v \in V$ consists of three numbers: LEVEL(v), INLABEL(v), and ASCENDANT(v)

LEVEL(v) is simply the distance of v from the root. Recall that the computation of LEVEL(v), for all $v \in V$ can be done in $O(\log n)$ time using $O(n/\log n)$ processors on an EREW PRAM using the Euler tour technique ([9, 10] and Chapter 2) together with the optimal logarithmic time parallel list ranking algorithm ([4, 2] and Chapters 2 and 3).

The INLABEL Numbers

To compute the INLABEL numbers, we need the preorder numbering of T.

The *preorder traversal* of a rooted tree is defined recursively as follows (cf. [1], pp. 54-55).

1. Visit the root r.
2. Visit in preorder all the subtrees of the root r (if such exist) one by one, from left to right.

The *preorder number* of a vertex $v \in V$ denoted PREORDER(v), is its number in the preorder traversal.

The computation of INLABEL(v), for each $v \in V$ is done in three steps.

Step 1 For each $v \in V$, compute PREORDER(v).

Step 2 For each $v \in V$, compute and SIZE(v), the number of vertices in the subtree rooted at v.

The parallel implementation of these two steps is left as an exercise.

The definition of the preorder numbering implies the following fact.

Fact: The preorder numbers of the vertices in the subtree rooted at v range between PREORDER(v) and PREORDER(v) + SIZE(v) − 1.

Motivated by this fact we define the closed interval [PREORDER(v), PREORDER(v) + SIZE(v) − 1] as the *the interval of v*.

In Step 3 we consider the binary representation of the integers in the interval of v.

Step 3 Find the integer that has the maximal number of rightmost "0" bits in the interval of v. This number is assigned to INLABEL(v). It will be clear from the implementation of this step that this number is unique.

Step 3 is implemented in two substeps.

Step 3.1 Compute the floor of the (base two) logarithm of (PREORDER(v) $-$ 1) $\oplus$ (PREORDER(v) + SIZE(v) $-$ 1) into i. This gives the index of the leftmost bit of difference between (the binary representations of) PREORDER(v) $-$ 1 and (PREORDER(v) + SIZE(v) $-$ 1). (Notice that the bit indexed i in PREORDER(v) $-$ 1 is "0" and the bit indexed i in (PREORDER(v) + SIZE(v) $-$ 1) is "1".)

In Step 3.2 the number INLABEL(v) is "composed". For this, we need two observations:

1. The $l-i+1$ leftmost bits of INLABEL(v) are the same as the $l-i+1$ leftmost bits in PREORDER(v) + SIZE(v) $-$ 1.
2. The i other bits in INLABEL(v) are "0"s.

Step 3.2 Compute $2^i \lfloor (\text{PREORDER}(v) + \text{SIZE}(v) - 1)/2^i \rfloor$ into IN-LABEL(v). This assigns the $l - i + 1$ leftmost bits in PREORDER(v) + SIZE(v) $-$ 1 to the $l - i + 1$ leftmost bits in INLABEL(v) and "0"s to the other bits of INLABEL(v).

For an example of the computations done in this section see Fig. 6.2.

The INLABEL numbers define a decomposition of T into chains. Each chain in the decomposition consists of all the vertices that have the same INLABEL number. To show that this is indeed a decomposition into vertex disjoint chains we prove the following lemma.

LEMMA 6.2
If vertices v and u have the same INLABEL *number, then:*

1. *v is either an ancestor or a descendant of u,*
2. *all the vertices in the path between v and u have the same* INLABEL *number as u and v.*

PROOF
From the definition of the INLABEL numbers it follows that INLABEL(v) is the preorder number of some vertex in the subtree rooted at v. Thus, if v

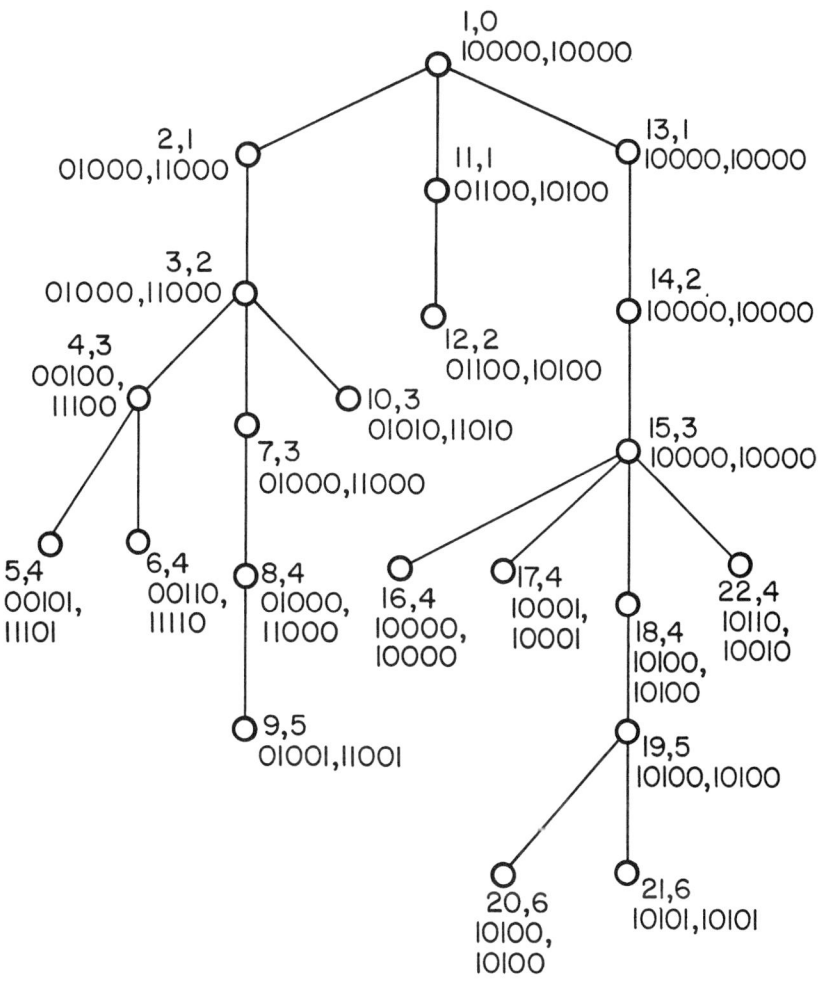

FIGURE 6.2
The preprocessing stage. The numbers PREORDER, LEVEL, INLABEL, and ASCENDANT are given at each vertex. (The last two numbers are given in binary representation.)

and u have the same INLABEL number, then the vertex whose preorder number is INLABEL(v) is in both the subtree rooted at v and the subtree rooted at u. This implies that v is either an ancestor or a descendant of u, proving part (i) of the Lemma. Without loss of generality assume

268   Chapter 6. Parallel Lowest Common Ancestor Computation

that v is an ancestor of u. Consider a vertex w in the path between v and u. Clearly, the vertex whose preorder number is INLABEL(v) is in the subtree rooted at w, and hence, INLABEL(v) is in the interval of w. Since INLABEL(v) has the maximal number of rightmost "0" bits in the interval of v, and the interval of w is contained in the interval of v, INLABEL(v) must also have the maximal number of rightmost "0" bits in the interval of w. Thus, INLABEL(v) = INLABEL(w), proving part (ii) of the Lemma.   ■

The INLABEL numbers define also the mapping of the chains into the vertices of the complete binary tree B with $2^{\lfloor \log_2 n \rfloor + 1} - 1$ vertices. The chain defined by the INLABEL number k is mapped to the vertex (whose inorder number is) k in B. Below, we prove that this mapping obeys the Ascendance-Preservation Property.

LEMMA 6.3
If v is an ancestor of u in T, then the vertex INLABEL(v) is an ancestor of the vertex INLABEL(u) in B.

PROOF
Let i be the number of rightmost "0" bits in INLABEL(v). Since INLABEL(u) belongs to the interval of v and INLABEL(v) has the maximal number of rightmost "0" bits in this interval, the number of rightmost "0" bits in INLABEL(u) is at most i. The $l - i$ leftmost bits are the same for all numbers in the interval of v. In particular, the $l - i$ leftmost bits in INLABEL(u) are the same as the $l - i$ leftmost bits in INLABEL(v). By Lemma 6.1 we get that INLABEL(v) is an ancestor of INLABEL(u) in B.   ■

The ASCENDANT Numbers

Observe that the opposite direction of Lemma 6.3 is not always true. That is, it may be that a vertex v is not an ancestor of a vertex u in T, while the vertex INLABEL(v) is an ancestor of the vertex INLABEL(u) in B. For example, in Fig. 6.2, the vertex with preorder number 18 is not an ancestor of the vertex with preorder number 22 in T, while the INLABEL number of vertex 18 is an ancestor of the INLABEL number of vertex 22 in the complete binary tree with 31 vertices.

For each $v \in V$, the single number ASCENDANT(v) records all those ancestors of INLABEL(v) in B that are also INLABEL numbers of ancestors of v in T. We can record all this information in *one* number in the range $[0, 2^{\lfloor \log_2 n \rfloor + 1}]$ because the INLABEL number of each of the ancestors of v can

6.3. The Algorithm for Arbitrary Trees

be fully specified by the index of its rightmost "1" bit. This is because the bits which are to the left of this "1" bit are the same as their respective bits in INLABEL(v).

Denote the binary representation of ASCENDANT(v) by the sequence $A_l(v), \ldots, A_0(v)$ (where $l = \lfloor \log_2 n \rfloor$). We set $A_i(v)$ to be 1 only if i is the index of a rightmost "1" bit in the INLABEL number of some ancestor of v in T. That is, each "1" bit in ASCENDANT(v) corresponds to the INLABEL number of some ancestor of v in T.

Computing the ASCENDANT numbers. Let r be the root of t. Since IN-LABEL(r) = 2^l and r has only one ancestor (which is r itself), ASCENDANT(r) = 2^l. Consider an internal vertex $v \neq r$ in T and let $P(v)$ be the parent of v in T. If INLABEL(v) = INLABEL($P(v)$) then ASCENDANT(v) = ASCENDANT($P(v)$); otherwise, ASCENDANT(v) = ASCENDANT($P(v)$) + 2^i, where i is the index of the rightmost "1" bit in INLABEL(v). This is because the only ancestor of v that is not an ancestor of $P(v)$ is v itself.

The computation is implemented using the Euler tour technique ([9, 10] and Chapter 2) together with the optimal logarithmic time parallel list ranking algorithm ([4, 2] and Chapters 2 and 3) in three steps.

Step 1 Assign a weight to each edge in the Euler tour defined by T as follows. For each internal vertex $v \neq r$, if INLABEL(v) $\neq$ INLABEL($P(v)$), then the weight of the edge incoming to v from $P(v)$ is set to 2^i and the weight of the edge outgoing from v to $P(v)$ is set to -2^i, where i be the index of the rightmost "1" bit in INLABEL(v). The weight of all other edges is set to zero.

Step 2 Compute the (weighted) distance of each edge in the Euler tour from the start of the tour.

Step 3 Set ASCENDANT(r) to be 2^l. For each internal vertex $v \neq r$, set ASCENDANT(v) to be the distance of the edge incoming to v from $P(v)$ plus 2^l.

In the preprocessing stage we also compute the look-up table HEAD. This table contains the highest vertex in each INLABEL path; that is, the entry HEAD(k) contains the vertex which is closest to the root among all vertices with INLABEL number k. We note that, given the INLABEL numbers, the table HEAD can be computed in constant time using n processors, and hence also in logarithmic time using $n/\log n$ processors.

6.3.2 Processing LCA Queries

In this subsection we show how to answer LCA queries using the outcome of the preprocessing stage.

Consider a query $LCA(x,y)$, for a pair of vertices x,y in T. (To illustrate the presentation the reader is referred to Fig. 6.2.) There are two cases:

1. INLABEL(x) = INLABEL(y). It must be that x and y are in the same INLABEL path. We conclude that $LCA(x,y)$ is x if LEVEL$(x) \leq$ LEVEL(y) and y otherwise.
2. INLABEL$(x) \neq$ INLABEL(y). Let z be $LCA(x,y)$. We find z in four steps:

Step 1 Find b, the LCA of INLABEL(x) and INLABEL(y) in the complete binary tree B.

The implementation of this step is described in the previous section.

In Step 2 we find INLABEL(z) (where z is $LCA(x,y)$). The Ascendance-Preservation property of the INLABEL numbers implies that INLABEL(z) is a common ancestor of INLABEL(x) and INLABEL(y). Notice that INLABEL(z) is not necessarily b, the *lowest* common ancestor of INLABEL(x) and INLABEL(y) in B. This is because the vertices on the chain that is mapped into b are not necessarily ancestors of x or y. However, it is not difficult to see that INLABEL(z) is the lowest ancestor of b in B which is the INLABEL number of an ancestor of both x and y in T. To find it we use the ASCENDANT numbers.

Step 2 Find the index of the rightmost "1" bit in INLABEL(z), denoted by j. Since z is a common ancestor of x and y in T, $A_j(x) = 1$ and $A_j(y) = 1$. Let i be the index of the rightmost "1" bit in b. Since INLABEL(z) is an ancestor of b the index j must be greater or equal to i. Since INLABEL(z) is the *lowest* ancestor of b which is a common ancestor of x and y, the index j must be the minimum index $j \geq i$ such that $A_j(x) = 1$ and $A_j(y) = 1$; that is, j is the index of the *rightmost* "1" bit in $A_l(x), \ldots, A_i(x) \wedge A_l(y), \ldots, A_i(y)$. Consequently, INLABEL$(z)$ consists of the $l-j$ leftmost bits of INLABEL(x) (or INLABEL(y)) followed by a single "1" and j "0"s.

The implementation of Step 2 is done in three substeps.

Step 2.1 Compute ASCENDANT$(x) \wedge$ ASCENDANT(y) into COMMON.

Step 2.2 Compute $2^i \lfloor \text{COMMON}/2^i \rfloor$ into COMMON$_i$. COMMON$_i$ lists all the "1"s in both $A_l(x), \ldots, A_i(x)$ and $A_l(y), \ldots, A_i(y)$.

Step 2.3 j is the index of the rightmost "1" bit in COMMON$_i$. To find it we compute the (base two) logarithm of COMMON$_i$ − [COMMON$_i \wedge$ (COMMON$_i$ − 1)].

In the next steps we find $z = LCA(x, y)$. The vertex z is the lowest vertex in the path defined by INLABEL(z) which is a common ancestor of x and y in T. To find it we first find x'', the lowest ancestor of x in the path defined by INLABEL(z) and y'', the lowest ancestor of y in this same path. z is the higher vertex among these two vertices.

Step 3 Find x'' and y''. We show how to find x''. y'' is found similarly.

If INLABEL$(x) =$ INLABEL(z) then $x'' = x$ and nothing has to be done. Suppose INLABEL$(x) \neq$ INLABEL(z). We set the following intermediate goal, as the main step towards finding x'': Find the child of x'' which is also an ancestor of x. Denote the vertex that we search by w. Since INLABEL$(w) \neq$ INLABEL(x'') and the index of the leftmost "1" bit in INLABEL(x'') is j, the index of the rightmost "1" bit in INLABEL(w) is the index of the leftmost "1" bit in $A_{j-1}(x), \ldots, A_0(x)$. Denote this index by k. Clearly, INLABEL(w) consists of the $l - k$ leftmost bits of INLABEL(x) followed by a single "1" and k "0"s. Observe that w is the head of its INLABEL path (since the INLABEL number of its parent x'' is different from INLABEL(w)). Therefore, w is HEAD(INLABEL(w)) and our intermediate goal is achieved. Finally, x'' is the parent of w.

Step 4 $LCA(x, y)$ is x'' if LEVEL$(x'') \leq$ LEVEL(y'') and y'' otherwise.

6.4
Exercises

6.1 Let $T(V, E)$ be a rooted tree with n nodes. Using the Euler tour technique, find a logarithmic time parallel algorithm that uses $n/\log n$ processors for computing the following for each vertex $v \in V$: LEVEL(v), PREORDER(v), and the size of the subtree rooted at v.

6.2 Find a logarithmic time parallel algorithm that uses $n/\log n$ processors for computing the inorder numbering of a binary tree. (*Hint:* Use the Euler tour technique. In assigning the weights for the edges in the Euler tour distinguish between edges touching a left child and edges touching a right child.)

6.3 Show how the bitwise exclusive OR of two l-bit numbers can be found using a look-up table that consists of the bitwise exclusive OR of all pairs of $\lfloor l/2 \rfloor$-bit numbers. (*Hint:* Use the look-up table to compute the bitwise exclusive OR of the $\lfloor l/2 \rfloor$ most significant bits, and the $\lceil l/2 \rceil$ least significant bits.) Conclude how look-up tables for the bitwise operations needed by the LCA algorithm can be precomputed in constant time using a linear number of processors.

The goal of the next exercises is to design a simple optimal logarithmic time preprocessing algorithm for answering LCA queries in *logarithmic* time. This algorithm is given in [10].

6.4 The *postorder traversal* of a rooted tree is defined by exchanging steps 1 and 2 in the definition of the preorder traversal. Denote the postorder number of a vertex v by POSTORDER(v). Let x and y be two vertices in T. Show that given PREORDER(x), POSTORDER(x), PREORDER(y) and POSTORDER(y), the test whether x is an ancestor of y can be done in *constant* time.

6.5 Consider the Euler tour defined by a rooted tree T. Let x and y be two vertices in T, such that x is neither an ancestor nor a descendant of y.

a) Prove that either the last occurrence of x on the Euler tour is before the first occurrence of y, or the last occurrence of y on the Euler tour is before the first occurrence of x.

b) Suppose that the last occurrence of x is before the first occurrence of y. Prove that $LCA(x,y)$ is the vertex with minimal level that is on the portion of the Euler tour from the last occurrence of x to the first occurrence of y.

c) Show how to preprocess the Euler tour such that the vertex with the minimal level on any portion of it can be found in logarithmic time.

6.6 Using the two previous exercises design an optimal logarithmic time preprocessing algorithm for answering LCA queries in logarithmic time.

Bibliography

[1] A.V. Aho, J.E. Hopcroft, and J.D. Ullman. *The Design and Analysis of Computer Algorithms*. Addison-Wesley, Reading, Ma, 1974.

[2] R.J. Anderson and G.L. Miller. Deterministic parallel list ranking. In *Proc. of AWOC 88, Lecture Notes in Computer Science*, No. 319, pages 81–90. Springer-Verlag, 1988.

[3] A. Apostolico, C. Iliopoulos, G.M. Landaun, B. Schieber, and U. Vishkin. Parallel construction of a suffix tree with applications. *Algorithmica*, Vol. 3, pp. 347–365, 1988.

[4] R. Cole and U. Vishkin. Approximate parallel scheduling. I. The basic technique with applications to optimal parallel list ranking in logarithmic time. *SIAM Journal on Computing*, 17(1):128–142, February 1988.

[5] D. Harel and R.E. Tarjan. Fast algorithms for finding nearest common ancestors. *SIAM Journal on Computing*, 13(2):338–355, May 1984.

[6] Y. Maon, B. Schieber, and U. Vishkin. Parallel Ear Decomposition Search (EDS) and *st*-numbering in graphs. *Theoretical Computer Science*, 47:277–298, 1986.

[7] B. Schieber and U. Vishkin. On finding lowest common ancestors: simplification and parallelization. *SIAM Journal on Computing*, 17(6):1253–1262, December 1988.

[8] Y.H. Tsin. Finding lowest common ancesors in parallel. *IEEE Transactions on Computers*, 35:764–769, 1986.

[9] R.E. Tarjan and U. Vishkin. An efficient parallel biconnectivity algorithm. *SIAM Journal on Computing*, 14:862–874, 1985.

[10] U. Vishkin. On efficient parallel strong orientation. *Information Processing Letters*, 20:235–240, 1985.

7

Parallel Open Ear Decomposition with Applications to Graph Biconnectivity and Triconnectivity

Vijaya Ramachandran

Department of Computer Science
University of Texas
Austin, TX 78712
vlr@cs.utexas.edu

7.1
Introduction

In this chapter we introduce *open ear decomposition*, which is a method for searching an undirected graph. We present an algorithm that either finds an open ear decomposition in an undirected graph or reports that no open ear decomposition exists. This algorithm runs in logarithmic time with a linear number of processors. A graph has an open ear decomposition if and only if it is biconnected. Hence this algorithm allows us to determine graph biconnectivity efficiently in logarithmic parallel time.

We use open ear decomposition to obtain a logarithmic time parallel algorithm using a linear number of processors to find the triconnected components of a graph. This algorithm is fairly complex and we present it in a top-down manner by first giving the high-level ideas leading to the algorithm and then giving efficient implementations of the various steps. In the last section we give some pointers towards obtaining optimal logarithmic time parallel algorithms for graph biconnectivity and triconnectivity.

Algorithmic Notation

The algorithmic notation in this chapter is from Tarjan [24]. We enclose comments between a pair of curly brackets with asterisks ('{*' and '*}'). We incorporate parallelism by use of the following statement that augments the **for** statement.

pfor iterator **in parallel** → statement list **rofp**

The effect of this statement is to perform the **pfor** loop in parallel for each value of the iterator.

7.2
Ear Decomposition and Two-Connectivity

In this section we define *ear decomposition* and *open ear decomposition* and relate these to graph *two-edge-connectivity* and *two-vertex-connectivity* (i.e., *biconnectivity*). We then describe efficient parallel algorithms to find these decompositions. We also relate these parallel algorithms to the classical sequential algorithm for testing graph biconnectivity, which is based on depth-first search.

7.2.1 Basic Definitions

An *undirected graph* G is a pair (V, E) where V is the set of *vertices* of G and E is the set of *edges* of G; an edge is an unordered pair of distinct vertices. We denote the undirected graph by $G = (V, E)$ and we sometimes refer to it as G. An edge (u, v) is *incident* on vertices u and v. Vertices u and v are *adjacent* in G if G contains edge (u, v). The *degree* of a vertex is the number of edges incident on the vertex. We will sometimes refer to an undirected graph as simply a *graph*.

A *directed graph* $G = (V, E)$ consists of a vertex set V and an edge set E containing ordered pairs of elements from V. An edge (u, v) in a directed graph is directed from u to v and is *outgoing* from u and *incoming* to v.

A *multigraph* G is a pair (V, E) where V is the set of vertices of G and E is the *multiset* of edges of G; an edge of a multigraph is an unordered pair of vertices. We allow edges of the form $(v, v), v \in V$ and we call such edges *self-loops*. An edge e in a multigraph may be denoted by (a, b, i) to distinguish it from other edges between a and b; in such cases the third entry in the triplet may be omitted for one of the edges between a and b.

A *path* P in G is a sequence of vertices $\langle v_0, \ldots, v_k \rangle$ such that $(v_{i-1}, v_i) \in E$, $i = 1, \ldots, k$; P is directed or undirected depending on whether G is directed or undirected. The path P *contains* the vertices $v_0, \ldots, v_k$ and the edges $(v_0, v_1), \ldots, (v_{k-1}, v_k)$ and has *endpoints* v_0, v_k, and *internal vertices* $v_1, \ldots, v_{k-1}$. The path P is a *simple path* if $v_0, \ldots, v_{k-1}$ are distinct and $v_1, \ldots, v_k$ are distinct, and all edges on P are distinct. A simple path $P = \langle v_0, \ldots, v_k \rangle$ is a *simple cycle* if $v_0 = v_k$; otherwise P is *noncyclic*. The path $\langle v \rangle$ is a *trivial path* with no edges.

A graph $G' = (V', E')$ is a *subgraph* of a graph $G = (V, E)$ if $V' \subseteq V$ and $E' \subseteq E$. The *subgraph of G induced by V'* is the graph $H = (V', F)$ where $F = \{(u, v) \in E \mid u, v \in V'\}$.

An undirected graph $G = (V, E)$ is *connected* if there exists a path between every pair of vertices in V. A *connected component* of a graph G is a maximal induced subgraph of G which is connected.

Let $G = (V, E)$ and $H = (W, F)$ be a pair of graphs. The graph $G \cup H$ is the graph $G' = (V \cup W, E \cup F)$. If $W \subseteq V$ then the graph $G - H$ is the graph $H' = (V, E - F)$.

A *tree* is a connected graph containing no cycle. A *leaf* in a tree is a vertex of degree 1. Let $T = (V, E)$ be a tree and let $r \in V$. The *out-tree* $T = (V, E, r)$ *rooted at* r (or simply the *tree T rooted at r*) is the directed graph obtained from T by directing each edge such that every path from r

to any other vertex is directed away from r. The *in-tree rooted at r* is the directed graph obtained from T by directing each edge such that the path from every vertex to r is directed towards r.

Let (x, y) be a directed edge in a rooted tree T. Then, x is the *parent* of y and y is a *child* of x in T. Vertex v is a *descendant* of vertex u (and equivalently, u is an *ancestor* of v) if there is a directed path from u to v in T. Vertex v is a *proper descendant* of u (and u a *proper ancestor* of v) if v is a descendant of u and $u \neq v$. Given a pair of vertices $u, v \in V$, the *least common ancestor of u and v*, denoted by $lca(u, v)$ is the vertex $w \in V$ that is an ancestor of both u and v with no child of w being an ancestor of both u and v. For an edge $e = (u, v)$ the *least common ancestor of e*, denoted by $lca(e)$, is the vertex $lca(u, v)$.

A *preorder* labeling of the vertices of a rooted tree T labels the root of T and then the vertices in the subtree rooted at each child of the root in turn.

Let $G = (V, E)$ be a connected graph. A *spanning tree T of G* is a subgraph of G with vertex set V such that T is a tree. An edge in $G - T$ is a *nontree edge with respect to T*.

Let T be a spanning tree of G. Any nontree edge e of G creates a cycle in the graph $T \cup \{e\}$, called the *fundamental cycle of e with respect to T*. Let $r \in V$, and let T be rooted at r.

Let $e = (u, v)$ be a nontree edge in $T = (V, E, r)$ and let $lca(e) = l$. The fundamental cycle of e with respect to T consists of the path from l to u, followed by edge e, followed by the path from v to l. Let (l, a) be the first edge on the path from l to u and (l, b) be the first edge on the path from l to v (it is possible for one of these edges to be missing). Then edges (l, a) and (l, b) are the *base edge(s) of the fundamental cycle of e* (when they exist) and the vertices a and b are the *base vertice(s) of the fundamental cycle of e* (when they exist).

An edge $e \in E$ in a connected graph $G = (V, E)$ is a *cutedge* if e does not lie on a cycle in G. A connected undirected graph $G = (V, E)$ is *2-edge connected* if it contains no cutedge. A *2-edge connected component of G* is a maximal induced subgraph of G which is 2-edge connected.

A vertex $v \in V$ is a *cutpoint* of a connected undirected graph $G = (V, E)$ if the subgraph induced by $V - \{v\}$ is not connected. A connected graph G is *biconnected* (or *two-vertex connected*) if it contains at least 3 vertices and has no cutpoint. A *biconnected component* (or *block*) of G is a maximal induced subgraph of G which is biconnected.

By *Menger's theorem* a graph is 2-edge connected if and only if there are at least two edge-disjoint paths between every pair of distinct vertices, and a graph is biconnected if and only if the graph is connected and has no more than two vertices or there are at least two vertex-disjoint paths between every pair of distinct vertices.

The *two-connectivity* problem is the problem of determining 2-edge connectivity and biconnectivity in a connected graph.

7.2.2 Ear Decomposition

An *ear decomposition* $D = [P_0, P_1, \ldots, P_{r-1}]$ of an undirected graph $G = (V, E)$ is a partition of E into an ordered collection of edge-disjoint simple paths $P_0, \ldots, P_{r-1}$ such that P_0 is an edge, $P_0 \cup P_1$ is a simple cycle, and each endpoint of P_i, for $i > 1$, is contained in some $P_j, j < i$, and none of the internal vertices of P_i are contained in any $P_j, j < i$. The paths in D are called *ears*. An ear is *open* if it is noncyclic and is *closed* otherwise. A *trivial ear* is an ear containing a single edge. D is an *open ear decomposition* if all of its ears are open.

Let $D = [P_0, \ldots, P_{r-1}]$ be an ear decomposition for a graph $G = (V, E)$. For a vertex v in V, we denote by $ear(v)$, the index of the lowest-numbered ear that contains v; for an edge $e = (x, y)$ in E, we denote by $ear(e)$ (or $ear(x, y)$), the index of the unique ear that contains e. A vertex v belongs to $P_{ear(v)}$.

LEMMA 7.1 *[27]*
An undirected graph $G = (V, E)$ has an ear decomposition if and only if G is 2-edge connected.

PROOF
We first prove the *if* part of the lemma. Assume G is 2-edge connected. We construct an ear decomposition for G as follows. To construct P_0 and P_1, we pick any edge $e = (u, v)$ in G. Since e is not a cutedge, there is a simple path between u and v in G that avoids e. Let P be such a path. We construct P_0 as $\langle e \rangle$ and P_1 as P. Then P_0 is an edge and $P_0 \cup P_1$ is a simple cycle as required.

Assume inductively that we have constructed $H_{i-1} = \cup_{j=0}^{i-1} P_j, i > 1$. To construct P_i, we pick an edge (x, y) that is not contained in H_{i-1} but with vertex x in H_{i-1}. We then find a simple path Q from y to x in G that avoids edge (x, y). Let z be the first vertex on path Q that is contained in H_{i-1}. We construct P_i as the edge (x, y) followed by the path Q from y to z. This path has each of its endpoints on some $P_j, j < i$, and none of its internal vertices on any $P_j, j < i$. Hence it is an ear.

We now prove the *only if* part. Let $D = [P_0, \ldots, P_{r-1}]$ be an ear decomposition for G. We will prove by induction on i for $i > 0$ that the graph $H_i = \cup_{j=0}^{i} P_j$ is 2-edge connected. For the base case, $P_0 \cup P_1$ is a simple cycle, and therefore H_1 is 2-edge connected.

Assume inductively that H_{i-1} is 2-edge connected and consider H_i. To show that H_i is 2-edge connected it suffices to show that every edge on P_i lies on a cycle. Let the endpoints of P_i be x and y and let Q be a path from x to y in H_{i-1}. The path Q exists since H_{i-1} is connected. Every edge on P_i lies on the cycle $P_i \cup Q$ in H_i and hence H_i is 2-edge connected. ∎

LEMMA 7.2 *[27]*
A graph has an open ear decomposition if and only if it is biconnected.

PROOF
Exercise 7.1. ∎

7.2.3 An Efficient Parallel Algorithm for Ear Decomposition

In this section we present an efficient parallel algorithm for finding an ear decomposition for a 2-edge connected graph. This algorithm is from [15] and [14], and is an efficient parallel implementation of an algorithm in [13].

ALGORITHM 7.1
Ear Decomposition Algorithm
Input: A 2-edge connected graph $G = (V, E)$, with $|V| = n$ and $|E| = m$.
Output: A numbering on the edges in E, specifying their ear number.

        vertex v, r; **edge** e;
1. {∗ Preprocess. ∗} find a spanning tree T for G, pick a root vertex r and number the vertices of T in preorder from 0 to $n-1$ with respect to root r;
2. {∗ Assign ear numbers to nontree edges in T. ∗}
   2a. label each nontree edge e in G by its least common ancestor $lca(e)$ in T;
   2b. sort the labels of nontree edges in nondecreasing order and relabel them in order as 1, 2, ...;
3. {∗ Extend the numbering assigned in step 2 to the tree edges by numbering each tree edge t by the label of the nontree edge with smallest label whose fundamental cycle contains t. ∗}
   3a. label each vertex with the label of the nontree edge incident on it with the minimum label;

3b. assign to each tree edge $(parent(v), v)$ in T, the label of the minimum label of any descendant of v (including v);

4. relabel the nontree edge labeled 1 by the label 0

end.

We now prove the correctness of Algorithm 7.1 and then provide implementation details.

LEMMA 7.3
Algorithm 7.1 obtains an ear decomposition of a 2-edge connected graph.

PROOF
We first observe that the label given to tree edge $t = (parent(v), v)$ in step 3b is the label of the nontree edge with smallest label whose fundamental cycle contains t. This is because any such nontree edge e must be incident on a descendant of v, and any nontree edge n incident on a descendant of v with $lca(n) \leq v$ must include edge t in its fundamental cycle.

We now prove by induction on i that the edges with label i form a simple path that satisfies the definition of ear P_i.

BASE: P_0 and P_1. Let e be the nontree edge given label 1 in step 2b. Then by step 3 every tree edge in the fundamental cycle of e will be assigned label 1. Further, any tree edge not on the fundamental cycle of e will be assigned a label greater than 1. Hence the edges labeled 1 at the end of step 3 are exactly the edges in the fundamental cycle of e and these form a simple cycle as required for $P_0 \cup P_1$. By step 4 the label of e is set to be 0. Hence $P_0 = \{e\}$ and P_1 becomes a simple noncyclic path with its two endpoints on e.

INDUCTION STEP: Assume the result is true for up to P_{i-1}, $i > 1$, and consider the nontree edge $f = (u, v)$ with label i. Let $lca(f) = l$. Hence the tree edges in the fundamental cycle of f are the edges on the tree path P from l to u and on the tree path Q from l to v.

Consider the tree path P. Assume that P contains at least one edge with label $j \neq i$ and let (x, y) be the first edge on $R = P \cup \{f\}$ that has label i. We claim that every edge on R from x to v has label i and every edge in P from l to x has label less than i. To see the first part of the claim we note that by step 3 f is the nontree edge with smallest label whose fundamental cycle contains tree edge (x, y). Every edge on P

from y to u lies on the fundamental cycle of f, so if any edge on this path does not have label i then it must have a label $j < i$. But then, the nontree edge g with label j has $lca(g) \leq l$ by the labeling in step 2b. But then, edge (x, y) would be in the fundamental cycle of g and would be labeled j rather than i, which is a contradiction. Hence every edge on P from x to u is labeled i. Finally, edge (u, v) is labeled i by assumption. Hence all edges on R from x to v have label i.

To see the second part of the claim, consider tree edge $s = (x, parent(x))$. Since by assumption the edge s has a label j that is different from i, we know that tree edge s lies on the fundamental cycle of a nontree edge h with label j and that $j < i$. Further since $j < i$ we must have $lca(h) \leq l$ and hence every edge on the path P from l to x lies on the fundamental cycle of h. Hence the label of every edge on P from l to x is at most j and hence is less than i.

A similar argument holds for the path Q for the case when Q contains at least one edge with label $j \neq i$. Hence the edges with label i form a simple path that consists of a portion of tree path P starting at some vertex x and extending up to u, followed by edge (u, v) followed by a portion of the tree path Q from v to some vertex $z > l$; further, the two endpoints of this path are contained in ears numbered lower than i.

Finally, if P or Q contains no edge with label $j \neq i$ then we note that the label of tree edge $(parent(l), l)$ is less than i since any nontree edge g whose fundamental cycle contains this tree edge has $lca(g) < l$. Further, such a nontree edge g must exist since the graph is 2-edge connected. Hence vertex l is contained in an ear P_k with $k < i$ and hence the endpoints of ear P_i are contained on an ear with label smaller than i. ∎

Let us analyze the complexity of Algorithm 7.1.

Step 1 requires the computation of a spanning tree T and its preorder numbering with respect to the root r [2].

Step 2a requires the computation of least common ancestors in T [22].

Step 2b requires sorting of integers in the range $[0..n - 1]$ [1].

Step 3a requires the computation of the minimum value in each adjacency list [11].

Step 3b can be performed efficiently in parallel by the following simple method using the Euler tour technique on trees [25]. Note that the vertices that are the descendants of a vertex v in the tree T lie between the first and last occurrences of v in the Euler tour of T. In step 3b we need to compute the minimum value in each such interval. For this we first build a table of such minimum values for all intervals of length 2^i, $0 \leq i \leq log n$. This table can be constructed in $O(\log n)$ time using n processors. Once we have this table, the minimum value for any other interval I can be computed from the precomputed minimum values of two overlapping intervals whose union gives I. This part of the computation can be performed in constant time using one processor for each interval.

Step 4 is trivial to implement.

As seen above, all of the steps in Algorithm 7.1 can be performed in logarithmic time with a linear number of processors using well-known algorithms that are described in other chapters in this book. We also leave it as an exercise for the reader to verify that Algorithm 7.1 runs in linear sequential time.

7.2.4 Ear Decomposition and Depth-First Search

Algorithm 7.1 of the previous section computes an ear decomposition of a graph in linear sequential time. The computation in Algorithm 7.1 can be simplified considerably in the sequential algorithm if the spanning tree T is a depth-first search tree rooted at r. In that case, the *lca* computation in step 2a is immediate, since every nontree edge in the depth-first search tree goes from a vertex to its ancestor, and this ancestor will be the *lca*. We defer step 2b to the end of the algorithm and to compute step 3, we define the following two functions on vertices. (We assume that the vertices are numbered in preorder, starting with 0, and that the input graph has n nodes.)

$$low(v) = \min(\{w \mid w \text{ lies on the fundamental cycle of a nontree edge incident on a descendant of } v\} \cup \{n\})$$

$$ear(v) = lexmin(\{(w, x) \mid (w, x) \text{ is a nontree edge with } x \text{ a descendant of } v\} \cup \{(n, n)\})$$

The values $low(v)$ and $ear(v)$ can be computed incrementally during the depth-first search of G that generates T. This is given in Algorithm 7.2 below. Note that Algorithm 7.2 is essentially the well-known linear time sequential algorithm for graph biconnectivity [23].

ALGORITHM 7.2
Sequential Ear Decomposition Algorithm
Input: A connected graph $G = (V, E)$ with a root $r \in V$, and with $|V| = n$.
Output: A depth-first search tree of G, together with a label on each edge in E, indicating its ear number.

set T **of edges**; **integer** *count*;

Procedure *dfs*(**vertex** v);

{* This is a recursive procedure. The call *dfs*(v) of the main program constructs a depth-first search tree T of G rooted at r; the recursive call *dfs*(w) constructs the subtree of T rooted at w. The depth-first search tree is constructed by placing the tree edges in the set T and labeling the vertices in the subtree rooted at vertex v in preorder numbering, starting with *count*. The procedure assigns ear labels to the edges of G while constructing the depth-first search tree. An edge that does not belong to any ear is given the label (∞, ∞). Initially, all vertices are unmarked. *}

    vertex w;
    'mark' v;
    $preorder(v) := count;\ count := count + 1$;
    $low(v) := n;\ ear(v) := (n, n)$;
    for each vertex w adjacent to $v \to$
    {* This **for** loop performs a depth-first search of each child of v in turn and assigns ear labels to the tree and nontree edges incident on vertices in the subtrees rooted at the children of v. *}
        if w is not marked $\to$
            add (v, w) to T; $parent(w) := v$; $dfs(w)$;
            if $low(w) \geq preorder(w) \to ear(parent(w), w) := (\infty, \infty)$
      0. $|low(w) < preorder(w) \to ear(parent(w), w) := ear(w)$
        fi;
      1. $low(v) := \min(low(v), low(w))$;
      2. $ear(v) := lexmin(ear(v), ear(w))$
    $|w$ is marked $\to$
        if $w \neq parent(v) \to$
      3. $low(v) := \min(low(v), preorder(w))$;
      4. $ear(w, v) := (preorder(w), preorder(v))$
      5. $ear(v) := lexmin(ear(v), ear(w, v))$;
        fi
    fi
    rof
end *dfs*;

{∗ Main program. ∗}

$T := \phi$; $count := 0$; $dfs(r)$;

sort the ear labels of the edges in lexicographically nondecreasing order and relabel distinct labels (except label (∞, ∞)) in order as 1, 2, ...;

relabel the nontree edge with label 1 as 0

end.

LEMMA 7.4
Tree edge $(parent(v), v)$ is a cutedge if and only if $low(v) \geq v$. If $low(v) < v$ for all $v \neq r$ then Algorithm 7.2 constructs an ear decomposition with each tree edge $(parent(v), v)$ contained in ear $P_{ear(v)}$.

PROOF
By the computation in steps 1 and 3 in Algorithm 7.2, $low(v)$ is the lowest numbered vertex w such that (x, w) is a nontree edge with x a descendant of v. Since nontree edges in a depth-first search tree go from a vertex to its ancestor, $low(v)$ is also the lowest numbered vertex in a fundamental cycle of a nontree edge incident on a descendant of v. If $low(v) \geq v$ then every nontree edge (y, z) incident on a descendant y of v has $z \geq v$. Hence tree edge $(parent(v), v)$ does not belong to any fundamental cycle and is a cutedge. Conversely, if $low(v) < v$ then there exists a nontree edge $f = (x, low(v))$ with x a descendant of v. Hence tree edge $(parent(v), v)$ lies on the fundamental cycle of f and is not a cutedge.

Each nontree edge $(w, v), w < v$, is labeled (w, v) in step 4. We have $lca(w, v) = w$ since nontree edges in a depth-first search go from a vertex v to an ancestor $w < v$. Hence the labels for the nontree edges are distinct and in nondecreasing order of their lca as required in step 2 of Algorithm 7.1.

By the computation in steps 2 and 5 in Algorithm 7.2, $ear(v)$ is set to be the lexicographic minimum among all nontree edges (u, w), with $u < w$ such that w is a descendant of v. In step 0 this label is assigned to tree edge $(parent(v), v)$. This is exactly the computation of step 3 of Algorithm 7.1 for assigning ear labels to tree edges. Hence by Lemma 7.3, Algorithm 7.2 constructs an ear decomposition for the input graph when it is 2-edge connected. ∎

While Algorithm 7.2 is an ear decomposition algorithm, it also gives an open ear decomposition in case G is biconnected. We establish this in the next lemma.

LEMMA 7.5

Algorithm 7.2 constructs an open ear decomposition if all of the following three conditions hold:

   a) *The root r has exactly one child c;*
   b) *$low(c) = r$;*
   c) *For all vertices v other than r and c, $low(v) < parent(v)$.*

Further, G is biconnected if and only if a), b) and c) hold.

PROOF

We first prove that conditions a) through c) imply that Algorithm 7.2 constructs an open ear decomposition. We prove this by establishing that the ear containing each tree edge is open. This suffices to establish this part of the lemma since any ear that contains no tree edge consists of a single nontree edge, and such an ear is guaranteed to be open.

Consider tree edge $t = (parent(i), i)$. Let $low(i) = w$ and $ear(i) = q$.

Case 1: $q = 1$. Then t is contained in ear P_1 which is an open ear.

Case 2: $q > 1$. The ear containing edge t consists of the nontree edge with label q, call it (w, v), followed by part of the tree path from v to w (this was shown in the proof of Lemma 7.3). Let the part of the tree path from v to w that is contained in ear P_q extend from v to u, where u is a descendant of w and a proper ancestor of i (see Figure 7.1). In order to show that ear P_q is open, it suffices to show that $u \neq w$.

Let (w, x) be the first tree edge on the path from w to i (Figure 7.1). If w is not the root, then $low(x) < w$ (by condition c) and hence $ear(x) < q$. Thus, edge (x, w) is not contained in ear P_q. Hence $u \geq x$, and since $x > w$, we are done. If w is the root then since $q > 1$, edge (w, x), which is equal to edge $(0, 1)$, has label 1, which is less than q. Hence $u \geq x$, and since $x > w$, we have $u > w$.

Hence the ear containing edge $(i, parent(i))$ is open. This concludes the proof of the statement that each tree edge is contained in an open ear. To complete the proof of the lemma we show that, if any one of conditions a) through c) is not satisfied, then G is not biconnected.

If condition a) is not satisfied, let c and d be two children of r with $c < d$. Then every path between c and d passes through r and hence r is a cutpoint and G is not biconnected.

7.2. Ear Decomposition and Two-Connectivity

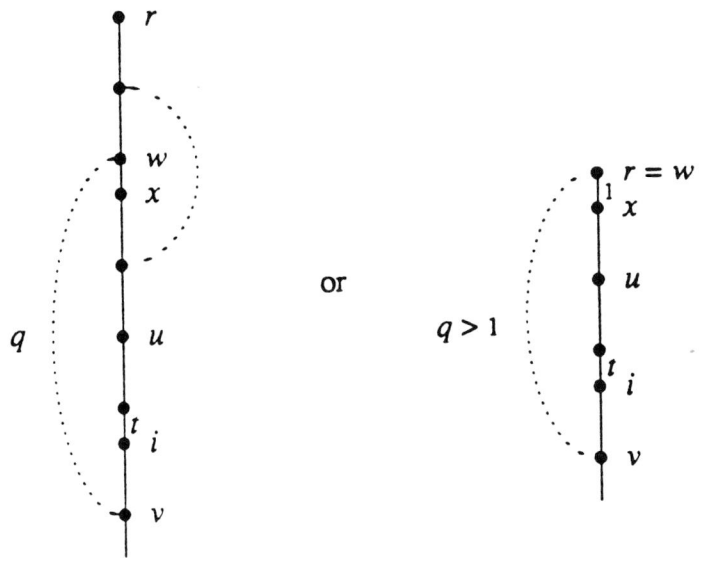

FIGURE 7.1
Illustrating case 2 in the proof of Lemma 7.5

If condition b) is not satisfied, then edge (r,c) is a cutedge and c is a cutpoint of G.

If condition c) does not hold, let v be a vertex for which it does not hold. The vertex v is neither the root nor the child of the root. If $low(v) > parent(v)$ then edge $(parent(v), v)$ is a cutedge (by proof of Lemma 7.4) and hence G is not biconnected. If $low(v) = parent(v) = w$ then any path between v and $parent(w)$ must pass through w. Hence w is a cutpoint of G. ∎

COROLLARY TO LEMMA 7.5
Algorithm 7.2 constructs an open ear decomposition for a biconnected graph.

Lemma 7.5 does not hold if we use Algorithm 7.1 in place of Algorithm 7.2. Figure 7.2 gives two different ear decompositions that are obtained using Algorithm 7.1 on a given input graph with the same spanning tree but with two different edge orderings. Of these, one is an open ear decomposition while the other is not.

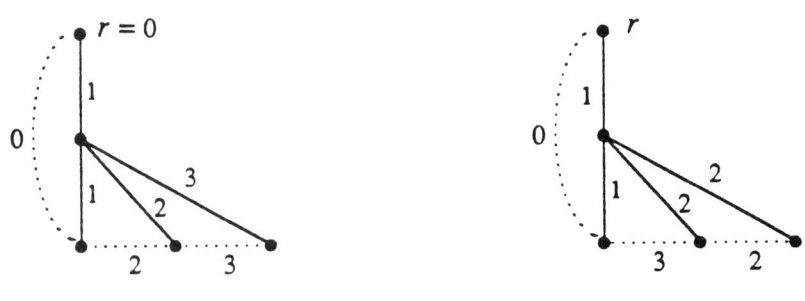

FIGURE 7.2
Examples of ear decompositions constructed by Algorithm 7.1

7.2.5 An Efficient Parallel Algorithm for Open Ear Decomposition

In the last section we noticed that Algorithm 7.1, when implemented using a depth-first search tree as the spanning tree for the input graph, serves as an algorithm to find an open ear decomposition of a biconnected graph; but if an arbitrary spanning tree is used, Algorithm 7.1 may no longer construct an open ear decomposition of a biconnected graph. Since no efficient parallel algorithm is known for finding a depth-first search tree in an undirected graph, we need to use a general spanning tree in an efficient parallel implementation of Algorithm 7.1.

Intuitively, the reason why a depth-first search tree is effective in finding an open ear decomposition is that all nontree edges go from a descendant to an ancestor. As a result the fundamental cycle of any nontree edge e contains only one base vertex v. Note that $lca(e) = parent(v)$. If the graph is biconnected, then there must be a path between v and some proper ancestor w (if it exists) of $lca(e)$ that avoids $lca(e)$. But this requires that edge $(parent(v), v)$ be contained in an ear that is incident on a proper ancestor of $lca(e)$.

When an arbitrary spanning tree is used in place of a depth-first search tree, the above property no longer need hold, and it is this that prevents Algorithm 7.1 from constructing an open ear decomposition for a biconnected

7.2. Ear Decomposition and Two-Connectivity

graph. In order to address this, we will modify step 2 of Algorithm 7.1 to introduce some ordering among nontree edges with the same lca. The modified version of step 2 is given below.

Step 2′

{* Assign ear numbers for an open ear decomposition to nontree edges in T. *}

pfor each vertex $v \in V - \{r\} \rightarrow$ compute $low(v)$ and 'mark' v if $low(v) < parent(v)$ **rofp**;

a. construct an auxiliary multigraph $H = (V', E')$ with $V' = V - \{r\}$ and for each nontree edge e in G place an edge in E' between the base vertices of its fundamental cycle;

{* In case e has only one base vertex u we place a self-loop at u. *}

pfor each connected component C of $H \rightarrow$

b. let a be any vertex in C and let b be the parent of a in T; $label(C) :=$ preorder number of b in T;

c. find a spanning tree S for C, root it at a 'marked' vertex if one exists, and number the vertices of S in preorder as $0, \ldots, k$;

d. label each tree edge $(parent(y), y)$ in S by the ordered pair $(label(C), y)$;

e. label the nontree edges in S (including multiple copies and self-loops) as $(label(C), k+1)$;

rofp;

pfor each nontree edge n in $G \rightarrow label(n) :=$ label of the edge in H that was placed in H by n **rofp**;

sort the labels of the nontree edges in G in lexicographically non-decreasing order and relabel them in order as $1, 2, \ldots$

end 2′;

LEMMA 7.6
Algorithm 7.1 with step 2 replaced by step 2′ constructs an ear decomposition if G is two-edge connected.

290 Chapter 7. Parallel Open Ear Decomposition

PROOF

Let C be any connected component in H. The value of $label(C)$ computed in step b is the lca of the fundamental cycle of every nontree edge that places an edge in C in step a. Hence the labels assigned to nontree edges of G by step $2'$ continue to be nondecreasing in the lca of their fundamental cycle and hence by Lemma 7.3 the modified algorithm constructs an ear decomposition for G. ∎

LEMMA 7.7

Let C be a connected component in H.

a) If $label(C) \neq 0$ and C contains no marked vertex then G is not biconnected;
b) If $label(C) = 0$ and there is another connected component C' with $label(C') = 0$ then G is not biconnected.

PROOF

The proof is similar to the proof of the converse of Lemma 7.5 and is left as an exercise. ∎

THEOREM 7.1

Algorithm 7.1 with step 2 replaced by step $2'$ constructs an open ear decomposition of G if G is biconnected.

PROOF

By Lemma 7.6, P_1 is an open ear.

Let n be the nontree edge of T with label $i, i > 1$. Then by Lemma 7.3 we know that the edges in G with label i form a simple path p that is part of the fundamental cycle c of n. We will show that $p \neq c$ thereby establishing that P_i is an open ear.

Let $lca(n) = l$ and let a and b be the base vertices of the fundamental cycle of n. (Let $b = a$ if there is only one base vertex.) Then a and b belong to the same connected component C in H. Let $a \leq b$ in the numbering of step c. We will show that edge (l, a) must belong to an ear numbered lower than i.

Consider $ear(l, a)$. If a is a 'marked' vertex then edge (l, a) belongs to the fundamental cycle of a nontree edge whose lca is less than l and hence $ear(l, a) < i$. If a is not 'marked' then if a has a parent p in S, the spanning tree of C, then consider the nontree edge n' in G that introduced edge (a, p) in C. By the labeling scheme in steps d and e we have $label(n') < label(n)$. Further the fundamental cycle of n' contains the edge (a, l). Hence $ear(a, l) \leq label(n') < i$.

Finally if a is neither 'marked' nor has a parent in S (i.e., a is the root of S) then $C = 0$ by Lemma 7.7 and hence $ear(a, l) = 1 < i$. ∎

Step $2'$ requires the computation of the *low* value for the vertices, the computation of connected components, spanning trees, preorder numbering, and sorting. All of these computations can be performed in logarithmic time using a linear number of processors using well-known algorithms. Hence the over-all open ear decomposition algorithm (i.e., Algorithm 7.1 with step 2 replaced by step $2'$) has the same processor-time bounds.

7.3 Graph Triconnectivity

In this section we describe an algorithm for testing three-vertex connectivity (or triconnectivity) of a biconnected graph using an open ear decomposition of the graph. We then extend this algorithm to one that decomposes the biconnected graph into certain pieces called triconnected components. This material is from Miller & Ramachandran [16].

We start by presenting several definitions in Section 7.3.1. Since our algorithm is fairly complex, we give a high-level description of the approach in Section 7.3.2. In Section 7.3.3 we give the details of the triconnectivity algorithm and prove its correctness. In Section 7.3.4 we extend this algorithm to finding triconnected components.

In this section we only establish the correctness of the algorithm to test triconnectivity and find triconnected components using open ear decomposition. In Section 7.4 we describe implementations of the various steps in the algorithm that run in logarithmic time with a linear number of processors. At the end of the chapter we provide some pointers towards achieving optimal performance of the algorithm in logarithmic parallel time.

7.3.1 Further Graph-theoretic Definitions

We first need to add to the graph-theoretic definitions given in Section 7.2.

Let G be a biconnected graph with an open ear decomposition $D = [P_0, \ldots, P_{r-1}]$. Two ears are *parallel to each other* if they have the same endpoints; an ear P_i is a *parallel ear* if there exists another ear P_j such that P_i and P_j are parallel to each other.

An *st-numbering* of a graph G is a numbering of the n vertices of G from $s = 1$ to $t = n$, such that every vertex v (other than s and t) has adjacent vertices u, w with $u < v < w$. An *st-graph* is a directed acyclic graph $G = (V, E)$ with $(s, t) \in E$ such that every vertex in V lies on a path from s to t.

Let $P = \langle v_0, \ldots, v_{k-1} \rangle$ be a simple path. The path $P(v_i, v_j)$, $0 \le i, j \le k - 1$ is the simple path connecting v_i and v_j in P, i.e., the path $\langle v_i, v_{i+1}, \ldots, v_j \rangle$, if $i \le j$ or the path $\langle v_j, v_{j+1}, \ldots, v_i \rangle$, if $j < i$. Analogously, $P[v_i, v_j]$ consists of the path (segments) obtained when the edges and internal vertices of $P(v_i, v_j)$ are deleted from P.

Given a noncyclic path $P = \langle v_0, \ldots, v_k \rangle$, the *innard of P* is the path $\langle v_1, \ldots, v_{k-1} \rangle$, i.e., the path obtained from P by deleting the first and last vertices.

Let $G = (V, E)$ be a biconnected graph, and let Q be a subgraph of G. We define the *bridges of Q in G* as follows: Let V' be the vertices in $G - Q$, and consider the partition of V' into classes such that two vertices are in the same class if and only if there is a path connecting them which does not use any vertex of Q. Each such class K defines a *nontrivial bridge* $B = (V_B, E_B)$ of Q, where B is the subgraph of G with $V_B = K \cup$ {vertices of Q that are connected by an edge to a vertex in K}, and E_B containing the edges of G incident on a vertex in K. The vertices of Q which are connected by an edge to a vertex in K are called the *attachments* of B on Q; the connecting edges are called the *attachment edges*. An edge (u, v) in $G - Q$, with both u and v in Q, is a *trivial bridge* of Q, with attachments u and v and attachment edge (u, v). The nontrivial and trivial bridges of Q together form the *bridges* of Q. The operation of *removing a bridge B of Q from G* is the removal from G of all edges and all nonattachment vertices of B.

Let $G = (V, E)$ be a graph and let $V' \subseteq V$ with the subgraph of G induced on V' being connected. The operation of *collapsing the vertices in V'* consists of replacing all vertices in V' by a single new vertex v, deleting all edges in G whose two endpoints are in V' and replacing each edge (x, y) with x in V' and y in $V - V'$ by an edge (v, y). In general this results in a multigraph even though G is not a multigraph.

Let $G = (V, E)$ be a biconnected graph, and let Q be a subgraph of G. The *bridge graph of Q*, $S = (V_S, E_S)$ is obtained from G by collapsing the nonattachment vertices in each nontrivial bridge of Q and by replacing each trivial bridge $b = (u, v)$ of Q by the two edges (x_b, u) and (x_b, v) where x_b is a new vertex introduced to represent the trivial bridge b. Note that in general the bridge graph is a multigraph.

Let $G = (V, E)$ be a biconnected graph with an open ear decomposition $D = [P_0, \ldots, P_{r-1}]$. We will denote the bridge graph of ear P_i by C_i. The *anchor bridges of P_i* are the bridges of P_i in G that contain nonattachment vertices belonging to ears numbered lower than i. For any two vertices x, y on P_i, we denote by $V_i(x, y)$, the internal vertices of $P_i(x, y)$, i.e., the vertices in $P_i(x, y) - \{x, y\}$; we denote by $V_i[x, y]$, the vertices in $P_i[x, y] - \{x, y\}$ together with the nonattachment vertices in the anchor bridges of P_i. Figure 7.3 illustrates some of our definitions relating to bridges.

A *star* is a connected graph in which exactly one vertex has degree greater than 1. The unique vertex of a star that has degree greater than 1 is called its *center*.

Let P be a simple noncyclic path in a graph G. If each bridge of P in G contains exactly one vertex not on P, then we call G the *star graph* $G(P)$. Each bridge of $G(P)$ is a star and is called a *star of* $G(P)$. Note that, in a connected graph G, the bridge graph X of any simple noncyclic path in G is a star graph $X(P)$. For example, in Figure 7.3, the bridge graph X of P_2 is a star graph $X(P_2)$. We will sometimes refer to a star graph $G(P)$ by G if the path P is clear from the context.

Let $G(P)$ be a star graph and let $P = \langle 0, 1, \ldots, k \rangle$. Given a star S of $G(P)$ with attachments $v_0 < v_1 < \ldots < v_r$ on P, we will call v_0 and v_r the *end attachments* of S and the remaining attachments the *internal attachments* of S; the vertex v_0 is the *leftmost attachment* of S, and the vertex v_r is its *rightmost attachment*; attachments v_i and v_{i+1} are *consecutive*, for $i = 0, \ldots, r-1$.

Two stars in a star graph $G(P)$ *interlace* if one of the following two hold:

1. There exist four distinct vertices a, b, c, d in increasing order on P such that a and c are attachments of one of the stars and b and d are attachments of the other star; or
2. There are three distinct vertices on P that belong to both stars.

The operation of *coalescing* two stars S_j and S_k is the process of forming a single new star S_l from S_j and S_k by combining the centers of S_j and S_k, and deleting S_j and S_k. Given a star graph $G(P)$, a *coalesced graph* G_c of G is the graph obtained from G by repeatedly coalescing a pair of interlacing stars in the current star graph until no pair of stars interlace; a *partially coalesced graph* of G is any graph obtained from G by performing this repeated coalescing at least once.

A *planar embedding* of a graph G is a mapping of each vertex of G to a distinct point on the plane and each edge of G to a curve connecting its endpoints such that no two edges intersect. A *face* of a planar embedding is a maximal region of the plane that is bounded by edges of the planar embedding.

294   Chapter 7.   Parallel Open Ear Decomposition

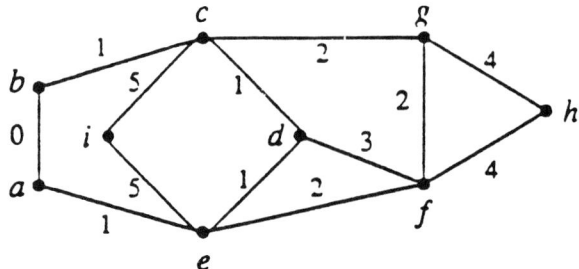

G with open ear decomposition $D = [P_0, P_1, P_2, P_3, P_4, P_5]$; $P_0 = \langle a, b \rangle$, $P_1 = \langle b, c, d, e, a \rangle$, $P_2 = \langle c, g, f, e \rangle$, $P_3 = \langle d, f \rangle$, $P_4 = \langle g, h, f \rangle$, $P_5 = \langle c, i, e \rangle$.

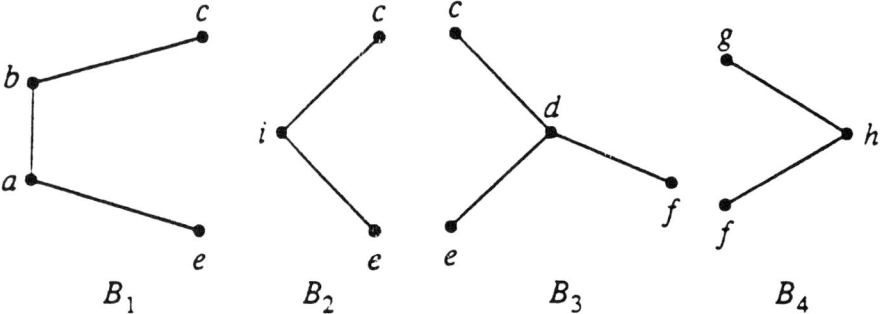

Bridges of P_2.

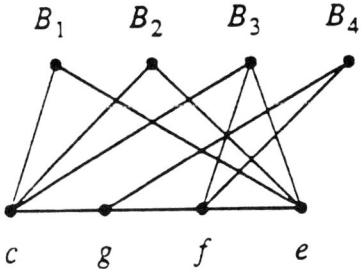

Bridge graph X of P_2. Anchor bridges are B_1 and B_3

FIGURE 7.3
Illustrating the definitions

The *outer face* of a planar embedding is the face with unbounded area. An *inner face* of a planar embedding is a face with finite area.

Let $G(P)$ be a star graph in which no pair of stars interlace. If $G(P)$ contains no star that has attachments to the endpoints x and y of P, then add a virtual star X to $G(P)$ with attachments to x and y. The *star embedding* $G^*(P)$ of $G(P)$ is the planar embedding of (the possibly augmented) $G(P)$ with P on the outer face. From some well-known results in planarity, it can be established that a star graph $G(P)$ has a planar embedding with P on the outer face if and only if no pair of stars interlace. We give some further definitions on planar combinatorial embeddings in the third part of Section 7.4.2.

Let $G(P)$ be a star graph with a star embedding $G^*(P)$. Let B and B' be two stars in $G(P)$. Then B is the *parent-star* of B' and B' is a *child-star* of B if there is a face in the star embedding $G^*(P)$ that contains the end attachment edges of B' as well as an attachment edge of B on either side of the end attachments of B'.

Let G be a biconnected graph with an open ear decomposition $D = [P_0, \ldots, P_{r-1}]$. Let $B_1, \ldots, B_l$ be the anchor bridges of ear P_i. The *ear graph of P_i*, denoted by $G_i(P_i)$, is the graph obtained from the bridge graph of P_i by

a) Coalescing all stars corresponding to anchor bridges;
b) Removing any multiple two-attachment bridges with the endpoints of the ear as attachments; and
c) Replacing all multiple edges by a single copy.

We will call the star obtained by coalescing all anchor bridges, the *anchoring star* of $G_i(P_i)$.

We conclude our list of definitions by defining the *triconnected components* of a biconnected multigraph (see, e.g., [26, 8]).

A pair of vertices a, b in a multigraph $G = (V, E)$ is a separating pair if and only if there are two nontrivial bridges, or at least three bridges, one of which is nontrivial, of $\{a, b\}$ in G. A biconnected graph with at least four vertices is triconnected if it has no separating pair. The pair a, b is a *nontrivial* separating pair if there are two nontrivial bridges of a, b in G. These definitions apply to a (simple) graph as well; in this case, all separating pairs are nontrivial. By Menger's theorem, a graph is triconnected if and only if it is biconnected and has at most 3 vertices or there are 3 vertex-disjoint paths between every pair of distinct vertices.

Let $\{a, b\}$ be a separating pair for a biconnected multigraph $G = (V, E)$. For any bridge X of $\{a, b\}$, let $\bar{X}$ be the induced subgraph of G on $(V - V(X)) \cup \{a, b\}$. Let B be a bridge of $\{a, b\}$ such that $|E(B)| \geq 2, |E(\bar{B})| \geq 2$ and either

B or $\bar{B}$ is biconnected. We can apply a *Tutte split* $s(a, b, i)$ to G by forming G_1 and G_2 from G, where G_1 is $B \cup \{(a, b, i)\}$ and G_2 is $\bar{B} \cup \{(a, b, i)\}$. Note that we consider G_1 and G_2 to be two separate graphs. Thus it should cause no confusion that there are two edges labeled (a, b, i) since one of these edges is in G_1 and the other is in G_2. The graphs G_1 and G_2 are called *split graphs of G with respect to a, b.* The *Tutte components* of G are obtained by successively applying a Tutte split to split graphs until no Tutte split is possible. Every Tutte component is one of three types: i) a triconnected simple graph; ii) a simple cycle (a *polygon*); or iii) a pair of vertices with at least three edges between them (a *bond*). The Tutte components of a biconnected multigraph G are the unique *triconnected components* of G.

7.3.2 Brief Overview of Results

In this section we give a high-level description of the results leading to our triconnectivity algorithm. Given a biconnected graph, our algorithm finds all separating pairs in the graph. The input graph is triconnected if and only if the algorithm finds no separating pair in the graph.

In Section 7.3.3 we show that if x, y is a separating pair in a biconnected graph G with an open ear decomposition D, then there exists an ear P_i in D that contains x and y as nonadjacent vertices, and further, every bridge of P_i has an empty intersection with either $V_i(x, y)$ or $V_i[x, y]$. This is the basic property that we use in our algorithm.

We further show that the above property is not altered by the operation of coalescing interlacing stars in the bridge graph $C_i(P_i)$ and thus applies to the ear graph of P_i as well as its coalesced graph. Finally we show that separating pairs satisfying the basic property with respect to P_i are simply those pairs of nonadjacent vertices on P_i that lie on a common face in the star embedding of this coalesced graph.

The above results lead to the following high-level algorithm for finding separating pairs in a biconnected graph G: Find an open ear decomposition D for G and for each nontrivial ear P_i in D, form the coalesced graph of its ear graph and extract separating pairs from its star embedding.

In Section 7.3.4 we build on the above results to give an efficient parallel algorithm to find the triconnected components of a graph. This algorithm finds the triconnected components using Tutte splits in contrast to the earlier algorithm based on depth-first search [8], which used certain other types of splits that required a clean-up phase at the end of the algorithm.

The definition of triconnected components given in Section 7.3.1 may appear contrived at first, but in reality it decomposes a biconnected graph into substructures that preserve the triconnected structure of G. In particular, questions relating to graph planarity and isomorphism between a pair of graphs can be mapped onto related questions regarding the triconnected components. Thus the problem of finding the triconnected components of a graph is an important one.

7.3.3 The Triconnectivity Algorithm

LEMMA 7.8

Let $D = [P_0, \ldots, P_{r-1}]$ be an open ear decomposition of a biconnected graph G and let x and y be the endpoints of ear P_i. Then every anchor bridge of P_i has attachments on x and y.

PROOF

Let B be an anchor bridge of P_i and let $H = \cup_{j=0}^{i-1} P_j$. By definition, the nonattachment vertices in B are the vertices in a connected component C of $G - \{P_i\}$ that contains a vertex in $H - \{x, y\}$.

The graph $(H - \{x, y\}) \cap P_i$ is empty since none of the internal vertices of P_i are contained in ears numbered lower than i. Hence C must contain all vertices in one or more connected component(s) of $H - \{x, y\}$. Let D be one such connected component contained in C. Since H has an open ear decomposition, it is biconnected by Lemma 7.2. Hence D contains vertices adjacent to x and y in H, since otherwise x or y would be a cutpoint of H. But this implies that C contains vertices adjacent to x and y in $G - \{P_i\}$, i.e., bridge B of P_i has attachments on x and y. ∎

LEMMA 7.9

Let $G = (V, E)$ be a biconnected undirected graph for which vertices x and y form a separating pair. Let $D = [P_0, \ldots, P_{r-1}]$ be an open ear decomposition for G. Then there exists a nontrivial ear P_i in D that contains x and y as nonadjacent vertices, such that every path from a vertex in $V_i(x, y)$ to a vertex in $V_i[x, y]$ in G passes through either x or y.

PROOF

Since x and y form a separating pair, the subgraph of G induced by $V - \{x, y\}$ contains at least two connected components. Let X_1 and X_2 be two such connected components.

Case 1. The ear P_1 contains no vertex in X_2 (see Figure 7.4):
Consider the lowest-numbered ear, P_i, that contains a vertex v in X_2. Since the endpoints of P_i are distinct and must be contained in ears numbered lower than i, P_i must contain x and y. Further, all vertices

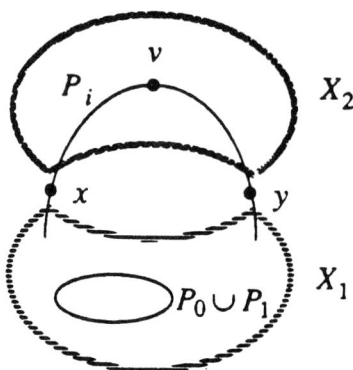

FIGURE 7.4
Case 1 in the proof of Lemma 7.9

in $V_i(x, y)$ lie in X_2, and none of the vertices in $V_i[x, y]$ lie in X_2. Hence every path from a vertex in $V_i(x, y)$ to a vertex in $V_i[x, y]$ in G passes through either x or y. Further, x and y are not adjacent on P_i since v lies between x and y.

Case 2. P_1 contains a vertex in X_2:
If P_1 contains no vertex in X_1, then case 1 applies to X_1. Otherwise P_1 contains at least one vertex from X_1, and one vertex from X_2. But then, since $P_0 \cup P_1$ is a simple cycle, and P_1 contains both vertices in P_0, we have the result that P_1 must contain x and y. Hence, by the argument of Case 1, every path from a vertex in $V_1(x, y)$ to a vertex in $V_1[x, y]$ must contain either x or y, and x and y are not adjacent on P_1. ∎

We will say that a separating pair x, y *separates* ear P_i if x and y are non-adjacent vertices on P_i, and the vertices in $V_i(x, y)$ are disconnected from the vertices in $V_i[x, y]$ in the subgraph of G induced by $V - \{x, y\}$. By Lemma 7.9, every separating pair in G separates some nontrivial ear. (Note that a separating pair may separate more than one nontrivial ear; for instance, in the graph G in Figure 7.2, the pair c, e is a pair separating ears P_1 and P_5—note that c, e does not separate P_2.)

LEMMA 7.10
Let $G = (V, E)$ be a biconnected graph with an open ear decomposition $D = [P_0, \ldots, P_{r-1}]$. Let ear P_i contain x and y as nonadjacent vertices.

Then x, y separates P_i if and only if every bridge of P_i has an empty intersection with either $V_i(x, y)$ or $V_i[x, y]$.

PROOF

Let every bridge of P_i have an empty intersection with either $V_i(x, y)$ or $V_i[x, y]$ and suppose x, y does not separate ear P_i. Hence, there exists a path $P = \langle a, w_1, \ldots, w_l, b \rangle$ in G, with a in $V_i(x, y)$ and b in $V_i[x, y]$, that avoids both x and y. This implies that there is a subpath P' of P with $P' = \langle w_r, \ldots, w_s \rangle$ such that w_r is in $V_i(x, y)$, w_s is in $V_i[x, y]$, and none of the intermediate w_k lie on P_i. Hence there is a bridge B of P_i containing w_r and w_s, i.e., B has a nonempty intersection with both $V_i(x, y)$ and $V_i[x, y]$, which is not possible by assumption. Hence x, y must separate ear P_i.

Conversely, suppose B is a bridge of P_i containing a vertex a in $V_i(x, y)$ and a vertex b in $V_i[x, y]$. Then we have a path from a vertex in $V_i(x, y)$ to a vertex in $V_i[x, y]$ that avoids both x and y. Hence x, y does not separate P_i. ∎

COROLLARY TO LEMMA 7.10

Let x and y be the endpoints of a nontrivial ear P_i in an open ear decomposition D of a graph G. Then x, y separates P_i if and only if no anchor bridge of P_i has an attachment in $V_i(x, y)$.

PROOF

Let x, y separate P_i. By Lemma 7.10, every bridge of P_i has an empty intersection with either $V_i(x, y)$ or $V_i[x, y]$. Since any anchor bridge of P_i has a nonempty intersection with $V_i[x, y]$, every anchor bridge must have an empty intersection with $V_i(x, y)$. Hence no anchor bridge can have an attachment in $V_i(x, y)$.

Conversely, suppose no anchor bridge of P_i has an attachment in $V_i(x, y)$. Then every anchor bridge has an empty intersection with $V_i(x, y)$. Since x and y are endpoints of P_i, every nonanchor bridge has an empty intersection with $V_i[x, y]$. Hence by Lemma 7.10, x, y separates P_i. ∎

We will call a pair of vertices x, y on an ear P_i a *candidate pair for P_i* if x, y is a pair separating P_i or (x, y) is an edge in P_i or x and y are endpoints of P_i. Clearly, if we can determine the set of candidate pairs for P_i, we can extract from it the pairs separating P_i by deleting pairs that are endpoints of an edge in P_i, and checking if the endpoints of P_i form a pair separating P_i using the criterion in the above Corollary.

More generally, let $G(P)$ be a star graph. A pair of nonadjacent vertices x, y on P will be called a *pair separating* P if the vertices in $P(x, y) - \{x, y\}$ are separated from the vertices in $P[x, y] - \{x, y\}$ when x and y are deleted from G. A pair of vertices x, y on P will be called a *candidate pair* for P in G if x, y is a pair separating P, or x and y are endpoints of P, or (x, y) is an edge in P.

The proof of the following claim is similar to the proof of Lemma 7.10 and is omitted.

CLAIM 7.1

Let $G(P)$ be a star graph. A pair x, y separates P in $G(P)$ if and only if every bridge of P in $G(P)$ has an empty intersection with either $P(x, y) - \{x, y\}$ or $P[x, y] - \{x, y\}$.

We now relate candidate pairs for P_i in G with candidate pairs for P_i in its bridge graph $C_i(P_i)$.

OBSERVATION 7.1

Let $G = (V, E)$ be a biconnected graph with an open ear decomposition $D = [P_0, \ldots, P_{r-1}]$. Then x, y is a candidate pair for P_i in G if and only if it is a candidate pair for P_i in the bridge graph $C_i(P_i)$.

PROOF

If (x, y) is an edge in P_i or if x and y are endpoints of P_i, then x, y is a candidate pair for P_i in both G and $C_i(P_i)$. So in the following we assume that x, y separates P_i and x and y are not both endpoints of P_i.

Let x, y separate P_i in G. By Lemma 7.10 every bridge of P_i in G has an empty intersection either with $V_i(x, y)$—and hence with $P_i(x, y) - \{x, y\}$—or with $V_i[x, y]$—and hence with $P_i[x, y] - \{x, y\}$. By construction this implies that every bridge of P_i in $C_i(P_i)$ has an empty intersection either with $P_i(x, y) - \{x, y\}$ or with $P_i[x, y] - \{x, y\}$. Hence by Claim 7.1, x, y separates P_i in $C_i(P_i)$.

Conversely, let x, y separate P_i in $C_i(P_i)$. By Claim 7.1, every bridge of P_i in $C_i(P_i)$ has an empty intersection either with $P_i(x, y) - \{x, y\}$ or with $P_i[x, y] - \{x, y\}$. Let $B_1, \ldots, B_k$ be the bridges of P_i in $C_i(P_i)$ corresponding to the anchor bridges of P_i in G. By Lemma 7.8, each B_j has attachments to the two endpoints e and f of P_i and by assumption either e or f is distinct from x and y. Assume without loss of generality that e is different from x and y. The vertex e is in $P_i[x, y] - \{x, y\}$ and each $B_j, j = 1, \ldots, k$ has an attachment on e. Hence each B_j has a nonempty intersection with $P_i[x,y] - \{x,y\}$ and therefore must have an empty intersection with $P_i(x,y) - \{x,y\}$.

The above implies that every anchor bridge of P_i in G has an empty intersection with $V_i(x,y)$ and every nonanchor bridge has an empty intersection either with $V_i(x,y)$ or with $V_i[x,y]$. Hence, by Lemma 7.10, x,y separates P_i in G. ∎

By Observation 7.1 we can work with the bridge graph of each ear in order to find the candidate pairs for that ear in G. We now develop results that will lead to an efficient algorithm to find candidate pairs in a star graph.

LEMMA 7.11

Let $G(P)$ be a star graph with stars $S_1, \ldots, S_k$. For $j = 1, \ldots, k$ let H_j be the subgraph of G consisting of $P \cup S_j$ and let H_j^* be the star embedding of H_j. Then a pair of vertices x, y on P is a candidate pair for P if and only if either x and y are the endpoints of P or x and y lie on a common inner face in each $H_j^*, j = 1, \ldots, k$.

PROOF

Let x, y be a candidate pair for P. If x and y are endpoints of P then the result follows immediately. If (x, y) is an edge on P then x and y must lie on a common inner face in each H_j^*. Otherwise, by Claim 7.1, each S_j has an empty intersection with either $P(x,y) - \{x,y\}$ or $P[x,y] - \{x,y\}$.

If S_j has an empty intersection with $P[x,y] - \{x,y\}$ then x and y belong to the unique inner face of H_j^* that contains the endpoints of P. If S_j has an empty intersection with $P(x,y) - \{x,y\}$, let $\langle a_1, \ldots, a_l \rangle$ be the attachments of S_j on P in the order that they are encountered on P from one endpoint of P to the other. The vertices x and y must lie between a_p and a_{p+1}, for some $1 \leq p < k$. Then x and y lie on the unique inner face of H_j^* containing a_p and a_{p+1}.

If x, y is not a candidate pair for P, then by Claim 7.1 there exists a star S_j with consecutive attachments a, b, with a in $P[x,y] - \{x,y\}$ and b in $P(x,y) - \{x,y\}$. Then, one of x and y, say x, lies in $P(a,b) - \{a,b\}$ and the other, y, lies in $P[a,b] - \{a,b\}$. Then x lies on the unique inner face containing a and b in H_j^* and y does not lies on this face. ∎

COROLLARY TO LEMMA 7.11

If G^* is the star embedding of $G(P)$, then a pair of vertices x, y on P is a candidate pair for P if and only if either x and y are the endpoints of P or x and y lie on a common inner face in G^*.

In general, this corollary may not apply, because $G(P)$ need not be planar. We now introduce the star coalescing property: namely, we establish that if we enforce the planarity required in the corollary by forming a coalesced graph G_c of $G(P)$ then the corollary applies to G_c.

The coalesced graph $G_c(P)$ of a star graph $G(P)$ is unique (exercise 7.3). Hence in the following we refer to G_c as 'the' coalesced graph of G (rather than 'any' coalesced graph of G).

THEOREM 7.2
Let $G(P)$ be a star graph and let $G_1(P)$ be obtained from $G(P)$ by coalescing a pair of interlacing stars S and T. Then a pair x,y on P is a candidate pair for $G(P)$ if and only if it is a candidate pair for $G_1(P)$.

PROOF
Let R be the star in $G_1(P)$ formed by coalescing S and T.

If (x,y) is an edge on P or if x and y are endpoints of P then x,y is a candidate pair for both $G(P)$ and $G_1(P)$.

Let x,y separate P in $G(P)$. Hence S and T have an empty intersection with either $P(x,y) - \{x,y\}$ or $P[x,y] - \{x,y\}$. Since S and T interlace, either both have empty intersection with $P(x,y) - \{x,y\}$ or both have empty intersection with $P[x,y] - \{x,y\}$. Hence R, which contains the union of the attachments of S and T, must have an empty intersection with either $P(x,y) - \{x,y\}$ or with $P[x,y] - \{x,y\}$. Hence by Claim 7.1, x,y separates P in $G_1(P)$.

Conversely, suppose x,y separates P in $G_1(P)$ and let R have an empty intersection with $P(x,y) - \{x,y\}$ ($P[x,y] - \{x,y\}$). Then both S and T have an empty intersection with $P(x,y) - \{x,y\}$ ($P[x,y] - \{x,y\}$) and hence x,y separates P in $G(P)$ by Claim 7.1. ∎

COROLLARY TO THEOREM 7.2
Let $G(P)$ be a star graph.

  a) *Let $G'(P)$ be any partially coalesced graph of $G(P)$. Then x,y is a candidate pair for $G(P)$ if and only if it is a candidate pair for $G'(P)$.*

  b) *A pair x,y is a candidate pair for $G(P)$ if and only if it is a candidate pair for the coalesced graph $G_c(P)$.*

Let $G(P)$ be a star graph and let $G_c(P)$ be its coalesced graph. Since no pair of bridges of P interlace in $G_c(P)$, Lemma 7.11 and its Corollary apply to this graph. Let us refer to the set of vertices on P that lie on a common inner face in G_c^* listed in the order they appear on P as a *candidate list for P*. A pair of vertices is a candidate pair for P if and only if it lies in a candidate list for P. A candidate list S for ear P is a *nontrivial candidate list* if it contains a pair separating P.

Let G be a biconnected graph with an open ear decomposition $D = [P_0, \ldots, P_{r-1}]$. Since every separating pair for G is a candidate pair for some nontrivial ear P_i (Lemma 7.9), any algorithm that determines the candidate lists for all nontrivial ears is an algorithm that finds all separating pairs for a graph. By the results we have proved above, we can find all candidate lists in G by forming the bridge graph for each nontrivial ear, and then extracting the nontrivial candidate lists from the coalesced graph of the bridge graph.

In order to obtain an efficient implementation of this algorithm, we will not use the bridge graph of each ear, but instead the closely related ear graph which we defined in Section 7.3.1.

LEMMA 7.12
A pair of vertices x, y separates ear P_i in G if and only if it separates P_i in the ear graph $G_i(P_i)$.

PROOF
By Claim 7.1, x, y separates ear P_i in G if and only if it separates P_i in the bridge graph $C_i(P_i)$.

Now consider the ear graph $G_i(P_i)$. The ear graph $G_i(P_i)$ is obtained from the bridge graph $C_i(P_i)$ by coalescing all anchor bridges, deleting multiple two-attachment bridges with the endpoints of the ear as attachments, and deleting all multiple edges by a single copy.

Deleting a star with attachments only to the endpoints of an ear can neither create nor destroy candidate pairs. Let $C_i'(P_i) = C_i(P_i) - \{2\text{-attachment bridges with endpoints of } P_i \text{ as attachments}\}$.

By Lemma 7.8, every anchor bridge of P_i has the two endpoints of P_i as attachments, and hence every pair of anchor bridges with an internal attachment on P_i must interlace. Hence $G_i(P_i)$ is the graph derived from $C_i'(P_i)$ by coalescing some interlacing stars. The lemma now follows from the Corollary to Theorem 7.2. ∎

LEMMA 7.13
Let $G = (V, E)$ be a biconnected graph with an open ear decomposition $D = [P_0, \ldots, P_{r-1}]$, and let $|V| = n$ and $|E| = m$. Then the total size of the ear graphs of all nontrivial ears in D is $O(m)$.

PROOF
Each ear graph consists of a nontrivial ear P_i together with a collection of stars on P_i. The size of all of the P_i is $O(m)$. So we only need to bound the size of all of the stars in all of the ear graphs.

Consider an edge (u, v) in G. This edge appears as an internal attachment edge in at most two ear graphs: once for the ear $P_{ear(u)}$ and once for ear $P_{ear(v)}$. Thus the number of internal attachment edges in all of the stars is no more than $2m$.

We now bound the number of attachment edges to endpoints of ears. Since we delete all stars with only the endpoints of an ear as attachments, every star in an ear graph $G_i(P_i)$ with an attachment to an endpoint of P_i also has an internal attachment in P_i. A star can contain at most two attachments to endpoints of an ear. Hence for each star that contains attachments to endpoints of its ear, we charge these attachments to an internal attachment. Since the number of internal attachment edges is no more than $2m$, the number of attachment edges to endpoints of ears is no more than $4m$. Hence the total size of all of the ear graphs is $O(m)$. ∎

The above results establish the validity of the following algorithm to find the nontrivial candidate lists in a biconnected graph.

ALGORITHM 7.3
Finding the Nontrivial Candidate Lists
Input: A biconnected graph $G = (V, E)$.
Output: The candidate lists for G.

>   **integer** j; **vertex** u, v;
>   1. find an open ear decomposition $D = [P_0, \ldots, P_{r-1}]$ for G;
>   **pfor** each nontrivial ear $P_j \rightarrow$
>   2. construct the ear graph $G_j(P_j)$;
>   3. coalesce all interlacing stars on $G_j(P_j)$ to form the coalesced graph G_{j_c};
>   4. construct the star embedding of $G_{j_c}^*$ of G_{j_c}, and identify each list of vertices on P_j on a common inner face in this embedding as a candidate list;
>   let u and v be the endpoints of P_j;
>   **if** $[u, v]$ is a candidate list for P_j and the anchoring star of P_j has an internal attachment on $P_j \rightarrow$ delete candidate list $[u, v]$
>   **fi**;

delete any candidate list for P_j that contains only the two endpoints of an edge in P_j

rofp

end.

In Section 7.2.3 we described a logarithmic time parallel algorithm with a linear number of processors on a CRCW PRAM for step 1 of Algorithm 7.1. In Section 7.4, we give algorithms with similar processor-time bounds to perform steps 2, 3 and 4 in parallel for all nontrivial ears. Clearly the remaining steps in the **pfor** loop are trivial to implement. Hence Algorithm 7.3 can be made to run in logarithmic time with a linear number of processors. However, before proceeding to an efficient implementation of Algorithm 7.3, we show in Section 7.3.4 how to obtain the triconnected components of a biconnected graph, given the nontrivial candidate lists.

7.3.4 Finding Triconnected Components

In this section we define a special type of split, called the *ear split* in a biconnected graph with an open ear decomposition. This split has the desirable property that the original open ear decomposition decomposes in a natural way into two open ear decompositions, one for each split graph. This also leads to a natural algorithm for finding triconnected components based on applying certain types of ear splits successively.

We also consider some issues that arise in a parallel implementation of the above algorithm. The obvious approach would be to perform all of the ear splits in parallel. However, this leads to complications when a vertex is shared by several Tutte pairs. We analyze some of the properties of ear splits in this section and we present a method for performing all of the relevant ear splits on a single ear. This method runs in logarithmic time with a number of processors linear in the size of the bridge graph of the ear. In Section 7.4.3 we apply this method to the 'local replacement graph' which is defined in Section 7.4.1 to obtain a logarithmic time algorithm using a linear number of processors to find the triconnected components of the input graph.

We start by defining a special type of split, called an *ear split*, on a biconnected graph G with an open ear decomposition $D = [P_0, \ldots, P_{r-1}]$. Let a, b be a pair separating ear P_i. Let $B_0, \ldots, B_k$ be the bridges of P_i with an attachment in $V_i(a, b)$, and let $T_i(a, b) = (\cup_{j=0}^{k} B_j) \cup P_i(a, b)$. It is easy to see that $T_i(a, b)$ is a bridge of a, b. Then the *ear split* $e(a, b, i)$ consists of forming the *upper split graph* $G_1 = T_i(a, b) \cup \{(a, b, i)\}$ and the *lower split* graph $G_2 = \bar{T}_i(a,b) \cup \{(a,b,i)\}$. Note that the ear split $e(a,b,i)$ is a Tutte split if one of $G_1 - \{(a,b,i)\}$ or $G_2 - \{(a,b,i)\}$ is biconnected.

Let S be a nontrivial candidate list for ear P_i. A pair u,v in S is an *adjacent separating pair for P_i* if S contains no vertex in $V_i(u,v)$. The pair u,v is a *nonvacuous adjacent separating pair for P_i* if u,v is an adjacent separating pair and there is a bridge of P_i with an attachment on $V_i(u,v)$. A pair a,b in S is an *extremal separating pair for P_i* if $|S| \geq 3$ and S contains no vertex in $V_i[a,b]$. We will refer to a nonvacuous adjacent or extremal separating pair as a *Tutte pair*.

We now prove the following theorem.

THEOREM 7.3
Let $G = (V,E)$ be a biconnected graph with an open ear decomposition $D = [P_0, \ldots, P_{r-1}]$. Let a,b be an adjacent (extremal) separating pair for P_i in G, and let G_1 and G_2 be, respectively, the upper and lower split graphs obtained by the ear split $e(a,b,i)$. Then,

a) $G_1 - \{(a,b,i)\}(G_2 - \{(a,b,i)\})$ *is biconnected.*
b) *The ear decomposition D_1 induced by D on G_1 by replacing P_i by the simple cycle formed by $P_i(a,b)$ followed by the newly added edge (b,a,i) is a valid open ear decomposition for G_1; likewise, the ear decomposition D_2 induced by D on G_2 by replacing $P_i(a,b)$ by the newly added edge (a,b,i) is a valid open ear decomposition for G_2.*
c) *Let c,d be a pair separating some $P_j, 0 \leq j \leq r-1$ in G. If $\{c,d\} \neq \{a,b\}$ or $i \neq j$ then c and d lie in one of G_1 or G_2, and c,d is a separating pair for P_j in the split graph in which $P_j, c,$ and d lie.*
d) *Every separating pair in G_1 or in G_2 is a separating pair in G.*

PROOF
a) Let a,b be an adjacent separating pair for P_i. If $G_1 - \{(a,b,i)\}$ is not biconnected then let c be a cutpoint in the graph. The vertex c cannot lie on $P_i(a,b)$ since this would imply that it is part of the candidate list for which a,b is an adjacent separating pair. But c cannot lie on a bridge of $P_i(a,b)$ since then c would be a cutpoint of G and this would imply that G is not biconnected.

Similarly $G_2 - \{(a,b,i)\}$ is biconnected if a,b is an extremal separating pair.

b) We establish by induction on ear number j, for $j \geq i$, that the graph $P_{0,j} = \cup_{k=0}^{j} P_k$ satisfies the property in part b) of the Theorem. The details are straightforward and are omitted.

c, d) If $i \neq j$ let P_j lie in G_k (where $k = 1$ or 2). We note that the ear graph of P_j in G_k is the same as the ear graph of P_j in G. Hence c, d is a pair separating P_j in G if and only if it is a pair separating P_j in G_k.

If $i = j$ we note that in G_1 the bridges of P_i are precisely those bridges of P_i in G that have attachments on an internal vertex of $P_i(a, b)$. Hence if c and d lie on $P_i(a, b)$ then c, d separates P_i in G if and only if it separates $P_i(a, b)$ in G_1. An analogous argument holds for G_2 in the case when c and d lie on $P_i[a, b]$. ∎

We now present the algorithm for finding triconnected components.

ALGORITHM 7.4
Finding Triconnected Components
Input: A biconnected graph $G = (V, E)$ with an open ear decomposition $D = [P_0, \ldots, P_{r-1}]$, and the nontrivial candidate lists for each ear.
Output: The triconnected components of G.

vertex u, v; **integer** i;

pfor each nontrivial candidate list S in each nontrivial ear $P_i \to$

   pfor each nonvacuous adjacent separating pair u, v in $S \to$

   form the upper split graph G_1 for the ear split $e(u, v, i)$ and replace G by the lower split graph G_2 for the ear split $e(u, v, i)$; replace D by the open ear decomposition D_2 for the lower split graph G_2 and form the open ear decomposition D_1 for the upper split graph G_1 as in part b) of Theorem 7.3

   rofp;

   if $|S| > 2 \to$

   form the upper split graph G_1 and replace G by the lower split graph G_2 for the extremal separating pair u, v in S;

   form the open ear decompositions D_1 and D_2 as in Theorem 7.3 and replace D by D_2. (if $i = 1$ and u and v are endpoints of ear P_1 then perform this ear split only if there are at least two edges between u and v)

   fi

rofp;

split off multiple edges in the remaining split graphs to form the bonds

end.

LEMMA 7.14
Algorithm 7.4 generates the Tutte components of G.

PROOF
By Theorem 7.3, each split performed in Algorithm 7.4 is a Tutte split, and at termination there is no separating pair in any of the generated graphs. ∎

For an efficient parallel implementation of Algorithm 7.4 we need a good method to perform all of the Tutte splits in the algorithm in parallel. This is quite simple if all of the Tutte pairs are disjoint. However, for the general case when the Tutte pairs are not necessarily disjoint, we need to specify a method to process the splits in parallel without causing conflicts between different splits that share a vertex in their Tutte pairs. In the rest of this section we develop a method to perform in parallel all of the splits on Tutte pairs in a single ear. This method is not necessarily efficient. However, it will be used in a general algorithm described in Section 7.4.3 that performs the splits corresponding to Tutte pairs in all ears in logarithmic time with a linear number of processors.

We start by associating a triconnected component with each ear split corresponding to a Tutte pair. Let $e(a,b,i)$ be such a split. Then by definition $T_i(a,b) \cup \{(a,b,i)\}$ is the upper split graph associated with the ear split $e(a,b,i)$. The *triconnected component of the ear split* $e(a,b,i)$, denoted by $TC(a,b,i)$, is $T_i(a,b) \cup \{(a,b,i)\}$ with the following modifications: Call a pair c,d separating an ear P_j in $T_i(a,b)$ a *maximal pair for* $T_i(a,b)$ if there is no e,f in $T_i(a,b)$ such that e,f separates some ear P_k in $T_i(a,b)$ and c,d is in $T_k(e,f)$. In $T_i(a,b) \cup \{(a,b,i)\}$ replace $T_j(c,d)$ together with all two-attachment bridges with attachments at c and d, for each maximal pair c,d of $T_i(a,b)$, by the edge (c,d,j) to obtain $TC(a,b,i)$. We denote by $TC(0,0,0)$, the unique triconnected component that contains P_0.

LEMMA 7.15
$TC(a,b,i)$ is a triconnected component of G.

PROOF
Each split of $T_i(a,b)$ in the above definition is a valid Tutte split, and the final resulting graph contains no unprocessed separating pair. Hence $TC(a,b,i)$ is a valid triconnected component of G. ∎

LEMMA 7.16
Every triconnected component of G is $TC(a,b,i)$ for some unique triplet (a,b,i).

PROOF
Straightforward. ∎

We note that if a,b is an extremal pair separating P_i then $TC(a,b,i)$ is a polygon and if a,b is a nonvacuous adjacent pair separating P_i then $TC(a,b,i)$ is a simple triconnected graph.

Let $G = (V, E)$ be a biconnected graph with an open ear decomposition $D = [P_0, P_1, \ldots, P_{r-1}]$. Let $C_i(P_i)$ be the bridge graph of P_i and let $D_i(P_i)$ be the coalesced graph of C_i. Note that D_i is closely related to $G_{i_c}(P_i)$, the coalesced graph of the ear graph of P_i in G, but is not exactly the same since D_i retains multiple attachment edges as well as multiple two-attachment bridges. (Note also that the sum of the sizes of the D_i over all nontrivial ears could be superlinear in the size of G.)

The proofs of the following two lemmas are left as exercises.

LEMMA 7.17
Algorithm 7.3 with G_{i_c} replaced by D_i will output the nontrivial candidate lists of G.

LEMMA 7.18
Let a,b be a nonvacuous adjacent separating pair for P_i in G and let (x,y) be an edge, not in P_i, which is incident on a vertex y on P_i. Then

a) *The edge (x,y) is in $T_i(a,b)$ if and only if it is in a star of D_i with an attachment on an internal vertex in $P_i(a,b)$;*

b) *D_i contains at most one star B with attachments on a, b, and an internal vertex in $P_i(a,b)$, and if edge (x,y) is in $TC(a,b,i)$ then it lies in B.*

We now give a lemma about two-attachment bridges.

LEMMA 7.19
Let B be a two-attachment bridge of P_i in D_i with attachments a and b. Then

a) If the span $[a,b]$ is degenerate (i.e., (a,b) is an edge in P_i) or if there is a bridge B' of P_i with attachments on a and b and at least one other vertex, then the graph $D_i - B$ defines the same set of polygons and simple triconnected components $TC(x,y,i)$, for i fixed, as $D_i(P_i)$.

b) If part a) does not hold then $\{a,b\}$ is an extremal pair separating P_i as well as an adjacent pair separating P_i.

PROOF

Let P_j be the lowest-numbered ear in B. Then $j > i$ and a and b are endpoints of P_j. Hence the ear split $e(a, b, j)$ separates B from P_i, and thus B is not part of $TC(x, y, i)$ for any pair $\{x, y\}$ separating P_i. So a two-attachment bridge of P_i in D_i is never part of a triconnected component associated with a pair separating P_i, though it may define some adjacent and extremal separating pairs as in case b) of the lemma.

We now prove parts a) and b) of the lemma.

Part a): Suppose span $[a, b]$ is degenerate. Then the triconnected component associated with split $e(a, b, i)$ is the single edge (a, b), which is a bond. Otherwise, if there is a bridge B' with attachments on a, b and at least one other vertex v, then the triconnected component associated with split $e(a, b, i)$ contains a portion of P_i between a and b, together with B' if v is in the interval (a, b) and is a polygon if v is not in $[a, b]$. Both of these situations can be inferred without the presence of B. Note that it is not possible for B' to have an attachment v in the interval (a, b) and another attachment w that is not in $[a, b]$, since the bridge B would interlace with B' in such a case.

Part b): Let the span $[a, b]$ be non-degenerate and let the portion of P_i between a and b be $\langle a = a_1, \ldots, a_n = b \rangle$. Since there is no k-attachment bridge, $k > 2$, with span $[a, b]$, there must exist an a_i, $1 < i < k$ such that a, a_i, and b are in the same candidate list C, and no vertex outside $[a, b]$ is in C. Hence $\{a, b\}$ is an extremal separating pair. Also, since there is no bridge with attachments on a, b and some other vertex c outside $[a, b]$, there must be some vertex c on P_i such that either $c < a < b$ or $a < b < c$, and a, b, and c are in the same candidate list C'. Further, no vertex in the interval (a, b) can belong to C'. Hence $\{a, b\}$ is an adjacent pair in the candidate list C'. ∎

Let us consider the case of a graph in which any pair of ear splits $e(a, b, i), e(c, d, j)$ with $i \neq j$ are disjoint. In this case we can perform the ear splits in Algorithm 7.4 corresponding to different ears in parallel. To process separating pairs on a single ear P_i we run the following algorithm.

ALGORITHM 7.5
Performing Ear Splits on a Single Ear
Input: A biconnected graph G together with $D_i(P_i)$, the coalesced graph of the bridge graph of a nontrivial ear P_i in an open ear decomposition of G, with $P_i = \langle 0, 1, \ldots, k \rangle$.

Output: The split graphs of G after all Tutte splits on P_i have been performed.

vertex j, u, v, w, x, y; {∗ These vertices may be subscripted. ∗}

delete redundant two-attachment bridges;

pfor each attachment vertex v of each star B in $D_i \to$ make a copy v_B of v **rofp**;

pfor each internal vertex v on $P_i \to$

    if there is no star with an internal attachment on $v \to$ make an additional copy v_P of v to represent the lower split graph formed when all adjacent separating pairs containing v have been processed **fi**;

rofp;

pfor $j = 0$ **to** $k - 1 \to$

    if there is no bridge with its leftmost attachment on $j \to$ replace edge $(j, j+1)$ on P_i by an edge incident on j_C, where C is B if there is a bridge B with an internal attachment on j and is P otherwise **fi**

rofp;

pfor $j = 1$ **to** $k \to$

    if there is no bridge with its rightmost attachment on $j \to$ replace edge $(j-1, j)$ on P_i by an edge incident on j_D, where D is B' if there is a bridge B' with an internal attachment on j and is P otherwise **fi**

rofp;

{∗ Process nonvacuous adjacent separating pairs. ∗}

pfor each star B in $D_i \to$

    let the end attachments of B on P_i be v and w, $v < w$;

    replace all edges in B incident on v by edges incident on v_B;

    replace all edges in B incident on w by edges incident on w_B;

    if B has no child-star B' with an attachment at $v \to$ replace edge $(v, v+1)$ on P by an edge incident on v_B **fi**;

    if B has no child-star B' with an attachment at $w \to$ replace edge $(w-1, w)$ by an edge incident on w_B **fi**;

    place a virtual edge (v_B, w_B, i) and another virtual edge (v_C, w_D, i), where C (resp. D) is the parent-star of B if the parent star of B has an attachment at v (resp. w) and is P otherwise;

replace each internal attachment edge of B on a vertex u in P_i by an edge incident on u_B

rofp;

{* Process extremal pairs. *}

pfor each star B in $D_i \to$

let the attachments of B on P_i be $v_0 < v_1 < \ldots < v_l$;

pfor each j in $\{0, \ldots, l-1\}$ for which (v_{j_B}, v_{j+1_B}) is not an edge in the current component containing $B \to$

for convenience of notation let x denote v_j and let y denote v_{j+1};

make a copy x_{B_r} of x and a copy y_{B_l} of y;

replace the edge on P_i connecting x_B to the next larger vertex in the current graph by an edge incident on x_{B_r};

replace the edge on P_i connecting y_B to the next smaller vertex in the current graph by an edge incident on y_{B_l};

place a virtual edge (x_B, y_B, i) and another virtual edge (x_{B_r}, y_{B_l}, i)

rofp

rofp

end.

Algorithm 7.5 is an implementation of Algorithm 7.4 on ear P_i using the results of Lemmas 7.17, 7.18 and 7.19. We leave the proof of correctness of the algorithm to the reader. We also leave it to the reader to verify that all steps in the algorithm can be performed in logarithmic time with a linear number of processors in the size of D_i.

There are two problems with using this approach in an efficient logarithmic time algorithm for forming the triconnected components of a graph. One is that we are working with the D_i and the total size of these graphs need not be linear in the size of G. The second is that this approach will not work if a vertex a appears in an ear split for two different ears. For instance, two-attachment bridges corresponding to nonvacuous adjacent separating pairs will be separated on two different ears and this would cause processor conflicts. In Section 7.4.3 we show how to overcome these two problems to obtain logarithmic time parallel algorithm using a linear number of processors for finding the triconnected components of a general biconnected graph.

7.4 Efficient Implementation of Triconnectivity Algorithm

This section deals with a logarithmic time, linear processor implementation of Algorithms 7.3 and 7.4.

Section 7.4.1 gives such an algorithm for constructing the ear graphs of the nontrivial ears in an open ear decomposition (step 2 of Algorithm 7.3). Section 7.4.2 gives an algorithm with these bounds for constructing the coalesced graph of a star graph, and for extracting the candidate lists from its star embedding (steps 3 and 4 of Algorithm 7.3). In Section 7.4.3 we show that the results in Sections 7.4.1 and 7.4.2 lead to a simple implementation of Algorithm 7.4 that runs in logarithmic time with a linear number of processors.

The algorithm in Section 7.4.1 for constructing the ear graphs is fairly intricate. A considerably simpler algorithm for this problem is given in Miller & Ramachandran [16] (Exercise 7.4). However, although the algorithm in [16] is efficient, it needs $\log^2 n$ parallel time.

7.4.1 Forming the Ear Graphs

In this section we develop a parallel algorithm to find the ear graph of each nontrivial ear. This algorithm is based on material from Fussell, Ramachandran & Thurimella [5], though the development here is somewhat different.

We begin by describing in the next section a simple linear processor, logarithmic time algorithm to find the bridge graph of each path in a collection of vertex-disjoint paths in a given graph. The set of nontrivial ears does not form a collection of vertex-disjoint paths since the endpoints of an ear are contained in other ears. Hence we cannot apply the algorithm in the next section to obtain the bridge graphs or ear graphs of nontrivial ears. However, in the following sections, we present a collection of results that allow us to transform the input graph G, together with an open ear decomposition $D = [P_0, \ldots, P_{r-1}]$, into a modified graph G_l, together with a collection of edge-disjoint paths $[P'_0, \ldots, P'_{r-1}]$ with the useful property that the innard of each P'_i is P_i and the ear graph of each nontrivial ear P_i in D can be derived from the bridge graph of P_i in G_l. This property allows us to use the simple technique of the next section on the innards of the P'_i, since these paths are vertex-disjoint. In the following section the local replacement technique is presented.

Bridges of Disjoint Collection of Paths

In this section we present an algorithm for constructing the bridge graph of each path in a collection of vertex-disjoint paths in a graph.

ALGORITHM 7.6
Forming the Bridge Graph of Each Path in a Collection of Vertex-Disjoint Paths
Input: Graph $G = (V, E)$, together with a collection of vertex-disjoint paths $\{Q_0, \ldots, Q_{k-1}\}$.
Output: The bridge graph of each $Q_i, i = 0, \ldots, k - 1$.

    integer i; **vertex** a, b, v; {* v will be subscripted by an integer. *}

    pfor each $i \to$ collapse the vertices in Q_i into a vertex v_i **rofp**;

    let the resulting graph be G^-;

    pfor each $i \to$

        pfor each block β of G^- with cutpoint $v_i \to$ form a nontrivial bridge B of Q_i with the edges of G^- in β that are incident on v_i as attachment edges **rofp**;

        pfor each edge (a, b) in $G - \{Q_i\}$ with a and b in $Q_i \to$ form a bridge of Q_i with attachments a and b **rofp**

    rofp

end.

It is straightforward to see that this algorithm correctly constructs the bridge graph of each of the Q_i, and that it runs in logarithmic time with a linear number of processors.

In the following sections we will use Algorithm 7.6 to find the ear graphs of the nontrivial ears in an open ear decomposition of a biconnected graph. We start by relating open ear decomposition to an *st*-graph in the next section.

The st-graph

Let $G = (V, E)$ be a biconnected graph with an open ear decomposition $D = [P_0, \ldots, P_{r-1}]$ with $P_0 = (s, t)$. Since G is biconnected, it has an *st* numbering (Exercise 7.2).

LEMMA 7.20
Let G be a biconnected graph with an open ear decomposition $D = [P_0, \ldots, P_{r-1}]$, where $P_0 = (s, t)$. Then it is possible to direct each ear in

7.4. Efficient Implementation of Triconnectivity Algorithm

D from one endpoint to the other such that the resulting directed graph G_d is an st graph.

PROOF

We prove the lemma by establishing, by induction on i, that the graph $P_{0,i} = \cup_{j=0}^{i} P_j$ satisfies the statement of the lemma.

BASE: $i = 0$. Direct (s,t) from s to t.

INDUCTION STEP: Assume that the result is true until $i-1$ and consider i.

Let D_{i-1} be the directed graph obtained from $P_{0,i-1}$ by directing its ears according to the statement of the lemma. Assume that the vertices in $P_{0,i-1}$ are numbered according to an st numbering consistent with D_{i-1}.

Let u and v be the endpoints of ear P_i and assume without loss of generality that u is numbered lower than v in the st numbering for $P_{0,i-1}$. Direct P_i from u to v.

We claim that $D_{i-1} \cup \{P_i$ directed from u to $v\}$ satisfies the statement of the lemma. This follows from the following construction. Number the internal vertices of P_i in order from u as $v, v+1, \ldots, v+k-1$, where k is the number of internal vertices of P_i. Replace the number of each vertex w in $P_{0,i-1}$ with $w \geq v$ by $w+k$. The resulting numbering is a valid st numbering for $P_{0,i}$ and $D_{i-1} \cup \{P_i$ directed from u to $v\}$ is its st graph. ∎

Given an open ear decomposition $D = [P_0, \ldots, P_{r-1}]$, Maon, Schieber & Vishkin [14] give a parallel algorithm to direct each ear in D as in Lemma 7.20 such that the resulting directed graph is an st graph. Let G_{st} be this graph, which we will call the *st-graph of D*. The graph T_{st}, the *st-tree of D*, is the directed spanning tree obtained from G_{st} by deleting the last edge in each ear except P_0. We can similarly construct G_{ts} and its directed spanning tree T_{ts} by considering P_0 to be directed from t to s. We will refer to G_{ts} as the *reverse directed graph of G_{st}* and vice versa.

We now state two simple but useful properties of open ear decomposition and the trees T_{st} and T_{ts}.

Property 7.1 Let P_i and P_j be two ears in an open ear decomposition D of graph G with $i < j$. Then, all vertices and edges of P_j belong to a single bridge of P_i in G.

Property 7.2 Let $p = \langle u_0, \ldots, u_i \rangle$ be a directed path in T_{st} or T_{ts}. Then the ear numbers of the vertices in p are nondecreasing when going from u_0 to u_i.

The Local Replacement Graph

In this section we describe a transformation of a biconnected graph G with an open ear decomposition $D = [P_0, \ldots, P_{r-1}]$ into a new graph G_l, called the *local replacement graph* of $(G; D)$. In the graph G_l, each ear P_i in G is converted into a path P'_i with the innard of P'_i being P_i and with the bridge graph of P_i in G_l corresponding to the ear graph of P_i in G.

Consider any vertex v in G. Let the degree of v be d $(d \geq 2)$. Of the d edges incident on v, two belong to $P_{ear(v)}$. Each of the remaining $d-2$ edges incident on v is an end edge of some ear P_j, with $j > ear(v)$. In the local replacement graph G_l we will replace v by a rooted tree with $d-1$ vertices, with one vertex for each ear containing v. The root of this tree will be the copy of v for the ear containing v. The actual form of the tree is computed from T_{st} and T_{ts} as in the algorithm below. The tree representing vertex v will be called the *local tree of v* and will be denoted by T_v.

ALGORITHM 7.7
Constructing the Local Replacement Graph
Input:
A biconnected graph $G = (V, E)$;
an open ear decomposition $D = [P_0, \ldots, P_{r-1}]$ for G, with $P_0 = (s, t)$;
the st-graph G_{st} with its spanning tree T_{st} and the ts-graph G_{ts} with its spanning tree T_{ts}.
Output: The local replacement graph G_l of $(G; D)$.

    integer i, j; {∗ These integers range in value from 0 to $r-1$. ∗}

    vertex a, q, u, v, w; {∗ q, u, v and w may be subscripted by an integer. ∗}

    edge a, e, f, n; {∗ e and f will be subscripted by an integer. ∗}

    rename each vertex v in G by v_j, where $ear(v) = j$;

    {∗ We will refer to the vertex $v_{ear(v)}$ interchangeably as either v or $v_{ear(v)}$. ∗}

1. **pfor** each outgoing ear P_i at each vertex v in G_{st} →

        let the edge in P_i incident on v be e_i and let the nontree edge in P_i be f_i;

        detach edge e_i from v and label the detached endpoint as v_i;

        let a be a base edge of the fundamental cycle created by f_i in T_{st} with $ear(a) \neq i$;

7.4. Efficient Implementation of Triconnectivity Algorithm 317

   if $ear(a) \leq ear(v) \rightarrow v_{ear(v)} := parent(v_i)$
   $|ear(a) > ear(v) \rightarrow v_{ear(a)} := parent(v_i)$ **fi** ;
   direct this edge from $parent(v_i)$ to v_i
rofp;
let the undirected version of the graph obtained in step 1 be G^1, the directed version be G^1_{st} and its associated spanning tree be T^1_{st} and the reverse directed graph be G^1_{ts} and its associated spanning tree be T^1_{ts};

2. repeat step 1 using G^1_{ts} and T^1_{ts} and let the resulting undirected graph be G^2, the resulting directed graph be G^2_{ts} and its associated spanning tree be T^2_{ts}, and the reverse directed graph be G^2_{st} and its associated spanning tree be T^2_{st};

{∗ In the following we process parallel ears by constructing a new graph H. ∗}

pfor each parallel ear $P_i \rightarrow$ **create** a vertex q_i **rofp** ;
pfor each nontree edge n in $T^2_{st} \rightarrow$
   if the base edges of the fundamental cycle of n belong to ears P_i and P_j, where P_i and P_j are parallel to each other $\rightarrow$ **create** an edge between q_i and q_j **fi**
rofp;
call the resulting graph H;
find a spanning tree in each connected component of H and root it at the vertex corresponding to the minimum numbered ear in the connected component;

3. **pfor** each vertex q_i in H that is not a root of a spanning tree $\rightarrow$
   let P_i be directed from endpoint u to endpoint w in G_{st}; let q_j be the parent of q_i in the spanning tree in H;
   replace the parent of u_i in T^2_{st} by u_j and the parent of w_i in T^2_{ts} by w_j
rofp;
denote the undirected version of the graph formed in step 3 by G_l, the directed graph from s to t by G'_{st} and its associated spanning tree by T'_{st} and the reverse directed graph by G'_{ts} and its associated spanning tree by T'_{ts}; call G_l the *local replacement graph* of G;
call the underlying undirected tree constructed in steps 1, 2 and 3 from each vertex v in G the *local tree* T_v; call $v_{ear(v)}$ the root of T_v, and consider T_v to be an out-tree rooted at $v_{ear(v)}$. Call the part of T_v constructed by assigning parents in T^2_{st} the *o-tree* OT_v of T_v and the part of T_v constructed by assigning parents in T^2_{ts} the *i-tree* IT_v of T_v;

{* In G^2_{st}, OT_v is an out-tree rooted at $v_{ear(v)}$ and IT_v is an in-tree rooted at $v_{ear(v)}$ and vice-versa in G^2_{ts}. *}

denote by P'_i the ear P_i, together with the edge connecting each endpoint of P_i to its parent in its local tree in G_l;

{* Note that the innard of P'_i (i.e., the path P'_i excluding its two end edges) is P_i. *}

denote the first vertex on P'_i when directed as in G'_{st} by $L(P'_i)$, *the left endpoint of* P'_i, and the last vertex on P'_i when directed as in G'_{st} by $R(P'_i)$, *the right endpoint of* P'_i.

end.

An example of the construction in Algorithm 7.7 is shown in Figure 7.5 We will prove the following:

1. All ears with endpoints as descendants of v_i in T_v must belong to the same bridge of P_i in G.

2. An ear P_j with v_j not a descendant of v_i in T_v must be part of an anchor bridge of P_i or of a bridge of P_i with attachments to only the endpoints of P_i in G.

We start with the following preliminary lemmas.

LEMMA 7.21
Let v_i be a proper ancestor of v_j in T_v, the local replacement tree of vertex v. Then either P_i and P_j are parallel to each other or $i < j$.

PROOF
Without loss of generality we assume that v_i and v_j belong to OT_v.

By the construction in Algorithm 7.7, either v_i is a proper ancestor of v_j in T^1_{st} or v_i and v_j are unrelated in T^1_{st} and v_i becomes a proper ancestor of v_j in step 3. In the latter case, v_i and v_j are parallel to each other and we are done. So for the rest of the proof we assume that v_i is a proper ancestor of v_j in T^1_{st}.

Let T^1_v be the out-tree for vertex v at the end of step 1 of Algorithm 7.7. The vertex set of T^1_v is $\{v_i|$ vertex v is contained in P_i in $G\}$. We claim that the subscripts of the vertices are strictly increasing in any directed path in T^1_v. To see this, let v_k be the parent of v_j in T^1_v. If $k = ear(v)$ then $k < j$ since one endpoint of P_j in G is v. If $k \neq ear(v)$ let w be the other endpoint of P_j. By the construction in step 1 of Algorithm 7.7, w

7.4. Efficient Implementation of Triconnectivity Algorithm

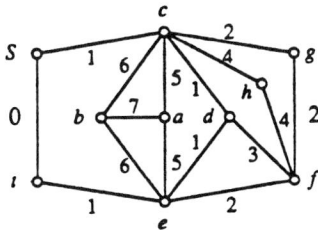

G with an open ear decomposition

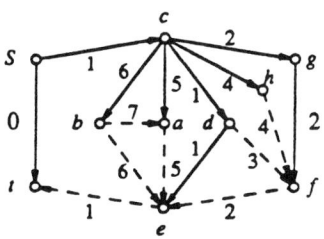

G_{st} and T_{st}

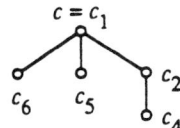

Out-tree at vertex c
after step 1 of
Algorithm 4.2

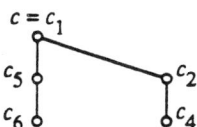

Out-tree at vertex c
after step 3 of Algorithm 4.2
(P_2, P_5 and P_6 are parallel ears)

FIGURE 7.5
Constructing G_l from G

is a proper descendant of v in T_{st}. Hence by Property 7.2, $k \leq ear(w)$ and since $ear(w) < j$ we have $k < j$.

Hence if v_i is a proper ancestor of v_j in T^1_{st} then $i < j$. ∎

DEFINITION
Let (v, w) be the first edge on P_i in G_{st}. Then $T_{st}(i)$ is the subtree of T_{st} rooted at w. Similarly if (x, y) is the first edge on P_i in G_{ts} then $T_{ts}(i)$ is the subtree of T_{ts} rooted at y.

LEMMA 7.22
Let v_i and v_j be vertices in T_v such that neither is a descendant of the other. Then in G, the following two properties hold.

   a) Either P_i and P_j are ears parallel to each other, or $P_i \cap P_j = \{v\}$;
   b) If $v_i \in OT_v$ then $P_j \cap T_{st}(i) = \{v\}$ and if $v_i \in IT_v$ then $P_j \cap T_{ts}(i) = \{v\}$.

PROOF

The proof of this lemma is straightforward and is left to the reader. ∎

LEMMA 7.23

Let v_i be a vertex in T_v and let $S_i = \{$ears P_j in $G \mid P_j$ contains v and v_j is not a proper descendant of v_i in $T_v\}$.

Let v_k be a child of v_i in T_v and let T_k be the subtree of T_v rooted at v_k. Then, all of the ears P_l in G such that v_l is in T_k belong to a single bridge of S_i in G.

PROOF

By induction on the height of T_k. We assume, without loss of generality that $v_i \in OT_v$.

BASE: Height of $T_k = 0$. Then T_k contains only one vertex and the claim is vacuously true since the corresponding ear P_k must belong to some single bridge of S_i (by Property 1 and Lemma 7.21 for those ears P_j in S_k with v_j an ancestor of v_i, and by Lemma 7.22, part a) for those P_j in S_k with v_j unrelated to v_i in T_v).

INDUCTION STEP: Assume that the lemma is true for height of T_k up to $h-1$ and let height of T_k be h. Let v_l be any child of v_k. Then T_l has height at most $h-1$ and hence by the induction hypothesis, all of the ears whose corresponding vertices lie in T_l belong to a single bridge of $S_i \cup \{P_k\}$ in G. Hence all of these ears belong to a single bridge B of S_i in G.

We now claim that bridge B contains ear P_k as well. The proof is a case analysis depending on whether v_k was made the parent of v_l in T_v in step 1 or in step 3 of Algorithm 7.7.

Case 1. v_k was made parent of v_l in step 1. Then P_k and P_l are not parallel to each other. Let (x,y) be the nontree edge (with respect to T_{st}) in ear P_l (Figure 7.6a). Then by construction, y is a descendant of w, where (v,w) is the first edge on P_k in G (since $v_k \neq v_{ear(v)}$). But by Property 1, Lemma 7.21 and Lemma 7.22, none of the vertices on the tree path from w to y can be contained in an ear in S_i. Hence all vertices and edges in ear P_k belong to bridge B of S_i in G.

Case 2. v_k was made parent of v_l in step 3. Then P_k and P_l are parallel to each other. Further since v_k was made parent of v_l in step 3, there is a nontree edge n (with respect to T_{st}^2) whose fundamental cycle C contains both v_k and v_l (Figure 7.6b). But none of the vertices in C other

7.4. Efficient Implementation of Triconnectivity Algorithm 321

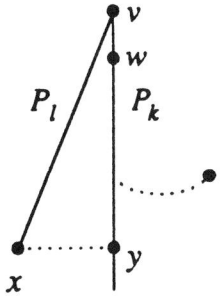
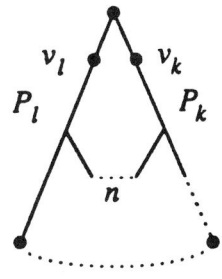

Case 1: Configuration of
P_l and P_k in G_{st}
(a)

Case 2: Configuration of
P_l and P_k in G_{st}^2
(b)

FIGURE 7.6
Illustrating the proof of Lemma 7.23

than the lca can belong to an ear in S_i by Lemma 7.22, part b, since all of these vertices are in either $T_{st}(k)$ or $T_{st}(l)$. Hence P_k is contained in bridge B of S_i in G.

This concludes the proof of the induction step and the lemma is proved. ∎

COROLLARY TO LEMMA 7.23
Let v_i be a vertex in T_v and let v_j be a child of v_i in T_v. Then all ears P_k in G with v_k in T_j belong to a single bridge of P_i in G.

PROOF
This follows immediately from Lemma 7.23 by observing that P_i is contained in S_i. ∎

LEMMA 7.24
Let v_i be a vertex in T_v and let v_j be another vertex in T_v which is not a descendant of v_i. Then in G, P_j either belongs to the anchoring star of P_i or belongs to a bridge of P_i that has attachments only to the endpoints of P_i.

PROOF
Without loss of generality we assume that $v_i \in OT_v$. If $v_j \in IT_v$ then the outgoing edge e of P_j in G_{st} cannot be a descendant of v (since in that

case G_{st} would contain a cycle). Further $P_i \cap P_j = \{v\}$ by Lemma 7.22. But then, there is a path from s to P_j in G that avoids ear P_i and hence P_j belongs to the anchoring star of P_i (since $ear(s) = 0$.)

For the rest of the proof we assume that $v_j \in OT_v$. Let $lca(v_i, v_j) = v_k$.

Case 1. $v_j = v_k$. If v_j is not parallel to v_i, then $i < j$ by Lemma 7.21, and hence P_j belongs to the anchoring star of P_i.
If P_j is parallel to P_i, let v_l be the root of the spanning tree of its connected component in H formed in step 2 of Algorithm 7.7. By construction, v_l must be an ancestor of v_j.

Case 1.1. If $v_l = v_j$ then $j < i$ (since the spanning tree is rooted at the vertex with minimum index) and hence P_j is part of the anchoring star of P_i.

Case 1.2. If v_l is a proper ancestor of v_j then consider a sequence of nontree edges that caused the edges on the path from v_j to v_l in T_v to be placed in H. None of the vertices in the fundamental cycle of any of these nontree edges in G lie on P_i. Hence in G, these nontree edges, together with appropriate tree edges, induce a path from a vertex in P_j to a vertex in P_l that avoids all vertices in P_i. Hence P_j is in the same bridge of P_i as P_l and hence belongs to the anchoring star of P_i (since l must be less than i).

Case 2. $v_j \neq v_k$.
In this case v_k is a proper ancestor of v_j. Let v_m be the child of v_k that is an ancestor of v_j. Then all ears with corresponding vertices in T_m lie on a single bridge of P_i (by Corollary to Lemma 7.23).
Let v_l be the nearest ancestor of v_i such that P_l is not parallel to P_i.

Case 2.1. v_l is a proper descendant of v_k.
In this case, P_m is not parallel to P_i, since otherwise, by step 3 of Algorithm 7.7, v_m would be a descendant of v_l. Also, by Lemma 7.22, $P_m \cap P_i = \{v\}$. Finally, the nontree edge in P_m completes a fundamental cycle in G, one of whose base edges belongs to some $P_q, q \leq k$. None of the vertices other than v in this fundamental cycle belongs to P_i, since by step 1 of Algorithm 7.7, the two base edges in the fundamental cycle of which P_i is part, belong to P_i and P_l. Hence, P_m (and thus P_j) belongs to the same bridge of P_i as P_q and is thus part of the anchoring star of P_i.

7.4. Efficient Implementation of Triconnectivity Algorithm

Case 2.2. $v_l = v_k$ (the nontrivial case).
Let y be the last vertex on P_i and let z be the child of v_k in T_{st}^2 that is an ancestor of y. By construction (step 1 of Algorithm 7.7), either $v_k = v_{ear(v)}$ or z lies in $T_{st}(k)$.

Case 2.2.1.   If v_m is not parallel to v_i then let (w, x) be the last edge in P_m in G. The vertex x is contained in P_q, for some $q \leq k$ and x is not contained in P_i (by Lemma 7.22, part a). If x lies on the path from v to y then P_m (and hence P_j) is part of the same bridge of P_i as P_k and hence is part of the anchoring star of P_i. Otherwise, x is not an ancestor of y and by the st-numbering property, there is a path from x to t (and hence to s) in G that avoids all vertices in P_i. Hence again we have the case that P_j is part of the anchoring star of P_i.

Now consider the case when P_m is parallel to P_i and assume that P_m (and hence P_j) is part of a bridge B of P_i with an internal attachment on P_i. We will show that B must be an anchor bridge of P_i.

Since B has an internal attachment on P_i, there is a path p in G_l from v_m to some vertex u that is internal to P_i that avoids all other vertices in P_i. The path p must contain at least one nontree edge whose lca is $\leq v_k$. Let n be the first such nontree edge encountered when traversing p from v_m to u.

Case 2.2.2. $lca(n) < v_k$ in T'_{st}. Then there is a path in G from a vertex in P_m to $lca(n)$ that avoids P_i and since $k < i$, $lca(n)$ belongs to an ear numbered less than i (by Property 1). Hence B is an anchor bridge of P_i.

Case 2.2.3. $lca(n) = v_k$ in T'_{st}. Let $e = (v_k, v_m)$ and $f = (v_k, v_n)$ be the base edges of the fundamental cycle of n in T_{st}^2. Then P_n cannot be parallel to P_m, since otherwise v_m and v_n would be in the same connected component of H and hence would be in a single subtree rooted at a child of v_k. But if P_n is not parallel to P_m, it is also not parallel to P_i, and we can use the analysis used with P_m in case 2.2.1 to deduce that P_n is part of an anchor bridge of P_i. Hence B is an anchor bridge of P_i.

Case 2.3. v_l is a proper ancestor of v_k.
The analysis of this case is similar to case 1.2 and we can deduce that P_j is part of the anchoring star of P_i.

This concludes the case analysis and the lemma is proved. ∎

THEOREM 7.4
Let G be a biconnected graph with an open ear decomposition $D = [P_0, \ldots, P_{r-1}]$ and let P_i be a nontrivial ear in D. Let B be a bridge of P_i in G_l and let a and b be any two edges of G that are in B. Then,

a) If B contains the endpoints of P_i' in G_l then a and b belong to the anchoring star of P_i in G.
b) If B does not contain the endpoints of P_i' in G_l then a and b are both part of a single nonanchor bridge of P_i in G.

Also, if c and d are two edges of G that do not belong to a single bridge of P_i in G_l then c and d belong to different bridges of P_i in G.

PROOF
The theorem follows from observing that any additional connectivity induced in G_l - $\{P_i\}$ that is not present in $G - \{P_i\}$ must occur at T_v, for some vertices $v \in P_i$, and by applying the Corollary to Lemma 7.23 and Lemma 7.24. ∎

COROLLARY TO THEOREM 7.4
Let G_i' be the bridge graph of P_i in G_l. Let G_i be obtained from G_i' by replacing all multiple edges in G_i' by a single copy. Then G_i is the ear graph of P_i.

By the above results, the following algorithm constructs the ear graph of each nontrivial ear in an open ear decomposition $D = [P_0, \ldots, P_{r-1}]$ of a biconnected graph G.

ALGORITHM 7.8
Constructing the Ear Graphs
Input: A biconnected graph $G = (V, E)$ together with an open ear decomposition $D = [P_0, P_1, \ldots, P_{r-1}]$.
Output: The ear graph of each nontrivial ear in D.

form the local replacement graph G_l of G, together with the associated paths $P_0', \ldots, P_{r-1}'$ using Algorithm 7.7;

apply Algorithm 7.6 to G_l with the nontrivial ears in D as the vertex-disjoint paths to obtain the bridge graph of each nontrivial P_i in G_l;

pfor each nontrivial ear $P_i \to$ obtain the ear graph G_i of P_i in G from the bridge graph of P_i in G_l by replacing all multiple edges by a single copy **rofp**

end.

7.4.2 Finding the Candidate Lists

In this section we describe an efficient algorithm to implement steps 3 and 4 in Algorithm 7.3. Given a star graph $G(P)$, steps 3 and 4 require us to find its coalesced graph $G_c(P)$ and extract the candidate lists from its star embedding.

We present the algorithm for forming the coalesced graph, which is based on material in Ramachandran & Vishkin [20], in two parts. In the next section we present an algorithm to find the coalesced graph when every star in $G(P)$ has exactly two attachments. Then in the following section we give an efficient reduction from a general star graph to this special case. The reduction presented here is somewhat different from the one in [20].

Our algorithm solves a more general problem than that of finding the coalesced graph of a star graph $G(P)$: It also provides an 'interlacing parity' (which is defined in the next section) for every pair of stars on P. This property is useful in determining planarity of $G(P)$ for the case when every star has to be embedded completely on one side of P. While this is not needed for the triconnectivity algorithm, it is an important step in the parallel algorithm for graph planarity given in Ramachandran & Reif [18].

In the last part of Section 7.4.2 we describe a simple efficient algorithm to form the star embedding of $G_c(P)$ and to extract from it the candidate lists of $G(P)$.

Determining Interlacings of Chords on a Path

Let $G(P)$ be a star graph in which each star has exactly two attachments on P. For simplicity we assume $P = \langle 0, \ldots, n \rangle$. Then G can be viewed as the simple path P together with a collection of chords (i, j) on P. We shall refer to such a graph as a *chord graph*.

Let $a = (u, v)$ and $b = (w, x)$ be two chords on P, where $u < v$ and $w < x$. If a and b interlace, then they cannot be placed on the same side of P in a planar embedding. If a and b do not interlace, then they can be placed in a planar embedding on the same (opposite) side of C if and only if there exists no sequence of chords $\langle a = a_0, a_1, \ldots, a_r = b \rangle$, with r odd (even) such that a_i interlaces with a_{i+1}, $0 \leq i \leq r - 1$. If there is such a sequence with r even then a and b have *even interlacing parity* and if there is such a sequence with r odd, then a and b have *odd interlacing parity*. If no such sequence exists for r either odd or even, then a and b have *null interlacing parity:* in this case a and b can be placed either in the same side or in opposite sides of P in a planar embedding. It is possible for a and b to have both odd and even parity—in this case, no planar embedding of G is possible if every chord is to be placed completely on one side of P.

We now present an efficient parallel algorithm for preprocessing the graph G so that an interlacing parity for any pair of chords can be determined in constant time. If a pair of chords a and b have both even and odd interlacing parities, then we will find only one of these parities. A high level description of the algorithm is as follows: We construct an auxiliary graph G_I called the *interlacing parity graph of* G with a vertex for each chord on P. We then place some edges in G_I. Each such edge connects a pair of vertices whose corresponding chords interlace. We do not put in an edge for every pair of interlacing chords, but only a subset of them, so that the size of G_I is linear in the size of G. In particular, if r is the number of chords on P then G_I will have at most $2r$ edges. Further, if a and b are two chords for which there exists some interlacing sequence then a and b will lie in the same connected component of G_I. Since each edge in G_I represents an actual interlacing, we can obtain an interlacing parity for each pair of chords with an interlacing sequence by finding a spanning tree in each connected component of G_I and two coloring the vertices of the spanning tree: Now, two chords, whose vertices are in the same connected component in G_I, have odd interlacing parity if they have different colors and even parity if they have the same color. Note that to form the coalesced graph of $G(P)$, we only need to find the connected components of G_I, and coalesce all chords that correspond to vertices in each connected component.

The algorithm is presented below. Since vertices in G_I correspond to chords on P, we will sometimes refer to a vertex in G_I as a chord; by this we mean the chord in G that this vertex represents.

ALGORITHM 7.9
Interlacing Parity for a Chord Graph
Input: Undirected graph G consisting of a simple path $P = \langle 0, \ldots, n \rangle$, together with a collection of chords on P.
Output: A label on each chord which allows an interlacing parity of any pair of chords to be determined in constant time by one processor.

    vertex u, v; **edge** c, l, r; {∗ u, v, l, r may be subscripted. ∗}

1. **pfor** each chord $c = (u, v)$, $u < v$, that interlaces with some other chord
$\rightarrow$

{* Left rule. *}

let u_l be the minimum numbered vertex on P such that c interlaces with a chord incident on u_l;

if $u_l < u$ → find the chord $l_c = (u_l, v_l)$ with maximum v_l that interlaces with c and place an edge (the *left edge*) in G_I between vertices c and l_c fi;

{* Right rule. *}

let v_r be the maximum numbered vertex on P such that c interlaces with a chord incident on v_r;

if $v_r > v$ → find the chord $r_c = (u_r, v_r)$ with minimum u_r that interlaces with c and place an edge (the *right edge*) in G_I between c and r_c fi

rofp;

find a spanning tree in each connected component of the interlacing parity graph G_I and two-color the spanning trees;

assign a label ⟨component number, color⟩ to each vertex;

pfor each chord in $G(P)$ → assign the label of the vertex in G_I corresponding to it **rofp**

end.

With this preprocessing we can determine an interlacing parity for any pair of chords c, d on P as follows: If component number of c is not equal to component number of d then c and d have null interlacing parity, otherwise they have even interlacing parity if they have the same color, and odd interlacing parity if they have different colors.

LEMMA 7.25
If a pair of chords α, β have an interlacing parity that is not null, then α and β appear in the same connected component of G_I.

PROOF
By induction on the number of chords r on P.

BASE: $r = 2$. This is immediate.

INDUCTION STEP: Assume that the claim is true for all simple paths with up to $r-1$ chords, and let G be a graph consisting of a simple path P, together with r chords. Let v be the lowest numbered vertex on P that has a chord incident on it, and let $a = (v, w)$ be the chord incident on v with maximum w. Delete a from G to form G'. By the induction hypothesis, every pair of chords that have an interlacing parity that is not null in G' appear in the same connected component in G'_I.

We will show two things:

A. Any pair of chords α, β that lie in the same connected component in G'_I continue to lie in the same connected component in G_I.

B. Any chord having an interlacing parity with a will be in the connected component of a in G_I.

Lemma 7.25 follows. To see this consider any pair of chords α, β, where $\alpha \neq a$ and $\beta \neq a$. Assume α and β have an interlacing parity in G and consider an interlacing sequence. If the interlacing sequence does not include a apply claim A. If the interlacing includes a, then we may assume that a appears only once. Apply claim B to show that a is in the same connected component as its predecessor and successor in the sequence. Finally, apply claim A.

We show A. For this, observe that the only edges in G'_I that are not present in G_I are the edges introduced by the left rule for chords that interlace with a: each such edge is replaced by an edge connecting the chord to vertex a. Let b be such a chord interlacing with a, let its left edge in G'_I be (b, c). Its left edge in G_I is (b, a). We claim that a, b and c lie in the same connected component in G_I.

Case 1. Chord c interlaces with chord a. Then edges (a, c) and (a, b) are present in G_1 and hence a, b and c belong to the same connected component in G_1.

Case 2. Chord c does not interlace with a. Consider the right edge (c, d) of c. Then chord d has its right endpoint at least as large as the right endpoint of b, and hence d interlaces with a. Hence edges $(a, d), (c, d)$ and (b, a) are present in G_I, i.e., a, b and c lie in the same connected component in G_I.

We show claim B. For each chord b having an interlacing parity with a, consider an interlacing sequence $a_0 = a, a_1, \ldots, a_k = b$. Chords a_1 and b are in the same connected component of G_I by claim A. It suffices to show that chords a and a_1, which actually interlace, are in the same connected component. But, there must be an edge connecting them in G_I by the left rule for a_1. Claim B and Lemma 7.25 follow. ∎

LEMMA 7.26
Algorithm 7.9 correctly finds an interlacing parity for each pair of chords on P.

PROOF
Since an edge (a, b) is placed in G_I only if chords a and b interlace, it follows that the algorithm finds a correct interlacing parity for every pair of chords that belong to the same connected component. By Lemma 7.25, vertices that belong to different connected components correspond to chords that have null interlacing parity. This establishes the correctness of the algorithm. ∎

To implement step 1 we determine at each vertex v, the chord with smallest attachment s_v and the chord with largest attachment l_v incident on v, and we store at v the ordered pairs $(s_v, -v)$ and (l_v, v). With this preprocessing it is a simple exercise to verify that all steps of Algorithm 7.9 can be performed in logarithmic time with a linear number of processors on a CRCW PRAM.

Determining Interlacings of Stars in a Star Graph

In this section we consider a general star graph $G(P)$. We replace each star in $G(P)$ with a collection of chords and construct the interlacing parity graph for this new graph as in the previous section. We then add in some additional vertices and edges to this graph and we establish that the resulting graph can be used to obtain the interlacing parity of each pair of stars in $G(P)$ efficiently.

We now describe our construction. Let $P = \langle 0, 1, \ldots, n \rangle$. We replace each star S on $G(P)$ by a collection of chords as follows: Let the attachments of S on P be $a_0, a_1, \ldots, a_k$ with $a_0 < a_1 < \ldots < a_k$. We replace S by the chords $(a_0, a_i), i = 1, \ldots, k$ and the chords $(a_i, a_k), i = 1, \ldots, k-1$. We will refer to these chords as the *chords of S*.

Let $H(P)$ be the graph obtained from $G(P)$ by replacing each star in $G(P)$ by a collection of chords as described above. Let $H_I = (V, E)$ be the interlacing parity graph of $H(P)$. We construct $G_I = (V', E')$, the *interlacing parity graph of $G(P)$* as follows:

$$V' = V \cup \{v_S | S \text{ is a star in } G(P)\};$$
$$E' = E_1 \cup \{(v_S, u) | S \text{ is a star in } G(P) \text{ and } u \in V \text{ represents a chord of } S\} \cup F,$$

where E_1 and F is defined as follows:

$E_1 = E - \{(u,v) \in E \mid u \text{ and } v \text{ are chords of the same star } S \text{ in } G(P)\};$

For each vertex i on P let

$$\begin{aligned}F_i = \{(v_S, v_T) \mid\ & S \text{ is a star in } G(P) \text{ with an internal attachment on } i \\ & \text{and } T \text{ ranges over all other stars in } G(P) \text{ with} \\ & \text{an internal attachment on } i\}\end{aligned}$$

Then $F = \cup_{i=1}^{n-1} F_i$.

LEMMA 7.27
Let S and T be two stars that interlace in $G(P)$. Then either S and T share an internal attachment on $G(P)$ or there exist a chord of S and a chord of T that interlace on P.

PROOF
Let S and T interlace on P. If they interlace by virtue of three common attachments then consider the middle attachment of these three. This attachment must be an internal attachment of both S and T.

If S and T interlace because there exist four vertices $a < b < c < d$ on P such that a and c are attachments of S and b and d are attachments of T, then the four vertices $a' < b < c < d'$, where a' is the first attachment of S and d' is the last attachment of T, also represent the interlacement of S and T. But (a', c) is a chord of S and (b, d') is a chord of T and (a', c) and (b, d') interlace.

A similar argument applies to the case when a and c are attachments of T and b and d are attachments of S. ∎

LEMMA 7.28
Let S and T be two stars that do not interlace on $G(P)$. Then S and T share no internal attachment on P, and for any pair of chords c and d, with c a chord of S and d a chord of T, c and d do not interlace on P.

PROOF
Exercise 7.8. ∎

LEMMA 7.29
Let G_I be the interlacing parity graph of a star graph $G(P)$. The following properties hold:

a) All vertices representing chords of a single star in $G(P)$ belong to a single connected component in $G(P)$.

b) *Two stars S and T in $G(P)$ have null interlacing parity if and only if the vertices corresponding to their chords lie in different connected components in G_I.*

PROOF
Exercise 7.8.   ∎

LEMMA 7.30
Let $G(P)$ be a star graph and let S and T be two stars on $G(P)$. Let c be a chord of S and d a chord of T and let their corresponding vertices lie in a single connected component C of G_I. Let $\mathbf{X}$ be a two-coloring of a spanning tree of C. Then if the vertices corresponding to c and d have the same color in $\mathbf{X}$ then stars S and T have even parity in $G(P)$ and if they have different colors then S and T have odd parity in $G(P)$.

PROOF
Exercise 7.8.   ∎

We now present the algorithm to determine interlacing parity for a general star graph.

ALGORITHM 7.10
Interlacing Parity for a Star Graph
Input: A star graph $G(P)$.
Output: A label on each star which allows an interlacing parity of any pair of stars to be determined in constant time by one processor.

   construct the interlacing parity graph G_I of G as described above;

   find a spanning tree in each connected component of G_I and two-color the spanning trees;

   assign a label ⟨component number, color⟩ to each vertex in G_I;

   pfor each star S in $G(P)$ → assign the label of a vertex in G_I corresponding one of its chords **rofp**

end.

To determine an interlacing parity for any pair of stars S, T in $G(P)$ we proceed as in the previous section: If component number of S is not equal to component number of T then S and T have null interlacing parity, otherwise they have even interlacing parity if they have the same color, and odd interlacing parity if they have different colors.

THEOREM 7.5
Algorithm 7.10 correctly determines the interlacing parity of any pair of stars in $G(P)$.

PROOF
The proof is a straightforward consequence of Lemmas 7.29 and 7.30. ∎

Finally, to form the coalesced graph $G_c(P)$ of $G(P)$, we replace all stars in each connected component of G_I by a new star whose attachments are the union of the attachments of these stars.

The Star Embedding and the Candidate Lists

In this section we give a method to find the candidate lists in a star graph $G(P)$, given its coalesced graph $G_c(P)$. The algorithm forms the star embedding of $G_c(P)$ and extracts the candidate lists as those sets of vertices on P that lie on a single face in the star embedding of $G_c(P)$.

We first give a combinatorial characterization of a graph embedding that was introduced by Edmonds [3]. Let $G = (V, E)$ be the graph to be embedded, and let D_G be the directed graph obtained from G by replacing each undirected edge (u, v) in E by two directed edges (u, v) and (v, u). A *combinatorial embedding*, $I(G)$, of the graph G is an assignment of a cyclic ordering to the set of outgoing edges from each vertex in the graph D_G. For each edge (u, v) in D_G let $next(u, v)$ be the edge (v, w) that follows edge (v, u) in the cyclic ordering of edges outgoing from vertex v. Then the graph defined by the next pointers is a collection of edge-disjoint cycles in D_G whose union is the edge set of D_G. Each cycle defined by the next pointers represents a *face* of $I(G)$. Let $n = |V|, m = |E|$, let c be the number of connected components in G and let f be the number of faces in $I(G)$. If Euler's formula $n - m + f = 1 + c$ is satisfied, then the combinatorial embedding $I(G)$ defines a planar embedding of G with the edges incident on each vertex embedded according to the cyclic ordering, and with each face in $I(G)$ representing a face in the planar embedding of G.

Our algorithm will find such a combinatorial embedding for $G_c(P)$ with the additional property that all stars in $G_c(P)$ lie on the same side of $G_c(P)$. This will give us a star embedding of $G_c(P)$. From this star embedding we can obtain the faces of $G_c^*(P)$ by following the next pointers as in the above definition, and we can obtain the candidate lists as those sequences of vertices of P that lie on a single face in the star embedding. The algorithm is given below.

7.4. Efficient Implementation of Triconnectivity Algorithm

ALGORITHM 7.11

Finding Candidate Lists
Input: The coalesced graph $G_c(P)$ of a star graph $G(P)$, with $P = \langle 0, 1, \ldots, n \rangle$.
Output: The candidate lists of $G(P)$.

    vertex c, u, v; {∗ c will be subscripted. ∗}

    edge e;

    interpret each edge (u, v) in $G_c(P)$ as a pair of directed edges (u, v) and (v, u);

      1. **pfor** each vertex i on $P \to$

          sort the attachment edges on vertex i that represent the last attachment edge of their star in nonincreasing order of the first attachment of the star containing the edge; break ties by placing an attachment edge of a two-attachment star *after* an attachment of a star with three or more attachments;
let the resulting array be A_i;
$B_i := \phi$;

          if there is a star with an internal attachment edge e on $i \to$
$B_i := e$ **fi**;
sort the attachment edges on i that represent the first attachment of their star in nonincreasing order of the last attachment edge of the star containing the edge; break ties by placing an attachment on i of a two-attachment star *before* an attachment of a star with three or more attachments;
let the resulting array be C_i;
rearrange the adjacency list for i by concatenating edge $(i, i-1)$ (if it exists), arrays A_i, B_i, C_i and edge $(i, i+1)$ (if it exists) in this order;
make this list cyclical by causing the first edge on the list to follow the last edge

      rofp;

      2. **pfor** each star $S \to$

          rearrange the adjacency list for its center c_S in nondecreasing order of the attachment vertices;
make this list cyclical by causing the first edge on the list to follow the last edge

      rofp ;

form the star embedding of $G_c(P)$ by embedding P and by embedding the edges incident on vertex $i, i = 0, \ldots, n$ and on the center of each star in $G(P)$ cyclically according to their order in the rearranged adjacency lists obtained in steps 1 and 2;

3. **pfor** each edge (u, v) in $G_c(P) \to next(u, v) :=$ the edge following (v, u) in the adjacency list of v **rofp**;

4. use pointer jumping on the next pointers to partition the directed edges into cycles;
   pfor each cycle formed in step 1 $\to$ find the list of vertices on P that are contained in the cycle and output this sequence as a candidate list **rofp**

end.

THEOREM 7.6
Algorithm 7.11 correctly finds the candidate lists of $G(P)$.

PROOF
We leave it as an exercise to verify that the combinatorial embedding in step 3 represents the star embedding of $G_c(P)$. The proof of correctness then follows from the properties of a combinatorial embeddings, together with the results in Section 7.3.3. ∎

7.4.3 Finding Triconnected Components Efficiently

In this section we give an efficient parallel algorithm to find the triconnected components of a biconnected graph using Algorithm 7.5. Algorithm 7.5, when used in parallel on all nontrivial ears, finds the triconnected components of a biconnected graph G with an open ear decomposition D when no vertex is part of pairs separating two different ears. This property need not hold for a general biconnected graph G but if we use G_l, the local replacement graph of $(G; D)$, all separating pairs in G_l corresponding to separating pairs of G are internal to their corresponding paths P_i' and hence this property holds. Further, the size of all of the bridge graphs of the nontrivial paths P_i in G_l is linear in the size of G since these paths are vertex-disjoint in G_l. Hence we can find the triconnected components of G using the following simple algorithm:

ALGORITHM 7.12
Efficient Triconnected Components Algorithm
Input: Biconnected graph $G = (V, E)$ together with an open ear decomposition $D = [P_0, \ldots, P_{r-1}]$ for G.

Output: The triconnected components of G.

form G_l from G using Algorithm 7.7;

form the bridge graph $D'_i(P_i)$ in G_l of each nontrivial path P_i in D using Algorithm 7.6;

pfor each nontrivial ear $P_i \to$ apply Algorithm 7.5 to D'_i to perform all of the ear splits on P_i **rofp**;

pfor each connected component in the resulting graph $\to$ collapse all vertices in each local tree to get back a triconnected component of G **rofp**

end.

7.5
Towards Optimality

Although we have stated the parallel bound for the various steps in our algorithms as $O(\log n)$ parallel time with a linear number of processors, most of the steps can, in fact, be performed optimally in $O(\log n)$ time. Only the need to find connected components in a graph and to perform bucket sort prevents us from obtaining true optimality. In this context we note the following.

1. *Finding connected components:* We need to find connected components at several places in our algorithms. At present there is no optimal $O(\log n)$ time parallel algorithm known for graph connectivity although the algorithm in [2] is 'almost optimal'. However, this algorithm assumes that the graph is represented by its adjacency lists. Even if we assume that the input graph is represented by its adjacency lists, we still need to ensure that the adjacency lists for the various derived graphs used in the algorithms can be obtained optimally. For the open ear decomposition algorithm, the details of this construction are worked out in [21]. The adjacency lists for some of the derived graphs used in the triconnectivity algorithm can also be obtained optimally. In the absence of an optimal algorithm for finding the adjacency list for a derived graph, we can use bucket sort in the range $[1..n]$, where n is the number of vertices in the graph, to rearrange an unordered edge list into adjacency lists for the individual vertices. This can be done in $O(\log n)$

time using $O((n + m) \cdot \log \log n / \log n)$ processors using the algorithm of Hagerup [7], where m is the number of edges in the graph. This is also the best bound known for testing graph connectivity if the input graph is specified as a list of edges that are not organized as adjacency lists for vertices.

The connected components of a graph can be obtained optimally in logarithmic time on a CRCW PRAM using the *randomized* algorithm of Gazit [6]. The required adjacency lists can also be obtained optimally in logarithmic time using a randomized algorithm for bucket sort in the range $[1..n]$ given in [17].

2. *Sorting:* Our algorithms use sorting in several places. All of the sorting that is needed can be performed using an algorithm for bucket sort in the range $[1..n^2]$ and this can be done using the algorithm of [7] with the performance bound stated in 1). Further, in some cases the sorting step can be replaced by a more sophisticated algorithm that runs optimally (see, e.g., [5]).

In the ear decomposition and open ear decomposition algorithms, sorting is needed only if we require consecutive ear numbers. In most applications (including triconnectivity, four-connectivity and planarity), any sequence of increasing labels from a totally ordered set suffices for the ear labels, and in such cases, no sorting is needed in either of these algorithms.

7.6 Conclusion

In this chapter we have presented efficient parallel algorithms for testing graph biconnectivity and triconnectivity and for finding the triconnected components of a biconnected graph. All of these algorithms run in linear sequential time and thus represent new linear time sequential algorithms for these problems.

The algorithms in this chapter differ significantly in structure from the earlier linear time algorithms for these problems which were based on depth-first search. For instance, our algorithms are very modular. Thus an implementer can choose a trade-off between efficiency of implementation and ease of programming in deciding which of several methods to use to implement each step in the various algorithms. This is in contrast to the earlier algorithms in which most of the steps were tied to a depth-first search of the input graph.

The algorithms in this chapter are somewhat more complex than the earlier sequential algorithms. Parallel algorithm design is a challenging task, and in asking for an efficient parallel logarithmic time algorithm whose sequential implementation runs in linear time, we are requiring much more of the algorithm designer than we do of a designer of linear sequential algorithms. Hence it is not surprising that the parallel algorithms in this chapter are rather complex.

Much of the additional complexity in the parallel algorithms of this chapter relative to the sequential ones is due to the fact that the parallel algorithms is unable to generate a depth-first search tree efficiently. We demonstrated this in the case of finding an open ear decomposition. However, due to the difference in the techniques used in the design of sequential and parallel algorithms, we sometimes obtain a simplified sequential algorithm from a parallel algorithm for a problem. For instance, in the case of finding triconnected components, the algorithm in this chapter is actually simpler than the earlier linear time sequential algorithm in that it directly performs Tutte splits and hence does not need to perform any recombinations.

A graph is *planar* if it can be embedded on the Cartesian plane so that no edges cross. The *planarity* problem is given a graph, determine if the graph is planar, and if so give a planar embedding. Ramachandran and Reif [19] recently developed a logarithmic time, linear processor algorithm for the planarity. The algorithm uses a reduction to open ear decomposition, LCA, and tree contraction.

Acknowledgement

I would like to thank the Austin Tuesday Afternoon Club for several comments on an earlier version of Section 7.2. I am especially grateful to Edsger W. Dijkstra and Jay Misra for their comments on algorithmic notation and specification of algorithms.

7.7 Exercises

7.1 Prove that a graph has an open ear decomposition if and only if it is biconnected.

7.2 Prove that a graph is biconnected if and only if it has an *st*-numbering.

7.3 Prove that the coalesced graph of a star graph is unique.

7.4 Let G be a biconnected graph with an open ear decomposition D and let G have n vertices and m edges. Use a divide and conquer approach to find the ear graph of each nontrivial ear in D in $O(\log^2 n)$ time with $O(n + m)$ processors.

7.5 Adapt the theorems and results in Section 7.3.3 to the problem of determining three-edge connectivity in an undirected graph. Simplify the results wherever possible.

7.6 Complete the proofs of lemmas in Section 7.4.1 by considering the case when vertices are in their corresponding in-trees.

7.7 Explain why it suffices to consider only T_{st} in step 3 of Algorithm 7.7.

7.8 Prove the correctness of Algorithm 7.10 by supplying the proofs of Lemmas 7.28, 7.29 and 7.30.

7.9 The *tree of triconnected components* of a biconnected graph G is a tree $T = (V, E)$ whose vertex set is the set of triconnected components of G, and which contains an edge $(x, y) \in E$ whenever the triconnected components x and y have the same copy of an edge introduced during a Tutte split.

Give a parallel algorithm that runs in logarithmic time with a linear number of processors to find the tree of triconnected components of a biconnected graph.

Bibliography

[1] R. Cole, "Parallel merge sort," *SIAM J. Comput.*, vol. 17, 1988, pp. 770-785.

[2] R. Cole, U. Vishkin, "Approximate and exact parallel techniques with applications to list, tree and graph problems," *Proc. 27th Ann. IEEE Symp. on Foundations of Comp. Sci.*, 1986, pp. 478-491.

[3] J. Edmonds, "A combinatorial representation for polyhedral surfaces," *Not. Am. Math. Soc.*, vol. 7, 1960, p. 646.

[4] S. Even, *Graph Algorithms*, Computer Science Press, Rockville, MD, 1979.

[5] D. Fussell, V. Ramachandran, R. Thurimella, "Finding triconnected components by local replacements," *Proc. ICALP 89*, Springer Verlag LNCS 372, 1989, pp. 379-393; *SIAM J. Comput*, to appear.

[6] H. Gazit, "An optimal randomized parallel algorithm for finding connected components in a graph," *Proc. 27th Ann. IEEE Symp. on Foundations of Comp. Sci.*, 1986, pp. 492-501.

[7] T. Hagerup, "Towards optimal parallel bucket sorting," *Inform. and Comput.*, vol. 75, 1987, pp. 39-51.

[8] J. E. Hopcroft, R. E. Tarjan, "Finding the triconnected components of a graph," TR 72-140, Computer Science Department, Cornell University, Ithaca, NY, 1972.

[9] J. E. Hopcroft, R. E. Tarjan, "Dividing a graph into triconnected components," *SIAM J. Comput.*, 1973, pp. 135-158.

[10] A. Kanevsky, V. Ramachandran, "Improved algorithms for graph four-connectivity," *Proc. 28th IEEE Symp. on Foundations of Comp. Sci*, 1987, pp. 252-259; and *Jour. Comput. Syst. Sci.*, vol. 42, 1991, pp. 288–306.

[11] R. M . Karp, V. Ramachandran, "Parallel algorithms for shared memory machines," *Handbook of Theoretical Computer Science*, J. Van Leeuwen, ed., North Holland, 1990, pp. 869-941.

[12] S. R. Kosaraju, A. L. Delcher, "Optimal parallel evaluation of tree-structured computations by raking," *Proc. 3rd Aegean Workshop on Computing*, Springer-Verlag LNCS 319, 1988, pp. 101-110.

[13] L. Lòvasz,"Computing ears and branchings in parallel," *Proc. 26th IEEE Ann. Symp. on Foundations of Comp. Sci.*, 1985, pp. 464-467.

[14] Y. Maon, B. Schieber, U. Vishkin, "Parallel ear decomposition search (EDS) and st-numbering in graphs," *Theoretical Comput. Sci.*, vol. 47, 1986, pp. 277-298.

[15] G. L. Miller, V. Ramachandran, "Efficient parallel ear decomposition with applications," manuscript, MSRI, Berkeley, CA, January 1986.

[16] G. L. Miller, V. Ramachandran, "A new graph triconnectivity algorithm and its parallelization," *Proc. 19th Annual ACM Symp. on Theory of Computing*, 1987, pp. 254-263, and *Combinatorica*, vol. 12, no. 1, 1992, pp. 53–76.

[17] S. Rajasekharan, J. H. Reif, "Optimal and sublogarithmic time randomized parallel sorting algorithms," *SIAM J. Comput.*, vol. 18, 1989, pp. 594-607.

[18] V. Ramachandran, J. H. Reif, "Planarity testing in parallel," TR-90-15, Dept. of Computer Sciences, The University of Texas, Austin, TX, 1990; preliminary version appears as "An optimal parallel algorithm for graph planarity," *Proc. 30th Ann. IEEE Symp. on Foundations of Comp. Sci.*, 1989, pp. 282-287.

[19] V. Ramachandran and J. Reif, *An optimal parallel algorithm for planarity*, 30th Annual Symposium on Foundations of Computer Science, Durham, NC, 1989, pp. 282–287. To appear in special issue of *Journal of Algorithms*, 1992.

[20] V. Ramachandran, U. Vishkin, "Efficient parallel triconnectivity in logarithmic time," *VLSI Algorithms and Architectures*, Springer Verlag LNCS 319, 1988, pp. 33-42.

[21] B. Schieber, *Design and Analysis of Some Parallel Algorithms*, Ph. D. thesis, Tel Aviv University, Tel Aviv, Israel, 1987.

[22] B. Schieber, U. Vishkin, "On finding lowest common ancestors: simplification and parallelization," *Proc. 3rd Aegean Workshop on Computing*, Springer-Verlag LNCS 319, 1988, pp. 111-123.

[23] R. E. Tarjan, "Depth first search and linear graph algorithms," *SIAM J. Computing*, vol. 1, 1972, pp. 146-160.

[24] R. E. Tarjan, *Data Structures and Network Algorithms*, SIAM Press, Philadelphia, PA, 1983.

[25] R. E. Tarjan, U. Vishkin, "An efficient parallel biconnectivity algorithm," *SIAM J. Computing*, vol. 14, 1984, pp. 862-874.

[26] W. T. Tutte, *Connectivity in Graphs*, University of Toronto Press, 1966.

[27] H. Whitney, "Non-separable and planar graphs," *Trans. Amer. Math. Soc.* 34, 1932, pp. 339-362.

8

Parallel Algorithms for Chordal Graphs

Philip Klein
Department of Computer Science
Brown University
Providence, RI 02912-1910
pnk@cs.brown.edu

8.1
Introduction

The algorithmic study of special kinds of graphs has yielded insights both combinatorial and algorithmic. In this chapter, we consider two classes of graphs, chordal graphs and interval graphs, that have been the subject of algorithmic study since the work of Lekkerkerker and Boland in 1962. These classes of graphs are rich enough to find application in diverse areas, from biology to VLSI, but are structured enough that efficient algorithms exist to analyze them. Problems such as maximum-weight clique and minimum coloring, which are NP-complete for arbitrary graphs, can be solved in linear time for chordal and interval graphs.

As motivation for studying chordal graphs, we start in Section 8.2 by considering the problem of recognizing *interval graphs*. The best algorithms for interval graph recognition involve finding the maximal cliques of the input graph. To do so, they make use of the fact that every interval graph is *chordal*. In Section 8.3, we show how to identify all the maximal cliques of a chordal graph, by using a node-ordering called a *perfect elimination ordering*, or peo. Next we show that the perfect elimination ordering can also be used to solve combinatorial-optimization problems on chordal graphs. In Section 8.4, we sketch the sequential algorithms for some of these problems and in Section 8.5 we describe some parallel algorithms.

Finally, we consider the problem of efficiently finding a perfect elimination ordering. In section 8.6, we describe a framework for solving this problem. In Section 8.7, we show how the classic sequential algorithm of Rose, Tarjan, and Lueker can be cast within this framework. In Section 8.8, we give an efficient parallel algorithm for the problem. The reader is referred to the subsection below for some elementary definitions.

DEFINITION

Let G be a graph. Throughout the paper, we let n and m denote the number of nodes and edges in G, respectively.

A path or cycle is simple *if each node occurs only once. A k-cycle is a cycle consisting of k nodes.*

Let H be a subgraph of G or a subset of the nodes of G. We let $G[H]$ denote the subgraph of G consisting of nodes of H and edges both of whose endpoints are in H. We call $G[H]$ the subgraph of G induced by nodes of H. A node-induced subgraph of G is a subgraph of the form $G[H]$. We let $G - H$ denote the subgraph of G induced by nodes not in H. The neighbors of a subgraph H of G are the nodes of $G - H$ connected to H by edges of G. In case the nodes of G have weights, the weight of a subgraph is the sum of the weights of its nodes.

A graph H is a clique *if there is an edge between every pair of nodes*, and an independent set *if it has no edges*. We say H is a clique of G if H is a clique and a subgraph of G. We say H is a maximal clique *of G if H is a clique of G and no clique H' of G strictly contains H. We say H is a* maximum clique *of G if H is a largest clique of G*. We call $H_1, \ldots, H_k$ a clique-cover *of G if each H_i is a clique of G and every node of G is in some H_i. A* minimum clique-cover *is one with a minimum number of cliques*. We can assume without loss of generality that the H_i's are node-disjoint. The above definitions also apply to independent sets. *An* independent set cover *of G is called a* coloring *of G if the independent sets are node-disjoint. A minimum coloring is also called an* optimal coloring.

Let P be a problem for which the best known sequential algorithm takes time $T(n)$ on instances of size n. A parallel algorithm to solve P is called efficient *if the product of its processor requirement with its time requirement is within a polylogarithmic factor of $T(n)$*. For example, if the parallel algorithm requires $T(n)$ processors and $O(\log^2 n)$ time, it is efficient.

8.2 Interval Graphs

In this section, we introduce interval graphs and one method for determining whether a given graph is an interval graph. A key component of this method is the identification of the maximal cliques of the given graph. In the next section, we show how the maximal cliques can be identified. We only touch on the other component of the recognition method; the reader is directed to other references for more details.

Interval graphs arise in modelling overlapping intervals of a line; each interval is represented by a node, and two nodes are adjacent if the represented intervals overlap. Figure 8.1 illustrates an example. The biologist Seymour Benzer proposed an examination of the structure of the gene to determine whether its subelements are arranged linearly; the analysis entailed

344 Chapter 8. Parallel Algorithms for Chordal Graphs

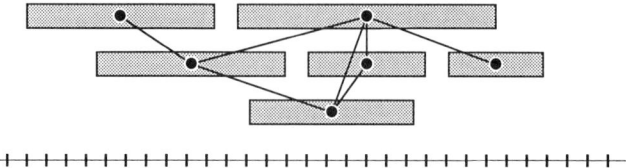

FIGURE 8.1
An example of an interval graph. For clarity, the intervals are drawn one above another. Two nodes are adjacent if the corresponding intervals share a point of the real line.

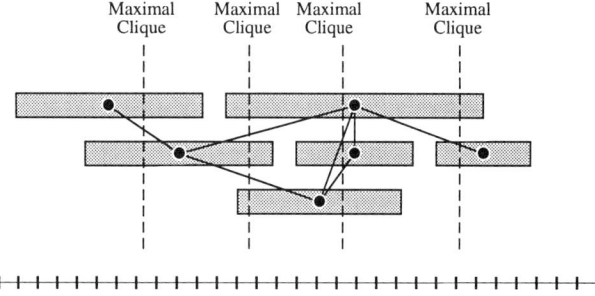

FIGURE 8.2
Each maximal clique corresponds to a point on the real line. The ordering of these points corresponds to an ordering of the maximal cliques such that for each node, the maximal cliques containing the node are consecutive.

determining whether an experimentally determined graph of correlations between mutations was in fact an interval graph (it was). Motivated by the biological application, Fulkerson and Gross took up the problem of determining whether a graph is an interval graph.

The intuitive basis for their interval graph recognition algorithm was the following observation. If a graph G corresponds to a collection of overlapping intervals of the real line R, distinct maximal cliques of G correspond to disjoint nonempty intervals of R. Hence the set of maximal cliques of G inherit a natural order from the interval representation of G. Based on this observation, it is not hard to prove the following theorem, illustrated in Figure 8.2.

THEOREM 8.1 *Fulkerson and Gross*
A graph is an interval graph if and only if there is an ordering of its maximal cliques so that for any node of the graph, the maximal cliques containing that node are consecutive.

Observe that such an ordering of maximal cliques of a graph actually yields an interval representation of the graph. Identify the maximal cliques with distinct points on the real line, and associate with each node v the interval spanning the maximal cliques that contain v. Then the intervals associated with two nodes overlap if and only if there is a maximal clique containing the two nodes, which holds if and only if the two nodes are adjacent.

In using this theorem as the basis for an algorithm, we come up against two problems: finding the maximal cliques, and finding a good ordering for them. The key to finding the maximal cliques is the following property, possessed by all interval graphs:

Chordality: Every cycle of length greater than three has a *chord*, an edge connecting two non-consecutive nodes of the cycle.

It is easy to see that every interval graph has this property. First, lay down three (or more) intervals to form a path, as shown in Figure 8.3. There is no way to lay down a fourth interval overlapping the end-intervals but not the middle one. We shall see in Section 8.3 how to find the maximal cliques of any graph possessing the chordality property. Of course, if the input graph did not have this property, Fulkerson and Gross could immediately conclude that the input graph was not an interval graph.

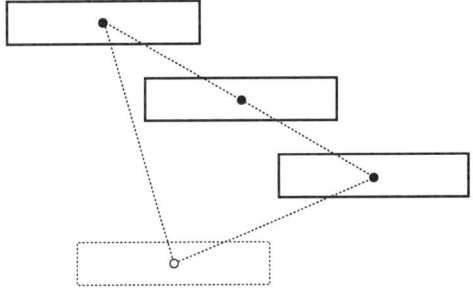

FIGURE 8.3
Three intervals are laid down to form a path. A fourth interval can be laid down to join the endpoints of the path only if it intersects all three intervals.

Once we have in hand the maximal cliques of the input graph, it remains to determine whether there existed a good ordering of the cliques. The abstract problem is to order some items $C_1, \ldots, C_k$ subject to consecutivity constraints, represented by subsets $S_1, \ldots, S_t$ of the set $\{C_1, \ldots, C_k\}$ of items to be ordered. For each subset S_i, we require that the elements of S_i form a consecutive subsequence of the ordered sequence of items. For example, suppose we wish to order the items A, B, C, D, E, F subject to the constraints $S_1 = \{A, C, E\}$, $S_2 = \{A, C, F\}$, and $S_3 = \{B, F, D\}$. Two consistent orderings are $BDFACE$ and $DBFCAE$; there are six others.

Fulkerson and Gross gave an algorithm based on linear algebra for solving the ordering problem. Their algorithm required $O(k^3)$ time. Years later, Booth and Lueker [6] developed a data structure, called the PQ-tree, for more efficiently solving this problem. The PQ-tree is a tree of size $O(k)$ for implicitly representing all orderings of $C_1, \ldots, C_k$ subject to consecutivity constraints. Given a PQ-tree T and a new constraint S_i, Booth and Lueker showed how to incorporate the new constraint into T, further restricting the set of orderings represented by T, in time proportional to $|S_i|$. Thus, starting from a PQ-tree representing all possible orderings, one can introduce the constraints $S_1, \ldots, S_t$ one by one, and obtain after $O(k + \sum_i^t |S_i|)$ time a PQ-tree representing the set of orderings consistent with all the constraints (possibly an empty set). (Klein [28] gave a parallel algorithm for introducing all these constraints at once into a PQ-tree.)

We can apply the PQ-tree to the ordering problem arising in recognizing interval graphs. If the input graph satisfies the chordality property, then, as we shall see, the number k of maximal cliques is at most the number n of nodes of the graph, and the sum of the sizes of the contraints S_i is at most $n + m$. Consequently, use of the PQ-tree yields a linear-time sequential algorithm (and a linear-processor parallel algorithm) for finding a good ordering, if one exists. In the next section, we examine more closely the chordality property, and show how it leads to efficient identification of the maximal cliques.

8.3
Chordal Graphs

The property of chordality seems quite innocent but in fact characterizes a rich and important class of graphs, *chordal* graphs. These graphs are also called *triangulated* graphs [41], because each minimal cycle is a triangle, and *rigid circuit* graphs, because every cycle is (fancifully) made "rigid" by its

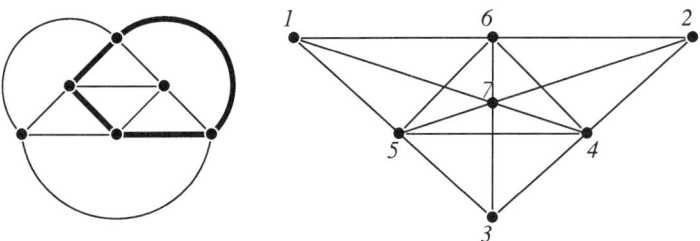

FIGURE 8.4
The first graph is nonchordal, and a chordless cycle is highlighted. The second graph is chordal, and a perfect elimination ordering is given. Note that the first graph has no simplicial nodes; in the second graph, nodes 1, 2, and 3 are all simplicial. The clique associated with node 1 is $C(1) = \{1, 5, 6, 7\}$. The clique associated with node 5 is $C(5) = \{5, 6, 7\}$. The first clique is maximal; the second is not—it is contained in the first.

chords. They are called *perfect elimination* graphs, because they represent the nonzero structure of symmetric matrices for which Gaussian elimination need not introduce a nonzero entry.[1] Figure 8.4 illustrates a nonchordal and a chordal graph.

Based on a theorem of Dirac on chordal graphs [15], Fulkerson and Gross (and also Rose [40]) gave a characterization of chordal graphs that remains today the most algorithmically useful characterization. Dirac showed that every chordal graph has a *simplicial* node, a node all of whose neighbors form a clique. It follows easily from the chordality property that deleting nodes of a chordal graph yields another chordal graph. That is, any node-induced subgraph of a chordal graph is chordal. In particular, deleting a single node yields a chordal graph. This observation leads to the following algorithm: find a simplicial node, delete it, and recurse on the resulting graph, until no nodes remain. By Dirac's theorem, this procedure successfully terminates on chordal graphs. Conversely, if the graph contains an unchorded k-cycle with $k \geq 4$ (e.g., the first graph of Figure 8.4), no node in that cycle will ever become simplicial. (For any node in the cycle, its neighbors in the cycle are not adjacent.) Thus the procedure succeeds for chordal graphs, and chordal graphs alone.

---

[1] Given an $n \times n$ symmetric matrix A, the corresponding graph has nodes $1, \ldots, n$, and an edge $\{i, j\}$ for each nonzero entry A_{ij}.

But the procedure provides more than a test for chordality; it unfolds the structure of the chordal graph. For example, it provides us with all the maximal cliques. For each node v_i, let $C(v_i)$ be the set consisting of v_i together with v_i's higher-numbered neighbors—those neighbors that are still in the graph when v_i is deleted by the procedure. The procedure ensures that $C(v_i)$ is a clique. It is easy to see that the set of cliques $C(v_i)$ for $1 \leq i \leq n$ include all the maximal cliques. Let C be any maximal clique, and let v_i be the first node of C to be deleted. Then every other node of C is one of the neighbors of v_i remaining in the graph, so C is contained in $C(v_i)$. Since C is maximal, it must be that $C = C(v_i)$. We have proved that every maximal clique occurs as $C(v_i)$ for some v_i. (The reader is encouraged to try applying this argument to any maximal clique in a chordal graph, e.g., the second graph in Figure 8.4.) Note that some cliques $C(v_i)$ are not maximal. However, it is easy to determine the set of v_i for which $C(v_i)$ is a maximal clique, and we discuss this step in Section 8.5.2.

The above argument shows also that there are at most n maximal cliques in a chordal graph. Moreover, we can infer that if S_i is the set of maximal cliques containing the node v_i, then the sum $\sum_i^n |S_i|$ is at most $n+m$. First, v_i is the minimum-numbered node in at most one maximal clique. Second, the number of maximal cliques of which v_i is a non-minimum member is bounded by the number of edges connecting v_i to a lower-numbered neighbor. The sum of these numbers over all nodes v_i is $n+m$.

The node-ordering provided by the procedure has many other uses as well. For example, Rose used it to establish a connection between chordal graphs and symmetric linear systems. For any $n \times n$ symmetric matrix, there is a corresponding n-node undirected graph—the graph contains the edge $\{i,j\}$ if and only if the ij entry of the matrix is nonzero. Rose showed that if the graph is chordal, a linear system associated with the matrix can be solved via Gaussian elimination *without introducing new nonzero entries*.[2] One simply eliminates variables in the order in which corresponding nodes are deleted by the procedure. For this reason, such an ordering is called a *perfect elimination ordering* (peo) or perfect elimination scheme.

The original peo algorithm seems to require $\Omega(nm)$ work; at each of n stages, a simplicial node must be located. Rose, Tarjan, and Lueker, however, came up with a linear-time algorithm [41] for constructing a peo. It follows that we can recognize interval graphs in linear time. We use Rose, Tarjan,

---

[2] Rose also required that the matrix be positive semidefinite, so that the elimination process is numerically stable regardless of the order of elimination.

and Lueker's linear time **peo** algorithm to help identify the maximal cliques of the input graph, and then use Booth and Lueker's PQ-tree to find a good ordering of the maximal cliques.

We discuss Rose, Tarjan, and Lueker's sequential **peo** algorithm and a parallel **peo** algorithm in Sections 8.7 and 8.8.

8.4
Sequential Optimization Algorithms for Chordal Graphs

With the maximal cliques provided by the perfect elimination ordering, Fulkerson and Gross were able to carry out interval graph recognition. Gavril [19] carried it a step further. He saw that the perfect elimination ordering was the basis for solving certain optimization problems on chordal graphs. First, observe that included among the maximal cliques is certainly the *maximum* clique. In fact, for any non-negative node-weights, a maximum-weight clique is also among the maximal cliques. Thus having a **peo** enables one to easily find the maximum-weight clique.

Gavril also gave algorithms, all based on the **peo**, for finding a maximum independent set, a minimum coloring (i.e., a minimum covering of the graph by independent sets), and a minimum covering by cliques. If one is given a perfect elimination ordering $v_1 \ldots v_n$, the algorithms are easy to carry out. For the remainder of this section, we assume a perfect elimination ordering has been computed. We call a node *low* or *high* according to its position in the ordering. For example, the first node in the ordering, v_1, is said to be the lowest node, and the other nodes are all higher than v_1.

To find a maximum independent set, use a greedy approach to build up an independent set I node by node as shown in Figure 8.5. Start by placing the lowest node v_1 in I, and at each subsequent step add to I the lowest remaining node that is not adjacent to any node in I. This procedure always yields a *maximal* independent set; in fact, Gavril proved that it yields a maximum independent set. For whenever a node v_i is added to I, the nodes it rules out—the neighbors of v_i not already ruled out—are higher neighbors, and hence form a clique $C(v_i)$ with v_i. When the procedure terminates, every node has either been placed in I or been ruled out by such a node; thus, we have covered the graph with cliques $C(v_i)$, one for each node v_i added to I. If there were a larger independent set than I, the pigeonhole principle would imply that two of its nodes would land in one of the cliques of our clique cover, a contradiction since no two nodes of an independent set are adjacent. Thus I

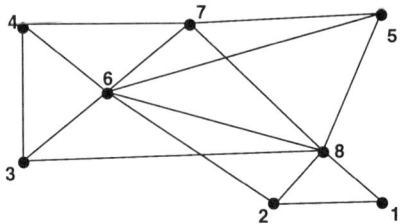

FIGURE 8.5
The independent set algorithm applied to the graph shown first places node 1 in I, and deletes 2 and 8. Then it places 3 in I, and deletes 4 and 6. Finally, it places 5 in I, and deletes 7. Thus the final independent set is $I = \{1, 3, 5\}$. The corresponding clique cover is $C(1) = \{1, 2, 8\}$, $C(3) = \{3, 4, 6, 8\}$, and $C(5) = \{5, 6, 7, 8\}$.

is a *maximum* clique. By the same token, if there were a smaller clique cover, two of the nodes of I would fall in one of the cliques, so the clique cover we have found is minimum.

This last bit of argument relies on a *min-max inequality*, one of the many that pervade combinatorics:

$$\text{maximum independent set} \leq \text{minimum clique cover} \tag{8.1}$$

Gavril's algorithm shows that for any chordal graph, (8.1) is satisfied with equality. In fact, if G is a chordal graph, then every node-induced subgraph of G is also chordal, so (8.1) holds with equality for every node-induced subgraph of G.

A graph all of whose node-induced subgraphs satisfy (8.1) with equality is called a *perfect* graph. Perfect graphs have long been a focus of combinatorial and algorithmic study. Although we know of no algorithmically useful characterization of perfect graphs, they contain as subclasses a variety of well-studied classes of graphs, including comparability graphs, line graphs, and, of course, chordal and interval graphs. Lovász's celebrated Perfect Graph Theorem states that the node-induced subgraphs of a graph all satisfy (8.1) with equality if and only if they satisfy the following inequality with equality.

$$\text{maximum clique} \leq \text{minimum covering by independent sets (coloring)} \tag{8.2}$$

Since a clique is the complement of an independent set, the Perfect Graph Theorem states that a graph is perfect if and only if its complement is perfect.

Returning to the subclass of interest, chordal graphs, we might be tempted by (8.2) to conjecture that there is a minimum coloring algorithm for chordal graphs. In fact, Gavril gives such an algorithm, again based on a greedy approach. We use the positive integers as colors. Starting at the last node v_n of the peo, and working backwards, assign to each v_i in turn the minimum color not assigned to its higher neighbors. Suppose the node receiving the highest color k is v_i. Then colors 1 through $k-1$ were already assigned to the higher neighbors of v_i. But then v_i and its higher neighbors form a clique $C(v_i)$ of size k. Since we have found a coloring using only k colors, the inequality (8.2) implies that our coloring is minimum (and k is the size of the largest clique).

Gavril's algorithms for maximum-weight clique, maximum independent set, minimum coloring, and minimum-clique cover can all be implemented in linear time, if a perfect elimination ordering is provided.

8.5
Parallel Optimization Algorithms

Naor, Naor, and Schäffer [35] first gave *parallel* algorithms for these optimization problems. They relied, largely, not on the peo, but on another characterization of chordal graphs. It turns out that chordal graphs are exactly the intersection graphs of subtrees of trees. That is, for a tree T and subtrees $T_1, \ldots, T_n$ of T there is a graph whose nodes correspond to subtrees T_i, and where two nodes are adjacent if the corresponding subtrees share a node of T. Such a representation is depicted in Figure 8.6.

The graphs arising in this way are exactly the chordal graphs, and interval graphs arise when the tree T happens to be a simple path. The *clique tree* of a chordal graph G is a special tree-and-subtrees representation of G, in which there is a node in T for each maximal clique, and, for each node v_i of G, the corresponding subtree T_i in the representation contains the nodes corresponding to the maximal cliques containing v_i.

Naor, Naor, and Schäffer's parallel approach was first to find all the maximal cliques of G, using a simple but expensive divide-and-conquer strategy, and then construct the clique tree representation using the fact that every tree has a good separator. Their algorithm for finding all maximal cliques used n^4 processors to achieve $O(\log^3 n)$ time or n^5 processors to achieve $O(\log^2 n)$ time. Given the maximal cliques, their algorithm for constructing the clique tree used n^3 processors to achieve $O(\log^2 n)$ time on a CREW PRAM. The

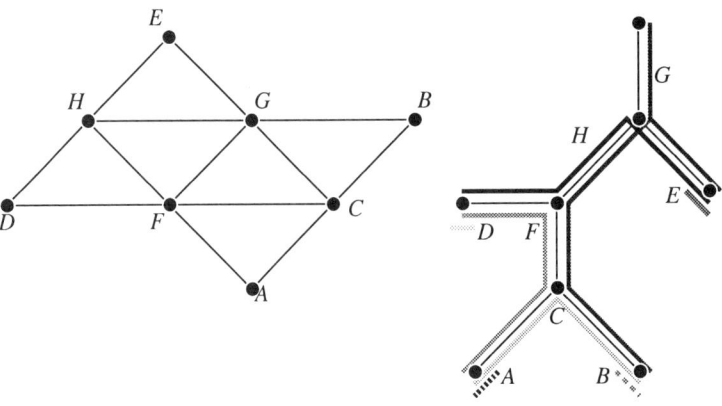

FIGURE 8.6
Illustrated are a chordal graph and a tree whose subtrees represent the chordal graph. Each highlighted subtree represents a node of the chordal graph; two nodes of the chordal graph are adjacent if and only if the corresponding subtrees share nodes of the tree.

disadvantage of this approach is that the number of processors required is quite large.

Naor, Naor, and Shäffer then showed how to use the clique tree as the basis for parallel algorithms to find a perfect elimination ordering, a maximum-weight clique, a minimum coloring, a minimum clique cover, a maximum independent set, and a maximum-weight independent set. The processor bounds range from n^2 for the first problem to n^4 for the fourth.

A key technique they developed was that of processing a tree by *pruning terminal branches*. A *terminal branch* of a tree is a maximal path in the tree consisting of degree-two nodes and a leaf; an example is illustrated in Figure 8.7. Naor, Naor, and Schäffer observe that removal of all terminal branches from a tree yields a tree with at most half the number of leaves. Thus $\lfloor \log n \rfloor + 1$ stages of terminal branch pruning reduces an n-node tree to nothing. The methodology of Naor, Naor, and Schäffer for solving a particular chordal graph problem was to develop a parallel subroutine for solving the subproblem in the case where the clique tree is a *path* (i.e., the graph is an interval graph), and repeatedly apply this subroutine to each of the terminal branches in parallel. The number of iterations is $O(\log n)$. The idea of pruning

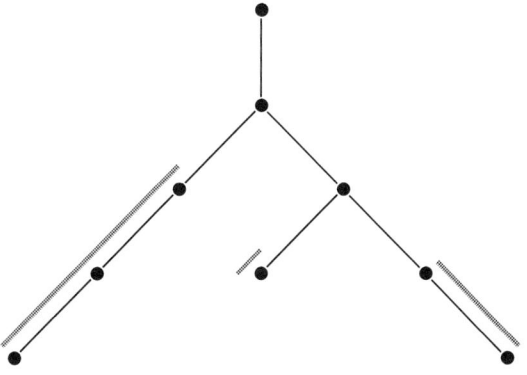

FIGURE 8.7
The rooted tree illustrated has three terminal branches, one per leaf, each consisting of the path from the leaf up to but not including the first ancestor with two or more children.

terminal branches is simple yet powerful, and will probably find its way into other parallel algorithms that operate on trees.

Klein [29] took a different approach to solving the optimization problems in parallel. The basis for these parallel algorithms was, as for the sequential algorithms, the perfect elimination ordering. In Section 8.8, we describe an *efficient* parallel algorithm for finding a peo—the algorithm uses only $(n+m)/\log n$ processors to achieve $O(\log^2 n)$ time on a CRCW PRAM. Thus the total work done by this algorithm is within a logarithmic factor of linear. In this light, the result can be viewed as a parallel analogue to Rose, Tarjan, and Lueker's result on finding a peo in linear time.

In the remainder of Section 8.5, we show that, if a peo is provided, the problems of maximum-weight clique, minimum coloring, maximum independent set, and minimum clique cover can all be solved efficiently in parallel, again in $O(\log^2 n)$ time using $(n+m)/\log n$ processors of a CRCW PRAM.[3] Thus Gavril's results also have analogues in the parallel realm.

There is also a parallel analogue of Booth and Lueker's PQ-tree data structure: given a PQ-tree T and some constraints $S_1, \ldots, S_t$, the constraints can be incorporated into T quickly and efficiently in parallel. Combining this

---

[3] In fact, the first problem is nearly trivial to solve, once the peo is obtained; the time bound is only $O(\log n)$.

8.5.1 The Elimination Tree

The algorithms we describe in this section, like those of Naor, Naor, and Schäffer, tend to rely on a tree to guide the algorithm. In the former case, however, the tree was the *elimination tree* determined by the peo. Given connected graph G and a peo $\sigma = v_1 \ldots v_n$ for G, the elimination tree $T(G_\sigma)$ is defined by choosing a parent $p(v_i)$ for each node v_i of G except the last, v_n. The choice of parent is given by

$$p(v_i) = \text{lowest neighbor } v_j \text{ such that } j > i \tag{8.3}$$

The parent has a higher number than the child, so the resulting directed graph is acyclic. Since each node but one has a parent, the directed acyclic graph is indeed a tree (see Figure 8.8).

It turns out that the elimination tree determined by a peo of a graph G was in fact a *depth-first search tree* for the graph G. For any rooted spanning tree T of G, a *cross-edge* of T is an edge of $G - T$ whose endpoints are not ancestors of each other in T. Figure 8.9 illustrates this definition. If T has no cross-edges, we call T a *depth-first search tree* for G.

THEOREM 8.2

Let G be a chordal graph. The elimination tree $T(G_\sigma)$ determined by a

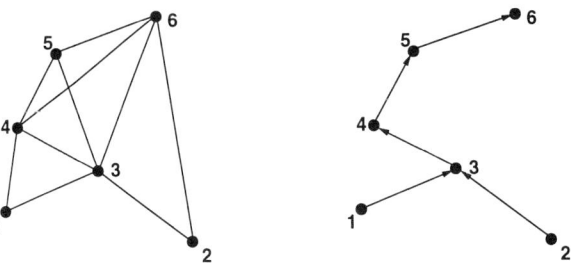

FIGURE 8.8
Given a graph G with a peo σ, we construct the elimination tree $T(G_\sigma)$ by choosing a parent for each node according to the rule (8.3): the parent of v is the lowest neighbor higher than v.

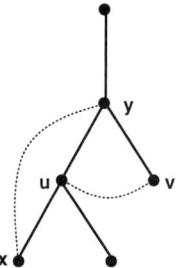

FIGURE 8.9
The edge $\{x, y\}$ is not a cross-edge because y is an ancestor of x. The edge $\{u, v\}$ is a cross-edge because neither u nor v is an ancestor of the other.

peo σ of G has no cross-edges.

To prove the theorem, we introduce the notion of an end-high path. For a graph G with a total ordering on its nodes, an *end-high path* is a path in G whose endpoints are higher in the total ordering than all internal nodes.

LEMMA 8.1   *Rose, Tarjan, and Lueker*
For a chordal graph G with the nodes ordered by a peo, *every end-high path has adjacent endpoints.*

PROOF
Suppose there is an end-high path with endpoints x and y. We must show that x and y are adjacent. Let P be the shortest end-high path with these endpoints. If P consists of a single edge between x and y, we are done. Therefore, assume for a contradiction that P contains internal nodes. Let v be the lowest internal node in P. The node that precedes v in P and the node that follows v in P are clearly neighbors of v, since P is a path. By choice of v, they are higher than v in the peo. Hence by definition of a peo, they are adjacent. The edge between them is a "short-cut" that allows us to skip v. That is, removing v from the path P yields a shorter end-high path. This contradicts the minimality of P. ∎

The proof of Theorem 8.2 follows easily from Lemma 8.1. The proof is based on an observation illustrated in Figure 8.10.

PROOF
Suppose for a contradiction that $\{v, y\}$ is a cross-edge. Let x be the

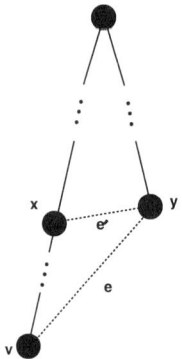

FIGURE 8.10
The existence of a cross-edge e connecting v and y implies the existence of a cross-edge e' connecting x and y.

highest ancestor of v that is less than y in the peo. There is an end-high path starting at y, going to v and then up through the tree to x. Hence by Lemma 8.1, x and y are adjacent. Since $\{v, y\}$ is a cross-edge, y is not an ancestor of v. In particular, x's parent $p(x)$ is not y. By choice of x, therefore, $p(x)$ is higher in the peo than y. But $p(x)$ is defined to be the lowest neighbor of x that is higher than x, and hence $p(x)$ must be lower than y. We have a contradiction. ∎

Another important structural property of the elimination tree is stated by the following simple corollary to Lemma 8.1.

COROLLARY 8.1
Suppose u is a descendent of v in the elimination tree determined by a peo. Then every neighbor of u that is a proper ancestor of v is also a neighbor of v.

PROOF
There is an end-high path starting each such neighbor of u, proceeding through u and up the tree to v. Hence by Lemma 8.1, each such neighbor is adjacent to v. ∎

The fact that an elimination tree is a depth-first search tree makes it a useful tool for choosing a chordal graph separator. In particular, we shall show that a node-separator for the elimination tree corresponds to a clique-

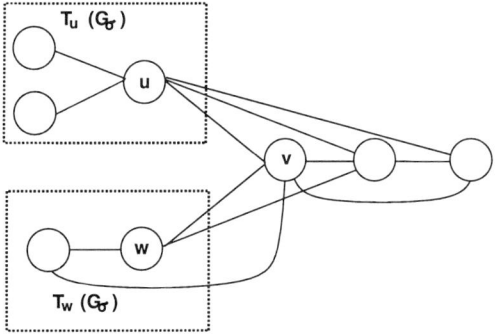

FIGURE 8.11
When the node v and its higher neighbors are removed, the subtrees rooted at children of v become separated from each other and from the remainder of the graph.

separator for the chordal graph. A *clique-separator* for a graph G is a clique K in G such that every component of $G - K$ has no more than half the nodes of G. The fact that every chordal graph has a clique-separator was first shown by Gilbert, Rose, and Edenbrandt [23]; they gave a linear-time sequential algorithm for finding such a separator. Using the elimination tree, Klein showed how to find a clique-separator efficiently in parallel.

The key lemma states that removing a node v and its higher neighbors separates each subtree rooted at a child of v from the rest of the graph. For a rooted tree T containing a node v, let T_v denote the subtree of T rooted at v. The idea for finding the clique-separator is illustrated in Figure 8.11 and is stated formally in the following lemma.

LEMMA 8.2
Let G be a chordal graph. Let $T = T(G_\sigma)$ be the elimination tree determined by a peo σ of G. Let $\hat{v}$ be a node of G, with children $v_1, \ldots, v_k$ in T. Let K be the clique of G consisting of $\hat{v}$ and its higher neighbors. Then $G[T_{v_i}]$ is a connected component of $G - K$, for $i = 1, \ldots, k$.

PROOF
To see that $G[T_{v_i}]$ is connected in $G - K$, note that edges in T_{v_i} are edges in G, and hence T_{v_i} is a spanning tree of $G[T_{v_i}]$. None of the nodes in T_{v_i} are in K, so $G[T_{v_i}]$ remains connected when K is removed from G.

Suppose there is an edge between a node v in T_{v_i} and a node w not in $K \cup$

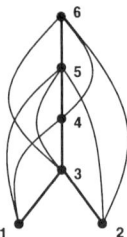

FIGURE 8.12
Each node v sends to its parent $p(v)$ a list of v's higher neighbors (excluding $p(v)$). Each node w verifies that it is a neighbor of every node on every list it received. For example, in the graph depicted, 1 sends to its parent 3 the list $\{4,5\}$, and 2 sends to its parent 3 the list $\{5,6\}$. The node 3 verifies that 4, 5, and 6 are among its neighbors.

T_{v_i}. The edge cannot be a cross-edge in T, so w must be an ancestor of v; since w is not in T_{v_i}, it must be an ancestor of $\hat{v}$ as well. Hence by Corollary 8.1, w must be adjacent to $\hat{v}$, so w belongs to K, a contradiction. ∎

As a corollary to Lemma 8.2, we can show that a chordal graph has a clique whose removal breaks the graph into pieces of at most half the size. Let the node $\hat{v}$ of Lemma 8.2 be the lowest node in the elimination tree having more than $n/2$ descendents. Then every component $G[T_{v_i}]$ has at most $n/2$ nodes, but together these components comprise at least $n/2$ nodes. Hence the clique consisting of $\hat{v}$ together with its higher neighbors is a separating clique.

8.5.2 Parallel Recognition of Chordal Graphs, and Identification of Maximal Cliques

As in the sequential case, the most efficient parallel algorithm for determining whether a graph G is chordal is to apply the parallel peo algorithm to G, and then determine whether the resulting node-ordering is in fact a peo. We carry out the second step using a parallel version of a technique from [41].

To determine whether a given node-ordering σ is a peo of G, we start by constructing the elimination tree $T(G_\sigma)$. Then each node v sends to its parent $p(v)$ in the elimination tree a list of v's higher neighbors (excluding $p(v)$). Finally, each node w sorts the elements of all the lists it received, together with w's own adjacency list, and verifies that it is a neighbor of every node on every list it received. An example is shown in Figure 8.12.

CLAIM 8.1

The node-ordering is a **peo** *if and only if no verification step fails.*

PROOF

Suppose the node-ordering is a **peo**. Then for every node v, the higher neighbors of v form a clique. In particular, the parent of v is adjacent to all v's other higher neighbors. Thus every verification step succeeds.

Suppose no verification step fails. We claim that for each node v, the higher neighbors of v form a clique. The proof is by reverse induction on the depth d of v in the tree $T(G_\sigma)$. The claim is trivial for $d = 0$, because the root has no higher neighbors. Suppose the claim holds for d, and let v be a node at depth $d + 1$. By the inductive hypothesis, $p(v)$ and its higher neighbors form a clique K. By the success of $p(v)$'s verification step, every higher neighbor of v is a neighbor of $p(v)$, and hence lies in K. This proves the induction step. ∎

The claim shows that we can determine whether a given ordering is a **peo** of G. The total number of items in all lists is $\sum_v (d(v) - 1)$, where $d(v)$ is the number of higher neighbors of v. Equivalently, $d(v)$ is the number of edges from v to higher neighbors; hence $\sum_v d(v) \leq m$. We can therefore allocate to each node w a number of processors equal to the number of items w must sort. All the sorts can then be carried out in $O(\log n)$ time. Thus the time for carrying out verification is $O(\log n)$ using $n + m$ processors using a CRCW PRAM. (In fact, a CREW PRAM is sufficient for this algorithm.)

Assuming the node-ordering is in fact a **peo**, we can use a similar algorithm to find the set of nodes v such that $C(v)$, the set of nodes consisting of v and its higher neighbors, is a maximal clique. Each node v computes the number $d(v)$ of higher neighbors it has, and sends the value $d(v)$ to its parent. Then each node w compares the values it received to $d(w)$, and concludes that $C(w)$ is a maximal clique only if none of the values it received exceeds $d(w)$. An example is shown in Figure 8.14.

The correctness of this procedure relies on the fact, immediate from Corollary 8.1, that every higher neighbor of a node v is either v's parent $p(v)$ or a higher neighbor of $p(v)$. If the number of higher neighbors of v exceeds the number of higher neighbors of $p(v)$, then the clique $C(p(v))$ is strictly contained in the clique $C(v) = \{v\} \cup C(p(v))$, and is therefore not maximal.

Conversely, suppose some clique C strictly contains $C(w)$; we shall show that w received a value exceeding $d(w)$. This argument is illustrated in Figure 8.13. Let u be the lowest node in C. If u was higher than w,

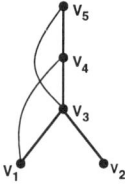

FIGURE 8.13
Identification of maximal cliques. For the graph depicted, v_3 has two higher neighbors, v_4 and v_5, so $d(v_3) = 2$. Similarly, $d(v_1)$ is 2, $d(v_2)$ and $d(v_4)$ are 1, and $d(v_5)$ is 0. The node v_3 receives the numbers 2 and 1 from its children v_1 and v_2; neither of these numbers exceeds $d(v_3)$, so we conclude that $C(v_3) = \{v_3, v_4, v_5\}$ is a maximal clique. The nodes v_1 and v_2 have no children, so the condition is trivially satisfied, and we conclude that the cliques $C(v_1) = \{v_1, v_3, v_4\}$ and $C(v_2) = \{v_1, v_3\}$ are maximal.

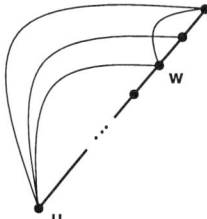
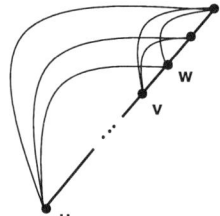

FIGURE 8.14
Proof of correctness of the procedure for identifying maximal cliques. If $C(w)$ is not a maximal clique, then w has a descendent u such that $C(u)$ contains $C(w)$. Let v be the child of w that has u as a descendent. Then $C(v)$ contains $\{v\} \cup C(w)$, so $d(v) > d(w)$. We conclude that w receives from one of its children (namely v) a number exceeding $d(w)$.

then C would not include w, and if u equaled w, C would be contained in $C(w)$, as we showed in Section 8.3. Hence u must be lower than w. Since u and w are together in a clique, there is an edge between them, and so by Lemma 8.2, u is a descendent of w. Let v be the child of w that is an ancestor of u. By Corollary 8.1, the neighbors of u that are proper ancestors of v (which include the nodes of $C(w)$) are also neighbors of v. Hence $d(v) \geq |C(w)| = 1 + d(w)$.

8.5.3 Maximum-Weight Clique

Suppose each node is assigned a non-negative weight. As Gavril observed, any maximum-weight clique is maximal, so finding all the maximal cliques is tantamount to finding a maximum-weight clique. For each clique $C(v)$, compute the total weight of its nodes, and compare the total weights to find the maximum. This procedure takes $O(\log n)$ time using $(\sum_v |C(v)|)/\log n \leq (n+m)/\log n$ processors of a CREW PRAM.

8.5.4 Breadth-First Search Trees

To obtain a breadth-first search tree of G, we construct a tree similar to the elimination tree, by choosing the parent of each node v (except the highest node) to be the highest neighbor of v. Let T be the resulting tree, rooted at the highest node, which we shall denote by r. It is easy to see that T can be computed in $O(\log n)$ time using $n+m$ processors of a CREW PRAM. Our proof that T is a breadth-first search tree relies on two claims.

CLAIM 8.2

For each node v the shortest path from v to r is monotonically increasing with respect to the peo.

PROOF
Any subpath whose internal nodes are lower than its endpoints can be replaced by a direct edge between the endpoints, by Lemma 8.1. ∎

For the second claim, let $s(v)$ denote the length of the shortest path in G from v to r.

CLAIM 8.3
If w is a descendent of v in the elimination tree, then $s(w) \geq s(v)$.

PROOF
The proof is by reverse induction on the position of w in the peo. The basis, in which $w = r$, is trivial. Otherwise, let w' be the second node on a shortest path from w to r, so $s(w) = 1 + s(w')$. By Claim 8.2, w' is higher than w, hence an ancestor of w by Lemma 8.2. If w' is a descendent of v, then $s(w') \geq s(v)$ by the inductive hypothesis. If w' is an ancestor of v, then there is an end-high path from v back along tree edges to w, and then forward to w', proving by Lemma 8.1 that v is adjacent to w', and hence that $s(v) \leq 1 + s(w')$. ∎

For each node $v \neq r$, let $q(v)$ be the highest neighbor of v. Any other neighbor w of v is a descendent of $q(v)$ in the elimination tree, so

$s(w) \geq s(q(v))$ by Claim 8.3. It follows that $q(v)$ is the second node in a shortest path from v to the highest node of G. Thus the tree defined by $q(\cdot)$ is a breadth-first search tree.

8.5.5 Maximum Independent Set

As we discussed in Section 8.4, Gavril showed that a maximum independent set $\mathcal{I}$ of the chordal graph G is obtained by the greedy maximal-independent-set algorithm when applied to nodes in order of the peo $\sigma = v_1 \ldots v_n$. We review his algorithm. First put v_1 into $\mathcal{I}$, and delete v_1 and its neighbors. Next, put the lowest remaining node in $\mathcal{I}$, and delete it and its neighbors, and so forth. Since in each iteration the lowest possible node v_i is placed in $\mathcal{I}$, all v_i's neighbors at that time are *higher* neighbors. Every deleted node, therefore, is either in $\mathcal{I}$ or is a *higher* neighbor of a node in $\mathcal{I}$. Recall that for each node x, $C(x)$ is a clique consisting of x together with its higher neighbors. Consequently, when the algorithm terminates, the family of cliques $\{C(x) : x \in \mathcal{I}\}$ is a clique cover (a set of cliques whose union contains all the nodes). Because any independent set has size at most that of any clique cover, it follows that the above procedure has identified a *maximum* independent set and a *minimum* clique cover.

We want to simulate Gavril's sequential greedy algorithm in parallel. First suppose that the elimination tree $T(G_\sigma)$ determined by the peo σ is a *path* with leaf x. In this case, we give a simple algorithm PMIS for simulating Gavril's algorithm. For each node v, let $b[v]$ be the lowest ancestor of v in $T(G_\sigma)$ that is not adjacent to v in G (or v, if no such ancestor exists).

CLAIM 8.4
The greedy independent set consists of $x, b[x], b[b[x]]$, and so on.

This set can be determined quickly in parallel using standard pointer-jumping techniques. The implementation shown in Figure 8.15 requires $O(\log n)$ time, m processors, and $O(m \log n)$ space using a CREW PRAM; use of more sophisticated techniques (e.g., [3], [11]) achieves the same time bound using only $m/\log n$ processors and $O(m)$ space on a EREW PRAM.

PROOF
Suppose we put x into $\mathcal{I}$ and delete the neighbors of x. The node $b[x]$ is by definition the lowest undeleted node. Moreover, we assert that for each undeleted node v, $b[v]$ is undeleted. If $b[v]$ were a neighbor of x, then by Corollary 8.1, $b[v]$ would be a neighbor of v, contradicting the

ALGORITHM
PMIS: Finding a maximum independent set.

Step 1 For each node v, let $b_0[v]$ denote the lowest ancestor of v that is not adjacent to v (or else v).

Step 2 For stages $k = 0, \ldots, \lceil \log n \rceil - 1$, for each node v, let $b_{k+1}[v] := b_k[b_k[v]]$.

Step 3 Mark the leaf x as being in the independent set.

Step 4 For stages $k = \lceil \log n \rceil - 1, \lceil \log n \rceil - 2, \ldots, 0$, for each marked node v, mark $b_k[v]$.

FIGURE 8.15
A simple implementation of the algorithm PMIS for finding a maximum independent set when the elimination tree is a path.

definition of $b[v]$. This argument proves the assertion; the claim follows by induction on the length of the elimination path. ∎

To generalize this procedure to the case in which $T(G_\sigma)$ is a tree, we use an idea of Naor, Naor, and Schäffer discussed at the beginning of Section 8.5: pruning terminal branches. A *terminal branch* of a tree is a maximal path of degree-two nodes ending in a leaf. Naor, Naor, and Schäffer observe that deletion of all terminal branches of a tree yields a new tree with half as many leaves. Therefore, $O(\log n)$ iterations of terminal branch elimination suffice to eliminate the entire tree. By applying this idea to the elimination tree, we obtain a parallel maximum independent set algorithm requiring only $m/\log n$ processors.

Before giving the algorithm, we introduce a bit of tree-surgery, called *splicing*. To *splice* a node v out of a tree T is to remove the node and reattach any children of v to v's parent in T, as illustrated in Figure 8.16. If a set of nodes are to be spliced from a tree, the resulting tree does not depend on the order in which the nodes are spliced out. In fact, they can all be spliced out

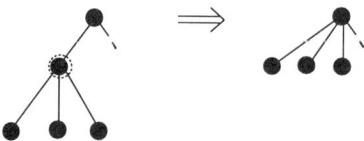

FIGURE 8.16
To splice a node out of a tree, remove the node and reattach the node's children to its parent.

at once; for each node v to be spliced out, the children of v are reattached to the lowest ancestor of v that is not to be spliced out.

We now describe the algorithm MIS, shown in Figure 8.17, for constructing a maximum independent set in a chordal graph G. Let $T(G_\sigma)$ be the elimination tree for G determined by a peo σ. The algorithm maintains a set $\mathcal{I}$, the independent set under construction, and a tree T, obtained from the elimination tree $T(G_\sigma)$ by splicing out nodes. We prove by induction that the following invariant holds before and after each iteration of the algorithm.

Invariant

1. $\mathcal{I}$ is an independent set.
2. Every neighbor of a node of $\mathcal{I}$ is in fact a *higher* neighbor of some node of $\mathcal{I}$.
3. T is obtained from $T(G_\sigma)$ by splicing out the nodes of $\mathcal{I}$ and their neighbors.

The algorithm terminates when T is empty, at which point $\mathcal{I}$ is an independent set such that every node of G is either in $\mathcal{I}$ or is a higher neighbor of some node in $\mathcal{I}$. Thus, as in Gavril's algorithm, the family of cliques $C(x) = \{x\} \cup \{\text{higher neighbors of } x\}$ for $x \in \mathcal{I}$ is a minimum clique cover, and $\mathcal{I}$ is a maximum independent set.

Initially, $\mathcal{I} = \emptyset$ and T is $T(G_\sigma)$, the elimination tree, so the invariant holds trivially. Suppose the invariant holds through the first k iterations of the algorithm, and consider the $k+1^{st}$ iteration. For each terminal branch $\mathcal{B}$, the algorithm finds a maximum independent set $\mathcal{I}_\mathcal{B}$ of the subgraph induced on $\mathcal{B}$, using PMIS as a subroutine. If two nodes of T lie in different terminal branches, neither is an ancestor of the other in $T(G_\sigma)$, and hence the two nodes are not adjacent, by Lemma 8.2. Thus $\bigcup_\mathcal{B} \mathcal{I}_\mathcal{B}$ is an independent set in G, where the union is over all terminal branches of T. Moreover, T contains

ALGORITHM
MIS: *Finding a maximum independent set.*
To initialize, let $\mathcal{I} = \emptyset$ and let T be the elimination tree $T(G_\sigma)$ for G determined by a peo σ.
While T is not empty,

Step 1 Use the algorithm PMIS to find the greedy maximum independent set $\mathcal{I}_B$ of the subgraph induced on each terminal branch $\mathcal{B}$ of T.

Step 2 Add the nodes $\bigcup_\mathcal{B} \mathcal{I}_B$ to $\mathcal{I}$.

Step 3 Splice out of T the nodes $\bigcup_\mathcal{B} (\mathcal{I}_B \cup \{\text{neighbors of } \mathcal{I}_B\})$.

FIGURE 8.17
The algorithm MIS to construct a maximum independent set $\mathcal{I}$ in the chordal graph G.

---

no neighbors of $\mathcal{I}$ (by part 3 of the invariant), so $\mathcal{I} \cup (\bigcup_\mathcal{B} \mathcal{I}_B)$ is an independent set of G. Thus when the nodes $\bigcup_\mathcal{B} \mathcal{I}_B$ are added to $\mathcal{I}$ in step 2, part 1 of the invariant remains true.

To show that part 2 remains true, we must prove that every node w that is newly a neighbor of a node in $\mathcal{I}$ is in fact a *higher* neighbor of a node in $\mathcal{I}$. Our simulation PMIS of Gavril's algorithm on terminal branches $\mathcal{B}$ guarantees this property when w lies on a terminal branch. Suppose therefore that w does not lie on a terminal branch, and let $v \in \mathcal{I}$ be a neighbor of w. Since there are no cross-edges by Lemma 8.2, v is either a descendent or an ancestor of w in $T(G_\sigma)$. The set $\mathcal{I}$ consists only of nodes in terminal branches of T and descendents of such nodes. Hence v must be a descendent of w, and a lower neighbor.

In the last step of an iteration of the algorithm, we splice out of T all nodes newly added to $\mathcal{I}$ and their neighbors. This step ensures that part 3 holds at the end of the iteration.

Having proved that the invariant continues to hold, we now consider the implementation of the algorithm MIS (Figure 8.17). In step 1, the algorithm must identify the nodes lying in terminal branches of T. An application of the Euler tree technique [45] suffices to determine, for each node v of T, the number of leaf descendents of v. The nodes for which this number is 1 are the nodes in terminal branches. Next, the algorithm must find a maximum independent set in each terminal branch. For each node v in T, $b_0[v]$ is assigned the lowest ancestor w of v in T that is not a neighbor of v, *if* w lies in a terminal branch. Otherwise, $b_0[v]$ is assigned v. As in PMIS, a pointer-jumping technique is then used to mark the nodes $x, b_0[x], b_0[b_0[x]]$, and so on, for all leaves x of T. The marked nodes are added to $\mathcal{I}$ in step 2.

To implement the splicing in step 3, we again use a pointer-jumping technique; for each node v, we compute the lowest ancestor of v in T that is not to be spliced out. Each step can be implemented in $O(\log n)$ time using $m/\log n$ processors and $O(m)$ space. Each iteration removes all nodes in terminal branches and hence reduces the number of leaves in T by a factor of two; consequently, $\lceil \log n \rceil + 1$ iterations suffice, for a total of $O(\log^2 n)$ time.

8.5.6 Minimum Coloring

Gavril showed that applying the greedy coloring algorithm to the nodes of G in reverse order of a peo yields a minimum coloring. In this section, we give a parallel algorithm for minimum coloring. The algorithm is efficient in its use of parallelism—it requires only $(n+m)/\log n$ processors—but takes $O(\log^2 n)$ time on a CRCW PRAM. It requires as input a chordal graph and a perfect elimination ordering.

The Color-Mapping Strategy

To gain intuition, let us consider the problem of coloring a graph G consisting of two overlapping subgraphs $\widehat{H_0}$ and $\widehat{H_1}$. We assume that each edge of G occurs in either $\widehat{H_0}$ or $\widehat{H_1}$—that is, no edge goes between $\widehat{H_0} - \widehat{H_1}$ and $\widehat{H_1} - \widehat{H_0}$. We would like to obtain a minimum coloring of G from minimum colorings of $\widehat{H_0}$ and $\widehat{H_1}$. The difficulty is ensuring that the separate colorings are consistent on the nodes $\widehat{H_0} \cap \widehat{H_1}$ common to the two subgraphs; if this were the case, merging the two colorings would be easy. We would take the color of a node of G to be the color assigned to that node in one of the two separate colorings. The resulting color assignment would not assign the same color to two adjacent nodes, because the nodes are also adjacent in either $\widehat{H_0}$ or $\widehat{H_1}$.

To make things easier, let us assume that the common nodes form a clique. The advantage is that there is only one way to color the nodes of a clique, up to renaming of colors—each node of the clique must receive a different color.

Our strategy for making the separate colorings consistent, which we call the *color-mapping strategy*, relies on the fact that colors are interchangeable. We start by independently choosing a coloring c_i for each subgraph $\widehat{H}_i$ ($i = 0, 1$), ignoring the fact that they overlap. We must then carry out a *repair step*, in which the coloring of one subgraph, say $\widehat{H}_1$, is modified to be consistent with the coloring of the other subgraph $\widehat{H}_0$ on the common nodes. The repair step consists in finding a one-to-one mapping of the colors used by c_1 satisfying the following condition:

> *Consistency condition*: the color assigned by c_1 to a common node $v \in \widehat{H}_1 \cap \widehat{H}_0$ is mapped to the color assigned to the same node by c_0.

Once we have found such a mapping, we can use it to transform the coloring c_1 of H_1; each color is replaced by its image under the mapping. By the consistency condition, the resulting coloring c'_1 assigns the same color to each common node as c_0 does. We can therefore combine the colorings c'_1 and c_0 to obtain a coloring c of the entire graph G.

What properties must the one-to-one mapping satisfy in order that the coloring of G be a *minimum* coloring? We must ensure that the transformed coloring c'_1 *re-uses* colors used by c_0 as much as possible. Let x_i be the number of colors used by c_i, for $i = 0, 1$. If $x_1 \leq x_0$, we can map every color used by c_1 to a color used by c_0 (while satisfying the consistency condition). In this case, a total of x_0 colors are used by the merged coloring c of G. If $x_1 > x_0$, we must map $x_1 - x_0$ of the colors used by c_1 to colors not used by x_0. In this case, a total of x_1 colors are used. Let us assume inductively that c_0 and c_1 are *minimum* colorings of $\widehat{H}_0$ and $\widehat{H}_1$, respectively. Then since these graphs are subgraphs of G, the minimum number of colors required for G is at least $\max(x_0, x_1)$. The merged coloring c of G achieves this lower bound; hence it is minimum.

A Divide-and-Conquer Approach to Coloring

To apply the color-permuting strategy to minimally coloring a chordal graph G, we use divide-and-conquer. For now, we outline the basic approach; we will later consider each step in more detail. Divide the graph G into node-disjoint subgraphs $H_0, H_1, \ldots, H_s$ such that the neighborhood of each H_i for $i = 1, \ldots, s$ is a clique K_i contained in $\widehat{H}_0$. Next, for $i = 1, \ldots, s$, let $\widehat{H}_i = G[H_i \cup K_i]$. The resulting subgraphs $\widehat{H}_0, \widehat{H}_1, \ldots, \widehat{H}_s$, illustrated in Figure 8.18, satisfy the conditions that make possible the color-mapping strategy.

1. The intersection of each $\widehat{H}_i$ ($i > 0$) with $\widehat{H}_0$ is a clique.
2. The union of $\widehat{H}_0, \ldots, \widehat{H}_s$ is the whole graph G.

368 Chapter 8. Parallel Algorithms for Chordal Graphs

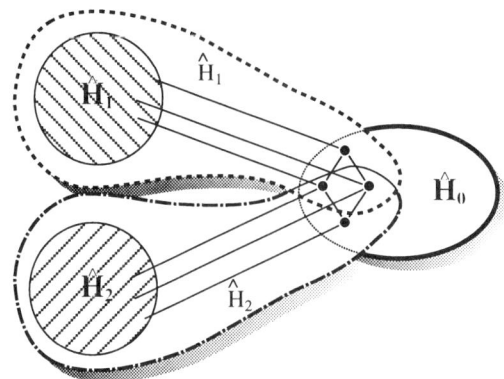

FIGURE 8.18
Divide G into disjoint subgraphs $\widehat{H}_0, H_1, H_2, \ldots, H_k$ (here $k = 2$) such that the neighborhood of each H_i ($i > 0$) is a clique contained in $\widehat{H}_0$.

In parallel, recursively find minimum colorings $c_0, c_1, \ldots, c_k$ for the subgraphs $\widehat{H}_0, \widehat{H}_1, \ldots, \widehat{H}_s$. For $i = 1, \ldots, s$ in parallel, find a transformed coloring c'_i of $\widehat{H}_i$. Derive the coloring of G from the colorings $c_0, c'_1, \ldots, c'_k$.

The questions that we need to address in order to turn this outline into an algorithm are:

1. How do we choose the decomposition of G into disjoint subgraphs?
2. Given that the subgraphs on which we recurse are not disjoint, how can we achieve a linear processor bound?
3. How do we guarantee small recursion depth?
4. How do we carry out the repair step, in which the colors are mapped?

We first address question 2. It seems that since the subgraphs $\widehat{H}_0, \widehat{H}_1, \ldots, \widehat{H}_s$ on which the algorithm recurses are not disjoint—$\widehat{H}_i$ shares with $\widehat{H}_0$ the edges and nodes in the cliques K_i—we cannot hope to make do with only one processor per edge and one processor per node.

To cope with this difficulty, we use a simple idea called *clique management*. Given the knowledge that K_i is a clique, the algorithm need not inspect the edges between nodes of K_i during the recursive call on $\widehat{H}_i = G[H_i \cup K]$; it "knows" without checking that every pair of nodes in K_i are adjacent. Hence for this recursive call, we do not need to assign processors to the edges of K_i, only to the "relevant edges"—those within H_i and those between H_i and K_i. For distinct subgraphs H_i and H_j, the relevant edge-sets are disjoint. Hence we can carry out the recursive calls while holding to our rule of assigning one processor per edge of the original graph.

The above idea helps us deal with edge duplications—how do we handle node duplications? Because every node of K_i is a neighbor of a node of H_i, every node of K_i is the endpoint of an edge relevant to H_i. We can have each processor assigned to a relevant edge do double-duty—it handles the edge and also handles the endpoint in the clique.

According to the idea of clique management, the recursive coloring procedure takes *two* inputs, the graph G to be colored and a clique K_0 contained in G. A sketch of the procedure COLOR(G, K_0) is given in Figure 8.19. Step 1 handles the base case of the recursion, where the graph G is a clique. In this case, it is easy to minimally color G—each node must receive a different color. Otherwise, step 2 finds a decomposition of $G - K_0$ into some number of subgraphs $H_0, \ldots, H_s$ such that for $i = 1, \ldots, s$, the neighborhood of H_i is a clique K_i. (Note that this clique K_i may share nodes with the clique K_0.) Step 3 recursively colors each of the subgraphs $\widehat{H_i}$; during the coloring of $\widehat{H_i}$, the clique K_i is managed.

In order to ensure that the recursion depth is small (question 4), the decomposition of $G - K_0$ is such that each subgraph H_i is at most half the size of $G - K_0$. Thus we measure progress not in terms of the size of the graph G being colored, but in terms of the number of nodes of the graph that are not in the clique being managed. On input G, K_0, the number of levels of recursion is $1 + \log|G - K_0|$. We shall show that each level can be implemented in $O(\log t)$ time using t processors, where t is the number of edges with at least one endpoint in $G - K_0$. To find a coloring in the original graph G, call COLOR$(G, \emptyset)$.

At last we address question 1: how is the decomposition chosen? This is where we use the special properties of the eliminination tree; in fact, this is the only place we use chordality. Inductively we assume we have an elimination tree T for G in which the nodes of K_0 are the highest nodes. Using the Euler-tour technique [45], choose the lowest node $\hat{v}$ in T that has more than $p/2$ descendents, where $p = |G - K_0|$. Let $v_1, \ldots, v_s$ be the children of $\hat{v}$ in T; then each subtree T_{v_i} has at most $p/2$ nodes. We let K be the clique $\{\hat{v}\} \cup$ {higher neighbors of $\hat{v}$}, and let $H_i = G[T_{v_i}]$. By Lemma 8.2, the subgraphs $H_1, \ldots, H_s$ are connected components of $G - K$. The neighborhood of each H_i in $G - H_i$ is contained in the clique K, and is therefore itself a clique K_i. Let $H_0 = G - K_0 - \bigcup_{i=1}^{s} H_i$. By choice of $\hat{v}$, H_0 has at most $p/2$ nodes.

ALGORITHM
COLOR(G, K_0)

Input: Connected graph G containing a clique K_0, such that every node of K_0 has a neighbor in $G - K_0$.

Output: Optimal coloring of G.

Step 1 If $G - K_0$ consists of a single node v, then G is a clique; assign the first $|V(G)|$ colors to its nodes, and end.

Step 2 Otherwise, break $G - K_0$ into subgraphs $H_0, \ldots, H_s$ such that
- each subgraph has size at most half that of $G - K_0$;
- $H_1, \ldots, H_s$ are distinct components of $G - K_0 - H_0$; and
- for $1 \leq i \leq s$, the neighborhood of H_i in $G - H_i$ is a clique K_i.

Step 3 For $i = 0, \ldots, s$ in parallel, call COLOR$(\widehat{H}_i, K_i)$, where $\widehat{H}_i = G[H_i \cup K_i]$, to get an optimal coloring c_i of $\widehat{H}_i$.

Step 4 For $i = 1, \ldots, s$ in parallel, modify the coloring c_i to be consistent with c_0 on the nodes $V(K_i)$ they have in common.

Step 5 Merge the colorings to obtain a coloring of G.

FIGURE 8.19
The recursive algorithm COLOR for finding an optimal coloring of a chordal graph G.

Step 4, in which we modify the colorings to be consistent, can be implemented using parallel prefix computation. We shall next describe this step in greater detail.

The Repair Step

It is convenient to use the positive integers as our colors. The repair step consists of finding a mapping ϕ_i from the colors used by c_i (for $i = 1, \ldots, s$)

into the set of all colors, and computing the transformed coloring $c'_i(v) = \phi_i(c_i(v))$, such that the following two conditions are satisfied:

Consistency condition: For each $1 \leq i \leq s$, the color assigned by c_i to a common node $v \in \widehat{H_i} \cap \widehat{H_0}$ is mapped to the color assigned to the same node by c_0.

Parsimony condition: The maximum color used by the transformed colorings $c'_1, \ldots, c'_s$ is no more than the maximum color used by the original recursive colorings $c_0, c_1, \ldots, c_s$.

The consistency condition ensures that each transformed coloring c'_i can be merged with c_0. The parsimony condition ensures that the resulting merged coloring is minimum (if the recursive colorings were minimum.)

To make the repair step easier, we inductively require that for $i = 0, \ldots, s$, the coloring c_i assigns colors 1 through $|K_i|$ to the nodes of the associated clique K_i. This requirement is easy to satisfy at the base of the recursion, step 1. It is automatically preserved in going from one level of recursion to the next higher level: the colors assigned by c to the nodes of K_0 are exactly those assigned by c_0.

Given the above requirement is satisfied, we now show how to transform a coloring c_i ($1 \leq i \leq s$) so that the consistency condition and the parsimony condition are satisfied. Let x_i be the number of colors used by c_i. By the requirement, the colors 1 through $|K_i|$ are assigned by c_i to the nodes of K_i; these are the colors that must be mapped to corresponding colors used by c_0 in order that the consistency condition be satisfied. We therefore partially define the mapping ϕ_i by

$$\forall v \in K_i : c_i(v) \longmapsto c_0(v) \qquad (8.4)$$

The other colors used by c_i, colors $|K_i| + 1$ through x_i, are assigned to the nodes of H_i. We must map these to other available colors used by c_0 (and use additional colors only when necessary). In order to help us identify the other available colors, we perform a preliminary computation. For each color q between $|K_i| + 1$ and x_i, we count the colors less than q that c_0 uses in coloring K_i: let

$$A_i[q] := |\{c_0(v) < q : v \in V(K_i)\}|.$$

(The values $A_i[\cdot]$ can be computed using a parallel prefix computation.) A color q between $|K_i| + 1$ and x_i is the $q - |K_i|\underline{{}^{th}}$ color used by c_i on H_i. We want to map it to the $q - |K_i|^{\underline{th}}$ color not already assigned by c_0 to a node of K_i. We therefore extend the definition (8.4) of the mapping ϕ_i as follows:

$$\forall q(|K_i| + 1 \leq q \leq x_i) : \quad q \longmapsto q - |K_i| + A_i[q] \qquad (8.5)$$

Thus the colors of nodes of H_i are mapped to colors starting at 1, with gaps only for colors already assigned to nodes of K_i. We use this mapping to transform the coloring c_i to obtain c'_i: for each node $v \in \widehat{H_i}$, define

$$c'_i(v) := \begin{cases} c_0(v) & \text{if } v \in K_i \\ c_i(v) - |K_i| + A_i[c_i(v)] & \text{if } v \in H_i \end{cases}$$

This assignment ensures that, for any node $v \in H_i$, colors 1 through $c'_i(v)$ all appear in the coloring c'_i of $\widehat{H_i}$. As a consequence, we claim, the parsimony condition is satisfied. If the maximum color q used by the transformed coloring c'_i is used for a node of K_i, then q is at most the maximum color used by c_0. Otherwise, colors 1 through q all appear in c'_i. Since c'_i uses the same number of colors as c_i, it follows that c_i uses at least q colors. The claim is proved.

Resource Requirements

The only nontrivial computation in implementing step 4 is computing the $A_i[q]$ values. For each $1 \leq i \leq s$, construct an array of length $x_i \leq |H_i|$, and initialize all entries to 1. Then set to zero the $q^{\underline{th}}$ entry for each color q that c_0 assigns to a node of K_i. Finally, perform a parallel prefix sum computation on the array to compute the values $A_i[q]$. The total work is proportional to $\sum_i |K_i| + |H_i|$.

Let t_i be the number of edges that either lie in H_i or connect H_i to K_i. Since H_i is connected, the number of nodes in H_i is at most one more than the number of edges in H_i. Since every node in K_i is a neighbor of some node in H_i, the number of nodes in K_i is at most the number of edges connecting H_i to K_i. Thus $|H_i \cup K_i| \leq t_i + 1$. So the total work is proportional to $\sum_i t_i$, which is just the number t of edges that either lie within $G - K_0$ or connect $G - K_0$ to G.

Assume inductively that $O(\log t_i \log |H_i|)$ time and $O(t_i / \log t_i)$ processors are sufficient to recursively color $G[H_i \cup K_i]$. Then $O(t/\log t)$ processors are sufficient to recursively color all the subgraphs $G[H_i \cup K_i]$ in $O(\log t(\log(|G - K_0|) - 1))$ time and to combine the colorings in $O(\log t)$ time, for a total of $O(\log t \log |G - K_0|)$ time.

8.6
The Framework for Finding a Perfect Elimination Ordering

8.6.1 Preliminaries

In this section, we develop two efficient algorithms for finding a perfect elimination ordering, one sequential and one parallel algorithm. The optimization algorithms of the previous section all assume that a perfect elimination ordering (or peo) is part of the input; hence the peo algorithms are crucial for finding a minimum coloring, a maximum independent set, etc., efficiently.

Recall the historically first peo algorithm, that of Fulkerson and Gross: repeatedly find a *simplicial* node (one whose neighbors form a clique), and delete it. This algorithm clearly consists of only n stages for an n-node graph; the disadvantage is that each stage seems fairly expensive. Indeed, even checking whether a given node is simplicial seems to require $\Omega(m)$ time in the worst case.

The theorem of Dirac guaranteeing the existence of a simplicial node in every chordal graph gives us a clue as to how to proceed in order to get an improved sequential algorithm. Dirac's theorem in fact ensures that every graph with at least two nodes has at least *two* simplicial nodes. For any node v, therefore, there is a simplicial node other than v; once it is eliminated, there is remaining another simplicial node other than v, and so on, until there is only v remaining. We see that we can initially set aside v to be eliminated last. This idea suggests we might be able to build a peo backwards, first choosing the last node of the peo, then choosing the second-to-last, and so on. The linear-time peo algorithm of Rose, Tarjan, and Lueker does exactly this.

Dirac's theorem ensures that the choice of last node is arbitrary. Once the last node of a peo is chosen, however, the choice of the second-to-last node is far from arbitrary, the subsequent choice of the third-to-last node is even more restricted, and so on. We need a condition that guides us in making these choices so that we never make a choice that rules out all peos.

We shall give a characterization of those partial solutions that can be extended to complete solutions. In order that our characterization yields not only a sequential algorithm but also a parallel algorithm, we use a general model of what constitutes a partial solution. A *semiorder* is a partition of a nodeset into an ordered sequence of classes: $\Psi = (C_1, C_2, \ldots, C_k)$. For example, a total ordering is a semiorder $(\{v_1\}, \{v_2\}, \ldots, \{v_n\})$ in which each class contains only one node. The partial solution in which we have chosen the last node v but have made no other decisions is represented by a

374   Chapter 8. Parallel Algorithms for Chordal Graphs

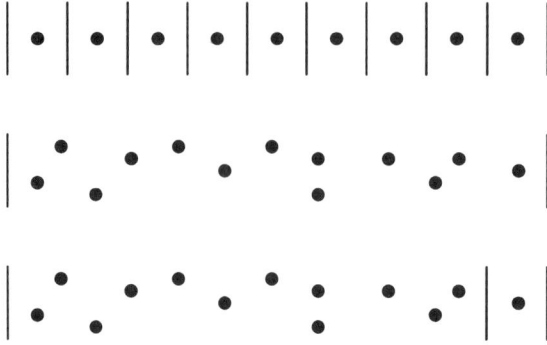

FIGURE 8.20
Example semiorders are depicted. The first is a semiorder in which every node is in a separate class, i.e., a total order. The second consists of only a single class; it is effectively a completely unordered set. The third is a semiorder in which the last node has been chosen but no other ordering decisions have been made.

semiorder $(C, \{v\})$ consisting of two classes: the second class contains v and the first class contains all other nodes. Figure 8.20 illustrates some examples of semiorders.

We make progress towards a complete solution by *refining* the semiorder, i.e., further dividing its classes into subclasses. Let $\Psi = (C_1, \ldots, C_k)$ be a semiorder on the nodeset V, and let $\Phi = (C_{i1}, C_{i2}, \ldots, C_{ir})$ be a semiorder on a class C_i of Ψ. To *refine* Ψ *by* Φ is to replace the class C_i in Ψ by the subclasses $C_{i1}, C_{i2}, \ldots, C_{ir}$ in order, yielding a new, *refined* semiorder. More generally, to *refine* a semiorder is to refine some of its classes, as illustrated in Figure 8.21.

We say one semiorder is *consistent* with a second if the second can be obtained by refining the first. For example, the semiorder with only one class is consistent with every semiorder, while a total order—a semiorder in which each class contains one node—is consistent only with itself.

For a graph G and a semiorder Ψ, we write G_Ψ for the graph G with each node labelled by the class containing it. For two nodes v and w, we write $\Psi(v) > \Psi(w)$ if v is higher than w. For example, a semiorder Ψ is consistent with another semiorder Φ if $\Psi(v) > \Psi(w)$ implies $\Phi(v) > \Phi(w)$ for all pairs of nodes v, w.

8.6. The Framework for Finding a Perfect Elimination Ordering   375

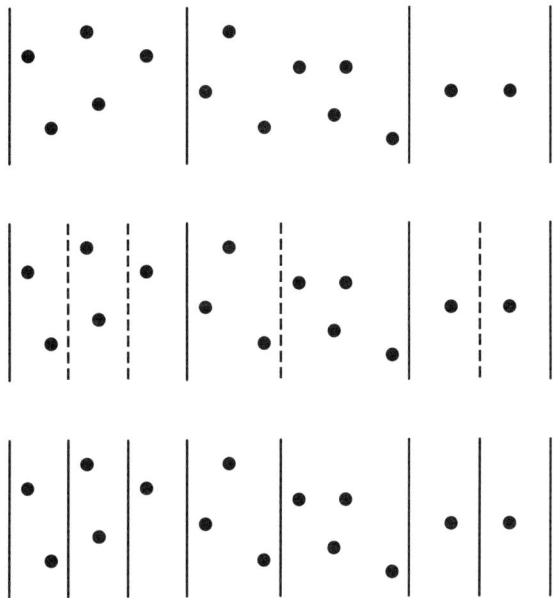

FIGURE 8.21
This figure illustrates the process of refining a semiorder. We start with a semiorder with three classes. Each class is divided into subclasses. Finally, these subclasses become the classes of the new semiorder.

Our notion of an end-high path, defined in Section 8.5.1 for graphs whose nodes are totally ordered, is equally applicable to graphs G_Ψ whose nodes are semiordered: an end-high path is a path in the graph whose endpoints are strictly higher than its internal nodes. We say a semiorder Ψ is *valid* for a graph G if every end-high path in G_Ψ has adjacent endpoints. The reason for this definition is made clear by the following theorem.

THEOREM 8.3   *End-high Theorem*
A semiorder on the nodes of a chordal graph is consistent with some **peo** *if and only if the semiorder is valid for the graph.*

Figure 8.22 gives examples of how this theorem is applied. We postpone the proof of the theorem until the end of this section; for now, we consider the special case in which the semiorder is restricted to be a total order.

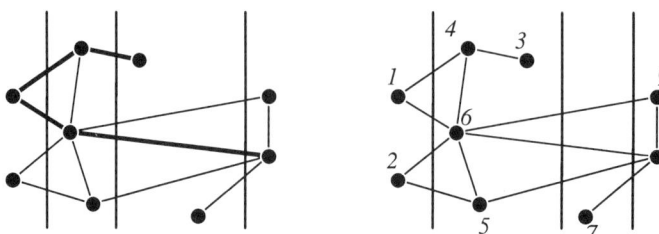

FIGURE 8.22
The first semiorder is not valid for the graph shown; the highlighted end-high path, for example, does not have adjacent endpoints. The second semiorder is valid, and the nodes are numbered in perfect elimination order.

LEMMA 8.3
A total ordering of the nodes of a chordal graph is a peo *if and only if it is valid.*

PROOF
It follows from Lemma 8.1 of Section 8.5.1 that a peo is valid. Suppose conversely that the total ordering Ψ is valid, and let v be a node. We must show that the higher neighbors of v form a clique. Let x and y be any two higher neighbors of v. Then xvy is an end-high path, so by validity, its endpoints x and y are adjacent. ∎

Terminology begets terminology. In order to prove that every end-high path has adjacent endpoints, it is easier to focus on a special kind of path, the *chordless* path. A *chord* of a path $P = v_1 \ldots v_k$ in a graph G is an edge $\{v_i, v_j\}$ in G, where $i < j-1$. We say a path is *chordless* if it has no chord. The following simple observation shows why we can focus exclusively on chordless paths.

LEMMA 8.4 *Shortcutting Lemma*
For any path P in a graph G, there is a chordless path P' with the same endpoints, such that $V(P') \subseteq V(P)$.

PROOF
If $P = v_1 \ldots v_k$ has a chord $\{v_i, v_j\}$, replace P with the path $v_1 \ldots v_i v_j \ldots v_k$, which has fewer chords; by iterating, we obtain the desired chordless path. ∎

For a graph G with a semiorder on its nodes, we say a path is *weakly end-high* if its endpoints are at least as high as all internal nodes. Contrast

ALGORITHM
ITERATED REFINEMENT
    Step 1   To initialize, let Ψ be the trivial semiorder consisting of one class.

    Step 2   While Ψ is not a total ordering.

    Step 2.1   Refine Ψ, maintaining validity

    Step 3   Output the total ordering Ψ.

FIGURE 8.23
The ITERATED REFINEMENT algorithm for finding a peo.

this notion with that of a (strictly) end-high path, whose endpoints must be strictly higher than the internal nodes. We say a path is *uniform* if all its nodes except possibly one endpoint belong to the same class.

LEMMA 8.5   *Uniformity Lemma*
If Ψ is valid for G, each chordless weakly end-high path P in G_Ψ is uniform.

PROOF
Suppose $P = v_1 \ldots v_k$ is not uniform, and let $v_i \ldots v_j$ be a maximal subpath consisting of nodes lower than the endpoints of P. Then $v_{i-1} \ldots v_{j+1}$ is an end-high path, so by validity of Ψ, $\{v_{i-1}, v_{j-1}\}$ is an edge, contradicting the chordlessness of P.   ∎

8.6.2   Iterated Refinement

We shall next lay out our basic framework for peo algorithms. In the following section, we describe the sequential algorithm of Rose, Tarjan, and Lueker, and show how it can be cast within the framework. In Section 8.8, we give the parallel algorithm of [29].

Our general framework for finding a peo, which we call *iterated refinement*, is illustrated in Figure 8.23. We assume at the outset that the input graph is chordal, and show how to construct a total ordering that is a peo if the assumption holds; once the total ordering has been constructed, the method of Section 8.5.2 can be used to verify that it is indeed a peo and hence that the input graph is indeed chordal.

The iterated refinement approach starts with the trivial semiorder, in which all nodes are in the same class, and iteratively refines the semiorder, maintaining validity throughout the process, until the semiorder becomes a total order. The initial, trivial semiorder is trivially valid. Since we maintain validity throughout, the final, total ordering is valid. But by Lemma 8.3, a valid total ordering is a peo.

The iterated refinement approach is the basis for proving the End-high Theorem.

> PROOF
> The "only if" direction is easy to prove; the proof resembles that of Lemma 8.1. Suppose Ψ is a semiorder on G that is consistent with the peo σ. We need to show that every chordless end-high path P in G_Ψ has adjacent endpoints. If P consists of a single edge $\{x, y\}$, we are done, so assume for a contradiction that P has internal nodes. Let u be the internal node with the minimum σ-number. Then the neighbors of u in P have a higher σ-number, so they are adjacent by definition of a perfect ordering, contradicting the chordlessness of P.
>
> To prove the "if" direction, suppose G is chordal. In Section 8.8, we give a procedure REFINE to validly refine any valid semiorder on G that is not a total order. By repeated refinements, we eventually obtain a valid total order, i.e., a peo. ∎

Even aside from the fact that we have not yet given the procedure RE-FINE or proved its correctness, the proof of the "if" direction may seem to be a cheat. How can we use an algorithm to prove a theorem "about" the algorithm? The answer is that the End-high Theorem is not used in proving the correctness of either the *iterated refinement* approach or the REFINE procedure; only the special case of Lemma 8.3 is needed.

The use of an algorithm to prove a combinatorial result is by no means new; illustrious examples include use of a maximum flow algorithm to prove the Max-flow Min-cut Theorem, and use of Edmonds' blossom-shrinking non-bipartite matching algorithm to prove the Gallai-Edmonds Structure Theorem. (For those readers still not satisfied, reference [28] contains a more direct proof of the End-high Theorem.)

A word about implementation is in order. For conceptual simplicity, we use the notion of semiorder refinement. For purposes of implementation, a semiorder must be represented in a way that can be efficiently updated when refinement takes place. In sequential computation, one can use doubly-linked lists, one for each class listing the elements of the class. This representation makes it easy to remove nodes from a class, one at a time, and form new subclasses. This is the representation used in implementing Rose, Tarjan, and Lueker's linear-time sequential algorithm for finding a peo, described in the next section.

In parallel computation, the same representation can be used, but removing nodes from a class is somewhat more involved, because a large collection of nodes may need to be removed all at the same time. This can be accomplished by use of list-ranking techniques. An alternative representation assigns distinct positive integers to each of the classes, in ascending order. Instead of assigning, say, 1 to the first class, 2 to the second, and so on, we can assign $1n$ to the first class, $2n$ to the second, and so on. This gives us enough room for refinement: we can split each class into subclasses and assign each subclass its own integer. After each class has been refined, we can use sorting to bring together all nodes in the same class, and then use a parallel prefix computation to renumber the classes in preparation for the next refinement. For brevity, we will not address these issues of representation when we discuss the parallel algorithm for finding a peo.

8.7 Rose, Tarjan, and Lueker's Sequential Algorithm for Finding a Perfect Elimination Ordering

The sequential peo algorithm of Rose, Tarjan, and Lueker used a method they called *lexicographic breadth-first search*. The idea is to construct the peo backwards, selecting each node in turn in an order dependent on the node's neighbors among previously selected nodes. The algorithm consists of n stages, one for each node, numbered from n down to 1. In stage i, a node is selected and assigned position number i in the peo being built. The selection is made from the set of as-yet-unnumbered nodes, based on the neighborhoods of these nodes among the numbered nodes. The node selected is one which has the *lexicographically highest* neighborhood. That is, given two distinct subsets S_1 and S_2 of the numbered nodes, let v_k be the highest-numbered node that is in one subset but not in the other; then the subset containing v_k is *lexicographically higher* than the subset not containing v_k. Figure 8.24 gives an example of this rule being applied.

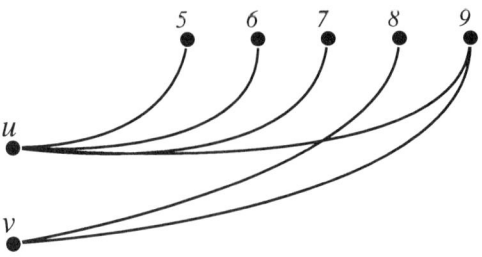

FIGURE 8.24
Both u and v are adjacent to 9, but only v is adjacent to 8. Therefore, the neighborhood of node v among the numbered nodes is lexicographically higher than that of u.

Linear-Time Implementation

Rose, Tarjan, and Lueker show that this version of lexicographic breadth-first search can be implemented in linear time. The key is to keep all the unnumbered nodes partitioned into classes according to their neighborhoods among the numbered nodes. That is, each class contains all the unnumbered nodes having a particular neighborhood, and the classes are ordered lexicographically by neighborhood. Each class is represented by a doubly-linked list.

Stage i consists of two phases. First, the algorithm selects an unnumbered node v belonging to the highest class, removes it from the linked list representing that class, and assigns it number i. Second, each class C is partitioned into two subclasses, a higher subclass C^+ and a lower subclass C^-. The higher subclass consists of the nodes in the class that are adjacent to the newly numbered node v, and the lower subclass consists of the remaining nodes in the class. The ordered sequence of subclasses is the sequence of classes used for the next stage. That is, if at the beginning of stage i, the sequence of classes was $C_1, C_2, \ldots, C_k$, then at the end the new sequence of classes is $C_1^-, C_1^+, C_2^-, C_2^+, \ldots, C_k^-, C_k^+$. (In general, some of these new classes may be empty; the empty classes are omitted from the sequence.)

In order to carry out the partitioning of the classes, the algorithm examines the adjacency list of the newly numbered node v to determine the set of unnumbered nodes having v as a neighbor. For each such unnumbered node w, the algorithm determines which class C currently contains w, removes w from the linked list representing C, and adds w to a new list representing the higher subclass C^+. Subsequently the elements not removed from C form the lower subclass C^-. Thus, work is only performed for the neighbors of w.

8.7. Sequential Algorithm for Finding a Perfect Elimination Ordering 381

Consequently, the stage in which v is assigned number i requires constant time to remove v from its class, plus time proportional to the number of edges connecting v to unnumbered nodes. Over all n stages, each node is examined once and each edge is examined once, so the total time required is $O(n + m)$.

Correctness

In order to prove that lexicographic breadth-first search yields a peo when the input graph is chordal, we cast the algorithm in the framework of Iterated Refinement, and show that each refinement maintains validity. At each stage in the algorithm, the semiorder consists of a sequence of classes containing the unnumbered nodes, followed by a sequence of single-node classes, one for each numbered node, in increasing order of number. An example is shown in Figure 8.25.

Stage i consists of two refinements, one for each phase. We must verify that these two refinements maintain validity. We inductively assume at the beginning of the stage that the current semiorder Ψ_i divides the unnumbered nodes into classes according to their neighborhoods among the numbered nodes. Thus all nodes in a class have the same neighborhood. We also assume that the current semiorder Ψ is valid. In the first refinement, we refine only the highest class C containing unnumbered nodes. We select a node v in C, and we partition C into a higher class containing only the node v—this corresponds to assigning the number i to v—and a lower class containing all

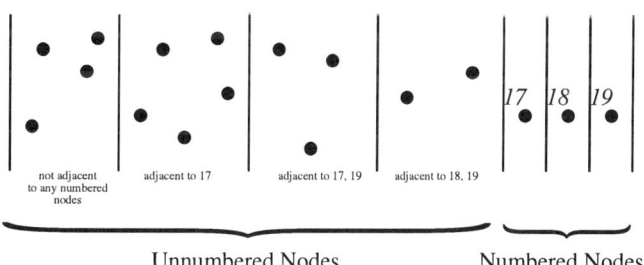

FIGURE 8.25
The semiorder consists of a sequence of classes of unnumbered nodes, followed by singleton classes, one for each numbered node. Each of the former classes contains all nodes with a specific neighborhood among the numbered nodes, and these classes are ordered according to the lexicographic ordering of the neighborhoods. The classes containing numbered nodes are ordered according to number.

the other nodes of C. Let Ψ' be the resulting refined semiorder. Figure 8.26 illustrates an example of this refinement step.

CLAIM 8.5
The first refinement $\Psi \to \Psi'$ maintains validity.

PROOF
Suppose Ψ is valid, and let P be a chordless end-high path in $G_{\Psi'}$ with endpoints x and y; we must show that x and y are adjacent. If neither of the endpoints is v, then P was an end-high path even before the refinement, in G_Ψ, so x and y are adjacent by the validity of Ψ. Assume therefore that the lower of the endpoints, say x, is v; the other endpoint y must be a previously numbered node.

By the uniformity lemma, the path P is uniform with respect to the old semiorder Ψ. Hence every node of P, except for the higher endpoint y, belonged to the same class C of Ψ as v. We assumed that all the nodes in each class of Ψ have the same numbered neighbors. In particular, the node of P adjacent to y has the same numbered neighbors as v. We have proved that v is adjacent to y. ∎

In the second refinement, we consider each class C containing unnumbered nodes, and we partition its nodes into a higher class C^+ and a lower class C^-, where the higher class contains the nodes of C adjacent to v. Let Ψ'' be the resulting refined semiorder.

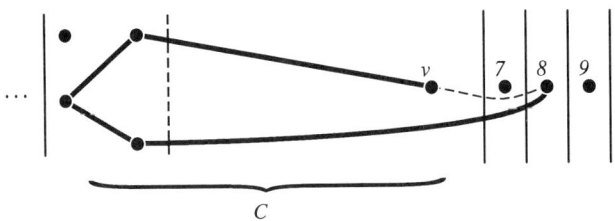

FIGURE 8.26
This figure shows the highest class C and the numbered nodes. The end-high path P has endpoints v, the node currently being numbered, and 8, a previously numbered node. Since all nodes of C have the same neighborhood among the previously numbered nodes, v must be adjacent to 8.

8.7. Sequential Algorithm for Finding a Perfect Elimination Ordering 383

CLAIM 8.6

The refinement $\Psi' \to \Psi''$ maintains validity.

PROOF

Suppose Ψ' is valid, and let P be a chordless end-high path in $G_{\Psi''}$ that was not end-high with respect to the previous semiorder Ψ'. By the uniformity lemma, all nodes of P except possibly the higher endpoint y, belonged to a single class C of Ψ'. There are four possibilities for y: it may be a node numbered in a previous iteration, the newly numbered node v, a node of C^+, or a node that belonged to a higher class C' than C in Ψ. All four cases are illustrated in Figure 8.27.

First suppose y is a previously numbered node. We assumed that all the nodes in each class of Ψ' have the same previously numbered neighbors; since the node of P adjacent to y belongs to the class C of Ψ', every node of C, including x, is adjacent to y.

Next suppose y is the newly numbered node v. The other endpoint x belonged to the class C of Ψ', along with the internal nodes of P. Since P is end-high with respect to Ψ'', the endpoint x must belong to the higher subclass C^+. Hence x is adjacent to y.[4]

Now suppose y is adjacent to v; this includes the case in which y is in C^+. By adding the edges $\{x, v\}$ and $\{y, v\}$ to the path P, we obtain a cycle θ. If θ has three edges then $\{x, y\}$ is the third, and we are done; otherwise by chordality θ has a chord. Since P is chordless, θ's chord cannot consist only of nodes of P; it must contain the node v. But the internal nodes of P belong to the lower subclass C^-, and hence are not adjacent to v, so we have a contradiction.

Suppose finally that y belongs to a higher class C' of Ψ' than C, and y is not adjacent to v. Then y has a numbered neighbor v_j that is not a neighbor of any node of C. By adding the edges $\{y, v_j\}$ and $\{x, v\}$ to P, we get a path P' that is end-high with respect to Ψ'. By Claim 8.5, its endpoints v and v_j are adjacent. By adding the edge $\{v, v_j\}$ to P', we obtain a cycle θ, which must have a chord. But v is not adjacent to any node of P except x, and v_j is not adjacent to any node of P except y, so the only possible chord is $\{x, y\}$. ∎

---

[4] In fact, this case cannot happen; the node of P adjacent to y would be in C^+, which would mean P was not an end-high path after all.

384 Chapter 8. Parallel Algorithms for Chordal Graphs

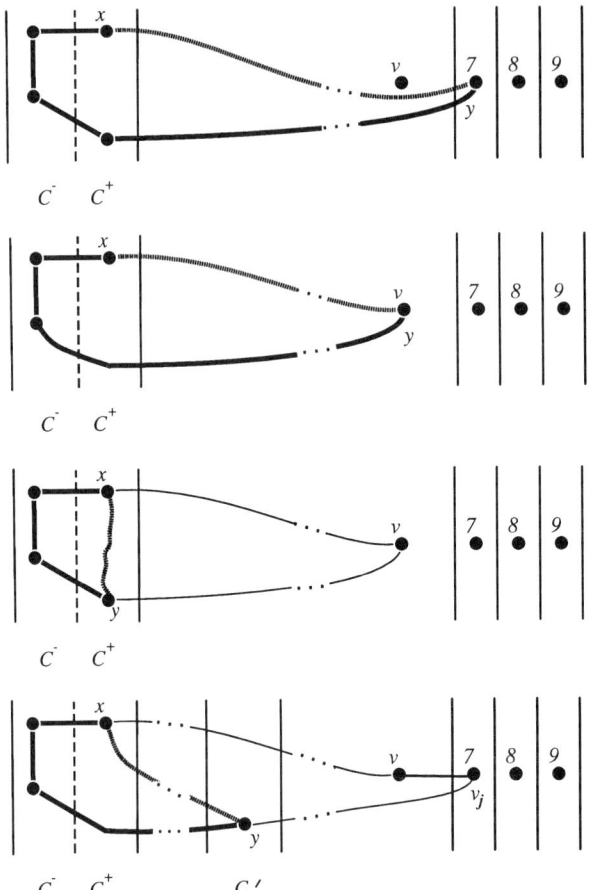

FIGURE 8.27

In the first semiorder, y is a previously numbered node; all the nodes in C have the same numbered neighbors, so x is adjacent to y. In the second semiorder, y is the newly numbered node v; since x is in C^+, x is adjacent to v. In the third semiorder, both endpoints are adjacent to v, but none of the internal nodes of the path is. In this case, chordality implies that x and y are adjacent. In the fourth semiorder, x is adjacent to v, and y is not. The node y is adjacent to a numbered node that is adjacent to v but not to x or any internal nodes of the path. Moreover, none of the internal nodes is adjacent to v. Chordality again implies that x and y are adjacent.

We have proved that the two refinements of stage i, $\Psi \to \Psi'$ and $\Psi' \to \Psi''$, maintain validity. The algorithm uses the semiorder Ψ'' for the next stage; hence the algorithm maintains validity throughout, and is therefore correct.

8.8 A Parallel Algorithm for Finding a Perfect Elimination Ordering

In this section, we show how to find a peo in $O(\log^2 n)$ time using $n+m$ processors.

8.8.1 The REFINE Procedure

In order for an *iterated refinement* algorithm to be a fast parallel algorithm, the number of iterations required must be small, much smaller than the $2n-1$ iterations required by the sequential algorithm of the previous section. We give an algorithm requiring only $O(\log n)$ iterations. The core of the algorithm is a subroutine REFINE that takes as input a graph G and a valid semiorder Ψ and returns a refined valid semiorder Ψ' whose largest nonsingleton class is four-fifths the size of that of Ψ. After t iterations, the largest nonsingleton class has size at most $(4/5)^t n$. Hence only $O(\log n)$ iterations are required before every class is singleton.

To achieve the refinement, the REFINE subroutine operates independently on each of the nonsingleton classes of the input semiorder Ψ, and refines each. *A priori*, we have no reason to believe that such a strategy could work—that we could refine each class independently and expect to thereby obtain a valid semiorder; perhaps some coordination between different classes is necessary. The key to proving that the strategy can be made to work is the Refinement Lemma, which we give presently. It states, intuitively, that a refinement of a class C preserves validity for the entire graph if it preserves validity for the subgraph consisting of C and its higher neighbors. To state the lemma formally, we need to introduce some notation.

Fix a graph G and a valid semiorder Ψ. For a class C, let $\Gamma(C)$ denote the set of higher neighbors of C, let $\widehat{C}$ denote the subgraph of G induced by $C \cup \Gamma(C)$, and let Ψ_C denote the semiorder $(C, \Gamma(C))$ for $\widehat{C}$.

LEMMA 8.6 Refinement Lemma

Suppose Ψ is a valid semiorder for G. Let Φ be a semiorder for a class C of G_Ψ. Then the following two conditions are equivalent:

1. The refinement of G_Ψ by Φ is valid.
2. The refinement of $\widehat{C}_{\Psi_C}$ by Φ is valid.

PROOF
Let Ψ' be the refinement of G_Ψ by Φ, and let λ be the refinement of $\widehat{C}_{\Psi_C}$ by Φ. We need to show that Ψ' is valid for G if and only if λ is valid for $\widehat{C}$. It suffices to show that the chordless end-high paths in $G_{\Psi'}$ that are not end-high with respect to Ψ are exactly the chordless end-high paths in $\widehat{C}_\lambda$ that are not end-high with respect to Ψ_C.

One direction is easy. Since $\widehat{C}$ is a subgraph of G, a path P in $\widehat{C}$ is also a path in G. Moreover, λ is consistent with the restriction of Ψ' to $\widehat{C}$, so if P is end-high with respect to λ, it is certainly end-high with respect to Ψ'.

Conversely, let P be a chordless end-high path in $G_{\Psi'}$ that was not end-high with respect to Ψ. By the Uniformity Lemma, P is uniform with respect to Ψ, so its lower endpoint x lies in the same class of Ψ as its internal nodes. Since P is end-high with respect to Ψ', x lies in a higher class of Ψ' than the internal nodes. Since Ψ' is the refinement of Ψ by Φ, it must be that x lies in a higher class of Φ that the internal nodes. Since Φ is only defined for the class C of Ψ, this class must contain x and all the internal nodes.

The other endpoint y of P is a neighbor of a node of C, and hence is either in C itself, or is a higher neighbor of C. In the first case, y is in a higher class of Φ than the internal nodes because P' is strictly end-high. In the second case, y is in a higher class of Ψ_C than the internal nodes. In either case, we conclude that P' is an end-high path in $\widehat{C}_\lambda$. ∎

The Refinement Lemma implies that to validly refine a class C in G_Ψ, we need only validly refine it in $\widehat{C}_{\Psi_C}$. In fact, we observe next that all the classes of G_Ψ may be thus refined simultaneously and independently, as far as validity is concerned. The reason is that for any class C, $\widehat{C}_{\Psi_C}$ remains unchanged when classes other than C are refined. More formally, fix a semiorder Φ on C. By applying the lemma to Ψ, we see that the following two conditions are equivalent:

1. The refinement of G_Ψ by Φ maintains validity.
2. The refinement of $\widehat{C}_{\Psi_C}$ by Φ maintains validity.

8.8. A Parallel Algorithm for Finding a Perfect Elimination Ordering

Now suppose Ψ' is obtained from Ψ by refining classes other than C. By applying the lemma to Ψ', we see that condition (2) is equivalent to the following condition:

3. The refinement of $G_{\Psi'}$ by Φ maintains validity.

We conclude that conditions (1) and (3) are equivalent, and hence that the refinement of C is independent of the refinement of other classes. Thus the REFINE subroutine can refine each class independently and still be assured of maintaining validity for the entire graph.

We make one more useful observation before giving the REFINE procedure. We shall see at the end of this subsection that it is easier to work with *connected* classes, i.e., classes C whose induced subgraphs $G[C]$ are connected. The following corollary shows that we can assume without loss of generality that this condition holds.

COROLLARY 8.2
Refining a class by dividing it into its connected components always maintains validity.

PROOF
No end-high path can be created by such a refinement, because an end-high path would have to cross components. More formally, let Ψ be a valid semiorder for G, and let C be a class of Ψ. Let Φ be a semiorder on C consisting of the connected components of $G[C]$ in any order. Suppose we refine $\widehat{C}_{\Psi_C}$ by Φ, obtaining $\widehat{C}_\lambda$. If we can prove that no new end-high paths result, the corollary will follow via the Refinement Lemma. Assume for contradiction that there is an end-high path P in $\widehat{C}_\lambda$ that was not end-high in $\widehat{C}_{\Psi_C}$. Then all the internal nodes and at least one endpoint x must be in C. Since they are connected by a path, these nodes all lie in the same connected component of $G[C]$. In Φ and λ, therefore, x is no higher than the internal nodes, contradicting our assumption that P was end-high. ∎

The procedure REFINE is given in Figure 8.28. The first step refines Ψ to ensure that each of its classes is connected. The validity of this refinement follows from Corollary 8.2. The main work of REFINE is carried out in the subroutine STRATIFY(G_Ψ, C), which validly refines a connected class C into subclasses such that each nonsingleton connected component of each subclass has size at most four-fifths the size of C. These connected components themselves become classes during the first step of the next iteration. Thus in

ALGORITHM
Procedure REFINE(G_Ψ)
For each nonsingleton class C, call STRATIFY(G_Ψ, C) to refine C.

(Follow-up step) For each resulting class C, refine C by dividing it into its connected components.

FIGURE 8.28
The procedure REFINE for refining a semiorder; it reduces the size of each class by a factor of four-fifths.

---

successive calls to STRATIFY, the maximum class size goes down by a factor of four-fifths.

Resource Requirements: The Return of Clique Management

There is an apparent obstacle to achieving an efficient algorithm. The second step of REFINE operates independently on each class. To refine a class C validly, one must examine the structure of $\widehat{C}$, the subgraph consisting of C and its higher neighbors; these higher neighbors belong to different classes. If we are to refine all classes simultaneously, we must operate on mutually *overlapping* subgraphs; how can we do this with only a linear number of processors?

The solution is *clique management*, the idea we used for minimum coloring in Section 8.5.6.

LEMMA 8.7
Suppose Ψ is a valid semiorder for G, and a class C forms a connected subgraph of G. Then the higher neighbors of C form a clique.

PROOF
For every pair of higher neighbors x and y of C, there is an end-high path through C with endpoints x and y; hence every pair of higher neighbors of C are adjacent. ∎

As in the minimum coloring algorithm, once we have identified a clique, we need not assign processors to the edges between its nodes. Consequently, we can carry out STRATIFY(G_Ψ, C) in work proportional to the number of edges with at least one endpoint in C; we need not do work for the edges between C's higher neighbors. Summing over all classes C, we find that each edge of G is charged at most twice, so the total work is $O(m)$.

The parallel time to carry out STRATIFY(G_Ψ, C) is $O(\log n)$ using t processors, where t is the number of edges with at least one endpoint in C. By applying STRATIFY to all classes in parallel, we achieve $O(\log n)$ time using $2m$ processors. The preliminary step in which we divide each class into its connected components can be carried out within the same bounds, using a standard connectivity algorithm.

8.8.2 Stratification of a Class

We will presently give a procedure for *stratifying* a class: breaking it into subclasses while maintaining validity. Given a validly numbered graph G_Ψ and a nonsingleton connected class C of G_Ψ, our goal is to refine C into subclasses such that each nonsingleton subclass has at most four-fifths the nodes of C.

Note that because of the follow-up step of the procedure REFINE, presented in Section 8.8.1, we need only refine C into subclasses such that each *connected component* of each subclass is a constant fraction smaller than C. We call such a refinement a *well-stratification* of C.

THEOREM 8.4
Suppose Ψ is a valid numbering of a chordal graph G, and C is a class of G_Ψ. Valid well-stratification of C can be done in $O(\log k)$ time using $O(k)$ processors of a CRCW P-RAM, where k is the number of edges with at least one endpoint in C. Hence, a peo *of the chordal graph G can be found in $O(\log^2 n)$ time using $O(m)$ processors.*

In this subsection, we present some results used in proving the correctness of the procedure. In subsequent subsections, we present parts of the procedure.

We start with a lemma of Dirac [15], illustrated in Figure 8.29.

LEMMA 8.8 *Dirac*
If S is a minimal set of nodes whose removal separates a chordal graph into exactly two connected components, then S is a clique.

The following corollary is illustrated in Figure 8.30.

COROLLARY 8.3
In a chordal graph, the common neighbors of two nonadjacent nodes form a (possibly empty) clique.

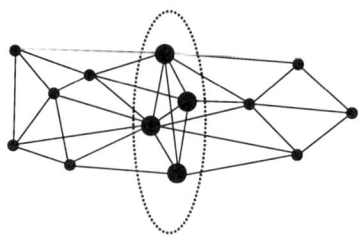

FIGURE 8.29
A minimal separator is a clique.

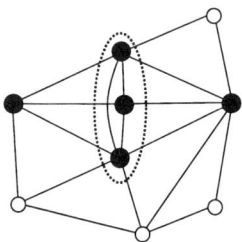

FIGURE 8.30
Common neighbors of nonadjacent nodes form a clique.

PROOF
In the induced subgraph consisting of the two nonadjacent nodes x and y, and their common neighbors, the common neighbors form a minimal separator between x and y. ∎

The following corollary is illustrated in Figure 8.31.

COROLLARY 8.4
Let H be a connected subgraph of a chordal graph G. Then the semiorder $\Phi = (G - H - \{neighbors\ of\ H\}, H \cup \{neighbors\ of\ H\})$ is valid.

PROOF
Let P be an end-high path in G_Φ, and let K be the component of the lower class that contains the internal nodes of P. Let D be the set of higher neighbors of nodes in K. In the subgraph induced on $K \cup D \cup H$, the nodes of D form a minimal separator between C and H, so D is a clique. The endpoints of P are in D, so they are adjacent. ∎

8.8. A Parallel Algorithm for Finding a Perfect Elimination Ordering

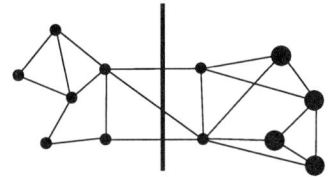

FIGURE 8.31
If H is connected, the semiorder $\Phi = (G - H - \{\text{neighbors of } H\}, H \cup \{\text{neighbors of } H\})$ is valid.

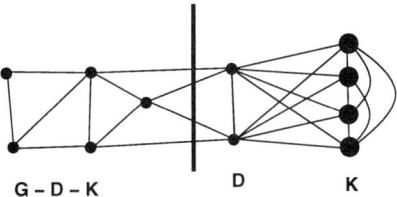

FIGURE 8.32
If K is a clique and D contains the nodes adjacent to every node of K, then the semiorder $\Phi = (G - K - D, K \cup D)$ is valid.

The following lemma is illustrated in Figure 8.32.

LEMMA 8.9
Let K be a clique in a chordal graph G. Let $D = \{v \in G : v \text{ adjacent to every node of } K\}$. Then the semiorder $\Phi = (G - K - D, K \cup D)$ is valid.

PROOF
The proof is by induction on $|K|$. The base case, in which $|K| = 1$, follows from Corollary 8.4. Suppose $|K| > 1$, and let v be a node of K. Let A consist of the nodes of $K - \{v\}$ and the nodes adjacent to all of $K - \{v\}$. Let α be the semiorder $(G-A, A)$. By the inductive hypothesis, α is valid. Because K is a clique, v is in A. Next we define a semiorder for A, $\beta = (A - \{v\} - \{\text{neighbors of } v\}, \{v\} \cup \{\text{neighbors of } v\})$. By Corollary 8.4, β is a valid semiorder for $G[A]$. Let G_γ be the refinement of G_α by β. The nodes of A have no higher neighbors in G_α, so $\hat{A} = A$. Hence by the Refinement Lemma, γ is a valid semiorder for G. But γ is

a refinement of the numbering Φ defined in the statement of the lemma, so Φ is also valid. This completes the inductive step. ∎

LEMMA 8.10

Let α be any valid semiorder of a graph G, and let K be a clique contained in the highest class C of G_α. Suppose γ is obtained from α by first refining C by $(C - K, K)$ and then refining K arbitrarily. Then γ is valid.

PROOF

The only end-high paths introduced have endpoints in the clique K. ∎

The following is a special case of the Refinement Lemma, where all nodes in the class C have the same higher neighbors. In this special case, there is no need to consider $\widehat{C}$.

LEMMA 8.11

Let Ψ be a valid semiorder for a graph G. Suppose C is a class of G_Ψ such that $G[C]$ is connected, and all nodes in C have the same higher neighbors in G_Ψ. Let Φ be a valid semiorder for C. Then the refinement of G_Ψ by Φ is valid.

PROOF

By the Refinement Lemma, we need only show that the refinement λ of $\widehat{C}_{\Psi_C}$ by Φ preserves validity. Assume Ψ is valid, so the nodes of $\widehat{C}$ not in C form a clique by Lemma 8.7. Let P be an end-high path in $\widehat{C}_\lambda$ with endpoints x and y. We must show that x and y are adjacent. If both endpoints are in C, they are already adjacent by the validity of Φ. If neither is in C, they belong to a clique, and hence are adjacent. Suppose therefore that x is in C and y is not. Since P is a weakly end-high path in $\widehat{C}_{\Psi_C}$ and an endpoint lies in C, all the internal nodes of P must also lie in C. Hence y is a higher neighbor of C in $\widehat{C}_\Psi$. Since all nodes in C have the same higher neighbors, it follows that y is a neighbor of x. ∎

8.8.3 The Procedure STRATIFY

In this subsection, we outline the procedure STRATIFY for well-stratifying a class. The main work of the procedure is done in subprocedures that are described in succeeding subsections.

The procedure STRATIFY(G_Ψ, C) appears in Figure 8.33. If C is a connected nonsingleton class of G_Ψ, the procedure well-stratifies C. The procedure takes $O(\log n)$ time using t processors, where t is the number of edges of G with at least one endpoint in C. To achieve this processor bound, the

ALGORITHM
Procedure STRATIFY(G_Ψ, C)

> **Step 1** For each node v in C, identify those edges connecting v to other nodes in C, and those edges connecting v to higher nodes.
>
> **Step 2** Let B be the set of higher neighbors of C.
>
> **Step 3** If B is empty, call NONE(G_Ψ, C), and end.
>
> **Step 4** If every node in B has at least $\frac{2}{5}|C|$ neighbors in C, call HIGHDEGREE(G_Ψ, C, B), and end.
>
> **Step 5** If there are nodes in B with fewer than $\frac{2}{5}|C|$ neighbors in C, call LOWDEGREE(G_Ψ, C, B).

FIGURE 8.33
The main procedure for finding a valid well-stratification.

procedure first identifies these edges by inspecting the adjacency lists of nodes in C, and subsequently never examines any other edges of G.

While inspecting the adjacency lists, the procedure also identifies the set B of higher neighbors of C in G_Ψ. The procedure uses the fact that if Ψ is a valid numbering of G, then the set of nodes B form a clique, by Lemma 8.7. The procedure assumes the existence of edges between nodes in B without ever checking for their presence. Specifically, in computing connected components of a graph involving nodes of B, the procedure uses an efficient connectivity algorithm such as that of Shiloach and Vishkin [44] or that of Gazit [21], suitably modified to take into account the fact that every two nodes of B are adjacent.

Namely, whenever we need to compute connectivity in a subgraph that includes nodes of B, we start by constructing a star graph containing all such nodes. The star graph connects up all these nodes using $k-1$ artificial edges, where k is the number of such nodes of B. Because we know that every two nodes of B are adjacent in the true graph, the artificial edges stand for true edges. The number of processors needed for handling these artificial edges is no more than the number of processors needed to handle the nodes of B. Hence the total number of processors needed for computing connected components in a node-induced subgraph of $G[C \cup B]$ is proportional to the number of nodes in the subgraph, plus the number of edges with at least one endpoint being a node of C. The processor bound does not depend on the number of edges with both endpoints in B.

Depending on the nodes in B, the procedure STRATIFY(G_Ψ, C) calls one of three subprocedures, in which most of the work is done. In each procedure, we make use of *parallel prefix computation*, due to Ladner and Fischer [32]. In the next subsection, we describe the simplest of these procedures, called NONE, which is applicable when the class has no higher neighbors (i.e., when B is empty). This procedure illustrates the techniques used in refinement, and should be sufficient for all but the most dedicated and thorough reader. For that reader, however, we have described the other two procedures in the subsequent subsection.

8.8.4 The Subprocedure NONE for Stratifying a Class with No Higher Neighbors

We now show that STRATIFY(G_Ψ, C) succeeds in finding a valid refinement that well-stratifies C. We shall refer to the nodes of C as *crimson* nodes, and the nodes of B as *blue* nodes. We consider three cases, corresponding to the three procedures. In this subsection, we consider only the first and simplest case, Case I, when there are no blue nodes.

Case I

In this case, procedure NONE(G_Ψ, C), shown in Figure 8.34, is called. The procedure first identifies the set D of high-degree nodes: $D = \{v \in C : v \text{ has } > \frac{3}{5}|C| \text{ neighbors in } C\}$. The procedure then branches according to the following cases:

Subcase (1): Some component H of $C - D$ has size $> \frac{4}{5}|C|$. In this case, as shown in Figure 8.35, the procedure finds a connected subgraph H' of H such that the set $A = H' \cup \{\text{neighbors of } H' \text{ in } C\}$ includes between $n/5$ and $4n/5$ nodes. Then C is refined by the semiorder $(C - A, A)$, resulting in a well-stratification of C. The validity of the refinement follows from Corollary 8.4.

The procedure chooses H' as follows. First, a spanning tree T of H is constructed. Next, the nodes of H are arranged in some order consistent with their distance in T from the root: $v_1, \ldots, v_k$. This order has the property that any initial subsequence $v_1, \ldots, v_j$ induces a connected subgraph of H.

Next, we carry out a parallel prefix computation. The goal is to identify the maximum j such that the nodes $v_1, \ldots, v_j$ and their neighbors comprise at most $\frac{4}{5}|C|$ nodes of C.

8.8. A Parallel Algorithm for Finding a Perfect Elimination Ordering

ALGORITHM
Procedure NONE(G_Ψ, C)

Step 1   Let $D = \{v \in C : v \text{ has } > \frac{3}{5}|C| \text{ neighbors in } C\}$.

Step 2   Find a spanning forest of $C - D$.

Step 3   If some component H of $C - D$ has at least $\frac{4}{5}|C|$ nodes,

Step 3.1   Let T be the spanning tree of H.

Step 3.2   Arrange the nodes of H in some order consistent with their distance in T from the root: $v_1, \ldots, v_k$.

Step 3.3   For $1 \leq j \leq k$, let A_j denote the set consisting of $v_1, \ldots, v_j$ and neighbors of these nodes in C.

Step 3.4   Using parallel prefix computation, choose $\hat{j} = \max\{j : |A_j| \leq \frac{4}{5}|C|\}$.

Step 3.5   Refine C by the semiorder $(C - A_{\hat{j}}, A_{\hat{j}})$.

Step 4   Otherwise (if every component of $C-D$ has fewer than $\frac{4}{5}|C|$ nodes),

Step 4.1   If D is a clique,

Step 4.2   Let $v_1, \ldots, v_k$ be the nodes of D.

Step 4.3   Refine C by the semiorder $(C - D, \{v_1\}, \ldots, \{v_k\})$.

Step 4.4   Otherwise,

Step 4.5   Let x and y be two nonadjacent nodes of D.

Step 4.6   Let $v_1, \ldots, v_k$ be their common neighbors. (They form a clique.)

Step 4.7   Refine C by the semiorder $(C - \{v_1, \ldots, v_k\}, \{v_1\}, \ldots, \{v_k\})$.

FIGURE 8.34
The subprocedure used to well-stratify C if C has no higher neighbors.

For each node $v \neq v_1$ in C, let *earliest-nbr*(v) be the minimum i such that v is adjacent to v_i, or *undefined* if v has no neighbors among $v_1, \ldots, v_k$. Let *earliest-nbr*$(v_1) = 1$.

Now let us consider the inverse of the relation *earliest-nbr*. For each node v_i in T, *earliest-nbr*$^{-1}(v_i)$ is the set of neighbors of v_i in C that are not neighbors of any lower-numbered node. Let $f(v_i) = |earliest\text{-}nbr^{-1}(v_i)|$,

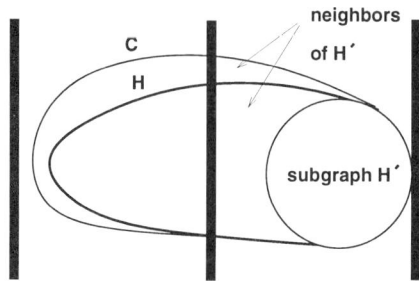

FIGURE 8.35
In Case I, subcase (1), we find a connected subgraph H' of H such that $H' \cup \{\text{neighbors of } H'\}$ has between $|C|/5$ and $4|C|/5$ nodes. Let A be $H' \cup \{\text{neighbors of } H'\}$. We refine C by the semiorder $(C - A, A)$. This refinement is valid because the neighborhood in A of any component of $C - A$ is a clique, by Corollary 8.4.

---

and let $g(\ell) = \sum_{i=1}^{\ell} f(v_i)$ for $\ell = 1, \ldots, k$. Then $g(\ell)$ is the number of nodes comprised by $v_1, \ldots, v_\ell$ and their neighbors in $\widehat{C}$. The function $g(\cdot)$ can be computed from $f(\cdot)$ by a parallel prefix computation.

After $g(\cdot)$ has been computed, it is easy to find the largest index j such that $g(j) \leq \frac{4}{5}|C|$. Let $\hat{j}$ denote that index. That is, $\hat{j}$ is the maximum j such that the nodes $v_1, \ldots, v_j$ and their neighbors comprise at most $\frac{4}{5}|C|$ nodes of C.

Finally, we let H' be the subgraph consisting of $v_1, \ldots, v_{\hat{j}}$, and we let A be the subgraph consisting of H' and its neighbors. As mentioned before, the class C is then refined by $(C - A, A)$.

We need to show that the result is a well-stratification, i.e., that each of the subclasses is significantly smaller than the class. It is clear that A has at most $\frac{4}{5}|C|$ nodes. We must show that it has at least $\frac{1}{5}|C|$ nodes, for then it will follow that $C - A$ has at most $\frac{4}{5}|C|$ nodes. Observe that by choice of $\hat{j}$, nodes $v_1, \ldots, v_{\hat{j}+1}$ and neighbors comprise *more than* $\frac{4}{5}|C|$ nodes. Not all these neighbors are neighbors of H'—some are neighbors of $v_{\hat{j}+1}$, which is not in H'. But $v_{\hat{j}+1}$ has at most $\frac{3}{5}|C|$ neighbors, since it is not in D. Hence leaving out the neighbors of $v_{\hat{j}+1}$, we get at least $\frac{1}{5}|C|$ nodes in A.

Subcase (2): Each component of $C - D$ has size at most $\frac{4}{5}|C|$, and D is a clique. Let $D = \{v_1, \ldots, v_k\}$. In this case, as shown in Figure 8.36, we refine C by the semiorder $(C - D, \{v_1\}, \ldots, \{v_k\})$, placing each node of D

8.8. A Parallel Algorithm for Finding a Perfect Elimination Ordering 397

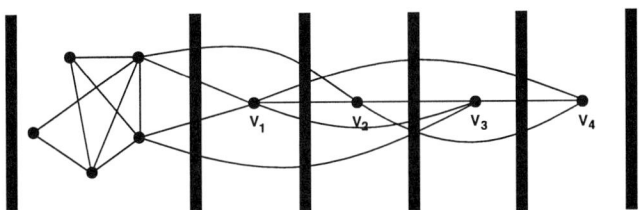

FIGURE 8.36
In Case I, subcases (2) and (3), we identify a clique $\{v_1, \ldots, v_k\}$. Then we place each node of the clique in a subclass by itself: we refine C by the semiorder $(C - \{v_1, \ldots, v_k\}, \{v_1\}, \{v_2\}, \ldots, \{v_k\})$. This refinement is valid because the neighborhood in $\{v_1, \ldots, v_k\}$ of any component of $C - \{v_1, \ldots, v_k\}$ is a clique.

in its own class. The components induced by the remaining nodes of C are small, so we have well-stratified C. The validity of the refinement follows from Lemma 8.10.

Subcase (3): D is not a clique. In this case, we choose two nonadjacent nodes x and y in D. At most $\frac{2}{5}|C|$ nodes of C are not neighbors of x, and so at least $\frac{1}{5}|C|$ neighbors of y are also neighbors of x. Let these common neighbors be $v_1, \ldots, v_k$. By Lemma 8.3, the common neighbors form a clique. We therefore proceed as in Subcase (2): we refine C by the semiorder $(C - \{v_1, \ldots, v_k\}, \{v_1\}, \ldots, \{v_k\})$, putting each of the common neighbors in its own class. Since there are so many common neighbors, the remaining nodes form a small subclass, and we have well-stratified C. The refinement is valid by Lemma 8.10.

8.8.5 The Other Subprocedures Used in STRATIFY

In this subsection, we describe the other two subprocedures. The subprocedures and their justification are considerably more complicated than that described in the previous subsection. It is therefore suggested that this subsection be skipped on first reading.

There are two remaining cases to consider, Cases II and III, corresponding to the two remaining subprocedures.

Case II

The set B of higher neighbors (called blue nodes) is not empty, and each blue node is adjacent to at least $\frac{2}{5}|C|$ crimson nodes. In this case, the procedure HIGHDEGREE(G_Ψ, C, B) of Figure 8.37 is called. The procedure

ALGORITHM
Procedure HIGHDEGREE(G_Ψ, C, B)
(Each node of B has at least $\frac{2}{5}|C|$ neighbors in C.)

Step 1 Arbitrarily order the nodes of B: $v_1, \ldots, v_k$.

Step 2 for $1 \leq j \leq k$, let F_j denote the set of nodes of C adjacent to *all* of the nodes $v_1, \ldots, v_j$.

Step 3 Using parallel prefix computation, choose $\hat{j} = \max\{j : |F_j| \geq \frac{1}{5}|C|\}$.

Step 4 Refine C by the semiorder $(C - F_{\hat{j}}, F_{\hat{j}})$.

Step 5 Let C' be the largest component of $G[F_{\hat{j}}]$.

Step 6 If $\hat{j} = k$, then call NONE(G_Ψ, C').

FIGURE 8.37
The procedure HIGHDEGREE used to stratify C in case every higher neighbor of C has high degree in C.

arbitrarily orders the blue nodes: $B = \{v_1, \ldots, v_k\}$. The procedure then proceeds as indicated in Figure 8.38. For $1 \leq j \leq k$, let F_j be the set of nodes $v \in C$ such that v is adjacent to *all* the nodes $v_1, \ldots, v_j$. Then $\hat{j}$ is chosen to be the maximum j such that F_j contains at least $|C|/5$ crimson nodes. (This step is similar to a step in the procedure NONE, described in the previous subsection, and can be implemented in essentially the same way.) The class C is then refined by the semiorder $(C - F_{\hat{j}}, F_{\hat{j}})$. The validity of the resulting numbering follows from Lemmas 8.9 and 8.10. At most $\frac{4}{5}|C|$ nodes of C remain in the lower subclass $C - F_{\hat{j}}$. However, the set $F_{\hat{j}}$ may be quite large. We consider two subcases.

Subcase (1): $\hat{j} < k$. In this case, by choice of $\hat{j}$, the set of crimson nodes adjacent to all the nodes $v_1, \ldots, v_{j+1}$ is less than $|C|/5$. Since there are at most $\frac{3}{5}|C|$ crimson nodes not adjacent to v_{j+1}, it follows that the number of nodes adjacent to all the nodes $v_1, \ldots, v_j$ is less than $\frac{1}{5}|C| + \frac{3}{5}|C| = \frac{4}{5}|C|$. Thus C has been well-stratified in this case.

8.8. A Parallel Algorithm for Finding a Perfect Elimination Ordering 399

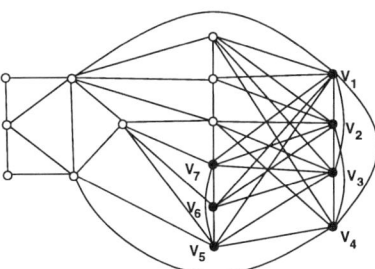

FIGURE 8.38
In Case II, we start by finding a subset $\{v_1, \ldots, v_j\}$ of blue nodes such that at least one-fifth of the nodes of C are adjacent to *all* the nodes of the blue subset. For the graph illustrated, the subset would be $\{v_1, v_2, v_3, v_4\}$.

Subcase (2): $\hat{j} = k$. In this case, every node in $F_{\hat{j}}$ is adjacent to every blue node. The procedure finds the largest component C' of $G[F_{\hat{j}}]$, and calls NONE(G_Ψ, C'), which stratifies C' as if C' had *no* higher neighbors. The validity of the resulting numbering of $\widehat{C}$ follows from Lemma 8.11 and the Refinement Lemma. Since C' is well-stratified and $|C - C'| \leq \frac{4}{5}|C|$, C is well-stratified.

Case III

The set B of higher neighbors is not empty, and Case II does not hold. In this case, the procedure LOWDEGREE(G_Ψ, C, B) in Figure 8.39 is called. The procedure defines D to be the set of nodes in $C \cup B$ having more than $\frac{3}{5}|C|$ crimson neighbors. The procedure then finds a spanning tree T of the (unique) component H of $G[(C \cup B) - D]$ containing blue nodes, and roots T at a blue node. The nodes of T are arranged in some order consistent with their distance from the root: $v_1, \ldots, v_k$. For $1 \leq j \leq k$, let A_j be the set consisting of $v_1, \ldots, v_j$ and neighbors of these nodes in $\widehat{C}$. Note that since v_1 is blue, and the blue nodes form a clique, every A_j contains all the blue nodes. The procedure chooses $\hat{j}$ to be the maximum j such that A_j includes at most $\frac{4}{5}|C|$ crimson nodes. (This step is similar to a step in the procedure NONE, described in the previous subsection, and can be implemented in essentially the same way.)

Next, C is refined by the semiorder $(C - A_{\hat{j}}, A_{\hat{j}} \cap C)$, in step 7. To see that the resulting semiorder θ of $\widehat{C}$ is valid, first consider the intermediate

ALGORITHM
Procedure LowDegree(G_Ψ, C, B)
(There exists a node in B having fewer than $\frac{2}{5}|C|$ neighbors in C.)

Step 1 Let $D = \{v \in C \cup B : v \text{ has} > \frac{3}{5}|C| \text{ neighbors in } C\}$.

Step 2 Let H be the connected component of $G[C \cup B - D]$ containing nodes of B.

Step 3 Find a spanning tree T of H, rooted at a node of B.

Step 4 Arrange the nodes of T in some order consistent with their distance in T from the root: $v_1, \ldots, v_k$.

Step 5 For $1 \leq j \leq k$, let A_j denote the set consisting of $v_1, \ldots, v_j$ and neighbors of these nodes in C.

Step 6 Using parallel prefix computation, choose $\hat{j} = \max\{j : |A_j \cap C| \leq \frac{4}{5}|C|\}$.

Step 7 Refine C by the semiorder $(C - A_{\hat{j}}, A_{\hat{j}} \cap C)$.

Step 8 Let C' be the largest component of $C - A_{\hat{j}}$.

Step 9 If $|C'| > \frac{4}{5}|C|$, then call Stratify(G_Ψ, C').

FIGURE 8.39
The procedure LowDegree used to stratify C in case some higher neighbor of C has low degree in C.

semiorder $(C - A_{\hat{j}}, A_{\hat{j}})$ of $\widehat{C}$. Since $G[\{v_1, \ldots, v_{\hat{j}}\}]$ is connected, the intermediate semiorder is valid by Corollary 8.4. To obtain the final semiorder θ of $\widehat{C}$ produced in step 7 from the intermediate numbering, we need only refine the class $A_{\hat{j}}$ by the semiorder $(A_{\hat{j}} - B, B)$; this refinement is valid by Lemma 8.10, since the blue nodes form a clique.

8.8. A Parallel Algorithm for Finding a Perfect Elimination Ordering

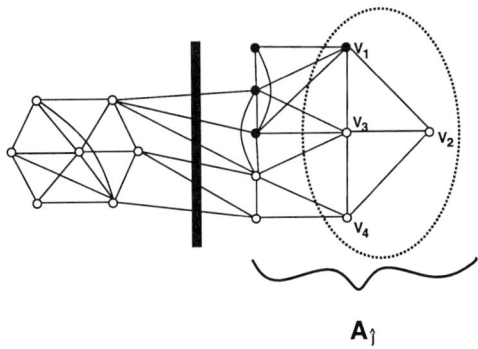

FIGURE 8.40

In Case III, we proceed similarly to Case I, subcase (1). We find a connected subgraph containing a blue node such that the subgraph together with its neighbors, collectively denoted $A_{\hat{\jmath}}$, contains at most four-fifths of the nodes of C. We next refine C by the semiorder $(C - A_{\hat{\jmath}}, A_{\hat{\jmath}} \cap C)$.

The higher subclass $A_{\hat{\jmath}} \cap C$ of C has size at most $\frac{4}{5}|C|$. However, the lower subclass $C - A_{\hat{\jmath}}$ may be quite large. The procedure finds the largest component C' of $C - A_{\hat{\jmath}}$, and proceeds according to the following two cases.

Subcase (1): $|C'| \leq \frac{4}{5}|C|$. In this case, we are done; C has been well-stratified.

Subcase (2): $|C'| > \frac{4}{5}|C|$. First, we observe that in this case, $\hat{\jmath} = k$. To see this, suppose $\hat{\jmath} < k$. Then the number of crimson nodes among $\{v_1, \ldots, v_{\hat{\jmath}+1}\}$ and neighbors is more than $\frac{4}{5}|C|$. Since $v_{\hat{\jmath}+1}$ is adjacent to at most $\frac{3}{5}|C|$ crimson nodes, it follows that the number of nodes whose numbers have increased is more than $\frac{4}{5}|C| - \frac{3}{5}|C| = \frac{1}{5}|C|$, which contradicts the fact that $|C'| > \frac{4}{5}|C|$.

Since $\hat{\jmath} = k$, every crimson node in T and every crimson neighbor of T ended up in the higher subclass. Therefore, C' contains no nodes of T and no neighbors of T. Every node in D has more than $\frac{3}{5}|C|$ crimson neighbors, and all but at most $\frac{1}{5}|C|$ of the crimson nodes are in C', so every node in D has more than $\frac{2}{5}|C|$ crimson neighbors in C'. Thus every higher neighbor of C' is adjacent to at least $\frac{2}{5}|C|$ nodes of C'. This shows that the recursive call to STRATIFY in step 9 results in a call to HIGHDEGREE, and not in a call to LOWDEGREE. Thus no further recursive calls occur. The recursive call well-stratifies C', and hence C as well, since $|C - C'| \leq \frac{1}{5}|C|$. The validity of the resulting numbering follows from the Refinement Lemma.

This completes the description of the algorithm STRATIFY for valid well-stratification of a class. At most one recursive call is made, as we have shown. The time for the algorithm is dominated by the time to compute connected components and find spanning trees, which is $O(\log |C \cup B|)$ using the algorithm of [44]; as discussed in Section 8.8.2, we need only $|C \cup B| + |E(C)| + |B| - 1$ processors, a number of processors that is bounded by at most twice the number of edges with at least one endpoint in C. We have thus proved Theorem 8.4.

Using our algorithm for valid well-stratification in the procedure ITERATED REFINEMENT, we can therefore find a peo of a graph G in $O(\log^2 n)$ time using $O(n + m)$ processors.

Acknowledgement

Thanks to Tina Cantor and Daniel Robbins for preparing the figures.

8.9 Exercises

The first 10 problems are fairly easy. The remaining ones are more difficult and more technical.

8.1   Show that the graph illustrated below is an interval graph by finding an interval representation for it.

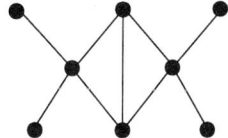

8.2   Prove that every tree is an interval graph.

8.3   Show that the graph illustrated below is *not* a chordal graph.

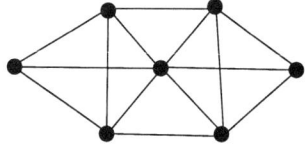

8.4   Find the tree-and-subtrees representation for the chordal graph illustrated below.

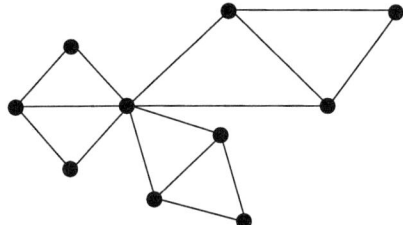

8.5 Apply Gavril's greedy algorithm to find a maximum independent set in the chordal graph of Exercise 8.4.

8.6 Construct the elimination tree for the numbered chordal graph depicted in Figure 8.4.

8.7 Construct the breadth-first search tree for the numbered chordal graph depicted in Figure 8.4, using the algorithm of Section 8.5.4.

8.8 Use lexicographic breadth-first search to construct a **peo** of the chordal graph of Exercise 8.4. Show the semiorder at each stage of the process.

8.9 Let G be the chordal graph consisting of four nodes, a, b, c, and d, in a path. There are 14 two-class semiorders for the set $\{a, b, c, d\}$. List them and, for each, say whether it is valid and, if not, why not.

8.10 For what kind of graphs is *every* possible semiorder valid? Why?

8.11 Give parallel algorithms requiring $O(\log n)$ time and $(n+m)/\log n$ processors to solve the following problems, given a graph G and a **peo** of G:

1. Find the connected components of G.
2. Construct a clique tree of G.

Hint: Use the concept of the elimination tree.

8.12 Give a linear-time greedy algorithm for maximum matching in an interval graph.

8.13 Tarjan and Yannakakis show that the following procedure yields a **peo** in a connected graph G:

- Start by assigning the number n to some node of G.
- For $i = n - 1$ down to 1, select an unnumbered node v with the most numbered neighbors, and assign i to v.

Prove that this algorithm works by showing that each numbering step preserves validity.

8.14 Show that in step 5 of the procedure HIGHDEGREE (Figure 8.37), the subgraph $G[F_j]$ is connected, so no computation need occur at that step.

8.15 Describe the parallel prefix computation for step 3 of procedure HIGHDEGREE (Figure 8.37) and for step 6 of LOWDEGREE (Figure 8.39).

Bibliography

[1] A. Aggarwal and R. Anderson, "A random NC algorithm for depth first search," *19th STOC* (1987), pp. 325-334.

[2] A. Aho, J. Hopcroft, and J. Ullman, *The Design and Analysis of Computer Algorithms*, Addison-Wesley, Reading, Massachusetts (1974).

[3] R. J. Anderson and G. L. Miller, "Optimal parallel algorithms for list ranking," extended abstract, 1986.

[4] C. Beeri, R. Fagin, D. Maier, and M. Yannakakis, "On the desirability of acyclic database schemes," *J. ACM 30* (1983), pp. 479-513.

[5] C. Berge, *Graphs and Hypergraphs*, North-Holland, Amsterdam (1973).

[6] K. S. Booth and G. S. Lueker, "Testing for the consecutive ones property, interval graphs, and graph planarity using PQ-tree algorithms," *J. Comp. Sys. Sci. 13:3* (1976), pp. 335-379.

[7] P. Buneman, "A characterization of rigid circuit graphs," *Discrete Math. 9* (1974), pp. 205–212.

[8] N. Chandrasekharan and S. S. Iyengar, "NC algorithms for recognizing chordal graphs and k-trees," Tech. Rept. #86-020, Dept. of Comp. Sci., Louisiana State Univ. (1986).

[9] C. J. Colbourn and K. S. Booth, "Linear time automorphism algorithms for trees, interval graphs, and planar graphs," *SIAM J. Comp. 10:1* (1981), pp. 203-225.

[10] R. Cole, "Parallel merge sort," *27th FOCS* (1986), pp. 511-516.

[11] R. Cole and U. Vishkin, "Approximate and exact parallel scheduling with applications to list, tree, and graph problems," *27th FOCS* (1986), pp. 478-491.

[12] D. Coppersmith and S. Winograd, "Matrix multiplication via arithmetic progressions," *19th STOC* (1987), pp. 1-6.

[13] E. Dahlhaus and M. Karpinski, "The matching problem for strongly chordal graphs is in NC," Tech. Rept. 855-CS, Institut für Informatik, Universität Bonn, 1986.

[14] E. Dahlhaus and M. Karpinski, "Fast parallel computation of perfect and strongly perfect elimination schemes," Tech. Rept. 8519-CS, Institut für Informatik, Universität Bonn, 1987, and IBM Research Report RJ5901, October 1987.

[15] G. A. Dirac, "On rigid circuit graphs," *Abh. Math. Sem. Univ. Hamburg 25* (1961), pp. 71-76.

[16] A. Edenbrandt, *Combinatorial Problems in Matrix Computation*, TR-85-695 (Ph.D. Thesis), Department of Computer Science, Cornell University (1985).

[17] A. Edenbrandt, "Chordal graph recognition is in NC," *Inf. Proc. Let. 24* (1987), pp. 239-241.

[18] D. Fulkerson and O. Gross, "Incidence matrices and interval graphs," *Pac. J. Math 15* (1965), pp. 835-855.

[19] F. Gavril, "Algorithms for minimum coloring, maximum clique, minimum covering by cliques, and maximum independent set of a chordal graph," *SIAM J. Comp. 1* (1972), pp. 180-187.

[20] F. Gavril, "The intersection graphs of subtrees in trees are exactly the chordal graphs," *J. Combin. Theory B 16* (1974), pp. 47-56.

[21] H. Gazit, "An optimal randomized parallel algorithm for finding connected components in a graph," *27th FOCS* (1986), pp. 492-501.

[22] H. Gazit and G. L. Miller, "An improved parallel algorithm that computes the BFS numbering of a directed graph," *Inf. Proc. Let. 28* (1988), pp. 61-65.

[23] J. R. Gilbert, D. J. Rose, and A. Edenbrandt, "A separator theorem for chordal graphs," *SIAM J. Alg. Disc. Meth. 5* (1984), pp. 306-313.

[24] P. C. Gilmore and A. J. Hoffman, "A characterization of comparability graphs and of interval graphs," *Can. J. Math 16* (1964), pp. 539–548.

[25] M. C. Golumbic, *Algorithmic Graph Theory and Perfect Graphs*, Academic Press, New York (1980).

[26] C.-W. Ho and R. C. T. Lee, "Efficient parallel algorithms for finding maximal cliques, clique trees, and minimum coloring on chordal graphs," *Inf. Proc. Let. 28* (1988), pp. 301-309.

[27] C.-W. Ho and R. C. T. Lee, "Counting clique trees and computing perfect elimination schemes in parallel," *Inf. Proc. Let. 31* (1989), pp. 61-68.

[28] P. N. Klein, *Efficient parallel algorithms for planar, chordal, and interval graphs*, TR-426 (Ph.D. thesis), Laboratory for Computer Science, MIT (1988).

[29] P. N. Klein, "Efficient parallel algorithms for chordal graphs," *Proc. 29th Symp. Found. of Comp. Sci.* (1989), pp. 150-161.

[30] P. N. Klein and J. H. Reif, "An efficient parallel algorithm for planarity," *J. Comp. Sys. Sci. 37* (1988), pp. 190-246. A preliminary version appeared in *27th FOCS* (1986), pp. 465-477.

[31] D. Kozen, U. Vazirani, and V. Vazirani, "NC algorithms for comparability graphs, and testing for unique perfect matching," *Proc. 5th Symp. Found. of Software Technology and Theor. Comp. Sci.*, published as *Lecture Notes in Computer Science 206*, Springer-Verlag, New York (1985), pp. 496-503.

[32] R. E. Ladner and M. J. Fischer, "Parallel Prefix Computation," *J. ACM 27:4* (1980), pp. 831-838.

[33] G. S. Lueker and K. S. Booth, "A linear time algorithm for deciding interval graph isomorphism," *J. ACM 26:2* (1979), pp. 183-195.

[34] G. L. Miller and J. H. Reif, "Parallel tree contraction and its application," *26th FOCS* (1985), pp. 478-489.

[35] J. Naor, M. Naor, and A. A. Schäffer, "Fast parallel algorithms for chordal graphs," *19th STOC* (1987), pp. 355-364; also submitted to *SIAM J. Comp.*

[36] M. B. Novick, personal communication (1987).

[37] M. B. Novick, personal communication (1988).

[38] M. B. Novick, "Logarithmic time parallel algorithms for recognizing comparability and interval graphs," manuscript.

[39] J. H. Reif, "An optimal parallel algorithm for integer sorting," *26th FOCS*, pp. 496-503.

[40] D. J. Rose, "Triangulated graphs and the elimination process," *J. Math Anal. Appl. 32* (1970), pp. 597-609.

[41] D. J. Rose, R. E. Tarjan, and G. S. Lueker, "Algorithmic aspects of vertex elimination on graphs," *SIAM J. Comp. 5* (1976), pp. 266-283.

[42] J. E. Savage and M. G. Wloka, "A parallel algorithm for channel routing," *Proceedings of WG '88, Graph-theoretic Concepts in Computer Science*, published as *Lecture Notes in Computer Science*, Springer-Verlag, New York (1988).

[43] R. Schreiber, "A new implementation of sparse Gaussian elimination," *ACM Trans. on Mathematical Software 8:3* (1982), pp. 256-276.

[44] Y. Shiloach and U. Vishkin, "An $O(\log n)$ parallel connectivity algorithm," *J. Algorithms 3* (1982), pp. 57-67.

[45] R. E. Tarjan and U. Vishkin, "Finding biconnected components and computing tree functions in logarithmic parallel time," *25th FOCS* (1984), pp. 12-22.

[46] J. R. Walter, *Representations of Rigid Circuit Graphs*, Ph.D. thesis, Wayne State Univ. (1972).

[47] J. R. Walter, "Representations of chordal graphs as subtrees of a tree," *J. Graph Theory 2* (1978), pp. 265-267.

Part III

Parallel Sorting and Computational Geometry

9

Random Sampling Techniques and Parallel Algorithm Design

Sanguthevar Rajasekaran

Department of Computer and Information Science
University of Pennsylvania
Philadelphia, PA 19104
raj@central.cis.upenn.edu

Sandeep Sen

Department of Computer Science
Duke University
Durham, NC 27706
ssen@cse.iitd.ernet.in

9.1 Introduction

9.1.1 Randomized Algorithms

The technique of randomizing an algorithm to improve its efficiency was first introduced in 1976 independently by Rabin and Solovay & Strassen. Since then, this idea has been used to solve myriads of computational problems successfully. Today randomization has become a powerful tool in the design of both sequential and parallel algorithms.

Even though the idea of randomization is at least as old as Hoare's quicksort algorithm, these previous approaches assume a distribution on the space of all possible inputs, which may not be a valid assumption at all. For example, Hoare's quicksort algorithm may run for a long period of time on certain inputs. However, the number of such bad input permutations is only a small fraction of the possible inputs. If we assume (which indeed Hoare does) that each input permutation is equally likely to occur, then quicksort algorithm is quite practical because with very high probability the given input permutation will not be a bad one and hence the algorithm will terminate quickly. But, Hoare's assumption of a uniform distribution on the input space is questionable, since the input distribution may vary quite unpredictably. Rabin and Solovay & Strassen rectify this problem by introducing randomization into the algorithm itself.

Informally, a randomized algorithm is one which bases some of its decisions on the outcomes of coin flips. We can think of the algorithm with one possible sequence of outcomes for the coin flips to be different from the same algorithm with a different sequence of outcomes for the coin flips. Therefore, a randomized algorithm can be viewed as a family of algorithms. For a given input, some of the algorithms in this family may run for indefinitely long periods of time. The objective in the design of a randomized algorithm is to ensure that the number of such bad algorithms in the family is only a small fraction of the total number of algorithms. If for *any* input we can show that at least $(1-\epsilon)$ (ϵ being very close to 0) fraction of algorithms in the family will run quickly on that input, then clearly, a random algorithm in the family will run quickly on any input with probability $\geq (1-\epsilon)$. In this case we say that this family of algorithms (or this randomized algorithm) runs quickly with probability at least $(1-\epsilon)$ where ϵ is called the error probability. Observe that this probability is independent of the input distribution.

To give a flavor for the above notions, we now give an example of a randomized algorithm. Given a polynomial of n variables $f(x_1, \ldots, x_n)$ over a field F, it is required to check if f is identically zero. We generate a random n-vector $(r_1, \ldots, r_n)$ $(r_i \in F, i = 1, \ldots, n)$ and check if $f(r_1, \ldots, r_n) = 0$. We repeat this for k independent random vectors. If there is even one vector for which f evaluates to a nonzero value, then clearly f is nonzero. If f evaluated to zero on all the k vectors tried, we conclude f is zero. It can be shown that the probability of error in our conclusion will be very small if we choose a sufficiently large k. In comparison, the best known deterministic algorithm for this problem is much more complicated and has a much higher time complexity.

Advantages of Randomization

Two of the most important advantages of using randomized algorithms are their simplicity and efficiency. A majority of the randomized algorithms found in the literature are simpler and easier to understand than the best deterministic algorithms for the same problems. The reader may have already got some feel for this from the above given example of testing if a polynomial is identically zero. Randomized algorithms have also been shown to yield better complexity bounds.

A skeptical reader at this point might ask: How dependable are randomized algorithms in practice? After all, there is a non zero probability that they may fail. Most readers are also aware that there is a probability (however small it might be) that the hardware itself might fail. So, if we design a fast algorithm for a problem with an error probability $< 2^{-k}$ for some integer k independent of the problem size, we can reduce the error probability far below the hardware error probability by making k large enough.

Different Types of Randomized Algorithms

Two types of randomized algorithms can be found in the literature: 1) algorithms that always produce the correct output but may run for an indefinite period (that is the running time is a random variable). These are called *Las Vegas* algorithms; and 2) those that run for a specified amount of time and whose output will be correct with a specified probability. These are called *Monte Carlo* algorithms. Primality testing algorithm of Rabin is of the second type.

The error of a randomized algorithm can either be 1-sided or 2-sided. Consider a randomized algorithm for recognizing a language. The output of the algorithm is either *yes* or *no*. There are algorithms which when saying *yes* are always correct, but when saying *no* may have produced a wrong answer. Such algorithms are said to have 1-sided error. Algorithms that have non zero error probability on both possible outputs are said to have 2-sided error.

Randomization in Parallel Algorithms

The ever-decreasing low cost of hardware nowadays has prompted computer scientists to design parallel machines and algorithms to solve problems very efficiently. In an early paper Reif proposed using randomization in parallel computation. In this paper he also solved many algebraic and graph theoretic problems in parallel using randomization. Since then a new area of research in computer science has evolved that tries to exploit the special features offered by both randomization and parallelization.

9.1.2 Model of Computation

A large number of parallel machine models have been proposed. Some of the more widely accepted models are: 1) fixed connection machines, 2) shared memory models, 3) the boolean circuit model, and 4) the parallel comparison trees. For all the algorithms in this chapter, we assume the shared memory model. In the randomized version of these models, each processor is equipped with a random number generator which can produce independent random bits in unit time. This is in addition to the computations allowed by the corresponding deterministic version. The *time complexity* of a parallel machine is a function of its input size. More precisely, time complexity is a function $g(n)$ that is the maximum over all inputs of size n of the time elapsed when the first processor begins execution until the time the last processor stops execution.

Just like big-O function serves to represent the complexity bounds of deterministic algorithms, $\widetilde{O}$ serves to represent the complexity bounds of randomized algorithms.

NOTATION

We say a randomized algorithm has a resource (time, space etc.) bound of $\widetilde{O}(g(n))$ if there exists a constant c such that the amount of resource used by the algorithm (on any input of size n) is no more than $c\alpha g(n)$ with probability $\geq 1 - 1/n^\alpha$. We shall refer to these bounds as *high probability* bounds.

We say a parallel algorithm is *optimal* if its processor bound P_n and time bound T_n are such that $P_n T_n = \widetilde{O}(S)$ where S is the time bound of the best known sequential algorithm for that problem.

In shared memory models, a number (say P) of processors work synchronously communicating with each other with the help of a common block of memory accessible by all processors. Each processor is a random access machine. Each step of the algorithm is an arithmetic operation, a comparison, or a memory access. Several conventions are possible to resolve read or write conflicts that might arise while accessing the shared memory. *EREW PRAM* is the shared memory model where no simultaneous read or write is allowed on any cell of the shared memory. *CREW PRAM* is a variation which permits concurrent read but not concurrent write. And finally, *CRCW PRAM* model allows both concurrent read and concurrent write. Each processor has a unique identifier (or *id* for short) which is a $\log P$ bit integer where P is the number of processors. We also assume that each processor has access to $O(\log n)$ bit random numbers in unit time.

9.1.3 Problems of Interest

In this chapter we consider the following problems: Selection, Sorting, and Convex hulls. These are very important fundamental problems in computer science which is reflected by the vast amount of literature devoted to these problems.

The problem of selection is to find the ith smallest key in a given sequence of keys (for some specified i). The problem of sorting is to rearrange the given sequence of keys in either descending order or ascending order.

DEFINITION 9.1
A convex hull of a set S of points is the smallest convex set containing S. A planar convex hull is the convex hull of a set of points in two dimensional space.

We shall describe a parallel algorithm for constructing planar convex hulls. The algorithm can actually be extended to three dimensions with some additional effort; however, we have confined ourselves to two dimensions to make the presentation simpler. The problem of sorting and construction of convex hulls are closely related. Specifically, any lower bound for sorting is also a lower bound for constructing convex hulls. Moreover, in the sequential context the algorithms for the problems share an interesting relationship. The techniques for sorting like mergesort and insertion sort can be extended almost directly for convex hulls.

9.2
Random Sampling Lemmas for Sorting and Selection
9.2.1 Selection

Let $X = \{k_1, k_2, \ldots, k_n\}$ be a set of n distinct keys. Let $<$ be a total ordering over X. The problem of selection is to find the ith smallest key in X

(for some specified $i \leq n$). Let the rank of any key in X be the number of keys less than this key plus one. A trivial sequential algorithm for selection computes the rank of each key and finds the one with rank i. Such an algorithm makes n^2 comparisons. But there are both deterministic and randomized algorithms that make only $O(n)$ comparisons. The constant in the time bound of the randomized algorithm is optimal, whereas no such deterministic algorithm is known. The power of randomization in algorithms design is also illustrated by the extremal selection algorithm given in the next section.

Chernoff Bounds

Frequently, in the design of randomized algorithms, many parameters of interest (like run time, space used, etc.) are random variables with a binomial distribution. For example, let's assume that the run time of an algorithm is binomial with mean m. Can we say the run time of the algorithm is $O(m)$ with high probability? The answer is yes if m is sufficiently large. Intuitively, if m is large, the area under the tail ends of a binomial distribution is negligible. Chernoff bounds provide fairly tight estimates for computing the area under the tail ends of a binomial. Before giving the Chernoff bounds, it is illustrative to consider the following simple example.

We toss an unbiased coin $10 \log n$ times. What is the probability P that $\leq \log n$ of the outcomes are tails? It is easy to see that

$$P = \sum_{i=0}^{\log n} \binom{10 \log n}{i} 2^{-10 \log n}.$$

i.e.,

$$P \leq \log n \binom{10 \log n}{\log n} 2^{-10 \log n}.$$

Using Stirling's approximation for factorials,

$$P \leq \log n 2^{(10 \log 10 - 9 \log 9) \log n} 2^{-10 \log n}.$$

Thus P is very, very small.

It turns out that if the mean of a binomial distribution X is $\Omega(\log n)$, then, the probability that X is greater than $O(\log n)$ is $\leq n^{-\alpha}$, for any $\alpha > 1$ (the constants in $\Omega()$ and $O()$ will depend on α). Very often it is easier to *bound* the random variable corresponding to the running time by a well-known distribution rather than analyze the exact distribution.

DEFINITION 9.2
We say a random variable X upper bounds another random variable Y (equivalently, Y lower bounds X) if for all x such that $0 \le x \le 1$, $Probability(X \le x) \le Probability(Y \le x)$.

A *Bernoulli trial* is an experiment with two possible outcomes, namely *success* and *failure*. The probability of success is p.

A *binomial variable* X with parameters (n, p) is the number of successes in n independent Bernoulli trials, the probability of success in each trial being p.

The *distribution function* of X can easily be seen to be

$$Probability(X \le x) = \sum_{k=0}^{x} \binom{n}{k} p^k (1-p)^{n-k}.$$

Chernoff and Angluin & Valiant have found ways of approximating the tail ends of a binomial distribution. In particular, their results can be summarized as

LEMMA 9.1 *Chernoff Bounds*
If X is binomial with parameters (n,p), and $m > np$ is an integer, then

$$Probability(X \ge m) \le \left(\frac{np}{m}\right)^m e^{m-np}. \quad (9.1)$$

Also,

$$Probability(X \le \lfloor(1-\epsilon)pn\rfloor) \le exp(-\epsilon^2 np/2) \quad (9.2)$$

and

$$Probability(X \ge \lceil(1+\epsilon)np\rceil) \le exp(-\epsilon^2 np/3) \quad (9.3)$$

for all $0 < \epsilon < 1$.

9.2.2 A Simple Algorithm for Extremal Selection

Consider the problem of selection when $i = n$, i.e., the problem of finding the largest (or smallest) key from among a given set of n keys. In this section we describe a randomized algorithm on an n processor $CRCW\ PRAM$ that solves this problem in $\widetilde{O}(1)$ time. In contrast there are known lower bounds which establish that any deterministic algorithm for maximal selection requires $\Omega(\log \log n)$ time on an n processor $CRCW\ PRAM$. Before we describe the algorithm we state and prove a lemma that will be used in the algorithm.

LEMMA 9.2
Maximum of n keys can be found in $O(1/\epsilon)$ time using $n^{1+\epsilon}$ CRCW PRAM processors.

PROOF

We prove this for the special case when $\epsilon = 1/2$ and leave the general case as an exercise problem. Let $k_1, k_2, \ldots, k_n$ be the input keys. Group the $n^{1+1/2}$ processors into $\sqrt{n}$ groups $G_1, G_2, \ldots, G_{\sqrt{n}}$ where each group has n processors. Each group G_i ($1 \leq i \leq \sqrt{n}$), in parallel, computes the maximum of $\sqrt{n}$ elements in constant time. This can be done by checking for each element k_i, if there is a larger element in the input than k_i. Using a processor for every pair and a special 'marker-cell' for every element this can be accomplished in $O(1)$ time. The processor comparing k_i and k_j writes (concurrently) into the 'marker-cell' i, if it finds that $k_i < k_j$. Only one 'marker-cell' will not be written into which corresponds to the maximum key. Subsequently we choose the maximum of the $\sqrt{n}$ maxima by another application of the previous strategy. (There are $n^{3/2}$ processors and only $\sqrt{n}$ keys in this phase). ∎

Now we present the maximal selection algorithm for n keys. In the beginning processor i is assigned the key k_i, and the set of 'surviving keys', Y, is the same as X.

Step 1
In this step roughly $n^{1/2}$ random keys are sampled (i.e. chosen randomly) from Y. Each processor decides to include its key in the sample with probability $\frac{1}{2n^{1-\frac{1}{2}}}$.

Step 2
Keys that were sampled in step 1 are assigned to unique locations in a memory block of size $n^{3/4}$. Call these memory cells $M(1), M(2), \ldots, M(n^{3/4})$. For each sampled key, a random cell is chosen. If that cell has not been assigned to any other key, assignment for the key is complete. If the chosen cell has been previously assigned to some other key, another random cell is chosen. This process is repeated until a unique assignment has been found for the key.

Step 3
n processors in parallel find the maximum of $M(1), M(2), \ldots, M(n^{3/4})$ in $O(1)$ time using the procedure described in the proof of Lemma 9.2.

(Note: All the $M(i)$'s will contain $-\infty$ at the beginning of step 2.) Let m be the maximum found.

Step 4
Each processor i, $(1 \leq i \leq n)$ (with a key in Y) in parallel compares k_i with m. If k_i is less than m, then k_i will be dropped from (Y and) future consideration as a potential candidate for maximum.

The surviving keys execute steps 2 and 3 once more and m found in step 3 is output as the *maximum*.

Analysis The number of keys that will be included in the sample in step 1 is a binomial variable with parameters $(n, 1/(2n^{1/2}))$. Thus using Chernoff bounds, this number is no more than $n^{1/2}$ with high probability. The number of keys surviving after step 4 is no more than $n^{1/2} \log n$ with high probability (Exercise 9.6). In the second step, assignment for sampled keys can be found in $O(1)$ time with high probability (Exercise 9.6).

From these assertions, it is easy to see that the algorithm runs in $\widetilde{O}(1)$ time.

9.2.3 A Selection Algorithm

In this section we give a sequential randomized selection algorithm and show how this can be modified to get a sorting algorithm also. We only perform expected time analysis, though the same time bounds also hold with high probability. Also, we can assume that the input keys are distinct. If they are not distinct, append i (as a $\log n$ bit binary number) to the right of k_i (for $1 \leq i \leq n$). This modification of the keys does not alter either the time or the processor bounds of the algorithms given in this chapter.

ALGORITHM 9.1
Select(i, X);
begin

    if $X = \{x\}$ **then return** x;
    Choose a random key $k \in X$ (called the splitter);
    Let $X_1 = \{x \in X | x < k\}$ and $X_2 = X - X_1$;
    if $|X_1| \geq i$ **then return** Select(i, X_1)
    else return Select$(i - |X_1|, X_2)$

end ;

Let $\bar{T}(i, n)$ be the expected (sequential) run time of Select. After selecting the splitter key, partitioning of X into X_1 and X_2 can be performed in n steps. Also, the randomly chosen splitter key can be any one of the n keys in X with equal probability. In particular, the splitter key will be the jth smallest key of X with probability $1/n$ (for $1 \leq j \leq n$). Thus, $\bar{T}(i,n)$ satisfies the following recurrence.

$$\bar{T}(i,n) = n + \frac{1}{n}\left[\sum_{j=1}^{i} \bar{T}(i-j, n-j) + \sum_{j=i+1}^{n} \bar{T}(i,j)\right].$$

By induction we can show $\bar{T}(i,n) \leq 2n + \min(i, n-i) + o(n)$. Notice that the expectation in the above equation is over the space of all outcomes of coin flips and not over the space of all possible inputs. It would be more desirable to show that the asymptotic time bound is the same with high probability, i.e., to show that the run time is no more than can with probability $> 1 - \frac{1}{n^\alpha}$ for some fixed constant c.

Markov's inequality asserts that if $\bar{T}$ is the expected value of a random variable T, then the probability that T exceeds $k\bar{T}$ is less than $1/k$. This implies for example that the run time of Select will not exceed $2\bar{T}(i,n)$ with probability $\geq 1/2$. This probability is not good enough when compared with $1 - \frac{1}{n^\alpha}$.

We can modify Select slightly so as to make the same time bound hold with high probability. The modification is to choose a random sample S of size s from X, to find the median of S, and to use this median as the splitter key. An exactly similar algorithm can be given for sorting also. The algorithm can be outlined as the following.

ALGORITHM 9.2
Sort(X);
begin

    if $|X| = 1$ **then return** X;
    Choose a random subset $S \subseteq X$ of size s;
    Let k =select($\lfloor s/2 \rfloor, S$);
    return Sort($\{x \in X | x < k\}$) . (k).
    Sort($\{x \in X | x > k\}$);

end;

The correctness of Sort is based on the following lemma:

LEMMA 9.3
Let $s = n/\log n$. Then,

$$\text{Prob.} \left[|\text{rank}(k, X) - n/2| > \sqrt{d\alpha n} \log n \right] < n^{-\alpha}$$

for some constant d.

The proof of Lemma 9.3 is left as an exercise.

Let $\bar{T}(n)$ be the expected number of comparisons made by Sort(X). Since selection on an input of n keys takes only $O(n)$ comparisons, we have for $s(n) = n/\log n$,

$$\bar{T}(n) \leq 2\bar{T}(n_1) + n^{-\alpha}\bar{T}(n) + O(n)$$

where

$$n_1 = n/2 + \sqrt{d\alpha n} \log n.$$

This solves to

$$\bar{T}(n) = O(n \log n),$$

which asymptotically approaches the optimal number of comparisons needed to sort n numbers.

It can be proven that the order of time bounds of both Sort and modified Select remains unaltered even with high probability.

9.3
General and Integer Sorting

Given a sequence of keys $k_1, k_2, \ldots, k_n$, the problem of sorting is to rearrange this sequence either in ascending order or descending order. Sorting is of vital importance in computer science since almost every application program calls for a sorting subroutine. There are a large number of optimal sequential algorithms for sorting. Examples include merge-sort, quick-sort, heap-sort etc. The run times of all these algorithms are $O(n \log n)$. All these algorithms assume no prior information about the keys being sorted. On the other hand, if we know some structure about the keys, sorting becomes easier. For example, if the keys to be sorted are integers with $O(\log n)$ binary bits each, sorting can be performed in $O(n)$ time, using the radix sort algorithm.

DEFINITION 9.3
Keys with no known structure will be called general keys, *and keys that are integers with at the most $O(\log n)$ bits will be called* integer keys.

Ajtai, Komlos and Szemeredi proposed the first non-trivial deterministic $O(\log n)$ depth parallel sorting network for sorting general keys. Subsequently Leighton improved their algorithm so that it ran on a sorting network with $O(n)$ processors in $O(\log n)$ time (or depth). The constant in this time bound is quite large. Following this, Cole gave a PRAM algorithm that was optimal with a small constant in the time bound. Cole's algorithm is less complicated than the previous algorithm but uses non-trivial pointer updating mechanism. For this reason it is not suitable for interconnection network models. In this section we describe a simple (and optimal) randomized algorithm for general-sort. A noteworthy feature of this algorithm is that it can be extended (with some variations) to run in the interconnection networks with similar asymptotic behavior; however the description of this implementation falls outside the scope of this chapter. We shall also describe an optimal algorithm for sorting integer keys in the range $[1, n]$. The algorithm INTEGER_SORT uses $n/\log n$ processors and runs in $\widetilde{O}(\log n)$ time.

9.3.1 Preliminary Results

Prefix Circuits

DEFINITION 9.4
Let Σ be a domain and let $\circ$ be an associative operation that takes $O(1)$ sequential time over this domain. The prefix computation *problem is defined as given input $(X(1), X(2), \ldots, X(n)) \in \Sigma^n$, compute outputs $(X(1), X(1) \circ X(2), \ldots, X(1) \circ X(2) \circ \ldots \circ X(n))$. When the $\circ$ operation is the ordinary summation, it is called the* prefix sum *problem.*

There are many optimal algorithms to solve this problem on various models. In particular,

LEMMA 9.4
Prefix computation can be performed using $n/\log n$ processors and $O(\log n)$ time on any PRAM.

Prefix sum can be computed optimally in time less than $\log n$ provided the elements are only integers of $O(\log n)$ bits. Cole and Vishkin have proved the following.

LEMMA 9.5
Prefix sum computation of n integers ($O(\log n)$ bits each) can be performed in $O(\log n / \log \log n)$ time using $n \log \log n / \log n$ CRCW PRAM processors.

9.3. General and Integer Sorting

DEFINITION 9.5
A sorting algorithm is said to be Stable if equal elements remain in the same relative order in the sorted sequence as they were in originally. In more precise terms, given input $k_1, k_2, \ldots, k_n$, the algorithm outputs a sorting permutation σ of $(1, 2, \ldots, n)$ such that for all $i, j \in [n]$, if $k_i = k_j$ and $i < j$ then $\sigma(i) < \sigma(j)$. A sorting algorithm that is not guaranteed to output a stable sorted sequence is called Non-Stable.

It is well known that Stable INTEGER_SORT of n keys can be done in time $O(n)$ by a deterministic sequential RAM.

NOTATION
Throughout this chapter we let $[m]$ stand for $\{1, 2, \ldots, m\}$.

An Assignment Problem

Let Q be a set $\{1, 2, \ldots, n\}$ of n indices where each index belongs to exactly one of m groups $G_1, G_2, \ldots, G_m$. Let g_i denote the number of indices belonging to group $G_i, i = 1, \ldots, m$. Given a sequence $N(1), N(2), \ldots, N(m)$ where $\sum_{i=1}^{m} N(i) = O(n)$ and $N(i)$ is an upper bound for $g_i, i = 1, 2, \ldots, m$. The problem is to find a permutation of $(1, 2, \ldots, n)$ in which all the indices belonging to G_1 appear first, all the indices belonging to G_2 appear next, and so on. (Assume that given an index i, the group $G_{i'}$ that i belongs to can be found in $O(1)$ time.)

For example, if $n = 5$, $m = 2$, $G_1 = \{2, 5\}$, $G_2 = \{1, 3, 4\}$, then $(5, 2, 1, 3, 4)$ and $(2, 5, 3, 1, 4)$ are (two of the) valid answers. It is a trivial task to design a linear time algorithm for the above problem.

LEMMA 9.6 *Assignment Lemma*
The above assignment problem can be solved in $\tilde{O}(\log n)$ parallel time using $n/\log n$ PRAM processors.

PROOF
The following algorithm achieves the bound stated in the lemma. We use a shared memory of size $2 \sum_{i=1}^{m} N(i)$ ($= L$, say). This memory is divided into m blocks $B_1, B_2, \ldots, B_m$ the size of B_i being $2N(i)$. A unique assignment for the indices belonging to G_i will be found in the block B_i, for $i = 1, 2, \ldots, m$.

Each one of the P ($= n/\log n$) processors is given $\log n$ successive indices. More precisely, processor π is given the indices $(\pi - 1) \log n + 1$, $(\pi - 1) \log n + 2, \ldots, \pi \log n$, for $\pi = 1, 2, \ldots, P$. There are three phases of the algorithm. In the first phase, boundaries of the m blocks are computed. In the second phase, every processor sequentially finds unique assignments for the $\log n$ indices given to it in their **respective** blocks. In the third phase, a prefix sum computation is done to eliminate the unused cells and the position of each index in the output is computed. Details follow.

Step 1
P processors collectively do a prefix sum of $(N(1), N(2), \ldots, N(m))$ and hence compute the boundaries of blocks in the common memory.

Step 2
Each processor π is given a total time of $d \log n$ (d being a constant to be fixed) to find assignments for all its indices sequentially.
π starts with its first index (call it) l. If $G_{l'}$ is the group that l belongs to, π chooses a random cell in $B_{l'}$ and tries to write its id in it. If the chosen cell did not contain the id of any other processor and π succeeds in writing, then that cell is assigned to l. The probability of success in one trial is $\geq 1/2$. If π has failed in this trial then it tries as many times as it takes to find an assignment for l and then it takes up the next index.
After $d \log n$ steps, even if there is a single processor that has not found assignments for all its keys, the algorithm is aborted and started anew.

Step 3
Each processor π writes a 1 in each of the cells that have been assigned to its indices. Unassigned cells in the common memory will have 0's. P processors perform a prefix sum computation on the contents of the memory cells $(1, 2, \ldots, L)$. Finally, every processor reads out from the prefix sum the position of each one of its indices in the output.

Analysis Steps 1 and 3 can be completed in $O(\log n)$ time in accordance with Lemma 9.4. In step 2, the probability that a particular processor π successfully finds an assignment for one of its keys in a single trial is $\geq 1/2$. Let Y be the random variable equal to the number of successes of π in $d \log n$ trials. We require Y to be $\geq \log n$ for every processor. Clearly Y is lower bounded by a binomial variable with parameters $(d \log n, 1/2)$. It follows from the Chernoff bounds (Equation 9.3) that the probability that there will be at least a single processor which has not found assignments for all of its indices after $d \log n$ trials can be made $\leq n^{-\alpha}$ for any $\alpha \geq 1$, if we choose a proper constant d.

Therefore the whole algorithm runs in time $\widetilde{O}(\log n)$. This completes the proof of Lemma 9.6. ∎

9.3.2 Preparata's GENERAL_SORT Algorithm

Preparata's algorithm is nearly optimal and runs on *CREW PRAM*. It uses $n \log n$ processors and runs in time $O(\log n)$. One of the subroutines used is an $O(\log \log n)$ time algorithm of Valiant for merging two n element sorted sequences using n processors. The problem of merging two sorted sequences is to obtain a sorted sequence of elements of both the sequences. A trivial sequential algorithm can achieve this task in linear time. Details of Preparata's algorithm follows.

Step 1
If the problem is of constant size, solve it directly and quit.

Step 2
Partition the given n keys into $\log n$ parts, with $n/\log n$ keys in each part. Sort each part recursively and separately in parallel, assigning n processors to each part. Let $S_1, S_2, \ldots, S_{\log n}$ be the sorted sequences.

Step 3
Merge S_i with S_j for $1 \leq i, j \leq \log n$ in parallel. This can be done by allocating $n/\log n$ processors to each pair (i, j). That is, using $n \log n$ processors this step can be accomplished in $O(\log \log n)$ time using Valiant's algorithm. As a by-product of this merging step, we have computed the rank of each key in each one of the S_i's ($1 \leq i \leq \log n$).

Step 4
Allocate $\log n$ processors to compute the rank of each key in the original input. This is done in parallel for all the keys by adding up the $\log n$ ranks computed (for each key) in step 3. This can be done in $O(\log \log n)$ time (Lemma 9.4). Finally, the keys are written out in the order of their ranks.

Analysis Let $T(n)$ be the run time of the above algorithm using $n \log n$ processors. Clearly, step 1 takes $T(n/\log n)$ time. Put together, steps 2 and 3 take $O(\log \log n)$ time. Thus we have,

$$T(n) = T(n/\log n) + O(\log \log n),$$

426 Chapter 9. Random Sampling Techniques and Parallel Algorithm Design

which solves to $T(n) = O(\log n)$. Also, the number of processors used in each step is $n \log n$.

We summarize as follows:

THEOREM 9.1
n general keys can be sorted in $O(\log n)$ time using $n \log n$ CREW PRAM processors.

COROLLARY 9.1
n general keys can be sorted in $O(t \log n)$ time using $n \log n / t$ CREW PRAM processors.

9.3.3 Integer Sorting

In this section we present an algorithm INTEGER_SORT for sorting integer keys in the range $[n]$. This algorithm employs $n / \log n$ processors and runs in time $\widetilde{O}(\log n)$.

Summary of the Algorithm

The main idea behind our algorithm is radix sorting. As an example of radix sorting, consider the problem of sorting a sequence of two-bit decimal integers. One way of doing this is to sort the sequence with respect to the least significant bits (LSB) of the keys and then to sort the resultant sequence with respect to the most significant bits (MSB) of the keys. This will work, provided that in the second sort keys with equal MSBs will remain in the same relative order as they were in originally. In other words, the second sort should be stable.

Given keys $k_1, k_2, \ldots, k_n \in [n]$, where each key is a $\log n$-bit integer, we first (non-stable) sort this sequence with respect to the $(\log n - 3 \log \log n)$ LSBs of the keys. (Call this sort *Coarse_Sort*). In the resultant sequence we apply a stable sort with respect to the $3 \log \log n$ MSBs of the keys. (Call this sort *Fine_Sort*).

Even though the sequential time complexity of stable sort is no different from that of non-stable sort, it seems that parallel stable sort is inherently more complex than parallel non-stable sort. This is the reason why we partition the bits of the keys unevenly.

In Coarse_Sort, we (non-stable) sort a sequence of n keys, each key being in the range $[1, n/\log^3 n]$ and, in Fine_Sort we (stable) sort n keys in the range $[1, \log^3 n]$. In more formal terms, algorithm INTEGER_SORT can be summarized as follows.

Let $D = n/\log^3 n$ and $k'_i = \lfloor k_i/D \rfloor$ and $k''_i = k_i - k'_i * D$ for all $i \in [n]$.
Coarse_Sort. Sort $k''_1, k''_2, \ldots, k''_n \in [D]$. Let σ be the resultant permutation.
Fine_Sort. Stable-sort $k'_{\sigma(1)}, k'_{\sigma(2)}, \ldots, k'_{\sigma(n)} \in [\log^3 n]$. Let ρ be the resultant permutation.
Output. The permutation $\rho.\sigma$, the composition of ρ and σ.

In the next two sections, we describe Fine_Sort and Coarse_Sort respectively.

Fine_Sort

We give a deterministic algorithm for Fine_Sort. First we will show how to stable-sort n keys in the range $[\log n]$ using $n/\log n$ processors in time $O(\log n)$ and then apply the idea of radix sorting to prove that we can stable-sort n keys in the range $[(\log n)^{O(1)}]$ within the same resource bounds.

LEMMA 9.7
n keys $k_1, k_2, \ldots, k_n \in [\log n]$ can be stable-sorted in $O(\log n)$ time using $P = n/\log n$ processors.

PROOF
In Fine_Sort algorithm, each processor π is given $\log n$ successive keys. Each one of the P processors starts by sequentially stable-sorting the keys given to it. Then, collectively, the P processors group all the keys with equal values. (There are $\log n$ groups in all.) Finally, they output a rearrangement of the given sequence in which all the 1's (i.e., keys with a value 1) appear first, all the 2's appear next, and so on. Throughout the algorithm the relative order of equal keys is preserved. More details follow.

To each processor $\pi \in [P]$ we assign the key indices $J(\pi) = \{j \mid (\pi - 1)\log n < j \leq min(n, \pi \log n)\}$. There are three steps in the algorithm.

Step 1
Each processor π sequentially stable-sorts the keys $\{k_j \mid j \in J(\pi)\}$ in time $O(\log n)$, and hence constructs $\log n$ lists $J_{\pi,k} = \{j \in J(\pi) \mid k_j = k\}$ for $k \in [\log n]$. Elements in $J_{\pi,k}$ are ordered in the same relative order as in the input.

Step 2
The P processors collectively perform the prefix sum of

$$(|J_{1,1}|, |J_{2,1}|, \ldots, |J_{P,1}|,$$
$$|J_{1,2}|, |J_{2,2}|, \ldots, |J_{P,2}|,$$
$$\ldots$$
$$|J_{1,q}|, |J_{2,q}|, \ldots, |J_{P,q}|)$$

where $q = \log n$. Call this sum

$$(S_{1,1}, S_{2,1}, \ldots, S_{P,1},$$
$$S_{1,2}, S_{2,2}, \ldots, S_{P,2},$$
$$\ldots$$
$$S_{1,q}, S_{2,q}, \ldots, S_{P,q}).$$

Step 3
Each processor π sequentially computes the position of each one of its keys in the output using the prefix sum. The position of keys in the list $J_{\pi,l}$ will be $S_{\pi-1,l} + 1, S_{\pi-1,l} + 2, \ldots, S_{\pi,l}$.

Analysis It is easy to see that steps 1 and 3 can be performed within the stated resource bounds. Step 2 also can be completed within the stated resource bounds as stated in Lemma 9.4.

This concludes the proof of Lemma 9.7. ∎

LEMMA 9.8
If n keys in the range $[R]$ (for any $R = n^{O(1)}$) can be stable-sorted in $O(\log n)$ time using $P = n/\log n$ processors, then n keys $k_1, k_2, \ldots, k_n \in [R^2]$ can be stable-sorted in time $O(\log n)$ using the same number of processors.

PROOF
Let $k_i' = \lfloor k_i/R \rfloor$ and $k_i'' = k_i - k_i' * R$ for every $i \in [n]$. First, stable-sort $k_1'', k_2'', \ldots, k_n''$ obtaining a permutation σ. Then stable-sort $k_{\sigma(1)}', k_{\sigma(2)}', \ldots, k_{\sigma(n)}'$ obtaining a permutation ρ. Output $\rho.\sigma$. Clearly both these sorts can be completed in time $O(\log n)$ using P processors. ∎

Lemmas 9.7 and 9.8 immediately imply the following.

LEMMA 9.9
n integer keys in the range $[(\log n)^{O(1)}]$ can be stable-sorted in time $O(\log n)$ using $n/\log n$ processors.

Coarse_Sort

In this sub-section we fix a key domain $[D]$ where $D = n/\log^3 n$. We assume for the sake of convenience, $\log^3 n$ divides n. Let the input keys be $k_1, k_2, \ldots, k_n \in [D]$. Define the *index sequence* for each key $k \in [D]$ to be $I(k) = \{i \mid k_i = k\}$. The randomized algorithm for Coarse_Sort to be presented in this sub-section employs $P = n/\log n$ processors and runs in time $\tilde{O}(\log n)$. The sorted sequence is non-stable.

The main idea is to calculate the cardinalities of the index sequences $I(k), k \in [D]$ approximately, and then to use the assignment algorithm in the proof of Lemma 9.6 to rearrange the given sequence in sorted order.

LEMMA 9.10 *Estimation Lemma*
Given as input $k_1, k_2, \ldots, k_n \in [D]$, *we can compute* $N(1), N(2), \ldots, N(D)$ *in* $\tilde{O}(\log n)$ *time using* $P = n/\log n$ *processors such that* $\sum_{k \in [D]} N(i) = O(n)$ *and furthermore, with very high likelihood,* $N(k) \geq |I(k)|$ *for each* $k \in [D]$.

PROOF
The following sampling algorithm serves as a proof.

Step 1
Each processor $\pi \in [D \log n]$ in parallel chooses a random index $s_\pi \in [n]$. Let S be the sequence $\{s_1, s_2, \ldots, s_{D \log n}\}$.

Step 2
The P processors collectively sort the keys with the chosen indices. That is, they sort $k_{s_1}, k_{s_2}, \ldots, k_{s_{D \log n}}$ and compute index sequences $I_S(k) = \{i \in S \mid k_i = k\}$ (for each $k \in [D]$).

Step 3
D of the P processors in parallel set $N(k) = d(\log^2 n) \max(|I_S(k)|, \log n)$ for $k \in [D]$, d being a constant to be fixed in the analysis. Output $N(1), N(2), \ldots, N(D)$.

Analysis Trivially, steps 1 and 3 can be performed in $O(1)$ time. Step 2 can be performed using Preparata's GENERAL_SORT algorithm in $O(\log n)$ time. (Notice that we have to sort only $n/\log^2 n$ keys in step 2). It remains to be shown that $N(i)$'s computed by the sampling algorithm satisfy the conditions in Lemma 9.10.

If $|I(k)| \leq d\log^3 n$, then always $N(k) \geq d\log^3 n \geq |I(k)|$. So suppose $|I(k)| > d\log^3 n$. Then it is easy to see that $|I_S(k)|$ is a binomial variable with parameters $(\frac{n}{\log^2 n}, \frac{|I(k)|}{n})$. The Chernoff bounds (see Lemma 9.1, Equation 9.2) imply that for all $\alpha \geq 1$, there exists a c such that

$$Prob\left(|I_S(k)| \leq c\alpha |I(k)|/\log^2 n\right) \leq \frac{1}{n^\alpha}.$$

Therefore, if we choose $d = (c\alpha)^{-1}$ then $N(k) \geq |I(k)|$ (for every $k \in [D]$) with probability $\geq 1 - n^{-\alpha}$. The Chernoff bounds (Lemma 9.1 Equation 9.3) also imply that for all $\alpha \geq 1$ there exists a h such that $N(k) \leq (h\alpha)|I(k)|$ (for every $k \in [D]$) with probability $\geq 1 - n^{-\alpha}$.

The bound on $\sum_{k \in [D]} N(k)$ clearly holds since

$$\sum_{k \in [D]} N(k) \leq \sum_{k \in [D]} d\log^2 n[|I_S(k)| + \log n]$$
$$= d\log^3 nD + d\log^2 n \sum_{k \in [D]} |I_S(k)|$$
$$= dn + d\log^2 nD \log n = 2dn$$

This concludes the proof of Lemma 9.10. ∎

Having obtained the approximate cardinalities of the index sets, we apply the assignment algorithm. The set Q is the set of key indices, namely $\{1, 2, \ldots, n\}$. An index i belongs to group $G_{\iota'}$ if the value of the key with index i is ι'. Under this definition, group G_j is the same as index sequence $I(j)$, $j = 1, 2, \ldots, D$. Since we can find approximate cardinalities of these groups (Lemma 9.10), we can use the assignment algorithm to rearrange the given sequence in sorted order. Thus we have the following:

LEMMA 9.11
n keys $k_1, k_2, \ldots, k_n \in [D]$ can be sorted in time $\tilde{O}(\log n)$ time using $n/\log n$ processors.

Lemmas 9.9 and 9.11, together with the algorithm summary in Section 9.3.3, prove the following:

THEOREM 9.2
n integer keys in the range $[n]$ can be sorted in $\tilde{O}(\log n)$ time using $n/\log n$ CRCW PRAM processors.

9.3.4 Reischuk's Algorithm for GENERAL_SORT

The original algorithm of Reischuk for sorting general keys was recursive and hence the analysis was tedious. We give a non-recursive version of his algorithm. This algorithm will use n processors and run in time $\widetilde{O}(\log n)$. The underlying constant is small. The basis for this algorithm is Preparata's sorting scheme.

Reischuk's algorithm runs in the same time bound as Preparata's (with high probability), but using only n processors. The idea is to randomly sample $N = \frac{n}{\log^4 n}$ keys from the input and sort these using a non optimal algorithm like Preparata's. The sorted sample partitions the original problem into N independent subproblems of nearly equal size, and hence all these subproblems can be solved easily. These ideas are made concrete in the following algorithm.

Step 1
$N = n/(\log^4 n)$ processors randomly sample a key (each) from $X = k_1, k_2, \ldots, k_n$, the given input sequence.

Step 2
Sort the N keys sampled in step 1 using Preparata's algorithm. Let $l_1, l_2, \ldots, l_N$ be the sorted sequence.

Step 3
Let $X_1 = \{k \in X \mid k \leq l_1\}$; $X_i = \{k \in X \mid l_{i-1} < k \leq l_i\}$, $i = 2, 3, \ldots, N-1$; $X_N = \{k \in X \mid k > l_N\}$. Partition the given input X into X_i's as defined. This is done by first finding the part each key belongs to (using binary search in parallel). Now partitioning the keys reduces to sorting the keys according to their part numbers.

Step 4
For $1 \leq i \leq N$ in parallel do: sort X_i using Preparata's algorithm.

Step 5
Output sorted(X_1), sorted(X_2),..., sorted(X_N).

Analysis Step 2 can be done using $N \log N \leq N \log n$ processors in $O(\log N) = O(\log n)$ time (theorem 9.1). In step 3, partitioning of X can be done using binary search and the INTEGER_SORT algorithms.

Thus step 3 can be performed in $\widetilde{O}(\log n)$ time, using $\leq n$ processors. With high probability there will be no more than $O(\log^5 n)$ keys in each of the X_i's ($1 \leq i \leq N$). Proof of this fact is left as an exercise. Within the same processor and time bounds, we can also count $|X_i|$ for each i. In step 4, each X_i can be sorted in $O(\log |X_i|)$ time using $|X_i| \log |X_i|$ processors. Also X_i can be sorted in $(\log |X_i|)^2$ time using $|X_i|$ processors (see corollary 9.1). Thus step 4 can be completed in $(\max_i \log |X_i|)^2$ time using n processors. If $\max_i |X_i| = O(\log^5 n)$, step 4 takes $O((\log \log n)^2)$ time.

Thus we have proved the following.

THEOREM 9.3
We can sort n general keys using n CRCW PRAM processors in $\widetilde{O}(\log n)$ time.

9.4
Recursive Parallel Divide and Conquer

Although we presented a non-recursive version of Reischuk's parallel sorting algorithm, it is instructive to analyze the recursive version of his algorithm. This will be used later for more general situations where the algorithms are recursive by nature. In Reischuk's original algorithm, $\lfloor \sqrt{n} \rfloor$ keys were chosen randomly such that each key was chosen with probability $\frac{1}{\sqrt{n}}$. These keys (called splitters) were then sorted using pairwise comparisons from which their ranks can be computed easily. The latter can be done in $O(\log n)$ time very easily and we can do the pairwise comparisons simultaneously in constant time by using one processor for each comparison. These splitters partition the n input keys into $\lfloor \sqrt{n} \rfloor + 1$ buckets. For each key we can determine the appropriate bucket by a simple binary search using one processor for each key (using simultaneous reads). If we let n_i denote the size of the ith bucket then we claim the following:

LEMMA 9.12
The probability that for any i, n_i is larger than $c\sqrt{n} \log n$ is less than $\frac{1}{n^{(c-1)}}$.

PROOF
We shall show that for any key (whether or not it is in the sample), the probability that it is more than $c\sqrt{n} \log n$ away (in rank) from the next sampled key on its right is less than $\frac{1}{n^{(c-1)}}$. This follows from the

fact that each key was chosen with probability $\frac{1}{\sqrt{n}}$ and hence the above event can happen with probability less than $(1 - \frac{1}{\sqrt{n}})^{c\sqrt{n}\log n}$. For large n this can be bounded by $\frac{1}{n^3}$. Hence the probability that it can happen for any element is less than $\frac{1}{n^2}$. Consequently the distance (in rank) between two sampled elements is less than $c\sqrt{n}\log n$ with probability at least $\frac{1}{n^{(c-1)}}$. Letting c be 3, we can write down the recurrence for the expected running time of the algorithm as:

$$\bar{T}(n) \leq (1 - \frac{1}{n^2})\bar{T}(3\sqrt{n}\log n) + \frac{1}{n^2}\bar{T}(n - \lceil\sqrt{n}\rceil + 1)$$

By induction it can be shown that $\bar{T}(n) \leq O(\log n)$ and we leave it as an exercise problem. ∎

To derive high probability bounds, we shall actually derive a more general bound. This will be used repeatedly for analyzing algorithms in future. We shall also make use of the fact that Reishchuk's algorithm would execute in the same asymptotic time bound even if the number of sampled keys is $\lfloor n^\epsilon \rfloor$ for any fixed ϵ, $0 < \epsilon \leq 1/2$. Note that in general, the probabilistic bound of Lemma 9.12 holds for partitions of sizes $O(n^{1-\epsilon}\log n)$ (the lemma was for the case $\epsilon = 1/2$). For convenience, we shall use the bound $O(n^{\epsilon_0})$ where $\epsilon_0 > 1 - \epsilon$. Thus at depth i from the root, the size of a sub-problem can be bounded from above by $n^{\epsilon_0^i}$.

It may be helpful to view the algorithm as a tree where a node represents a subproblem and its children represent the recursive calls made by this node. For example, the root represents the procedure Sort[1..n] which has $\lfloor n^\epsilon \rfloor + 1$ children procedure calls each of size at most $n^{1-\epsilon}\log n$. The leaves of this tree represent problems of size less than a pre-determined threshold, say $\log^r n$ for some fixed integer r. At this stage the problem size is so small that we can use a direct sorting procedure like Batcher's sort to sort in time $O(\log\log n)$ thereby adding a factor of $o(\log n)$. Our objective is to show that all the leaf-level procedures are completed in $O(\log n)$ time with high probability. For this it suffices to show that a particular leaf-level procedure is completed in $O(\log n)$ time. This leaf-node defines a fixed path from the root to the leaf such that the problem sizes at successive nodes of this path are decreasing. For this let us denote the node at depth i from the root as N_i, the problem-size as n_i and the time taken at N_i by T_i. We claim the following:

LEMMA 9.13
$$Prob[T_i \geq k \cdot \epsilon_0^i c\alpha \log n] \leq 2^{-c\alpha\epsilon_0^i \log n}$$

where c and α are integers and k is a constant.

PROOF

From the previous claim and the comment in the previous paragraph, $Prob[n_{(i+1)} > n_i^{\epsilon_0}] < \frac{1}{n_i^2}$ for an appropriately chosen constant $0 < \epsilon_0 < 1$. We can verify this in $O(\log n_i)$ time (using prefix sum) and we repeat the sampling until $n_{(i+1)} \leq n_i^{\epsilon_0}$. If $k \log n_i$ is the time for each iteration (of the sampling algorithm), then we can immediately conclude the following:

$$Prob[T_{(i+1)} > kc\alpha \log n_i] < 2^{-c\alpha 2 \log n_i}$$

If $n_i = 2^{\epsilon_0^i \log n}$, then we arrive at the required inequality. From our resampling scheme we have guaranteed that $n_{(i+1)} \leq 2^{\epsilon_0^i \log n}$ so we have to prove that the claim is true when n_i is strictly less than $2^{\epsilon_0^i \log n}$. Let $n_i = 2^{(1/a)\epsilon_0^i \log n}$ where $a > 1$. Substituting this value of n_i in the previous inequality we get:

$$Prob[T_{(i+1)} > kc\alpha(1/a)\epsilon_0^i \log n] < 2^{-c\alpha 2(1/a)\epsilon_0^i \log n_i}$$

The above inequality implies that

$$Prob[T_{(i+1)} > kc\alpha(\lfloor a \rfloor/a)\epsilon_0^i \log n] < 2^{-c\alpha 2(\lfloor a \rfloor/a)\epsilon_0^i \log n_i}$$

Since $2\lfloor a \rfloor/a \geq 1$ for any $a > 1$, the lemma follows. ∎

We are now ready to prove the main result of this section:

THEOREM 9.4

Given a process-tree which has the property that a procedure at depth i from the root takes time T_i such that

$$P[T_i \geq kc\alpha \log n(\epsilon_0)^i] \leq 2^{-(\epsilon_0)^i c\alpha \log n}$$

then, all the leaf-level procedures are completed in $\tilde{O}(\log n)$ time.

PROOF

Setting $t_i = k(\epsilon_0)^i \log n\alpha(c - c_o)$, where c_o is some constant, we obtain

$$Prob[T_i \geq k\alpha c \log n(\epsilon_0)^i + t_i] \leq 2^{-(\epsilon_0)^i c\alpha \log n} \leq 2^{-t_i/k}$$

If T is the total time for this worst case chain of nested calls and $m = 1/(1-\epsilon_0)$, the probability that it takes more than $mk\alpha \log nc_o + t$, is less than the sum of the probability of events where $\sum_i t_i = t+j$, $0 \leq j \leq \mu$. Here $\mu = mk\alpha \log nc_o$. We shall compute the probability that $\sum_i t_i = t$ and multiply by μ.

$$\prod_{\sum t_i = t} 2^{-t_i} \leq \sum 2^{-t/k} \text{ over } t^{O(\log \log n)} \text{ tuples.}$$

Thus

$$Prob[T > km\alpha \log nc_o + t] < \mu 2^{-t/k + O(\log t \log \log^2 n)}$$

Using $t \geq km\alpha(c - c_o)\log n$, for large values of n and $m > 1$, we can rewrite the above expression as

$$Prob[T > km\alpha c \log n] < \mu 2^{-\alpha m(c-c_o)\log n}$$

For $c > 4c_o$, i.e. $c - c_o > 3/4c$, we have the following required bound,

$$Prob[T > \alpha log n] \leq \mu 2^{-(3/4)c\alpha \log n} \leq n^{-c_1 \alpha}.$$

assuming that k, m and c are larger than 1. ∎

9.5 Higher Dimensional Problems: Convex Hulls

We shall now try to apply the techniques developed in the previous section to a specific geometric problem on the plane. For this purpose we have chosen the problem of constructing two-dimensional convex hull of point sites. Although optimal $O(\log n)$ time parallel algorithms have been known for some time, its close relation to sorting makes it a natural candidate for the methods developed in the earlier part of this chapter. Moreover, the additional methods that will be developed for this problem are applicable to more general situations and the reader will be referred to the relevant literature at the end of this chapter.

9.5.1 A Straightforward Extension

We shall actually look at its dual problem, namely, the intersection of n half-planes. For readers unfamiliar with this transform, a brief sketch is included under Notes and References at the end of the chapter. Without loss of generality we can assume that the origin lies in its interior (we can ensure that the origin lies in the interior of the primal problem from which this property will hold trivially). Following on the lines of the sorting algorithm, we choose a random subset of $\lfloor n^\epsilon \rfloor$ half-planes where $0 < \epsilon < 1$. We can construct their intersection in $O(\log n)$ time using a brute-force method like checking for each pairwise intersection, if it is a vertex of the convex hull. Let $h_0, h_1, ..$ be the vertices of the convex-hull in a cyclic order (there can be at most $\lfloor n^\epsilon \rfloor$ of them). Consider the triangles (will also be referred to as sectors) of the form O,h_1, h_2. These will be intersected by a number of half-planes that were not chosen in the sample. The output convex hull is the union of the boundaries formed by the intersection of the half-planes inside each of the sectors. This gives a recursive procedure for constructing the convex hull. For each sector, we determine the half-planes intersecting the region and then call the algorithm recursively for each of the sectors.

To determine the sectors that a half-plane intersects we can use a very simple procedure due to Chazelle and Dobkin which uses *Fibonacci search* on a bimodal sequence to determine extremal points. The distances of the vertices of a convex n-gon from a line form a bimodal sequence and the closest vertex (including the intersecting points) can be determined using *Fibonacci search* which is similar to a binary search on the cyclic sequence of the vertices. The interested reader is encouraged to look up the reference mentioned in the last section for further details of this search procedure. Using one processor for every half-plane, in $O(\log n)$ time we can determine the edges of the convex hull that the half-plane intersects which gives us the information of the bounding sectors that this half-plane intersects. Because of convexity, it intersects all the sectors that lie in between. Note that we can very easily determine the number of sectors that it intersects in the same time (without explicitly listing the sectors).

In essence, the algorithm appears to be identical to that of parallel sorting. To prove any interesting result we have to determine how quickly the subproblem sizes are decreasing. However, there is an obvious difference, namely the total size of the subproblems may not be bound by the parent's problem size. This happens because a half-plane can intersect more than one sector which results in fragmentation. This is crucial for the processor's bound and large fragmentation could ruin the possibility of an optimal algorithm (i.e., one in which the *PT* product is $O(n \log n)$). Below we prove some results on the problem size and obtain a bound on fragmentation. For simplicity of the arguments we shall assume that no three lines have a common intersection point.

LEMMA 9.14
The probability that the maximum number of half-planes intersecting any sector exceeds $2(c+2)n^{1-\epsilon} \log n$ is less than n^{-c}.

PROOF
Consider all the $O(n^2)$ pairwise intersections defined by the lines bounding the half-planes. Draw the segments joining O and each of these

intersections and consider the ordered intersections of lines (representing the boundaries of half-planes) on this segment. For any given segment the probability that the number of intersections exceeds $(c+2) \times n^{1-\epsilon} \log n$ before the first (counting from O) sampled half-plane is less than $(1 - \frac{1}{n^{1-\epsilon}})^{(c+2)n^{1-\epsilon}\log n}$ which is less than $n^{-(c+2)}$ for large n. Thus the probability that this event happens for any segment is less than n^{-c}. This implies the lemma, since any half-plane intersecting the sector intersects at least one of the two bounding segments. ∎

LEMMA 9.15
The expected value of the sum taken over all the sectors of all the half-planes intersecting a sector is $O(n)$.

PROOF
From the proof of the previous lemma, it is clear that if the number of halfplanes intersecting a segment is greater than $n^{1-\epsilon}$, then the expected number of half-planes obeys a geometric distribution. The probability of success is $1/n^{1-\epsilon}$ which is the probability of being selected in the sample. So the expected number of half-planes that we do not select in the ordered list of half-planes (starting from O) before we select the first half-plane is $n^{1-\epsilon}$. Using the property that the expectation of the sum of random variables is the sum of the individual expectations, we arrive at the required bound. In our case we are interested in the sum of n^ϵ random variables. Note that in the proof there is no conditioning on the set of half-planes chosen; rather it is on the number of such planes chosen. Technically, the number of such planes chosen is a random variable whose expected value is n^ϵ. The reader is encouraged to solve Exercise 9.14. ∎

In the parallel setting, if we use as an abstract representation of a recursive algorithm the tree as described in the previous section, the running time of the algorithm is proportional to the longest path (in time) from the root to a leaf. Thus at any given node we are interested in the recursive call that takes the the longest time. Given that we only have a bound for the *expected* time taken by the child-procedures, and there are n^ϵ of them, we cannot derive any useful tail estimates. The reason that we could derive interesting bound for the sorting algorithm is because we were able to get tail estimates (of the problem size exceeding a certain size). However, for sequential algorithms one may obtain bounds on the *expected* running time by the using the

438 Chapter 9. Random Sampling Techniques and Parallel Algorithm Design

linearity property of *expectations*. We shall come back to this issue at the end of the chapter.

The bound on the total size of the subproblems is not known to hold with high-probability. However, we can claim the following:

LEMMA 9.16
For some suitable constant k_{total} and large n, the following conditions hold with probability at least 1/2:

(i) The maximum number of half-planes intersecting any sector is less than $2n^{1-\epsilon} \log n$
(ii) The sum of half-planes taken over all the sectors of the number of half-planes intersecting a sector is less than $k_{total} n$.

DEFINITION 9.6
We shall call a random sample good if the above conditions are satisfied and bad otherwise.

PROOF
From Lemma 9.15 and Markov's inequality we can choose k_{total} such that the probability that (ii) fails is at most 1/3 (i.e., kn is thrice the mean). For sufficiently large n, $1/n + 1/3$ is less than 1/2. Thus the probability that both (i) and (ii) are satisfied is at least 1/2. ∎

9.5.2 Resampling and Polling

As a consequence of the previous claim, if we repeat the sampling algorithm $r \log n$ times, the probability that the conditions are not satisfied during all the tries is less than n^{-r}. That is, if we choose independently $p(n) = O(\log n)$ sets of samples, one of them is good with very high likelihood. However, to determine if a sample is 'good', we would have to carry out the search procedure $O(\log n)$ times, each of which requires $O(\log n)$ time (such a method was described earlier). Instead, we try to estimate the number of half-planes intersecting a sector C_i using only a fraction of the input half-planes. For example, we can choose $c_0 \cdot n/\log^d n$ for some fixed integer $d > 2$ and a constant c_0 (the actual value will be determined from the required success probability of the algorithm) of the input half-planes randomly for the jth sample, R_j. Let X_i^j be the number of half-planes intersecting sector C_i corresponding to sample R_j, $1 \le j \le b \log n$ where b is fixed integer greater than 0 which is determined from the success probability of the algorithm. A_i^j be the number of half-planes intersecting C_i out of the $n/\log^d n$ randomly chosen input half-planes for the same sample. Clearly, A_i^j is a binomial random variable with parameters $c_0 \cdot n/\log^d n$ (number of trials) X_i^j/n (probability of success). Assuming that X_i^j is greater than $\bar{c} \cdot \log^{d+1} n$, for some constant $\bar{c}$, we will apply Chernoff bounds to tightly bound the estimates within a constant multiplicative factor. Since we do it only for $1/\log^d n$ of the input half-planes, the total number of operations for the $O(\log n)$ random subsets can be bounded by $O(n \log n)$ (as we show in the next section). Note that $X_i^j < \bar{c} \log^{d+1} n$, is an easy case since $n^\epsilon \cdot \bar{c} \log^{d+1} n = o(n)$.

9.5.3 Probabilistic Analysis of Polling

More formally, by invoking Chernoff bounds, for any $\alpha > 0$ (α is a function of c_0), there exists a c_1, independent of n, $Prob(A_i^j \leq \alpha c_1 X_i^j/\log^d n) \leq 1/n^\alpha$ and $Prob(A_i^j \geq c_2 \alpha c_0 \cdot X_i^j/\log^d n) < 1/n^{c_0 \alpha} < 1/n^\alpha$ (for $c_0 > 1$).

From the last two inequalities, X_i^j is bounded by $L^j = A_i^j \log^d n/c_0 c_2 \alpha$ from below, and by $U^j = A_i^j \log^d n/c_1 \alpha$ from above. With appropriate changes in the constants, this condition holds with high likelihood (as defined in section 9.2.1) for all X_i^j simultaneously. We do the procedure (described in the next section) simultaneously for all the samples R_j and choose the sample R^{j_o} using the following simple test:

ALGORITHM 9.3
Finding a good sample using Polling

Input: Samples $R_1 \ldots R_m$ where $m = O(\log n)$.
Output: A good sample R^{j_o}.
Notation: Let $N^j = \sum A_i^j$ and let the actual number of intersections be denoted by T^j and the upper and lower bounds obtained from N^j by U^j and L^j respectively.

(clearly good)

if $k_{total}n > U^j$ then accept sample R^j (since $k_{total}n \geq U^j \geq T^j$),

(clearly bad)

if $k_{total}n \leq L^j$ then the sample is 'bad' (since $k_{total}n \leq L^j \leq T^j$),

(choose the best)

if $L^j \leq k_{total}n \leq U^j$, then accept the sample R^{j_o} for which N^{j_o} is minimum. Since both $k_{total}n$ and T^{j_o} lie in this interval, this guarantees that $T^{j_o} \leq c_3 \cdot k_{total}n$ where $c_3 = U^j/L^j$ which is a constant.

Recall that from our earlier discussion at least one of the samples would satisfy conditions 1 or 3 with very high likelihood. We summarize as following:

LEMMA 9.17 *Polling lemma*
Using the previous procedure we can obtain a sample that is 'good' with high probability.

The more careful reader would have noticed there is a technical inconsistency with the above claim. From our definition of a good sample, the sum of the half-planes should be less than $k_{total} \cdot n$, whereas the output of the Polling algorithm could be larger by a factor of c_3. Strictly speaking, one needs to modify the definition to accommodate an extra factor c_3. However, it should be clear that we have succeeded in our objective of choosing a sample for which the sum of subproblems is $O(n)$ with high probability. The above procedure can be used in a more general situation where we need 'good' samples with very high likelihood from samples that only expect to be 'good'. Moreover, according to our previous discussion, the extra amount of overhead does not affect the asymptotic work done by the algorithm, because it uses only a fraction of the input to test the samples.

9.5.4 Bounding the Number of Processors

Until now we had focused on getting a 'good' sample with high probability which will ensure that the sum of the sizes of sub-problems is within a constant factor of size of the parent problem. But this only guarantees that over $O(\log \log n)$ levels of recursive calls, the sum of the sizes of the subproblems will be $O(n \log^b n)$ for some constant b. This implies that either we use $n \log^b n$ processors or settle for a corresponding trade-off in the running time. Clearly, we need some stronger observations to prevent this proliferation over every level of recursive calls.

For this we shall use geometric properties of the specific problem. After getting a 'good' sample we would do some further processing to bound the sum of sizes of the sub-problems by the exact size of the parent problem. This will prevent proliferation in the problem size over successive recursive calls. This *filtering* procedure is a kind of post-processing step after random-sampling. The input size is at most a constant factor times the input size (guaranteed by the polling algorithm) of the problem while the output is no more than the input size.

In case of two-dimensional convex hulls, we make use of the following filtering scheme. For any sector, we identify the half-planes that intersect it.

9.5. Higher Dimensional Problems: Convex Hulls

Some of them may be part of the output while some of them may not show up in the output (in that sector). Since the output size is bounded by the input size, our objective is to eliminate all the half-planes from a sector that do not show up in the output. There are the following cases to consider.

Case 1. If a half-plane is occluded completely by another half-plane within a sector (see Figure 9.1a), then we can discard these half-planes by the following strategy. For every half-plane intersecting a sector, consider the points of intersection with the two boundaries of the sector. Sort these intersection points in increasing order starting from O. At the end of this step we have two sorted lists. For every half-plane consider a tuple of the form (x_i, y_i) where x_i and y_i are respectively the ranks of the the sorted intersection on the boundaries. A half-plane H_i is completely occluded by another half-plane H_j if and only if $x_i > x_j$ and $y_i > y_j$. In other words, if we compute the maximal elements of the tuples, then the elements that are not a part of this set can be left out from further calls of recursion. This is the easy case.

Case 2. A half-plane may be occluded due to the combined action of two half-plane (see Figure 9.1b). In this case notice that in all other sectors, that this half-plane intersects it will be eliminated by the previous case.

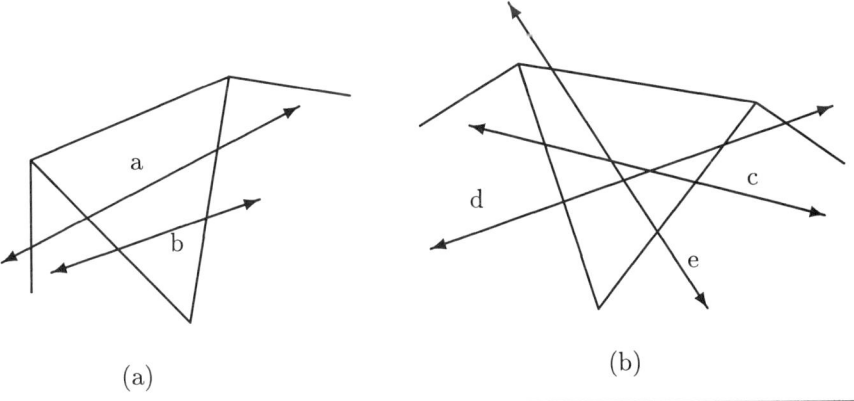

(a)   (b)

FIGURE 9.1
(a) illustrates case (a)—line a is completely occluded by b. (b) illustrates case (b). Line c is occluded by lines d and e but not by any one of them completely. Also it is clear that in all other cones line c will be eliminated by case (a).

Case 3. If a half-plane is visible in at most one sector, it will clearly be eliminated from all other sectors by case 1 (by convexity arguments).

Case 4. If a half-plane is visible in more than one sector, then in all but two sectors it will eliminate all other half-planes by case 1 and moreover we do not have to call the algorithm in these sectors (since we know the output). This half-plane can contribute to vertices of the output hull in at most two sectors. During subsequent recursive calls on these two sectors, at most one copy will be retained in each.

From a global view-point, at any stage of the algorithm, we retain at most two copies of a half-plane. If we let an output vertex be represented by the two intersecting lines, a half-plane can appear in at most two such tuples, which we can denote by the left-vertex (l_v) and the right-vertex (r_v). The *output size* is defined as twice the number of vertices. (Thus if a sector does not have a vertex the output size is 0, implying that an edge traverses the sector, and hence we have already determined the convex hull in that sector). Assume that we have $2n$ processors for the algorithm. Then our processor-allocation strategy is as follows:

For each half-plane that we know to be visible in more than one sector, allocate one processor each to the bounding sectors on the left and right. These processors can be charged to l_v and r_v respectively. Note that in subsequent recursive calls, at most one processor will be allocated for l_v and r_v.

For each half-plane whose visibility in the current recursive call we have not been able to determine, allocate two processors. From our previous observations, these half-planes can appear in at most one sector.

To summarize, we have achieved the following:

- The total number of processors is always bounded by $2n$.

- For any subproblem (at any recursive call), the number of processors is greater than or equal to the output-size of that sector. This ensures that we do not have to rebalance processors between these sub-problems as the algorithm proceeds recursively.

We now have a parallel algorithm for constructing a convex hull in two dimensions which satisfies the properties of parallel sorting algorithm, namely Lemma 9.12 (although we had to work much harder to get to it). Hence by invoking Theorem 9.1, we can claim that the above algorithm runs in $\tilde{O}(\log n)$ using n processors in a *CREW PRAM* model.

Let us recapitulate the main steps of the algorithm for 2-D convex hulls described in the last section in a more general context of divide-and conquer strategy

1. Select $O(\log n)$ subsets of random objects (in case of 2-D hulls these were half-planes) each of size $\lfloor n^\epsilon \rfloor$ for some $0 < \epsilon < 1$.
2. Select a 'good' sample using *Polling*
3. Divide the original problem into smaller sub-problems (the maximum size can be bounded by $O(n^{1-\epsilon} \log n)$) using the 'good' sample.
4. Use a *Filtering* scheme to bound the sum of the sub-problem size by some fixed measure like the output size or input size. This step is problem dependent and uses the specific geometry properties of a problem. The purpose is to bound the number of processors.
5. If the size of a sub-problem is more than a threshold (usually it is chosen to be $O(\log^k n)$ for some constant k), then call the algorithm recursively, else solve the problem using some direct method.

The above general strategy has been used successfully to obtain efficient algorithms for a number of fundamental problems like triangulation and convex-hulls in three dimensions. However, the implementation of some of these steps depends heavily on the specific problem. The probabilistic bounds used in step 3 have to be proved for the specific problem. In this regard, Clarkson presented bounds for very general situations which are applicable to a certain extent; however, we have chosen to present alternate arguments which are simpler. Moreover, the procedure used for dividing the subproblems would naturally depend on the problem at hand. Perhaps the step that is most specific to a problem is the *Filtering* step where we have to use the geometry of the problem.

9.6
Exercises

9.1 Prove Lemma 9.2 for any fixed ϵ.

9.2 Prove Lemma 9.3.

9.3 Given a two-sided biased coin (i.e., the probability of a *head* is not $\frac{1}{2}$). How will you use it to simulate an unbiased coin? Also, how will you simulate an n-sided coin using a 2-sided coin?

9.4 (Parallel Selection). Using the bounds of Lemma 9.3, design a simple algorithm for choosing the k-th largest element for any $1 \leq k \leq n$. Prove that your algorithm runs in $\widetilde{O} \log n)$ time using $n/\log n$ EREW PRAM processors, i.e., it is PT optimal.

Hint: In the first phase, select two elements such that the required element is smaller than one and larger than the other. Show that the total number of elements that lie between these two elements is $O(n^\epsilon)$ for some $\epsilon < 1$ with high probability.

9.5 Let $A[1..n]$ be an array of n elements. An element x of A is said to be a *semi-majority element* if $|\{i : A(i) = x\}| \geq \frac{n}{4}$. Give an $\widetilde{O}(\log n)$ time $n/\log n$ EREW PRAM processor algorithm to find all the semi-majority elements of A.

9.6 Prove the claims about the size bounds after steps 3 and 4 of the algorithm for maximal selection. Do the same for step 3 of the General sorting algorithm.

9.7 Given a coin for which the probability of getting a head is at least α, $0 < \alpha < 1$, prove that there is a constant c ($c > 1$) such that with high probability, there are at least $\log n$ heads if the coin is tossed $\frac{c \log n}{\alpha}$ times.

9.8 You roll an N sided dice. If you get n ($1 \leq n \leq N$), you roll an n-sided dice. What is the expected number of times you have to roll to get a 1? Use this result to derive an expected time bound for Quicksort.

9.9 Let $S_1, S_2, \ldots, S_k$ (where $k = (\log n)^{O(1)}$) be sets of integers in the range $[1, n(\log n)^{O(1)}]$. Given also that $\sum_{i=1}^{k} |S_i| = n$. Present an $\widetilde{O}(\log n)$ time CRCW PRAM algorithm that sorts all the k sets using $n/\log n$ processors.

9.10 Given n integers in the range $[1, n^{O(1)}]$. If the computer word length is n^ϵ, for some fixed $0 < \epsilon < 1$, how will you sort these keys in $O(\log n)$ time using $n/\log n$ CREW PRAM processors?

9.11 If each one of n integers is picked randomly and uniformly from the range $[1, n^{O(1)}]$, how will you sort them using $n/\log n$ CRCW PRAM processors such that for a large fraction of all possible inputs the algorithm terminates in $\widetilde{O}(\log n)$ time?

Hint: Make use of the integer sorting algorithm.

9.12 Given n keys (not necessarily integers) with many duplications such that the number of distinct keys is $(\log n)^{O(1)}$, present an $\widetilde{O}(\log n)$ time, $\frac{n \log \log n}{\log n}$ CRCW PRAM processor algorithm to sort this input.

9.13 Consider a probabilistic experiment where there are N events of interest each of which has a success probability of $\frac{1}{2}$ independent of each other. We consider the experiment to be successful if all the N events are successful.

What can you say about the success probability of the experiment if you have $O(\log N)$ independent runs of the experiment? Can you obtain any meaningful bound as we could do with *Polling?*

9.14 Let X_i, be a family of random variables (not necessarily independent) such that the mean of each X_i is less than μ. Suppose $Y = \sum_{i=1}^{n} X_i$ where n is a random variable (integral valued) with mean N. Show that the expectation of Y can be bound by $N \cdot \mu$. Y is called a *random sum*; it is the sum of a random number of random variables.
Hint: Use the method of conditional expectation.

9.15 Present an $\widetilde{o}(n)$ algorithm to compute an *approximate rank* of a given element x in a set of n keys. An approximate rank of any element x is an integer in the range $[r - \delta n, r + \delta n]$, where r is the true rank of x and δ is a fixed number $(0 < \delta < 1)$.

9.16* An *approximate median* of a given set of n elements is an element of the set whose rank is γn for some fixed $0 < \gamma < 1$ (γ is independent of n). Describe a parallel algorithm to choose such an element that runs in $O(1)$ time using n *CRCW* processors. Your algorithm should succeed with probability $> \frac{1}{2}$, i.e., the output element of your algorithm should satisfy the property of an *approximate median* with this probability.

Notes and References

Selection and Sorting were among the first problems that captured the attention of researchers in parallel algorithms. One of the earliest significant results was obtained by Valiant [34]. He proved a lower bound of $\Omega(\log \log n)$ for extremal selection which we overcame by use of randomization. Beame and Hastad [4] proved a lowerbound of $\Omega(\log n / \log \log n)$ for general selection as long as one uses a polynomial number of *CRCW PRAM* processors. For some special cases (Exercise 9.16), the lower bound can be circumvented [32]. The first optimal $O(\log n)$ time algorithm for selection was presented by Reischuk [31] (Exercise 9.4) adopting the sequential algorithm of Floyd and Rivest [13]. These techniques were extended by Rajasekaran to obtain an optimal randomized selection algorithm for the hypercube [22].

The first optimal $O(\log n)$ time PRAM sorting algorithm was obtained by Reischuk [31] using random-sampling. The *flashsort* algorithm of Reif and Valiant [30] gives an optimal logarithmic time randomized sort for the n size *butterfly network*, a regularly connected network which has degree 4. This algorithm uses a combination of randomized routing techniques and advanced randomized sampling techniques. Let S be the set of input keys. *Flashsort* executes as follows:

1. Take a random sample of n^ϵ elements of S to form a sample set S' of size n^ϵ, for $0 < \epsilon < 1/2$.
2. Sort S' in logarithmic time within the n size network using a less efficient deterministic algorithm.
3. Form a set S'' by choosing every $(\log n)$th element from S'. A result in Reif and Valiant [30] shows with high probability, S'' splits S into the subsets of expected size of size at most $(1+\mu)\, n^{1-\epsilon} c \log n$ where μ is of the order of $\frac{d}{\log n}$ and where $c, d \geq 2$.
4. Separate S into the sets $S_0, S_1, \ldots S_t$ on the basis of S'', where t is in the range from $\frac{n^\epsilon}{(1+\mu)c\log n} + 1$ to $\frac{n^\epsilon}{c\log n} + 1$.
5. The algorithm is applied recursively for each S_i in parallel.
6. The recursion terminates when the sub-problems have polylog size. At that point a less efficient deterministic algorithm can be used.

This algorithm also uses *pipelining*, which is a sequence of parallel processors, with a stream of data passing through the processors, where each data item is processed in sequence by successive processors. The *flashsort* algorithm pipelines the logarithmic recursive levels of sorts. The total time is logarithmic with a network of linear size. A recent efficient (considerably faster than the system sort) implementation of *flashsort* on a massively parallel machine, the MASSPAR, is described in [14]. The basic methodology was adapted for the *Flashsort* algorithm by Reif and Valiant [30] around the same time as the celebrated AKS network was proposed by Ajtai, Komlos and Szemeredi [3]. Leighton [19] reduced the processor complexity in AKS to obtain a truly optimal deterministic algorithm. However, the AKS has suffered due to astronomical constants involved in the construction of the underlying network which is an expander graph. More recently Cole [7] designed an elegant $O(\log n)$ time sorting algorithm which has virtually settled the problem of sorting on PRAM models; however, Flashsort continues to remain the most practical algorithm for networks. The optimal sub-logarithmic algorithm for prefix-sum stated in Lemma 9.5 was discovered by Cole and Vishkin [11]. The first optimal sub-logarithmic time algorithms for General sorting and integer sorting were provided by Rajasekaran and Reif [24].

The presentation of our general sorting algorithm uses ideas drawn from a lot of the earlier work and was given in Rajasekaran and Reif [23]. [23] also provides a survey of parallel selection and sorting algorithms. The first optimal algorithm for integer sorting in the range $[n]$ was given in [24]. Integer sorting in range $[n^2]$ continues to remain a challenging problem although significant progress has been achieved recently (by Bhatt et al. [5], and by Rajasekaran and Sen [25]).

The algorithms for general-sorting and convex hulls presented in this chapter are not only optimal in PT bounds but are also optimal in the time bounds for the model ($CREW$) used. However, for the stronger $CRCW$ model one can actually improve some of the algorithms to obtain an optimal time bound of $\Theta(\frac{\log n}{\log \log n})$. This is discussed in [24].

Use of randomized methods for parallel computational geometry was introduced by Reif and Sen [27]. Given a set of n non-intersecting line segments in the Cartesian plane, the *trapezoidal decomposition* is a decomposition of the Cartesian plane into trapezoids, where

1. the top and bottom (with respect to the y-axis) of each trapezoid is a segment of an input segment, and
2. on each side of each trapezoid is a line segment parallel with the y-axis which ends or begins at an endpoint of one of the input segments.

These optimal randomized PRAM algorithms for 2-D convex hull were extended by Reif and Sen [27] to many other 2-D problems, such as trapezoidal decomposition, and also extended by Reif and Sen [29] to give optimal logarithmic time algorithms for 2-D convex hull and trapezoidal decomposition on regular, bounded degree fixed connection networks, such as the *butterfly*. The *3-dimensional(D) convex hull* is the minimal convex polyhedron containing n input points in $\mathbf{R}^3$. Given a set of n points in the Cartesian plane, the *Voronoi diagram* is a polygonal map (a planar embedded graph with straight line edges) that decomposes the Cartesian plane into n regions each consisting of the set of points closest to a given input point. In Reif and Sen [28], they provide further applications of these methods to three-dimensional convex hulls and 2-D Voronoi diagrams. Random sampling in computational geometry has proved to be very useful especially in a sequential context. Clarkson [10] introduced the use of random sampling to computational geometry and derived similar bounds for a more general setting using more involved techniques. He had also given numerous applications of these very general probabilistic bounds. In his case, however he was dealing with the sequential algorithms and so he could derive bounds on the expected running time of the algorithms as the sum of the expected time for individual steps.

Another direction for research in randomized algorithms is to minimize the use of random bits in the algorithms. This is especially crucial in practical situations where it is often difficult to generate *truly* random bits (as opposed to *pseudo-random bits*). Using techniques of Chor and Goldreich [9], Karloff and Raghavan [18] were able to show that Reischuk's algorithm can be made to run in the same asymptotic bounds using only $O(\log n)$ *purely* random bits. Their methods were further extended by Reif and Sen [28] to show that some of the algorithms in computational geometry (including the convex-hull algorithm) can be implemented using $O(\log^2 n)$ bits.

The Dual Transform: The convex-hull problem has a very interesting dual problem, namely the intersection of half-spaces. This dual transformation $\mathcal{D}$ maps a point in E^d to a non-vertical hyperplane in E^d and vice-versa. Let $p = (\pi_1, \pi_2, \ldots, \pi_d)$ be a point in E^d. Then $\mathcal{D}(p)$ is the hyperplane $1 = \pi_1 x_1 + \pi_2 x_2 + \ldots \pi_d x_d$ and vice-versa such that a hyperplane h not containing the origin is mapped to a point p for which $\mathcal{D}(p) = h$.

The transform $\mathcal{D}$ is extended to sets of points (hyperplanes) in a natural way. Let $\mathcal{P}$ be a convex polytope with non-empty interior $int\mathcal{P}$ and assume that the origin O is contained in $\mathcal{P}$. Then $\mathcal{D}(\mathcal{P})$ is an infinite set of hyperplanes that avoid some convex region around O. The dual of $\mathcal{P}$ is defined as

$$\bar{\mathcal{P}} = closure(\bigcap_{h \in \mathcal{D}(\mathcal{P})} h^{pos})$$

where h^{pos} denotes the half-space containing the origin.

It can be verified that, given a set of points S, the vertices of the convex hull are the dual transform of the facets of the intersection of the half-spaces $\mathcal{D}(S)$ This property has been exploited very often so that the same algorithm can be used for both convex-hulls and intersection of half-spaces.

Bibliography

[1] A. Aho, J. Hopcroft and J. Ullman, *The Design and Analysis of Computer Algorithms*, Addison-Wesley, 1974.

[2] D. Angluin and L.G. Valiant, *Fast Probabilistic Algorithms for Hamiltonian Paths and Matchings*, J. Comp. Syst. Sci., 18 (1979), pp. 155–193.

[3] M. Ajtai, J. Komlós and E. Szemerédi, *An $O(n \log n)$ Sorting Network*, Proc. 15th ACM Symposium on Theory of Computing, 1983, pp. 1–9.

[4] P. Beame and J. Hastad, *Optimal Bounds for Decision Problems on the CRCW PRAM*, 19th ACM Symposium on Theory Of Computing, 1987, pp. 83–93.

[5] P. Bhatt, K. Diks, T. Hagerup, V. Prasad, T. Radzik, and S. Saxena, *Improved Deterministic Parallel Integer Sorting*, Unpublished Manuscript, 1989.

[6] H. Chernoff, *A Measure of Asymptotic Efficiency for Tests of a Hypothesis Based on the Sum of Observations*, Annals of Math. Statistics 23, 1952, pp. 493–507.

[7] R. Cole, *Parallel Merge Sort*, Proc. 27th IEEE Symposium on Foundations of Computer Science, 1986, pp. 511–516.

[8] B. Chazelle and D. Dobkin, *Intersection of convex objects in two and three dimensions*, Journal of the ACM, Volume 34(1), 1987, pp. 1–27.

[9] B. Chor and O. Goldreich, *On the power of two-point based sampling*, Journal of Complexity, Volume 5, 1989, pp. 96–106.

[10] K. Clarkson, *Applications of random random sampling in computational geometry II*, Proc. of the 4th Annual Symp. on Computational Geometry, 1988, pp. 1–11.

[11] R. Cole and U. Vishkin, *Approximate and Exact Parallel Scheduling with Applications to List, Tree, and Graph Problems*, Proc. 27th IEEE Symposium on Foundations of Computer Science, 1986, pp. 478–491.

[12] W. Feller, *An Introduction to Probability Theory and Its Applications*, vol.1, Wiley, New York, 1950.

[13] Floyd and Rivest, *Expected Time Bounds for Selection*, Communications of the ACM, vol. 18, no. 3, 1975, pp. 165–172.

[14] W.L. Hightower, J. Prins and J. Reif, Randomized Sorting on Large Parallel Machines Implementations of Randomized Sortings on Large Parallel Machines, *4th Annual ACM Symposium on Parallel Algorithms and Architectures*, San Diego, CA, July 1992.

[15] Hoare, *Quicksort*, Computer Journal 5, 1962, pp. 10–15.

[16] W. Hoeffding, *On the Distribution of the Number of Successes in Independent Trials*, Annals of Math. Stat. 27, 1956, pp. 713–721.

[17] D.E. Knuth, *The Art of Computer Programming, Vol.3: Sorting and Searching*, Addison-Wesley Publishing Company, Massachusetts, 1973.

[18] H. Karloff and P. Raghavan, *Randomized algorithms and pseudorandom numbers*, Proc. 20th ACM Symposium on Theory of Computing, 1988, pp. 310–321.

[19] T. Leighton, *Tight Bounds on the Complexity of Parallel Sorting*, 16th ACM Symposium on Theory of Computing, Washington, D.C., 1984, pp. 71–80.

[20] F. Preparata, *New Parallel Sorting Schemes*, IEEE Transactions on Computers, vol. C27, no. 7, 1978, pp. 669–673.

[21] M. O. Rabin, *Probabilistic Algorithms*, in *Algorithms and Complexity, New Directions and Recent Results*, edited by J. TRAUB, Academic Press, 1976, pp. 21–36.

[22] S. Rajasekaran, *Randomized Parallel Selection*, 10th Annual Conference on Foundations of Software Technology and Theoretical Computer Science, 1990. Springer-Verlage Lecture Notes in Computer Science 472, pp. 215–224.

[23] S. Rajasekaran, and J.H. Reif, *Derivation of Randomized Algorithms for Sorting and Selection*, Technical Report, Aiken Computing Lab., Harvard University, 1987.

[24] S. Rajasekaran, and J.H. Reif, *Optimal and Sub-Logarithmic Time Randomized Parallel Sorting Algorithms*, SIAM Journal on Computing, vol. 18, no. 3, 1989, pp. 594–607.

[25] S. Rajasekaran, and S. Sen, *On Parallel Integer Sorting*, Technical Report, Department of Computer Science, Duke University, 1987. To appear in ACTA INFORMATICA.

[26] J.H. Reif, *On the Power of Probabilistic Choice in Synchronous Parallel Computations*, SIAM J. Computing 13(1), 1984b, pp. 46–56.

[27] J.H. Reif and S. Sen, Optimal Randomized Parallel Algorithms for Computational Geometry. *16th International Conference on Parallel Processing*, St. Charles, IL, August 1987, pp. 270–276. Also in Algorithmica, Vol. 7, pp. 91–117, January 1992.

[28] J.H. Reif and S. Sen, Polling: A New Randomized Sampling Technique for Computational Geometry. *21st Annual ACM Symposium on Theory of Computing*, Seattle, WA, May 1989, pp. 394–404. Revised as Optimal Parallel Randomized Algorithms for Three-Dimensional Convex Hulls and Related Problems, *SIAM Journal on Computing*, Vol. 21, No. 3, pp. 466–485, June 1992.

[29] J.H. Reif and S. Sen, Randomized Algorithms for Binary Search and Load Balancing on Fixed Connection Networks with Geometric Applications. *2nd Annual ACM Symposium on Parallel Algorithms and Architectures*, Crete, Greece, July 1990, pp. 327–337. To appear in *SIAM Journal of Computing*, 1992.

[30] J. Reif and L. Valiant, A Logarithmic Time Sort on Linear Size Networks. *Journal of the Association for Computing Machinery*, 34(1):60–76, January 1987.

[31] R. Reischuk, *A Fast Probabilistic Sorting Algorithm*, Proc. 22nd IEEE Symposium on Foundations of Computer Science, 1981, pp. 88–102.

[32] S. Sen, *Finding an approximate median with high probability in constant parallel time*, Information Processing Letters, 34, 1990, pp. 77–80.

[33] Y. Shiloach and U. Vishkin, *Finding the Maximum, Merging, and Sorting in a Parallel Computation Model*, J. Algorithms 2, 1981, pp. 212–219.

[34] L. Valiant, *Parallelism in Comparison Problems*, SIAM Journal of Computing, vol. 4, 1975, pp. 348–355.

[35] Wilks, *Mathematical Statistics*, John Wiley and Sons, New York 1962.

10

Parallel Merge Sort

Richard Cole
*Computer Science Department
Courant Institute of Mathematical Sciences
New York University
251 Mercer Street
New York, NY 10012
cole@cs.nyu.edu*

10.1
Introduction

This chapter describes three deterministic parallel algorithms based on the merge sort paradigm. The first algorithm is Batcher's odd-even sorting circuit; it has depth $\frac{1}{2}\log n(\log n + 1)$. The second algorithm is an optimally efficient CREW PRAM implementation of Batcher's odd-even sorting circuit; it runs on $n/\log n$ processors in $O(\log^2 n)$ time. The third algorithm is a refinement of the second algorithm; it runs on n processors in $O(\log n)$ time.

The problem is to sort a set of records. In general a record comprises several fields, one of which is designated the *key field*, or *key* for short. The key field is ordered; that is, there is an order relation, denoted "<", such that for any two keys, k_1 and k_2, one of the following three relations holds: either $k_1 < k_2$, or $k_1 = k_2$, or $k_1 > k_2$. The problem is to rearrange the records so that following the reordering, for every pair of records, R_1, R_2, if $R_1.key < R_2.key$, then R_1 precedes R_2.

EXAMPLE 10.1
Consider a telephone directory. Here a record is an entry in the directory; it comprises a person's name, the person's address, and the person's telephone number. The key used is the person's name and the ordering is the alphabetic ordering of the keys. (Of course, considerably more detail is required to specify the exact form of allowable names and addresses.) The telephone directory is a listing of these records in sorted order, that is in the alphabetic ordering by name.

The basic step in most sorting algorithms and in all the sorting algorithms presented in this chapter is an ordering step; it takes two records, R_1 and R_2 as input and places them in sorted order, naming the ordered records R_3 and R_4, respectively. Formally, $Order(R_1, R_2, R_3, R_4)$ is defined as follows.
  if $R_1.key \leq R_2.key$
    then $R_3 := R_1; R_4 := R_2$
    else $R_3 := R_2; R_4 := R_1$

Henceforth, in order to simplify the presentation, all of the record apart from its key is ignored; the reduced record, comprising a key, is called an *item*.

This chapter considers two classes of algorithms, namely sorting circuits and PRAM algorithms; more specifically, Concurrent Read, Exclusive Write (CREW) PRAM algorithms.

10.1. Introduction   455

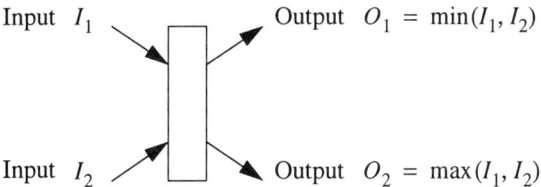

FIGURE 10.1
A comparator unit.

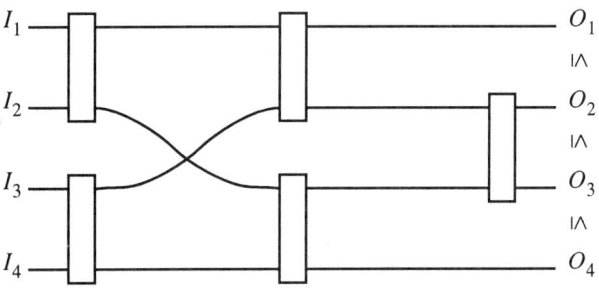

FIGURE 10.2
A sorting circuit for sets of size 4.

A sorting circuit is a device built of comparator units. A comparator unit takes as input two keys, I_1 and I_2, and produces two outputs, O_1 and O_2; it computes the function $Order(I_1, I_2, O_1, O_2)$. (See Figure 10.1.)

A sorting circuit for n items has n input wires and n output wires. The inputs arrive one on each input wire. The task of the sorting circuit is to reorder the inputs so that they appear on the output wires in sorted order (the output wires are labeled $O_1, O_2, \cdots$; the smallest input appears on wire O_1, the second smallest on wire O_2, and so on). The input wires are connected to the output wires through a collection of comparator units. Each comparator unit receives two inputs; an input can come either from an input wire or be the output of another comparator unit. An example of a sorting circuit for sets of size 4 is shown in Figure 10.2.

A sorting circuit comprises a number of *levels* of comparators, a level containing the comparators which perform their computation simultaneously. A level can contain at most $\lfloor \frac{n}{2} \rfloor$ comparators. The *depth* of a circuit is the number of levels it comprises. For example, the circuit of Figure 10.2 has three levels. So the depth of a circuit is exactly the running time of the circuit, where a unit of running time is the time a comparator takes to order its two inputs and provide its outputs to the output destinations. A sorting circuit must have $\Omega(n \log n)$ comparators (see Exercise 10.1). In Section 10.2, a family of sorting circuits for sorting sets of n items, for $n \geq 1$, is given; the circuits in this family are called the *odd-even sorting circuits*. The odd-even circuit for sets of size n has depth $\frac{1}{2}\lceil \log n \rceil (1 + \lceil \log n \rceil)$, and the circuit comprises $\Theta(n \log^2 n)$ comparators. Asymptotically more efficient circuits, of depth $O(\log n)$ and comprising $O(n \log n)$ comparators, are known to exist; but as the constants hidden by the big "Oh" notation are of the order of several thousand, the odd-even circuits are more efficient in practice.

In Sections 10.3 and 10.4 optimal CREW PRAM algorithms for sorting sets of size n are described; they have parallel running times $O(\log^2 n)$ and $O(\log n)$, respectively.

The odd-even sorting circuits and the PRAM sorting algorithms given here are all based on the sequential merge sort algorithm. This is reviewed briefly. The sequential merge sort algorithm is built on a merge procedure. This is described by way of an example.

EXAMPLE 10.2

Suppose two piles of playing cards in sorted order are given; the task is to merge the two piles into one new sorted pile of cards. The ordering for cards is defined as follows: they are first ordered by suit (♣ < ◊ < ♡ < ♠) and within each suit by value ($A < 2 < \cdots < 10 < J < Q < K$). The merge procedure is straightforward: the two piles are placed face up; repeatedly, the smaller of the two visible cards is placed face down on the new pile; when one of the old piles is emptied, the remainder of the other old pile is transfered, card by card, to the new pile. Turning the new pile face up provides the desired sorted pile of cards.

The standard merge procedure proceeds in the same way, but its input is provided in two arrays and its output is produced in a third array; of course, it is not restricted to using card values for its keys.

The merge procedure is used to produce a sorting algorithm as follows. From now on, the set of n items to be sorted is assumed to be provided in an array $SortMe[1:n]$. If the array $SortMe$ contains a single item it is already in sorted order. Otherwise, the following two subarrays are recursively sorted: $First = SortMe[1 : \lfloor \frac{n}{2} \rfloor]$ and $Last = SortMe[1 + \lfloor \frac{n}{2} \rfloor, n]$. Then the arrays $SortFirst$ and $SortLast$, resulting from sorting $First$ and $Last$, respectively, are merged. This algorithm has running time $O(n \log n)$. A proof of this running time and a more detailed description of the sequential merge sort algorithm can be found in most textbooks on algorithms and algorithm analysis.

There is one further important classification of sorting algorithms: *stability*. It is significant if distinct records may have identical keys. A sorting algorithm is stable if it maintains the original order of records with equal keys. The sequential merge sort algorithm is stable as are the algorithms of Sections 10.3 and 10.4; but the circuits of Section 10.2 are not stable.

10.2 The Odd-Even Sorting Circuit

In this section, the *odd-even sorting circuit*, one of the sorting circuits discovered by Batcher, is presented. While the circuit correctly sorts sets including items with equal keys, the circuit is more easily described and understood if all the keys are imagined to be distinct.

10.2.1 Overview

The odd-even sorting circuit is based on the merge sort algorithm described in Section 10.1. The key component of the sorting circuit is a subcircuit for merging two sorted sets, called an *odd-even merging* circuit.

A description of the odd-even sorting circuit for sorting an array *SortMe* of n items follows (note that there is a distinct circuit for each value of n). If $n = 1$, *SortMe* is already sorted. Otherwise, for $n > 1$, proceed as follows. Let $First = SortMe[1 : \lfloor \frac{n}{2} \rfloor]$ and let $Last = SortMe[1 + \lfloor \frac{n}{2} \rfloor : n]$. The arrays *First* and *Last* are sorted recursively using odd-even sorting circuits for sets of size $\lfloor n/2 \rfloor$ and $\lceil n/2 \rceil$, respectively. Let the outputs of these two circuits be stored in arrays *SortFirst* and *SortLast*, respectively. The sort is completed by merging arrays *SortFirst* and *SortLast*, using an odd-even merging circuit.

Next, in Section 10.2.2, the odd-even merging circuit is described. The complexity of the odd-even sorting circuit is analyzed in Section 10.2.3. Following the description of the odd-even merging circuit, the odd-even sorting circuit for sets of size 8 is shown, in Figure 10.4.

10.2.2 The Odd-Even Merging Circuit

An example helps to provide some intuition. Following the example, a more formal treatment is presented.

458   Chapter 10.   Parallel Merge Sort

Suppose the following two sorted arrays, L and R, of integers are to be merged (L and R are short for $Left$ and $Right$, respectively):

$$L = 1, 7, 12, 15, 18, 22, 26, 29$$

and

$$R = 3, 5, 8, 20, 21, 23, 25, 27$$

The output is to be placed in the array M (short for $Merge$). A natural approach is to merge the items in the even index positions in each of L and R into array EM (short for $EvenMerge$) and simultaneously to merge the items in the odd index positions in L and R into array OM (short for $OddMerge$). Following this, the items in arrays EM and OM are combined. Returning to the example, the merge of the items in odd index positions in L and R comprises the array

$$OM = 1, 3, 8, 12, 18, 21, 25, 26$$

and the merge of the items in even index positions comprises the array

$$EM = 5, 7, 15, 20, 22, 23, 27, 29$$

For each item, based on its current position in OM or EM, its final position in array M can be predicted fairly accurately, in fact, up to a choice of two adjacent locations in array M, as is shown below.

Associate with each item in L and R, apart from the first item in each array, its *predecessor*, the item in the preceding position in its original array, be it L or R (for instance, the predecessor of 12 is 7). Likewise, associate with each item in L and R, apart from the last item in each array, its *successor*, the item in the succeeding position in its original array (for instance, the successor of 15 is 18). Each item is smaller than its successor and larger than its predecessor (when these exist). Consider item 15, in EM; it is in position 3. The predecessors of the first three items in EM are smaller than 15 and the successors of the items of ranks 3–7 are larger than 15. All these predecessors and successors are stored in OM. Hence the first three items in OM are smaller than 15 (namely, items $1, 3, 8$). Likewise, the last four items in OM are all larger than 15. There is only one item in OM about whose size relative to item 15 there might be any doubt: the item in position 4 in OM.

Similarly, for $1 \leq i \leq 8$, for the item e_i in position i in array EM, the first i items in OM must be smaller than e_i and the last $8 - i - 1$ items must be larger than e_i. The only item about which there might be doubt is the item in position $i+1$ in OM (for $i \neq 8$). So, for $i \neq 8$, e_i belongs in either position $2i$ or $2i+1$ in array M, while for $i = 8$, e_i belongs in position $2i$ in array M. The indeterminancy, for $i \neq 8$, can be resolved by a comparison between e_i and the item in position $i+1$ in OM. Likewise, the item in position $i+1$ in OM belongs in position $2i$ or $2i+1$ in M.

So the merge is completed as follows: The item in position 1 in OM provides the item in position 1 in M; the item in position 8 in EM provides the item in position 16 in M; for $1 \leq i < 8$, the smaller of the items in position i in EM and position $i+1$ in OM provides the item in position $2i$ in M, while the larger of these two items provides the item in position $2i+1$ in M.

The above construction for sets of size 8 is generalized to sets of size 2^k, for arbitrary integer k, as follows. Suppose that the arrays, L and R, to be merged, are both of size 2^k, for some $k \geq 1$. The merged set is placed in array M, of size 2^{k+1}. If $k = 0$, the two arrays (both of size 1) are merged using a single comparator unit. Otherwise, the following arrays are merged recursively: array EL (short for $EvenL$), comprising the items in the even index positions in L, is merged with array ER (short for $EvenR$), comprising the items in the even index positions in R; the result of this merge is stored in array EM. Simultaneously, array OL (short for $OddL$), comprising the items in the odd index positions in array L, is merged with array OR (short for $OddR$), comprising the items in the odd index positions in array R; the result of this merge is stored in array OM. Then OM and EM are merged, the result being placed in M.

Formally, the following lemma proves the correctness of the odd-even merge algorithm. The proof follows the intuitive sketch given above.

LEMMA 10.1
For $1 \leq i < 2^k$, the smaller of the items in position i in EM and position $i+1$ in OM provides the items in position $2i$ in M and the larger provides the item in position $2i+1$ in M. The item in position 1 in OM provides the item in position 1 in M, while the item in position 2^k in EM provides the item in position 2^{k+1} in M.

PROOF
Associate with each item in L and R, apart from the first item in each array, its *predecessor*, the item in the preceding position in its original array, be it L or R. Each item is larger than its predecessor. Let e_i be the item in position i in EM. The predecessors of the first i items in EM must all be smaller than e_i; they are all stored in OM. Hence the first i items in OM are smaller than e_i. Likewise, associate with each

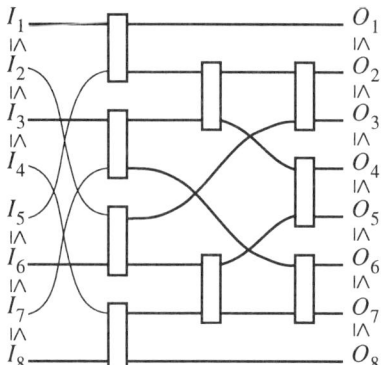

FIGURE 10.3
The Odd-Even Merging Circuit for Two Sets of Size 4.

item in L and R, apart from the last item in each array, its *successor*, the item in the succeeding position in its original array. With the last item from each of L and R associate the imaginary successor ∞. Each item is smaller than its successor. Consider item e_i again; as it is in position i in EM, the successors of items in positions i through 2^k in EM are all larger than e_i; all of these successors, apart from the two instances of ∞, are stored in OM; this is $2^k - i - 1$ successors stored in OM, all larger than e_i. So the last $2^k - i - 1$ items in OM must all be larger than e_i. The only item in OM whose size relative to e_i is uncertain is the item in position $i+1$, and then only if $i+1 \leq 2^k$. It follows that e_i can only be in positions $2i$ or $2i+1$ in M; similarly, e_k must be in position 2^{k+1} in M.

A similar argument shows that item o_{i+1}, the $(i+1)$th item in OM must also be in positions $2i$ or $2i+1$ in M, for $1 \leq i < 2^k$. Additionally, item o_1 is in position 1 in M.

So the smaller of e_i and o_{i+1} occupies position $2i$ in M and the larger occupies position $2i+1$. ∎

The merging circuit is called an *odd-even merging circuit*. The odd-even merging circuit for merging two sorted sets of size 4 is shown in Figure 10.3.

It may not always be the case that the sets to be merged are both of the same size, nor need these sets have size a power of two. Both these problems are easily handled. So suppose the task is to merge two sorted arrays L and R

10.2. The Odd-Even Sorting Circuit

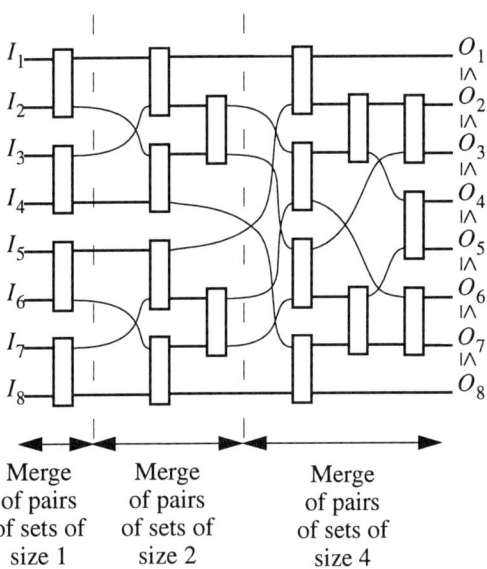

Merge of pairs of sets of size 1   Merge of pairs of sets of size 2   Merge of pairs of sets of size 4

FIGURE 10.4
The Odd-Even Sorting Circuit for a set of size 8.

of sizes n and m, respectively, where $n \leq m$. Further suppose that $2^{k-1} < m \leq 2^k$, for some integer $k > 0$. Then L and R are padded with dummy items of value ∞, to bring each of L and R to size 2^k. The previous merging circuit is used, except that there is no need to have a comparator when one or both of the items being compared is a dummy item. The details of specifying the merging circuit in such cases are left as an exercise (see Exercise 10.2).

To illustrate how the merging circuits are put together to form a sorting circuit, the odd-even sorting circuit for sorting a set of size 8 is shown in Figure 10.4.

10.2.3 The Analysis

LEMMA 10.2

The depth of the odd-even merging circuit for merging two sets of size 2^k, for integer k, is $k+1$.

PROOF

The odd-even merging circuit comprises the following levels: a level at which the merge of sets of size $1 = 2^0$ is completed, a level at which the merge of sets of size $2 = 2^1$ is completed, a level at which the merge of sets of size $4 = 2^2$ is completed, $\cdots$, and finally a level at which the merge of sets of size 2^k is completed. This is a total of $k+1$ levels. ∎

This lemma is readily generalized to the following result; its proof is left to the reader (see Exercise 10.4).

LEMMA 10.3
The depth of the odd-even merging circuit for merging two sets of size n and m, $n \leq m$, where $2^{k-1} < m \leq 2^k$, for some integer k, is $k+1$.

LEMMA 10.4
The depth of the odd-even sorting circuit for a set of size n is $\frac{1}{2}\lceil \log n \rceil (\lceil \log n \rceil + 1)$.

PROOF
Let $k = \lceil \log n \rceil$. The sorting circuit comprises the following series of subcircuits. A collection of circuits for merging pairs of sets each of size $1 = 2^0$, followed by a collection of circuits for merging pairs of sets each of size at most $2 = 2^1$, followed by a collection of circuits for merging pairs of sets each of size at most $4 = 2^2$, $\cdots$, and finally a circuit for merging two sets each of size at most 2^{k-1}. These collections of circuits require the following number of levels: $1, 2, 3, \cdots, k$, respectively. This is a total depth of $\frac{1}{2}k(k+1)$. ∎

10.3
An Efficient $O(log^2 n)$ Time Sorting Algorithm

The sorting algorithm of the previous section is readily converted to a PRAM algorithm using $n/2$ processors and running in parallel time $O(\log^2 n)$ time (each processor simulates a distinct comparator at each level of the sorting circuit). Actually, this is not trivial; attempt Exercise 10.9. However, this algorithm performs $\frac{1}{2}n\lceil \log n \rceil(\lceil \log n \rceil + 1)$ comparisons on a set of size n. This compares unfavorably to the $O(n \log n)$ comparisons performed by efficient sequential algorithms. It is possible to obtain this $O(n \log n)$ performance with parallel algorithms, both in the PRAM and the comparator circuit models. The harder task is to provide a sorting circuit which performs only $O(n \log n)$ comparisons in small depth. In fact, as indicated in Section 10.1, an $O(\log n)$ depth sorting circuit is known. This circuit is not considered further for several reasons:

1. It has very large depth for realistic values of n (the constant hidden by the big-Oh notation is very large).
2. It is beyond the scope of this text.
3. It is not based on the merge sort paradigm.

In this chapter, algorithms in the more permissive PRAM model are considered. In particular, in this section, the sorting circuit of the previous section is implemented in a more efficient manner as a PRAM algorithm, so that the total number of operations performed is $O(n \log n)$.

10.3.1 Overview

Why might an improvement be anticipated? Again, consider the problem of merging two sorted arrays, L and R, each of 2^k items, for some integer $k \geq 1$. The result of the merge is to be placed in array M. Suppose the items in the even index positions in L are merged with the items in the even index positions in R, the result being placed in array EM, as before. Also, as before, let the array OM comprise the merge of the items in the odd index positions in L and R. The key observation is that given the array EM it is fairly easy to compute the array OM also. This would appear to save a factor of two in the complexity; by proceeding recursively, a factor of $\log n$ savings is achieved.

The following problem instance is used as a running example throughout this section. Consider the problem of merging arrays L and R, where

$$L = 1, 4, 8, 22, 24, 26, 29, 31$$

$$R = 2, 5, 10, 12, 15, 16, 19, 27$$

Then
$$EM = 4, 5, 12, 16, 22, 26, 27, 31$$

$$OM = 1, 2, 8, 10, 15, 19, 24, 29$$

After computing EM, what is known about an item in OM? For instance, consider item 24 in L. It appears between items 22 and 26 in L. In EM, 22 is preceded by item 16 from R; so item 16 and every item preceding item 16 in R must precede item 24 in M. Likewise, in EM, item 27 from R follows item 26. So item 27 and every item following item 27 in R must follow item 24 in M. This leaves just one item from R whose location in M relative to item 24 is uncertain: item 19 in R. However, the uncertainty is not always as limited. For instance, consider item 8 in L. It is straddled by items 4 and 22 from L which also appear in EM. But the only item from R in EM, outside the interval $[4, 22]$, is item 27; so the position of item 8 in M is still very uncertain. The large uncertainty arises because it is not known which two items from EM straddle item 8, an item belonging in OM.

This suggests the following three stage algorithm. In Stage 1, for each item, e, belonging in (unsorted) OM, compute the largest item, em, from EM smaller than e, if any. Actually, rather than computing em, the index, iem, of em in EM is computed; if em does not exist, then $iem := 0$. In Stage 2, compute the position of e in array OM. In Stage 3, merge EM and OM as before. In fact, Stages 2 and 3 can be compressed into a single stage, called Stage 2 henceforth.

10.3.2 Details of the Merging Algorithm

To simplify the exposition, the array indices are described as pointers; so often, where an item e is written, strictly speaking, item e's index in an appropriate array is intended. In addition, to avoid exceptional cases for out of bound array references, array EM is extended at front and rear by items $-\infty$ (in position index 0) and $+\infty$ (in position index $2^k + 1$). The values $\pm\infty$ are never referenced; all that will be needed are pointers associated with these locations. Arrays L and R are extended in the same way. The following notation is helpful: A^+ denotes the array A extended by $+\infty$, A^- denotes A extended by $-\infty$, and $A^\pm$ denotes the array A extended by $\pm\infty$. Again, on a first reading, it may be helpful to assume all the items being sorted are distinct; in fact, as before, the algorithm is correct even if some of the items are equal.

The computation is performed by processors associated with each of the items in arrays $EM^\pm$, L and R.

At the start of the step, besides array EM, the following information is assumed to be at hand. First, each item, e, in EM knows which array (L or R) it came from; this is stored in array *Origin*. Second, e has a pointer to its *predecessor* in its original array, the largest item smaller than e; this is stored in array L_Ptr or R_Ptr, as appropriate. Note that these pointers are pointing to items of OM. In addition, the items $\pm\infty$ in $EM^\pm$ have pointers to their predecessors in both L^- and R^-, where the predecessor of $-\infty$ is redefined to be $-\infty$. Figure 10.5 shows the initial state for the running example.

Stage 1 proceeds in two substages. The computation for items originally in L is discussed; the procedure for items originally in R is entirely analogous. In Stage 1.1, for each item, e, present in both L and EM, compute the largest item, r, in R^- smaller than e; r is called the *predecessor* of e in R^-; a pointer

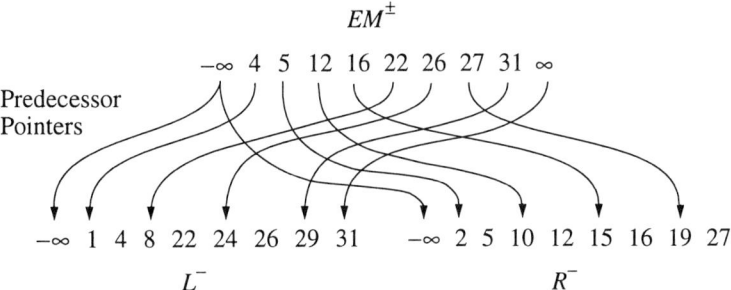

FIGURE 10.5
Information available at the start of Stage 1.1; the running example.

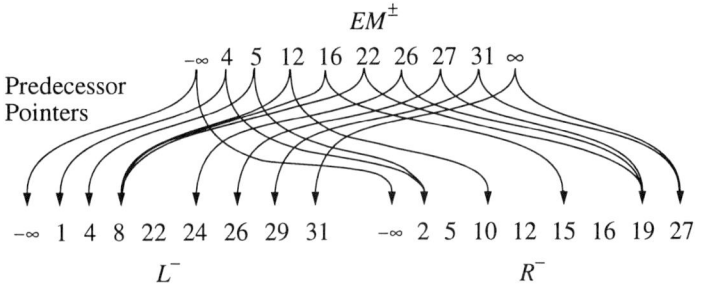

FIGURE 10.6
The effect of Stage 1.1; the running example.

to r is stored in array R_Ptr. In Stage 1.2, for each item, l, in L, but not in EM, compute the largest item, e, in EM^- smaller than l; a pointer to e is stored in array L_EM_Ptr. Figure 10.6 shows the effect of Stage 1.1 on the running example, and Figure 10.7 shows the effect of Stage 1.2 on the running example.

The computations are performed as follows.

Stage 1.1 At the end of Stage 1.1, for each item, e, in $EM^\pm$, $R\_Ptr(e)$ will point to its predecessor, the largest item in R^- smaller than e. L_Ptr is defined analogously.

Without loss of generality, suppose that e is from L. See Figure 10.8.

466   Chapter 10.   Parallel Merge Sort

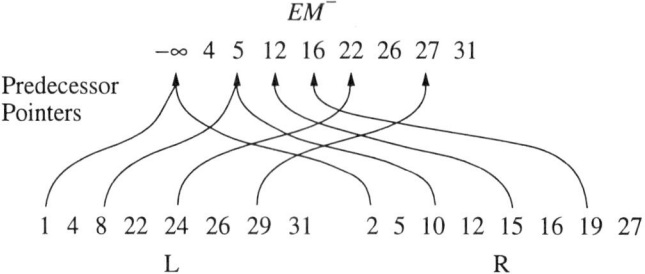

FIGURE 10.7
The Effect of Stage 1.2; the running example.

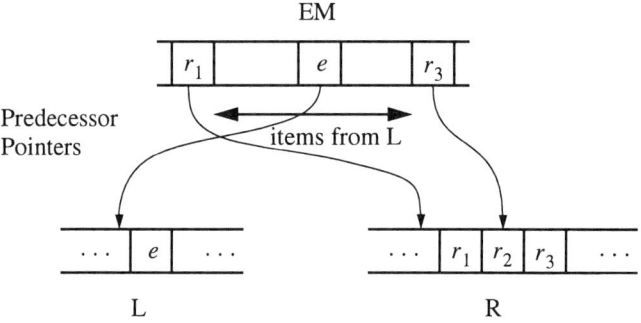

FIGURE 10.8
Notation for Stage 1.1.

Let $r1$ be the largest item from R^- in EM^- smaller than e ($\pm\infty$ are assumed to be from both $L^\pm$ and $R^\pm$). The procedure for computing a pointer to the copy of $r1$ in R^- is explained in the next paragraph. Let $r2$ be the right neighbor of r_1 in $R^\pm$. If $r2 < e$ then the predecessor of e in R^- is $r2$; otherwise, the predecessor of e in R^- is $r1$.

One way of computing a pointer to $r1$ is as follows. Here we need to mention array indices explicitly. Let e have index ie in EM; its index in L is $L\_Ptr(e) + 1$. So the index, $ir1$, of $r1$ in R^- is $ir1 = 2 \cdot ie - L\_Ptr(e) - 1$.

Pseudo-code for Stage 1.1 is given in Figure 10.9.

In the programs, the index for an item e is denoted ie; the notation is consistent with the text.

for all $1 \leq ie \leq 2^k$ **pardo**
    if $Origin(ie) = L$
        then begin
            let $ir1(ie)$ be the index in R^- of the predecessor of $EM(ie)$
            from R^- in EM^-, if any, and 0 otherwise.
                $ir1(ie) := 2 \cdot ie - L\_Ptr(ie) - 1$;
            if $ir1(ie) < 2^k$ **cand** $R(ir1(ie) + 1) < EM(ie)$
                then $R\_Ptr(ie) := ir1(ie) + 1$
                else $R\_Ptr(ie) := ir1(ie)$
        end
        else begin ($Origin(ie) = R$)
            let $il1(ie)$ be the index in L^- of the predecessor of $EM(ie)$
            from L^- in EM^-, if any, and 0 otherwise.
                $il1(ie) := 2ie - R\_Ptr(ie) - 1$;
            if $il1(ie) < 2^k$ **cand** $L(il1(ie) + 1) \leq EM(ie)$
                then $L\_Ptr(ie) := il1(ie) + 1$
                else $L\_Ptr(ie) := il1(ie)$
        end
end

FIGURE 10.9
Pseudo-code for Stage 1.1.

The computation of Stage 1.1 requires one comparison per item in EM, and $O(1)$ other operations per item in EM.

Stage 1.2 In Stage 1.2, the arrays L_EM_Ptr and R_EM_Ptr are computed. L_EM_Ptr stores, for each item, l, in L, which is not in EM, a pointer to the largest item in EM^-, smaller than l. R_EM_Ptr is defined analogously. The computation of L_EM_Ptr is described; the computation of R_EM_Ptr is entirely analogous.

The computation is performed by the items in EM^-, or rather the associated processors. See Figure 10.10. Let e be an item in EM^-, and let e' be its right neighbor in EM^+. Item e is responsible for setting the L_EM_Ptr for the odd index items in the range $(L\_Ptr(e), L\_Ptr(e')]$.

468   Chapter 10.   Parallel Merge Sort

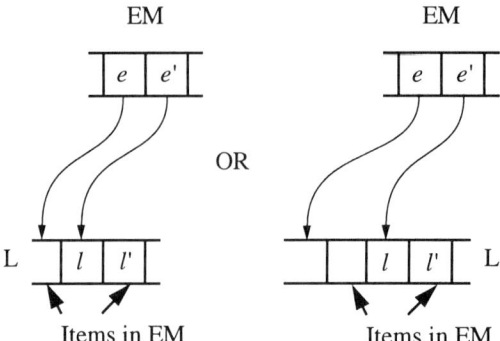

FIGURE 10.10
Notation for Stage 1.2.

There is at most one such item l (for every even index item in L is in EM, and hence every odd index location in L occurs as an L_Ptr value). (In fact, for the same reason, the only item in the range $(L\_Ptr(e), L\_Ptr(e')]$ that might have odd index in L is $L\_Ptr(e')$.) $L\_EM\_Ptr(l)$ is set to point to e.

It remains to show that for every odd index item, l, in L, $L\_EM\_Ptr(l)$ is initialized. Let l' be the right neighbor of l in L; l' is in EM and $L\_Ptr(l') = l$. Let e' be the leftmost item in EM with $L\_Ptr(e') = L\_Ptr(l')$ (possibly $e' = l'$). Let e be the left neighbor of e' in EM^- (e exists for $e' > -\infty$). $L\_EM\_Ptr(l)$ is initialized to e.

Pseudo-code for Stage 1.2 is given in Figure 10.11. Stage 1.2 requires $O(1)$ computations for each item in EM^-. No comparisons of keys are needed in Stage 1.2, however.

Stage 2   Each item in EM obtains its location in M by adding its L_Ptr (an array index) to its R_Ptr. An item, l, in L but not in EM, or rather the associated processor, proceeds as follows. See Figure 10.12. Item l follows its L_EM_Ptr pointer from L to item e in EM^- and then the R_Ptr pointer from e's right neighbor, e', in EM^+ to R^-. Let r be the item $R\_Ptr(e)$ in R^- and let r' be the item $R\_Ptr(e')$. Since $e < l$ and $r \leq e$ (equality holds only if $e = -\infty$), r and all the items preceding r in R are smaller than l; likewise, as $e' > l$ and all the items to the

Computation for array L
    for all $0 \leq ie < 2^k$ **pardo**
        if $L\_Ptr(ie) < L\_Ptr(ie+1)$ and $L\_Ptr(ie+1)$ is odd
            then $L\_EM\_Ptr(L\_Ptr(ie+1)) := ie$
    end;
Computation for array R
    for all $0 \leq ie < 2^k$ **pardo**
        if $R\_Ptr(ie) < R\_Ptr(ie+1)$ and $R\_Ptr(ie+1)$ is odd
            then $R\_EM\_Ptr(R\_Ptr(ie+1)) := ie$
    end;

FIGURE 10.11
Pseudo-code for Stage 1.2.

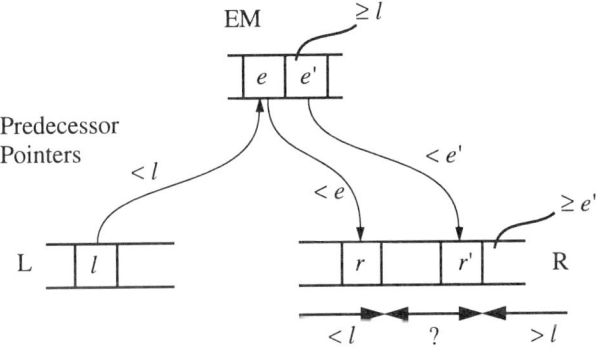

FIGURE 10.12
Notation for Stage 2.

right of r' in R are at least as large as e', all items to the right of r' in R are larger than l. Note that the range $(r, r']$ includes at most two items in R since every second item in R is in EM. Further, the size relative to l of the even index item from R in the range $(r, r']$, if any, is known, since this item would be present in EM. So by means of a comparison with the odd index item in R in the range $(r, r']$, if any, the position of l in M is determined. Further note that there is an R_Ptr to every odd index item in R, so only r' can have odd index; the other item from R in the range $(r, r']$, if any, must have even index; in addition it must be smaller than l. The computation is finished by placing each odd index item in L and R in its correct location in M.

The computation of Stage 2 requires one comparison for each odd index item in L and R, and a constant number of other operations for each odd index item in arrays L and R, and each item in array EM.

The array $EM^{\pm}$ is computed recursively (except if EM comprises two items, in which case $EM^{\pm}$ is computed directly by means of a single comparison). When computing $EM^{\pm}$, arrays $Origin$, L_Ptr and R_Ptr must also be computed; their computation takes a constant number more operations for each item placed in $EM^{\pm}$.

Pseudo-code for Stage 2 is given in Figure 10.13. Note the presence of additional arrays M_Origin, M_L_Ptr, and M_R_Ptr. This is to indicate the additional computations needed in the recursive computation of $EM^{\pm}$.

10.3.3 Complexity of the Merging Algorithm

LEMMA 10.5

The algorithm of this section requires at most $4n$ comparisons between keys to merge two sorted arrays of n items, for $n = 2^k$, for any integer $k \geq 0$. It takes parallel time $O(1 + \log n)$ using $n + 2$ processors.

PROOF

The computation in Stage 1 performs $|EM| = n$ comparisons between keys. Assuming there is one processor for each item in array $EM^{\pm}$, each processor performs its computation in $O(1)$ time. Stage 2 performs n comparisons between keys; again, assuming one processor for each item in array $EM^{\pm}$, and each odd index item in arrays L and R, each processor performs its computation in $O(1)$ time.

The array EM is of size n; it is obtained by recursively merging two arrays of size $\frac{n}{2}$. The following recursive equation bounds the number of key comparisons, $C(n)$, used to compute the array M of size $2n$:

$$C(n) \leq C(\frac{n}{2}) + 2n \quad for \ n > 1$$

$$C(n) = 1 \quad for \ n = 1$$

These equations have solution $C(n) \leq 4n$.

Using $n + 2$ processors (one processor for each odd index item in arrays L and R, and for each item in array $EM^{\pm}$), the merge is completed in

10.3. An Efficient $O(\log n)$ Time Sorting Algorithm

pardo
   Computation for array L
      for all odd $1 \leq il \leq 2^k$ **pardo**
         $ir'(il) := R\_Ptr(L\_EM\_Ptr(il) + 1);$
         if $L(il) > R(ir'(il))$ **then** $M\_Ptr(il) := il + ir'(il)$
            else $M\_Ptr(il) := il + ir'(il) - 1;$
         $M(M\_Ptr(il)) := L(il);\ M\_Origin(M\_Ptr(il)) := L;$
         $M\_L\_Ptr(M\_Ptr(il)) := 2il - 1$
      end
   Computation for array R
      for all odd $1 \leq ir \leq 2^k$ **pardo**
         $il'(ir) := L\_Ptr(R\_EM\_Ptr(ir) + 1);$
         if $R(ir) \geq L(il'(ir))$ **then** $M\_Ptr(ir) := ir + il'(ir)$
            else $M\_Ptr(ir) := ir + il'(ir) - 1;$
         $M(M\_Ptr(ir)) := R(ir);\ M\_Origin(M\_Ptr(ir)) := R;$
         $M\_R\_Ptr(M\_Ptr(ir)) := 2ir - 1$
      end
   Computation for array EM
      for all $1 \leq ie \leq 2^k$ **pardo**
         $M\_Ptr(ie) := L\_Ptr(ie) + R\_Ptr(ie);$
         $M(M\_Ptr(ie)) := EM(ie);$
         $M\_Origin(M\_Ptr(ie)) := Origin(ie);$
         if $Origin(ie) = L$
            then $M\_L\_Ptr(M\_Ptr(ie)) := 2 \cdot L\_Ptr(ie) + 1$
            else $M\_R\_Ptr(M\_Ptr(ie)) := 2 \cdot R\_Ptr(ie) + 1$
      end
   Computation for the remainder of array $EM^{\pm}$
      pardo
         $-\infty$ entry
            $M\_L\_Ptr(0) := 0;\ M\_R\_Ptr(0) := 0$
         $+\infty$ entry
            $M\_L\_Ptr(2^{k+1} + 1) := 2^{k+1} + 1;$
            $M\_R\_Ptr(2^{k+1} + 1) := 2^{k+1} + 1$
      end
end

FIGURE 10.13
Pseudo-code for Stage 2.

$O(1 + \log n)$ time, since the computation at each recursive level takes $O(1)$ time and there are $\log n + 1$ levels. ∎

COROLLARY 10.1
The algorithm of this section requires at most $4n$ comparisons between keys to merge two sorted arrays of n items, for $n = 2^k$, for any integer $k \geq 0$. It takes parallel time $O(\log n)$ using $(n+2)/\log n$ processors.

Verifying the corollary is a straightforward application of Brent's Theorem. It is left as an exercise (see Exercise 10.14).

Again, the above algorithm is readily modified so as to merge two arrays, of sizes n and m, respectively, where $n \leq m$, using $(n+m+2)/(2\log n)$ processors in time $O(\log n)$ and performing at most $2(n+m)$ key comparisons (see Exercise 10.16).

10.3.4 The Sorting Algorithm

An $O(\log^2 n)$ sorting algorithm for an array *SortMe* of n items is readily created from the above merging procedure. The algorithm has the following form. If the array holds one item then it is already sorted. If it holds 2 items then a single comparison suffices. Otherwise, if it is larger, of size n, then the following two arrays of items are recursively sorted in parallel: *First* = $SortMe[1 : \lfloor \frac{n}{2} \rfloor]$ and *Last* = $SortMe[1+\lfloor \frac{n}{2} \rfloor, n]$. Let the results of these two recursive sorts be stored in arrays *SortFirst* and *SortLast*, respectively. The algorithm is completed by merging the arrays *SortFirst* and *SortLast*, using the merging procedure of this section.

The performance of the sorting algorithm is given by the following theorem.

THEOREM 10.1
The above sorting algorithm performs $2n\lceil \log n \rceil$ key comparisons. It runs in $O(\log^2 n)$ time using $2n$ processors.

PROOF
The merge of a sorted array of size n with a sorted array of size m takes $2(n+m)$ key comparisons and $O(\log n)$ time, for $n \leq m$. The sorting algorithm comprises a series of $\lceil \log n \rceil$ phases. In Phase 1, arrays of size $1 = 2^0$ are merged; in Phase 2, arrays of size at most $2 = 2^1$ are merged; in Phase 3, arrays of size at most $4 = 2^2$ are merged; $\cdots$; in Phase $\lceil \log n \rceil$, arrays of size at most $2^{\lceil \log n \rceil - 1} \geq \lceil n/2 \rceil$ are merged. This is a total of $\lceil \log n \rceil$ phases. Each phase performs a number of key comparisons equal to at most 2 times the number of items being merged. But in each

phase, each of the n items is involved in at most one merge computation. So the total number of key comparisons per phase is at most $2n$, which yields a bound of $2n\lceil \log n \rceil$ key comparisons for the whole algorithm.

For the processor count, note that $p+2$ processors are needed for a merging problem of size p. The smallest such problem is of size 3, so at most $2p$ processors are needed for a merging subproblem (more accurately, $\frac{5}{3}p$). So the total number of processors needed is at most $2n$ since the maximum total size of the merging problems being solved simultaneously is n.

For $1 \leq i \leq \lceil \log n \rceil$, Phase i takes parallel time $O(i+1)$. So the parallel time used by the whole algorithm is $O(\log^2 n)$. ∎

COROLLARY 10.2
The above sorting algorithm performs $2n\lceil \log n \rceil$ key comparisons. It can be implemented to run in $O(\log^2 n)$ time on $n/\log n$ processors.

This is a straightforward application of Brent's theorem, as with Corollary 10.1 (see Exercise 10.17).

COMMENT 10.1
The sorting algorithm is stable. The reader should note the slight differences in dealing with equalities in the comparisons of keys in the code for Stages 1.1 and 2. We leave it as an exercise to demonstrate that this does indeed produce a stable algorithm (see Exercise 10.18).

The reader should note the equally efficient $O(\log^2 n)$ time, $n/\log n$ processor algorithm of Exercise 10.20. The virtue of the sorting algorithm of this section, from the perspective of the present chapter, is that it facilitates the presentation of the sorting algorithm of Section 10.4.

10.4
An Efficient log$n$ Time Sorting Algorithm

This section shows how to speed up the sorting algorithm of the previous section so as to achieve a logarithmic running time, but with only a constant factor loss in efficiency (i.e., the number of comparisons and other operations performed remains $O(n \log n)$).

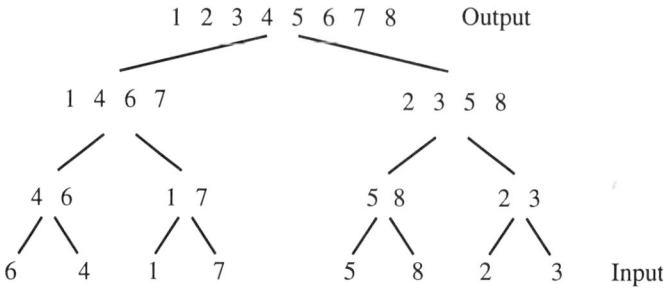

FIGURE 10.14
The Merge Tree for the Sorting Algorithm.

10.4.1 Overview

Why might a speed-up of the previous algorithm be possible? A slightly different view of the sorting algorithm of Section 10.3 is helpful. The sorting is considered to occur on a complete binary tree. The n items to be sorted are initially distributed one each to the n leftmost leaves of the tree. At each internal node of the tree the task is to compute the sorted array of the items at the leaves of its subtree. For each internal node, u, of the tree this sort is accomplished by merging the items in the arrays at u's children, once these arrays have been computed (see Figure 10.14).

A little notation is helpful. Internal node u has left and right children v and w, respectively, while internal node v has left and right children x and y, respectively. As the merge of the arrays at x and y proceeds at v, larger and larger arrays are computed at v; these, in some imprecise sense, provide better and better approximations of the final array sought at v. The arrays computed at v are not exactly the arrays needed for the merge computation at u; nonetheless, it is tempting to use these arrays, as they are computed, to start the merge computation at u.

The intended computation is illustrated in Figures 10.15 and 10.16. The new algorithm proceeds in phases. At phase 0, the array at each leaf contains a single item. At each phase $t > 0$, at each internal node u, array $M_t(u)$ is computed. In this paragraph, to simplify the exposition, it is assumed that $n = 2^k$, for some integer $k > 0$. There are an initial series of phases in which $M_t(u)$ remains empty. Let t' be the first phase in which the arrays at u's children become non-empty; in phase $t = t' + 1$, $M_t(u)$ acquires size 2;

10.4. An Efficient log n Time Sorting Algorithm

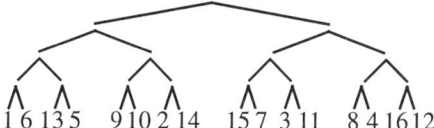

Start of Phase 1

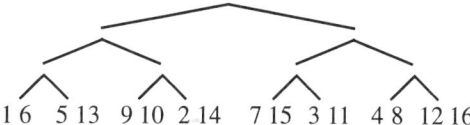

Start of Phase 2

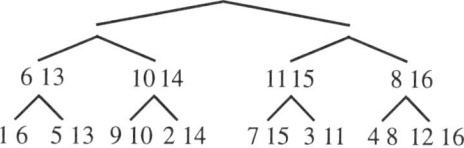

Start of Phase 3

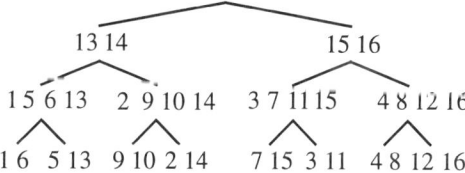

Start of Phase 4

FIGURE 10.15
Data movement for the sort—Part I.

thereafter, at each phase, the size of $M_t(u)$ doubles, until it reaches its final maximum size, namely the number of items at the leaves of its subtree. Let t'' be the phase at which $M_t(u)$ acquires its maximum size. For the moment, we assume the algorithm has the following form (some modification will be

476 Chapter 10. Parallel Merge Sort

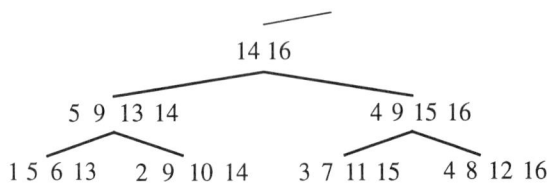

Start of Phase 5

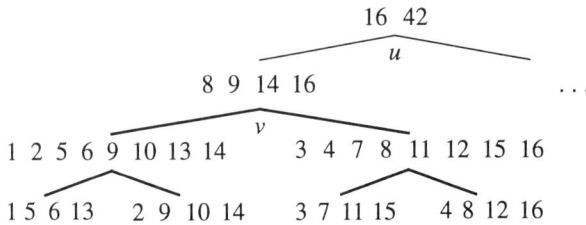

Start of Phase 6

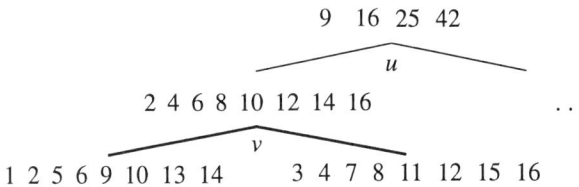

Start of Phase 7

FIGURE 10.16
Data movement for the sort—Part II.

needed). For phases $t < t''$, $M_t(u)$ comprises the merge of the even index items in $M_{t-1}(v)$ and $M_{t-1}(w)$; $M_{t''}(u)$ comprises the merge of $M_{t''-1}(v)$ and $M_{t''-1}(w)$; for $t > t''$, $M_t(u) = M_{t''}(u)$.

It is not possible to use exactly the procedure of Section 10.3 for performing the merges of the new algorithm. The data flow in the new algorithm has several features not present before. It is helpful to refer to

10.4. An Efficient log n Time Sorting Algorithm 477

Figure 10.16. Look at the arrays present at the left child, v, of the root, in Phases 6 and 7; notice that item 9 present in v's array in Phase 6 is no longer present in v's array in Phase 7; that is, the computation of array M, from phase to phase, is not simply a matter of "filling in" more items. This can have surprising effects; for instance, in Phase 7, at the root, node u, the array contains item 9, but this item is not present in the array at node v.

So the state assumed at the start of a phase, as given in Section 10.3, needs to be modified. As before, three additional arrays are maintained: $Origin_t(u)$, $L\_Ptr_t(u)$, and $R\_Ptr_t(u)$. The definitions of $L\_Ptr_t(u)$ and $R\_Ptr_t(u)$ are modified; the new definitions are given below. Some new notation is needed. The entry for item e in an array, $M_t(u)$ for example, is denoted $M_t(u)[e]$. As before, to avoid special cases in the discussion that follows, each array is extended by items $-\infty$ and $+\infty$ at the front and back, respectively. If an item, e, in $M_t(u)$ came from v's subtree, then $Origin_t(u)[e] = L$; $L\_Ptr_t(u)[e]$ is defined below. Three more arrays are introduced. First, $E_{t+1}(u)$ (E is short for $Extract$); it comprises the items from $M_t(u)$ to be placed in $M_{t+1}(parent(u))$. For the moment, suppose that $E_{t+1}(u)$ comprises every second item in $M_t(u)$, unless $M_{t-1}(u)$ has received all its elements, in which case $E_{t+1}(u) = M_t(u)$. (As we will see later, this definition leads to difficulties, so it will need to be modified.) $M_{t+1}(u)$ comprises the merge of $E_{t+1}(v)$ and $E_{t+1}(w)$. Now, $L\_Ptr_t(u)[e]$ can be defined; it exists if e is from u's left subtree, and then it comprises a pointer to the largest item in $E_{t+1}^-(v)$ smaller than e. The second array is $B_t(u)$ (B is short for $Before$); for each item, e, in $M_t(u)$, if e is from v's subtree, $B_t(u)[e]$ is a pointer to the following item in $M_t^-(u)$: the largest item from w's subtree smaller than e (in this definition, the items $\pm\infty$ are treated as if they came from the subtrees of both v and w). The third array is $A_t(u)$ (A is short for $After$); if e is from v's subtree, $A_t(u)[e]$ is a pointer to the following item in $M_t^+(u)$: the smallest item from w's subtree larger than e. Analogous definitions apply if e came from w's subtree. See Figure 10.17 for an example.

We need to introduce a further definition. Array C is said to 3-cover array D, if for each two adjacent items e_1 and e_2 in C, the range $[e_1, e_2)$ contains at most 3 items from D. Let us assume for the time being that $E_t^{\pm}(u)$ is a 3-cover of $E_{t+1}(u)$ (we will show later how to enforce this property). The 3-cover property is needed in order to carry out the computation of $M_{t+1}^{\pm}(u)$ in constant time. See Figure 10.18 for an example of this property. Note that as $M_t^{\pm}(u)$ contains $E_t^{\pm}(u)$, $M_t^{\pm}(u)$ is a 3-cover of $E_{t+1}(u)$.

478   Chapter 10.   Parallel Merge Sort

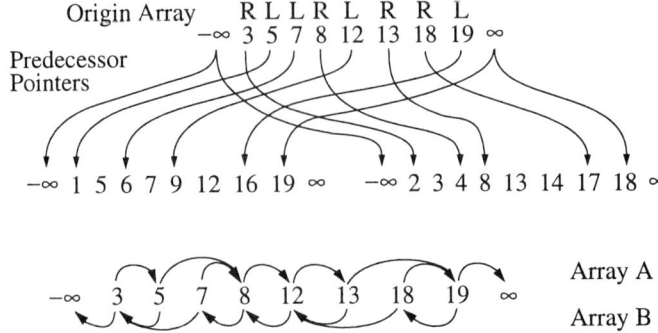

FIGURE 10.17
The initial configuration at step $t+1$; an example.

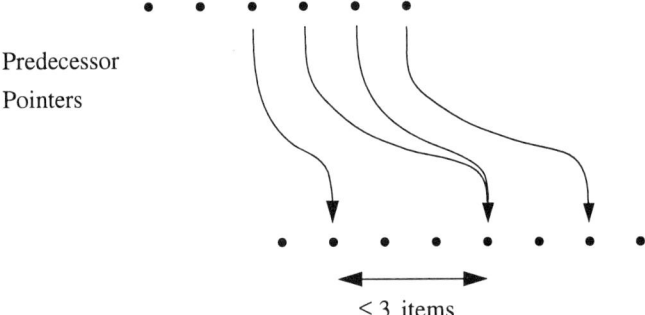

FIGURE 10.18
Example of 3-Cover Property.

10.4.2   Details of the Merge Computation

The merge computation has the same essential structure as in Section 10.3. The details need some modification, however. The new version of each Stage is specified below.

Stage 1.1 For each item, e, in $M_t(u)$, if e came from v's subtree, compute its *predecessor* in $E^-_{t+1}(w)$; e's predecessor is the largest item smaller than e in $E^-_{t+1}(w)$. A pointer to e's predecessor is stored in $R\_Ptr_t(u)[e]$. An analogous procedure is applied if e came from w's subtree.

10.4. An Efficient log n Time Sorting Algorithm

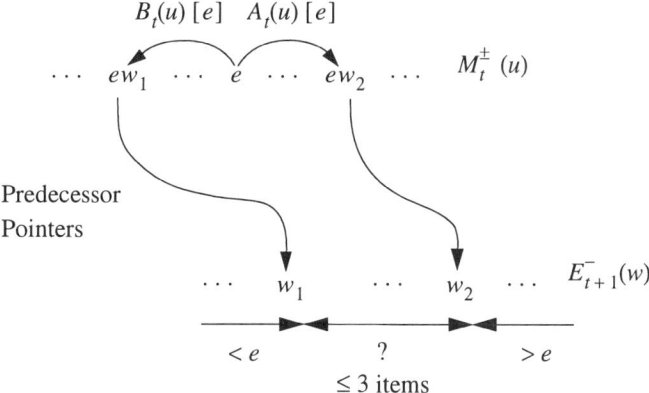

FIGURE 10.19
Notation for Stage 1.1.

See Figure 10.19. Let $ew1 = B_t(u)[e]$ and $ew2 = A_t(u)[e]$ be the two items from w's subtree in $M_t^{\pm}(u)$, preceding and following e, respectively. Let $w1$ and $w2$ be the following items in $E_{t+1}^-(w)$: $R\_Ptr_t(u)[ew1]$ and $R\_Ptr_t(u)[ew2]$, respectively. Then e is larger than $w1$ and all the items to its left in $E_{t+1}(w)$; likewise, e is smaller than all the items to the right of $w2$ in $E_{t+1}(w)$. Note that $ew1$ and $ew2$ are neighboring items in $E_t^{\pm}(u)$. By the 3-cover property, there are at most 3 items in $E_{t+1}(w)$ whose size relative to e is uncertain (these 3 items are those in the range $(w_1, w_2]$, which in $E_{t+1}(w)$ is also the range $[ew1, ew2)$). e, or rather the associated processor, can determine which of these 3 (or fewer) items are smaller than e by means of two comparisons. This determines $R\_Ptr_t(u)[e]$.

Stage 1.2 For each item, ev, in $E_{t+1}(v)$, its *predecessor* in $M_t^-(u)$ is determined; ev's predecessor is the largest item in $M_t^-(u)$ smaller than ev. A pointer to ev's predecessor is stored in array $P\_Ptr_{t+1}(v)$ (P_Ptr is short for *Parent_Pointer*). Array $P\_Ptr_{t+1}(w)$ is defined and computed analogously.

The computation is performed by the processors associated with items in $M_t^-(u)$. See Figure 10.20. Let $e1$ be an item in $M_t^-(u)$, and let $e2$ be its right neighbor in $M_t^+(u)$. Let $ev1$ and $ev2$ be the following items in $E_{t+1}^-(v)$: $L\_Ptr_t(u)[e1]$ and $L\_Ptr_t(u)[e2]$, respectively. Then $e1$, or

480   Chapter 10.   Parallel Merge Sort

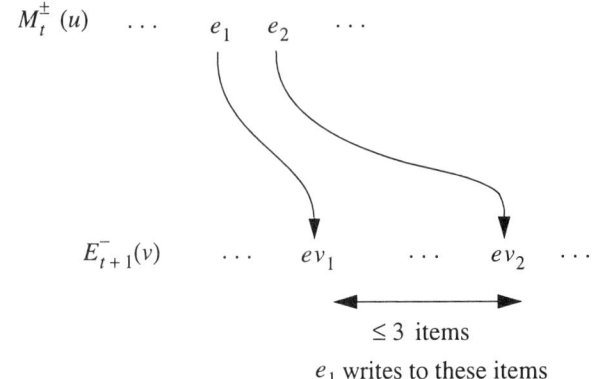

FIGURE 10.20
Notation for Stage 1.2.

rather the associated processor, writes the entries in $P\_Ptr_{t+1}(v)$ for the items of $E_{t+1}(v)$ in the range $[e1, e2)$, which is exactly the items in the range $(ev1, ev2]$. Each entry is a pointer to item e_1 in $M_t^-(u)$. The 3-cover property ensures that there are at most 3 items in the range $[e1, e2)$. So this computation can be performed in constant time. It requires no comparisons of keys.

Stage 1.3 The items in $E_{t+1}(v)$ and $E_{t+1}(w)$ are merged to produce $M_{t+1}(u)$. The computation for items in $E_{t+1}(v)$ is described; the computation for $E_{t+1}(w)$ is entirely analogous. Each item, ev, in $E_{t+1}(v)$, or rather its associated processor, determines ev's predecessor in $E_{t+1}^-(w)$, the largest item smaller than ev; ev can then deduce its index in $M_{t+1}(u)$, which is stored in array $F\_Ptr_{t+1}(v)$ (F_Ptr is short for *Future_Pointer*). It will also be useful to set $F\_Ptr_{t+1}(v)[+\infty]$ to point to the item $+\infty$ in $M_{t+1}(u)$.

See Figure 10.21. The computation is performed by the processors associated with the items in $E_{t+1}^-(v)$. Let ev be an item in $E_{t+1}(v)$. Let $e1$ be the item $P\_Ptr_{t+1}(v)[ev]$ in $M_t^-(u)$, and let $e2$ be $e1$'s right neighbor in $M_t^+(v)$. Let $we1$ and $we2$ be the following items in $E_{t+1}^-(w)$: $R\_Ptr_t(u)[e1]$ and $R\_Ptr_t(u)[e2]$, respectively. Then the only items whose size relative to ev is uncertain are the items in the range $(we1, we2]$;

10.4. An Efficient $\log n$ Time Sorting Algorithm

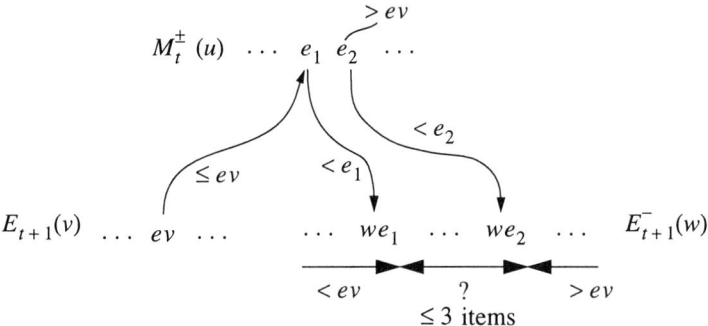

FIGURE 10.21
Notation for Stage 2.1.

this range contains the same items in $E_{t+1}(w)$ as range $[e1, e2)$. By the 3-cover property, range $[e1, e2)$ contains at most 3 items in $E_{t+1}(w)$. So by means of two comparisons, the processor associated with ev can determine its predecessor in $E^-_{t+1}(w)$. Finally, the processor for $ev = -\infty$ initializes $F\_Ptr_{t+1}(v)[+\infty]$ (this processor is not performing any computation for the item $ev = -\infty$ at this point).

Stage 2.2. It remains to compute the arrays $Origin_{t+1}(u)$, $L\_Ptr_{t+1}(u)$, $R\ Ptr_{t+1}(u)$, $A_{t+1}(u)$ and $B_{t+1}(u)$. These arrays need be computed only if $M_{t+1}(u)$ has not attained the maximum possible size. Computing $Origin_{t+1}(u)$ is straightforward. Below, the computation of $L\_Ptr_{t+1}(u)$ for items from v's subtree is described (it is computed for items from w's subtree in Stage 1.1 of Phase $t + 2$). The computation of $R\_Ptr_{t+1}(u)$ is analogous. Then, the computation of $A_{t+1}(u)$ is described; the computation of $B_{t+1}(u)$ is similar.

See Figure 10.22. Let ev be an item in $E_{t+1}(v)$. ev is also an item in $M_t(v)$.

Case 1. $M_t(v) \neq M_{t+1}(v)$. Then $M_{t+1}(v)$ comprises the merge of $E_{t+1}(x)$ and $E_{t+1}(y)$. There are $L\_Ptr_t(v)[ev]$ items in $E_{t+1}(x)$ smaller than ev (viewing this pointer an as array index) and $R\_Ptr_t(v)[ev]$ items in $E_{t+1}(y)$ smaller than ev. So $L\_Ptr_{t+1}(u)[ev] = L\_Ptr_t(v)[ev] + R\_Ptr_t(v)[ev]$.

482 Chapter 10. Parallel Merge Sort

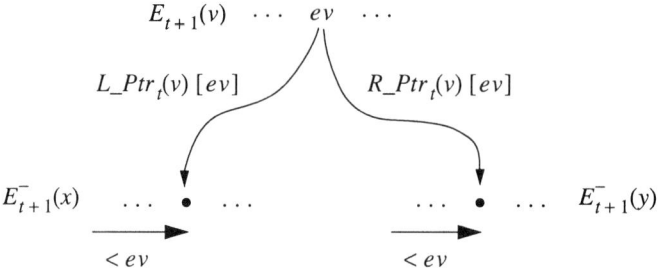

FIGURE 10.22
Notation for Stage 2.2.

Case 2. $M_t(v) = M_{t+1}(v)$. Then $L\_Ptr_{t+1}(u)[ev]$ is set to point to the predecessor of ev in $E^-_{t+2}(v) = M_t(v)$, whose location is known to item ev in $M_t(v)$.

Now, the computation of $A_{t+1}(u)$ is described. Let ev be an item in $E_{t+1}(v)$. In Stage 2.1, ev learned its predecessor ew' in $E^-_{t+1}(w)$. Let ew'' be the right neighbor of ew' in $E^+_{t+1}(w)$. Recall that $F\_Ptr_{t+1}(w)[ew'']$ stores a pointer to the new location of ew'' in $M_{t+1}(u)$. So $A_{t+1}(u)[ev] := F\_Ptr_{t+1}(w)[ew'']$. An analogous computation is performed for items ew in $E_{t+1}(w)$.

Finally, and in parallel with the remainder of Stage 2.2, the extension by $\pm \infty$ of $M_{t+1}(u)$ and associated arrays must be created; this can be done by the processors associated with the extension of $M_t(u)$.

The computation of Stage 2.2 requires no key comparisons.

COMMENT 10.2
The first stage, t, in which $M_t(u)$ becomes non-empty can use a simpler computation to determine $M_t(u)$: a single comparison suffices since $|M_t(u)| \leq 2$. However, $M^{\pm}_t(u)$ must still be created.

This completes the description of the merging procedure. Its constant time performance depends critically on the 3-cover property. Unfortunately, as was alluded to earlier, it need not hold (see Exercise 10.21). However, a simple modification of the algorithm restores the 3-cover property. Instead of comprising every second item in $M_t(u)$, $E_{t+1}(u)$ will comprise every fourth item in $M_t(u)$. Again, for each internal node u, after a number of phases in which $M_t(u)$ remains empty, $M_t(u)$ acquires size 2; it then doubles in size from phase to phase until it attains its maximum size. Again, a little care is needed in the definition of $E_t(u)$. Suppose that $M_t(u)$ first acquires its maximum size in Phase t'. Then, $E_{t'+1}$ comprises every fourth item in $M_{t'}(u)$, as specified. But in order to ensure the continued doubling in size of $M_t(parent(u))$, $E_{t'+2}(u)$ comprises every second item in $M_{t'+1}(u) = M_{t'}(u)$ and $E_{t'+3}(u)$ comprises every item in $M_{t'+2}(u) = M_{t'}(u)$.

Once more, the above description of the growth of $M_t(u)$ applies only if $n = 2^k$, for some integer $k > 0$.

10.4.3 Correctness Proofs for the Algorithm

In general, the merges proceed according to the just described pattern, but it need not be the case that the size of $M_t(u)$ doubles from phase to phase. $M_t(u)$ is specified as follows. Suppose internal node u is at height h in the tree (the leaves are at height 0). Then in phases t, $2h+1 \leq t \leq 3h-2$, $E_t(u)$ comprises every fourth item in $M_{t-1}(u)$; in phase $3h-1$, $E_t(u)$ comprises every second item in $M_{t-1}(u)$; and in phase $3h$, $E_t(u)$ comprises every item in $M_{t-1}(u)$.

LEMMA 10.6
Let u be an internal node at height h. $M_{3h}(u)$ contains every item originally in u's subtree.

PROOF
The result follows by an induction on h. ■

LEMMA 10.7
Let u be an internal node at height h. If u is not the rightmost node at its level in the tree, then $|M_t(u)|$ doubles from phase to phase, for $2h+1 < t \leq 3h$; in addition, $|M_{2h+1}(u)| = 2$. While if u is the rightmost node at its level in the tree, let $t = k$ be the first phase at which $|M_t(u)| > 0$; $k \geq 2h+1$. For phases $k < t \leq 3h$, $|M_t(u)|$ is either $2|M_{t-1}(u)|$ or $2|M_{t-1}(u)| + 1$; in addition, $|M_k(u)|$ is 1 or 2.

PROOF
An induction on t is used. First suppose that u is not the rightmost node at its level in the tree. For $t = 2h+1$, $|M_t(u)| = \lfloor |M_{t-1}(v)|/4 \rfloor + \lfloor |M_{t-1}(w)|/4 \rfloor$; by the inductive hypothesis, $|M_{t-1}(v)| = |M_{t-1}(w)| = 4$, since v and w are at height $h-1$ and neither of them is the rightmost node at its level in the tree. So $|M_{2h+1}(u)| = 2$ as claimed.

Next suppose that u is the rightmost node at its level, and consider phase $t = k$. Then $|M_{k-3}(v)|, |M_{k-3}(w)| \leq 1$, for if not and $k > 2h+1$, this contradicts the definition of $t = k$ being the first phase for which $|M_t(u)| > 0$, while if $k = 2h+1$, by definition, $M_{2h-2}(v)$ and $M_{2h-2}(w)$ are empty (v and w are nodes at height $h-1$). So, by the inductive hypothesis, $|M_{k-1}(v)|, |M_{k-1}(w)| \leq 7$; it follows that $|M_k(u)| \leq 2$.

Now consider phase t, $\max\{2h+1, k\} < t \leq 3h-2$.

Case 1. w is present in the tree. $|M_t(u)| = |M_{t-1}(v)|/4 + \lfloor |M_{t-1}(w)|/4 \rfloor$ (for note that v is not the rightmost node at its level and hence $|M_{t-1}(v)|$ is a multiple of four). As $t - 3 \geq 2h - 1$, by the inductive hypothesis, $|M_{t-1}(v)| = 4|M_{t-3}(v)|$; likewise, either $|M_{t-3}(w)| = 0$ or $\lfloor |M_{t-1}(w)|/4 \rfloor = |M_{t-3}(w)|$. So $|M_t(u)| \leq |M_{t-3}(v)| + \max\{1, |M_{t-3}(w)|\}$ (the 1 handles the case that $|M_{t-3}(w)| = 0$). Similarly, $|M_{t-1}(u)| = |M_{t-3}(v)|/2 + \lfloor |M_{t-3}(w)|/2 \rfloor$. So $|M_t(u)|$ is either $2|M_{t-1}(u)|$ or $2|M_{t-1}(u)| + 1$.

Case 2. w is not present in the tree. $|M_t(u)| = \lfloor |M_{t-1}(v)|/4 \rfloor$. By the inductive hypothesis, either $|M_{t-3}(v)| = 0$ or $\lfloor |M_{t-1}(v)|/4 \rfloor = |M_{t-3}(v)|$; but the second possibility always applies for $t > k$. So $|M_t(u)| = |M_{t-3}(v)|$. Similarly, $|M_{t-1}(u)| = \lfloor |M_{t-3}(v)|/2 \rfloor$. So $|M_t(u)|$ is either $2|M_{t-1}(u)|$ or $2|M_{t-1}(u)| + 1$.

The result for $t = 3h-1, 3h$ follows readily from the fact that E_{3h-2}, E_{3h-1}, E_{3h} of v and w comprise, respectively, every fourth, second and consecutive item in M_{3h-3} of v and w. ∎

Next, the 3-cover property is shown (Corollary 10.3 to Lemma 10.8). Let A be an array of integers. Let $e_1, e_2, \cdots, e_{k+1}$ be a sequence of $k+1$ items in $A^{\pm}$. Then the k intervals defined by these $k+1$ items comprise the range $[e_1, e_{k+1})$; these intervals are said to be in $A^{\pm}$.

LEMMA 10.8
Let $k \geq 1$. Any k adjacent intervals in $E_t^{\pm}(u)$ contain at most $2k+1$ items from $E_{t+1}(u)$.

PROOF
The result is proved by induction on t. The claim is true initially, for when $E_t(u)$ is empty, $E_{t+1}(u)$ contains at most one item, while at that time $t = t'$ at which $E_t(u)$ first becomes non-empty, $E_t(u)$ contains one item, so $M_{t-1}(u)$ contains at most seven items, hence $M_t(u)$ contains at most fifteen items (by Lemma 10.7), and so $E_{t+1}(u)$ contains at most three items.

10.4. An Efficient log n Time Sorting Algorithm 485

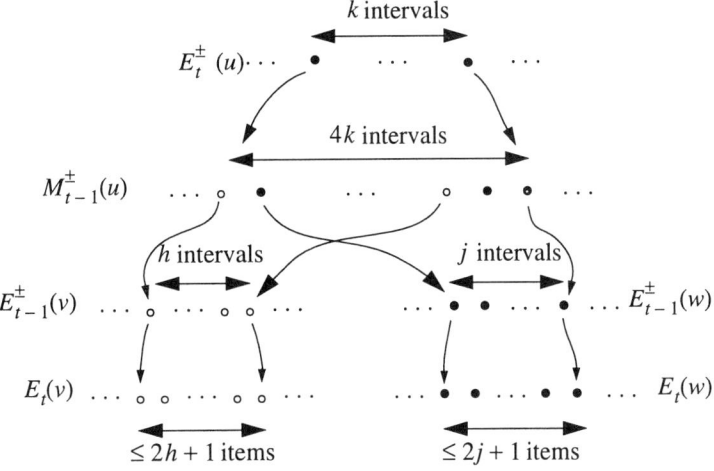

FIGURE 10.23
Cover Properties of k Intervals in $E_t(u)$.

Suppose $M_{t-1}(u)$ has not yet attained its maximum size. See Figure 10.23. Consider a sequence of k adjacent intervals in $E_t^\pm(u)$; they cover the same range as some sequence of $4k$ adjacent intervals in $M_{t-1}^\pm(u)$. Recall that $M_{t-1}(u)$ comprised the merge of $E_{t-1}(v)$ and $E_{t-1}(w)$. The $4k$ intervals in $M_{t-1}^\pm(u)$ overlap some $h \geq 1$ adjacent intervals in $E_{t-1}^\pm(v)$ and some $j \geq 1$ intervals in $E_{t-1}^\pm(w)$, with $h+j = 4k+1$. The h intervals in $E_{t-1}^\pm(v)$ contain at most $2h+1$ items from $E_t(v)$ by the inductive hypothesis; likewise, the j intervals in $E_{t-1}^\pm(w)$ contain at most $2j+1$ items from $E_t(w)$. Recall that $M_t(u)$ comprises the merge of $E_t(v)$ and $E_t(w)$. Thus the $4k$ intervals in $M_{t-1}^\pm(u)$ contain at most $2h + 2j + 2 = 8k + 4$ items from $M_t(u)$. But $E_{t+1}(u)$ comprises every fourth item in $M_t(u)$; so the k adjacent intervals in $E_t^\pm(u)$ contain at most $2k + 1$ items from $E_{t+1}(u)$.

It remains to prove the lemma for times t for which $M_{t-1}(u)$ has attained its maximum size. Clearly, once $E_t(u)$ has attained its maximum size ($E_t(u) = M_{t-1}(u)$) there is no need for subsequent computations of $E_t(u)$; formally, as regards the lemma, $E_t(u) = E_{t+1}(u)$. So only two phases need be considered: phases t' and $t'+1$, where $t'-1$ is the first phase in which $M_t(u)$ attains its maximum size. For these two phases the following stronger claim applies:

CLAIM 10.1
k adjacent intervals in $E_t^{\pm}(u)$ contain exactly $2k$ items from $E_{t+1}(u)$, and every item in $E_t(u)$ occurs in $E_{t+1}(u)$.

This is readily seen. Consider time t'. $E_{t'}(u)$ comprises every fourth item in $M_{t'-1}(u)$, while $E_{t'+1}(u)$ comprises every second item in $M_{t'-1}(u)$; clearly the claim is true for phase t'. The argument is similar for phase $t'+1$. ∎

COROLLARY 10.3
$E_t^{\pm}(u)$ is a 3-cover of $E_{t+1}(u)$.

10.4.4 Complexity of the Algorithm

For the computation of $M_{t+1}^{\pm}(u)$ and associated arrays one processor is assigned to each item in $M_t^{\pm}(u)$, $E_{t+1}^{-}(v)$ and $E_{t+1}^{-}(w)$. Each of Stages 1.1, 1.2, 2.1, and 2.2 require a constant number of operations per processor. Stage 1.1 requires at most 2 key comparisons for each item in $M_t(u)$ and Stage 2.1 requires at most 2 key comparisons for each item in $E_{t+1}(v)$ and $E_{t+1}(w)$.

LEMMA 10.9
Suppose the subtree at node u originally contained $L(u)$ items. Then at most $\frac{15}{4} \cdot L(u)$ key comparisons are performed in computing the arrays $M_t(u)$.

PROOF
Let $t = t'$ be the phase in which $M_t(u)$ acquires size s. In Stage 1.1, $2 \cdot \lfloor s/2 \rfloor$ key comparisons are performed, and in Stage 2.1, $2s$ key comparisons are performed. This is at most $3s$ key comparisons in Phase t'. Summing over all t, since the size of $M_t(u)$ at least doubles from phase to phase, this is a total of at most $3 \cdot \sum_{i \geq 0} L(u)/2^i$ key comparisons; this is bounded by $6L(u)$.

This is an overcount, however. In phase $t = t''$, the phase in which $M_t(u)$ attains its maximum size, and in the preceding phase Claim 10.1 applies. So, in these two phases, only one comparison need be performed for each item in $M_t(u)$, $E_{t+1}(v)$, and $E_{t+1}(w)$. This reduces the number of comparisons performed at node u to at most $3 \cdot \sum_{i \geq 2} L(u)/2^i + \frac{3}{4}L(u) + \frac{3}{2}L(u) = \frac{15}{4}L(u)$. ∎

REMARK 10.1
By using the algorithm of Section 10.3 for the last two phases of computation at each node, the total number of comparisons performed at a node u can be reduced to $3 \cdot L(u)$.

10.4. An Efficient $\log n$ Time Sorting Algorithm

Next, the number of phases in the algorithm is bounded.

LEMMA 10.10
The algorithm terminates in $3\log n$ phases.

PROOF
Suppose $M_t(v)$ first attains its maximum size at phase $t = t'$. Then, $M_t(u)$ first attains its maximum size at phase $t' + 3$. But the tree comprises $\log n$ levels of internal nodes. The lemma follows. ∎

The algorithm uses a linear number of processors, as is shown in the following lemma.

LEMMA 10.11
The algorithm uses $\frac{12}{7}n + (4n + 4\lceil \log n \rceil)$ processors.

PROOF
Consider the computation of array $M^{\pm}_{t'+1}(u)$ of size s. It uses $\lfloor \frac{3}{2} \cdot s \rfloor + 4$ processors (one for each item in $M^{\pm}_{t'}(u)$, $E^{-}_{t'+1}(v)$, and $E^{-}_{t'+1}(w)$). No processors are required for those nodes z for which $M_t(z)$ has attained its maximum size in some phase $t \leq t'$; such nodes z are said to be *inactive*; all other nodes are said to be *active*. So the active nodes at the deepest level in the tree require at most $\lfloor \frac{3}{2}n \rfloor + 4r$ processors, where r is the number of nodes at this level; note that $r \leq \lceil n/2 \rceil$. The (active) nodes one level higher in the tree require at most $\lfloor \frac{1}{8}\frac{3}{2}n \rfloor + 4\lceil r/2 \rceil$ processors, and so on. (For if $M_{t'+1}(v)$ has size s, $M_{t'+1}(u)$ has size $2\lfloor s/8 \rfloor$; in addition, the number of nodes at u's level is $\lceil r/2 \rceil$). Summing over all the active nodes, this is a total requirement of at most $\frac{12}{7}n + (4n + 4\lceil \log n \rceil)$ processors. ∎

The following theorem has been shown.

THEOREM 10.2
There is a CREW PRAM sorting algorithm that runs in time $O(\log n)$ on $5\frac{5}{7}n + 4\lceil \log n \rceil$ processors, performing at most $\frac{15}{4}n$ comparisons.

COMMENT 10.3
The $4n + 4\lceil \log n \rceil$ term can be considerably reduced by initially performing, in parallel, sequential sorts on sets of size 4; the term is then reduced to $n + 4\lceil \log n \rceil - 8$. Also, the processors for $M^{\pm}_t$ are never needed simultaneously with the processors for $E^{-}_{t'+1}(v)$ and $E^{-}_{t'+1}(w)$; exploiting this reduces the processor requirement further to $\frac{8}{7}n + \frac{n}{2} + 2\lceil \log n \rceil - 4$ processors.

REMARK 10.2

The algorithm needs only $O(n)$ space. A simple implementation uses $O(n \log n)$ space. It provides $O(L(u))$ space to each node u whose subtree originally contained $L(u)$ items. However, the nodes at different levels do not need their maximum space simultaneously. In phase $t + 1$, only the arrays for phases t and $t+1$ are needed; further, at the end of phase $t + 1$, the phase t arrays can be discarded and the space reused. The details of this implementation are left to the reader (see Exercise 10.23).

REMARK 10.3

The algorithm of this section is less efficient than the sorting algorithm of Section 10.3 by roughly a factor of 2. It is worthwhile to use the Section 10.4 algorithm only if the full exploitation of the parallelism of the machine at hand requires the use of $\Omega(n)$ virtual processors, rather than $\Omega(n/\log n)$ processors; note that the number of virtual processors needed to fully exploit a given machine is presumably fixed, while n, the size of the set to be sorted, may vary for different problem instances. It is worth remarking that the number of virtual processors needed to fully exploit a given machine may be larger than the number of physical processors in the machine. Also, see Exercise 10.25.

REMARK 10.4

The algorithm of this section can be made stable in the same way as the sorting algorithm of Section 10.3. The details are left as an exercise (Exercise 10.26).

10.5 Exercises

10.1 A comparison based algorithm may perform only two operations on the keys: comparing and copying. (Hence all branch points in the algorithm are explicitly or implicitly controlled by key comparisons.) Show that any comparison based sorting algorithm performs $\Omega(n \log n)$ key comparisons.

10.2 Draw the Odd-Even merging circuit for merging two sorted sets of sizes 5 and 6, respectively.

10.3 Draw the Odd-Even sorting circuit for a set of size 16.

10.4 Prove Lemma 10.3.

10.5 Recall that a sorting algorithm is stable if it preserves the relative order of items with equal keys.

a) Show that the odd-even sorting circuit is not stable.

b) Can this circuit be made stable by interchanging the order of the inputs to some of the comparators? If yes, describe how the circuit needs to be changed and prove that it then sorts stably. If not, show that no such set of interchanges will produce a stable sorter.

10.6 The following result is known as the Zero-One Law for Sorting Circuits: If a circuit, intended to sort sets of n integers, correctly sorts all sets comprising n zeros and ones then it correctly sorts all sets of n integers.

a) Prove the Zero-One Law for Sorting Circuits.

b) Does this law also apply to comparison based algorithms? (See Exercise 10.1.)

10.7 An alternative merging circuit, also developed by Batcher, is based on bitonic sequences. Before defining a bitonic sequence a few other definitions are useful. An *up sequence* is a sequence in increasing order (or more strictly, non-decreasing order); likewise, a *down sequence* is a sequence in non-increasing order. An up-down sequence comprises an up sequence followed by a down sequence. A bitonic sequence is a circular shift of an up-down sequence. (A circular shift by s positions of an n item sequence places the item in position i in the original sequence in position $(i + s)$ mod n in the new sequence.)

Show that a bitonic sequence comprises one of the following.

i. An up sequence, followed by a down sequence, followed by a second up sequence; any of these sequences may be empty. In addition, the last item of the second up sequence is no larger than the first item of the first up sequence.

ii. A down sequence, followed by an up sequence, followed by a second down sequence; any of these sequences may be empty. In addition, the last item of the second down sequence is no smaller than the first item of the first down sequence.

10.8 a) Consider the following algorithm applied to a $2n$-item array A.
for $i = 1$ **to** n **do** $Order(A(i), A(i+n), A(i), A(i+n))$ **end**
(Recall that $Order$ is the function that places its first two arguments in sorted order in its last two arguments.) Show that if A stores a bitonic sequence (see Exercise 10.7) then the algorithm rearranges the sequence so that the smallest n items are in the first n positions and the largest n items are in the last n positions. In addition, show that the first n items form a bitonic sequence, as do the last n items.

b) Hence devise a depth $\log n$ sorting circuit for $n = 2^k$ item bitonic sequences, for integer $k > 0$.

c) Hence devise a depth $O(\log^2 n)$ sorting circuit for sets of size n. What is the exact depth of your circuit?

10.9 Provide a description of the PRAM version of Batcher's odd-even sorting circuit; that is, specify, for each processor, what comparison, if any, it performs at each time step. (Hint: First provide a description for sets of size $n = 2^k$, for integer $k > 0$. In addition, suppose that the algorithm compares and exchanges pairs of items in the input array of items; this gives a different apparent pattern to the comparisons than that shown in Figure 10.4 for instance.)

10.10 In the comparison model of parallel computation, the only computations to be performed on keys are comparisons and copyings. Any computations on non-keys (for instance, computations on array indices) are for free. In parts (i)-(v) below, give algorithms in the comparison model of computation.

a) Give an algorithm to compute the maximum of n items stored in an array in $O(\max\{1, \log \log n - \log \log(p/n)\})$ time using $p \geq n$ processors.

b) Modify the algorithm of (a) to obtain an $O(\log \log n)$ running time with $n/\log \log n$ processors.

c) Give an algorithm to merge arrays of n and m items, respectively, where $m \geq n$, in $O(\log \log n)$ time using $n + m$ processors.

d) Modify the algorithm of (c) to obtain an $O(\log \log n)$ running time with $(n + m)/\log \log n$ processors.

e) Generalize the algorithm of (c) to obtain an $O(\max\{1, \log \log n - \log \log(p/(m + n))\})$ running time with $p \geq m + n$ processors.

10.11 This exercise is concerned with a linear time sequential algorithm for selection. (Input: An array A of n items, plus another integer k, $1 \leq k \leq n$. Output: The kth smallest item in A.) Let $A = A_0$. Consider applying the following extraction process repeatedly, yielding, in turn, sets $A_1, A_2, \cdots A_h$ of sizes $\lfloor n/2 \rfloor, \lfloor n/4 \rfloor, \cdots, \lfloor n/2^h \rfloor$, respectively. For $1 \leq i < h$, A_{i+1} is obtained from A_i by partitioning A_i into sets of size $8 \cdot 2^i$ (there may be one undersize set), sorting each set, and taking every second item in each sorted set. h is chosen to be the smallest i such that $n < 4^{i+2}$.

a) Show that the rth smallest item in A_{i+1} is at least the $2r$th smallest item in A_i and at most the $(2r + n/(8 \cdot 4^i))$th smallest item in A_i, for $0 \leq i < h$.

b) Show that the $\lceil (k - n/4)/2^h \rceil$th item in A_h is no larger than the kth smallest item in A and is no smaller than the $k - n/4$th item in A. Similarly, show that the $\lceil k/2^h \rceil$th item in A_h is no smaller than the kth smallest item in A and is at most the $k + n/4$th item in A.

c) Show that the extraction algorithm takes linear time.

d) Hence obtain a linear time selection algorithm. Obtain a bound of the form $cn + o(n)$ on the number of key comparisons performed by your algorithm. Try to implement the algorithm so as to keep c small.

10.12 In the comparison model of parallel computation (see Exercise 10.10) design an n processor, $O((\log \log n)^2)$ time parallel algorithm for selection (see Exercise 10.11). (Hint: Design an n processor, $O(\log \log n)$ time algorithm for approximate selection; it should find two items straddling the kth smallest item, e_k, both within $n/4$ items of e_k. Exercise 10.11 may be suggestive. Generalize this to a $p \geq n$ processor, $O(\log \log n)$ time algorithm for finding an item within $n^2/4p$ items of the sought item. Apply appropriate approximate selection algorithms repeatedly.)

10.13 Design a CREW PRAM algorithm for selection that runs in $O(\log n \log^* n)$ time on n processors (see Exercise 10.11). Modify your algorithm so that it performs only $O(n)$ operations while still achieving the $O(\log n \log^* n)$ running time. $\log^* n$ is the least i such that $\log^{(i)} n \leq 1$, where $\log^{(0)} n = n$ and, for $i \geq 0$, $\log^{(i+1)}(n) = \log(\log^{(i)} n)$. (Hint: Design an approximate selection algorithm that runs in $O(\log n)$ time; using $p \geq n$ processors, it should find two items straddling the kth smallest item, e_k, both within $n^2/4p$ positions of e_k. Exercise 10.12 may be helpful.)

10.14 Show how to implement the algorithm of Section 10.3 for merging two arrays each of n items using $(n+2)/\log n$ processors so that it still runs in $O(\log n)$ time.

10.15 Design a CRCW PRAM algorithm for merging two arrays of n and m items, respectively, where $n \leq m$. Your algorithm should run in time $O(\log \log n)$ on $(n+m)/\log \log n$ processors.

10.16 Generalize the merging algorithm of Section 10.3 so that it can merge two arrays of sizes n and m respectively, while having the following performance: It is to run in $O(\log n)$ time on $(n+m)/\max\{1, \lfloor \log n \rfloor\}$ processors, and to perform at most $2(n+m)$ key comparisons.

10.17 Show how to implement the algorithm of Section 10.3 for sorting an array of n items using $(n+2)/\log n$ processors so that it still runs in $O(\log^2 n)$ time.

10.18 Show that the sorting algorithm of Section 10.3 is stable. (See Exercise 10.4 for the definition of a stable sort.)

10.19 Write a program that implements the sorting algorithm of Section 10.3. (Comment. If a parallel machine is not available, first write a simulator that implements the **pardo** statement.)

10.20 The following exercise develops an $n/\log n$ processor, $\log^2 n$ time algorithm for sorting based on the bitonic merge algorithm of Exercise 10.8. In parts (a)-(e), below, assume that $n = 2^k$, for some integer $k > 0$. Consider the first step of a sort of a bitonic sequence: Placing the smaller and larger items, respectively, in the first and second halves of the array.

a) Show that the items placed in the first half comprise an initial portion of the array plus a final portion of the array.

b) Hence devise a 1 processor, $O(\log n)$ time EREW PRAM algorithm for carrying out this swap (you may assume the bitonic sequence is stored in left to right order in the nodes of a complete binary tree plus one extra node to the right of the binary tree).

c) Hence, using pipelining, provide an n processor, $O(\log n)$ time EREW PRAM algorithm for sorting a bitonic sequence.

d) Observe that the algorithm for (c) performs $O(n)$ operations (if need be, modify your algorithm). Conclude that this algorithm can be implemented to run in $O(\log n)$ time on $n/\log n$ processors. How many key comparisons does your algorithm perform? (A bound of $2n$ can be achieved.)

e) Hence obtain an EREW PRAM algorithm for sorting an n item set; your algorithm should run in $O(\log^2 n)$ time on $n/\log n$ processors.

f) Generalize the algorithm of (e) to work for arbitrary n; it should perform no more than $2n\lceil \log n \rceil$ key comparisons.

10.21 a) Give an example of an array to be sorted by the algorithm presented at the start of Section 10.4 for which the 3-cover property of $E_t^\pm(u)$ with respect to $E_{t+1}(v)$ does not hold. (This is the algorithm in which $E_{t+1}(v)$ comprises every second item in $M_t(v)$.)

b) The 3-cover property can be generalized to a c-cover property, as follows. Array $E_t^\pm(u)$ c-covers array $E_{t+1}(v)$ if for each two adjacent items, e_1 and e_2 in $E_t^\pm(u)$, there are at most c items in $E_{t+1}(v)$ in the range $[e_1, e_2)$. For this same algorithm, give a second example for which no c-cover property holds for any constant c.

c) For this same algorithm, for what values of c can a c-cover property be proved? Give a tight bound on c, up to constant multiplicative factors.

10.22 a) Modify the merging algorithm of Section 10.3 so as to merge two arrays of 2^k items using $O(2^k)$ processors in $O(k+1)$ time while performing just $\frac{3}{2} \cdot 2^k$ key comparisons. (Hint: The odd index items in L should compute their location in M as before. However, the odd index items in R should have their locations in M determined without any further key comparisons.)

b) In a similar way, reduce the number of key comparisons performed in the sorting algorithm of Section 10.4 to $\frac{17}{8} n \log n$. The algorithm should still run in $O(\log n)$ time on n processors. (See Remark 10.1.)

10.23 Provide an implementation of the sorting algorithm of Section 10.4 that uses linear space.

10.24 This problem is concerned with the sorting algorithm of Section 10.4.
   a) Write pseudo-code for the procedure of Stage 1.1.
   b) Write pseudo-code for the procedure of Stage 1.2.
   c) Write pseudo-code for the procedure of Stage 2.1.
   d) Write pseudo-code for the procedure of Stage 2.2.
   e) Write a program to implement the sorting algorithm. (Comment: If a parallel machine is not available, first write a simulator that implements the **pardo** statement.)

10.25
   a) Counting just the steps in which key comparisons are performed, how many parallel steps does the sorting algorithm of Section 10.3 perform?
   b) Counting just the steps in which key comparisons are performed, how many parallel steps does the sorting algorithm of Section 10.4 perform?
   c) For what values of n does the algorithm of Section 10.4 perform fewer parallel steps than the sorting algorithm of Section 10.3?

10.26 Explain how to make the sorting algorithm of Section 10.4 stable. (See Exercise 10.4 for the definition of a stable sort.)

10.27 Research problem. Can the sorting algorithm of Exercise 10.20 be modified to run in time $O(\log n)$ on n processors, possibly in the spirit of the sorting algorithm of Section 10.4? Before attempting this problem, the reader is advised to examine the articles of Goodrich and Kosaraju and of Bilardi and Nicolau.

Notes and References

The sorting circuits of Section 10.2 and Exercise 10.8 are due to Batcher [4]. Further information on sorting circuits can be found in Knuth's [11] text. An $O(\log n)$ depth sorting circuit is given by Ajtai, Komlos and Szemeredi [1]; a simplified version of this circuits is given by Paterson [12]. The sorting algorithm of Section 10.4 is due to Cole [7], though the presentation is based in part on the generalization of this algorithm by Atallah et al [3]. Cole also gives an EREW PRAM version of this sorting algorithm. A related sorting algorithm for pointer machines, having the same asymptotic complexity, is given by Goodrich and Kosaraju [10]. The first work on parallel computation in the comparison model is due to Valiant [14]; Exercise 10.10 is drawn from this work. Valiant considered the problems of merging, sorting, and finding the maximum. In this same model, the selection problem is studied by Cole and Yap [9]; Exercise 10.12 is drawn from this work. A tight bound of $\Theta(\log \log n)$ time on n processors for selection in the comparison model is due to Ajtai, Komlos, Steiger and Szemeredi [2]. Exercise 10.13, on selection in the PRAM model, is due to Cole [8]. Exercise 10.15, on merging in the CRCW PRAM model, is due to Borodin and Hopcroft [6]. Exercise 10.20, which gives another efficient $O(\log^2 n)$ time, $n/\log n$ processor sorting algorithm is due to Bilardi and Nicolau [5]. Exercise 10.22 is due to Cole [7]. Richards [13] gives a substantial bibliography on parallel sorting. Valiant [15] also considers the issue of parallel slackness: how many processes are needed to fully exploit a given parallel machine.

Bibliography

[1] Ajtai, M., Komlos, J., Szemeredi, E. An $O(n \log n)$ sorting network. *Combinatorica*, 3(1983), 1–19.

[2] Ajtai, M., Komlos, J., Steiger, W., Szemeredi, E. Deterministic selection in $O(\log \log n)$ parallel time. *Proceedings ACM Symposium on Theory of Computing*, 188–195, 1986.

[3] Atallah, M., Cole, R., Goodrich, M. Cascading divide and conquer: a technique for designing parallel algorithms. *SIAM J. Computing*, 3(1989), 499–532.

[4] Batcher, K. Sorting networks and their applications. *AFIPS Spring Joint Computing Conference*, 32(1968), 307–314.

[5] Bilardi, G., Nicolau, A. Adaptive bitonic sorting: An optimal parallel algorithm for shared memory machines. *SIAM J. Computing*, 2(1989), 216–228.

[6] Borodin, A., Hopcroft, J.E. Routing, merging and sorting on parallel models of computation. *J. Computer and Systems Sciences*, 30(1985), 130–145.

[7] Cole, R. Parallel Merge Sort. *SIAM J. Computing*, 4(1988), 770–785.

[8] Cole, R. An optimally efficient parallel selection algorithm. *Information Processing Letters*, 26(1987/88), 295–299.

[9] Cole, R., Yap, C.K. A parallel median algorithm. *Information Processing Letters*, 20(1985), 137–139.

[10] Goodrich, M., Kosaraju, S.R. Sorting on a parallel pointer machine with applications to set expression evaluation. *Proceedings IEEE Symposium on Foundations of Computer Science*, 1989, 190–195.

[11] Knuth, D. Searching and Sorting. *Addison-Wesley*, 1973.

[12] Paterson, M.S. Improved sorting networks with $O(\log n)$ depth. *Algorithmica*, 1(1990), 75–92.

[13] Richards, D. Parallel sorting – a bibliography. *ACM SIGACT News*, Summer 1986, 28–48.

[14] Valiant, L. Parallelism in comparison problems. *SIAM J. Computing*, 4(1975), 348–355.

[15] Valiant, L. General Purpose Parallel Architectures. *Harvard University Technical Report*, No. 07-89, 1989.

11

Deterministic Parallel Computational Geometry

Mikhail J. Atallah

Department of Computer Science
Purdue University
West Lafayette, IN 47907
mja@nadia.cs.purdue.edu

Michael T. Goodrich

Department of Computer Science
The Johns Hopkins University
Baltimore, MD 21218-2694
goodrich@blaze.cs.jhu.edu

11.1
Introduction

Computational Geometry is concerned with the design and analysis of algorithms for solving geometric problems. These problems generally deal with collections of simple geometric objects, such as lines, points, planes, circles, etc., about which we are asked to answer some basic questions. Many of the problems in computational geometry come from applications in pattern recognition, computer graphics, statistics, operations research, computer-aided design, robotics, etc. The problems that arise from these areas are likely to come from real-time applications or situations in which the input consists of a large number of geometric objects. This implies that these problems need to be solved as fast as possible. For many of these problems, however, we already are at the limits of what can be achieved through sequential computation. Thus, it is natural to study what kinds of speed-ups can be achieved through parallel computing. As an indication of the importance of this research direction, we note that four of the eleven problems used as benchmark problems to evaluate parallel architectures for the DARPA Architecture Workshop Benchmark Study of 1986 were computational geometry problems.

Unfortunately, many of the techniques used to find efficient sequential algorithms for computational geometry problems do not translate well into a parallel setting. While providing elegant paradigms for designing sequential algorithms, these techniques use methods that seem to be inherently sequential. Therefore, one needs to develop new paradigms for computational geometry, ones better suited for finding efficient parallel algorithms. This chapter gives a number of algorithms for solving computational geometry problems efficiently in parallel. The algorithms all have linear or "almost" linear speed-ups over the best known sequential algorithms for these problems.

The geometric problems that we address in this chapter all deal with planar objects. The only exception is in Section 11.5, where we consider a problem dealing with 3-dimensional points. In each section in this chapter we address an important computational geometry problem and show how one can solve this problem efficiently in parallel. In some cases the parallel efficiency comes primarily from new geometric insights, and in other cases it comes primarily from new algorithmic techniques, which is an interesting aspect of parallel computational geometry. The specific problems we address include computing convex hulls, constructing the intersection of half planes, computing the visible region from a point in the presence of opaque obstacles, finding the separation distance between two polygons, and computing three-dimensional maxima points. Each of these problems is considered fundamental in computational geometry; we have tried to motivate their significance where appropriate.

11.2
Convex Hull

This section deals with the problem of computing the convex hull of a set S of points in the plane. We begin with a few definitions. A *polygonal chain* is a sequence $(p_1, s_1, p_2, s_2, \ldots p_n, s_n, p_{n+1})$, where each s_i is a line segment with endpoints p_i and p_{i+1}. This notational convention implies that s_i and s_{i+1} share a common endpoint, p_{i+1}, since each s_i can alternately be denoted $\overline{p_i p_{i+1}}$. The s_i's are called the *edges* of the polygonal chain, and the p_i's are called its *vertices*. A polygonal chain is *simple* if for every segment s_i in it, the only other segments that intersect s_i are s_{i-1} and s_{i+1}, and their respective intersections with s_i are p_i and p_{i+1} (see Figure 11.1a). Note, to simplify our discussions we sometimes use the the term *n-gon* as a shorthand for "simple polygon with n edges." Unless otherwise stated, every polygonal chain we consider in this chapter is simple. A polygonal chain is the *boundary of a simple polygon* if $p_1 = p_{n+1}$, in which case its sequence of vertices and edges can be viewed as being *circular*, in the sense that one can write the

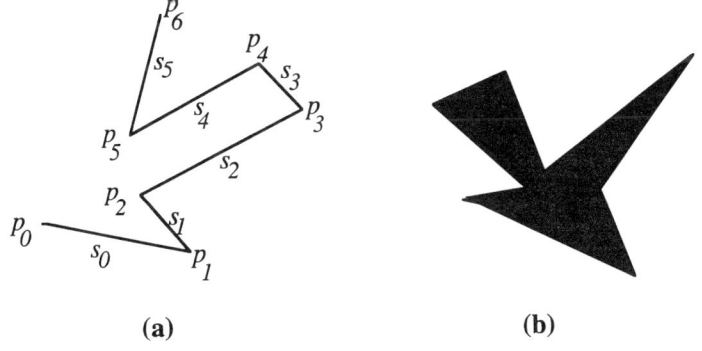

(a) **(b)**

FIGURE 11.1
Examples of (a) a simple polygonal chain and (b) a polygon.

sequence beginning at any vertex. In this case the polygonal chain partitions the plane into *two* regions—one finite region, the other infinite (this is due to a topological fact known as the Jordan Curve Theorem). By convention, the finite region is called the *interior* of the polygon, the polygonal chain is called the *boundary* of the polygon, and the infinite region is called the *exterior* of the polygon (see Figure 11.1b).

Let $P = (p_1, s_1, p_2, \ldots, p_n, s_n, p_{n+1})$ be a polygon. A polygon $P' = (p_{i_1}, s'_{i_1}, p_{i_2}, \ldots, p_{i_k}, s'_{i_k}, p_{i_{k+1}})$ is a *subpolygon* of P if $1 \leq i_1 < i_2 < \cdots < i_{k+1} \leq n+1$, and each s'_{i_j} is the edge from p_{i_j} to $p_{i_{j+1}}$. This definition is analogous to the notion of a subsequence of a sequence of numbers.

A planar region R is *convex* if, given any two points p and q in R, the line segment $\overline{pq}$ is also (entirely) in R. If R is a simple polygon, then this is equivalent to requiring that the intersection of R with any line is either empty or a line segment. The following lemma establishes an interesting property of subpolygons of convex polygons.

LEMMA 11.1
Any subpolygon of a convex polygon is itself a convex polygon.

PROOF
See Exercise 11.5. ∎

As we show in this section, this simple geometric observation has important consequences for the design of an efficient parallel algorithm for constructing the *convex hull* of a set S of n points in the plane. Given such a set of points, the convex hull of S is the smallest convex polygon that contains no points of S in its exterior, i.e., each point of S is either in its interior or on its boundary. The *convex hull problem* is to produce a polygonal representation of the convex hull of S, with the vertices listed in clockwise order (so the interior of the convex hull is always to the right of each convex hull edge). Intuitively, we can imagine the plane to be a large wooden board with a nail sticking up from the position of each point in S. If we were to stretch a rubber band so that it surrounds all the nails, and let the rubber band shrink to a resting state, then it would assume the contour of the boundary of the convex hull of S, with a nail at each vertex. Convex hulls have applications in a host of problem domains, including computer vision, computer graphics, and statistics. They are often useful any time one wishes to capture the "shape" of a set of points.

The problem of constructing the convex hull of a set S of n points in the plane is known to have an $O(n \log n)$ time sequential solution (see

11.2. Convex Hull

Exercise 11.1), together with an $\Omega(n \log n)$ lower bound in the comparison model (see Exercise 11.2). In this section we give an n-processor CREW PRAM algorithm that runs in $O(\log n)$ time.

Let $S = \{p_1, p_2, \ldots, p_n\}$. The convex hull of S consists of two portions: the *upper hull* and the *lower hull*. The upper (resp., lower) hull consists of the edges of the hull such that the interior of the hull polygon is below (resp., above) them. See Figure 11.2. We shall focus on the problem of computing the upper hull $UH(S)$. The problem of computing the lower hull $LH(S)$ is symmetrical and therefore omitted. We assume that the recursive procedure returns an array $UH(S)$ containing the points of the upper hull in left to right order. It also returns the size of the upper hull (i.e., $|UH(S)|$).

To simplify the exposition, we assume that n is a power of 2, that no two points in S have same x (respectively, y) coordinate, and that no three points in S lie on the same line. We also assume that the points have already been sorted by x coordinate, so that $x(p_i) < x(p_{i+1})$ for all $i \in \{1, 2, \ldots, n-1\}$. This can be accomplished in $O(\log n)$ time using n processors in the CREW PRAM model by the parallel mergesort procedure.

The main idea is to recursively solve the problem for the two point sets $S_1 = \{p_1, p_2, \ldots, p_{(n/2)-1}\}$ and $S_2 = \{p_{(n/2)}, p_{(n/2)+1}, \ldots, p_n\}$. As we will show, when these two recursive calls return $UH(S_1)$ and (respectively) $UH(S_2)$, we can combine their answers and obtain $UH(S)$ in constant time and using a linear number of processors on a CREW PRAM. Thus, the

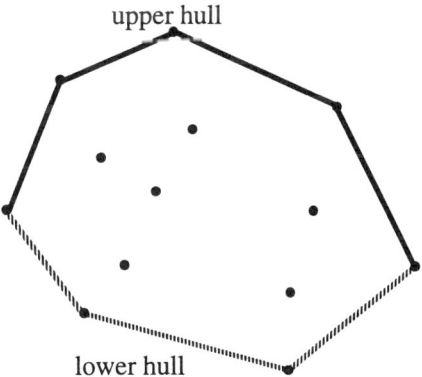

FIGURE 11.2
Illustrating the upper and lower hulls.

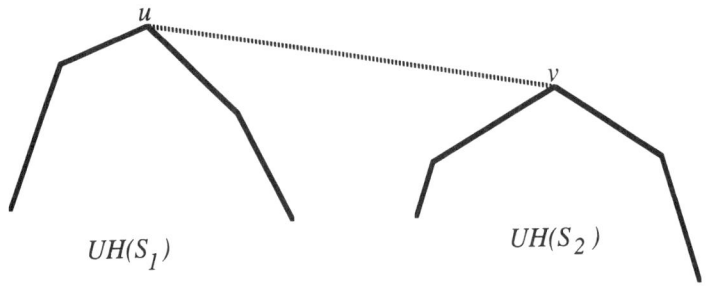

FIGURE 11.3
Illustrating the definition of u and v.

time and processor recurrences will be $T(n) = T(n/2) + c_1$ and $P(n) = \max\{n, 2P(n/2)\}$, respectively with boundary conditions $T(1) = c_2$ and $P(1) = 1$. Their solutions are $T(n) = O(\log n)$ and $P(n) = n$, respectively. We now show that the "combine" stage can indeed be done in constant time with a linear number of processors.

Consider the line T that is above and tangent to both $UH(S_1)$ and $UH(S_2)$. Let u (resp., v) be the point at which this tangent touches $UH(S_1)$ (resp., $UH(S_2)$). See Figure 11.3. The line segment $\overline{uv}$ is called the *upper common tangent* to $UH(S_1)$ and $UH(S_2)$. Observe that $UH(S)$ consists of the portion of $UH(S_1)$ to the left of u, followed by u, the edge $\overline{uv}$, v, and the portion of $UH(S_2)$ to the right of v. This observation implies that, if we somehow knew the points u and v, then we could obtain $UH(S)$ in constant time with $O(n)$ processors, as follows. Let α (resp., β) be the rank of u (resp., v) in the array $UH(S_1)$ (resp., $UH(S_2)$). We copy the first α entries of $UH(S_1)$ into the first α positions of $UH(S)$, add the edge $\overline{uv}$, and then fill the remainder of $UH(S)$ with the last $|UH(S_2)| - \beta + 1$ entries of $UH(S_2)$. Thus the problem of obtaining $UH(S)$ from $UH(S_1)$ and $US(S_2)$ is essentially that of identifying these two points u and v.

Let $UH(S_1) = (p_1, s_1, p_2, \ldots, p_l)$ and $UH(S_2) = (q_1, t_1, q_2, \ldots, q_m)$, where $l \leq n/2$ and $m \leq n/2$. The procedure we now give for locating u and v uses $l + m$ ($\leq n$) processors. Before we describe this method, however, note that if we had lm processors available, then it would be trivial to find the desired upper common tangent in constant time by the following "brute force" algorithm: assign a processor to each possible pair (p_i, q_j). This processor tests whether this pair is the one we seek (i.e., whether $p_i = u$ and $q_j = v$) by checking, in constant time, whether the segment $\overline{p_i q_j}$ is tangent to $UH(S_1)$ at p_i and to $UH(S_2)$ at q_j. This test is accomplished just by looking locally around p_i and q_j. Our goal, however, is to use only $l + m$ processors.

One may be tempted to give the following constant time, n processor "solution" (which doesn't work):

1. Consider the subpolygon P of $UH(S_1)$ obtained by choosing every $(k\sqrt{l})$-th vertex of $UH(S_1)$, $1 \leq k \leq \sqrt{l}$. That is, P is the $\sqrt{l}$-gon $P = (p_{\sqrt{l}}, s'_{\sqrt{l}}, p_{2\sqrt{l}}, \ldots, p_l)$.
2. Consider the subpolygon Q of $UH(S_2)$ obtained by choosing every $(k\sqrt{m})$-th vertex of $UH(S_2)$, $1 \leq k \leq \sqrt{m}$. That is, Q is the $\sqrt{m}$-gon $Q = (q_{\sqrt{m}}, t'_{\sqrt{m}}, q_{2\sqrt{m}}, \ldots, q_m)$.
3. Use the above-mentioned brute force approach to find the common tangent to P and Q in constant time. (It requires $\sqrt{lm} \leq l + m$ processors.) Say it is the line joining $p_{i\sqrt{l}} \in P$ to $q_{j\sqrt{m}} \in Q$.
4. The vertices of P divide $UH(S_1)$ (resp., $UH(S_2)$) into $\sqrt{l}$ portions, call them $P_1, P_2, \ldots, P_{\sqrt{l}}$. Similarly, the vertices of Q divide $UH(S_2)$ into $\sqrt{m}$ portions, call them $Q_1, Q_2, \ldots, Q_{\sqrt{m}}$. Use the brute force algorithm between the $2\sqrt{l}$ points in $P_i \cup P_{i+1}$ and the $2\sqrt{m}$ points in $Q_j \cup Q_{j+1}$ (i.e., between the two portions of P adjacent to $p_{i\sqrt{l}}$ and the two portions of Q adjacent to $q_{j\sqrt{m}}$).

The reason the above approach fails is that the "locality" property needed for Step 4 need not hold. Namely, we do not necessarily have $u \in P_i \cup P_{i+1}$ and $v \in Q_j \cup Q_{j+1}$. Indeed, the portion of $UH(S_1)$ containing u might be quite far from $p_{i\sqrt{l}}$, as might the portion of $UH(S_2)$ containing v be quite far from $q_{j\sqrt{m}}$. Figure 11.4 gives an example of how this might happen. The correct solution to the common tangent problem makes a more judicious use of the basic idea of the above (erroneous) steps 1–4. It also makes use of the next two propositions.

PROPOSITION 11.1
Let p be any point of $UH(S_1)$. Then the upper tangent to $UH(S_2)$ passing through p can be computed in time $O(k)$ by an $m^{1/k}$ processor CREW PRAM, where k is any integer such that $2 \leq k \leq \log m$. Similarly, if p is any point of $UH(S_2)$, then the upper tangent to $UH(S_1)$ passing through p can be computed in time $O(k)$ by an $l^{1/k}$ processor CREW PRAM, where k is any integer such that $2 \leq k \leq \log l$.

504   Chapter 11.   Deterministic Parallel Computational Geometry

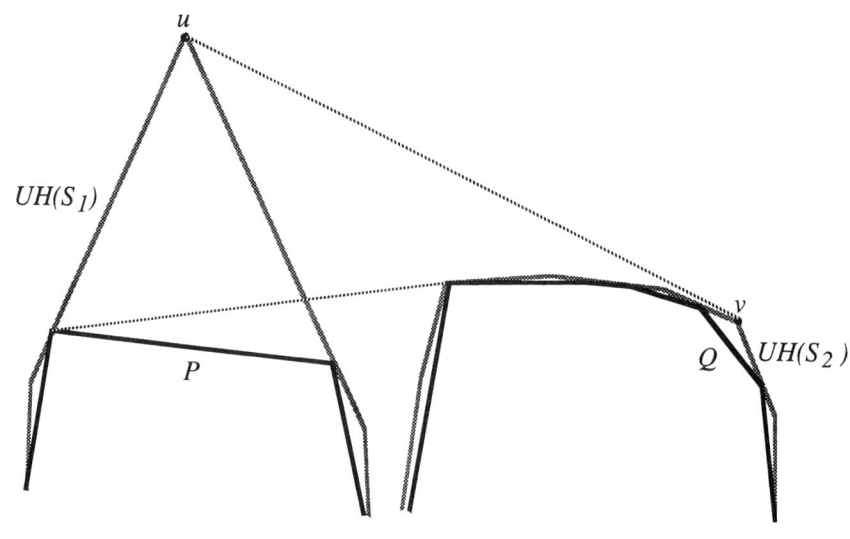

FIGURE 11.4
Counterexample.

PROOF
We give the proof for $p \in UH(S_1)$ (the proof for the $p \in UH(S_2)$ case is similar). Let $m' = m^{1-1/k}$. Let Q' be the subpolygon of $UH(S_2)$ consisting of every m'-th vertex of $UH(S_2)$, i.e., $Q' = (q_{m'}, q_{2m'}, \ldots, q_m)$. Since Q' has $m^{1/k}$ vertices and we have $m^{1/k}$ processors, it is easy to find in constant time the upper tangent to Q' passing through p, say this tangent touches Q' at $q_{im'}$ (the tangent is found in constant time by the same method as for searching for x in an array of n elements using n CREW PRAM processors). Let q_j be the vertex of $UH(S_2)$ at which the desired tangent touches $UH(S_2)$. We can easily test whether $q_j = q_{im'}$ in constant time and one processor, just by looking in the vicinity of $q_{im'}$ in $UH(S_2)$. If the test is positive, and $q_j = q_{im'}$, then we are done. So suppose $q_j \neq q_{im'}$. We then test whether q_j is to the left of $q_{im'}$ or to the right of $q_{im'}$, as follows. Let L be the line through p and $q_{im'}$. Also, let L_{left} the portion of L strictly to the left of $q_{im'}$, and let L_{right} be the portion of L strictly to the right of $q_{im'}$ (so neither L_{left} nor L_{right} contain $q_{im'}$). If L_{left} is above segment $q_{im'-1}q_{im'}$ and L_{right} is below segment $q_{im'}q_{im'+1}$ (see Figure 11.5a), then q_j is to the right of $q_{im'}$. Otherwise, if L_{left} is below segment $q_{im'-1}q_{im'}$

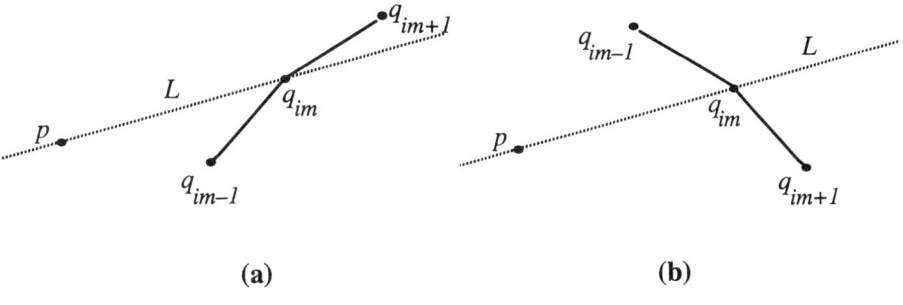

FIGURE 11.5
Illustrating the proof of Proposition 11.1.

and L_{right} is above segment $q_{im'}q_{im'+1}$, then q_j is to the right of $q_{im'}$ (see Figure 11.5b). Without loss of generality, assume the test reveals that q_j is to the left of $q_{im'}$, i.e., $j < im'$ (the case $im' < j$ is symmetrical). Then it is not hard to prove that we have $(i-1)m' \le j$ (see Exercise 11.8). Therefore, it suffices to find the upper tangent to the subpolygon $(q_{im'-m'}, s'_{im'-m'}, q_{im'-m'+1}, \ldots, q_{im'-1})$ passing through p. Thus, by doing a constant amount of work, we have reduced the polygon size by a factor of $m^{1/k}$. Doing this at most k times finds the desired point of tangency. ∎

PROPOSITION 11.2
Given a vertex, p, on $UH(S_1)$, one can determine if u is to the left of p, to the right of p, or at p in $O(k)$ time using $m^{1/k}$ processors, where k is any integer such that $2 \le k \le \log m$. Similarly, given a vertex, q, on $UH(S_2)$, one can determine if v is to the left of q, to the right of q, or at q in $O(k)$ time using $l^{1/k}$ processors, where k is any integer such that $2 \le k \le \log l$.

PROOF
We only consider the case $p \in UH(S_1)$, since the case $q \in UH(S_2)$ is similar. Use the previous proposition to find the tangent to $UH(S_2)$ passing through point p; let T be this tangent. If T is tangent to P then $u = p$. Otherwise, let γ be the vertex of P just to the left of p. Observe that u is to the left of p on $UH(S_1)$ if and only if γ is above line T. ∎

We can now give the algorithm **TANGENTS** for finding the upper common tangent to $UH(S_1)$ and $UH(S_2)$ (and hence u and v). Recall that both

$UH(S_1)$ and $UH(S_2)$ are monotone in the x direction, i.e., the x-coordinate of p_i is smaller than that of p_{i+1} and the x-coordinate of q_i is smaller than that of q_{i+1}.

1. Set $P' := UH(S_1)$, $Q' := UH(S_2)$.

   Comment. The algorithm will iteratively decrease the size of P' and/or Q', maintaining the property that the upper common tangent between $UH(S_1)$ and $UH(S_2)$ is the same as the one between P' and Q'.

2. Repeat the following steps 3–7 until either P' is a single point or Q' is a single point. Without loss of generality, assume that it is P' that ends up becoming a single point (call it p_u); now use Proposition 11.1 to find, in constant time, the tangent to Q' passing through p_u, and output the tangent thus found (this is the desired upper common tangent between $UH(S_1)$ and $UH(S_2)$).

3. Let $P'' = (a_1, s'_1, a_2, \ldots, a_{\sqrt{l}})$ be the subpolygon obtained by considering every $\sqrt{l}$-th vertex of P', i.e., the $\sqrt{l}$ vertices of P'' divide P' into $\sqrt{l}$ equal portions. Call these portions $A_1, A_2, \ldots, A_{\sqrt{l}}$, so that a_i is adjacent in P' to portions A_i and A_{i+1}. By convention, a_i belongs to A_i but not to A_{i+1}. Let $Q'' = (b_1, t'_1, b_2, \ldots, b_{\sqrt{m}})$ be analogously defined for Q', and let the resulting portions of Q' be $B_1, B_2, \ldots, B_{\sqrt{m}}$. Use the already mentioned brute force method for finding the common tangent between P'' and Q'' (this is possible and takes constant time because we have $l + m \geq \sqrt{lm}$ processors). Say the tangent thus found joins $a_i \in P''$ to $b_j \in Q''$. (See Figure 11.6.)

4. Test whether the common tangent to P' and Q' touches P' in A_i. (This is done in constant time by using Proposition 11.2 twice, once at vertex p_{i-1} and once at vertex p_i.) If the answer is "yes" then do $P' := A_i$, otherwise P' remains unchanged.

   Implementation Note. The assignment $P' := A_i$ is done in constant time simply by remembering the new first and last vertex of P'.

5. Test whether the common tangent to P' and Q' touches P' in A_{i+1}. If it does then do $P' := A_{i+1}$, otherwise P' remains unchanged.

6. Test whether the common tangent to P' and Q' touches Q' in B_j. If it does then do $Q' := B_j$, otherwise Q' remains unchanged.

7. Test whether the common tangent to P' and Q' touches Q' in B_{j+1}. If it does then do $Q' := B_{j+1}$, otherwise Q' remains unchanged.

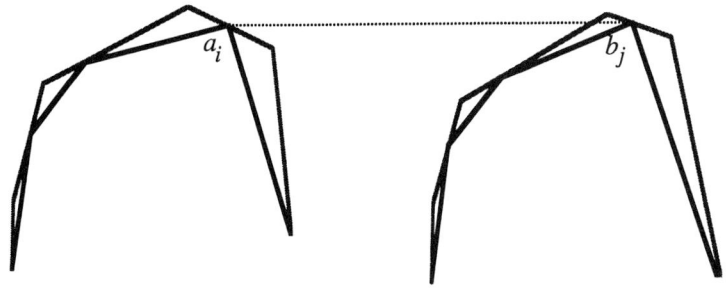

FIGURE 11.6
Illustrating algorithm **TANGENTS**.

Since every usage of Proposition 11.2 takes constant time, the time complexity of the algorithm is equal to the number of times that steps 3–7 get executed. We now bound the number of times steps 3–7 are executed.

LEMMA 11.2
*Let a_i, b_j, P', Q', P'', and Q'' be as in Step 3 of the algorithm **TANGENTS**. Also, let $p_u q_v$ be the common tangent to P' and Q' ($p_u \in P'$, $q_v \in Q'$). Then at least one of the following statements (1), (2), (3), or (4) is true:*

1. $p_u \in A_i$;
2. $p_u \subset A_{i+1}$;
3. $q_v \in B_j$;
4. $q_v \in B_{j+1}$.

PROOF
If $p_u = a_i$ or $q_v = b_j$ then the lemma holds, so suppose that $p_u \neq a_i$ and $q_v \neq b_j$. By its definition, the line through p_u and q_v is above both a_i and b_j. Therefore at least one of p_u or q_v is above the line through a_i and b_j. If p_u is above the line through a_i and b_j, then we prove that (1) or (2) must hold by the following case analysis:

Case 1: In P', p_u is to the left of a_i. Then we claim that $p_u \in A_i$ (and hence (1) holds). Suppose to the contrary that $p_u \notin A_w$ where $w < i$. By the definition of a_i and b_j, the vertex $a_w \in P''$ must lie on or below the line $a_i b_j$. The three vertices p_u, a_w, a_i occur in that order on P' (see Figure 11.7). Consider the positions of these three vertices relative to

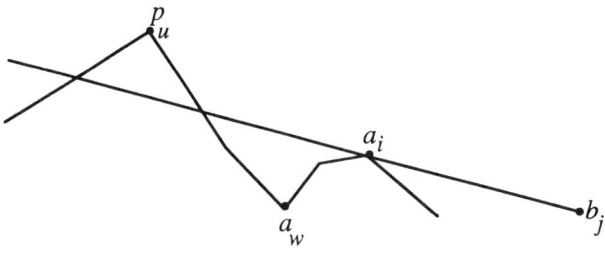

FIGURE 11.7
Illustrating Case 1.

the line $a_i b_j$: The first vertex is (by hypothesis) above that line, the second is (as we have just argued) on or below it, and the third is (by definition) on it. This contradicts the convexity of P'. Thus, (1) holds.

Case 2: In P', p_u is to the right of a_i. An argument similar to that for Case 1 shows that $p_u \in A_{i+1}$; hence, (2) holds.

If q_v is above line $a_i b_j$, then an argument similar to that above shows that one of (3) or (4) must hold. ∎

COROLLARY 11.1
Steps 3–7 of algorithm **TANGENTS** *are executed at most three times.*

PROOF
Lemma 11.2 implies that, every time we execute steps 3–7, at least one of the statements $P' := A_i$, $P' := A_{i+1}$, $Q' := B_j$, $Q' := B_{j+1}$ is executed. This implies that either P' decreases in size by a factor of $\sqrt{l}$, or Q' decreases in size by a factor of $\sqrt{m}$. This proves the corollary. ∎

We have thus established the following:

THEOREM 11.1
Algorithm **TANGENTS** *correctly computes the upper common tangent between* $UH(S_1)$ *and* $UH(S_2)$, *in constant time with* $l + m$ *processors.*

COROLLARY 11.2
The convex hull of a planar set of n points can be computed in $O(\log n)$ time using n processors on a CREW PRAM.

The next section uses the convex hull algorithm to solve the problem of computing the intersection of n half-planes. That a convex hull algorithm can be used in this way may seem surprising, since this problem is defined on a set of half-planes and any convex hull algorithm assumes it is given a set of points.

11.3
Intersections of Half-Planes

Let $L = \{l_1, l_2, \ldots, l_n\}$ be a set of n planar lines. We let $H(l_i)$ denote the half-plane that is to the left of l_i, assuming a given orientation of l_i. One could, for example, specify each line l_i by two points p_i and q_i that determine it, with the convention that the orientation of l_i is from p_i to q_i. In this section we address the problem of computing the intersection of the n half-planes, i.e., in computing the region $F = \cap_{i=1}^{n} H(l_i)$. For the reader familiar with the terminology of operations research, this problem is equivalent to computing the feasible region defined by a two-variable linear program. It has a number of applications, one of which we will explore (in Section 11.3.1).

Without loss of generality, we assume that no l_i is parallel to the y-axis. If this is not the case, then one can easily change the coordinate axes so that it is true. In this section we show how to compute F in $O(\log n)$ time and using n processors. This region is convex (see Exercise 11.4) and is described by a polygonal chain that may be closed or may begin and end at semi-infinite edges.

Rather than give a new algorithm for solving this problem, we shall show how it can actually be solved by using the convex hull algorithm of the previous section. The main idea is based on a geometric transform f that maps a point into a line and a line into a point: a point $p = (a, b)$ is mapped into the line $f(p)$ whose equation is $y = ax + b$, and the line l whose equation is $y = ax + b$ is mapped into the point $f(l) = (-a, b)$. It is easy to verify (Exercise 11.14) that the transform f maintains the relative positions of lines and points: if one of $\{\alpha, \beta\}$ is a line and one is a point, then α is below β if and only if $f(\alpha)$ is below $f(\beta)$. If A is a set of lines or points, then we use $f(A)$ to denote the set $\{f(a) : a \in A\}$.

To simplify the description of our method, we partition L into two sets, L^+ and L^-, where L^+ is the subset of L containing the lines l_i such that $H(l_i)$ is the half-plane above l_i, and L^- is the subset of L containing the lines l_i such that $H(l_i)$ is the half-plane below l_i. Let $F^+ = \cap_{l \in L^+} H(l)$ and $F^- = \cap_{l \in L^-} H(l)$. Then, clearly $F = F^+ \cap F^-$. Therefore, since computing the intersection of two convex polygonal chains can easily be done in $O(\log n)$ time

510   Chapter 11.   Deterministic Parallel Computational Geometry

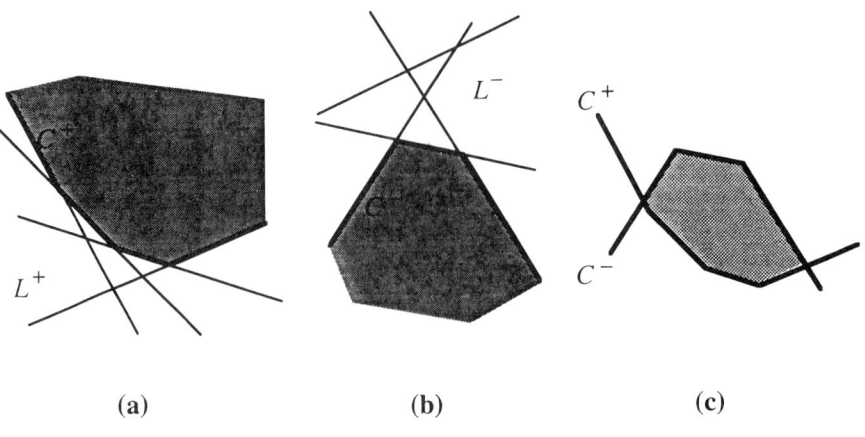

(a)   (b)   (c)

FIGURE 11.8
Examples of (a) C^+, (b) C^-, and (c) their intersection.

with $O(n/\log n)$ processors (see Exercise 11.4), we can restrict our attention to the computation of F^+ and F^- separately. (See Figure 11.8.)

Before we describe our algorithm, however, we must generalize our definition of polygonal chains to allow for chains that begin and end with a ray (or "semi-infinite" line), rather than requiring that chains begin and end with a point. Specifically, we allow a polygonal chain P to be given as an alternating sequence of points and edges, where each point is given by its coordinates and each edge is specified by the line that contains it. This is essentially the same as the definition of the previous section, except we now allow a polygonal chain to have an edge at each end, as well as allowing it to have a point at each end, as before. An "edge" that is at the end of a polygonal chain has only one endpoint, hence, is a ray. The beauty of using this convention is that, for any polygonal chain P, $f(P)$ is also a polygonal chain (see Exercise 11.15).

Since the computations of F^+ and F^- are symmetric, let us concentrate on the construction of F^+. Let $C^+ = (s_1, p_1, s_2, p_2 \ldots, s_{m-1}, p_{m-1}, s_m)$ denote the polygonal chain defining F^+, which, of course, starts and edges with a ray, one emanating from p_1 along the line specified by s_1 and one emanating from p_{m-1} along the line specified by s_m. Our algorithm for constructing C^+ is simply the following:

- Construct $f(C^+)$ and apply f^{-1} to get C^+.

But what is $f(C^+)$?

11.3. Intersections of Half-Planes

LEMMA 11.3
$f(C^+)$ is the upper hull of the points in $f(L^+)$, i.e., $f(C^+) = UH(f(L^+))$.

PROOF
As we have already observed, since C^+ is a polygonal chain having a ray at each end, $f(C^+)$ is a polygonal chain having a point at each end. Let I^+ be the set of intersections between lines in L^+, and observe that $f(I^+)$ is the set of lines determined by all pairs of points of $f(L^+)$. Note that the vertices of C^+ are exactly those points in I^+ that are on or above all the lines in L^+, listed from left to right. Thus, the edges of $f(C^+)$ are determined by the *lines* of $f(I^+)$ that are above all the *points* of $f(L^+)$, listed by increasing slopes. In other words, $f(C^+)$ is the list of lines that contain edges of the upper hull of $f(L^+)$, listed by increasing slopes. Therefore, $f(C^+)$ is, in fact, the upper hull of $f(L^+)$, listed right to left. ∎

Since $f(L^+)$ and its upper hull can be computed in $O(\log n)$ time by n CREW PRAM processors, it follows that $f(C^+)$, and hence C^+, can be computed within the same bounds. A symmetrical argument shows that C^- can be computed within the desired complexity bounds. Combining this with the observation that computing F is easy given F^+ and F^-, gives us the following theorem:

THEOREM 11.2
Given a set $L = \{l_1, l_2, \ldots, l_n\}$ of oriented lines in the plane, one can compute $F = \bigcap_{i=1}^{n} H(l_i)$ in $O(\log n)$ time using n processors in the CREW PRAM model.

We consider an application of this theorem in the next subsection.

11.3.1 The Kernel of a Simple Polygon

Let $P = (p_1, s_1, p_2, s_2, \ldots, s_{n-1}, p_n)$ be a simple polygon P (so $p_1 = p_n$). We consider each edge s_i of P to be *opaque*, i.e., an observer in the plane cannot see through it. We define the *kernel of P*, denoted $K(P)$, to be the set of points from each of which all of P's boundary is visible; that is, point p is in $K(P)$ if for every point q on the boundary of P, the segment joining p to q does not intersect any edge e of P except possibly at e's endpoints. Note that this implies that $K(P)$ is a subset of the interior of P. Also note that the kernel may be empty. If the kernel of a polygon P is not empty, then P is said to be *star-shaped*. The kernel of a convex polygon is, of course, the polygon itself. Figure 11.9 shows a star-shaped polygon and its kernel (shaded in the

FIGURE 11.9
A polygon and its kernel.

figure). Intuitively, if we imagine the polygon P to be the floor plan of an art gallery, then the kernel of P can be viewed as the region where a guard could stand and be able to see every painting on the walls of P.

It turns out that, as a consequence of the algorithm for computing the intersection of n half-planes, the kernel of a polygon P can be computed in $O(\log n)$ time with n processors. We consider each edge s_i of P to be an oriented line segment such that the interior of P is on its right. We let $H(s_i)$ denote the half-plane to the right of the oriented line containing the edge s_i. It is not hard to show (see Exercise 11.6) that the kernel of P is the intersection of the n half-planes determined by the edges in P, i.e., it is $\cap_{i=1}^{n} H(s_i)$. This implies that the kernel can be computed in $O(\log n)$ time by n processors on a CREW PRAM.

The next section deals with the problem of computing the distance between two convex polygons. There, too, success will hinge on a careful exploitation of convexity.

11.4
The Distance Between Two Convex Polygons

In the problem we address in this section the input consists of two convex polygons $P = (p_1, s_1, p_2, s_2, \ldots, s_{n-1}, p_n)$ and $Q = (q_1, t_1, q_2, t_2, \ldots, t_{n-1}, q_n)$, where the p_i's (resp., q_i's) are given in clockwise cyclic order, that is, the interior of P is to the right of each oriented segment $\overline{p_i p_{i+1}}$, and the similar property holds for Q. We are interested in computing the shortest distance between P and Q. This distance is formally defined as follows:

$$d(P,Q) = \min_{u \in P, w \in Q} d(u,w)$$

where $d(u,w)$ denotes the Euclidean distance between points u and w, and the notation "$u \in P$" means that u is a point of P (either on the boundary of P or interior to P). This problem often arises in machine learning problems where one wishes to determine if two sets of points A and B can be separated by a line, and if they can be so separated, then one wishes to determine how "close" A is to B. Constructing the convex hulls of A and B immediately reduces this learning problem to that of computing the distance between two convex polygons, or determining if they intersect.

The algorithm we present in this section actually returns a pair of points u, w such that $d(P,Q) = d(u,w)$. It runs in $O(c^2)$ time with $O(n^{1/c})$ processors on a CREW PRAM. We give the algorithm assuming that P and Q are disjoint, so that there is a line separating them. Exercise 11.17 deals with the case of possibly zero distance, i.e., when P and Q intersect, in which case there is no separating line. Of course, once we have these points, u and w, any perpendicular to the line segment joining u and w is a line separating P from Q. Therefore the algorithm given below for the closest distance can also be used to give us a separating line.

To simplify the exposition, we assume that no three successive vertices of either polygon are collinear, and that no edge of P is parallel to an edge of Q. The assumption that no edge of P is parallel to an edge of Q implies that the desired points u and w are unique. The algorithm can easily be modified for the case when edges of P might be parallel to edges of Q, e.g., by adopting a suitable convention for returning a unique u, w pair in case $d(P,Q)$ is the distance between two parallel edges of P and Q respectively. In this latter case there is an infinite number of choices for u and w, and this is the only case where u and w are not unique.

Let p be a point, which is not necessarily a vertex, on the boundary of P, and define q similarly for Q. Let T_p (resp., T_q) be the line perpendicular to the line segment $\overline{pq}$ at point p (resp. q). It is easy to show (see Exercise 11.16) that $d(P,Q) = d(p,q)$ if and only if (i) T_p and T_q are tangent to P and Q respectively, and (ii) P and Q are on opposite sides of the region between T_p and T_q. Note, conditions (i) and (ii) are "local" and can thus be tested by one processor in constant time for a given pair of points p and q.

The above simple observation implies that, with $O(n^2)$ processors and in constant time, it is possible to compute the closest distance between P and Q and a pair of points achieving it. The brute force procedure for doing this assigns a processor to each pair (a, b), where a is a vertex or edge of P, and b is a vertex or edge of Q (but not when both a and b are edges—that case need not be considered, as it is subsumed by one of the cases in which one of $\{a, b\}$ is a vertex). Such a processor then computes, in constant time, the points $a' \in a$ and $b' \in b$ such that $d(a', b') = d(a, b)$. It then tests, also in constant time, whether (i) $T_{a'}$ and $T_{b'}$ are tangent to (respectively) P and Q, and (ii) P and Q are on opposite sides of the region between $T_{a'}$ and $T_{b'}$ (i.e., this region separates them). If so, then the processor decides that $u = a'$ and $w = b'$, and furthermore no other processor reaches such a decision about the identity of u and w (by the uniqueness of u and w). Thus, there are no "write conflicts" when that processor writes the names of u and w in the specified registers for them.

The algorithm we shall give still takes constant time, but it uses far fewer processors than n^2: in fact it uses $O(n^{1/k})$ processors for any constant k of our choice. It relies on the above simple observations as well as on the next two propositions.

PROPOSITION 11.3
Let p be a point external to Q. Then the point $q \in Q$ such that $d(p, q) = d(p, Q)$ can be computed in time $O(k)$ by an $n^{1/k}$ processor CREW PRAM, where k is any integer of our choice.

PROOF
Let $l = n^{1-1/k}$. Let Q' consist of every l-th vertex of Q, i.e., Q' is the subpolygon $(q_l, s'_l, q_{2l}, s'_{2l}, \ldots, q_n)$. Since Q' has $n^{1/k}$ vertices and we have $n^{1/k}$ processors, it is trivial to find in constant time the point $q' \in Q'$ such that $d(p, q') = d(p, Q')$. (Note that q' need not be a vertex of Q'.) If the perpendicular to the line segment $\overline{pq'}$ at point q' is tangent to Q, then we can stop and declare point q' as the desired point q. Otherwise, let α (resp. β) be the vertex of Q' that immediately precedes (resp. follows) point q' when the boundary of Q' is traced in a clockwise manner (see Figure 11.10). Note that in Q, there are $2l + 1$ vertices between α and β (inclusive) if q' is a vertex, and there are $l + 1$ vertices between α and β otherwise. We leave it to the reader to prove that, in Q, the desired point q occurs between α and β (inclusive). Let γ be the median of the (at most $2l+1$) vertices between α and β (inclusive): Test whether the desired point q is at γ, between α and γ, or between γ

11.4. The Distance Between Two Convex Polygons

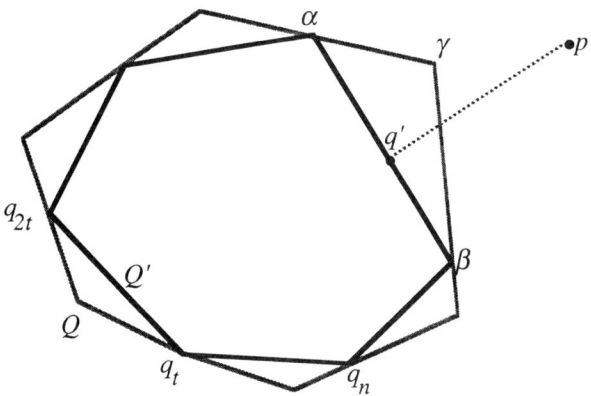

FIGURE 11.10
Illustrating the proof of Proposition 11.3.

and β. This test trivially takes constant time with one processor. If $q = \gamma$ then we're done, so suppose (without loss of generality) that the test reveals that q is between α and γ. If this occurs, then we can focus our search for q to the section of Q between α and γ (excluding γ), which contains at most l vertices. Therefore, by doing a constant amount of work, we have reduced the polygon size by a factor of at least $n^{1/k}$. Doing this at most k times finds the desired point q. ∎

PROPOSITION 11.4
Let p_i and p_j be any two vertices of P, $i < j$, and let p_u be the vertex of P such that $d(p_u, Q) = d(P, Q)$. Then for any integer k of our choice, an $n^{1/k}$ processor CREW PRAM can, in $O(k)$ time, locate where p_u occurs with respect to p_i and p_j in the sequence $p_1, p_2, \ldots, p_n$ (i.e., it can determine whether $u = i$, $u = j$, $i < u < j$, or none of these).

PROOF
For any two indices $1 \leq a, b \leq n$, let $\sigma_{a,b}$ denote the sequence

$$\sigma_{a,b} = d(p_a, Q), d(p_{a+1}, Q), \ldots, d(p_b, Q)$$

(assuming index $n+1$ equals 1). For example,

$$\sigma_{9,2} = d(p_9, Q), \ldots, d(p_n, Q), d(p_1, Q), \ldots, d(p_2, Q).$$

Observe that, because of convexity, there exist two indices a and b, $1 \leq a \leq b \leq n$, such that $\sigma_{a,b}$ and $\sigma_{b,a}$ are both sorted, one in increasing order and the other in decreasing order. This implies that we can locate where p_u occurs with respect to any pair p_i, p_j in the sequence $p_1, p_2, \ldots, p_n$ by performing a constant number of distance computations of the type $d(p_l, Q)$. By Proposition 11.3, each such distance computation can be done within the desired time and processor bounds. ∎

The following algorithm shows that, for any integer c of our choice, an $n^{1/c}$ processor CREW PRAM can find, in $O(c^2)$ time, the points $u \in P$ and $w \in Q$ such that $d(u, w) = d(P, Q)$.

ALGORITHM 11.1
D for computing distance:
Input: Two disjoint convex polygons $P = (p_1, s_1, p_2, s_2, \ldots, p_n)$ and $Q = (q_1, t_1, q_2, t_2, \ldots, q_n)$. The p_i's (resp., q_j's) are given in clockwise cyclic order.
Output: Points u, w such that $d(u, w) = d(P, Q)$.

1. Set $\hat{P} := P$, $\hat{Q} := Q$, $s := n^{1/2c}$.
2. Repeat the following steps 3–5 until either $\hat{P}$ is a single point or $\hat{Q}$ is a single point. Without loss of generality, assume it is $\hat{P}$ that ends up becoming a single point (call it x): Use Proposition 11.3 to find, in $O(c)$ time, the point $y \in Q$ such that $d(x, y) = d(x, Q)$. Output the points x and y (these are the desired points u, w).
3. Let $P' = (a_1, s'_1, a_2, s'_2, \ldots, a_s)$ be the subpolygon obtained by considering every $(|\hat{P}|/l)$-th vertex of $\hat{P}$, i.e., the l vertices of P' divide $\hat{P}$ into l equal portions. Call these portions $A_1, A_2, \ldots, A_l$, so that a_i is adjacent in $\hat{P}$ to portions A_i and A_{i+1}. By definition, a_i belongs to A_i but not to A_{i+1}. Let $Q' = (b_1, t'_1, b_2, t'_2, \ldots, b_l)$ be analogously defined for $\hat{Q}$, and let the resulting portions of $\hat{Q}$ be $B_1, B_2, \ldots, B_l$. Use the already mentioned brute force method for finding the points $a \in P'$ and $b \in Q'$ such that $d(a, b) = d(P', Q')$. Since we have l^2 processors, this takes constant time.
Let α_P be the vertex of P' that immediately precedes a on the boundary of P', and let β_P be the vertex of P' that immediately follows a on the boundary of P'. (Figure 11.11 illustrates the case when a is not a vertex of P'.) If a is a vertex of P' then α_P and β_P are (respectively) its predecessor and successor vertices on P', and hence there are then $(2|\hat{P}|/l) + 1$ vertices of $\hat{P}$ between α_P

11.4. The Distance Between Two Convex Polygons

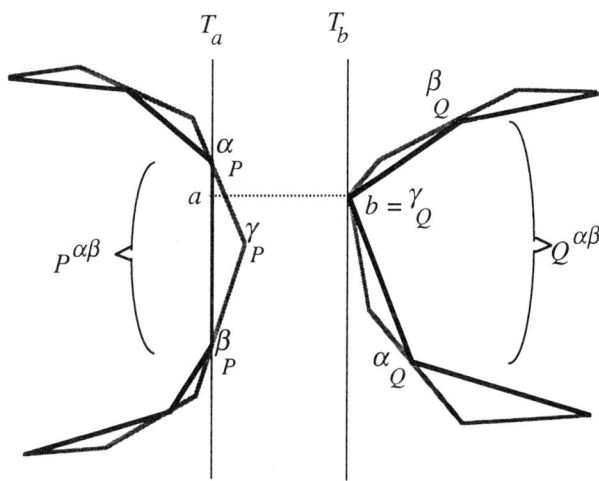

FIGURE 11.11
Illustrating the algorithm for the distance.

and β_P (inclusive). If a is not a vertex of P' then α_P and β_P are consecutive vertices of P', point a is on the segment of P' that joins α_P to β_P, and there are $(|\hat{P}|/l) + 1$ vertices of $\hat{P}$ between α_P and β_P (inclusive). Let γ_P be the median of the (at most $(2|\hat{P}|/l) + 1$) vertices of $\hat{P}$ that are between α_P and β_P (inclusive) (Note that, if a is a vertex of P', then $\gamma_P = a$.) We use $P^{\alpha\beta}$ to denote the portion of P that is between α_P and β_P (excluding α_P and β_P). $P^{\alpha\gamma}$ and $P^{\gamma\beta}$ are analogously defined.

Let α_Q, β_Q, γ_Q, $Q^{\alpha\beta}$, $Q^{\alpha\gamma}$ and $Q^{\gamma\beta}$ be similarly defined for b, Q' and $\hat{Q}$. (Figure 11.11 illustrates the case when b is a vertex of Q'.)

4. Use Proposition 11.4 to detect whether $u = \alpha_P$, $u = \beta_P$, $u = \gamma_P$, $u \in P^{\alpha\gamma}$, $u \in P^{\gamma\beta}$, or none of these. If u equals α_P (resp. γ_P, β_P) then set $\hat{P}$ equal to α_P (resp., γ_P, β_P) and go to Step 5. Otherwise, if $u \in P^{\alpha\gamma}$ then do $\hat{P} := P^{\alpha\gamma}$ and go to Step 5. Otherwise, if $u \in P^{\gamma\beta}$ then do $\hat{P} := P^{\gamma\beta}$ and go to Step 5. Otherwise leave $\hat{P}$ unchanged. (An assignment like $\hat{P} := P^{\alpha\gamma}$ is done in constant time simply by remembering the new first and last vertex of $\hat{P}$.)

5. Use Proposition 11.4 to detect whether $w = \alpha_Q$, $w = \beta_Q$, $w = \gamma_Q$, $w \in Q^{\alpha\gamma}$, $w \in Q^{\gamma\beta}$, or none of these. If w equals α_Q (resp. γ_Q, β_Q) then set $\hat{Q}$ equal to α_Q (resp. γ_Q, β_Q) and go to Step 3. Otherwise,

if $w \in Q^{\alpha\gamma}$ then do $\hat{Q} := Q^{\alpha\gamma}$ and go to Step 3. Otherwise, if $w \in Q^{\gamma\beta}$ then do $\hat{Q} := Q^{\gamma\beta}$ and go to Step 3. Otherwise leave $\hat{Q}$ unchanged.

Since every usage of Proposition 11.4 takes $O(c)$ time, the time complexity of the algorithm is equal to c multiplied by the number of times that steps 2–4 get executed. We now bound the number of times steps 2–4 are executed.

LEMMA 11.4
Let a, b, $P^{\alpha\beta}$, $Q^{\alpha\beta}$, u, and w be as in algorithm D. Assume that $u \notin \{\alpha_P, \beta_P\}$ and $w \notin \{\alpha_Q, \beta_Q\}$. Then at least one of the following statements (1) or (2) is true:

1. $u \in P^{\alpha\beta}$,
2. $w \in Q^{\alpha\beta}$.

PROOF
Let T_a be the line perpendicular at a to the segment ab, and let T_b be the line perpendicular at b to the segment ab (see Figure 11.11). By definition, T_a is tangent to P', and T_b is tangent to Q'. Without loss of generality, T_a and T_b are vertical, P' is to the left of T_a, and Q' is to the right of T_b. If $u = a$ or $w = b$ then the lemma holds, so suppose that $u \neq a$ and $w \neq b$. By the definition of u and w, we must have $d(u,w) \leq d(a,b)$. This implies that u is to the right of T_a or w is to the left of T_b (possibly both). Hence, it suffices to show that if u is to the right of T_a, then (1) holds, and if w is to the left of T_b, then (2) holds. We prove this by contradiction. Suppose that u is to the right of T_a and (1) does not hold, i.e., $u \notin P^{\alpha\beta}$. Without loss of generality, assume that u is below γ_P. Now, by walking from vertex γ_P clockwise along the boundary of $\hat{P}$, one encounters vertex β_P before reaching u. Since γ_P is on or to the right of T_a, β_P on or to the left of T_a, and u to the right of T_a, this contradicts the convexity of P. A similar argument shows that if w is to the left of T_b, then (2) holds. ∎

COROLLARY 11.3
Steps 3–5 of algorithm D are executed a total of at most $4c - 1$ times.

PROOF
Lemma 11.2 implies that, every time we execute Steps 3–5, at least one of $\hat{P}$ or $\hat{Q}$ decreases in size by a factor of at least $s = n^{1/2c}$, thus proving the corollary. ∎

11.4. The Distance Between Two Convex Polygons

We have thus established the following:

THEOREM 11.3

Algorithm D correctly finds points $u \in P$ and $w \in Q$ such that $d(u, w) = d(P, Q)$. It uses $n^{1/c}$ processors and it runs in time $O(c^2)$, where c is any integer such that $2 \leq c \leq \log n$.

Before proceeding to our next problem (3-dimensional maxima), we note that there is another version of the definition of the distance: the *boundary-to-boundary* distance, in which "$u \in P$" means that u is a point of the *boundary* of P. Interestingly, if we use this notion of distance, then determining the distance between two convex polygons has an $\Omega(\log n)$ *lower bound*, as the following shows.

THEOREM 11.4

The problem of computing the shortest boundary-to-boundary distance between two convex polygons P and Q has an $\Omega(\log n)$ lower bound on a CREW PRAM having a polynomial number of processors.

PROOF

We use the fact that there is a known $\Omega(\log n)$ lower bound for the problem of computing the logical OR of n bits $x_1, x_2, \ldots, x_n$. Therefore it suffices to show that a CREW PRAM can compute the logical OR of n bits in time equal to a constant plus the time for computing the distance between the boundaries of P and Q. We do this by converting the problem of computing the OR of $x_1, x_2, \ldots, x_n$ to that of computing the distance between the boundaries of the following two convex n-gons P and Q. Let P be any regular convex n-gon, so that all its edges have the same length. Let Q_0 consist of the regular convex n-gon whose vertices are the midpoints of the n boundary edges of P. The convex n-gon Q is obtained from Q_0 by "deforming" Q_0 very slightly so as to move some of its vertices slightly into the interior of P, and others just outside of P. This deformation is governed by the Boolean values $x_1, x_2, \ldots, x_n$: we associate the i-th vertex, q_i, of Q_0 with x_i, and move q_i slightly inside of P if $x_i = 0$ and move q_i slightly outside of P if $x_i = 1$. The polygon Q so obtained has the property that its boundary is at a distance of zero from the boundary of P if and only if the logical OR of the x_i's is 1. The construction of P and Q from $x_1, x_2, \ldots, x_n$ can clearly be done in constant time with n processors. This establishes the lower bound. ∎

The rest of this chapter illustrates how the technique presented in Chapter 10 on parallel mergesorting can be used to solve geometric problems.

11.5
3-Dimensional Maxima

Let $V = \{p_1, p_2, \ldots, p_n\}$ be a set of points in $\Re^3$. For simplicity, we assume that no two input points have the same x (resp., y, z) coordinate. We denote the x, y, and z coordinates of a point p by $x(p)$, $y(p)$, and $z(p)$, respectively. We say that a point p_i *1-dominates* another point p_j if $x(p_i) > x(p_j)$, *2-dominates* p_j if $x(p_i) > x(p_j)$ and $y(p_i) > y(p_j)$, and *3-dominates* p_j if $x(p_i) > x(p_j)$, $y(p_i) > y(p_j)$, and $z(p_i) > z(p_j)$. A point $p_i \in V$ is said to be a *maximum* if it is not 3-dominated by any other point in V. The 3-dimensional maxima problem, then, is to compute the set, M, of maxima in V. We show how to solve the 3-dimensional maxima problem efficiently in parallel in the following algorithm.

11.5.1 Maximal Elements

The method is based on a divide-and-conquer strategy in which the subproblem merging step involves the computation of *two* labeling functions for each point, but is otherwise similar to the parallel mergesort algorithm described earlier in this book. We call such a divide-and-conquer scheme a *cascading divide-and-conquer*, since it involves a cascading merge. Specifically, for each point p_i we compute the maximum z-coordinate from among all points that 1-dominate p_i and use that label to also compute the maximum z-coordinate from among all points that 2-dominate p_i. We can then test if p_i is a maximum point by comparing $z(p_i)$ to this latter label. The details follow.

Without loss of generality, we assume the input points are given sorted by increasing y-coordinates, i.e., $y(p_i) < y(p_{i+1})$, since if they are not given in this order we can sort them in $O(\log n)$ time using n processors. Let T be a complete binary tree with leaf nodes $v_1, v_2, \ldots, v_n$ (in this order). In each leaf node v_i we store the list $B(v_i) = (-\infty, p_i)$, where $-\infty$ is a special symbol such that $x(-\infty) < x(p_j)$ and $y(-\infty) < y(p_j)$ for all points p_j in V. Initializing T in this way can be done in $O(\log n)$ time using n processors. We then perform a cascading merge from the leaves of T upwards, basing comparisons on increasing x-coordinates of the points (not their y-coordinates). Let $U(v)$ denote the sorted array of the points stored in the descendants of $v \in T$ sorted by increasing x-coordinates. For each point p_i in $U(v)$ we store two labels: $zod(p_i, v)$ and $ztd(p_i, v)$, where $zod(p_i, v)$ is the largest z-coordinate of the points in $U(v)$ that 1-dominate p_i, and $ztd(p_i, v)$ is the largest z-coordinate of the points in $U(v)$ that 2-dominate p_i. Initially, *zod* and *ztd* labels are only defined for the leaf nodes of T. That is, $zod(p_i, v_i) = ztd(p_i, v_i) = -\infty$ and $zod(-\infty, v_i) = ztd(-\infty, v_i) = z(p_i)$ for all leaf nodes v_i in T (where $U(v_i) = (-\infty, p_i)$). In order to be more explicit in how we refer to various ranks, we let $\text{pred}(p_i, v)$ denote the predecessor of p_i in $U(v)$ (which would be $-\infty$ if the x-coordinates of the input points are all larger than $x(p_i)$). (See Figure 11.12.) As we are performing the merge, we update the labels zod and ztd based on the equations in the following lemma:

LEMMA 11.5
Let p_i be an element of $U(v)$ and let $u = lchild(v)$ and $w = rchild(v)$. Then we have the following:

$$zod(p_i, v) = \begin{cases} \max\{zod(p_i, u), zod(\text{pred}(p_i, w), w)\} & \text{if } p_i \in U(u) \\ \max\{zod(\text{pred}(p_i, u), u), zod(p_i, w)\} & \text{if } p_i \in U(w) \end{cases} \quad (11.1)$$

$$ztd(p_i, v) = \begin{cases} \max\{ztd(p_i, u), zod(\text{pred}(p_i, w), w)\} & \text{if } p_i \in U(u) \\ ztd(p_i, w) & \text{if } p_i \in U(w) \end{cases} \quad (11.2)$$

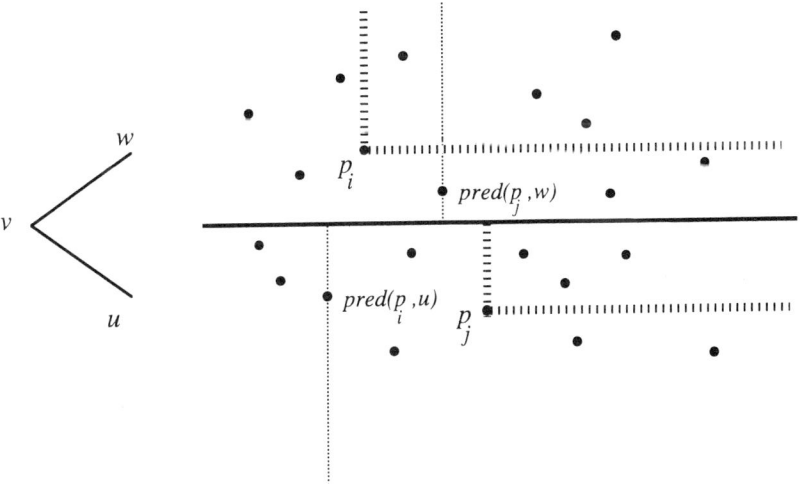

FIGURE 11.12
The combining step for 3-dimensional maxima points to the right of the thin dotted line 1-dominate p_i (resp. p_j), and points enclosed in the thick dotted lines 2-dominate p_i (p_j).

PROOF

Consider Equation (11.1). If $p_i \in U(u)$, then every point that 1-dominates p_i's predecessor in $U(w)$ also 1-dominates p_i, since p_i's predecessor in $U(w)$ is the point with largest x-coordinate less than $x(p_i)$ (or $-\infty$ if every point in $U(w)$ has larger x-coordinate than p_i). Thus $zod(p_i, v)$ is the maximum of $zod(p_i, u)$ and $zod(\text{pred}(p_i, w), w)$ in this case. The case when $p_i \in U(w)$ is similar. Next, consider Equation (11.2). We know that every point in $U(w)$ has y-coordinate greater than every point in $U(u)$, by our construction of T. Therefore, if $p_i \in U(u)$, then every point in $U(w)$ that 1-dominates p_i's predecessor in $U(w)$ must 2-dominate p_i. Thus, $ztd(p_i, v)$ is the maximum of $ztd(p_i, u)$ and $zod(\text{pred}(p_i, w), w)$. On the other hand, if $p_i \in U(w)$ then no point in $U(u)$ can 2-dominate p_i; thus, $ztd(p_i, v) = ztd(p_i, w)$. ∎

We use these equations during the cascading merge to maintain the labels for each point. By Lemma 11.5, when v becomes full (and we have $U(u)$, $U(w)$, and $U(v) = U(u) \cup U(w)$ available), we can determine the labels for all the points in $U(v)$ in $O(1)$ additional time using $|U(v)|$ processors. Thus, the running time of the cascading merge algorithm, even with these additional label computations, is still $O(\log n)$ using n processors. Moreover, after v's parent becomes full we no longer need $U(v)$ any more, and can deallocate the space it occupies, resulting in an $O(n)$ space algorithm. After we complete the merge, and have computed $U(root(T))$, along with all the labels for the points in $U(root(T))$, note that a point $p_i \in U(root(T))$ is a maximum if and only if $ztd(p_i, root(T)) \leq z(p_i)$ (there is no point that 2-dominates p_i and has z-coordinate greater than $z(p_i)$). Thus, after completing the cascading merge we can construct the set of maxima by compressing all the maximum points into one contiguous list using a simple parallel prefix computation. We summarize in the following theorem:

THEOREM 11.5
Given a set V of n points in $\Re^3$, one can construct the set M of maxima points in V in $O(\log n)$ time and $O(n)$ space using n processors in the CREW PRAM model.

11.6
Visibility from a Point

Given a set of line segments $S = \{s_1, s_2, \ldots, s_n\}$ in the plane that do not intersect, except possibly at endpoints, and a point p, the visibility from

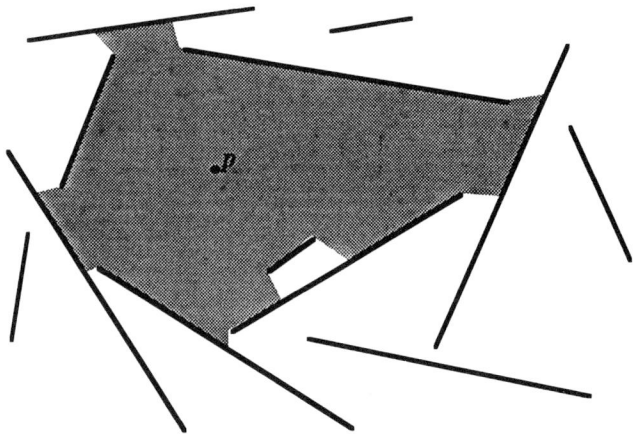

FIGURE 11.13
The shaded region is visible from p.

a point problem is to determine the part of the plane that is visible from p assuming every s_i is opaque. Intuitively, one can think of the point p as a specular light source, the segments as walls, and the problem to determine all the parts of the plane that are illuminated (see Figure 11.13).

We can use the cascading divide-and-conquer technique to solve this problem in $O(\log n)$ time and $O(n)$ space using n processors. Without loss of generality, we assume that the point p is at negative infinity below all the segments. The algorithm is essentially the same if p is a finite point, except that the notion of segment endpoints being ordered by x-coordinate is replaced by the notion that they are ordered radially around p. In other words, it suffices to compute the *lower envelope* of the n segments to give a method for computing the visibility from a point. For simplicity of expression, we also assume that the x-coordinates of the endpoints are distinct.

In the previous section the set of objects consisted of points, but in the visibility problem we are dealing with line segments. The method is slightly different in this case. In this case we store the segments in the leaves of a binary tree and perform a cascading merge of the x-coordinates of intervals of the x-axis determined by segment endpoints. We maintain a single label for each interval that represents the segment that is visible from $-\infty$ on that interval. The details follow.

524   Chapter 11.   Deterministic Parallel Computational Geometry

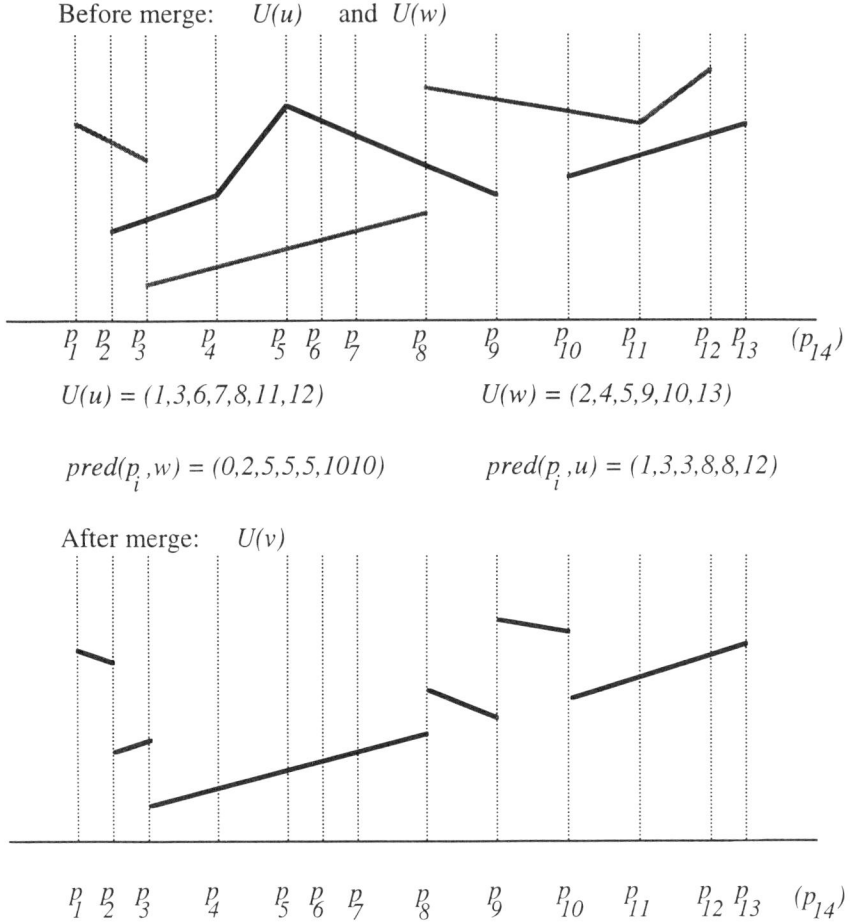

FIGURE 11.14
An example of visibility merging. The dashed segments correspond to the visible region for $x(u)$ and the solid segments correspond to the visible region for $x(w)$. For simplicity, we store the pointers $pred(p_i, u)$, and $pred(p_i, w)$ in arrays and denote each point p_i by its index i. Note that points are never removed, even if the same segment defines the visible region for many consecutive intervals (e.g., p_3 through p_7).

Let T be a complete binary tree with leaf nodes $v_1, v_2, \ldots, v_n$ ordered from left to right. We associate the segment s_i with the leaf v_i and at v_i store the list $U(v_i) = (-\infty, p_1, p_2)$, where p_1 and p_2 are the two endpoints of s_i, with $x(p_1) < x(p_2)$, and $-\infty$ is defined such that $x(-\infty) < x(p)$ and $y(-\infty) < y(p)$ for all points p. We then perform a generalized cascading-merge from the leaves of T, basing comparisons on increasing x-coordinates of the points. For each internal node v we let $U(v)$ denote an array of the points stored in the descendants of $v \in T$ sorted by increasing x-coordinates. For each point p_i in $U(v)$ we store a label $vis(p_i, v)$, which stores the segment with endpoints in $U(v)$ that is visible in the interval $(x(p_i), x(\text{succ}(p_i, v)))$, where $\text{succ}(p_i, v)$ denotes the successor of p_i in $U(v)$ (based on x-coordinates). Initially, the vis labels are only defined for the leaf nodes of T. That is, if $U(v) = (-\infty, p_1, p_2)$, where $s_i = p_1 p_2$, then $vis(-\infty) = +\infty$, $vis(p_1) = s_i$, and $vis(p_2) = +\infty$. We use $\text{pred}(p_i, v)$ to denote the predecessor of p_i in $U(v)$. As we are performing the cascading-merge, we update the vis labels based on the equation in the following lemma (see Figure 11.14):

LEMMA 11.6
Let p_i be an element of $U(v)$ and let $u = lchild(v)$ and $w = rchild(v)$. Then we have the following (if two segments s_i and s_j are comparable by the "above" relation, then we let $\min\{s_i, s_j\}$ denote the lower of the two):

$$vis(p_i, v) = \begin{cases} \min\{vis(p_i, u), \ vis(\text{pred}(p_i, w), w)\} & \text{if } p_i \in U(u) \\ \min\{vis(\text{pred}(p_i, u), u), \ vis(p_i, w)\} & \text{if } p_i \in U(w) \end{cases}$$

PROOF
If we restrict our attention to the segments with an endpoint in $U(u)$, then for any point $p_i \in U(u)$ the segment visible (from $-\infty$) on the interval $(x(p_i), x(\text{succ}(p_i, v)))$ is the minimum of the segment visible on the interval $(x(p_i), x(\text{succ}(p_i, u)))$ and the segment that is visible on the interval $(x(\text{pred}(p_i, w)), x(\text{succ}(\text{pred}(p_i, w), w)))$. This is because the interval $(x(p_i), x(succ(p_i, v)))$ is exactly the intersection of the interval $(x(p_i), x(succ(p_i, u)))$ and the interval $(x(\text{pred}(p_i, w)), x(\text{succ}(\text{pred}(p_i, w), w)))$, and there is no segment in $U(v)$ with an endpoint interior to the interval $(x(p_i), x(succ(p_i, v)))$. Thus, $vis(p_i, v)$ is equal to the minimum of $vis(p_i, u)$ and $vis(\text{pred}(p_i, w), w)$. The case when $p_i \in U(v)$ is similar. ∎

By Lemma 11.6, after merging the lists $U(lchild(v))$ and $U(rchild(v))$ we can determine the labels for all the points in $U(v)$ in $O(1)$ additional time using $|U(v)|$ processors. Thus, the running time of this generalized cascading-merge algorithm is still $O(\log n)$ using n processors. After we complete the merge, and have computed $U(root(T))$, along with all the vis labels for the points in $U(root(T))$, then we can compress out duplicate entries in the list $(vis(p_1, root(T)), vis(p_2, root(T)), \ldots, vis(p_{2n}, root(T)))$ using a parallel prefix computation to construct a compact representation of the visible portion of the plane. We summarize in the following theorem:

THEOREM 11.6
Given a set S of n non-intersecting segments in the plane, one can find the lower envelope of S in $O(\log n)$ time and $O(n)$ space using n processors in the CREW PRAM model, and this is optimal.

PROOF
The correctness and complexity bounds follow from the discussion above. Since we require that the points in the description of the lower envelope be given by increasing x-coordinates, we can reduce sorting to this problem, and thus can do no better than $O(\log n)$ time using n processors (in the comparison model). ∎

11.7 Exercises

11.1 Prove an $O(n \log n)$ time sequential upper bound for the convex hull problem.

11.2 Prove an $\Omega(n \log n)$ time sequential lower bound for the convex hull problem in the comparison model. (Hint: Find a way to reduce the sorting problem to the convex hull problem.)

11.3 Prove an $\Omega(\log n)$ time lower bound for the convex hull problem and a CREW PRAM with a polynomial number of processors.

11.4 Prove that the intersection of two convex regions is also convex. Also, give an $O(\log n)$ time algorithm, using $O(n/\log n)$ processors in the CREW PRAM model, for computing the intersection of two convex regions. Your method should work even if either of the regions is unbounded. (Hint: Use the fact that one can merge two n-element sorted lists in $O(\log n)$ time using $O(n/\log n)$ processors in the CREW PRAM model.)

11.5 Let $P = (p_1, s_1, p_2, \ldots, p_n)$ be a simple polygon, and let $P' = (p_{i_1}, s'_{i_1}, p_{i_2}, \ldots, p_{i_k})$ be a subpolygon of P, i.e., $1 \leq i_1 < i_2 < \cdots < i_k \leq n$. Prove or disprove the following statements:

1. If P is simple and convex, then P' is simple and convex.
2. If P is simple, then P' is simple.
3. If P is simple and star-shaped, then P' is simple and star-shaped.

11.6 Prove that the two definitions of "kernel" given in Section 11.3.1 are equivalent.

11.7 Let P be a star-shaped polygon, P' be a subpolygon of P. Prove or disprove the following statement: "the kernel of P' is inside the kernel of P."

11.8 Let p, Q', q_j and q_{im} be as in the proof of Proposition 11.1, with $j < im$. Prove that $(i-1)m \leq j$.

11.9 Generalize Proposition 11.2 to arbitrary convex polygons P and Q, rather than just upper hulls.

11.10 Generalize the algorithm for finding the upper common tangent between $UH(S_1)$ and $UH(S_2)$ so that it runs in $O(k^2)$ time and uses $O(n^{1/k})$ processors, where k is a constant.

11.11 Design an algorithm for determining whether a given line L and the boundary of a given convex polygon P intersect or not. If they do intersect, your algorithm should give both points of this intersection, and your algorithm should run in $O(k)$ time with $O(n^{1/k})$ processors, where $2 \leq k \leq n$ is any integer.

11.12 Let S be a set of points in the plane. We wish to "mark" the points of S that belong to the convex hull of S. Give a constant-time CRCW algorithm for this problem, using a polynomial number of processors.

11.13 Let S be a set of points in the plane, given in sorted order by their x-coordinates. We wish to "mark" each point of S that belongs to the upper hull of S as "on hull," and also mark each each point p that is not on the upper hull of S with the name of the upper hull edge that is directly above p. Derive an $O(\log \log n)$ time algorithm that uses only n processors. (Hint: use the fact that the maximum of n items can be found in constant time using n^2 processors in the CRCW PRAM model.)

11.14 Let f be the transform defined in Section 11.3. Prove that f maintains the relative positions of lines and points: if one of $\{\alpha, \beta\}$ is a line and one is a point, then α is below β if and only if $f(\alpha)$ is below $f(\beta)$.

11.15 Let f be the transform defined in Section 11.3, as in the previous exercise. Prove or disprove the following statement: "If P is a simple polygon, then $f(P)$ is a simple polygon."

11.16 Let p be a point on the boundary of a convex polygon P, and define q similarly for a convex polygon Q. Note, p and q need not be vertices. Let T_p (resp. T_q) be the line perpendicular to the line segment $\overline{pq}$ at p and q, respectively. Prove that $d(P,Q) = d(p,q)$ if and only if (i) T_p and T_q are tangent to P and Q respectively, and (ii) P and Q are on opposite sides of the region between T_p and T_q.

11.17 Modify the algorithm for computing the distance between two convex polygons so that it also works if they are not disjoint. That is, the distance between them may be zero (this happens either by one of them being contained inside the other, or by their two boundaries intersecting).

11.18 Show that 2-dimensional maxima problem can be solved in $O(\log n)$ time and $O(n)$ space by a sorting step followed by a parallel prefix step.

Notes and References

The convex hull problem is probably the most-studied of all computational geometry problems. See the book by Edelsbrunner [33] or the book by Preparata and Shamos [63] for further references on this and other sequential algorithms for computational geometry. A proof of the Jordan Curve Theorem can be found in the book by Munkres [58]. The convex hull algorithm presented in this chapter is due to Atallah and Goodrich [12], who also give the algorithm for computing the distance between convex polygons. The solutions to Exercises 11.10 and 11.17 can be found in [12]. Dadoun and Kirkpatrick [31] show how to achieve $O(k)$ time using $O(n^{1/k})$ processors for the polygon separation problem, which improves the running time for larger values of k. Other efficient parallel convex hull algorithms can be found in [1, 11, 57, 81], and a solution to Exercise 11.13 can be found in [17]. Although it is a good example of applying geometric duality in parallel computational geometry, the algorithm given in this chapter for computing the kernel of a simple polygon is not the best possible, for Cole and Goodrich [25] show how to solve this problem in $O(\log n)$ time algorithm for using only $O(n/\log n)$ processors in the EREW PRAM model. For more information about this problem and other "art gallery" problems, see the excellent book by O'Rourke [59]. The algorithm for the 3-dimensional maxima and visibility from a point is due to Atallah, Cole, and Goodrich [10]; they also give a number of other geometric applications of the cascading divide-and-conquer technique. Recently, Atallah and Chen [9] show how one can improve the processor bound for visibility from a point to $O(n/\log n)$ for the case when the segments form a simple polygon. Exercise 11.12 is from Akl [4].

The field of parallel computational geometry is relatively young; hence, the body of literature in this field is rather sparse. Still, the bibliography given below attempts to cover most of the papers covering this area to date. It is by no means exhaustive, however, and, in fact, omits a number of papers dealing with geometric algorithms for network models (these network algorithms are outside the scope of this chapter, but are covered in a book by Miller and Stout [56], which also contains an extensive bibliography of geometric algorithms for network models). Recently, deterministic parallel algorithms have been developed by Cole, Goodrich, and Dunlaing for Voronoi diagram [26] and Goodrich for 3-D convex hull [40]. These algorithms are quite efficient, requiring $O(\log(n) \log \log n)$ time using $O(n \log^2 n)$ work, or alternatively time $O(\log^2 n)$ with optimal $O(n \log n)$ work.

Bibliography

[1] A. Aggarwal, B. Chazelle, L. Guibas, C. Ó'Dúnlaing, and C. Yap, "Parallel Computational Geometry," *Algorithmica*, Vol. 3, No. 3, 1988, pp. 293–328.

[2] A. Aggarwal, D. Kravets, J. K. Park, and S. Sen, "Parallel Searching in Generalized Monge Arrays with Applications," *Proc. 2nd ACM Symp. on Parallel Algorithms and Architectures*, 1990, pp. 259–267.

[3] A. Aggarwal and J. Park, "Parallel Searching in Multidimensional Monotone Arrays," *J. Algorithms*, 1989.

[4] S.G. Akl, "A Constant-Time Parallel Algorithm for Computing Convex Hulls," *BIT*, Vol. 22, 1982, pp. 130–134.

[5] N. Alon and N. Megiddo, "Parallel Linear Programming in Fixed Dimension Almost Surely in Constant Time," *Proc. 31st IEEE Symp. on Foundations of Computer Science*, 1990, pp. 574–582.

[6] R. Anderson, P. Beame, and E. Brisson, "Parallel Algorithms for Arrangements," *Proc. 2nd ACM Symp. on Parallel Algorithms and Architectures*, 1990, pp. 298–306.

[7] A. Apostolico, M. J. Atallah, L. L. Larmore, and H. S. McFaddin, "Efficient Parallel Algorithms for String Editing and Related Problems," *Proc. 26th Allerton Conf. on Communication, Control, and Computing*, 1988, pp. 253–263.

[8] M. J. Atallah, P. Callahan, and M. T. Goodrich, "P-Complete Geometric Problems," *Proc. 2nd ACM Symp. on Parallel Algorithms and Architectures*, 1990, pp. 317–326.

[9] M. J. Atallah and D. Z. Chen, "Optimal Parallel Algorithms for Visibility of a Simple Polygon From a Point," *Proc. 5th ACM Symp. on Computational Geometry*, 1989, pp. 114–123.

[10] M. J. Atallah, R. Cole, and M. T. Goodrich, "Cascading Divide-and-Conquer: A Technique for Designing Parallel Algorithms," *SIAM Journal on Computing*, Vol. 18, No. 3, 1989, pp. 499–532.

[11] M. J. Atallah and M. T. Goodrich, "Efficient Parallel Solutions to Some Geometric Problems," *Journal of Parallel and Distributed Computing*, Vol. 3, No. 4, 1986, pp. 492–507.

[12] M. J. Atallah and M. T. Goodrich, "Parallel Algorithms for Some Functions of Two Convex Polygons," *Algorithmica*, Vol. 3, 1988, pp. 535–548.

[13] M. J. Atallah and J. J. Tsay, "On the Parallel Decomposability of Geometric Problems," *Proc. 5th ACM Symp. on Computational Geometry*, 1989, pp. 104–113.

[14] M. Ben-Or, "Lower Bounds for Algebraic Computation Trees," *Proc. 15th ACM Symp. on Theory of Computing*, 1983, pp. 80–86.

[15] J. L. Bentley and M. I. Shamos, "Divide-And-Conquer in Multidimensional Space," *Proc. 8th ACM Symp. on Theory of Computing*, 1976, pp. 220–230.

[16] J. L. Bentley and D. Wood, "An Optimal Worst Case Algorithm for Reporting Intersections of Rectangles," *IEEE Trans. on Computers*, Vol. C-29, No. 7, 1980, pp. 571–576.

[17] O. Berkman, D. Breslauer, Z. Galil, B. Schieber, U. Vishkin, "Highly Parallizable Problems," *Proc. 21st ACM Symp. on Theory of Computing*, 1989, pp. 309–319.

[18] G. Bilardi and A. Nicolau, "Adaptive Bitonic Sorting: An Optimal Parallel Algorithm for Shared Memory Machines," TR 86-769, Dept. of Computer Science, Cornell Univ., August 1986.

[19] A. Borodin and J. E. Hopcroft, "Routing, Merging, and Sorting on Parallel Models of Computation," *Journal of Computer and System Sciences*, Vol. 30, No. 1, 1985, pp. 130–145.

[20] R. P. Brent, "The Parallel Evaluation of General Arithmetic Expressions," *J. ACM*, Vol. 21, No. 2, 1974, pp. 201–206.

[21] B. M. Chazelle and D. P. Dobkin. Detection is Easier than Computation, *Proc. 12th ACM Annual Symp. on Theory of Computing*, 1980, pp. 146–153.

[22] B. Chazelle and L. J. Guibas, "Fractional Cascading: I. A Data Structuring Technique," *Algorithmica*, Vol. 1, No. 2, pp. 133–162.

[23] A. Chow, "Parallel Algorithms for Geometric Problems," Ph.D. thesis, Computer Science Dept., Univ. of Illinois at Urbana-Champaign, 1980.

[24] R. Cole, "Parallel Merge Sort," *SIAM J. Comput.*, Vol. 17, No. 4, August 1988, pp. 770–785.

[25] R. Cole and M. T. Goodrich, "Optimal Parallel Algorithms for Point-Set and Polygon Problems", to appear in *Algorithmica* (appeared in preliminary form in *Proc. 4th ACM Symp. on Computational Geometry*, 1988, pp. 201–210).

[26] R. Cole, M. T. Goodrich, and C. O. Dunlaing, "Merging Free Trees in Parallel for Efficient Voronoi Diagram Construction," *International Colloquium on Automata, Languages, and Programming*, Warwich, England, July 1990, 432–445.

[27] R. Cole and O. Zajicek, "An Optimal Parallel Algorithm for Building a Data Structure for Planar Point Location," *J. Parallel and Distributed Computing*, Vol. 8, 1990, 280–285.

[28] S. Cook and C. Dwork. Bounds on the Time for Parallel RAM's to Compute Simple Functions, *Proc. 14th ACM Annual Symp. on Theory of Computing*, 1982, pp. 231–233.

[29] N. Dadoun and D. Kirkpatrick, "Parallel Construction of Subdivision Hierarchies," *Journal of Computer and System Sciences*, Vol. 39, 1989, pp. 153–165.

[30] N. Dadoun and D. Kirkpatrick, "Cooperative Subdivision Search Algorithms with Applications," *Proc. 27th Allerton Conf. on Communication, Control, and Computing*, 1989.

[31] N. Dadoun and D. Kirkpatrick, "Optimal Parallel Algorithms for Convex Polygon Separation," Technical Report 89-21, Dept. of Computer Science, Univ. of British Columbia, 1989.

[32] H. Edelsbrunner. Computing the Extreme Distances between Two Convex Polygons, *J. Algorithms*, Vol. 6, 1985, pp. 213–224.

[33] H. Edelsbrunner, *Algorithms in Combinatorial Geometry*, Springer-Verlag, NY, 1987.

[34] H. Edelsbrunner and M. H. Overmars, "On the Equivalence of Some Rectangle Problems," *Information Processing Letters*, Vol. 14, No. 3, 1982, pp. 124–127.

[35] H. ElGindy, "A Parallel Algorithm for Triangulating Simplical Point Sets in Space with Optimal Speed-up," *Proc. 24th Allerton Conf. on Communication, Control, and Computing*, 1986.

[36] H. ElGindy and M. T. Goodrich, "Parallel Algorithms for Shortest Path Problems in Polygons," *The Visual Computer: International Journal of Computer Graphics*, Vol. 3, No. 6, 1988, pp. 371–378.

[37] A. Fournier and Z. Kedem, "Comments on the All Nearest-Neighbor Problem for Convex Polygons," *Info. Proc. Letters*, Vol. 9, No. 3, 1979, pp. 105–107.

[38] M. T. Goodrich, "Efficient Parallel Techniques for Computational Geometry," Ph.D. thesis, Department of Computer Science, Purdue University, 1987.

[39] M. T. Goodrich, "Finding the Convex Hull of a Sorted Point Set in Parallel," *Information Processing Letters*, Vol. 26, 1987, pp. 173–179.

[40] M. T. Goodrich, "Geometric Partitioning Make Easier, Even in Parallel," manuscript, 1992.

[41] M. T. Goodrich, "Intersecting Line Segments in Parallel with an Output-Sensitive Number of Processors," *SIAM Journal on Computing*, to appear (appeared in preliminary form in *Proc. 1989 ACM Symp. on Parallel Algorithms and Architectures*, 127–137).

[42] M. T. Goodrich, "Triangulating a Polygon in Parallel," *J. Algorithms, Journal of Algorithms*, Vol. 10, 1989, pp. 327–351.

[43] M. T. Goodrich, "Constructing Arrangements Optimally in Parallel," Technical Report 90/06, Dept. of Computer Science, Johns Hopkins Univ., 1990.

[44] M. T. Goodrich, M. Ghouse, and J. Bright, "Generalized Sweep Methods for Parallel Computational Geometry," *Proc. 2nd ACM Symp. on Parallel Algorithms and Architectures*, 1990, pp. 280–289.

[45] M. T. Goodrich, C. Ó'Dúnlaing, and C. Yap "Computing the Voronoi Diagram of a Set of Line Segments in Parallel," *Algorithmica*, to appear (appeared in preliminary form in *Lecture Notes in Computer Science: 382, Algorithms and Data Structures, WADS '89*, Springer-Verlag, 1989, pp. 12–23).

[46] M. T. Goodrich, S. Shauck, and S. Guha, "Parallel Methods for Visibility and Shortest Path Problems in Simple Polygons," *Proc. 6th ACM Symp. on Computational Geometry*, 1990, pp. 73–82.

[47] L. Guibas, L. Ramshaw, and J. Stolfi, "A Kinetic Framework for Computational Geometry," *Proc. 24th IEEE Symp. on Found. of Computer Science*, 1983, pp. 100–111.

[48] T. Hagerup, H. Jung, and E. Welzl, "Efficient Parallel Computation of Arrangements of Hyperplanes in d Dimensions," *Proc. 2nd ACM Symp. on Parallel Algorithms and Architectures*, 1990, pp. 290–297.

[49] H. T. Kung, F. Luccio, F. P. Preparata, "On Finding the Maxima of a Set of Vectors," *J. ACM*, Vol. 22, No. 4, 1975, pp. 469–476.

[50] C. P. Kruskal, L. Rudolph, and M. Snir, "The Power of Parallel Prefix," *Proc. 1985 IEEE Int. Conf. on Parallel Processing*, St. Charles, IL., pp. 180–185.

[51] R. E. Ladner and M. J. Fischer, "Parallel Prefix Computation," *J. ACM*, 1980, pp. 831–838.

[52] D. T. Lee and F. P. Preparata, "The All Nearest-Neighbor Problem for Convex Polygons," *Info. Proc. Letters*, Vol. 7, No. 4, 1978, pp. 189–192.

[53] D. T. Lee and F. P. Preparata, "An Optimal Algorithm for Finding the Kernel of a Polygon," *J. ACM*, Vol. 26, No. 3, 1979, pp. 415–421.

[54] D. T. Lee and F. P. Preparata, "Computational Geometry—A Survey," *IEEE Trans. on Computers*, Vol. C-33, No. 12, 1984, pp. 872–1101.

[55] E. Merks, "An Optimal Parallel Algorithm for Triangulating a Set of Points in the Plane," Technical Report TR 86-9, Simon Fraser Univ., Burnaby, British Columbia, 1986.

[56] R. Miller and Q.F. Stout, *Parallel Algorithms for Regular Architectures*, The MIT Press, Cambridge, Massachusetts, 1989.

[57] R. Miller and Q.F. Stout, "Efficient parallel convex hull algorithms," *IEEE Trans. Computers* **C-37** (1988), pp. 1605-1618.

[58] J. R. Munkres, *Topology: A First Course*, Prentice-Hall, Inc. (Englewood Cliffs, NJ: 1975).

[59] J. O'Rourke, *Art Gallery Theorems and Algorithms*, Oxford University Press (New York: 1987).

[60] M. H. Overmars and J. Van Leeuwen, "Maintenance of Configurations in the Plane," *Journal of Computer and Systems Sciences,* Vol. 23, 1981, pp. 166–204.

[61] F. P. Preparata and S. J. Hong, "Convex Hulls of Finite Sets of Points in Two and Three Dimensions," *CACM,* Vol. 20, No. 2, 1977, pp. 87–93.

[62] F. P. Preparata and D. E. Muller, "Finding the Intersection of n Halfspaces in Time $O(n \log n)$," *Theoretical Computer Science*, Vol. 8, 1979, pp. 45–55.

[63] F. P. Preparata and M. I. Shamos, *Computational Geometry: An Introduction*, Springer-Verlag, 1985.

[64] J. H. Reif, "An Optimal Parallel Algorithm for Integer Sorting," *Proc. 26th IEEE Symp. on Foundations of Comp. Sci.*, 1985, pp. 496–504.

[65] J. H. Reif and S. Sen, "Optimal Parallel Algorithms for Computational Geometry," *Proc. 1987 IEEE Int. Conf. on Parallel Processing*, pp. 270–277.

[66] J. H. Reif and S. Sen, "An Efficient Output-Sensitive Hidden-Surface Removal Algorithm and Its Parallelization," *Proc. 4th ACM Symp. on Computational Geometry*, 1988, pp. 193–200.

[67] J. H. Reif and S. Sen, "Polling: A New Randomized Sampling Technique for Computational Geometry," *Proc. 21st ACM Symp. on Theory of Computing*, 1989, pp. 394–404.

[68] J. H. Reif and S. Sen, "Randomized Algorithms for Binary Search and Load Balancing with Geometric Applications," *Proc. 2nd ACM Symp. on Parallel Algorithms and Architectures*, 1990, pp. 327–337.

[69] C. Rüb, *Parallele Algorithmen zum Berechnen der Schnittpunkte von Liniensegmenten*, Doctoral Dissertation, Universität des Saarlandes, 1990.

[70] S. Sen, *Random Sampling Techniques for Efficient Parallel Algorithms in Computational Geometry*, Ph.D. thesis, Computer Science Dept., Duke Univ., 1989.

[71] M. I. Shamos, "Geometric Complexity," *Proc. 7th ACM Symp. on Theory of Computing*, 1975, pp. 224–233.

[72] M. I. Shamos and D. Hoey, "Closest-Point Problems," *Proc. 15th IEEE Symp. on Foundations of Computer Science*, 1975, pp. 151–162.

[73] Y. Shiloach and U. Vishkin, "Finding the Maximum, Merging, and Sorting in a Parallel Computation Model," *J. Algorithms*, Vol. 2, 1981, pp. 88–102.

[74] Q. F. Stout, "Constant-time geometry on PRAMs", *Proc. 1988 Int'l. Conf. on Parallel Computing*, vol. III, IEEE, pp. 104-107.

[75] R. Tamassia and J. S. Vitter, "Optimal Cooperative Search in Fractional Cascaded Data Structures," *Proc. 2nd ACM Symp. on Parallel Algorithms and Architectures*, 1990, pp. 307–316.

[76] R. Tamassia and J. S. Vitter, "Optimal Parallel Algorithms for Transitive Closure and Point Location in Planar Structures," *Proc. 1989 ACM Symp. on Parallel Algorithms and Architectures*, pp. 399–408.

[77] R. E. Tarjan and C. J. Van Wyk, "An $O(n \log \log n)$-Time Algorithm for Triangulating a Simple Polygon," *SIAM J. Comput.*, Vol. 17, 1988, pp. 143–178.

[78] G. T. Toussaint, "Solving Geometric Problems with Rotating Calipers," *Proc. IEEE MELECON '83*, Athens, Greece, May 1983.

[79] P. M. Vaidya, "Reducing the Parallel Complexity of Certain Linear Programming Problems," *31st IEEE Symp. on Foundations of Computer Science*, 1990, pp. 583–589.

[80] L. Valiant, "Parallelism in Comparison Problems," *SIAM Journal on Computing*, Vol. 4, No. 3, 1975, pp. 348–355.

[81] H. Wagener, "Optimally Parallel Algorithms for Convex Hull Determination," manuscript, 1985.

[82] C. A. Wang and Y. H. Tsin, "An $O(\log n)$ Time Parallel Algorithm for Triangulating a Set of Points in the Plane," *Information Processing Letters*, Vol. 25, 1987, pp. 55–60.

[83] C. C. Yang and D. T. Lee, "A Note on the All Nearest-Neighbor Problem for Convex Polygons," *Info. Proc. Letters*, Vol. 8, No. 4, 1979, pp. 193–194.

[84] C. K. Yap, "Parallel Triangulation of a Polygon in Two Calls to the Trapezoidal Map," *Algorithmica*, Vol. 3, 1988, pp. 279–288.

Part IV

Fundamental Parallel Algebraic Algorithms

12

Newton Iteration and Integer Division

Stephen R. Tate

Department of Computer Science
Duke University
Durham, NC 27706
srt@cs.duke.edu

12.1
Introduction

At the heart of all numerical computations are the basic operations of addition, subtraction, multiplication, and division (also, to a lesser extent, more complex operations such as powering, finding square roots, computing logarithms, etc.). It is vital to study these problems from the standpoint of parallel algorithms, because even in commonly used single processor machines the basic operations are done in parallel (by parallel paths through low-level logic circuits).

Early in school, a student learns how to perform the functions of addition, subtraction, multiplication, and division. In fact, these topics are usually presented in this order due to the increasing difficulty of the operations. Studies in parallel algorithms support the sense of difficulty assigned by our elementary school teachers—optimal algorithms exist for addition and subtraction, while good algorithms exist for multiplication (even the best of which is not known to be optimal), and division seems to be even harder. For more information on the operations of addition, subtraction, and multiplication, see the references section at the end of this chapter.

The common computational model used when examining arithmetic problems is the *bounded fan-in Boolean circuit*. A bounded fan-in circuit is simply a directed acyclic graph with each node having bounded in-degree. There are two sets of special nodes: the input nodes (which have in-degree zero) and the output nodes (which have out-degree zero). Furthermore, since the circuits we consider are Boolean, each non-input node is labeled with a Boolean function (AND, OR, or NOT). By applying Boolean values to the input nodes and computing all nodes as the labeled function from their predecessor values, circuits can be regarded as computing functions. To compute a function, *circuit families* are considered (one circuit for each possible input size). The *size* of a circuit family is a function $S(n)$ that gives (for each $n \geq 1$) the number of nodes in the circuit for inputs of length n. The *depth* of a circuit family is a function $D(n)$ that gives (for each $n \geq 1$) the length of the longest path from an input node to an output node in the circuit for length n inputs.

In this chapter, the parallel complexity of division is compared with the complexity of the other elementary operations. The problem of integer division is defined to be a function that takes a pair of input values (y, x), and produces the pair of values (q, r) such that $y = qx + r$, where $0 \leq r < x$ (i.e., a quotient and a remainder). Reduction of division to multiplication

12.1. Introduction

via Newton approximation is shown to provide good sequential results, but these results do not translate well to parallel algorithms. In this chapter, we describe a parallel algorithm due to Reif and Tate [12] which is a modified version of Newton approximation (called *high order* Newton approximation for reasons that will become clear). This algorithm obtains parallel results that are almost optimal. On the way to the results for division, it will be discovered that finding limited integer powers is vital for division, so ways of accomplishing this are discussed.

As is standard practice when comparing the complexity of algorithms, the focus of this chapter will be on reductions to other problems. It is sufficient to consider the problem of finding reciprocals in place of division. As we are considering only integer operations, the idea of a reciprocal is not clear—in general, the real reciprocal of an integer will *not* be an integer. Therefore, given an n-bit input integer x, the integer reciprocal is defined as the value

$$\left\lfloor \frac{2^{2n}}{x} \right\rfloor.$$

Notice that this is simply the shifted binary fixed point approximation to the real reciprocal; it should be obvious how the reciprocal can be used with a constant number of multiplications to solve the division problem.

The notation $\leq_{sd}$ denotes a constant size and depth reduction; in other words, if f and g are two functions, then $f \leq_{sd} g$ if, given any circuit family computing g in size $S(n)$ and depth $D(n)$, a circuit family can be constructed which computes f in size $O(S(n))$ and depth $O(D(n))$. Letting SQ denote the function that squares an n-bit integer and MULT denote the problem of multiplying two n-bit integers, it is easy to see that SQ $\leq_{sd}$ MULT. It is also true, but not quite as obvious, that MULT $\leq_{sd}$ SQ since $xy = \frac{1}{2}[(x+y)^2 - x^2 - y^2]$ (addition is easily accomplished, and the multiplication by $\frac{1}{2}$ is simply a binary shift by one bit). The notation $\equiv_{sd}$ is used for two problems that are constant size-depth reducible to each other, so SQ $\equiv_{sd}$ MULT as just shown.

Re-examining our rather arbitrary hierarchy of difficulty for arithmetic problems, a good candidate for a reduction of division would be multiplication. In fact, letting REC denote the integer reciprocal problem and using the new notation, it can be shown that SQ $\leq_{sd}$ REC by

$$x^2 = \frac{1}{\frac{1}{x} - \frac{1}{x+1}} - x.[1] \tag{12.1}$$

---

[1]This is actually an abuse of notation, since equation (12.1) uses real reciprocals instead of the defined integer reciprocal; however, it is not hard to see how the integer reciprocal can be used to approximate real reciprocals in the calculation of equation (12.1).

Noting that the $\leq_{sd}$ relation is transitive, this also means that MULT $\leq_{sd}$ REC, so finding reciprocals is at least as hard (in the sense of constant size-depth reductions) as multiplication. This verifies the fact that multiplication is a good candidate when trying to reduce division.

Throughout this chapter, the notation $M(n)$ will be used to represent the smallest size required by any circuit family that multiplies two n-bit numbers in $O(\log n)$ depth. As there are no known optimal algorithms for multiplication at this time, the exact value of $M(n)$ is unknown; however, the value is easily lower-bounded by $M(n) = \Omega(n)$ and upper-bounded by $M(n) = O(n \log n \log \log n)$ (the upper bound is due to an algorithm by Schönhage and Strassen—see the references at the end of the chapter for more information). It is assumed that $M(n)$ satisfies the equation

$$M(cn) \leq cM(n) \qquad (12.2)$$

for all positive $c \leq 1$. Almost all complexity measures that are $\Omega(n)$ satisfy this bound, so the assumption is not too great.

In the text that follows, the notation RECIPROCAL(x, n) refers to the *function* of integer reciprocal, without reference to a particular algorithm; the arguments x and n denote the input value and the size of the input, respectively. When referring to specific algorithms that compute the reciprocal function, the notation used will be RECIP1(x, n), RECIP2(x, n), etc.

12.2
Newton Approximation

Newton approximation is a tool commonly used by numerical analysts to find the zeros of a function. In numerical analysis terms, Newton approximation (in general) has quadratic convergence—what this means to the division problem will become clear shortly.

Consider a differentiable function $f(x)$ that has first derivative $f'(x)$ and has a zero at x_0 (so $f(x_0) = 0$). Assuming that $f'(x)$ is non-zero in a reasonable neighborhood of x_0, we can make an initial guess for x_0 (call the initial guess y_1) and use the slope $f'(y_1)$ to estimate how far y_1 is from the zero. This produces a new estimate for x_0 (call it y_2) and the process can be repeated producing a sequence of estimates $y_1, y_2, y_3, \ldots$ that converges to x_0 for all well-behaved functions and good initial approximations. In mathematical terms, this becomes

$$y_{i+1} = y_i - \frac{f(y_i)}{f'(y_i)}. \qquad (12.3)$$

The convergence rate for the general case is beyond the scope of interest of this chapter—the interested reader can consult any introductory numerical analysis text.

Consider the function $f(y) = 1 - \frac{1}{xy}$. Obviously, $\frac{1}{x}$ is a zero of f, and the derivative $f'(y) = \frac{1}{xy^2}$ is non-zero for all $y \neq 0$. Using this function f, equation (12.3) gives a sequence defined by

$$y_{i+1} = 2y_i - xy_i^2. \tag{12.4}$$

This equation will take a good initial estimate and converge to $\frac{1}{x}$. A word of warning is appropriate here—notice how easily we slipped into solving the problem of real reciprocals instead of integer reciprocals. Fortunately, the problem is not too great—as was noted before, the integer reciprocal is simply a scaled representation of the fixed point binary approximation to the real reciprocal. Re-writing the above equation with this scaling in mind, the following equation generates a sequence that converges to the integer reciprocal using only integer operations.

$$y_{i+1} = \left\lfloor \frac{2^{2n+1} y_i - xy_i^2}{2^{2n}} \right\rfloor \tag{12.5}$$

This formula works quite well, and direct implementation yields a circuit that computes integer reciprocals in size $O(M(n) \log n)$ and depth $O(\log^2 n)$. The $\log n$ multiplier in the size comes from the fact that $\Theta(\log n)$ iterations of equation (12.5) are needed, each of which requires a multiplication of n bit values.

Noticing that the approximation y_i is very inaccurate in the early stages, it seems pointless to do calculations with all the erroneous bits of y_i. In fact, this observation produces a new algorithm which removes the $\log n$ multiplier from the size bound above; the algorithm that accomplishes this is shown in figure 12.1, and proofs of correctness and complexity are given in theorem 12.1. The approximation formula used in figure 12.1 looks different from that in equation (12.5), but the only difference is due to the new scaling required by having only $\frac{n}{2}$ bits for y_i.[2] The "for loop" in algorithm RECIP1 is also a new addition; it is present to overcome errors induced by using fixed point approximation to real numbers. The usefulness of this adjustment stage will become apparent from the proof of theorem 12.1.

---

[2] One way to view this is that now the precision of the fixed point representation is changed at each stage; at the smallest stage, the fractional precision is only $\pm\frac{1}{2}$, but at the next stage the precision is $\pm\frac{1}{4}$, and then $\pm\frac{1}{16}, \pm\frac{1}{256}, \ldots$

Algorithm RECIP1(x, n);
   if $n = 1$
      then begin
         $y \leftarrow 4$;
      end;
      else begin
         $t \leftarrow$ RECIP1$(\lfloor \frac{x}{2^{n/2}} \rfloor, \frac{n}{2})$;
         $y \leftarrow \left\lfloor \frac{2^{\frac{3}{2}n+1}t - xt^2}{2^n} \right\rfloor$;
         for $i \leftarrow 3$ **downto** 0 **do**
            if $(x(y + 2^i) \le 2^{2n})$
               then begin
                   $y \leftarrow y + 2^i$;
               end;
      end;
   return (y);
end.

FIGURE 12.1
Algorithm RECIP1.

For the remainder of this section, as well as in sections 12.4 and 12.5, the n-bit input x is assumed to satisfy $2^{n-1} \le x < 2^n$ (i.e., the high-order bit is set). The algorithms may be modified so they do not require this assumption by simply shifting x (by bits) into the appropriate range, performing the algorithms found in this chapter, and shifting the results back into the proper range. The complexity of the shifting stages is negligible compared to the complexity of the algorithms discussed. Similarly, it is assumed that n is a power of 2.

THEOREM 12.1
Algorithm RECIP1 *in figure 12.1 correctly computes the integer reciprocal of x, and is realized with a circuit family of size $O(M(n))$ and depth $O(\log^2 n)$.*

PROOF
The following proof of the correctness is rather tedious; this comes from the fact that fixed point approximations to real numbers are used, so

small errors (from rounding or truncating) are introduced at various points. A very simple way to get a feeling for why this method works is to examine how the error of an approximation is affected by equation (12.4); while this is not a proof that algorithm RECIP1 is correct, it does provide insight that is useful if the following proof is found to be confusing.

To simplify notation, let r represent the value returned by RECIP1(x, n). To prove the correctness of RECIP1, it is necessary to show that $r = \lfloor \frac{2^{2n}}{x} \rfloor$; in other words, $xr = 2^{2n} - s$ where $0 \le s < x$. The proof is by induction on n; the correct value for $n = 1$ is stated explicitly in the algorithm.

Assume that the algorithm returns a correct value for inputs of size $\frac{n}{2}$. Let t be the value of RECIP1$(\lfloor \frac{x}{2^{n/2}} \rfloor, \frac{n}{2})$ as in figure 12.1, and let $d = 2^{\frac{3}{2}n+1}t - xt^2$. Also, denote the most significant $\frac{n}{2}$ bits by x_1 and the least significant $\frac{n}{2}$ bits by x_0, so $x = x_1 2^{n/2} + x_0$. The value d can now be written as

$$d = 2^{\frac{3}{2}n+1}t - t^2(x_1 2^{n/2} + x_0).$$

The value of interest in this proof is xr, so first we will find xd and then bound the difference between this and xr.

$$xd = 2^{2n+1}x_1 t + 2^{\frac{3}{2}n+1}x_0 t - t^2(x_1 2^{n/2} + x_0)^2$$

Using the induction hypothesis (that $x_1 t = 2^n - s'$, where $0 \le s' < x_1$), this can be simplified to

$$xd = 2^{3n} - (2^{n/2}s' - tx_0)^2.$$

Dividing by 2^n, the result is

$$\frac{xd}{2^n} = 2^{2n} - (s' - \frac{tx_0}{2^{n/2}})^2.$$

Noting that s' and $\frac{tx_0}{2^{n/2}}$ are both positive and that the difference of these two is squared, it is possible to bound

$$\left(s' - \frac{tx_0}{2^{n/2}}\right)^2 \le \max\left\{(s')^2, \left(\frac{tx_0}{2^{n/2}}\right)^2\right\}.$$

By the induction hypothesis, $s' < x_1 < 2^{n/2}$, so $(s')^2 < 2^{n/2}x_1 \le x$. Furthermore,

$$\left(\frac{tx_0}{2^{n/2}}\right)^2 \le \left(\frac{2^{n/2}x_0}{x_1}\right)^2 < \left(2^{n/2+1}\right)^2 = 2^{n+2} \le 8x,$$

so $(s' - \frac{tx_0}{2^{n/2}})^2 < 8x$. In other words,

$$\frac{xd}{2^n} > 2^{2n} - 8x.$$

Now, considering the value y calculated by the Newton approximation equation,

$$xy = x\left\lfloor \frac{d}{2^n} \right\rfloor > x(\frac{d}{2^n} - 1) = \frac{xd}{2^n} - x > 2^{2n} - 9x$$

The adjustment stage of RECIP1 will adjust the least significant four bits of y to the correct value, as long as RECIPROCAL$(x, n) - y \le 15$ entering the adjustment stage. It has just been shown that, in fact, RECIPROCAL$(x, n) - y \le 9$, so RECIP1 correctly returns the integer reciprocal.

The complexity of the circuit is very straightforward to calculate. To calculate the size, notice that RECIP1 performs only a constant number of multiplications and simpler operations on $O(n)$ bit numbers in addition to the recursive call. In other words, the recurrence

$$S(n) \le S(\frac{n}{2}) + cM(n) \qquad (12.6)$$

$$S(1) = 1$$

describes the size of the circuit for RECIP1. The solution to equation (12.6) is given by

$$S(n) \le c \sum_{i=0}^{\log n} M(\frac{n}{2^i}).$$

The earlier assumption that $M(n)$ satisfies equation (12.2) implies that $M(\frac{n}{2^i}) \le \frac{1}{2^i} M(n)$, so the resulting size is $S(n) = O(M(n))$.

The depth of each level of recursion is bounded by $O(\log n)$, and since there are $\log n$ stages, the total depth is bounded by $O(\log^2 n)$. This is a rather simplistic depth analysis, but closer examination shows that this is the tightest upper bound possible. ∎

12.3
Integer Powering

The seemingly unrelated problems of integer reciprocal and integer powering are actually very closely related. In fact, it has been shown by Beame,

Algorithm REPEATSQ(x, m);
{Consider m in its binary representation:
$$m = m_{\lfloor \log m \rfloor} 2^{\lfloor \log m \rfloor} + \cdots + m_2 2^2 + m_1 2^1 + m_0 2^0\}$$
$\quad i \leftarrow 0$;
$\quad p \leftarrow x$;
$\quad y \leftarrow 1$;
$\quad$ while $i \leq \log n$ do begin
$\quad\quad$ if $m_i = 1$
$\quad\quad\quad$ then begin
$\quad\quad\quad\quad y \leftarrow yp$;
$\quad\quad\quad$ end;
$\quad\quad p \leftarrow p^2$;
$\quad$ end;
end.

FIGURE 12.2
Repeated squaring method of taking powers.

Cook, and Hoover [3] that the two problems are equivalent with respect to constant depth reductions.[3] A survey of the research on integer division shows that all known efficient reciprocal algorithms use powering as an integral part (see the references for information on other reciprocal algorithms).

As an introduction to powering, consider a simple powering algorithm; the problem is to raise an n-bit number x to the m-th power, where $m \leq n$. Now write m in its binary notation, so $m = m_{\lfloor \log m \rfloor} 2^{\lfloor \log m \rfloor} + \cdots + m_2 2^2 + m_1 2^1 + m_0 2^0$. An algorithm (called repeated squaring) that takes advantage of the binary representation of m is shown in figure 12.2.

The complexity analysis of this algorithm is particularly easy, resulting in a circuit family with size $O(M(nm))$ and depth $O(\log n \log m)$. Note that this is considerably better than simply multiplying x by itself m times which takes size $O(mM(nm))$ and depth $\Theta(m \log n)$.

With this algorithm in mind, consider the reciprocal algorithm of the previous section; at first glance, RECIP1 doesn't seem to take any powers greater than squaring y_i. However, if a more global view is invoked, this

---

[3] A constant depth reduction is similar to the constant size and depth reduction mentioned earlier in this chapter, except that the size can increase by a polynomial amount.

squared term is again squared in the next stage, and repeatedly squared until the end of the algorithm. In other words, the algorithm actually takes large powers using the repeated squaring algorithm! An observant reader would have noticed that the depth of the algorithm RECIP1 is the same as the depth of the algorithm REPEATSQ (with $m = n$). Now it can be seen that this is no coincidence—RECIP1 was actually performing operations almost identical to REPEATSQ.

An interesting question now arises: Can reciprocals be computed in depth smaller than $\Omega(\log^2 n)$ if there were an algorithm for computing powers in small depth? Indeed, this is the case (more information on this will be presented in following sections); unfortunately, finding small depth circuits for powering seems to be as hard as looking at the reciprocal problem directly. What follows is a description of a powering algorithm that only requires $O(\log n \log \log n)$ depth (for $m = n$); the algorithm is rather confusing to people who haven't seen anything like it before. A good "warm-up" exercise would be to read and understand the multiplication algorithm of Schönhage and Strassen (see the references). The algorithm presented here consists of two parts: reducing the size of the input number x, and reducing the power.

12.3.1 Bit Reduction

Again, we wish to raise an n-bit number x to a power m, where $m \leq n$. The number x has at most $d = \left\lfloor \frac{n}{\log b} \right\rfloor + 1$ digits in base b notation and can be written as

$$x = x_{d-1}b^{d-1} + \cdots + x_2 b^2 + x_1 b + x_0. \qquad (12.7)$$

If an indeterminate z is substituted for the occurrences of b that are raised to a power, then x can be represented by a polynomial $p(z) = x_{d-1}z^{d-1} + \cdots + x_2 z^2 + x_1 z^1 + x_0$, where $p(b) = x$.

Operations with such polynomials mirror the same operations performed on the numbers themselves, so for example, if x is represented by $p(z)$ and y is represented by $q(z)$, then the product of the two polynomials has the property that $p(z)q(z)|_{z=b} = p(b)q(b) = xy$.[4] The current interest is in powering, and it can be noticed that if x is represented by $p(z)$ and m is an integer, then $[p(z)]^m|_{z=b} = [p(b)]^m = x^m$. Efficient polynomial arithmetic is made possible by a domain change through Fourier transforms; we now see how this is done.

---

[4]The notation $p(x)|_{x=a}$ means the polynomial $p(x)$ evaluated at $x = a$; in other words, $p(a)$. Similarly, $p(z)q(z)|_{z=b}$ means to multiply the polynomials $p(z)$ and $q(z)$, and evaluate the resulting polynomial at $z = b$.

Returning to the original problem, let x be an n-bit number, where n is a power of 2—say $n = 2^p$. The input x can be broken into $k = 2^r$ blocks of $l = 2^{p-r}$ bits each, so letting $b = 2^l$, equation (12.7) becomes

$$x = x_{k-1} 2^{l(k-1)} + \cdots + x_2 2^{2l} + x_1 2^l + x_0.$$

The polynomial representation of x (as described above) is therefore $p(z) = x_{k-1} z^{k-1} + \cdots + x_2 z^2 + x_1 z + x_0$; notice that $p(2^l) = x$.

To raise x to the mth power, simply find the polynomial $[p(z)]^m$ and evaluate at $z = 2^l$. Unfortunately, the polynomial $[p(z)]^m$ has degree $m(k-1)$, which is too large for an efficient powering algorithm (it is an interesting exercise to follow the development of the powering algorithm using all terms of $[p(z)]^m$ to see exactly where things go amiss).

Consider calculating $[p(z)]^m (\bmod z^k - 1)$. When the value $z = 2^l$ is inserted, the result is $x^m (\bmod 2^n - 1)$; by padding the input with zeros and increasing n to insure that $2^n - 1 > x^m$, this method produces the exact answer. Furthermore, polynomials modulo $z^k - 1$ never have degree greater than $k - 1$, so the problem of growing polynomial degrees has disappeared. With this in mind, the subject of most of the remainder of this section will be the problem of modular powering.

Let $b(z) = b_{m(k-1)} z^{m(k-1)} + \cdots + b_2 z^2 + b_1 z + b_0$ be the exact value of $[p(z)]^m$, and let $d(z) = d_{k-1} z^{k-1} + \cdots + d_1 z + d_0$ be the reduction of $b(z)$ modulo $z^k - 1$ so that $d(z)$ has degree less than k. Since $z^k \equiv 1 (\bmod z^k - 1)$, it is easy to see that for $i = 0, 1, \ldots, k - 1$,

$$d_i = \sum_{j=0}^{m-1} b_{jk+i}.$$

All b_i with $i > m(k - 1)$ are assumed to be zero. Let $D = (d_0, d_1, \ldots, d_{k-1})$ and $X = (x_0, x_1, \ldots, x_{k-1})$ denote vectors of the coefficients of $d(z)$ and $p(z)$, respectively. The following lemma demonstrates an efficient way of computing the modular power polynomial $d(x)$ using Discrete Fourier Transforms (DFTs).[5]

LEMMA 12.1
Let $\mathsf{DFT}_k(X) = (t_0, t_1, \ldots, t_{k-1})$. Then $\mathsf{DFT}_k^{-1}((t_0^m, t_1^m, \ldots, t_{k-1}^m)) = D$.

---

[5] If the reader is unfamiliar with the Fourier transform or the Fast Fourier Transform algorithm, an introductory level discussion can be found in (Aho et al., [1]).

PROOF
By the definition of the DFT,

$$t_i = \sum_{j=0}^{k-1} x_j \omega^{ij} = p(\omega^i)$$

for all $i = 0, 1, \ldots, k-1$, where ω is a principal kth root of unity. Raising each t_i to the mth power gives

$$t_i^m = [p(\omega^i)]^m = [p(z)]^m|_{z=\omega^i} = \sum_{j=0}^{m(k-1)} b_j \omega^{ij} = \sum_{p=0}^{m-1} \sum_{q=0}^{k-1} b_{pk+q} \omega^{i(pk+q)}.$$

But $\omega^{i(pk+q)} = (\omega^k)^{ip} \omega^{iq} = \omega^{iq}$ since ω is a kth root of unity, so

$$t_i^m = \sum_{p=0}^{m-1} \sum_{q=0}^{k-1} b_{pk+q} \omega^{iq} = \sum_{q=0}^{k-1} \left(\sum_{p=0}^{m-1} b_{pk+q} \right) \omega^{iq} = \sum_{q=0}^{k-1} d_q \omega^{iq}.$$

By the definition of the DFT, this is simply the ith term of $\text{DFT}_k(D)$. As this holds for all $i = 0, 1, \ldots, k-1$, then $\text{DFT}_k^{-1}((t_0^m, t_1^m, \ldots, t_{k-1}^m)) = D$.
■

The Fourier transform of a k-vector (representing a degree $k-1$ polynomial) requires a principal kth root of unity ω. The polynomials that represent integers have integer coefficients, and to avoid doing computations over the complex field, it is possible to use finite rings as the basis of our computation. The ring of integers modulo $2^k - 1$ has a principal kth root of unity of $\omega = 2$, giving this ring the further nice property that multiplication by powers of ω is easily accomplished by bit shifts. Since computations on each element of X are now done modulo $2^k - 1$ it is clear how the original problem (powering an n-bit number modulo $2^n - 1$) is reduced to smaller subproblems (powering k-bit numbers modulo $2^k - 1$). This reduction can be repeated until the size of the subproblems is trivial. Furthermore, $k^{-1} (\bmod 2^k - 1)$ exists by insuring that k is a power of 2, so the inverse DFT is possible.

A problem arises from the fact that the previous discussion of powering assumes that the *exact* values for the coefficients of $[p(z)]^m (\bmod z^k - 1)$ are known, and the previous paragraph refers to only finding the coefficients modulo $2^k - 1$. The following lemma addresses this problem by showing how large to make k to insure that the coefficients are uniquely represented in this ring (i.e., the coefficients are less than $2^k - 1$).

12.3. Integer Powering

LEMMA 12.2
The coefficients of $[p(z)]^m (\bmod\ z^k - 1)$ are less than $2^k - 1$ if

$$2^r - r(m-1) - lm > 0 \qquad (12.8)$$

(where r, l, and m are defined in the preceding text).

PROOF
First, it is proved by induction that the coefficients of $[p(z)]^m (\bmod\ z^k - 1)$ are less than or equal to $k^{m-1}(2^l - 1)^m$ for $m = 1, 2, \ldots$. The basis of the induction is easy; simply let $m = 1$ and the claimed bound becomes $2^l - 1$. The coefficients of $p(z)$ are all less than or equal to $2^l - 1$ since each coefficient is l bits long.

Now assume the claim is true for $m - 1$ (that is, the coefficients of $[p(z)]^{m-1} (\bmod\ z^k - 1)$ are less than or equal to $k^{m-2}(2^l - 1)^{m-1}$). Let the expansion of $p(z)$ and $[p(z)]^{m-1} (\bmod\ z^k - 1)$ be as follows:

$$p(z) = x_{k-1}z^{k-1} + x_{k-2}z^{k-2} + \cdots + x_2 z^2 + x_1 z + x_0$$
$$[p(z)]^{m-1} = y_{k-1}z^{k-1} + y_{k-2}z^{k-2} + \cdots + y_2 z^2 + y_1 z + y_0$$

Notice that since $z^i \equiv z^{i(\bmod\ k)} (\bmod\ z^k - 1)$,

$$[p(z)]^m = [p(z)][p(z)]^{m-1} \equiv \sum_{i=0}^{k-1} \left(\sum_{j=0}^{k-1} x_j y_{i-j(\bmod\ k)} \right) z^i (\bmod\ z^k - 1).$$

Regardless of the particular values of i and j, it must be true that $x_j \leq 2^l - 1$ and $y_{i-j(\bmod\ k)} \leq k^{m-2}(2^l - 1)^{m-1}$ (by the induction hypothesis), so $x_j y_{i-j(\bmod\ k)} \leq k^{m-2}(2^l-1)^m$. Since there are k terms like this added together for each coefficient of $[p(z)]^m (\bmod\ z^k - 1)$, each coefficient must be less than or equal to $k^{m-1}(2^l - 1)^m$, and the proof by induction is finished.

Returning to the lemma, condition (12.8) states that $2^r > r(m-1) + lm$. In other words, taking each side as an exponent, $2^{(2^r)} > 2^{r(m-1)}2^{lm}$, and since $k = 2^r$ this implies that $k^{m-1}2^{lm} < 2^k$. Loosening the inequality slightly, this implies that (for $m \geq 1$)

$$k^{m-1}(2^l - 1)^m < 2^k - 1. \qquad (12.9)$$

The previous inductive proof showed the coefficients of $[p(z)]^m (\bmod\ z^k - 1)$ must be less than or equal to the left hand side of inequality (12.9), so each coefficient must also be less than $2^k - 1$, completing the proof of the lemma. ∎

Algorithm REDUCE1(x, m, n);
  $p \leftarrow \log n$;
  $q \leftarrow \lceil \log m \rceil$;
  $r \leftarrow \lceil \frac{p}{2} + \frac{2q}{3} \rceil$;
  $k \leftarrow 2^r$;
  Divide x into k blocks of $l = 2^{p-r}$ bits each as $(x_0, x_1, \ldots, x_{k-1})$;
  $(t_0, t_1, \ldots, t_{k-1}) \leftarrow \text{DFT}_k(x_0, x_1, \ldots, x_{k-1})$;
  for all $i = 0, 1, \ldots, k-1$ **pardo begin**
    $u_i \leftarrow \text{MODPOWER}(t_i, m, k)$;
  end;
  $(y_0, y_1, \ldots, y_{k-1}) \leftarrow \text{DFT}_k^{-1}(u_0, u_1, \ldots, u_{k-1})$;
  $y \leftarrow y_0 + y_1 2^l + y_2 2^{2l} + \cdots + y_{k-1} 2^{(k-1)l} (\bmod\, 2^n - 1)$;
  return (y);
end.

FIGURE 12.3
Powering reduction style 1.

As an example of the reduction technique just described, consider a single stage of bit reduction as shown in figure 12.3. The value for k comes from calculations involving lemma 12.2; lemma 12.3 shows how this works. Notice the call on MODPOWER in REDUCE1—this is a recursive call that is left unspecified for the moment. As it turns out, a second type of reduction will be needed for efficient powering, and the recursive call (named MODPOWER here) may be on a *different* type of reduction. Notice the new assumption that $m \leq n^{\frac{3}{8}}$. This assumption simply makes the lemma easier to prove, and will not affect the final powering result at all (in fact, it will become apparent that this is the result of passing the assumption $m \leq n$ down through several lemmas).

LEMMA 12.3
Let $m \geq 16$ and $m \leq n^{\frac{3}{8}}$. Then assuming that MODPOWER(t_i, m, k) correctly returns $t_i^m (\bmod\, 2^k - 1)$, the reduction REDUCE1 shown in figure 12.3 correctly returns $x^m (\bmod\, 2^n - 1)$. Furthermore, if the call on MODPOWER requires size $S(m, k)$ and depth $D(m, k)$, then REDUCE1 requires total size $kS(m, k) + O(nm^{\frac{4}{3}} \log n)$ and total depth $D(m, k) + O(\log n)$.

12.3. Integer Powering

PROOF

The correctness of REDUCE1 follows directly from the previous discussion with the important points being lemma 12.1 and lemma 12.2. The only verification that needs to be done is that the condition (12.8) of lemma 12.2 holds; that is, that $2^r - r(m-1) - lm > 0$. What follows is basically an exercise in minimizing the function on the left hand side.

From figure 12.3, let $r = \lceil \frac{p}{2} + \frac{2q}{3} \rceil$. To avoid the ceiling function write r as $\frac{p}{2} + \frac{2q}{3} + \epsilon$, where ϵ is some value satisfying $0 \leq \epsilon < 1$. Obviously, if condition (12.8) holds for all ϵ in this interval, then the condition must also hold with the ceiling. Substituting this value for r and letting $m = 2^q$ and $l = 2^{p-r}$, the left hand side of condition (12.8) becomes

$$f(p,q,\epsilon) = 2^{\frac{p}{2}+\frac{2q}{3}+\epsilon} - \left(\frac{p}{2} + \frac{2q}{3} + \epsilon\right)(2^q - 1) - 2^{\frac{p}{2}+\frac{q}{3}-\epsilon}.$$

This formula is quite messy, but can be simplified greatly just by taking the partial derivative with respect to ϵ (which will reveal a lot of useful information).

$$\frac{\partial f}{\partial \epsilon}(p,q,\epsilon) = 2^{\frac{p}{2}+\frac{q}{3}} \ln 2 \left(2^{\epsilon+\frac{q}{3}} + 2^{-\epsilon}\right) - (2^q - 1)$$

This function is easily minimized for a given p and q (for an easy trick, substitute $t = 2^\epsilon$ and minimize with respect to t) when $\epsilon = -\frac{q}{6}$. Substituting this value into $\frac{\partial f}{\partial \epsilon}$, for any p and q the minimum value of the partial derivative with respect to ϵ is

$$2^{\frac{p}{2}+\frac{q}{2}+1} \ln 2 - (2^q - 1). \tag{12.10}$$

In terms of p and q, the assumption that $m \leq n^{\frac{3}{8}}$ translates to $q \leq \frac{3p}{8}$ (or equivalently, that $p \geq \frac{8q}{3}$). We wish to show that equation (12.10) is greater than zero for all valid p and q. For any given q, equation (12.10) is minimum when p is at its minimum—in other words, when $p = \frac{8q}{3}$. Making this substitution, equation (12.10) becomes

$$2^{\frac{11q}{6}+1} \ln 2 - (2^q - 1),$$

which is easy to show greater than zero for all $q \geq 0$.

The past few paragraphs have shown that for all valid p, q, and ϵ, the derivative $\frac{\partial f}{\partial \epsilon}(p,q,\epsilon) > 0$. In other words, for all valid p and q, $f(p,q,\epsilon)$ is increasing in ϵ; therefore, for all valid p, q, and ϵ,

$$f(p,q,\epsilon) \geq f(p,q,0) = 2^{\frac{p}{2}+\frac{2q}{3}} - 2^{\frac{p}{2}+\frac{q}{3}} - \left(\frac{p}{2} + \frac{2q}{3}\right)(2^q - 1).$$

Differentiating the right hand side with respect to p gives

$$2^{\frac{p}{2}}\frac{\ln 2}{2}\left(2^{\frac{2q}{3}}-2^{\frac{q}{3}}\right)-\frac{2^q-1}{2}.$$

This is obviously increasing in p, so is minimized when p is minimum; after making the substitution $p = \frac{8q}{3}$ and doing some rearranging, the above becomes

$$\frac{1}{2}\left[\left(2^{2q}-2^{\frac{5q}{3}}\right)\ln 2-(2^q-1)\right],$$

which is easily shown to be greater than zero for all $q \geq 4$. In other words, for a given $q \geq 4$, $f(p,q,0)$ is increasing in p, so to minimize $f(p,q,0)$, again set p to $\frac{8q}{3}$. Therefore,

$$f(p,q,0) \geq f(\frac{8q}{3},q,0) = 2^{2q} - 2^{\frac{5q}{3}} - 2q(2^q-1),$$

which is greater than zero for all $q \geq 4$.

Summarizing, it has been shown that for all valid p, q, and ϵ,

$$f(p,q,\epsilon) \geq f(p,q,0) \geq f(\frac{8q}{3},q,0) \geq 0,$$

so condition (12.8) must hold, and REDUCE1 gives the correct answer by lemma 12.2, lemma 12.1, and the properties of polynomials discussed in the text before lemma 12.1.

The complexity of REDUCE1 relies on two results beyond the scope of this chapter: namely, the DFT_k and DFT_k^{-1} can be computed in size $O(k^2 \log k)$ and depth $O(\log n)$, and the evaluation of $[p(z)]^m|_{z=2^l}$ can be done in size $O(k^2)$ and depth $O(\log n)$. In other words, all steps except the call on MODPOWER can be done in size $O(k^2 \log k)$ and depth $O(\log n)$. Furthermore, since

$$k = 2^{\lceil \frac{p}{2}+\frac{2q}{3}\rceil} \leq 2^{\frac{p}{2}+\frac{2q}{3}+1} = 2n^{\frac{1}{2}}m^{\frac{2}{3}},$$

the above size can be written as $O(nm^{\frac{4}{3}} \log n)$. Including the size for the k calls on MODPOWER, the resulting size is $kS(m,k) + O(nm^{\frac{4}{3}} \log n)$. All recursive calls are done in parallel, so the total depth is $D(m,k) + O(\log n)$. ∎

The problem with repeatedly applying REDUCE1 is that the requirements of lemma 12.3 make reduction to a trivial problem size impossible (since n must be at least $m^{\frac{8}{3}}$); however, it is possible to reduce the power as well as the number of bits.

12.3.2 Power Reduction

Consider raising a number x to the mth power. If m is a perfect square with $w = \sqrt{m}$, it is easy to see that $x^m = (x^w)^w$; unfortunately, m is usually not a perfect square. To handle the more common case, let $v = \lfloor \sqrt{m} \rfloor$ and calculate $(x^v)^v$. Of course, this is not the desired answer, but notice that if $e = m - v^2$ is the error in the exponent of this approximation, e can be easily bounded by

$$e = m - v^2 \leq ((v+1)^2 - 1) - v^2 = 2v = 2\lfloor \sqrt{m} \rfloor.$$

Letting $e' = \lfloor \frac{e}{2} \rfloor$, $x^{e'}$ can be computed, squared, and multiplied by x (if e is odd) to achieve x^e. Notice that this computation of x^e can be done in parallel with the computation of $(x^v)^v$, so the original problem has been reduced to 3 smaller powerings (each of which raises a number to a power less than or equal to $\sqrt{m}$) and a constant number of multiplications. This reduction is called REDUCE2 and is shown in figure 12.4.

The correctness of REDUCE2 follows easily from the above discussion, and the complexity analysis is simple, so the following lemma is stated without proof.

LEMMA 12.4
Assuming MODPOWER(t, m, n) *correctly returns* $t^m (\mod 2^n - 1)$ *for all t, m, and n, the reduction* REDUCE2 *shown in figure 12.4 correctly returns* $x^m (\mod 2^n - 1)$. *Furthermore, if the call on* MODPOWER(t, m, n) *requires size $S(m, n)$ and depth $D(m, n)$, then* REDUCE *requires total size* $3S(\sqrt{m}, n) + O(M(n))$ *and total depth* $2D(\sqrt{m}, n) + O(\log n)$.

Again, there is a problem with using just REDUCE2—while the correct answer is returned, the number of subproblems grows too rapidly, and the depth of the powering circuit using just REDUCE2 is $\Theta(\log n \log m)$. Fortunately, in the design of REDUCE1 and REDUCE2 there were some subtle adjustments made (such as the choice for r in REDUCE1) that allow the two reductions to work very well together. Combining the two reductions is addressed in the following section.

12.3.3 Putting the Pieces Together

The final modular power algorithm consists of an initial reduction using REDUCE2 followed by a test to see if the power has been reduced to smaller than 16. If the power is less than 16, then the result can be computed using the REPEATSQ algorithm presented at the beginning of this section (taking

Algorithm REDUCE2(x, m, n);
  $p \leftarrow \lfloor \sqrt{m} \rfloor$;
  In Parallel do part1, part2
    part1: **begin**
        $t \leftarrow$ MODPOWER(x, p, n);
        $u \leftarrow$ MODPOWER(t, p, n);
      end;
    part2: **begin**
        $e \leftarrow m - p^2$;
        $e' \leftarrow \lfloor \frac{e}{2} \rfloor$;
        $v \leftarrow$ MODPOWER(x, e', n);
        if $(2e' = e)$
          then begin
            $w \leftarrow v^2 (\bmod\, 2^n - 1)$;
          end;
          else begin
            $w \leftarrow xv^2 (\bmod\, 2^n - 1)$;
          end;
      end;
  $y \leftarrow uw (\bmod\, 2^n - 1)$;
  return (y);
end.

FIGURE 12.4
Powering reduction style 2.

---

size $O(M(n))$ and depth $O(\log n)$); otherwise, the subproblems are further reduced by two applications of REDUCE1. All three of these reductions can be viewed together as a single "composite reduction" that produces subproblems with reduced size (i.e., number of bits) and reduced power. A proof of the correctness of this algorithm, along with the complexity analysis, is given in the following theorem.

THEOREM 12.2
Let x be an n-bit integer, and m be an integer with $m^2 \leq n$. The algorithm just described computes $x^m (\bmod\, 2^n - 1)$ in $O(nm^4 \log n \log \log n)$ size and $O(\log n + \log m \log \log m)$ depth.

12.3. Integer Powering

PROOF

The correctness of the above algorithm is proved by induction on the number of complete composite reductions required before the power is reduced below 16. If no reductions are required, the result is correct by the correctness of algorithm REPEATSQ. Assume that $R \geq 1$ reductions are required—by the condition of the theorem, $m \leq n^{\frac{1}{2}}$, so after the first reduction using REDUCE2, each subproblem of raising an n-bit number to the m'th power is such that $m' \leq n^{\frac{1}{4}}$. (Note that this means the condition for lemma 12.3 is satisfied.)

After the first reduction via REDUCE1, each resulting subproblem has $k \geq n^{\frac{1}{2}}(m')^{\frac{2}{3}}$ bits. (Notice that

$$m' = (m')^{\frac{3}{4}}(m')^{\frac{1}{4}} \leq n^{\frac{3}{16}}(m')^{\frac{1}{4}} = \left(n^{\frac{1}{2}}(m')^{\frac{2}{3}}\right)^{\frac{3}{8}} \leq k^{\frac{3}{8}},$$

so the condition for lemma 12.3 is again satisfied.)

Following the second reduction via REDUCE1, each subproblem has $k' \geq k^{\frac{1}{2}}(m')^{\frac{2}{3}}$ bits; using the previous bounds for k, $(m')^2$ can be bounded as

$$(m')^2 = (m')(m') \leq n^{\frac{1}{4}}(m') = \left(n^{\frac{1}{2}}(m')^{\frac{2}{3}}\right)^{\frac{1}{2}}(m')^{\frac{2}{3}} \leq k^{\frac{1}{2}}(m')^{\frac{2}{3}} \leq k'.$$

In other words, after one composite reduction each subproblem of raising a k'-bit number to the m'th power satisfies $(m')^2 \leq k'$. Only $R-1$ composite reductions are required for these subproblems (since R reductions were required for the original problem), and since $(m')^2 \leq k'$, the induction hypothesis applies to say that all these subproblems are correctly solved.

Going backwards through each individual reduction in the composite reduction, it has been noted that the conditions for lemmas 12.3 and 12.4 have been satisfied, so the correctness of the algorithm follows directly from these lemmas.

Now examine the size required for this algorithm. Let $S(m,n)$ denote the size of raising an n-bit number to the mth power modulo $2^n - 1$. The result of applying the size of REDUCE2 (from lemma 12.4) to the size of REDUCE1 (from lemma 12.3) which is again applied to itself gives the size for one composite reduction. The result is (using k, k', and m' as defined above)

$$S(m,n) = 3kk'S(m',k') + O(k^2(m')^{\frac{4}{3}} \log k)$$
$$+ O(n(m')^{\frac{4}{3}} \log n) + O(M(n)).$$

Using the bounds $k \leq 2n^{\frac{1}{2}}(m')^{\frac{2}{3}}$ (see the proof of lemma 12.3) and $m' \leq m^{\frac{1}{2}}$, in addition to the new bound $k' \leq 2k^{\frac{1}{2}}(m')^{\frac{2}{3}} = 2^{\frac{3}{2}}n^{\frac{1}{4}}m'$, gives a size of

$$S(m,n) = 3kk'S(m',k') + O(nm^{\frac{4}{3}}\log n) + O(nm^{\frac{2}{3}}\log n) + O(M(n)).$$

Using the Schönhage and Strassen algorithm, we know that $M(n) = O(n \log n \log \log n)$, so this can be simplified greatly to

$$S(m,n) = 3kk'S(m',k') + O(nm^{\frac{4}{3}}\log n \log \log n).$$

Removing the big-O notation, the above size bound can be expressed (for some constant c) as

$$S(m,n) \leq 3kk'S(m',k') + cnm^{\frac{4}{3}}\log n \log \log n.$$

Notice that this size only applies if a complete composite reduction is performed (i.e., $m' \geq 16$ or $m \geq 256$). For $m < 256$, only a constant number of multiplications are required, so $S(m,n) = O(M(n))$.

The claim is that $S(m,n) \leq c'nm^4 \log n \log \log n$ for some c', and is proved by induction on m. For $m < 256$ and the appropriate c' and c'',

$$S(m,n) \leq c''M(n) \leq c'nm^4 \log n \log \log n,$$

so this serves as a basis for the induction. Now assume $m \geq 256$, and the induction hypothesis states that

$$S(m',k') \leq c'k'(m')^4 \log k' \log \log k'$$

for $m' < m$. Using the bound $k' \leq 2^{\frac{3}{2}}n^{\frac{1}{4}}m^{\frac{1}{2}}$ and noticing that $2^{\frac{3}{2}}m^{\frac{1}{2}} \leq m^{\frac{11}{16}}$ for $m \geq 256$, k' can now be bounded as $k' \leq n^{\frac{1}{4}}m^{\frac{11}{16}} \leq n^{\frac{19}{32}}$. This means that $\log k' \leq \frac{19}{32}\log n$, so using all the upper bounds,

$$3(kk')S(m',k') \leq 3\left(2^{\frac{5}{2}}n^{\frac{3}{4}}m^{\frac{5}{6}}\right)\left(c'2^{\frac{3}{2}}n^{\frac{1}{4}}m^{\frac{1}{2}}m^2 \frac{19}{32}\log n \log \log n\right)$$
$$= \frac{57}{2}c'nm^{\frac{10}{3}}\log n \log \log n,$$

so

$$S(m,n) \leq \frac{57}{2}c'nm^{\frac{10}{3}}\log n \log \log n + cnm^{\frac{4}{3}}\log n \log \log n$$
$$\leq \left(\frac{57}{2}c'm^{-\frac{2}{3}} + cm^{-\frac{8}{3}}\right)nm^4 \log n \log \log n.$$

Since $m \geq 256$, this can be loosely upper bounded by

$$S(m,n) \leq \left(\frac{3}{4}c' + c\right) nm^4 \log n \log\log n,$$

and for $c' \geq 4c$ this becomes

$$S(m,n) \leq c' nm^4 \log n \log\log n,$$

proving the claimed size bound.

Turning to the depth, let $D(m,n)$ represent the depth of raising an n-bit number to the mth power modulo $2^n - 1$, and the depth of a composite reduction can be expressed as

$$D(m,n) = 2D(m',k') + O(\log n)$$

for $m \geq 256$ (i.e., $m' \geq 16$), and $D(m,n) = O(\log n)$ for $m < 256$. A depth bound of $D(m,n) \leq c'(\log n + \log m \log\log m)$ can be proved by induction; the basis follows easily for $m < 256$.

For $m \geq 256$, the induction hypothesis states that

$$D(m',k') \leq c'(\log k' + \log m' \log\log m').$$

Since $m' \leq m^{\frac{1}{2}}$, we can bound $\log m' \log\log m' \leq \frac{1}{2}\log m(\log\log m - 1)$, so

$$D(m',k') \leq c'(\frac{3}{2} + \frac{1}{4}\log n + \frac{1}{2}\log m + \frac{1}{2}\log m(\log\log m - 1)$$

$$= c'(\frac{3}{2} + \frac{1}{4}\log n + \frac{1}{2}\log m \log\log m).$$

In other words, for some constant c,

$$D(m,n) \leq 2D(m',k') + c\log n$$
$$\leq \left(\frac{3c'}{\log n} + \frac{c'}{2} + c\right) \log n + c' \log m \log\log m.$$

Since $\frac{3c'}{\log n} \leq \frac{3c'}{2\log m} \leq \frac{3c'}{16}$ for $m \geq 256$,

$$D(m,n) \leq \left(\frac{11}{16}c' + c\right) \log n + c' \log m \log\log m.$$

For $c' \geq \frac{16}{5}c$, this can be simplified to

$$D(m,n) \leq c'(\log n + \log m \log\log m),$$

proving the claimed depth bound. ∎

Returning to the original (exact) powering problem, the following easy corollary completes the study of integer powering.

COROLLARY 12.1
If x is an n-bit integer and m is an integer satisfying $m \leq n$, then x^m can be computed by a circuit of size $O(nm^5 \log n \log \log n)$ and depth $O(\log n + \log m \log \log m)$.

PROOF
Let $N = nm$. Since $m \leq n$, multiplying both sides of the inequality by m shows that $m^2 \leq nm = N$. By theorem 12.2, after padding x with zeros in the most significant $n(m-1)$ places, $x^m (\bmod\, 2^N - 1)$ can be computed in size $O(Nm^4 \log N \log \log N) = O(nm^5 \log n \log \log n)$ and depth $O(\log N + \log m \log \log m) = O(\log n + \log m \log \log m)$. Since x^m must be less than $2^{nm} - 1$, the modular computation actually gives the exact value of x^m. ∎

12.4
High Order Convergence with Newton Approximation

Given that repeated application of the Newton approximation formula given in Section 12.2 computes powers in a depth-inefficient way, it is worthwhile to examine how efficient powering methods can be incorporated to reduce the complexity of finding reciprocals.

Recall the approximation formula for finding *real* reciprocals given in equation (12.4). The initial ideas here are presented in terms of real reciprocals, and then the simple changes to the integer reciprocal problem are examined. Some algebraic manipulation shows that applying the approximation formula twice, the approximation refinement becomes

$$y_{i+2} = y_i(1 + (1 - xy_i) + (1 - xy_i)^2 + (1 - xy_i)^3).$$

In fact, the original equation can be rewritten as

$$y_{i+1} = y_i(1 + (1 - xy_i)),$$

with the basic pattern emerging of

$$y_{i+m} = y_i \sum_{j=0}^{2^m - 1} (1 - xy_i)^j. \tag{12.11}$$

12.4. High Order Convergence with Newton Approximation

(Of course, we haven't *proven* that this is the general form of repeated application of equation (12.4)—this is left to the interested reader. A proof that this equation, after scaling, gives the correct answer will be given in theorem 12.5.)

A nice property of equation (12.11) is that the upper limit of the sum does not necessarily have to be of the form $2^m - 1$ in order to work correctly. We wish to view an application of equation (12.11) as a single approximation step, so the new approximation formula can be written as

$$y_{i+1} = y_i \sum_{j=0}^{k-1} (1 - xy_i)^j. \tag{12.12}$$

This equation is called the kth order Newton approximation formula; the name comes from the fact that convergence is of order k. Desirable convergence properties can be proven for equation (12.12), but as we are interested in integer reciprocals, the scaled version should be examined first. Performing fixed point scaling exactly as was done for the second order formula of Section 12.2 gives a fixed-point equation; however, as before, only a small number of bits of y_i need to be considered in the calculation of y_{i+1}. If we let $y_i = \mathsf{RECIPROCAL}(\lfloor \frac{x}{2^{n-d}} \rfloor, d)$ (i.e., the integer reciprocal of the d most significant bits of x), and $x' = \lfloor \frac{x}{2^{n-dk}} \rfloor$ (the dk most significant bits of x) then the resulting equation is

$$y_{i+1} = \left\lfloor \frac{y_i \sum_{j=0}^{2k-1} 2^{d(k+1)(2k-j-1)} (2^{d(k+1)} - x'y_i)^j}{2^{2dk^2}} \right\rfloor \tag{12.13}$$

Notice that here the upper limit on the sum is $2k - 1$ instead of $k - 1$—the upper limit has been raised to overcome the same type of problem that required the adjustment stage of RECIP1; however, equation (12.13) is still referred to as the kth order Newton approximation formula.

To construct an algorithm using equation (12.13), the exact order of each approximation step must be considered; this schedule of approximations depends on complexity considerations and will be addressed in the next section. The following lemma shows how equation (12.13) affects an approximation.

LEMMA 12.5
If $d \geq 2$ and $y_i = \mathsf{RECIPROCAL}(\lfloor \frac{x}{2^{n-d}} \rfloor, d)$, then equation (12.13) gives y_{i+1} that satisfies

$$0 \leq \mathsf{RECIPROCAL}(\lfloor \frac{x}{2^{n-dk}} \rfloor, dk) - y_{i+1} \leq 2.$$

Furthermore, equation (12.13) can be evaluated by a circuit family with size $O(dk^7 \log dk \log \log dk)$ and depth $O(\log dk + \log k \log \log k)$.

PROOF

This proof closely parallels the proof of theorem 12.1. Writing x' in two parts as $x' = x_1 2^{d(k-1)} + x_0$, the assumption on y_i states that $y_i = \text{RECIPROCAL}(x_1, d)$, or that $x_1 y_i = 2^{2d} - s$, where $0 \leq s < x_1$. This implies that $x' y_i = (x_1 2^{d(k-1)} + x_0) y_i = 2^{d(k+1)} - (2^{d(k-1)} s - x_0 y_i)$. To simplify notation, let $w = 2^{d(k+1)}$ and $z = (2^{d(k-1)} s - x_0 y_i)$, so $x' y_i = w - z$.

Let

$$d = y_i \sum_{j=0}^{2k-1} 2^{d(k+1)(2k-j-1)} (2^{d(k+1)} - x' y_i)^j = y_i \sum_{j=0}^{2k-1} w^{2k-j-1} z^j.$$

The quantity of interest is $x' y_{i+1}$, so first compute $x' d$ as

$$x' d = (w - z) \sum_{j=0}^{2k-1} w^{2k-j-1} z^j = \sum_{j=0}^{2k-1} w^{2k-j} z^j - \sum_{j=0}^{2k-1} w^{2k-j-1} z^{j+1}$$

$$= w^{2k} - z^{2k} = 2^{2dk(k+1)} - (2^{d(k-1)} s - x_0 y_i)^{2k}.$$

Dividing by 2^{2dk^2} gives

$$\frac{x' d}{2^{2dk^2}} = 2^{2dk} - \left[\frac{s}{2^d} - \frac{x_0 y_i}{2^{dk}} \right]^{2k}.$$

Since $\frac{s}{2^d}$ and $\frac{x_0 y_i}{2^{dk}}$ are both positive, we can bound

$$\left| \frac{s}{2^d} - \frac{x_0 y_i}{2^{dk}} \right| \leq \max \left\{ \frac{s}{2^d}, \frac{x_0 y_i}{2^{dk}} \right\}. \quad (12.14)$$

The first of these terms is easy to bound: $\frac{s}{2^d} < 1$ since $s < x_1 < 2^d$. To bound the second term, notice that $y_i = \left\lfloor \frac{2^{2d}}{x_1} \right\rfloor \leq \frac{2^{2d}}{x_1}$, so

$$\frac{x_0 y_i}{2^{dk}} \leq \frac{x_0}{2^{d(k-2)} x_1} < \frac{2^{d(k-1)}}{2^{d(k-2)} 2^{d-1}} = 2.$$

Therefore, using equation (12.14),

$$\left(\frac{s}{2^d} - \frac{x_0 y_i}{2^{dk}} \right)^{2k} = \left(\left| \frac{s}{2^d} - \frac{x_0 y_i}{2^{dk}} \right| \right)^{2k} < 2^{2k}. \quad (12.15)$$

Since $d \geq 2$, this can be further bounded as

$$2^{2k} \leq 2^{dk} < 2 \cdot 2^{dk-1} \leq 2x'.$$

Notice that since the power $2k$ on the left hand side of equation (12.15) is even, the error term in equation (12.15) must be positive; in other words, $\frac{x'd}{2^{2dk^2}} \leq 2^{2dk}$. It follows that $2^{2dk} - 2x' < \frac{x'd}{2^{2dk^2}} \leq 2^{2dk}$.

The formula in equation (12.13) actually uses $\left\lfloor \frac{d}{2^{2dk^2}} \right\rfloor$, so

$$x'y_{i+1} = x' \left\lfloor \frac{d}{2^{2dk^2}} \right\rfloor > x' \left(\frac{d}{2^{2dk^2}} - 1 \right) = \frac{x'd}{2^{2dk^2}} - x' > 2^{2dk} - 3x'$$

If $\text{RECIPROCAL}(x', dk) - y_{i+1} \geq 3$, then $x'y_{i+1} \leq 2^{2dk} - 3x'$. As just shown, this is impossible, so $\text{RECIPROCAL}(x', dk) - y_{i+1} \leq 2$.

To evaluate equation (12.13), a circuit has to compute the jth power of $d(k+1)$ bit numbers, for $0 \leq j < 2k$. Noticing that for each j the size of this powering is $O(dkj^5 \log dk \log \log dk)$ from corollary 12.1, the total size required to take all the powers necessary is asymptotically upper-bounded by

$$\sum_{j=0}^{2k-1} cdkj^5 \log dk \log \log dk = cdk \log dk \log \log dk \sum_{j=0}^{2k-1} j^5$$

$$< cdk^7 \log dk \log \log dk.$$

As the reader can easily verify, the cost of adding these powers, multiplying by y_i, and scaling back down are all negligible compared the cost of powering, so the size of the circuit to evaluate equation (12.13) is $O(dk^7 \log dk \log \log dk)$.

All the powers are done in parallel, each having depth at most $O(\log dk + \log k \log \log k)$, and every other operation (the large sum and the rescaling) in the evaluation of equation (12.13) can be shown to have depth $O(\log dk)$; therefore, the total depth of evaluating equation (12.13) is $O(\log dk + \log k \log \log k)$. ∎

12.5
An Efficient Parallel Reciprocal Circuit

The results of the previous section can be used to design a parallel algorithm for finding reciprocals in depth $O(\log n \log \log n)$. In essence, lemma 12.5 says that an approximation to the reciprocal that is accurate to d bits can be extended to an accuracy of dk bits in $O(dk^7 \log dk \log \log dk)$ size and $O(\log dk + \log k \log \log k)$ depth.

Chapter 12. Newton Iteration and Integer Division

To design a reciprocal algorithm, we need to come up with a sequence of approximation accuracies $d_1, d_2, d_3, \ldots$ such that after doing i approximation refinements, the result is accurate to d_i bits; eventually, all n bits should be known. In searching for criteria to design such a sequence, a desirable feature of parallel algorithms is that the work is spread out evenly across time. Looking at the form of the size bound from lemma 12.5, a good candidate is to set the size of each stage to $O(n \log n \log \log n)$. Setting $d_1 = 2$ (so two bit are known initially), the schedule then works out as

$$d_i k_i^7 \log d_i k_i \log \log d_i k_i \leq n \log n \log \log n$$

$$\Longrightarrow d_i \left(\frac{d_{i+1}}{d_i}\right)^7 \log d_{i+1} \log \log d_{i+1} \leq n \log n \log \log n$$

$$\Longrightarrow d_{i+1}^7 \log d_{i+1} \log \log d_{i+1} \leq n d_i^6 \log n \log \log n$$

Noticing that $d_{i+1} \leq n$ at all times (otherwise, the whole answer would be known!), the above inequality is satisfied with

$$d_{i+1} = n^{\frac{1}{7}} d_i^{\frac{6}{7}}.$$

Solving this recurrence (with the initial condition $d_1 = 2$) reveals that the sequence of accuracies is

$$d_i = 2n^{1-\left(\frac{6}{7}\right)^{i-1}}.$$

Unfortunately, this schedule does not produce just integers for accuracies (in fact, not necessarily even rational numbers!), so instead, let $m = \log n$ (recall that n is a power of 2 by assumption) and define the function

$$f(i) = \left\lfloor m\left(1 - \left(\frac{6}{7}\right)^{i-1}\right) \right\rfloor. \tag{12.16}$$

Then the schedule can be defined by

$$d_i = 2^{f(i)}. \tag{12.17}$$

The result is the algorithm shown in figure 12.5.

LEMMA 12.6

Algorithm RECIP2 *shown in figure 12.5 correctly computes the reciprocal of an n-bit number, and can be realized with a circuit family of size $O(n \log n (\log \log n)^2)$ and depth $O(\log n \log \log n)$.*

Algorithm RECIP2(x, n);
    $m \leftarrow \log n$;
    $d_1 \leftarrow 2$;
    $i \leftarrow 2$;
    if $(x \geq 3 \cdot 2^{n-2})$
      then begin
        $y_1 \leftarrow 5$;
      end;
      else begin
        $y_1 \leftarrow 8$;
      end;
    while $i \leq \left\lceil \frac{\log \log n}{\log \frac{7}{6}} \right\rceil$ **do begin**
      $t \leftarrow \left\lfloor m \left(1 - \left(\frac{6}{7}\right)^{i-1}\right) \right\rfloor$;
      $d_i \leftarrow 2^t$;
      $k_i \leftarrow \frac{d_i}{d_{i-1}}$;
      $x' \leftarrow \left\lfloor \frac{x}{2^{n-d_i}} \right\rfloor$;
      $y_{i+1} \leftarrow \left\lfloor \frac{y_i \sum_{j=0}^{2k_i - 1} 2^{d_{i-1}(k_i+1)(k_i-j-1)} \left(2^{d_{i-1}(k_i+1)} - x'y_i\right)^j}{2^{2d_{i-1}k_i^2}} \right\rfloor$;
      for $j \leftarrow 1$ **downto** 0 **do**
        if $(x'(y_{i+1} + 2^j)) \leq 2^{2d_i}$
          then begin
            $y_{i+1} \leftarrow y_{i+1} + 2^j$;
          end;
      $i \leftarrow i + 1$;
    end;
    return (y_i);
end.

FIGURE 12.5
Algorithm RECIP2.

PROOF

The fact that algorithm RECIP2 meets the schedule of equation (12.17) is a very simple proof by induction. The basis of the induction is trivial—the integer reciprocals of the two possible two-bit numbers are hard-wired into the algorithm. The induction step is proved by lemma 12.5 (notice the adjustment step in figure 12.5 that takes up the slack in possible error from lemma 12.5). The final answer after $p = \left\lceil \frac{\log \log n}{\log \frac{7}{6}} \right\rceil + 1$ steps is d_p bits. Computing $f(p)$ (where f is defined in equation (12.16)) shows that $f(p) = m$; in other words, $d_p = 2^m = n$.

By lemma 12.5, the size of stage i is $O(d_{i-1} k_i^7 \log d_{i-1} k_i \log \log d_{i-1} k_i)$. Examining k_i, the order of approximation at stage i is $k_i = 2^{f(i)-f(i-1)}$. Focusing on the exponent,

$$f(i) - f(i-1) = \left\lceil m \left(\frac{6}{7}\right)^{i-2} \right\rceil - \left\lceil m \left(\frac{6}{7}\right)^{i-1} \right\rceil$$

$$\leq \left\lceil m \left(\frac{6}{7}\right)^{i-2} \right\rceil - \frac{6}{7} \left\lceil m \left(\frac{6}{7}\right)^{i-2} \right\rceil + 1$$

$$= \frac{1}{7} \left\lceil m \left(\frac{6}{7}\right)^{i-2} \right\rceil + 1.$$

This means that

$$d_{i-1} k_i^7 \leq 2^{m - \lceil m(\frac{6}{7})^{i-2} \rceil + \lceil m(\frac{6}{7})^{i-2} \rceil + 7} = O(n).$$

Furthermore, since $d_{i-1} k_i < n$, the total size of stage i (regardless of i) is $O(n \log n \log \log n)$. Over all $O(\log \log n)$ stages, the total size becomes $O(n \log n (\log \log n)^2)$.

By lemma 12.5, the depth of stage i is $O(\log d_{i-1} k_i + \log k_i \log \log k_i)$. Examining each term separately, the first term is $O(\log n)$ for all i, which produces a total depth of $O(\log n \log \log n)$ over all stages. In the second term, $\log \log k_i$ can be bounded by $\log \log n$ to obtain a depth over all stages of

$$\sum_{i=2}^{p} \log k_i \log \log k_i \leq \log \log n \sum_{i=2}^{p} \log k_i$$

$$= \log \log n \sum_{i=2}^{p} [f(i) - f(i-1)]$$

$$= \log \log n \, [f(p) - f(1)] = O(\log n \log \log n).$$

12.5. An Efficient Parallel Reciprocal Circuit

Combining both terms of the depth, the total depth can be seen to be $O(\log n \log \log n)$. ∎

The algorithm RECIP2 just described is certainly an efficient reciprocal algorithm (in terms of both size and depth), but it does not clearly specify a relationship between the complexity of multiplication and that of division. (The similarity of the size bound with the size of the Schönhage-Strassen multiplication algorithm is mere coincidence.) In this sense, algorithm RECIP1 was better, since the size was closely tied to the size of multiplication (in fact, the size was $O(M(n))$). Can the good qualities of both algorithms (the size bound of RECIP1 and the small depth of RECIP2) be combined? Fortunately, the answer to this question is yes.

The new algorithm is RECIP3 shown in figure 12.6; the value N is the number of bits of the *original* problem (before any reductions). The basic idea behind algorithm RECIP3 is to use RECIP2 to find a sufficiently accurate initial estimate of the integer reciprocal so that only $O(\log \log n)$ stages of second order approximations are needed.

THEOREM 12.3
Algorithm RECIP3 *in figure 12.6 correctly computes the reciprocal of an n-bit number, and can be realized with a circuit family of size $O(M(n))$ and depth $O(\log n \log \log n)$.*

PROOF
Algorithm RECIP3 is a hybrid of RECIP1 and RECIP2, and the correctness follows directly from the correctness of those algorithms (see theorem 12.1 and lemma 12.6).

After i steps of recursion in RECIP3, $n = \frac{N}{2^i}$, so it only takes $\log(\log^2 N) = O(\log \log N)$ steps of second order reduction before $n \leq \frac{N}{\log^2 N}$. The complexity analysis of the second order stages is identical to theorem 12.1, but with only $O(\log \log N)$ stages. In other words, the size of the second order approximations (not counting the call on RECIP2) is $O(M(N))$ and the depth is $O(\log N \log \log N)$.

The size of the call on RECIP2 is easily computed from lemma 12.6 to be
$$O(\frac{N}{\log^2 N} \log \frac{N}{\log^2 N} \log \log \frac{N}{\log^2 N}) = O(N),$$
and the depth is $O(\log N \log \log N)$.

Algorithm RECIP3(x, n);
    if $n \leq \frac{N}{\log^2 N}$ {N is the size of the original problem.}
        then begin
            $y \leftarrow$ RECIP2(x, n);
        end;
    else begin
        $t \leftarrow$ RECIP3($\lfloor \frac{x}{2^{n/2}} \rfloor, \frac{n}{2}$);
        $y \leftarrow \lfloor \frac{2^{\frac{3}{2}n+1}t - xt^2}{2^n} \rfloor$;
        if $(x(y+1) \leq 2^{2n})$
            then begin
                $y \leftarrow y + 1$;
            end;
    end;
    return (y);
end.

FIGURE 12.6
Algorithm RECIP3.

Combining the complexity of the second order stages with the complexity of the call on RECIP2, the final result is that the circuit for RECIP3 has size $O(M(N))$ and depth $O(\log N \log \log N)$. ∎

12.6
Summary

This chapter examined the most complex of the basic arithmetic problems—division. The algorithm presented in this chapter is essentially that of Reif and Tate [12]; some minor changes have been made to clarify the presentation. While it can be shown that division is at least as hard as the other arithmetic problems (addition, subtraction, and multiplication), it is unknown whether division is strictly *harder* than the other operations. In comparison with multiplication (the second hardest problem), the results of theorem 12.3 show that while it may still be possible that division is harder than multiplication, the difference is not all that great (in terms of asymptotic growth).

There is potential for future research on division, either in finding a lower bound or in finding a better upper bound that could possibly match that of multiplication (as is the case sequentially).

For further information, the interested reader can consult Pippenger [9] for an excellent summary of computing arithmetic functions with various circuit models. For more in-depth treatment of multiplication, consult Schönhage and Strassen [13]. Sequential treatment of the basic arithmetic problems (including Schönhage and Strassen's multiplication algorithm), as well as an introduction to the Fast Fourier Transform, is covered in Aho et al. [1]. For a historical tour through the development of division algorithms (most of which use variations on the method described in Exercise 12.6), consult Cook [5], Reif [10], Beame et al. [3], Reif [11], Hastad and Leighton [7] Melhorn and Preparata [8], and Shankar and Ramachandran [14]. Treatment of more complex functions (such as square root and logarithms) can be found in Alt [2]. For various approaches to the polynomial reciprocal problem, see Bini and Pan [4] and Eberly [6] as well as Reif and Tate [12] which describes the algorithm hinted at by Exercise 12.7.

12.7
Exercises

12.1   Prove that the integer division problem can be solved by finding a single integer reciprocal and performing a constant number of multiplications and divisions.

12.2   Derive an equation that describes how the error is affected by

   a) equation (12.4)

   b) equation (12.12).

   What assumptions must be made on the value of x in order for the iteration equations to converge to the appropriate answer?

12.3   The depth of any bounded fan-in circuit family for multiplication is $\Omega(\log n)$. Calculate the depth of algorithm RECIP1 more exactly than was done in theorem 12.1 to show that it is $\Omega(\log^2 n)$.

12.4   Since two reductions by REDUCE1 give subproblems with approximately $n^{1/4}m$ bits, why couldn't we define r in REDUCE1 at $\frac{1}{4}p + q$ so as to only require *one* reduction? How does using two reductions overcome this problem (what is the number of subproblems produced in each case)?

12.5 How would the error estimate of lemma 12.5 be affected if the upper limit on the sum in equation (12.13) was $k-1$ instead of $2k-1$?

12.6 There are, of course, other ways of computing reciprocals than with the algorithm presented in this chapter. The purpose of this exercise is to derive a different algorithm for computing reciprocals. For simplicity, assume we are solving the real reciprocal problem using fixed point binary representation. The input x takes n bits to represent and is assumed to be in the range $(0,1)$. We want the answer z such that $|z - \frac{1}{x}| < 2^{-n}$.

   a) Derive the Maclaurin series expansion for $f(u) = \frac{1}{1-u}$. For what values of u does this series converge?
   b) Use this series to design an algorithm that computes the reciprocal of x. The range of values for u should be restricted so that only $O(n)$ terms of the series need to be computed to achieve the desired accuracy.
   c) Using the powering complexity from corollary 12.1, what is the complexity of this reciprocal algorithm?

12.7 Let $R = \{D, +, \cdot, 0, 1\}$ be a ring, and let $p(x)$ be a degree n polynomial in $R[x]$. The polynomial reciprocal problem on input $p(x)$ is defined as computing the unique polynomial $q(x)$ such that

$$x^{2n-2} = q(x)p(x) + r(x),$$

where the degree of $r(x)$ is less than n. The floor notation can be used to equivalently state

$$\mathsf{PRECIP}(p(x)) = q(x) = \left\lfloor \frac{x^{2n-2}}{p(x)} \right\rfloor.$$

The model of computation for this problem is a circuit where each node can perform addition, multiplication, or reciprocation (when it is defined) in the ring R.

A form of Newton iteration can be used to solve this problem, and provides good sequential results. Consider the polynomial $p(x)$ in "halves", so $p(x) = p_1(x)x^{n/2} + p_0(x)$, and let $q_1(x) = \mathsf{PRECIP}(p_1(x))$. Then the reciprocal of $p(x)$ can be calculated by

$$\mathsf{PRECIP}(p(x)) = \left\lfloor \frac{2q_1(x)x^{(3/2)n-2} - p_0(x)(q_1(x))^2}{x^{k-2}} \right\rfloor.$$

Notice how similar this formula is to the second-order integer iteration formula of algorithm **RECIP1**. Decide what is meant by a "high-order" formula for this problem, and derive the iteration formula. Use this formula to design an efficient parallel algorithm for the polynomial reciprocal problem.

In analyzing the complexity of this algorithm, let $PM(n)$ be the size complexity of multiplying two degree n polynomials in depth $O(\log n)$. Also assume that the mth power of a degree n polynomial can be computed in size $O(nm \log n)$ and depth $O(\log n)$. The algorithm for this problem should have size $O(PM(n))$ and depth $O(\log n \log \log n)$.

Bibliography

[1] A. V. Aho, J. E. Hopcroft, and J. D. Ullman. *The Design and Analysis of Computer Algorithms*. Addison-Wesley, Reading, MA, 1974.

[2] H. Alt. Comparing the combinatorial complexities of arithmetic functions. *J. Assoc. Comput. Mach.*, 35(2):447–460, April 1988.

[3] P. W. Beame, S. A. Cook, and H. J. Hoover. Log depth circuits for division and related problems. *SIAM J. Comput.*, 15(4):994–1003, November 1986.

[4] D. Bini and V. Pan. Polynomial division and its computational complexity. *J. of Complexity*, 2:179–203, 1986.

[5] S. A. Cook. *On The Minimum Computation Time of Functions*. PhD thesis, Harvard University, Cambridge, MA, 1966.

[6] W. Eberly. Very fast parallel matrix and polynomial arithmetic. *Proc. 25th Annual IEEE Symposium on Foundations of Computer Science*, pages 21–30, 1984.

[7] J. Hastad and T. Leighton. Division in $O(\log n)$ depth using $O(n^{1+\epsilon})$ processors, 1986. Unpublished note.

[8] K. Melhorn and F. P. Preparata. Area-time optimal division for $T = \Omega((\log n)^{1+\epsilon})$. *Symposium on Theoretical Aspects of Computer Science*, pages 341–352. Lecture Notes in Computer Science 210, Springer-Verlag, 1986.

[9] N. Pippenger. The complexity of computations by networks. *IBM J. Res. Dev.*, 31(2):235–243, March 1987.

[10] J. H. Reif. Logarithmic depth circuits for algebraic functions. *Proc. 24th Annual IEEE Symposium on Foundations of Computer Science*, pages 138–145, 1983.

[11] J. H. Reif. Logarithmic depth circuits for algebraic functions. *SIAM J. Comput.*, 15(1):231–241, February 1986.

[12] J. H. Reif and S. R. Tate. Optimal size integer division circuits. *21st STOC*, pages 264–273, 1989.

[13] A. Schönhage and V. Strassen. Schnelle multiplikation grosser zahlen. *Computing*, 7:281–292, 1971.

[14] N. Shankar and V. Ramachandran. Efficient parallel circuits and algorithms for division. *Inform. Process. Lett*, 29:307–313, 1988.

13

Parallel Linear Algebra

Joachim von zur Gathen

Department of Computer Science
University of Toronto,
Toronto, Ontario M5S 1A4, Canada
gathen@theory.toronto.edu

13.1
Introduction

A natural model of algebraic computation (for polynomials over a field, say) is the *arithmetic circuit*, using the arithmetic operations $+, -, *, /$. Sections 13.2 and 13.3 give some easy parallel algorithms, and Section 13.4 deals with the first fundamental problem: how to solve systems of linear equations fast in parallel (and: how to compute the determinant or characteristic polynomial of a matrix). As usual in this volume, "fast parallel" means parallel time $(\log n)^{O(1)}$ (in fact, $O(\log^2 n)$ for our problems), and $n^{O(1)}$ processors, where n is the input size.

Section 13.5 introduces the important tool of *reductions* within our framework. This allows us to consider the "relative difficulty" of one problem with respect to another, without requiring the knowledge of the "absolute difficulty" of these problems.

Section 13.6 gives the second fundamental algorithm: computing the rank of a matrix. With this machinery, the elementary problems of linear algebra can be classified into very few groups, with the same parallel complexity within each group (although we may not know what exactly this complexity is).

In Section 13.8, the model is extended to include *tests for zero* and *selection*; this is necessary in order to deal with problems like general (possibly singular) systems of linear equations, or the rank of matrices. Furthermore, the *parallel complexity classes* NC_F^k and SAC_F^k are introduced, analogous to the Boolean complexity classes NC^k and SAC^k.

The pervasion of the theoretical tool of reductions gives the following hint for the design of parallel computers: implement one of the problems in dedicated hard/software, and then use subroutine calls to this one problem to solve all others.

The last section mentions several other topics in parallel arithmetic complexity which cannot be discussed in detail here: the exponentiation problem in finite fields, which shows that for certain tasks, our arithmetic circuits are *not* the right model, the surprising fact that "$NC^2 = P$" for polynomial computations, optimal algorithms for division with remainder of polynomials, computing normal forms of matrices, general lower bounds in terms of the degree, and permutation group algorithms.

The theory of parallel algebraic computation as presented here is a younger cousin of the more classical sequential algebraic complexity theory, which has well-established models of computation and fundamental results.

Surveys of this sequential theory are in Strassen [51] and von zur Gathen [25]. The present chapter is largely based on von zur Gathen [23].

The main prerequisite for our subject is basic linear algebra, plus the material typically presented in a one-semester introductory algebra course. Only the last section requires more algebraic background.

The material has been used in courses and seminars at University of Toronto (Canada), Université Laval (Québec, Canada), Universität Zürich (Switzerland), Universität des Saarlandes (Saarbrücken, Germany), and Universidad Católica de Santiago (Chile).

13.2
Arithmetic Circuits

The simplest case of algebraic computation is provided by an *arithmetic circuit* (called a *straight-line program* in Strassen [49]) such as in Figure 13.1.

This arithmetic circuit computes the two polynomials $3x_1 + \sqrt{2}\,x_2$ and $3x_2 - \sqrt{2}\,x_1$; 3 and $\sqrt{2}$ are constants belonging to the ground domain $\mathbb{R}$, and x_1 and x_2 inputs.

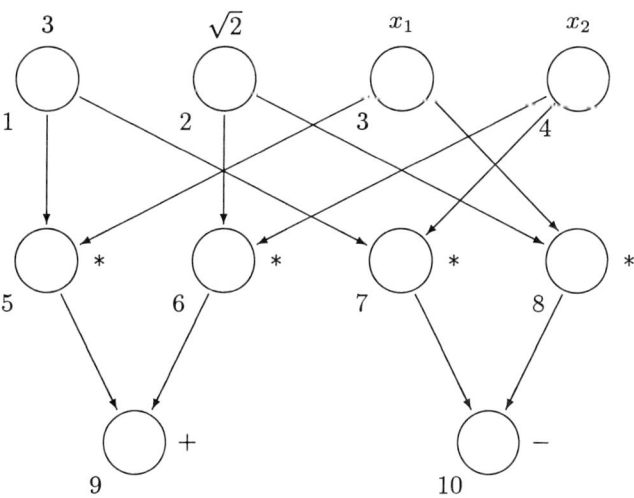

FIGURE 13.1
An arithmetic circuit over $\mathbb{R}$.

In general, we have a ground ring F, and an arithmetic circuit over F is a labelled directed acyclic graph. (Recall that a *ring* is equipped with two binary operations + and ∗ and elements 0 and 1 having the usual properties; the ring $\mathbb{Z}$ of integers is an example.) The label of each gate (or node) has two components. The first component is either a constant from F, an input, or an arithmetic operation $+, -, *, /$. The second component is a numbering of the gates. This numbering is only needed to distinguish the two inputs of the non-commutative operations "$-$" and "$/$": the lower numbered input is the first operand. (In Section 13.8, we will have to introduce more operations, such as testing for zero.) This model of computation forms the theoretical basis for much of *computer algebra*, where all operations are exact, in contrast to numerical computations of limited precision.

In Figure 13.1, we have the two constants 3 and $\sqrt{2}$ from the ground field $F = \mathbb{R}$ of real numbers, and two inputs x_1 and x_2. Interchanging the two second components 7 and 8 of labels, the arithmetic circuit would compute $\sqrt{2}\, x_1 - 3x_2$ at gate 10. In further examples, we will usually leave out this second component of labels; in the figures, the "left" input is the first operand.

If F is a field and $x_1, \ldots, x_n$ are the inputs, then at each gate a rational function in $F(x_1, \ldots, x_n)$ is computed. (A *field* is a ring with the further property that each nonzero element has a multiplicative inverse; an example is the field $\mathbb{Q}$ of rational numbers.) A technical requirement is that no division by the rational function zero may occur. (In the general case, where F is a ring, but not necessarily a field, each division must be executable in the ring; this is the case, e.g., when the denominator has an inverse.) We say that an arithmetic circuit *computes* any of the rational functions computed at any of its gates.

In Figure 13.2, the graph to the left is not an arithmetic circuit, since division by $x - x = 0$ occurs. The arithmetic circuit to the right computes the rational function $1/(x^2 - x)$. Although a division by zero occurs for the special inputs 0 and 1 for x, it is still a legal arithmetic circuit—even in the extreme case that the ground field $F = \mathbb{Z}_2$ contains only 0 and 1.

Two measures of an arithmetic circuit α are of interest here. The *depth* $D(\alpha)$ (= parallel time) is the number of arithmetic operations on a longest path in the graph of α. The *size* $S(\alpha)$ (or sequential time) is the total number of arithmetic operations. (It equals the "number of processors" when processors are not re-used.) For the α of Figure 13.1, we have $D(\alpha) = 2$ and $S(\alpha) = 6$.

Since all operations are binary (i.e., with two inputs), increasing the depth by one can at most double the number of inputs, so that depth $\log_2 n$ is optimal.

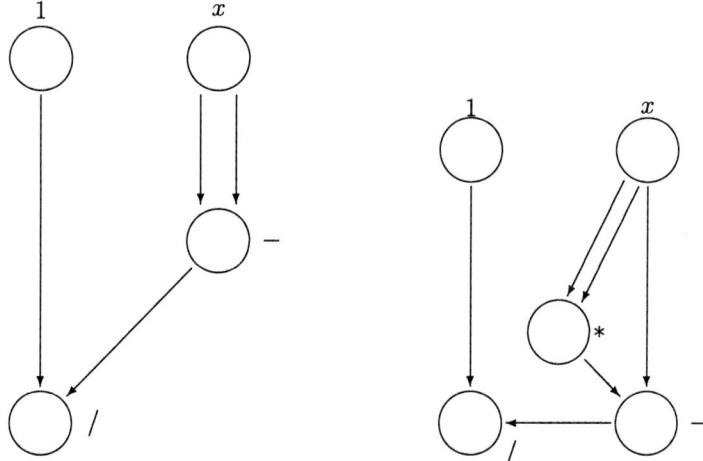

FIGURE 13.2
Two examples.

THEOREM 13.1
Let α be an arithmetic circuit. Then

1. $D(\alpha) \leq S(\alpha)$.
2. If there exists a single "output gate", with a directed path from every gate to it, then $S(\alpha) \leq 2^{D(\alpha)} - 1$.
3. If there exists an "output gate" with a directed path from each of n inputs to it, then $D(\alpha) \geq \log_2 n$.

PROOF
(1) is trivial. To prove (2), for any gate v in α, we consider the maximal length $D(v)$ of paths (where the length of a path is the number of arithmetic gates on it) leading from inputs or constants to v; thus $D(v)$ is the depth of v. We show by induction on $D(v)$ that v is connected to at most $2^{D(v)} - 1$ arithmetic gates. (We do not count the input or constant gates.) This is sufficient, since by assumption, there is some gate v connected to all $S(\alpha)$ gates, so that then $2^{D(\alpha)} - 1 \geq 2^{D(v)} - 1 \geq S(\alpha)$.

If $D(v) = 0$, then v is an input or constant gate. If $D(v) > 0$, let w_1 and w_2 be the two input gates to v. Then

$$D(v) = \max\{D(w_1), D(w_2)\} + 1,$$

and from the induction hypothesis it follows that the number of arithmetic gates connected to v is at most

$$(2^{D(w_1)} - 1) + (2^{D(w_2)} - 1) + 1 \leq 2^{D(v)-1} + 2^{D(v)-1} - 1 = 2^{D(v)} - 1, \quad (13.1)$$

where the $+1$ comes from the fact that v is connected to v.

The proof of (3) is similar (Exercise 13.1.a). ∎

Note that this is a purely graph-theoretic proof, independent of the types of gates we use, and thus valid for any directed acyclic graphs with fan-in two.

As an example, we consider the "iterated" sum $f = x_1 + \cdots + x_n$ of n indeterminates. A binary tree α of additions computes f with $D(\alpha) = \lceil \log_2 n \rceil$ and $S(\alpha) = n - 1$. Figure 13.3 shows the case $n = 7$. The same size and depth works also for the product $x_1 \cdots x_n$.

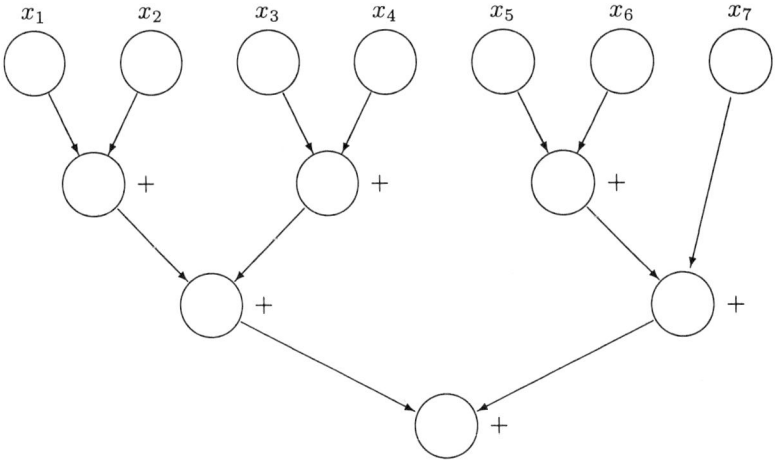

FIGURE 13.3
A binary addition tree, for $n = 7$.

Theorem 13.1 (3) says, in particular, that the binary trees for iterated sum or product cannot be improved in depth; the same is true for the size (see Exercise 13.1.b).

Only for rather simple problems like iterated sum are such optimal circuits known; Section 13.9 mentions other examples. In the next few sections we pursue a more modest but more realistic goal. For a variety of problems we will present algorithms which are not too far off from the trivial lower bounds of $\log_2 n$ and n for depth and size, resp. More precisely, their depth should be polynomial in $\log n$ (also called *polylogarithmic in n*, $O((\log n)^k)$ for some fixed k, or $(\log n)^{O(1)}$), and their size polynomial in n, i.e., $O(n^k)$ for some fixed k, or $n^{O(1)}$. Implicit in such asymptotic notions is the assumption that we are not dealing with a single computational problem, but with an infinite family $(f_n)_{n \in \mathbb{N}}$ of rational functions indexed by a parameter n, say $f_n \in F(x_1, \ldots, x_n)$, and a circuit family $(\alpha_n)_{n \in \mathbb{N}}$ with α_n computing f_n.

EXERCISE 13.1
Let α be a directed acyclic graph with indegree at most two, n "inputs" (= vertices with indegree 0), and a special "output" vertex (with outdegree zero) to which every vertex is connected.
   a) Prove that the depth of α is at least $\log_2 n$.
   b) Prove that the size of α is at least $n - 1$.

13.3
Multiplication

In this section, we consider multiplication of matrices and polynomials, and inversion of polynomials modulo a power of the indeterminate.

Given two square matrices $A, B \in F^{n \times n}$ over a ring F, their product $C = A \cdot B \in F^{n \times n}$ has entries

$$C_{ik} = \sum_{1 \leq j \leq n} A_{ij} B_{jk}$$

for $1 \leq i, k \leq n$. ($F^{n \times n}$ is the ring of $n \times n$-matrices with entries from F.) An arithmetic circuit is obvious from the formula:

1. For all i, j, k ($1 \leq i, j, k \leq n$) compute $A_{ij} \cdot B_{jk}$.
2. For all i, k ($1 \leq i, k \leq n$) compute C_{ik} as the above sum.

The depth of this circuit is $1 + \lceil \log_2 n \rceil = O(\log n)$, and the size is $n^3 + n^2(n-1) = O(n^3)$. By Theorem 13.1 (3) the circuit is depth-optimal; is it also size-optimal? It is not obvious, but in fact the size of matrix multiplication circuits may be drastically improved. The first surprising improvement was by Strassen [48], to $O(n^{2.81})$, and the smallest size known today is $O(n^{2.376})$, with depth still $O(\log n)$ (Coppersmith & Winograd [13]).

An extension is the problem of *iterated matrix multiplication*, where we are given matrices $A_1, \ldots, A_n \in F^{n \times n}$, and want to compute their product $A_1 \cdots A_n$. We can form a binary tree, of depth $\lceil \log_2 n \rceil$ and size $n-1$, with each "gate" being a multiplication of two matrices. The resulting depth is $O(\log^2 n)$, and the size $O(n^4)$.

The ring $F[x]$ of polynomials in x over a ring F consists of formal expressions of the form

$$f = a_0 + a_1 x + \cdots + a_n x^n \in F[x],$$

with $n \in N$ and $a_0, \ldots, a_n \in F$. This f has degree at most n; if $a_n \neq 0$, then the degree $\deg f$ is equal to n. Given a second polynomial $g = \sum_{0 \leq j \leq n} b_j x^j$, their product $h = \sum_{0 \leq k \leq 2n} c_k x^k = f \cdot g$ has coefficients

$$c_k = \sum_{\substack{0 \leq i,j \leq n \\ i+j=k}} a_i b_j.$$

The obvious circuit:

1. For all i, j ($0 \leq i, j \leq n$) compute $a_i \cdot b_j$.
2. For all k ($0 \leq k \leq 2n$) compute c_k as the above sum.

has depth $O(\log n)$ and size $O(n^2)$.

We note that addition of two matrices or polynomials can be done in depth 1, and iterated addition of n such items in depth $O(\log n)$.

Iterated polynomial product is the problem of computing $f_1 \cdots f_n$, where $f_1, \ldots, f_n \in F[x]$ have degree at most n. A binary tree of polynomial multiplications solves this problem in depth $O(\log^2 n)$. What is the resulting size? Let us assume for simplicity that n is a power of 2, and consider the levels 0 (inputs), 1, 2, $\ldots, k = \log_2 n$ (output) of the tree. At level i, a total of $n/2^i$ multiplications of polynomials of degree at most $2^{i-1}n$ are performed, in depth $O(\log n)$ and size at most

$$n/2^i \cdot O((2^{i-1}n)^2) = O(2^i n^3).$$

Summing these sizes over i, we get $O(n^4)$, and overall depth $O(\log^2 n)$. We will mention in Section 13.9 that this problem can even be solved in optimal depth $O(\log n)$.

Given three polynomials $f_1, f_2, g \in F[x]$, f_1 and f_2 are called *congruent modulo g* if their difference is divisible by g:

$$f_1 \equiv f_2 \bmod g \iff \exists h \in F[x] \ \ f_1 - f_2 = gh.$$

If $f \in F[x]$ and $n \in \mathbb{N}$, then f is *invertible modulo* x^{n+1} if there exists some $g \in F[x]$ such that $fg \equiv 1 \bmod x^{n+1}$; such a g is called a (modular) *inverse* of f. If $f = a_0 + \cdots$ is invertible modulo x^{n+1}, then it is invertible $\bmod x$, so that $a_0 = f(0) \in F$ is invertible; if F is a field, this is equivalent to $a_0 \neq 0$. The algorithm below shows that also the converse is true.

As an example, let $n = 3$ and $f = 1 - x$. Then $g = 1 + x + x^2 + x^3$ satisfies

$$fg = 1 - x^4 \equiv 1 \bmod x^4,$$

and g is an inverse of f modulo x^4.

The following algorithm generalizes this *geometric series*.

ALGORITHM 13.1
Polynomial inversion
Input: $f \in F[x]$ with $f(0) \in F$ invertible.

Output: $g \in F[x]$ with $fg \equiv 1 \bmod x^{n+1}$.

1. Compute $b = f(0)^{-1} \in F$,
2. compute $h = (f(0) - f) \cdot b \in F[x]$,
3. for all i, $0 \leq i \leq n$, compute h^i,
4. return $g = b \cdot \sum_{0 \leq i \leq n} h^i$.

EXAMPLE 13.1
Let $n = 3$ and $f = 1 + 7x + 29x^2 + 80x^3 \in \mathbb{Q}[x]$. Then $f(0) = 1$ and $h = -7x - 29x^2 - 80x^3$. The algorithm calculates

$$g = 1 + h + h^2 + h^3 \equiv 1 - 7x + 20x^2 - 17x^3 \bmod x^4.$$

Note that we have left out the terms of order 4 or higher, since they are irrelevant modulo x^4. The reader might check that indeed $fg \equiv 1 \bmod x^4$.

We first convince ourselves that algorithm **Polynomial inversion** works correctly. Note that x divides h (which we write as "$x \mid h$"), so that $x^{n+1} \mid h^{n+1}$, or $h^{n+1} \equiv 0 \bmod x^{n+1}$. Thus

$$fg = (1-h)f(0) \cdot b \sum_{0 \le i \le n} h^i = 1 - h^{n+1} \equiv 1 \bmod x^{n+1}, \qquad (13.2)$$

and g is indeed a modular inverse of f.

Making use of iterated polynomial product for the powers in step 2, the depth is $O(\log^2 n)$, and the size $O(n^5)$.

The algorithm displays the technique of "reduction" that we will use profusely in the next sections: solving one problem (here: modular inversion) by appealing to another one (here: iterated polynomial product). Although conceptually important and convenient, it has the disadvantage of blowing up the size (and sometimes the depth) more than necessary. In our case, we observe that—as in Example 13.1—we only need all polynomials modulo x^{n+1}, i.e., only the first $n+1$ coefficients. If we truncate all results modulo x^{n+1}, we only have to perform $n-1$ multiplications of polynomials of degree at most n, resulting in size $O(n^3)$. We have proved the following result.

THEOREM 13.2
Let F be a ring. Polynomials in $F[x]$ with constant term invertible in F can be inverted modulo x^{n+1} in depth $O(\log^2 n)$ and size $O(n^3)$.

13.4 The Determinant

The parallel algorithms discussed so far were straightforward. We now turn to a fundamental problem for which a good parallel solution is not obvious: the solution of systems of linear equations.

The problem is of central importance, and many sequential algorithms for it are well-studied. Suppose we want to solve

$$Ax = b,$$

where an $n \times n$-matrix $A \in F^{n \times n}$ over the ground field F and an n-vector $b \in F^n$ are given, and we are looking for a vector $x \in F^n$ satisfying the equation. Such a solution exists if and only if b is a linear combination of the n columns of A, and if the determinant $\det A$ is nonzero, there exists a unique solution x.

The classical algorithm of *Gaussian elimination* consists of n stages. In each stage, appropriate scalar multiples of a "pivot row" are subtracted from other rows to introduce zero entries in one column. The end result is an upper triangular system of linear equations with the same solutions as the original one. It can now easily be solved by "back-substitution". All the row operations of one stage can easily be performed in three parallel operations. However, the execution of the stages looks inherently sequential, and it is not clear how to obtain parallel time less than n, say.

We now discuss a very different algorithm, invented by the Leningrad mathematician Chistov [11]. Csanky [14] had presented the first parallel algorithm for the determinant using depth $O(\log^2 n)$ and size $n^{O(1)}$ (Exercise 13.4). It has the merit of being the first nontrivial parallel algorithm in linear algebra, within the framework of this chapter. Unfortunately, Csanky's algorithm only works over fields F of characteristic zero, i.e., if $\mathbb{Q} \subseteq F$, and this excludes the important case of finite fields. Next, Borodin et al. [8] gave an (admittedly awful) solution for the general case. Soon after that, Berkowitz [6]—then a student at University of Toronto—found an algorithm that competes with Chistov's in cost and clarity.

If $Ax = b$, $x = (x_1, \ldots, x_n) \in F^n$, and $\det A \neq 0$, then *Cramer's rule* says that $x_i = \det A^{[i]} / \det A$, where $A^{[i]} \in F^{n \times n}$ is obtained by substituting b for the ith column vector of A. Thus it is sufficient to compute determinants of matrices. Note that after performing Gaussian elimination, $\det A$ is the product of the diagonal entries of the resulting upper triangular matrix, and thus easy to compute.

We will actually solve the seemingly harder problem of computing the characteristic polynomial

$$\chi(A) = \det(xI_n - A) = c_0 + c_1 x + \cdots + c_{n-1}x^{n-1} + x^n \in F[x]$$

of a matrix A, where F is a ring, $I_n \in F^{n \times n}$ the identity matrix (with ones on the diagonal, and zeroes elsewhere), and x an indeterminate. Then $\det A = (-1)^n c_0$ can be read off (and $-c_{n-1}$ is the sum of the diagonal entries of A).

EXAMPLE 13.2

Let us take

$$A = \begin{bmatrix} 2 & -3 & 0 \\ 1 & 2 & -1 \\ -2 & 1 & 3 \end{bmatrix} \in \mathbb{Q}^{3 \times 3}.$$

Then

$$\chi(A) = \det \begin{vmatrix} x-2 & 3 & 0 \\ -1 & x-2 & 1 \\ 2 & -1 & x-3 \end{vmatrix} = -17 + 20x - 7x^2 + x^3,$$

and $\det A = -17$. ∎

If F is a field, $A = (a_{ij})_{1 \leq i,j \leq n} \in F^{n \times n}$ and $1 \leq r \leq n$, we consider the lower right submatrix

$$A_r = (a_{ij})_{r \leq i,j \leq n} \in F^{r' \times r'}$$

of A, where $r' = n - r + 1$ (see Figure 13.4). (Thus the rows and columns of A_r are indexed by $r, r+1, \ldots, n$.) If we let

$$d_r = \det(I_{r'} - xA_r) \in F[x],$$

then

$$\begin{aligned} \chi(A) &= \det(xI_n - A) = \det\left(xI_n \cdot (I_n - x^{-1}A)\right) \\ &= \det(xI_n) \cdot \det(I_n - x^{-1}A) = x^n \det(I_{1'} - x^{-1}A_1) = x^n d_1(x^{-1}). \end{aligned}$$

The polynomial $x^n d_1(x^{-1})$ is called the *reversal* (for degree n) of d_1, since its coefficient sequence is the reversed coefficient sequence of d_1.

The matrix $I_{r'} - xA_r \in F[x]^{r' \times r'}$ is invertible over $F(x)$, since its determinant d_r is a nonzero polynomial, with value 1 at $x = 0$. We denote by

$$B^{(r)} = (b_{ij}^{(r)})_{r \leq i,j \leq n} = (I_{r'} - xA_r)^{-1}$$

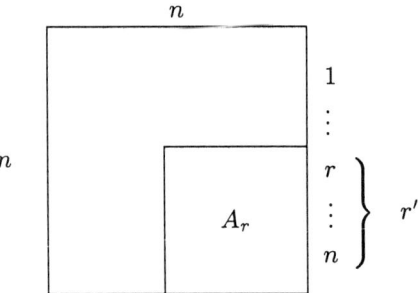

FIGURE 13.4
The lower right submatrix A_r.

its inverse. Thus each $b_{ij}^{(r)} \in F(x)$ is a rational function in x, and $b_{ij}^{(r)} \cdot d_r \in F[x]$. Let us further denote by $b_{r*}^{(r)} = (b_{rj}^{(r)})_{r \leq j \leq n}$ the leftmost column, and by $b_r = b_{rr}^{(r)} \in F(x)$ the top left entry of $B^{(r)}$. Then by definition of the inverse

$$(I_{r'} - xA_r) \cdot b_{r*}^{(r)} = (1, 0, \ldots, 0)^t.$$

The determinant of the matrix obtained by substituting $(1, 0, \ldots, 0)^t$ for the leftmost column of $I_{r'} - xA_r$ equals

$$\det(I_{(r+1)'} - xA_{r+1}) = d_{r+1}.$$

Thus expressing the top entry b_r of $b_{r*}^{(r)}$ according to Cramer's rule, we find

$$b_r = d_{r+1}/d_r. \tag{13.3}$$

(With $d_{n+1} = 1$, we note that $d_r(0) = 1$ for all r, so that in particular all denominators are nonzero.) Multiplying all equations (13.3) together, we obtain

$$\prod_{1 \leq r \leq n} b_r = d_1^{-1}.$$

This is an equation between rational functions in x (which again evaluate to 1 at $x = 0$).

A special case of the geometric series for the inverse (13.2) is

$$(1 - xa) \cdot \sum_{0 \leq k \leq n} x^k a^k \equiv 1 \bmod x^{n+1}. \tag{13.4}$$

This equation holds when a is in any ring and x an indeterminate over this ring commuting with a. We apply (13.4) with $a = A_r$, and obtain

$$B^{(r)} = (I_{r'} - xA_r)^{-1} \equiv \sum_{0 \leq k \leq n} x^k A_r^k \bmod x^{n+1}.$$

If we define $\tilde{B}^{(r)}$ as the sum on the right hand side, and $\tilde{b}_r$ as its top left entry, then each $\tilde{b}_{ij}^{(r)} \in F[x]$ is a polynomial in x of degree at most n, the denominator d_r of b_r is invertible modulo x, and $b_r \equiv \tilde{b}_r \bmod x^{n+1}$. Thus we have

$$\prod_{1 \leq r \leq n} \tilde{b}_r \equiv d_1^{-1} \bmod x^{n+1}.$$

Putting things together, we have the following algorithm.

ALGORITHM 13.2
Characteristic polynomial
Input: A matrix $A \in F^{n \times n}$, where F is a commutative ring with 1, and $n \in N$.
Output: The coefficients $c_0, \ldots, c_n$ of $\chi(A) \in F[x]$.

1. For all k, r $(1 \leq k, r \leq n)$, compute A_r^k and $\tilde{b}_r = \left(\sum_{0 \leq k \leq n} x^k A_r^k \right)_{rr}$.
2. Compute $b \in F[x]$ with $\deg b \leq n$ and
$$b \equiv \prod_{1 \leq r \leq n} \tilde{b}_r \mod x^{n+1}.$$
[Then $b \equiv d_1^{-1} \mod x^{n+1}$.]
3. Compute $c \in F[x]$ with $\deg c \leq n$ and $c \equiv b^{-1} \mod x^{n+1}$, using the algorithm **Polynomial inversion**. [Then $c = d_1$.]
4. Return the coefficients of the reverse $\chi(A) = x^n c(x^{-1})$ of c.

We have already seen that the algorithm works correctly over a field F. The cost follows from the estimates of the subroutines used:

Step	Subroutine	Depth	Size
1	iterated matrix product	$O(\log^2 n)$	$O(n^6)$
2	iterated polynomial product	$O(\log^2 n)$	$O(n^3)$
3	polynomial inversion	$O(\log^2 n)$	$O(n^3)$
4		0	0

For the estimates of steps 2 and 3, we use the truncating algorithm of Theorem 13.2. Step 4 is free, since only the coefficient sequence is reversed.

The algorithm actually works over an arbitrary commutative ring with 1. For this, we note that the algorithm has no divisions (the "division" in step 3 is by 1), and computes the characteristic polynomial of a matrix over $\mathbb{Z}$ with indeterminate entries. Therefore it computes $\chi(A)$ for any square matrix A over a commutative ring F.

EXAMPLE 13.3
We first trace the algorithm on the matrix of Example 13.1, and then check one of the equations used in deriving the algorithm.

$$A_1 = A = \begin{array}{|ccc|} \hline 2 & -3 & 0 \\ 1 & 2 & -1 \\ -2 & 1 & 3 \\ \hline \end{array},$$

$$A_1^2 = \begin{bmatrix} 2 & -12 & 3 \\ 6 & 0 & -5 \\ -9 & 11 & 8 \end{bmatrix}, \quad A_1^3 = \begin{bmatrix} -16 & -24 & 21 \\ 22 & -23 & -15 \\ -23 & 57 & 13 \end{bmatrix},$$

$$A_2 = \begin{bmatrix} 2 & -1 \\ 1 & 3 \end{bmatrix}, \quad A_2^2 = \begin{bmatrix} 3 & -5 \\ 5 & 8 \end{bmatrix}, \quad A_2^3 = \begin{bmatrix} 1 & -18 \\ 18 & 19 \end{bmatrix},$$

$$A_3 = \boxed{3}, \quad A_3^2 = \boxed{9}, \quad A_3^3 = \boxed{27},$$

$$\begin{aligned}
\tilde{b}_1 &= 1 + 2x + x^2 - 16x^3 \;(\equiv b_1 \bmod x^4), \\
\tilde{b}_2 &= 1 + 2x + 3x^2 + x^3 \;(\equiv b_2 \bmod x^4), \\
\tilde{b}_3 &= 1 + 3x + 9x^2 + 27x^3 \;(\equiv b_3 \bmod x^4), \\
b &= 1 + 7x + 29x^2 + 80x^3, \\
c &= 1 - 7x + 20x^2 - 17x^3 \;\text{(see Example 13.1)}, \\
\text{reverse of } c &= -17 + 20x - 7x^2 + x^3,
\end{aligned}$$

which is indeed $\chi(A)$. Here is the special case $b_1 = d_2/d_1$ of equation (13.3):

$$\begin{aligned}
d_1 \cdot b_1 &= \det(I_3 - xA_1) \cdot b_1 \equiv (1 - 7x + 20x^2 - 17x^3) \cdot \\
&\quad (1 + 2x + x^2 - 16x^3) \\
&= 1 - 5x + 7x^2 = \det(I_2 - xA_2) = d_2 \bmod x^4.
\end{aligned}$$

With a little care, we can improve the size of the circuit to $O(n^4)$, in fact, to even less. The following notation is convenient. We consider all families $\alpha = (\alpha_n)_{n \in \mathbb{N}}$ of arithmetic circuits, where α_n computes the product of two $n \times n$-matrices over F; then α *computes matrix multiplication*. We define the (parallel) matrix multiplication size as the smallest size sufficient to multiply matrices with logarithmic depth:

$$M(n) = \min\{S(\alpha_n)\colon \alpha \text{ computes matrix multiplication} \\ \text{and } D(\alpha_n) = O(\log n)\}.$$

The standard algorithm used in Section 13.3 shows $M(n) \leq 2n^3$. Clearly $M(n) \geq n^2$, since n^2 outputs have to be computed; the best lower bound known today is $M(n) \geq 2n^2 - 1$ (Brockett & Dobkin [9]). The best upper bound known today is by Coppersmith & Winograd [13]:

$$2n^2 - 1 \leq M(n) = O(n^{2.376}).$$

The reader misses nothing essential in the following if she thinks of $M(n) = 2n^3$ and the standard matrix multiplication algorithm. The following observation is from von zur Gathen & Eberly [28].

LEMMA 13.1
If $A \in F^{n \times n}$ and $b \in F^n$, then all vectors $A^0 b, A^1 b, A^2 b, \ldots, A^n b \in F^n$ can be computed in depth $O(\log^2 n)$ and size at most $2M(n)(1 + \log_2 n)$.

PROOF
Let $l = \lceil \log_2 n \rceil < 1 + \log_2 n$, so that $n \leq 2^l$. In a first stage we compute

$$A^{2^0}, A^{2^1}, A^{2^2}, \ldots, A^{2^l},$$

in depth $O(\log^2 n)$ and size $lM(n)$. In the second stage, we compute successively $B_0, B_1, B_2, \ldots, B_l \in F^{n \times n}$ as follows. B_0 has b as its first column, and zeroes elsewhere. B_i's first 2^{i-1} columns equal those of B_{i-1}, the next 2^{i-1} columns equal the first 2^{i-1} columns of $A^{2^{i-1}} \cdot B_{i-1}$, and the other columns are zero. (In B_l, only the next $n - 2^{l-1}$ columns are new.) One sees inductively that the jth column of B_i is $A^{j-1} b$ if $1 \leq j \leq 2^i$, and zero otherwise. (In B_l, this is valid for $1 \leq j \leq n$.) The cost is the same as for the first stage. ∎

In step 1 of the algorithm **Characteristic polynomial**, we do not really need all A_r^k, but only all top left entries $e_r A_r^k e_r^t$, where $e_r = (1, 0, \ldots, 0) \in F^{r'}$. For any $r \leq n$, all vectors $A_r^k e_r^t$ ($0 \leq k \leq n$) can be computed in size at most $2M(n) \log_2 n$ by the lemma. Thus we have the following result.

THEOREM 13.3
The characteristic polynomial of $n \times n$-matrices can be computed on an arithmetic circuit of depth $O(\log^2 n)$ and size at most $2nM(n)(1 + \log_2 n)$, or size $O(n^4 \log n)$.

A better size bound $O(n^{1/2} M(n))$ can be obtained in characteristic zero (Preparata & Sarwate [45]), and a slight improvement in the general case is in Galil & Pan [20]. Kaltofen & Pan [37] show how to solve $Ax = b$ by a probabilistic circuit of depth $O(\log^2 n)$ and size $O(M(n) \log n)$ if A is nonsingular, the characteristic of the field F is zero or larger than n, and the field is sufficiently large (say, $\#F \geq 6n^2$).

EXERCISE 13.2 (Inversion of triangular matrices)
Let F be a field, and

$$A = \begin{pmatrix} B & 0 \\ C & D \end{pmatrix} \in F^{2 \times 2}$$

a non-singular lower triangular matrix. Show that

$$A^{-1} = \begin{pmatrix} B^{-1} & 0 \\ -D^{-1}CB^{-1} & D^{-1} \end{pmatrix}.$$

Generalize this fact to arbitrary non-singular lower triangular matrices $A \in F^{n \times n}$ and use your results to construct a recursive parallel algorithm for computing the inverse of such matrices in $O(\log^2 n)$ time using $O(n^3)$ processors.

EXERCISE 13.3 (Linear recurrences)
Assume you are given a system of linear recurrences

$$\begin{aligned} x_1 &= c_1, \\ x_2 &= a_{21}x_1 + c_2, \\ &\vdots \\ x_n &= a_{n1}x_1 + a_{n2}x_2 + \cdots + a_{nn-1}x_n + c_n, \end{aligned}$$

where $a_{ij}, c_i \in F$, and F is a field. Letting $A = (a_{ij}) \in F^{n \times n}$, $x = (x_i) \in F^n$ and $c = (c_i) \in F^n$, this can be rewritten

$$Ax + c = x.$$

Give an efficient parallel algorithm to solve this system of linear recurrences. [Hint: Use your solution to Exercise 13.2.]

EXERCISE 13.4 (Csanky's algorithm)
Let F be a field of characteristic zero (or of characteristic larger than n), $A \in F^{n \times n}$ and write

$$\chi = \det(Ix - A) = x^n - s_1 x^{n-1} + s_2 x^{n-2} - \cdots + (-1)^n s_n \in F[x]$$

for the characteristic polynomial of A. Let $\lambda_1, \ldots, \lambda_n$ be the eigenvalues of A (say in an algebraic closure of F); these are just the roots of χ.

a) Show that
$$s_n = \det(A) = \prod_{i=1}^{n} \lambda_i.$$

The trace $\text{tr}(A) \in F$ of A is defined to be the sum of the diagonal entries of A:
$$\text{tr}(A) = \sum_{1 \leq i \leq n} A_{ii}.$$

b) Show that
$$s_1 = \text{tr}(A) = \sum_{1 \leq i \leq n} \lambda_i.$$

In other words, the trace of such a matrix is also the sum of its eigenvalues. Show also that $\text{tr}(A^k) = \sum_{i=1}^{n} \lambda_i^k$ for all $k \in \mathbb{N}$.

c) (Newton identities) Prove that
$$s_k = \frac{1}{k}(s_{k-1} \cdot \text{tr}(A) - s_{k-2} \cdot \text{tr}(A^2) + \cdots$$
$$+ (-1)^{k-2} s_1 \cdot \text{tr}(A^{k-1}) + (-1)^{k-1} \cdot \text{tr}(A^k)).$$

[Hint: Use a) and the fact that
$$s_k = \sum_{1 \leq i_1 < i_2 < \cdots < i_k \leq n} \lambda_{i_1} \lambda_{i_2} \cdots \lambda_{i_k}.$$

In other words, s_k is the *kth elementary symmetric polynomial* in the λ_i's.]

d) Now apply c) and Exercise 13.3 above to give an efficient parallel algorithm for computing the coefficients s_i of the characteristic polynomial $\chi \in F[x]$ of $A \in F^{n \times n}$.

EXERCISE 13.5 (Inversion of non-singular matrices)
The Cayley-Hamilton theorem states that any matrix A satisfies its characteristic polynomial: if $\chi = x^n - s_1 x^{n-1} + \cdots \pm s_{n-1} x \mp s_n x^0$ is the characteristic polynomial of $A \in F^{n \times n}$, then
$$\chi(A) = A^n - s_1 A^{n-1} + \cdots \pm s_{n-1} A \mp s_n I = 0.$$

Use this fact, together with Exercise 13.4 above, to show that if $A \in F^{n \times n}$ is non-singular, then the entries of A^{-1} can be computed from A in $O(\log^2 n)$ parallel arithmetic steps using $O(n^4)$ processors.

13.5
Polynomial and Matrix Problems

The goal of this section is a set of *reductions* between polynomial and matrix problems. A reduction $f \leq g$ is an arithmetic circuit for f of depth $O(\log n)$ that uses g; precise definitions are in Section 13.8. The following construction is useful for our purpose. We let F be any ring, possibly noncommutative, x an indeterminate over F, $d \in \mathbb{N}$, and consider the mapping

$$\tau_d \colon F[x] \to F^{d \times d}$$

$$a_0 + a_1 x + \cdots \mapsto \begin{pmatrix} a_0 & a_1 & \cdots & a_{d-1} \\ & \ddots & & \vdots \\ & & \ddots & a_1 \\ 0 & & & a_0 \end{pmatrix}.$$

Thus the image of τ_d is the set of Toeplitz matrices A in upper triangular form: $A_{ij} = A_{i+k, j+k}$ for all appropriate values of i, j, k, and $A_{ij} = 0$ if $i > j$. The proof of the following lemma is left as Exercise 13.6.

LEMMA 13.2
τ_d *is a ring homomorphism with kernel* (x^d).

It is convenient to have a standard language to describe our computational problems, such as the following.

PROD = $(\text{PROD}_n)_{n \in \mathbb{N}}$ with $\text{PROD}_n = x_1 \cdots x_n$ is the product problem. DETERMINANT = $(\text{DETERMINANT}_n)_{n \in \mathbb{N}}$ with

$$\text{DETERMINANT}_n = \det\big((x_{ij})_{1 \leq i,j \leq n}\big)$$

is the determinant problem.

We define further computational problems:

POLYPROD: product of two polynomials.

This is shorthand for defining a family POLYPROD = $(\text{POLYPROD}_n)_{n \in \mathbb{N}}$ of sequences of polynomials $\text{POLYPROD}_n = (c_0, \ldots, c_{2n})$, where $c_k = \sum_{i+j=k} a_i b_j \in F[a_0, \ldots, a_n, b_0, \ldots, b_n]$, and $a_0, \ldots, b_n$ are indeterminates over F.

Similarly, we have

ITPOLYPROD$_n$: $f_1 \cdots f_n$ for $f_1, \ldots, f_n \in F[x]$ of degree at most n,
POLYINV$_n$: $f^{-1} \bmod x^n$ for $f \in F[x]$, $f(0) \neq 0$,
MATPROD$_n$: $A \cdot B$ for $A, B \in F^{n \times n}$,
ITMATPROD$_n$: $A_1 \cdots A_n$ for $A_1, \ldots, A_n \in F^{n \times n}$,
MATINV$_n$: A^{-1} for $A \in F^{n \times n}$ invertible.

THEOREM 13.4
1. POLYPROD $\leq_F$ MATPROD,
2. POLYINV $\leq_F$ MATINV,
3. POLYINV $\leq_F$ ITPOLYPROD $\leq_F$ ITMATPROD.

PROOF
For (1), suppose we want to compute the product of $f, g \in F[x]$ with degree at most n. By Lemma 13.2 we have

$$\tau_{2n+1}(fg) = \tau_{2n+1}(f) \cdot \tau_{2n+1}(g) = \text{MATPROD}_{2n+1}(\tau_{2n+1}(f), \tau_{2n+1}(g)).$$

Thus the reduction has three (trivial) steps: 1. produce the matrices $\tau_{2n+1}(f)$ and $\tau_{2n+1}(g)$, 2. form their product, by calling MATPROD, and 3. read off the required output. More formally, the reduction circuit has $2n + 2$ input gates for the coefficients of f and g, the constant zero, and a single computation gate MATPROD$_{2n+1}$, with $2(2n+1)^2$ inputs and $(2n+1)^2$ outputs. The input gates and zero are connected to the MATPROD gate according to τ_{2n+1}, and the functions required for POLYPROD are among those computed by the MATPROD gate (in fact, the first row). The depth is 1, and the size is $2(2n+1)^2 - 1$.

The reduction for (2), and the second one in (3) are similar. The first reduction in (3) is given by algorithm **Polynomial inversion**. ∎

We now define further problems:

DETERMINANT$_n$: $\det A$ for $A \in F^{n \times n}$,
CHARPOLY$_n$: $\chi(A)$ for $A \in F^{n \times n}$,
MATPOWERS$_n$: $A^2, A^3, \ldots, A^n$ for $A \in F^{n \times n}$,
NONSINGEQ$_n$: x with $Ax = b$ for $A \in F^{n \times n}$ invertible and $b \in F^n$.

We write "$f \leq_F g + h$" for a reduction computing f that makes oracle calls both to g and h, and observe that $f \leq_F g + h$ and $g \leq_F h$ imply that $f \leq_F h$. We say that f is *equivalent* to g ($f \equiv g$) if and only if $f \leq g$ and $g \leq f$.

THEOREM 13.5
DETERMINANT, CHARPOLY, ITMATPROD, MATPOWERS, MATINV, and NONSINGEQ *are equivalent.*

PROOF
We will exhibit a complete circle of six reductions. The claim then follows from the transitivity of $\leq_F$ (Theorem 13.10 (1)).

1. DETERMINANT $\leq_F$ CHARPOLY: the determinant is the constant term of the characteristic polynomial, up to the sign.
2. CHARPOLY $\leq_F$ ITMATPROD: The algorithm **Characteristic polynomial** of Section 13.4 has shown that

$$\text{CHARPOLY} \leq_F \text{MATPOWERS} + \text{ITPOLYPROD} + \text{POLYINV}.$$

Together with Theorem 13.4 (3) and the trivial
MATPOWERS $\leq_F$ ITMATPROD, the required reduction follows. This is by far the most challenging reduction in this proof.

3. ITMATPROD $\leq_F$ MATPOWERS: Given $A_1, \ldots, A_n \in F^{n \times n}$, we may consider

$$B = \begin{bmatrix} I & A_1 & & & 0 \\ & & \ddots & & \\ & & & \ddots & A_n \\ 0 & & & & I \end{bmatrix} \in (F^{n \times n})^{(n+1) \times (n+1)}$$

as an $(n^2 + n) \times (n^2 + n)$-matrix. Then

$$B^n = \begin{bmatrix} I & * & \cdots & A_1 \cdots A_n \\ & I & \ddots & * \\ & & \ddots & * \\ 0 & & & I \end{bmatrix}$$

and ITMATPROD$_n(A_1, \ldots, A_n)$ can be read off MATPOWERS$_{n^2+n}(B)$.

4. MATPOWERS $\leq_F$ MATINV: Given $A \in F^{n \times n}$, we consider

$$B = \tau_{n+1}(1 - Ax) \in (F^{n \times n})^{(n+1) \times (n+1)}$$

as an $(n^2 + n) \times (n^2 + n)$-matrix. By Lemma 13.2,

$$\tau_{n+1}((Ax)^{n+1}) = 0,$$

$$B^{-1} = (1 - \tau_{n+1}(Ax))^{-1}$$

$$= \sum_{0 \leq k \leq n} \tau_{n+1}(Ax)^k$$

$$= \tau_{n+1}\left(\sum_{0 \leq k \leq n} A^k x^k \right)$$

$$= \begin{bmatrix} I & A & \cdots & A^n \\ & I & \ddots & \vdots \\ & & \ddots & A \\ 0 & & & I \end{bmatrix}.$$

Again, MATPOWERS$_n(A)$ can be read off MATINV$_{n^2+n}(B)$.

5. MATINV $\leq_F$ NONSINGEQ: Given an invertible $A \in F^{n \times n}$, find $x_1, \ldots, x_n \in F^n$ satisfying $Ax_i = e_i^t$, where $e_i = (0, \ldots, 0, 1, 0, \ldots, 0) \in F^n$ has a 1 in position i, and zeroes elsewhere. Then x_i is the ith column of A^{-1}.

6. NONSINGEQ $\leq_F$ DETERMINANT: follows with Cramer's rule. ∎

We define the "complexity class"

$$DET_F = \{f \colon f \leq \text{DETERMINANT}\}$$

of problems reducible to the determinant. (This is not an honest complexity class, since it is not defined just by explicit constraints on computational resources like depth and size.) As usual, we call a problem $f \in DET_F$ complete if $g \leq f$ for all $g \in DET_F$. Theorem 13.5 can then be stated as follows.

THEOREM 13.6

Let F be a field. DETERMINANT, CHARPOLY, ITMATPROD, MATPOWERS, MATINV, *and* NONSINGEQ *are complete for* DET_F.

EXERCISE 13.6
Prove Lemma 13.2.

13.6
Rank of Matrices

In order to solve general (possibly singular) systems of linear equations, we start with a related problem: the *rank* of matrices, and present a fast parallel algorithm, due to Mulmuley [41], in this section. Before that result, Ibarra et al. [31] had found a very simple algorithm which works over a "real field" F such as $F = \mathbb{Q}$ or $F = \mathbb{R}$. The first shallow circuit for arbitrary F was in Borodin et al. [8]; it has the drawback of requiring random choices in the algorithm. All these methods use depth $O(\log^2 n)$ and size $n^{O(1)}$.

The rank $r = \text{rank}(A)$ of a matrix $A \in F^{n \times n}$ over the ground field F is the maximal size of nonsingular minors of A. If

$$\ker A = \{x \in F^n : Ax = 0\}$$

denotes the nullspace of A, then

$$r + \dim_F \ker A = n. \tag{13.5}$$

If $F \subseteq K$ are fields and $A \in F^{n \times n} \subseteq K^{n \times n}$, then A has the same rank whether considered as a matrix over F or K; in other words, the rank is invariant under field extensions.

Section 13.8 discusses in more detail the model in which the computations of this section are performed (see Example 13.4).

The rank $r = \text{rank} A$ is sometimes called the *geometric rank*. A related quantity is the *algebraic rank* $t = \text{rank}_{\text{alg}} A$ of A, defined by

$$t + \mu_0(A) = n,$$

where $\mu_0(A)$ is the multiplicity of 0 as a root of $\chi(A)$. Note the analogy with (13.5); the "geometric multiplicity" $\dim \ker A$ of 0 in A is replaced by the algebraic multiplicity μ_0. Since $\mu_0 \geq \dim \ker A$, we have $t \leq r$. Using the algorithm **Characteristic Polynomial**, we can compute $\chi(A)$ and t quickly. The idea now is to reduce the computation of $\text{rank} A$ to that of $\text{rank}_{\text{alg}} A$.

What is the relation between rank and algebraic rank? Let $s = n - \operatorname{rank} A$, so that $s = \dim \ker A$, and suppose that $u_1, \ldots, u_n$ is a basis of F^n, with $u_1, \ldots, u_s$ being a basis of $\ker A$. Such a basis always exists, and in this basis A has the form

$$A = \begin{array}{c} \\ \\ \end{array} \begin{array}{|cc|} \hline 0 & * \\ \hline 0 & B \\ \hline \end{array} \begin{array}{l} 1 \\ \vdots \\ s \\ s+1 \\ \vdots \\ n \end{array} \qquad (13.6)$$

Clearly $\chi(A) = x^s \cdot \chi(B)$, and thus

$$\operatorname{rank} A = r = n - s \leq n - t = \operatorname{rank}_{\mathrm{alg}} A.$$

We can calculate $\chi(A)$ and $\operatorname{rank}_{\mathrm{alg}} A$ fast in parallel (Section 13.4), and would like to use this to compute $\operatorname{rank} A$. Here is a sufficient criterion.

LEMMA 13.3
If $\operatorname{rank} A = \operatorname{rank} A^2$, then $\operatorname{rank} A = \operatorname{rank}_{\mathrm{alg}} A$.

PROOF
We clearly have $\ker A \subseteq \ker A^2$, so that the hypothesis implies that $\ker A = \ker A^2$. In (13.6), it is sufficient to have B nonsingular, since then $\chi(B)$ does not have 0 as a root, and t is the multiplicity of 0 as a root of $\chi(A)$, and hence $\operatorname{rank} A = \operatorname{rank}_{\mathrm{alg}} A$.

So suppose $a = (a_{s+1}, \ldots, a_n) \in F^{n-s}$ with $Ba = 0$, and let $\bar{a} = (0, \ldots, 0, a_{s+1}, \ldots, a_n) \in F^n$. Then $A\bar{a} \subset \ker A$, and hence $A^2 \bar{a} = 0$. Thus $\bar{a} \in \ker A^2 = \ker A$. The special form of $\bar{a}$ implies that $\bar{a} = 0$, hence $a = 0$, and indeed B is nonsingular. ∎

We now try to get into this favorable case by constructing from A a matrix B with $\operatorname{rank} B = \operatorname{rank} B^2$, and such that $\operatorname{rank} A$ is easy to compute from $\operatorname{rank} B$. We first replace $A \in F^{n \times n}$ by

$$A' = \begin{array}{|cc|} \hline 0 & A \\ A^t & 0 \\ \hline \end{array} \in F^{2n \times 2n}.$$

Then $\operatorname{rank} A = \frac{1}{2} \operatorname{rank} A'$, and A' is symmetric. Writing A for A' now, we may assume that A is symmetric.

Let y be an indeterminate over F, $F(y)$ the field of rational functions in y over F,

$$Y = \mathrm{diag}(1, y, \ldots, y^{n-1}) = \begin{bmatrix} 1 & & & 0 \\ & y & & \\ & & \ddots & \\ 0 & & & y^{n-1} \end{bmatrix} \in F(y)^{n \times n},$$

and $B = YA$. Since Y is nonsingular, we have $\mathrm{rank}\, B = \mathrm{rank}\, A$. (We use the fact that $\mathrm{rank}\, A$ is invariant under the field extension $F \subseteq F(y)$.)

LEMMA 13.4
$\mathrm{rank}\, B^2 = \mathrm{rank}\, B$.

PROOF
It is sufficient to show that

$$\mathrm{rank}\, AYA = \mathrm{rank}\, A,$$

since Y is nonsingular and

$$\mathrm{rank}\, B^2 = \mathrm{rank}\, YAYA = \mathrm{rank}\, AYA = \mathrm{rank}\, A = \mathrm{rank}\, YA = \mathrm{rank}\, B.$$

Since $\mathrm{rank}\, AYA \leq \mathrm{rank}\, A$, we only have to show $\mathrm{rank}\, AYA \geq \mathrm{rank}\, A$. We prove this by showing $\ker AYA \subset \ker A$.

So let $u \in F(y)^n$ with $AYAu = 0$. We want to show that $Au = 0$. After multiplying up the denominators (in $F[y]$) of the coordinates of u, we may assume that $u \in F[y]^n$. Set $v = Au \in F[y]^n$. Let z be a new indeterminate over F, $w = v(z) = Au(z) \in F[z]^n$, and

$$s = \sum_{1 \leq i \leq n} w_i v_i y^{i-1} = w^t Yv = v(z)^t Yv = u(z)^t A^t Y Au = 0,$$

where we have used that A is symmetric: $A^t = A$. Suppose that $v \neq 0$. Let $m_i = \deg v_i$ (with $\deg 0 = -\infty$), $m = \max\{m_i \colon 1 \leq i \leq n\}$, $k = \max\{i \colon 1 \leq i \leq n, m_i = m\}$. Terms containing z^{m_k} only occur in summands of $\sum w_i v_i y^{i-1} \in F[y, z]$ with $\deg w_i = m = m_k$, and when $i < k$, then such a summand has degree less than $m_k + k - 1$ in y. Therefore $z^{m_k} y^{m_k} y^{k-1}$ has nonzero coefficient in the above sum. Thus $s \neq 0$. This contradiction shows that indeed $v = 0$, and thus $\ker A = \ker AYA$. ∎

ALGORITHM 13.3
Matrix rank
Input: A symmetric matrix $A \in F^{n \times n}$.
Output: rank A.

1. Compute $B = YA$, where Y is defined above, using an indeterminate y.
2. Compute $\chi(B) = \det(xI - B)$.
3. Return $r = \text{rank}_{\text{alg}} B$.

THEOREM 13.7
Over any field F, MATRANK $\leq$ ITMATPROD.

PROOF
The algorithm **Matrix rank** reduces the rank of $n \times n$-matrices to the computation of the characteristic polynomial of matrices in $F[y]^{n \times n}$, with each entry of degree less than n. Each coefficient of such a characteristic polynomial has degree less than n^2. Algorithm **Characteristic polynomial** in Section 13.4 reduces this in turn to the iterated product of $n \times n$-matrices over $F[y]$, again with degree in y less than n.

Thus suppose we want to compute $E = D_1 \cdots D_n$, with $D_1, \ldots, D_n \in F[y]^{n \times n}$. Write $D_i = \sum_{0 \leq j < n} D_{ij} y^j$, with all $D_{ij} \in F^{n \times n}$. Then the entries of E have degree less than n^2 and can be read off the iterated product of all

$$\phi_{n^2}(D_i) \in (F^{n \times n})^{n^2 \times n^2} \cong F^{n^3 \times n^3},$$

by Lemma 13.2. Overall, we have reduced MATRANK to ITMATPROD. ∎

Note that just saying "a product of n $n \times n$-matrices, each entry a polynomial of degree less than n" would only yield depth $O(\log^3 n)$.

COROLLARY 13.1
Let F be a field. The rank of $n \times n$-matrices can be computed in depth $O(\log^2 n)$ and size $n^{O(1)}$.

13.7
Linear Algebra Classes

In this section, we show that most elementary problems from linear algebra are complete for one of two complexity classes: DET_F or $RANK_F$.

We define the "complexity class"

$$RANK_F = \{f : f \leq \text{MATRANK}\},$$

and the further problems:

BASIS: compute an n-bit vector marking a maximal set of linearly independent columns of $A \in F^{n \times n}$ (i.e., a basis for the column space of A),

SOLVABILITY: compute the bit ($\exists x \in F^n \; Ax = b$),

MAXMINOR: mark rows and columns of $A \in F^{n \times n}$ forming a maximal nonsingular submatrix.

These problems are in general not functions, but relations with several possible answers. This issue is discussed in greater detail in Section 13.8.

THEOREM 13.8
Let F be a field. Then

1. $RANK_F \subseteq DET_F$.
2. MATRANK, MAXMINOR, BASIS, *and* SOLVABILITY *are complete for* $RANK_F$.

PROOF
(1) follows from MATRANK $\leq$ ITMATPROD $\in DET_F$.

(2) We give a circle of four reductions:

1. MATRANK $\leq$ MAXMINOR: The rank equals the number of columns of a maximal nonsingular minor.
2. MAXMINOR $\leq$ BASIS: Given $A \in F^{n \times n}$, mark a basis $A_{i_1}, \ldots, A_{i_r}$ of columns of A for the column space of A. Append $n - r$ zero columns to these to get $B \in F^{n \times n}$. Mark a basis for the row space of B (obtained from a column basis for B^t). Then the marked columns and rows of A form a maximal nonsingular minor.
3. BASIS $\leq$ SOLVABILITY: Suppose we are given $A \in F^{n \times n}$, with columns $A_1, \ldots, A_n \in F^n$. For all i, $1 \leq i \leq n$, check whether the system

$$\sum_{1 \leq j < i} A_j x_j = A_i$$

of n linear equations in $i-1$ indeterminates has a solution $(x_1, \ldots, x_{i-1}) \in F^{i-1}$. (To bring this into the required square format, append $n-i+1$ zero columns, and check for $x \in F^n$.) A basis is formed by those A_i for which no solution exists.

4. SOLVABILITY $\leq$ MATRANK: $Ax = b$ has a solution if and only if
$$\operatorname{rank} A = \operatorname{rank}(A|b),$$
where $A|b \in F^{n \times (n+1)}$ is A with column b appended. ∎

We define further problems:

INDEPENDENCE: on input $x_1, \ldots, x_i \in F^n$, decide whether they are linearly independent,

SINGULAR: decide whether $A \in F^{n \times n}$ is singular,

EQ: given $A \in F^{n \times n}$ and $b \in F^n$, compute the bit $c = (\exists y \in F^n \; Ay = b)$, and if $c = $ true, compute $x \in F^n$ with $Ax = b$.

NULLSPACE: compute a basis for the nullspace $\{b \in F^n : Ab = 0\}$ of $A \in F^{n \times n}$.

THEOREM 13.9
Let F be a field. EQ and NULLSPACE are complete for DET_F.

PROOF
We show
$$\text{EQ} \leq \text{NULLSPACE} \leq \text{NONSINGEQ} \leq \text{EQ}.$$

1. EQ $\leq$ NULLSPACE: Let $A \in F^{n \times n}$, $b \in F^n$. Then
$$\forall x \in F^n \left(Ax = b \iff \boxed{A \mid b} \cdot \boxed{\begin{array}{c} x \\ y \end{array}} = 0 \text{ with } y = -1 \right).$$

Determine a basis $z_1, \ldots, z_k \in F^{n+1}$ of the nullspace of
$$\boxed{\begin{array}{cc} A & b \\ 0 \ldots 0 & 0 \end{array}} \in F^{(n+1) \times (n+1)}.$$

Then return $c = $ true if and only if $z_{i,n+1} \neq 0$ for some i, $1 \leq i \leq k$. If $c = $ true, let i be the smallest index such that $z_{i,n+1} \neq 0$, and return also
$$x = \frac{-1}{z_{i,n+1}} (z_1, \ldots, z_n).$$

2. NULLSPACE $\leq$ NONSINGEQ: Given $A \in F^{n \times n}$, find a maximal nonsingular minor M of A, using MAXMINOR $\leq$ NONSINGEQ. For simplicity, we may assume that M is the upper left $r \times r$-submatrix, where $r = \operatorname{rank} A$. For all i, $r < i \leq n$, solve the nonsingular system of linear equations

$$Mx_i = y_i,$$

where $y_i \in F^r$ consists of the top r entries of the ith column of A:

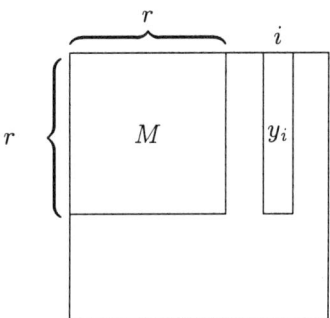

For $r < i \leq n$, let

$$z_i = (x_i, 0, \ldots, 0, \ -1, \ 0, \ldots, 0) \in F^n.$$
$$\uparrow$$
$$i$$

We now claim that $z_{r+1}, \ldots, z_n$ form a basis of the nullspace $\ker A$ of A. Let $r < i \leq n$. We first show that $z_i \in \ker A$. For $1 \leq j \leq r$, the jth entry of Az_i is

$$\sum_{1 \leq k \leq n} A_{jk} z_{ik} = \sum_{1 \leq k \leq r} A_{jk} x_{ik} - A_{ji} = 0.$$

For $r < j \leq n$, the jth row A_{j*} of A is a linear combination of the first rows $A_{1*}, \ldots, A_{r*}$, since M is a maximal nonsingular minor. Therefore again $(Az_i)_j = A_{j*} z_i = 0$. Combining these, we have $Az_i = 0$.

On the other hand, $\dim_F \ker A = n - r$ and $z_{r+1}, \ldots, z_n \in \ker A$ are linearly independent, because z_i has a -1 in position i, and all the other z_k's have a zero there. Therefore these form a basis of $\ker A$.

3. NONSINGEQ $\leq$ EQ is trivial. ∎

The following problems are all unsolved.

OPEN QUESTION 13.1
1. Is $RANK_F \neq DET_F$?
2. Is INDEPENDENCE complete for $RANK_F$?
3. Is INDEPENDENCE $\leq$ SINGULAR?

13.8
Arithmetic Boolean Circuits

In the previous sections, we have derived (exact) algorithms using parallel time $O(\log^2 n)$ for the basic problems of linear algebra. In this section, we describe the elements of a theory of parallel algebraic computation, and where the above results fit into that general framework.

We start with *arithmetic Boolean circuits*, a generalization of our arithmetic circuits necessary to deal with decision problems, which we have already used for the rank of matrices. Then we define some *parallel complexity classes*, in analogy with well-studied Boolean complexity classes, and finally formalize *reductions* between two problems. In Sections 13.5 and 13.7, many problems from linear algebra turned out to be equivalent either to the determinant or to the rank problem, so that any depth improvement for one of them would automatically improve the depth for all of them.

How can we solve a general system $Ax = b$ of linear equations, where A may be singular? Even for a single equation $ax = b$, with $a, b \in F$, all we can do with an arithmetic circuit is to return $x = b/a$. However, we would also like to output the information "no solution" if $a = 0$ and $b \neq 0$. So we now extend the model to allow such tests.

First recall that a *Boolean circuit* is a labelled directed acyclic graph, similar to an arithmetic circuit. The difference is that the values manipulated are not from an algebraic domain, but the two Boolean values **T** (for "true", or 1) and **F** (for "false", or 0). Accordingly, the operations are the Boolean $\neg$ (negation "not"), $\wedge$ (conjunction "and"), and $\vee$ (disjunction "or"). Boolean circuits are a model of the electronic circuits, the innards of digital computers.

We now define an *arithmetic Boolean circuit* (over a ring F) to be a labelled directed acyclic graph, where both arithmetic and Boolean labels are allowed. Thus we have arithmetic inputs and constants from F, and Boolean inputs and constants from the Boolean universe $\mathbb{B} = \{\mathbf{T}, \mathbf{F}\}$, and the seven operations $+, -, *, /, \neg, \wedge, \vee$. Each gate has a *type*—either arithmetic or

13.8. Arithmetic Boolean Circuits

Boolean—and the appropriate number of inputs with the right type. As an example, a ∧-gate has two inputs, both from a gate with type "Boolean".

Two further gates provide the interface between the arithmetic and the Boolean parts: test gates and selection gates. A *test gate* "$x \stackrel{?}{\neq} 0$" has an arithmetic input x and a Boolean output y:

$$y = \begin{cases} \mathbf{T} & \text{if } x \neq 0, \\ \mathbf{F} & \text{if } x = 0. \end{cases}$$

A *selection gate* has two arithmetic inputs x_1 and x_2, a Boolean input y, and an arithmetic output z:

$$z = \begin{cases} x_1 & \text{if } y = \mathbf{T}, \\ x_2 & \text{if } y = \mathbf{F}. \end{cases}$$

For simplicity, we assume in the sequel that the ground domain F is a field. Since we now have zero-tests, we insist that every division in α is by a nonzero field element, for any specific input supplied for the variables.

Figure 13.5 shows an arithmetic Boolean circuit α with two inputs x_1 and x_2 (at gates 2 and 3), the arithmetic constant 1 at gate 1, and one

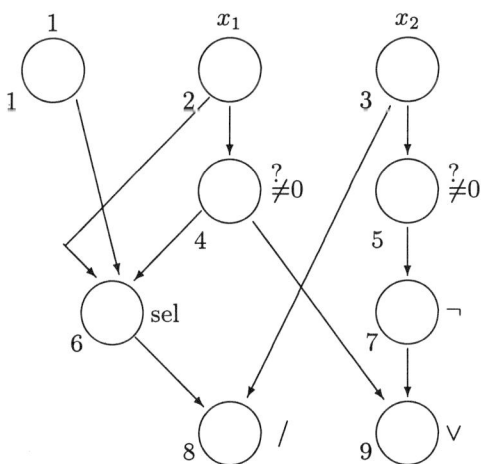

FIGURE 13.5
An arithmetic Boolean circuit for a linear equation $x_1 t = x_2$.

arithmetic plus one Boolean output, at gates 8 and 9. The value of gate 9 is
T if $z = (x_1 \neq 0 \vee x_2 = 0)$ is true, and the value y of gate 8 is

$$y = \begin{cases} x_2/x_1 & \text{if } x_1 \neq 0, \\ x_2 & \text{if } x_1 = 0. \end{cases}$$

Thus α solves the linear equation $x_1 t = x_2$ in a satisfactory way: the output z is **T** if a solution exists, and **F** otherwise, and if $z = \mathbf{T}$, then y is indeed a solution.

The arithmetic Boolean circuit α splits the input space $S = F^2$ into three regions: $S_1 = F^2 \setminus (\{0\} \times F) = \{(x_1, x_2) \in F^2 : x_1 \neq 0\}$, $S_2 = (\{0\} \times F) \setminus \{(0,0)\}$, and $S_3 = \{(0,0)\}$. These form a *partition* of S: $S = S_1 \cup S_2 \cup S_3$, and $S_i \cap S_j = \emptyset$ if $i \neq j$. In each region, two functions are computed:

$$\begin{aligned} y &= x_2/x_1 \quad \text{and} \quad z = \mathbf{T} \quad \text{on } S_1, \\ y &= x_2 \quad \text{and} \quad z = \mathbf{F} \quad \text{on } S_2, \\ y &= 0 \quad \text{and} \quad z = \mathbf{T} \quad \text{on } S_3. \end{aligned}$$

We will call the collection f_α of all these

$$f_\alpha = \big((S_1, x_2/x_1, \mathbf{T}), (S_2, x_2, \mathbf{F}), (S_3, 0, \mathbf{T})\big)$$

a *piecewise rational function*. The problem of determining whether a linear equation $x_1 t = x_2$ has a solution t, and computing a solution if one exists, corresponds to "the" piecewise rational function

$$g = \big((S_1, x_2/x_1, \mathbf{T}), (S_2, \mathbf{F}), (S_3, w, \mathbf{T})\big),$$

where w is any element of F. Thus in fact g consists of not one, but many piecewise rational functions, one for each $w \in F$.

We say that α *computes* g since for any input $(x_1, x_2) \in F^2$, one of the piecewise rational functions making up g can be read off the gates of α. If $(x_1, x_2) \in S_1$, the outputs y and z contain the unique correct answer to g. If $(x_1, x_2) \in S_2$, z has the correct value, and y is irrelevant. If $(x_1, x_2) \in S_3$, the answer $(y, z) = (0, \mathbf{T})$ is among the correct ones for g (with $w = 0$); the value $y = 0$ was chosen arbitrarily.

In general, we call any of the objects computed by an arithmetic Boolean circuit α a piecewise rational function, and each *computational problem* is a family $g = (g_n)_{n \in \mathbb{N}}$ of collections g_n of such piecewise rational functions. A family $\alpha = (\alpha_n)_{n \in \mathbb{N}}$ of arithmetic Boolean circuits *computes* g if for each n, α_n computes one of the piecewise rational functions in g_n.

13.8. Arithmetic Boolean Circuits

EXAMPLE 13.4
The problem of computing the rank of matrices would be formalized as follows: MATRANK $=$ (MATRANK$_n$)$_{n\in \mathbb{N}}$ with

$$\text{MATRANK}_n(A) = \underbrace{\mathbf{T}\cdots\mathbf{T}}_{r}\underbrace{\mathbf{F}\cdots\mathbf{F}}_{n-r} \text{ for } A \in F^{n\times n} \text{ and } r = \text{rank}\, A.$$

EXAMPLE 13.5
Let us see how the computational problem MAXMINOR $=$ (MAXMINOR$_n$)$_{n\in \mathbb{N}}$ of determining a maximal nonsingular minor of matrices, discussed in Section 13.7, fits into this framework. Intuitively, for each $n \in \mathbb{N}$ and matrix $A \in F^{n\times n}$ one should find some sets $I, J \subseteq \{1,\ldots,n\}$ such that the minor ($=$ submatrix) of A with row indices from I and column indices from J is a maximal nonsingular minor; i.e., it is a nonsingular square matrix, and A has no nonsingular minors of larger size. Formally, MAXMINOR$_n$ is the set of all functions

$$g\colon F^{n\times n} \longrightarrow \mathbb{B}^N \times \mathbb{B}^N$$

(with $N = \{1,\ldots,n\}$) where $g(A)$ are the row and column indices of a maximal nonsingular minor of A, for all $A \in F^{n\times n}$. We have to agree on some coding of $g(A)$ over $\mathbb{B}$: one possibility is a string $(y_1,\ldots,y_n, z_1,\ldots,z_n) \in \mathbb{B}^{2n}$ with $i \in I \iff y_i = \mathbf{T}$, and $j \in J \iff z_j = \mathbf{T}$. As a specific example, consider

$$A = \begin{bmatrix} 1 & 1 & 2 \\ 1 & 1 & -2 \\ 2 & 2 & 0 \end{bmatrix} \in \mathbb{Q}^{3\times 3}.$$

The last row of A is the sum of the first two rows. In our formalism, the fact that $\begin{bmatrix} 1 & 2 \\ 1 & -2 \end{bmatrix}$ is a maximal nonsingular minor of A is expressed as follows:

$$(A, \mathbf{TTFFTT}) \in \text{MAXMINOR}_3.$$

An arithmetic Boolean circuit solving MAXMINOR would compute, on any input $A \in F^{n\times n}$, some output $f_\alpha(A) = (\eta_1,\ldots,\eta_n, \zeta_1,\ldots,\zeta_n) \in \mathbb{B}^{2n}$, and the requirement is that indeed $f_\alpha(A)$ describe a maximal nonsingular minor of A. Natural algorithms, such as the ones to be discussed in Section 13.7 for this and similar problems, will compute a "natural" candidate among the $g(A)$'s, e.g., the "lexicographically first maximal nonsingular minor".

A powerful tool in the theory of computation is to collect problems of the "same" cost in *complexity classes*. Before we discuss these within our framework, we first need two rather technical notions; we do not give detailed definitions here. The first is *uniformity* of arithmetic circuits. Uniformity of Boolean circuits is discussed in Ruzzo [47] and Cook [12]. Among the various notions we choose P-uniform families of arithmetic Boolean circuits $\alpha = (\alpha_n)_{n \in \mathbb{N}}$ as our standard: there must exist a Turing machine which, on input n in unary, generates in polynomial time a description of the labels and connections of α_n. A further complication for us is that we have to define *uniformity of field constants*. For this, we require a polynomial-time Turing machine that, on input n, produces a polynomial-size arithmetic circuit β_n, whose only inputs or constants are the inputs to α_n and 1, and whose outputs are the constants used in α_n. (Often, 1 will be the only constant used.)

Furthermore, we have to use the *degree* $\deg g_n$ of a piecewise rational function g_n. If $g_n = (F^n, h_n)$ with $h_n \in F[x_1, \ldots, x_n]$ consists of a single polynomial, then its degree is simply the degree of h_n. In the general case, we require a deep generalization of this notion from algebraic geometry; for discussions, see Strassen [50] and Heintz [30].

Here are now some complexity classes of importance to parallel arithmetic computation.

$$\begin{aligned}
P_F &= \{g = (g_n)_{n \in \mathbb{N}} \colon \exists\ P\text{-uniform } \alpha = (\alpha_n)_{n \in \mathbb{N}} \text{ computing } g \text{ with} \\
&\qquad S(\alpha_n) = n^{O(1)}, \text{and } \deg g_n = n^{O(1)}\}, \\
NC_F^k &= \{g = (g_n)_{n \in \mathbb{N}} \colon \exists\ P\text{-uniform } \alpha = (\alpha_n)_{n \in \mathbb{N}} \text{ computing } g \\
&\qquad \text{with } S(\alpha_n) = n^{O(1)} \text{and } D(\alpha_n) = O(\log^k n), \\
&\qquad \text{and } \deg g_n = n^{O(1)}\}, \text{ for any } k \in \mathbb{N}, \\
NC_F &= \bigcup_{k \in \mathbb{N}} NC_F^k.
\end{aligned}$$

Here, g stands for a computational problem, and α for a family of arithmetic Boolean circuits.

The arithmetic complexity classes defined above are analogues of the Boolean classes NC^k, defined as the set of Boolean functions computed by *log-space uniform* Boolean circuits of depth $O(\log^k n)$ and size $n^{O(1)}$, for input size n. Also, $NC = \bigcup_{k \in \mathbb{N}} NC^k$, and P is defined by polynomial size $n^{O(1)}$ only. Nick Pippenger [44] introduced these Boolean classes NC^k—an acronym for "Nick's class", coined at Toronto where he was then working. $NC^k(P$-uniform) and $NC(P$-uniform) are obtained by the more generous notion of P-uniformity, which we use for our arithmetic classes. Then $NC \subseteq NC(P$-uniform), and Cook [12] conjectures that inequality holds. For any field F, we have $NC(P$-uniform$) \subseteq NC_F$, and similarly for the other classes.

13.8. Arithmetic Boolean Circuits

In Boolean circuit complexity, another model of importance is the *unbounded fan-in Boolean circuit*, where ∨- and ∧-gates may have any number of inputs. Restricting the depth to $O(\log^k n)$, for some $k \in \mathbb{N}$, the size to $n^{O(1)}$, and requiring uniformity, one obtains the complexity class AC^k. This acronym stands for "alternating class" and comes from the fact that, for $k \geq 1$, it can also be characterized by *alternating Turing machines* using space $O(\log n)$ and alternation depth $O(\log^k n)$. We clearly have $NC^k \subseteq AC^k \subseteq NC^{k+1}$. One of the few separation theorems in Boolean complexity theory is the breakthrough result $AC^0 \subsetneq NC^1$ of Furst et al. [19] and Ajtai [1]; the parity function is in NC^1_F, and they show that it is not in AC^0_F.

If we allow unbounded fan-in ∧-gates, but only ∨-gates with fan-in two, we obtain the classes SAC^k and SAC (for "semi-AC"). Borodin et al. [7] show that one obtains the same classes with unbounded fan-in ∨ and bounded ∧.

We take the latter model as our template for the arithmetic case. A *semi-unbounded fan-in arithmetic Boolean circuit* α is like an ordinary arithmetic Boolean circuit, except that the + and ∨-gates are allowed to have arbitrary fan-in. The depth of a gate with fan-in k is 1, and its size is $k-1$. Then, as usual, $D(\alpha)$ is the depth of a deepest path in α, and $S(\alpha)$ the sum of the sizes of all gates in α. These circuits lead to the following complexity classes.

$$SAC^k_F = \{g = (g_n)_{n \in \mathbb{N}} \colon \exists \alpha = (\alpha_n)_{n \in \mathbb{N}} \text{ of semi-unbounded fan-in}$$
$$\text{computing } g \text{ with } S(\alpha_n) = n^{O(1)}, D(\alpha_n) =$$
$$O(\log^k n), \text{ and } \deg g_n = n^{O(1)}\}, \text{ for any } k \in \mathbb{N},$$

$$SAC_F = \bigcup_{k \in \mathbb{N}} SAC^k_F.$$

The circuit families have to be *P*-uniform. We clearly have the following *hierarchy* of complexity classes:

$$NC^0_F \subseteq SAC^0_F \subseteq NC^1_F \subseteq \cdots$$
$$NC^k_F \subseteq SAC^k_F \subseteq NC^{k+1}_F \subseteq \cdots NC_F = SAC_F \subseteq P_F.$$

Whenever one has such a hierarchy, one of the first (and usually most difficult) questions is: Does it collapse or not? I.e., is $NC^k_F \subsetneq SAC^k \subsetneq NC^{k+1}_F$ for all k? Is $NC_F \subsetneq P_F$? Throughout this chapter we have worked at the low end of this hierarchy, within NC^2_F.

Only one difference is easy to see: $NC^0_F \neq SAC^0$. We have SUM $\in AC^0_F$, and by Theorem 13.1 (3), $\text{SUM}_n = x_1 + \cdots + x_n$ requires depth at least $\log_2 n$ on arithmetic Boolean circuits with fan-in two, and so is not in the trivial class NC^0_F.

We define a *reduction* between two problems $f = (f_n)_{n \in \mathbb{N}}$ and $g = (g_n)_{n \in \mathbb{N}}$ to be a family $(\alpha_n)_{n \in \mathbb{N}}$ of arithmetic Boolean circuits α_n of constant depth and polynomial size $n^{O(1)}$, with α_n computing f_n. As gates we allow in α_n the usual arithmetic and Boolean gates, as above, plus gates computing some g_j. Furthermore, both the +-gates and the ∨-gates may have an arbitrary number of inputs. If such an "arbitrary fan-in"-gate or an "oracle gate" for g_j has k inputs, we define its size to be $\max\{j, k\} - 1$, and its depth to be 1. (This agrees with the previous definitions for the "binary" +, with only two inputs; the size comes from the lower bound of Exercise 13.1.) If such a reduction exists, we write $f \leq_F g$: f is *reducible* to g. Furthermore, $f \equiv_F g$ means that $f \leq_F g$ and $g \leq_F f$; then f and g are *equivalent*. One ambiguity is that g_m—as a relation—may not have a single, but many possible answers. We stipulate that for any correct answer to any g_m a correct answer for f must result.

EXAMPLE 13.6
PROD $\leq_F$ DETERMINANT. Consider the circuit α_n with input $x_1, \ldots, x_n$ and the constant 0, which produces an $n \times n$-matrix A with $x_1, \ldots, x_n$ on the diagonal and zeroes elsewhere, and then calls DETERMINANT$_n$ with input A. The output is PROD$_n$. The depth of this reduction is 1, and the size $n^2 - 1$.

Intuitively, "$f \leq_F g$" means that f is not harder than g. If we find a good algorithm for g, then we automatically obtain a good algorithm for f. On the other hand, a lower bound on the complexity of f translates into one for the complexity of g.

THEOREM 13.10
Let F be an integral domain.

1. Reducibility is a partial order.
2. Equivalence is an equivalence relation.
3. If $f \leq_F g$ and g can be computed in depth $O(\log^k n)$, for some $k \geq 1$, and size $n^{O(1)}$, then f can be computed at the same (asymptotic) cost.

The theorem is proven by taking a circuit family $\alpha = (\alpha_n)_{n \in \mathbb{N}}$ for g, plugging it into the oracle nodes calling g_j in the reduction, and thus obtaining a circuit family for f; we forego the details.

If $\mathcal{C}$ is one of our complexity classes (or any set of problems), we say that a problem f is *complete for* $\mathcal{C}$ if

1. $f \in \mathcal{C}$,
2. $\forall g \in \mathcal{C} \; g \leq f$.

Thus a complete problem is hardest within its complexity class.

An important result is that the iterated product of 3 × 3-matrices (see Section 13.3) is complete for NC_F^1 (Ben-Or & Cleve [4]).

The two main algorithmic results in this chapter, namely the computations for determinant and rank of matrices, can be summarized as follows.

THEOREM 13.11
Let F be a field. Then

$$RANK_F \subseteq DET_F \subseteq SAC_F^1 \subseteq NC_F^2.$$

PROOF
The first inclusion is in Theorem 13.8 (1). For the second one, we check that MATPROD $\in SAC_F^0$ and ITMATPROD $\in SAC_F^1$. The claim then follows from Algorithm **Characteristic Polynomial**, using Theorem 13.4. ∎

The following problems are unsolved.

OPEN QUESTION 13.2
1. Is $DET_F = SAC_F^1$?
2. Is $SAC_F^1 \neq NC_F^2$?

REMARK 13.1
Our arithmetic circuits are the arithmetic analogues of Boolean circuits, and one reason for choosing them for this chapter is their conceptual simplicity. Another highly popular model of parallel Boolean computation is the PRAM. Its analogue, the arithmetic PRAM, *has arithmetic values in its memory cells, and each processor can perform an arithmetic operation on two of those values in one time step. (Similarly, one defines* arithmetic Boolean PRAMs, *with the same instruction set as our arithmetic Boolean circuits.) There are various possibilities to regulate the read/write conflicts on Boolean PRAMs. Similarly, the arithmetic PRAMs come in several flavors; we do not discuss this here.*

For a comparison of the two models, it is easiest to use levelled *arithmetic circuits, where each gate has an integer associated to it, its* level, *and inputs to a gate come only from previous levels. The* width *of a levelled arithmetic circuit is the maximum number of arithmetic gates at each level. Then an arithmetic circuit can be simulated by an arithmetic PRAM, and, for a given input size n, an arithmetic PRAM by an arithmetic circuit, with circuit depth corresponding to PRAM time, circuit size corresponding to PRAM memory, and circuit width corresponding to the number of PRAM processors.*

The simulation question is not so clear, however, when we consider circuit families, complexity classes, and the various notions of circuit uniformity.

13.9
Further Results

In this section, we give pointers to various results in parallel algebraic complexity theory, without proofs.

Exponentiation turned out to be a very interesting problem for parallel computation: computing a^b in parallel, where $a \in F$ is in the ground domain, and $b \in \mathbb{N}$. This problem and related tasks are used in many algorithms, e.g., factoring integers, primality test, cryptographic protocols, and factoring polynomials over finite fields.

The standard sequential algorithm of "repeated squaring" has linear depth n when b is an n-bit integer. The problem looks unamenable to parallelization, and Kung [39] shows with a degree agrument that indeed over an infinite field no arithmetic circuits with less than the disappointing linear depth are possible. The argument seems to fail over finite fields, where one can use Fermat's Little Theorem ($a^q = a$ for all a in $\mathbb{F}_q$, the field with q elements) to compute the values of large powers at no cost. However, von zur Gathen [24] shows that this is the only obstacle: $D(\alpha) \geq \min\{\log_2 b, \log_2(q-b)\}$ if $1 \leq b < q$ and α computes bth powers in $\mathbb{F}_q$. It was a big surprise when Fich & Tompa [17] proved in a slightly different—yet perfectly reasonable—model that the problem does have a fast parallel solution in an important special case (large finite fields of small characteristic). This leads to the rather shocking observation that for this (and some other) problem arithmetic circuits are *not* the appropriate model of computation (von zur Gathen & Seroussi [29]). The use of "normal bases" in finite fields leads to a natural setting in which the parallel complexity of exponentiation can be determined exactly (von zur Gathen [27]).

Eberly [16] solves various problems (such as the determinant, characteristic polynomial, and solution of systems of linear equations) for *banded* $n \times n$-*matrices* of bandwidth b in depth $O(\log n \log b)$ and size $n^{O(1)}$; in particular, for constant bandwidth he has optimal depth $O(\log n)$. Kaltofen et al. [35, 36] prove that the *Hermite* and *Smith normal forms* of polynomial matrices can be computed in *probabilistic NC*. These normal forms contain much information about the (geometric) structure of the linear mapping associated with a matrix.

Many problems in *polynomial arithmetic* can be solved in NC_F^2, for a field F, such as the gcd (Borodin et al. [8]), more generally all entries of the Extended Euclidean Scheme of two polynomials, various interpolation problems (rational, Hermite), partial fraction decomposition (for a given factorization of the denominator), Chinese remainder algorithm, and Padé approximation (von zur Gathen [22]). One of the most important unresolved issues is the status of the Boolean analogue:

OPEN QUESTION 13.3
Is the gcd *of integers in (Boolean) NC?*

It is widely conjectured that $NC \neq P$. However, Valiant et al. [52] showed that for polynomials over a field F we have "$NC_F^2 = P_F$": polynomial families with polynomial degree and polynomial-size arithmetic circuits can be computed on arithmetic circuits of depth $O(\log^2 n)$. Miller et al. [40] give a different version of that result, and Kaltofen [34] extends it to rational functions; see Kaltofen's Chapter 16 in this book.

Reif [46] and Beame et al. [3] showed that (Boolean) problems like division with remainder and iterated product of n-bit integers can be solved in optimal (P-uniform) depth $O(\log n)$ on Boolean circuits. Eberly [16] shows similar results for polynomials over a field. This leads to optimal-depth solutions for the exponentiation problem in finite fields of small characteristic (von zur Gathen & Seroussi [29]), for inversion in finite fields (von zur Gathen [26]), and for the Boolean exponentiation problem of computing $a^b \mod 2^n$, where $a, b \in \mathbb{N}$ are n-bit integers (von zur Gathen [24]).

A central problem in *computer algebra* is the *factorization of polynomials*. Over finite fields of *small characteristic*, the problem is in NC^2, but in general it is at least as hard as exponentiation (von zur Gathen [21]). Kaltofen [32] has an algorithm for absolute irreducibility. The polynomial-time sequential method over $\mathbb{Q}$ uses "short vectors in $\mathbb{Z}$-modules", which is conjectured to be P-complete; the (possibly "easy") integer gcd problem is reducible to a special case of this (von zur Gathen [21]).

A vast generalization of factoring polynomials is the question of determining the roots of a system of polynomial equations, or, more generally, of deciding first-order sentences in the theory of fields. Ben-Or et al. [4], Davenport & Heintz [15], and Fitchas et al. [18] contain parallel results for algebraically closed fields (like $\mathbb{C}$) and real closed fields (like $\mathbb{R}$).

Strassen [50] introduced the *degree* as an important tool in algebraic complexity theory (see Section 13.8). For an arithmetic circuit α computing a single polynomial f, one has $D(\alpha) \geq \log_2 \deg f$ (Kung [39]); this generalizes to a set of polynomials or rational functions, with the notion of degree from algebraic geometry. Unfortunately, the argument breaks down for our piecewise rational functions, and one can show that the best result here is $D(\alpha) \geq \log_2 \log_2 \deg f$ (see von zur Gathen [23]).

Throughout this chapter, we have implicitly assumed that an almost unbounded (namely, polynomial) number of processors is available. In practice today, however, one can count only on a limited number of processors, and *processor-efficient parallel algorithms* are important, which use small parallel time (say, $(\log n)^{O(1)}$) and not (many) more processors than the best sequential algorithm known; this question is briefly addressed at the end of Section 13.4. Kaltofen [34] has a result in this spirit on the gcd of polynomials, and Kaltofen & Pan [37] achieve this goal (up to logarithmic factors) for the solution of systems of linear equations. Pan & Reif [42, 43] started an interesting line of work for linear algebra; they work in a different "numerical" model where one has access to the individual bits of the real (or rational) inputs.

Let $M(n)$ be the number of PRAM processors required to multiply an $n \times n$ dense matrix in $O(\log n)$ time using $M(n)$ processors. For well-conditioned matrices, Pan and Reif use a Newton iteration to efficiently compute the matrix inverse within high accuracy in $O(\log^2 n)$ time using $M(n)$ PRAM processors. Kaltofen and Pan developed similarly efficient PRAM algorithms for matrix inverse, determinant, and rank over general fields.

Cheriyan and Reif [10] give output sensitive (where the complexity depends on the output) PRAM algorithms for various classes of algebraic problems. They present a randomized algorithm for computing the rank r of an $n \times n$ matrix which runs in parallel time $O(\log n + \log^3 r)$ using $(n^2 + M(r)) \log^{O(1)} n$ processors. They also present randomized algorithms for finding a maximum linearly independent subset of rows that run either in parallel time $O((\log n) \log^2 r)$ using $(n^2 + rM(r)) \log^{O(1)} n$ processors, or in parallel time $O(\log n + \log^2 r)$ using $(n^2 + nM(r)) \log^{O(1)} n$ processors. As an application, they give an output sensitive algorithm for computing greatest common divisors (GCD) of polynomials. Given two polynomials of degree n, the degree r of the polynomial GCD is computed in randomized parallel time $O(\log n + \log^3 r)$ using $(n^2/r + r^2) \log^{O(1)} n$ processors, and the GCD as well as the extended GCD are computed in randomized parallel time $O((\log n) \log^2 r)$ using $(n^2/r + r^3) \log^{O(1)} n$ processors.

A different set of problems concerns *permutation groups*. As an example, the *membership problem* is: given some permutations $\pi_1, \ldots, \pi_k, \sigma$ on n letters, in some standard representation, is σ in the subgroup generated by $\pi_1, \ldots, \pi_k$? A substantial line of research, culminating in Babai et al. [2], shows that the membership problem and related questions are in (Boolean) NC.

Bibliography

[1] M. Ajtai, Σ_1^1-Formulae on Finite Structures. Ann. of Pure and Applied Logic **24** (1983), 1–48.

[2] L. Babai, E.M. Luks, and Á. Seress, Permutation groups in NC. Proc. 19th Ann. ACM Symp. Theory Comput., New York NY, 1987, 409–420.

[3] P.W. Beame, S.A. Cook and H.J. Hoover, Log depth circuits for division and related problems. SIAM. J. Comput. **15** (1986), 994–1003.

[4] M. Ben-Or and R. Cleve, Computing Algebraic Formulas Using a Constant Number of Registers. Proc. 20th Ann. ACM Symp. Theory of Computing, Chicago IL, 1988, 254–257. SIAM. J. Comput., to appear.

[5] M. Ben-Or, D. Kozen, and J. Reif, The Complexity of elementary algebra and geometry. J. Computer System Sciences **32** (1986), 251–264.

[6] S.J. Berkowitz, On computing the determinant in small parallel time using a small number of processors. Information Processing Letters **18** (1984), 147–150.

[7] A. Borodin, S. A. Cook, P. W. Dymond, W. L. Ruzzo, and M. Tompa, Two applications of inductive counting for complementation problems. SIAM J. Comput. **18** (1989), 559–578.

[8] A. Borodin, J. von zur Gathen, and J. Hopcroft, Fast parallel matrix and GCD computations. Information and Control **52** (1982), 241–256.

[9] R.W. Brockett and D. Dobkin, On the number of multiplications required for matrix multiplication. SIAM J. Comput. **5** (1976), 624–628.

[10] J. Cheriyan and J. Reif, Parallel and Output Sensitive Algorithms for Combinatorial and Linear Algebra Problems, manuscript, 1992.

[11] A.L. Chistov, Fast parallel calculation of the rank of matrices over a field of arbitrary characteristic. Proc. Int. Conf. Foundations of Computation Theory, Springer Lecture Notes in Computer Science **199**, 1985, 63–69.

[12] S.A. Cook, A taxonomy of problems with fast parallel algorithms. Information and Control **64** (1985), 2–22.

[13] D. Coppersmith and S. Winograd, Matrix multiplication via arithmetic progressions. J. Symb. Comp. **9** (1990), 251–280.

[14] L. Csanky, Fast parallel matrix inversion algorithms. SIAM J. Comput. **5** (1976), 618–623.

[15] J. H. Davenport and J. Heintz, Real quantifier elimination is doubly exponential. J. Symb. Comp. **4** (1988).

[16] W. Eberly, Very fast parallel matrix and polynomial arithmetic. SIAM J. Comput. **18** (1989), 955–976.

[17] F. Fich and M. Tompa, The parallel complexity of exponentiating polynomials over finite fields. J. Assoc. Comput. Mach. **53** (1988), 651–667.

[18] N. Fitchas, A. Galligo, and J. Morgenstern, Algorithmes rapides en séquentiel et en parallèle pour l'élimination de quantificateurs en géometrie élémentaire, Séminaire Structures Algébriques Ordonnées, UER de Mathématiques : Université de Paris VII (1987).

[19] M. Furst, J. B. Saxe, and M. Sipser, Parity, Circuits, and the Polynomial Time Hierarchy. Math. Systems Theory **17** (1984), 13–28.

[20] Z. Galil and V. Pan, Parallel evaluation of the determinant and of the inverse of a matrix. Inform. Process. Letters **30** (1989), 41–45.

[21] J. von zur Gathen, Parallel algorithms for algebraic problems. SIAM J. Comput. **13** (1984), 802–824.

[22] J. von zur Gathen, Representations and parallel computations for rational functions. SIAM J. Comput **15** (1986a), 432–452.

[23] J. von zur Gathen, Parallel arithmetic computations: a survey. Proc. 12th Int. Symp. Math. Foundations of Computer Science, Bratislava, Springer Lecture Notes in Computer Science **233**, 1986b, 93–112.

[24] J. von zur Gathen, Computing powers in parallel. SIAM J. Computing **16** (1987), 930–945.

[25] J. von zur Gathen, Algebraic complexity theory. Annual Review of Computer Science **3** (1988), 317–347.

[26] J. von zur Gathen, Inversion in finite fields using logarithmic depth. J. Symb. Comp. **9** (1990), 175–183.

[27] J. von zur Gathen, Efficient and optimal exponentiation in finite fields. Comput. Complexity 1 (1991), 360–394.

[28] J. von zur Gathen and W. Eberly, Course notes for CSC 2408 "Parallel arithmetic computations", University of Toronto, January 1984.

[29] J. von zur Gathen and G. Seroussi, Boolean circuits versus arithmetic circuits. Information and Computation **91** (1991), 142–154.

[30] J. Heintz, Definability and fast quantifier elimination in algebraically closed fields. Theor. Computer Science **24** (1983), 239–277.

[31] O.H. Ibarra, S. Moran, and L.E. Rosier, A note on the parallel complexity of computing the rank of order n matrices. Inform. Process. Letters **11** (1980), 162.

[32] E. Kaltofen, Fast parallel absolute irreducibility testing. J. Symb. Computation **1** (1985), 57–67.

[33] E. Kaltofen, Greatest common divisors of polynomials given by straight-line programs. J. Assoc. Comput. Mach. **35** (1988), 231–264.

[34] E. Kaltofen, Parallel Algebraic Algorithm Design, Lecture Notes, Dept. Comput. Sci., Rensselaer Polytechnic Inst., November 1989.

[35] E. Kaltofen, M.S. Krishnamoorthy, and B.D. Saunders, Fast parallel computation of Hermite and Smith forms of polynomial matrices. SIAM J. Algebraic and Discrete Methods **8** (1987), 683–690.

[36] E. Kaltofen, M. Krishnamoorthy, and B.D. Saunders, Parallel algorithms for matrix normal forms. Linear Algebra and Applications **136** (1990), 189–208.

[37] E. Kaltofen and V. Pan, Processor efficient parallel solution of linear systems over an abstract field, Proc. 3rd Annual ACM Symposium on Parallel Algorithms and Architectures, (1991), 180–191.

[38] E. Kaltofen and V. Pan, Processor-efficient parallel solution of linear systems II. The positive characteristic and singular cases, Proc. 33rd Annual IEEE Symposium on F.O.C.S. (1992), 714–723.

[39] H.T. Kung, New algorithms and lower bounds for the parallel evaluation of certain rational expressions and recurrences. J. Assoc. Comput. Mach. **23** (1976), 252–261.

[40] G.L. Miller, V. Ramachandran, and E. Kaltofen, Efficient parallel evaluation of straight-line code and arithmetic circuits. SIAM J. Comput. **17** (1988).

[41] K. Mulmuley, A fast parallel algorithm to compute the rank of a matrix over an arbitrary field. Combinatorica **7** (1987), 101–104.

[42] V. Pan and J. Reif, Efficient Parallel Solution of Linear Systems, Proceedings of the 17th Annual ACM Symposium on Theory of Computing, Providence, RI, May 1985, ACM-SIGACT, pp. 143–152.

[43] V. Pan and J. Reif, Fast and Efficient Parallel Solution of Dense Linear Systems. Computers and Mathematics with Applications, Vol. 17, no. 11, pages 1481–1491, 1989.

[44] N. Pippenger, On simultaneous resource bounds. Proc. IEEE 20th Annual Symp. Foundations of Computer Science, 1979, 307–311.

[45] F.P. Preparata and D.V. Sarwate, An improved processor bound in fast matrix inversion. Inform. Process. Letters **7** (1978), 148–150.

[46] J. Reif, Logarithmic depth circuits for algebraic functions. SIAM J. Comput. **15** (1986), 231–242.

[47] W.L. Ruzzo, On uniform circuit complexity. J. Computer System Sciences **22** (1981), 365–383.

[48] V. Strassen, Gaussian elimination is not optimal. Numer. Mathematik **13** (1969), 354–356.

[49] V. Strassen, Berechnung und Programm. I. Acta Inf. **1** (1972), 320–335.

[50] V. Strassen, Die Berechnungskomplexität von elementarsymmetrischen Funktionen und von Interpolationskoeffizienten. Numer. Math. **20** (1973), 238–251.

[51] V. Strassen, Algebraic complexity theory. In Handbook of Theoretical Computer Science, ed. by J. van Leewen, Volume A (1990), 633–672. Elsevier, Amsterdam.

[52] L. Valiant, S. Skyum, S. Berkowitz, and C. Rackoff, Fast parallel computation of polynomials using few processors. SIAM J. Comput. **12** (1983), 641–644.

Part V

Advanced Parallel Algebraic Algorithms

14

Parallel Solution of Sparse Linear and Path Systems

Victor Pan

Department of Mathematics
Lehman College
CUNY
Bronx, NY 10468
vpan@lcvax.bitnet

14.1
Introduction

A parallel algorithm has no practical value if its computational time is dominated by the time required for processor communication and synchronization. Therefore, economization of processors used in fast parallel algorithms is a highly important practical problem. Its solution is also a theoretical challenge; the solution techniques are frequently not simpler than for devising fast parallel NC algorithms, in particular, in the case of computations with matrices and polynomials (compare Bini and Pan [11]). In this chapter we devise a fast (NC) *and* processor efficient parallel algorithm for a sparse linear system of equations (Section 14.2) and extend the solution to various practically and theoretically important computations of paths in graphs (Section 14.3).

The topic of Section 14.2 may seem narrow, but in fact, solving sparse linear systems practically dominates the scientific and engineering computations.

A modified version of the algorithm of Section 14.2 has been implemented on two massively parallel SIMD computers (the Connection Machine CM-1 and MPP), as this is described in Leiserson et al. [53] and Opsahl and Reif [62].

In Section 14.2 we combine the generalized nested dissection techniques, which are fundamental techniques for computations for sparse graphs and for the solution of the associated linear systems of equations, with the techniques of recursive factorization of the coefficient matrix of a given linear system.

As a result, some material of Section 14.2 is quite advanced, particularly in Sections 14.2.7 and 14.2.8, where we use some sophisticated techniques and results of Lipton, Rose and Tarjan [54]. To avoid further complication, we restricted our study in Section 14.2 to symmetric positive definite linear systems, most important in practice, and left out the derivation of the parallel complexity estimates for matrix multiplication and inversion, which we just cite.

We hope that Section 14.3 is easier to read than Section 14.2, and it can be read independently of Section 14.2, except that in Section 14.3 we use a recursive factorization of the inverse of the input matrix as a black box subroutine, described in Section 14.2.

14.2
Sparse Linear Systems

14.2.1 Model of Computing and Some Basic Estimates

In this section, we will first specify the model of our computations, and then define the parallel complexity of matrix multiplication and inversion, which are the basic blocks of the main algorithm.

In Section 14.2, we assume the customary EREW PRAM algebraic model of computing (see Eppstein and Galil [21], Karp and Ramachandran [48], Kaltofen [42]), where each processor in each step performs at most one arithmetic operation. We write $O_A(t,p)$ to show that the algorithm uses $O(t)$ arithmetic steps and $O(p)$ processors. We adopt a variant of *Brent's scheduling principle* (Karp and Ramachandran [48] and Pan and Preparata [69]), according to which we may slow down such an algorithm and use O(st) steps and at most p/s processors for any $s \leq p$, that is, $O_A(t,p)$ implies $O_A(tp,1)$ (but not vice versa in general).

Our algorithm is reduced to performing matrix multiplications and inversions, for which the known parallel algorithms use $O(\log n)$ and $O(\log^2 n)$ parallel arithmetic steps for $n \times n$ input matrices, respectively. Let $M(n)$ and Inv(n) denote the associated number of processors and let us abbreviate by writing MULT and INVERT for the problems of $n \times n$ matrix multiplication and inversion.

Allowing computations with real or rational constants (and actually even under milder assumptions) and randomization, we obtain from Pan [68], Bini and Pan [11], and Kaltofen and Pan [45, 46] that $\text{Inv}(n) = O(M(n))$.

Similarly, $\text{Inv}(n) = O(M(n))$ if we adjust our model for numerical computations with finite precision and restrict the input class to the practically important well-conditioned matrices (Pan and Reif [70, 76], Pan and Schreiber [79]).

We refer the reader to the cited papers or to Bini and Pan's book [11] for details and for the definition of well-conditioned matrices, and in this chapter will just apply the parallel complexity estimates,

$$(t_M, p_M) = O_A(\log n, M(n)) \tag{14.1}$$

for MULT, and

$$(t_{\text{inv}}, p_{\text{inv}}) = O_A(\log^2 n, M(n)) \tag{14.2}$$

for INVERT.

For theoretical purposes we may assume that

$$M(n) \leq n^{2.376}. \tag{14.3}$$

This bound follows since

$$M(n) \leq n^\omega,$$

as long as $\omega > \beta$ and $O(n^\beta)$ arithmetic operations suffice for MULT (Pan and Reif [70, 76], Pan [66]) and since we may set $\beta < 2.376$ (Coppersmith and Winograd [18]). The straightforward matrix multiplication algorithm only supports the bound

$$M(n) \leq n^3/\log n, \qquad (14.4)$$

but outperforms the algorithms of Coppersmith and Winograd [18] unless n is immense (much larger than, say, 10^{100}).

14.2.2 Preliminaries for Solving Sparse Linear Systems; Reduction to Matrix Factorization; The Associated Graphs, Fill-in and Elimination Ordering

To solve a nonsingular linear system

$$A\mathbf{x} = \mathbf{b}, \qquad (14.5)$$

we may first compute A^{-1} and then $\mathbf{x} = A^{-1}\mathbf{b}$, thus arriving at the complexity bound (14.2). In practice, large scale linear systems are usually sparse and structured, and for such systems we will present a much better algorithm. This is the subject of Section 14.2.

In this section we will start with restricting the input class, then reduce our problem to factorization of the input matrix A and then define a graph associated to A and the concepts of fill-in and elimination ordering.

Specifically, we restrict the class of linear systems (14.5) to the case where the real input matrix $A = [a_{ik}]$ is s.p.d., that is, symmetric ($a_{ik} = a_{ki}$ for all i and k) and positive definite ($\mathbf{v}^T A \mathbf{v} > 0$ for any vector $\mathbf{v} \neq \mathbf{0}$; here and hereafter W^T denotes the transpose of a vector or matrix W). In principle, we may easily relax this restriction on the system (14.5) and extend our algorithm respectively (see Remark 14.6), except that we would then lose its practically important property of numerical stability. Namely, for s.p.d. matrices A, the main (factorization) stage of our algorithm only involves the values bounded by $a = \max\{\|A\|_2, \|A^{-1}\|_2\}$ (see Lemma 14.3) and can be performed numerically with a bounded precision (of $\log_2(a/\epsilon) + O(\log n)$ binary digits, where ϵ defines the tolerance to the output errors). Here $\|W\|_2$ is the 2-norm of a matrix W (see Golub and van Loan [34]) and

$$w \leq \|W\|_2 \leq nw, w = max_{i,k}|w_{ik}|$$

for any $n \times n$ matrix $W = [w_{ik}]$. If the input matrix A is not s.p.d., the algorithm may involve arbitrarily large values, and thus may require an arbitrarily high precision of computations. We refer the reader to Golub and van Loan [34] and Bini and Pan [11] on this issue and to Young [97] on the importance of the class of s.p.d. linear systems.

Next, we will make our first auxiliary step by reducing the solution of an s.p.d. linear system (14.5) to computing the following *Cholesky factorization:*

$$PAP^T = LL^T, \tag{14.6}$$

such that $A = P^{-1}LL^T(P^{-1})^T$; L is a nonsingular lower triangular matrix with positive diagonal entries; P is a permutation matrix, such that premultiplication by P amounts to a certain row interchange. (In fact, $W = LL^T$ for some nonsingular triangular matrix L if and only if W is an s.p.d. matrix.)

Actually, our parallel solution of the system (14.5) will be presented in Sections 14.2.4–14.2.8, where we will rely on a recursive version of (14.6). In this section, for demonstration, we will only consider the evaluation of (14.6) and the sequential complexity of reduction of (14.5) to (14.6).

Computing the factorization (14.6) amounts to the elimination stage of *Gaussian elimination* (in the s.p.d. case) *with ordering* of the elimination of the variables defined by the matrix P. Assuming for simplicity that $P = I$ is the identity matrix or replacing PAP^T by A, we may write this stage as follows:

ALGORITHM 14.1
Cholesky factorization
Input. $n \times n$ s.p.d. matrix $A = [a_{ik}]$, $a_{ik} = a_{ki}$.
Output: the $n \times n$ matrix $L = [l_{ik}]$ of (14.6) where $P = I, l_{ik} = 0$ for $i < k$.
For $k = 1, 2, \ldots, n$

$$l_{kk} := (a_{kk} - \sum_{j=1}^{k-1} a_{kj}^2)^{1/2}$$

For $i = k+1, \ldots, n$

$$l_{ik} := (a_{ik} - \sum_{j=1}^{k-1} a_{ij}a_{kj})/l_{kk}$$

The permutation matrix P and Algorithm 14.1 completely define the elimination stage. The subsequent solution of the linear system (14.5) is reduced to solving two triangular linear systems

$$L\mathbf{y} = P\mathbf{b},$$

$$L^T\mathbf{z} = \mathbf{y},$$

which immediately gives $\mathbf{x} = P^T\mathbf{z} = A^{-1}\mathbf{b}$.

To estimate the complexity of the latter reduction, we will associate a graph to the matrix A.

DEFINITION 14.1
Let $G(W) = (V, E), V = V(W), E = E(W)$, denote the undirected graph associated with a symmetric matrix $W = [w_{ij}]$ such that the vertex set V and the edge set E are in the one-to-one correspondence with the row set of W and with the set of nonzero subdiagonal entries of W, respectively.

The numbers of nonzero subdiagonal entries of A and L are given by $|E(A)|$ and $|E(L)|$, respectively, where $|S|$ denotes the cardinality of a set S.

DEFINITION 14.2
The set $F = E(L) - E(A)$ of the edges of $E(L)$ not lying in $E(A)$ is called the fill-in of the factorization (14.6) of A.

REMARK 14.1
Hereafter, we will assume that F and $E(L)$ only depend on $G(A)$ and P; the entries of A, associated with $E(A)$, are indeterminates, and no entries of L vanish unless this takes place for all the instances of A.

FACT 14.1
(See Exercise 14.2.) $|E| = O(n^{1/2})$ in the applications where:

a) $G(A)$ is a d-dimensional grid for $d \geq 2$,
b) $G(A)$ is a planar graph,
c) $G(A)$ is a finite element graph, that is, the graph associated with a linear system that arises in solving partial differential equations by the finite element method (George and Liu [31], Rice [86]).

The reduction of the system (14.5) to two triangular linear systems immediately implies the following estimate:

FACT 14.2
Given the factorization (14.6), $O(n+|F|+|E(A)|)$ arithmetic operations suffice for the solution of the linear system (14.5).

REMARK 14.2
Algorithm 14.1 involves n evaluations of square roots of positive numbers. As an exercise, the reader may modify this algorithm to avoid such operations but still to compute a modified factorization,

$$PAP^T = \tilde{L}D\tilde{L}^T, \tag{14.7}$$

where $\tilde{L}$ is a unit triangular matrix (whose diagonal is filled with ones) and D is the diagonal matrix, $D = diag(l_{11}^2, l_{22}^2, \ldots, l_{nn}^2)$.

REMARK 14.3
A^{-1} is frequently dense even where the matrices L and A are sparse (see Example 14.1 below), so that using the factorization (14.6) is by far more effective than the reduction of solving a sparse linear system (14.5) to matrix inversion.

14.2.3 Elimination Graphs and the Minimum Degree Ordering; Separatable Associated Graphs and a Sequential Nested Dissection Algorithm

In this section, we recall and analyze the estimates of Lipton, Rose and Tarjan [54] for the sequential complexity of computing the factorization (14.6). We also recall some definitions and techniques involved in such computation. We will use them (some with modifications) in Sections 14.2.4 to 14.2.8.

Our first problem, suggested by Fact 14.2, is to choose the elimination ordering given by the matrix P to decrease the upper bounds on the fill-in and on the complexity of computing the factorization (14.6). The next example shows a dramatic impact of the choice of P on these bounds:

EXAMPLE 14.1
Let all nonzero entries of the matrix A lie on its diagonal and in its first column and row. Then $|F| = n^2 - 3n + 2$ if $P = I$ whereas $|F| = 0$ if P just interchanges the first and last rows. Similarly, the storage space and the number of arithmetic operations for computing (14.6) change from at least n^2 for $P = I$ to O(n). (Also note that $|E(A^{-1})| = n^2$, so that A^{-1} is dense.)

The elimination process of Algorithm 14.1 generates a sequence of smaller auxiliary matrices, whose associated graphs are called *elimination graphs*. Every elimination step updates the elimination graph according to the following rules (compare Rose [88] and George and Liu [31]).

Rules for Updating Elimination Graphs

Compare Figures 14.1 through 14.5:
The rules for updating elimination graphs are:
a) the eliminated vertex (variable) v is deleted from the vertex set;
b) for every pair of edges $\{u_1, v\}$ and $\{u_2, v\}$, the fill-in edge $\{u_1, u_2\}$ is included into the edge set unless it is already there;

628   Chapter 14. Parallel Solution of Sparse Linear and Path Systems

c) when the stage b) has been completed, all the edges adjacent to v are deleted from the edge set.

The latter observations lead us to the following result, which will be refined in Lemma 14.4:

LEMMA 14.1
The edge $\{u, w\} \in F \cup E(PAP^T)$ if and only if there is a path $u = v_0, v_1, \ldots, v_k = w$ in the graph $G(PAP^T)$ with $v_i < \min\{u, w\}$ for $0 < i < k$.

The *minimum degree ordering*, which is the most effective rule for the elimination ordering for general sparse symmetric matrices A (compare Exercise 14.4), requires that each elimination step eliminate the variable corresponding to the vertex of the associated graph that has the minimum degree, that is, to the vertex incident to the minimum number of edges (ties can be broken arbitrarily).

We will not use the minimum degree elimination ordering because we will deal with an important special class of linear systems, for which there exists an even more effective policy, namely, with the class of linear systems associated with graphs $G = (V, E)$ that satisfy the following properties (Schreiber [89], Liu [58], Gilbert and Tarjan [33]).

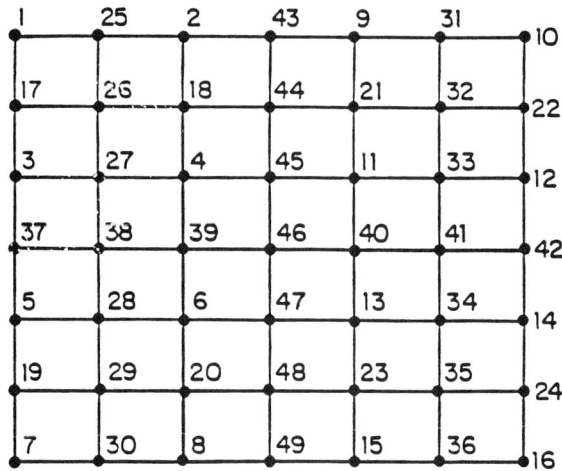

FIGURE 14.1
The 7×7 grid graph G_0, with elimination numbering.

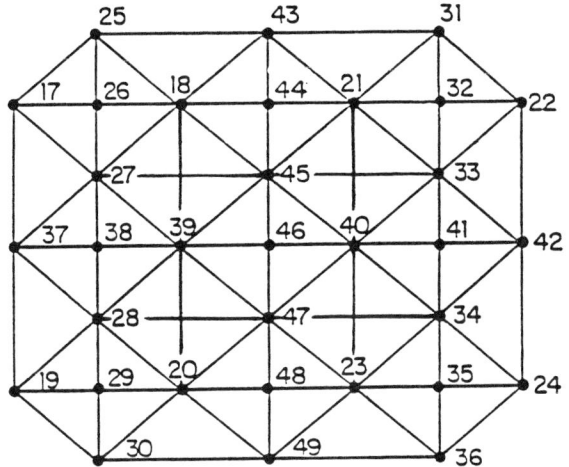

FIGURE 14.2
The graph G_1, derived from G_0 by simultaneous elimination of $R_0 = \{1, 2, \ldots, 16\}$.

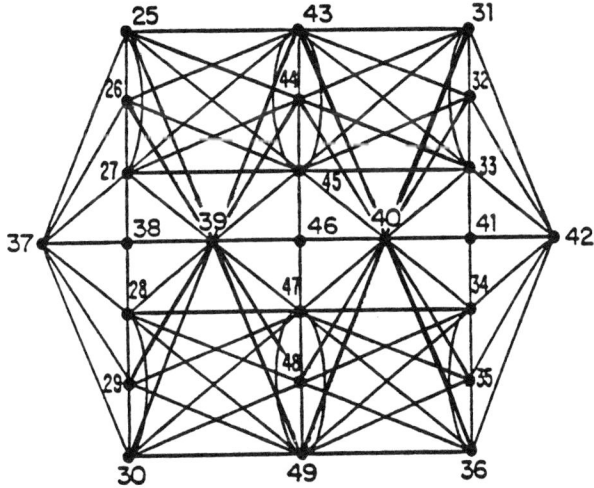

FIGURE 14.3
The graph G_2 derived from G_1 by simultaneous elimination of $R_1 = \{17, 18, \ldots, 24\}$.

630 Chapter 14. Parallel Solution of Sparse Linear and Path Systems

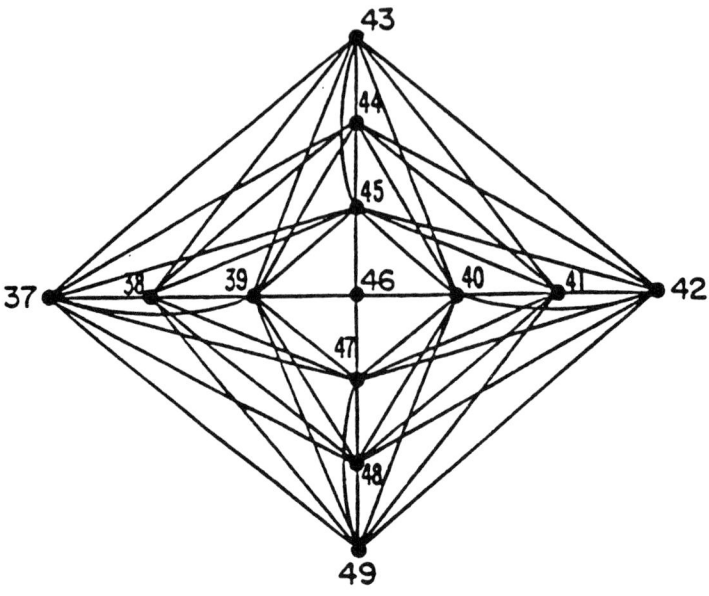

FIGURE 14.4
The graph G_3 derived from G_2 by simultaneous elimination of $R_2 = \{25, 26, \ldots, 36\}$.

DEFINITION 14.3
A graph $G = (V, E)$ is said to have an $s(n)$-separator family (with respect to two constants, α and n_0) if either $|V| \leq n_0$ or, deleting some separator set S of vertices, such that $|S| \leq s(|V|)$, we may partition G into two disconnected subgraphs with the vertex sets V_1 and V_2, such that $|V_i| \leq \alpha |V|$, $i = 1, 2$, and if, furthermore, each of the two subgraphs of G defined by the vertex sets V_i, $i = 1, 2$, also has an $s(n)$-separator family (with respect to the same constants α and n_0). The resulting recursive decomposition of G is known as the $s(n)$-separator tree, so that each partition of the subgraph of G in the decomposition defines its children in the tree. The vertices of the tree can thus be interpreted as subgraphs of G or as their vertex sets (we will assume the latter interpretation), and the edges of the tree can be interpreted as the separator sets. Then the vertex set V equals the union of all the vertex subsets in V associated with the edges of the $s(n)$-separator tree and with its leaves. We call a graph $s(n)$-separatable if it has an $s(n)$-separator family and its $s(n)$-separator tree is available.

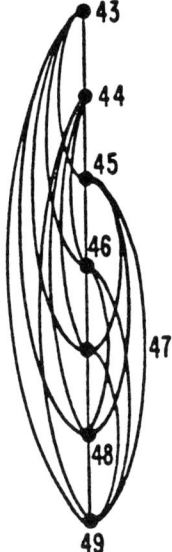

FIGURE 14.5
The graph G_4 derived from G_3 by simultaneous elimination of $R_3 = \{37, 38, \ldots, 42\}$.

In particular, a d-dimensional grid has an $s(n)$-separator family (for $s(n) = n^{1-(1/d)}$ and with respect to the constants $\alpha = 1/2$ and $n_0 = 0$). The separators are defined by the secant hyperplanes passing through the center of the vertex sets of the graph G or its subgraphs and parallel to the coordinate hyperplanes. Then each vertex of the original grid $G = (V, E)$ lies in a certain separator, so that we may recursively define the *tree of the separators*, such that the root S is given by the separator of G, its two children S_1 and S_2 are the separators of the two graphs induced by V_1 and V_2, and so on. The leaves of the tree are the singletons. (We will slightly refine this definition in Section 14.2.7.)

The elimination order is converse, starting with the elimination of the variables associated with such singletons and continuing toward the vertices of the preceding level of the tree of separators.

This process is called *nested dissection* in the case of grid graphs, *generalized nested dissection* for $s(n)$-separatable graphs and is illustrated in Figures 14.1–14.5 in the case of the 7 × 7 grid.

In Gilbert and Tarjan [33] it is proven that the overall number of arithmetic operations required to compute the factorization (14.6) in this way is $O(n^{3/2})$, and the fill-in has cardinality $O(n \log n)$ in the cases of grid graphs, planar graphs and two-dimensional finite element graphs.

Actually, instead of graphs satisfying Definition 14.3, we will study the more general class of graphs that have $s(n)$-separator families according to a modification of Definition 14.3 where the $s(n)$-separator sets are recursively defined, not for the vertex sets V_1 and V_2, but for $V_1 \cup S$ and $V_2 \cup S$, that is, the set S is included in each induced subgraph (see Remark 14.4 and Definition 14.6). This modified version will be called **Definition 14.4** and will be hereafter assumed whenever we cite $s(n)$-separator families of trees or $s(n)$-separatable graphs. We will also assume that

$$s(n) = cn^\sigma, 1/2 \leq \sigma < 1, c \text{ is a constant.} \qquad (14.8)$$

THEOREM 14.1 *(Lipton, Rose and Tarjan [54])*
Let $G(A)$ be an $s(n)$−separatable graph where $s(n)$ satisfies (14.8). Then $M^*(s(n))$ arithmetic operations suffice to compute the factorization (14.6), and $|F| = O(s^2(n) \log n)$.

Clearly, this result exploits the property of $s(n)$-separatable graphs $G(A)$ in order to greatly decrease the upper bound $M^*(n)$ on the sequential complexity of solving a linear system (14.5) with a general input matrix A.

The chief assumption that we need in order to arrive at these improved estimates, as well as at the parallel complexity estimates of Theorem 14.2 in Section 14.2.5, and of "Computational Cost Estimates" in Section 14.3.2, is that the $s(n)$-separator tree is given or readily computable. *For a planar graph, its $O(\sqrt{n})$-separator family can be computed in $O(n)$ sequential time* (Lipton and Tarjan [55] and Djidjev [20]) or, on a PRAM, in $O((\log n)\sqrt{n})$ parallel time using $(\sqrt{n})/\log n$ processors (Gazit and Miller [28]) with small overhead constants, in both algorithms of Lipton and Tarjan [55] and Gazit and Miller [28]; *or in $O(\log^2 n)$ time with $n + f^{1+\epsilon}$ processors for any positive ϵ* where f is the number of faces of the graph (Gazit and Miller [29]); the latter algorithm requires randomization at its auxiliary stage of computing a maximal independent set; for a d-dimensional grid graph, a desired separator tree (with $s(n) = n^{1-(1/d)}$) is immediately available, as well as a $(4\lfloor k/2 \rfloor \sqrt{n})$-separator tree for an n-vertex finite element graph with at most k boundary vertices per element. (See Miller, Teng, and Vavasis [58] on further substantial progress.)

Furthermore, we may assume the separator tree preprocessed, particularly, in the extensions to many graph computations, say, for several computations for a graph with fixed sets of edges of variable lengths. The latter assumption is very reasonable in the case of many large scale operations research problems, where the underlying graph G is fixed, but the costs associated with its edges (that is, the entries of the matrix A) may vary dynamically with the input. In particular, such a dynamic situation arises in an important case of a large computer network where the costs of the links may vary in time and moreover may grow to the infinity, but where no new links are dynamically created. Another example is the commodity transportation problems where the transportation links are known and fixed, but may have dynamically changing costs, and, furthermore, these costs may grow so exorbitant that, for some of these links, they may be essentially infinite. In both of these examples, it is certainly useful to preprocess the underlying structure of the network in order to increase the efficiency of the resulting dynamic algorithm. In such cases we shall presume a precomputation stage where we shall find separators for the underlying graph and for a certain family of its subgraphs. This will lead us to a significant improvement of the dynamic path algebra computations reduced to sparse linear systems (see Section 14.3).

REMARK 14.4
In Gilbert and Tarjan [33], the asymptotic estimates of Theorem 14.1 with smaller overhead constants can be proven (based on a modified nested dissection algorithm) for sparse linear systems (14.5) where $G(A)$ is an $s(n)$-separatable graph under (14.3) and Definition 14.3; this result holds for a class of graphs $G(A)$ which is broad, although not as general as the class referred to in Theorem 14.1.

Next, we will study the parallel cost of a modification of the nested dissection algorithm and will compare it with the sequential algorithm of Lipton, Rose, Tarjan [54].

14.2.4 Recursive Factorization of Matrices and an Outline of the Parallel Nested Dissection Algorithm

Our parallel generalized nested dissection algorithm follows the pattern of Lipton, Rose, and Tarjan [54] but relies on a distinct recursive factorization of the matrix $A_0 = PAP^T$, which plays the same role as (14.6), that is, having such a factorization simplifies the solution of the linear system $A\mathbf{x} = \mathbf{b}$. (The shift from (14.6) will slightly change the meaning of the fill-in set F and some other concepts.)

DEFINITION 14.5
A recursive s(n)-factorization of a symmetric matrix A (associated with a graph $G = G(A)$ that has an $s(n)$-separator family with respect to two

constants $\alpha, \alpha < 1$, and n_0) is a sequence of matrices $A_0, A_1, \ldots, A_d$ such that $A_0 = PAP^T$, P is an $n \times n$ permutation matrix,

$$A_h = \begin{bmatrix} X_h Y_h^T \\ Y_h Z_h \end{bmatrix}, Z_h = A_{h+1} + Y_h X_h^{-1} Y_h^T \text{ for } h = 0, 1, \ldots, d-1, \quad (14.9)$$

and X_h is a symmetric block diagonal matrix consisting of square blocks of sizes at most $s(n_h) \times s(n_h)$, associated with the separators of the graph G or, for $h = d$, with the leaves of the $s(n)$-separator tree. Here,

$$n_d = n, n_{h-1} \le \alpha n_h + s(n_h), h = 1, \ldots, d. \quad (14.10)$$

(The constant n_0, associated with the separator family, is also the order of the diagonal blocks of the matrix X_d.)

Our definition of a recursive $s(n)$-factorization relies on the matrix identities

$$A_h = \begin{bmatrix} I & O \\ Y_h X_h^{-1} & I \end{bmatrix} \begin{bmatrix} X_h & O \\ O & A_{h+1} \end{bmatrix} \begin{bmatrix} I & X_h^{-1} Y_h^T \\ O & I \end{bmatrix} \quad (14.11)$$

and

$$A_h^{-1} = \begin{bmatrix} I & -X_h^{-1} Y_h^T \\ O & I \end{bmatrix} \begin{bmatrix} X_h^{-1} & O \\ O & A_{h+1}^{-1} \end{bmatrix} \begin{bmatrix} I & O \\ -Y_h X_h^{-1} & I \end{bmatrix}. \quad (14.12)$$

Here, the matrix A_{h+1} defined by (14.9) is known in linear algebra as Schur's complement of X_h, h ranges from 0 to $d-1$, I denotes the identity matrices, and O denotes the null matrices of appropriate sizes.

The recursive decomposition (14.9)–(14.12) is intimately related to the recursive decomposition of the associated graph $G = G(A)$ defined by (and defining) the s(n)-separator tree. Specifically, we will see that the matrices X_h are block diagonal matrices whose blocks for $h = 0, 1, \ldots, d-1$ are associated with the separator sets of level h from the root of the tree, whereas the blocks of X_d are associated with the leaves of the tree.

Given a symmetric $n \times n$ matrix A associated with an $s(n)$-separatable graph $G(A)$, we may compute the recursive $s(n)$-factorization (14.9)–(14.12) by performing the following stages:

Stage 0: Compute an appropriate permutation matrix P, matrix $A_0 = PAP^T$ and the decreasing sequence of positive integers $n = n_d, n_{d-1}, \ldots, n_0$ satisfying (14.10) and defined by the sizes of the separators in the

$s(n)$-separator family of $G = G(A)$ (as specified in Section 14.2.7). The permutation matrix P and the integers $n_d, n_{d-1}, \ldots, n_0$ completely define the order of the elimination of the variables (vertices), so that we first eliminate the vertices of G corresponding to the leaves of the $s(n)$-separator tree, then we eliminate the vertices of G corresponding to the edges adjacent to the leaves of the tree (that is, the vertices of the separators used at the final partition step), then the vertices of G corresponding to the next edge level of the tree (separators of the previous partition step) and so on, as we formally analyze this in Sections 14.2.7 and 14.2.8.

Stage $h+1$ ($h = 0, \ldots, d-1$): Compute the matrices $X_h^{-1}, -Y_h X_h^{-1}$ (which also gives us the matrix $-X_h^{-1} Y_h^T = (-Y_h X_h^{-1})^T,$) and $A_{h+1} = Z_h - Y_h X_h^{-1} Y_h^T$ satisfying (14.9), (14.11), (14.12) and such that A_{h+1} has size $n_{d-h-1} \times n_{d-h-1}$. (Each of these stages amounts to inversion, two multiplications, and subtraction of some matrices.)

When the recursive factorization (14.12) has been computed, it will remain to compute the vector $\mathbf{x} = A^{-1}\mathbf{b} = P^T A_0^{-1}(P\mathbf{b})$ for a column vector $\mathbf{b}$. This can be done by means of recursive premultiplications of some subvectors of the vector $P\mathbf{b}$ by the matrices

$$\begin{bmatrix} I & O \\ -Y_h X_h^{-1} & I \end{bmatrix}, \begin{bmatrix} X_h^{-1} & O \\ O & A_{h+1}^{-1} \end{bmatrix} \text{ and } \begin{bmatrix} I & -X_h^{-1} Y_h^T \\ O & I \end{bmatrix}$$

for $h = 0, 1, \ldots, d$. At the stage of the premultiplication by

$$\begin{bmatrix} X_h^{-1} & O \\ O & A_{h+1}^{-1} \end{bmatrix},$$

the premultiplication by X_h^{-1} is done explicitly, and the premultiplication by A_{h+1}^{-1} is performed by means of recursive application of (14.12). Thus, (14.12) defines a simple recursive algorithm for computing $A^{-1}\mathbf{b}$ for any column vector $\mathbf{b}$ of length n, provided that a recursive $s(n)$-factorization (14.9) is given.

It is instructive to compare the recursive $s(n)$-factorization (14.9)–(14.12) with the factorization of A_h of the form (14.7), used in Lipton, Rose, and Tarjan [54]. Both factorizations rely on the matrix identities (14.11), (14.12), which in fact represent the block Jordan elimination algorithm for a 2×2 block matrix A_h of (14.9). The factorization $PAP^T = \tilde{L}D\tilde{L}^T$ is obtained in Lipton, Rose, and Tarjan [54] by the application of the Jordan elimination to the matrix PAP^T, which is equivalent to the recursive application of (14.11) to *both* submatrices X_h and A_{h+1}. (This defines $\tilde{L}$ and $\tilde{L}^T$ in factorized form, but the entries of the factors do not interfere with each other, so that all the entries of $Y_h X_h^{-1}$ coincide with the respective entries of $\tilde{L}$.) Fast and processor efficient parallelization of this algorithm (yielding $O(\log^3 n)$ parallel time) is straightforward, except for the stage of the factorization of the matrices X_h, which, by Lemma 14.5 of Section 14.2.7, are block diagonal with dense diagonal blocks of sizes of an order of $s(n_h) \times s(n_h)$. Actually, even for this stage a good parallel algorithm has been finally obtained (Hafsteinson [36]), but at that time it has already been observed that for the purpose of solving the systems $A\mathbf{x} = \mathbf{b}$, we do not have to factorize the matrices X_h. It suffices to invert them, and this can be efficiently done according to the estimate (14.2). Thus we arrive at the recursive s(n)-factorization (14.9)–(14.12), where we recursively factorize only the matrices A_{h+1} in (14.11) and A_{h+1}^{-1} in (14.12), but not the matrices X_h and X_h^{-1}. This modification of the factorization scheme turned out to be useful in some important combinatorial computations (see Section 14.3 of this chapter), to which the cited alternate approach of Hafsteinson [36] does not apply.

14.2.5 Parallel Generalized Nested Dissection: The Main Theorem

For simplicity, we will assume hereafter that

$$M(n) = n^{\bar{\omega}} \text{ for some constant } \bar{\omega} > 2 \geq 1/\sigma, \qquad (14.13)$$

where σ is defined by (14.8) and consequently that

$$M(ab) \leq M(a)M(b). \qquad (14.14)$$

Next, we will state the Main Theorem of Section 14.2 (compare Remark 14.4).

THEOREM 14.2
Let $G = (V, E)$ be an $s(n)$-separator graph for $s(n)$ satisfying (14.8). Then, given an $n \times n$ symmetric positive definite matrix A such that $G = G(A)$, we may compute a recursive $s(n)$-factorization of A in time $O(\log^3 n)$ using at most $M(s(n))/\log n$ processors (provided that with $M(s(n))$ processors we may multiply a pair of $s(n) \times s(n)$ matrices in time $O(\log n)$ and that $M(n)$ satisfies (14.13)). Whenever such a recursive $s(n)$-factorization of A is available, $O(\log^2 n)$ time and at most

$(|E|/\log n) + s(n)^2$ processors suffice in order to solve a system of linear equations $A\mathbf{x} = \mathbf{b}$ for any given column vector $\mathbf{b}$ of dimension n.

REMARK 14.5
It is possible to extend Theorem 14.2 to the case where (14.13) does not hold, by using Theorems 7-9 of Lipton, Rose, and Tarjan [54].

REMARK 14.6
The total work, that is, the product of the parallel time and processor bounds is the same, $TP = (PARALLEL\ TIME) * PROCESSOR = O(M(s(n))\log^2 n)$, both for computing the whole recursive factorization (14.11), (14.12) and its proper stage of inverting X_{d-1}. The total work can be decreased by the factor of $\log n$ (at the cost of a minor slow-down of the computation) using the techniques of Pan and Preparata [69].

Let us demonstrate some consequences of the complexity bounds of our algorithm. We will first assume the weak bound $M(n) = n^3/\log n$ for matrix multiplication. It is significant that already under this assumption our parallel nested dissection algorithm, for polylog time bounds, has processor bounds that substantially improve the previously known bounds. Let G be a fixed planar graph with n vertices given with its $O(\sqrt{n})$-separator tree. (For example, G might be a $\sqrt{n} \times \sqrt{n}$ grid graph.) Then, for any $n \times n$ matrix A such that $G = G(A)$, our parallel nested dissection algorithm takes $O(\log^3 n)$ time and $n^{1.5}/\log^2 n$ processors to compute the special recursive factorization of A and then $O(\log^2 n)$ time and n processors to solve any linear system $Ax = b$ with A fixed. We have the time bounds $O(\log^3(kn))$ and $O(\log^2(kn))$ and the processor bounds $k^3 n^{1.5}/\log^2(kn)$ and kn^2, respectively, if $G(A)$ is an n-vertex finite element graph with at most k vertices on the boundary of each face. In yet another example, $G(A)$ is a 3-dimensional grid, so it has an $n^{2/3}$-separator family. In this case, we have the same asymptotic time bounds as for planar graphs, and our processor bounds are $n^2/\log^2 n$ and $n^{1.33}$, respectively. Furthermore, if we use theoretical bounds for matrix multiplication, say, $M(n) = n^{2.4}$, then our processor bound, for computing the special recursive factorization are further decreased to $n^{1.2}$ in the planar case, to $k^{2.4}n^{1.2}$ in the case of n-vertex finite element graphs with at most k vertices per face and to $n^{1.6}$ for the 3-dimensional grid.

14.2.6 Proof of the Main Theorem: Initial Comments and Partition into Two Main Stages

We will start with some observations that we need for the proof of the Main Theorem. We will first show that the parallel algorithm uses only

$d = O(\log n)$ stages. Let $\delta = \delta(n)$ denote $cn^{\sigma-1} = s(n)/n$, and let n_0 be large enough, so that

$$\alpha k + s(k) = (\alpha + \delta)k = \beta k; \quad \beta = \alpha + \delta < 1 \text{ if } k > n_0. \tag{14.15}$$

(14.8) and (14.10) together imply that

$$n_h \leq (\alpha + \delta)^{d-h} n = \beta^{d-h} n, \quad h = 0, 1, \ldots, d. \tag{14.16}$$

LEMMA 14.2
$d = O(\log n)$ for fixed n_0 and $\alpha < 1$.

PROOF
(14.16) for h=0 implies that $d \leq \log(n/n_0)/\log(1/(\alpha + \delta)) = O(\log n)$.
∎

Next, we will point out that all the arithmetic operations required in our parallel nested dissection algorithm (except for those needed in order to invert X_h for all h) are also involved in the sequential algorithm of Lipton, Rose, and Tarjan [54]. (As in the latter paper, we ignore the arithmetic operations where at least one operand is zero; we assume that no random cancellations of nonzero entries takes place, for if there are such cancellations, we would only arrive at more optimistic bounds; and we note that both X_h and X_h^{-1} are block diagonal matrices having nonzero blocks in the same places.)

Next, we will indicate the two main stages of the proof of Theorem 14.2. For each h, we group all the arithmetic operations involved, to reduce them to a pair of matrix multiplications, $U_h = Y_h X_h^{-1}$ (which also gives $X_h^{-1} Y_h^T = U_h^T$) and $W_h = U_h Y_h^T$, and to a (low cost) matrix subtraction, $A_{h+1} = Z_h - W_h$ (assuming here that the matrix X_h^{-1} has been precomputed). Theorem 14.3 of the next section will provide a bound on the complexity of computing the inverse of the auxiliary matrices, $X_0, \ldots, X_d$, and Theorem 14.4 of Section 14.2.8 will provide a bound on the cost of parallel multiplication of the auxiliary matrices and will imply the time bound of $O(\log^2 n)$ for the entire computation, excluding the stage of the inversion of the matrices X_h. The product of the number of processors used times log n is bounded from above by the number of arithmetic operations used in the algorithm of Lipton, Rose, and Tarjan [54], that is, by $O(s(n)^3)$ or in fact by $O(M(s(n))\log n)$, (compare (14.3), (14.4)). (See Theorem 3 there and Remark 14.6 above.)

The estimates of Theorem 14.2 for the cost of computing the recursive factorization (14.9)–(14.12) immediately follow from Theorems 14.3 and 14.4.

Finally, we will discuss the parallel complexity of solving the linear system $A\mathbf{x} = \mathbf{b}$ given the recursive factorization. As we have already pointed out, when the recursive factorization (14.9)–(14.12) has been computed, we evaluate $\mathbf{x} = A^{-1}\mathbf{b}$ by means of successive premultiplications of some subvectors of $\mathbf{b}$ by the matrices $Y_h X_h^{-1}$, X_h^{-1} and $X_h^{-1} Y_h^T$ for $h = d, d-1, \ldots, 0$. The parallel time bounds are $O(\log n)$ for each h and $O(\log^2 n)$ for all h. The obvious processor bound is $(|E|+|F|)/\log n$ where $|E|+|F|$ denotes the number of the entries of an $n \times n$ array that are filled with nonzeros for at least one of the submatrices $Y_h X_h^{-1}$, A_h for $h = 0, 1, \ldots, d-1$ (for each h, A_h occupies the lower right corner of the array, and $Y_h X_h^{-1}$ occupies the lower left corner of the array). The nonzeros of A_0 form the set E of the edges of $G(A)$; other nonzeros, introduced in the process of computing the $s(n)$-factorization (14.9)–(14.12), form a set F called *fill-in*. (Here we extend Definition 14.2.)

It remains to estimate $|F|$. By Theorem 2 of Lipton, Rose, and Tarjan [54], $|F| = O(n + s(n)^2 \log n)$, and this bound can be applied to our algorithm as well. The proofs in Lipton, Rose, and Tarjan [54] are under the assumption that $s(n) = O(\sqrt{n})$, but the extension to any s(n) satisfying (14.8) is immediate. Likewise, Lipton, Rose and Tarjan only estimated the number of multiplications involved, but including the additions and subtractions would increase the upper estimates yielded in their and our algorithms only by a constant factor. We will recall this proof in Appendix 14.A.

REMARK 14.7
The algorithms and the complexity estimates of this chapter can be immediately extended to the case of nonsymmetric linear systems associated with undirected graphs, if, for all h, we replace the matrices Y_h^T by matrices W_h (which are not generally transposes of Y_h) and remove the assumption that the matrices X_h are symmetric. Then all the results and proofs can be extended. The algorithm remains powerful for the computations over semirings (dioids) with interesting combinatorial applications (see Section 14.3 of this chapter). On the other hand, if A is an s.p.d. matrix, the algorithm can be successfully performed with a finite precision. The next lemma bounds the magnitude of the values involved in the computation of (14.11), (14.12) and therefore, bounds the precision of this computation. This is crucially important for practical application of the algorithm, although in our proof of the Main Theorem, we only use the first part of this lemma. The lemma follows from the interlacing property of the eigenvalues of a symmetric matrix (see Golub and van Loan [34] or Parlett [80]).

LEMMA 14.3

The matrix A_{h+1} is symmetric positive definite if the matrix A_h is, and furthermore, $\max\{\|X_h\|_2, \|A_h\|_2\} \leq \|A\|_2$, $\max\{\|X_h^{-1}\|_2, \|A_h^{-1}\|_2\} \leq \|A^{-1}\|_2$ *for all h.*

14.2.7 First Stage of Proof: the Cost of Parallel Inversion of the Auxiliary Matrices X_h

In this section, we will specify the elimination order of our nested dissection algorithm and then will prove the following result:

THEOREM 14.3

Let A be an $n \times n$ symmetric positive definite matrix, having a recursive $s(n)$-factorization. Then $O(\log^3 n)$ parallel time and $M(s(n))/\log n$ processors suffice in order to invert the auxiliary matrices $X_0, \ldots, X_d$ that appear in the recursive $s(n)$-factorization $A_0, A_1, \ldots, A_d$ of A.

We will first reexamine the rule b) for updating the elimination graphs when the variable (vertex) v is eliminated (see Section 14.2.3) and will extend it to operations with matrix blocks. We first recall that adding a fill-in edge $\{u_1, u_2\}$ to the edge set corresponds to four arithmetic operations of the form $z - y_1 x^{-1} y_2$ where x, y_1, y_2, z represent the edges $\{v, v\}, \{u_1, v\}, \{v, u_2\}, \{u_1, u_2\}$, respectively (compare Figures 14.1–14.5). Now, if a block of variables is eliminated, then a set S, representing this block, should replace a vertex in the above description. Specifically, for every pair of edges $\{u_1, s_1\}, \{u_2, s_2\}$ with the endpoints s_1 and s_2 in S, we add the edge $\{u_1, u_2\}$ to the set of edges, and then, when all such pairs of edges have been scanned, we delete all the edges with one or two endpoints in S. This corresponds to the matrix operations of the form $Z - Y_1 X^{-1} Y_2^T$, where X, Y_1, Y_2^T, Z represent the blocks of edges of the form $\{s_1, s_2\}, \{u_1, s_1\}, \{s_2, u_2\}, \{u_1, u_2\}$, respectively, where $s_1, s_2 \in S$ and where u_1, u_2 denote two vertices connected by edges with S. For symmetric matrices, we may assume that $Y_1 = Y_2 = Y$.

In the nested dissection algorithm, the elimination is ordered so as to decrease the fill-in and the (sequential and parallel) arithmetic cost, specifically so that every eliminated block of vertices is connected by edges only to relatively few vertices, that is, to the vertices of an $s(n)$-separator that separates the vertices of the eliminated block from all other vertices.

In this process we use a *separator tree* T_G for the graph $G = G(A)$ that we define as follows (this refines our earlier definition of Section 14.2.3):

14.2. Sparse Linear Systems 641

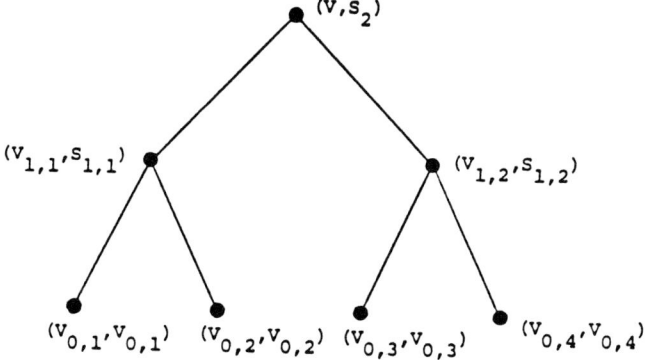

FIGURE 14.6
Tree T_G in the case $d = 2$.

DEFINITION 14.6
(Compare Figure 14.6 and Definition 14.4.) Suppose that the graph $G = (V, E)$ has n vertices. If $n \leq n_0$, let T_G be the trivial tree with no edges and with single leaf, (V, S), where $S = V$. If $n > n_0$, we know an $s(n)$-separator S of G, and obtain a partition $V(1), V(2), S$ of the vertex set V, such that there exists no edge in E between the sets $V(1)$ and $V(2)$, and furthermore, $|V(1)| \leq \alpha n, |V(2)| \leq \alpha n$, and $|S| \leq s(n)$. Then we define T_G as the binary tree with the root (V, S) having exactly two children, that are the roots of the two subtrees T_{G_1}, T_{G_2} of T_G, where G_j is the subgraph of G induced by the vertex set $S \cup V(j)$ for $j = 1, 2$. (Note that T_G is not equivalent to the $s(n)$-separator tree of Definition 14.3, which does not include the separator in the induced subgraphs.)

DEFINITION 14.7
Let the height of a node v in T_G equal d minus the length of the path from the root to v where d, the height of the root, is the length of a longest path from the root to a leaf. Let N_h be the number of nodes of height h in T_G, so that $N_h \leq 2^{d-h}$ since T_G is a binary tree. Let $(V_{h,1}, S_{h,1}), \ldots, (V_{h,N_h}, S_{h,N_h})$ be a list of all nodes of height h, and let $S_h = \cup_{k=1}^{N_h} S_{h,k}$.

Let $n = |V|$ denote the number of vertices of G. Since G has an $s(n)$-separator family, $|V_{h,k}| \leq n_h$ and $|S_{h,k}| \leq s(n_h)$ for each $h \geq 1$ and for $k = 1, \ldots, N_h$ (see (14.10)); furthermore $|V_{0,k}| \leq n_0$ and $S_{0,k} = V_{0,k}$ for $k = 1, \ldots, N_0$, by the definition of the tree T_G.

DEFINITION 14.8
For each $k = 1, \ldots, N_h$, let $R_{h,k}$ denote the set of all the elements of $S_{h,k}$ that are not in S_{h^*} for $h^* > h$. (Actually, $R_{h,k} = S_{h,k} - \cup S_{h^*,k^*}$ where the union is over all the ancestors $(V_{h^*,k^*}, S_{h^*,k^*})$ of $(V_{h,k}, S_{h,k})$ in T_G. Furthermore, since G has an $s(n)$-separator family, $R_{h,k_1} \cap R_{h,k_2} = \phi$ if $k_1 \neq k_2$.) Let $R_h = \cup_{k=1}^{N_h} R_{h,k}$.

Observe that, by definition, $R_{h,k_1} \cap R_{h,k_2} = \phi$ if $k_1 \neq k_2$, $R_h \cap R_h^* = \phi$ if $h \neq h^*$ and $V = \cup_{h=0}^{d} R_h$. Furthermore, for distinct k, the subsets $R_{h,k}$ of R_h are not connected by edges with each other; moreover, the vertices of each set $R_{h,k}$ can be connected by edges only with the vertices of the set $R_{h,k}$ itself and of the separator sets $S_{h+g,q}$ in the ancestor nodes of $(V_{h,k}, S_{h,k})$ of the tree T_G.

We will eliminate the vertices and the associated variables in the following order: at first, the vertices of R_0, then the vertices of R_1, of R_2, and so on, thus processing all the vertices of the set $V = \cup_h R_h$. (For each h, the elimination of the vertices of R_h is done in parallel for all the disjoint subsets $R_{h,1}, R_{h,2}, \ldots, R_{h,N_h}$.) In particular, the rows and columns of A_h associated with the vertices of $R_{h,k}$ form an $|R_{h,k}| \times |R_{h,k}|$ diagonal blocks of X_h for $k = 1, 2, \ldots, N_h$; X_h is the block diagonal submatrix of A_h with these N_h diagonal blocks, and $n_{h+1} = n_h - |R_h|$.

Let us formally define the elimination process and present the formal proofs and estimates. Let $\pi : \{1, \ldots, n\} \to \{1, \ldots, n\}$ be an enumeration of the n vertices of G such that $\pi(v) < \pi(v^*)$ if $v \in R_h$, $v^* \in R_{h^*}$, $h^* > h$, and furthermore, let π order the vertices of $R_{h,k}$ consecutively for each h and k. Then the elements of $\pi(R_h)$ are in the range from $\delta_h + 1, \ldots, \delta_{h+1}$ where $\delta_h = \sum_{gh} |R_g|$. Such an enumeration can be easily computed directly from the separator tree in parallel time $O(\log^2 n)$ using n/log n processors, by first enumerating the vertices of $R_d = S_d$ by $n, n-1, \ldots$ and then enumerating (also in the decreasing order) all the previously unenumerated vertices of R_h of height h for each h where $h = d-1, d-2, \ldots, 0$.

Let $P = [p_{ij}]$ denote the permutation matrix such that $p_{ij} = 1$ if $j = \pi(i)$ and $p_{ij} = 0$ otherwise. Then define the initial matrix $A_0 = PAP^T$. Recursively, for $h = 0, 1, \ldots, d-1$, let

$$A_h = \begin{bmatrix} X_h & Y_h^T \\ Y_h & Z_h \end{bmatrix}$$

denote the $(n-\delta_h) \times (n-\delta_h)$ symmetric matrix where X_h is the $|R_h| \times |R_h|$ upper left submatrix of A_h, Y_h is the $(n-\delta_h-|R_h|) \times |R_h|)$ lower left submatrix of A_h, and Z_h is the $(n-\delta_h-|R_h|) \times (n-\delta_h-|R_h|)$ lower right submatrix of A_h. Then compute $A_{h+1} = Z_h - Y_h X_h^{-1} Y_h^T$. This is elimination stage h, in which we eliminate the variables (vertices) associated with R_h. By Lemma 14.3, A_{h+1} is symmetric positive definite if A_h is.

We now claim that, given X_h, Z_h, and Y_h for a fixed h, we can compute A_{h+1} in time $O(\log^2 s(n))$ using at most $M(s(n))/\log n$ processors. To prove this, we will investigate the structure of A_h.

Let $A_h = [a_{ij}^{(h)}]$. Let $G_h = (V_h, E_h)$ denote the associated graph with the vertex set $V_h = \{\delta_h + 1, \delta_h + 2, \ldots, n\}$ and the edge set $E_h = \{\{i+\delta_h, j+\delta_h\} | a_{ij}^{(h)} \neq 0\}$; that is, we derive G_h from $G(A_h)$ by adding δ_h to each vertex number (see Figures 14.1–14.5). Note that $i, j \in V_h$ if the edge $\{i, j\}$ belongs to E_h. (The *fill-in in stage h* is the set of edges that are in E_h but not in E_{h-1}.)

Now, we are ready to establish a lemma, that refines Lemma 14.1 and provides some useful information about the fill-in, about the connectivity of G_h, and consequently about the sparsity of X_h (compare George and Liu [31], Liu [57], Gilbert and Tarjan [33]).

LEMMA 14.4
Let $h \geq 0$. Then

    a) if p is a path in G_h between two vertices $i \notin V_{h^*,k}$ and $j \in R_{h^*,k}$ for some $h^* \geq h$ and some k, then p visits some vertex v such that $\pi(v) > \delta_{h^*+1}$, that is, $v \notin R_q$ for $q \leq h^*$;
    b) $E_{h+1} \subseteq \{\{i,j\} \in E_h \mid i, j \notin R_h\} \cup \{\{i,j\} \mid \exists k \exists \{i, j_1\}, \{j_1, j_2\}, \ldots, \{j_{l-1}, j_l\}, \{j_l, j\} \in E_h\}$ where $j_1, \ldots, j_l \in R_{h,k}$ and $\pi(i) > \delta_h$, $\pi(j) > \delta_h$.

PROOF
Part a). Observe that if there is a path p in G_h between some $i \notin V_{h^*,k}$ and some $j \in R_{h^*,k}$ for some $h^* \geq h$, then p must contain a vertex, say, j^* of the separator set S_{h^*+1,k^*}, where $(V_{h^*+1,k^*}, S_{h^*+1,k^*})$ is the parent node of $(V_{h^*,k}, S_{h^*,k})$ in T_G. This vertex j^* has number $\pi(j^*) > \delta_{h^*+1}$, as required. This proves the part a) of Lemma 14.4.

Part b). By definition, $A_{h+1} = Z_h - Y_h X_h^{-1} Y_h^T$. Let $X_h = (x_{ij}), X_h^{-1} = \bar{X}_h = (\bar{x}_{ij}), Y_h = (y_{ij}), Z_h = (z_{ij})$, and $W_h = Y_h X_h^{-1} Y_h^T = (w_{ij})$. If $\{i, j\} \in E_{h+1}$, then $i, j \notin R_h$, by the definition of R_h, so $\pi(i) > \delta_{h+1}$, $\pi(j) > \delta_{h+1}$. Furthermore we have $z_{i-\delta_h, j-\delta_h} \neq 0$ or otherwise

$w_{i-\delta_h,j-\delta_h} \neq 0$. If $z_{i-\delta_h,j-\delta_h} \neq 0$, then $\{i,j\} \in E_h$, as claimed. On the other hand, if $w_{i-\delta_h,j-\delta_h} \neq 0$, then there exist vertices $j^* + \delta_h, j^{**} + \delta_h \in R_h$ such that $y_{ij^*} \neq 0$, $\bar{x}_{j^*,j^{**}} \neq 0$ and $y_{jj^{**}} \neq 0$. It follows that $\{i, j^* + \delta_h\}, \{j, j^{**} + \delta_h\} \in E_h$. Furthermore the Cayley-Hamilton theorem implies that $X_h^{-1} = c_0 I + c_1 X_h + \ldots + c_{r-1} X_h^{r-1}$ for some scalars $c_0, c_1, \ldots, c_{r-1}$ where $r \leq n$ (and in fact $r \leq s(n_h)$ since X_h is a block diagonal matrix with $s(n_h) \times s(n_h)$ diagonal blocks). Hence $\bar{x}_{j^*j^{**}} \neq 0$ implies that the (j^*, j^{**}) entry of X_h^l is not 0 for some $0 < l < r-1$, unless $j^* = j^{**}$ (we will omit the latter trivial case). Thus there exists a path $\{j_1, j_2\}, \{j_2, j_3\}, \ldots, \{j_{l-1}, j_l\}$ in E_h visiting only the vertices $j_1, \ldots, j_l$ in R_h where $j_1 = j^* + \delta_h$, $j_l = j^{**} + \delta_h$; but the part (a) of the lemma implies that there can be no edge in E_h between $R_{h,k}$ and R_{h,k^*} for $k \neq k^*$, so $j_1, \ldots, j_l \in R_{h,k}$ for some unique k. Thus we have established the part (b) of the lemma for E_{h+1}. ■

Lemma 14.4 gives the desired restriction on the edge connections in $G(A_h)$. In particular, Lemma 14.4(a) implies that E_h contains no edge between $R_{h,k}$ and R_{h,k^*} for $k \neq k^*$.

The next result follows since π groups the vertices of $R_{h,k}$ together and since $\max_k |R_{h,k}| \leq \max_k s(|V_{h,k}|) \leq s(n_h)$:

LEMMA 14.5
X_h *is a block diagonal matrix consisting of* $N_h \leq 2^{d-h}$ *square blocks of sizes* $|R_{h,k}| \times |R_{h,k}|$, *so that each block is of size at most* $s(n_h) \times s(n_h)$.

Lemma 14.5 implies that for $h > 0$ the inversion of X_h can be reduced to $N_h \leq 2^{d-h}$ parallel inversions of generally dense matrices, each of size at most $s(n_h) \times s(n_h)$. Therefore, one dense matrix is associated with each $R_{h,k}$, so, its size is at most $|R_{h,k}| \times |R_{h,k}|$. (14.2) implies that this can be done in $O(\log^2 n)$ time, using $N_h M(s(n_h)) \leq 2^{d-h} M(s(n_h))$ processors.

The next lemma is from Lipton and Tarjan [50].

LEMMA 14.6 *(compare (14.8), (14.15), (14.16) and Definition 14.3) For any triple of constants α, α^* and n_0 such that $1/2 \leq \alpha^* < \alpha < 1$ and $n_0 > 0$, if a graph has an $s(n)$-separator family with respect to α and n_0, then this graph has an $s^*(n)$-separator family with respect to α^* and n_0, where $s^*(n) \leq \sum_{h=0}^{d} s(n_h)$; in particular, $s^*(n) \leq cn^\sigma/(1-\beta^\sigma)$ if $s(n) \leq cn^\sigma$ for some positive constants c, β and σ, where $\beta < 1$.*

14.2. Sparse Linear Systems 645

PROOF (Pan and Reif [64])
Let us first assume that $\alpha^* > 1/2$ (in fact we will need Lemma 14.6 only under this assumption). Let a graph $G = (V, E)$ have $s(n)$-separator family with respect to α and n_0 and let $n = |V| \geq n_0$. Then we can partition V into three disjoint sets V_1^*, V_2^*, S^* such that $|V_1^*| \leq \alpha^* n, |V_2^*| \leq \alpha^* n, |S^*| \leq O(s(n))$, and V_1^* and V_2^* are not connected by E. Such a partition can be found by appropriately combining the $N_h \leq 1 + (\alpha^* - 1/2)\log_2 \beta = O(1)$ vertex subsets $V_{h,k}$ and $S_{h,k}$ found at depth $h = d - \lceil \log((\alpha^* - 1/2)/\log \beta) \rceil$ of T_G (compare (14.15), (14.16)). Specifically, the separator S_h^* is defined to be the union of all the separators of depths at least h in T_G; since there is only a constant number of such separators, $|S_h^*| = O(s(n))$. Further, $V_1^* = V(h) - S_h^*$, where V(h) is any maximal collection of the vertex subsets of depth h in T_G and where $n/2 \leq |V_1^*| \leq \alpha^* n$. [Note that $|V_{h,k}| \leq n\beta^h$ for all k (see (14.16)), and $n\beta^h \leq (\alpha^* - 1/2)n$.] Letting $V_2^* = V - V_1^* - S_h^*$, we obtain that $|V_2^*| = n - (n/2) = n/2 \leq \alpha^* n$. Such a subdivision is recursively repeated for the subgraphs with the vertex sets $V_i^* \cup S^*$, $i = 1, 2$, and so on, until we finally arrive at the desired $s^*(n)$-separator family of G with respect to α^*, where $s^*(n) = O(s(n))$, provided that $\alpha^* > 1/2$. The extension to the case $\alpha^* = 1/2$ and the specific upper bound on $s^*(n)$ of Lemma 14.6 can be obtained by extending the proof of Corollary 3 of Lipton and Tarjan [50]. ∎

LEMMA 14.7
$2^{d-h} M(s(n_h)) \leq M(s(n))\eta^{d-h}$ for some $\eta < 1$.

PROOF
(14.8) and (14.16) imply that $s(n_h) \leq c(\alpha+\delta)^{\sigma(d-h)} n^\sigma$, so that $M(s(n_h)) \leq c^{\bar\omega}(\alpha+\delta)^{\sigma\bar\omega(d-h)} n^{\sigma\bar\omega}$ (see (14.13)). We choose n_0 sufficiently large so as to make δ sufficiently small and then apply Lemma 14.6 to make sure that $\alpha + \delta$ lies as close to $1/2$ as we like. Since $\sigma\bar\omega > 1$ (see (14.13)), we may assume that $\eta = 2(\alpha+\delta)^{\sigma\bar\omega} < 1$. Then

$$2^{d-h} M(s(n_h)) \leq \eta^{d-h} c^{\bar\omega} n^{\sigma\bar\omega} = \eta^{d-h} M(s(n)). \qquad \blacksquare$$

Due to (14.2) and to Brent's scheduling principle of Section 14.2.1, we may invert X_h by using $O(k \log^2 n)$ arithmetic steps and $\lceil 2^{d-h} M(s(n_h))/k \rceil \leq \lceil M(s(n))\eta^{d-h}/k \rceil$ processors for some $\eta < 1$ and for any k such that $1 \leq k = k(h)$. Choosing the minimum $k = k(h) \geq 1$ such that $M(s(n))\eta^{d-h}/k(h) \leq M(s(n))/\log n$ (so that $k(h) = \eta^{d-h} \log n$ if $h \leq d + \log\log n/\log \eta$, $k(h) = 1$

otherwise), we simultaneously yield the time bound $O(\log^3 n)$ (see Lemma 14.2) and the processor bound $M(s(n))/\log n$, as this is required in Theorem 14.3.

14.2.8 Estimating the Cost of Parallel Multiplication of Auxiliary Matrices

THEOREM 14.4

The 2d matrix multiplications,

$$U_h = Y_h X_h^{-1}, W_h = U_h Y_h^T, h = 0, 1, \ldots, d-1, \quad (14.17)$$

involved in the recursive $s(n)$-factorization (14.9)–(14.12), can be performed by using $O(\log^2 n)$ parallel time and $M(s(n))$ processors or (if we slow down the computations by a factor of $\log n$) by using $O(\log^3 n)$ parallel time and $M(s(n))/\log n$ processors.

In this section, we will prove Theorem 14.4 by estimating the complexity of parallel evaluation of the matrix products of (14.17) (given Y_h and X_h^{-1}) for $h = 0, 1, \ldots, d - 1$. We will first rearrange the matrix multiplications of (14.17), reducing them to several matrix multiplications of the form

$$U_{h,k} = Y_{h,k} X_{h,k}^{-1}; \; W_{h,k} = U_{h,k} Y_{h,k}^T,$$
$$k = 1, 2, \ldots, N_h, \; h = 0, 1, \ldots, d-1. \quad (14.18)$$

For this, partition Y_h into N_h submatrices $Y_{h,k}$ associated with the row sets $R_{h,k}$ and having the sizes $m_{h,k} \times |R_{h,k}|$ where $m_{h,k} \leq n - \delta_h$ for $k = 1, \ldots, N_h$. Let $X_{h,k}^{-1}$ denote the respective dense diagonal blocks of X_h^{-1}. By the definition of G_h and T_G and by virtue of Lemma 14.4, the matrix $Y_{h,k}$ may have nonzero entries in row i only if i lies in one of the sets $R_{h+g,q}$ (for $1 < g \leq d - h$; q=q(g,h,k)) corresponding to an ancestor $(V_{h+g,q}, S_{h+g,q})$ of the node $(V_{h,k}, S_{h,k})$ in T_G.

To deduce the estimates of Theorem 14.4, examine the cost of the matrix multiplications (14.18), grouping them not in the *horizontal* order, where k ranges from 1 to N_h for a fixed h, but in the *vertical* order of Definition 14.3, that is, going from the root of the tree T_G to its leaves.

Slightly abusing the notation, denote $n = |R_{h,k}|$, $m = m_{h,k}$ for a fixed pair h and k and consider the matrix multiplications of (14.18) associated with the node $(V_{h,k}, S_{h,k})$ and with its descendants in the tree T_G. Surely, these matrix multiplications can be performed in $O(\log^2 n)$ time; let $P(n, m)$ denote the associated processor bound. For the two children of the node $(V_{h,k}, S_{h,k})$, the two associated processor bounds will be denoted by $P(n_1, m_1)$

14.2. Sparse Linear Systems

and $P(n_2, m_2)$, where, by virtue of Lemma 14.5 and Definition 14.3,

$$m_1 + m_2 \leq m + 2s(n),$$
$$n \leq n_1 + n_2 \leq n + s(n), \qquad (14.19)$$
$$(1 - \alpha)n \leq n_i \leq \alpha n + s(n) \text{ for } i = 1, 2.$$

Let M(p,q,r) denote the number of processors defined up to within a constant factor and required in order to multiply $p \times q$ by $q \times r$ matrices in $O(\log(pqr))$ parallel steps; then $M(p,q,r) \leq M(q)\lceil \frac{p}{q} \rceil \lceil \frac{r}{q} \rceil$ if $p \geq q$, $r \geq q$. For fixed h and k (and, therefore, for a fixed separator $S_{h,k}$), the matrix multiplications (14.18) can be performed using $O(\log n)$ parallel steps and $M(s(n)+m, s(n), s(n)+m) \leq \lceil (1+m/s(n))^2 \rceil M(s(n))$ processors. Recursively, it follows that

$$P(n,m) \leq (1 + (1+m/s(n))^2)M(s(n)) + P(n_1,m_1) + P(n_2,m_2) \qquad (14.20)$$

for some n_1, n_2, m_1, m_2 satisfying (14.19).

Using (14.20), we will prove the following claim, which in its special case where m=0 amounts to Theorem 14.4 (recall that we already have the parallel time bound $O(\log^2 n)$, and we will use the slowdown by the factor of log n to arrive at the time and processor bounds of this theorem):

CLAIM 14.1
$P(n,m) \leq (c_0 + c_1(m/s(n)) + c_2(m/s(n))^2)M(s(n))$ for all m and n and for some constants c_0, c_1, c_2.

PROOF
If $n \leq n_0$, then $P(n,m) \leq M(n) \leq c_0$, provided that $c_0 \geq M(n_0)$. Thus let $n \geq n_0$ and prove the claim by induction on n. We will assume that n_0 is large enough, so that (14.16) implies that $n_i < n$ for $i = 1, 2$. By the induction hypothesis, the claim holds if n is replaced by n_i for $i = 1, 2$. Consequently,

$$P(n_1, m_1) + P(n_2, m_2) \leq \sum_{i=1}^{2} (c_0 + c_1(m_i/s(n_i)) + c_2(m_i/s(n_i))^2) M(s(n_i)).$$

Therefore,

$$\sum_i P(n_i, m_i) \leq c_0 \sum_i M(s(n_i))$$
$$+ c_1 \sum_i m_i M(s(n_i))/s(n_i) \qquad (14.21)$$
$$+ c_2 \sum_i m_i^2 M(s(n_i))/s(n_i)^2.$$

Next we deduce from (14.19) that for g=1 and g=2,

$$\sum_i m_i^g M(s(n_i))/s(n_i)^g \le \left(\sum_i m_i{}^g\right) \max_i (M(s(n_i))/s(n_i)^g)$$
$$\le (m+2s(n))^g M(s(\alpha n + s(n)))/s(\alpha n + s(n))^g$$
$$\le (m+2s(n))^g M(s(\beta n))/s(\beta n)^g \text{ for } \beta < 1$$

(apply (14.15) to deduce the latter inequality). Applying here (14.8) and (14.13), we obtain that

$$\sum_i m_i^g M(s(n_i))/s(n_i)^g \le \gamma(m+2s(n))^g M(s(n))/s(n)^g, \qquad (14.22)$$

where $\gamma = \beta^{(\bar{\omega}-g)\sigma}$ is a constant, $\gamma < 1$, $g = 1, 2$, $g < \bar{\omega}$.

Furthermore, (14.8) and (14.19) imply that the sum $M(s(n_1))+M(s(n_2))$ takes its maximum value where one of n_1, n_2 is as large as possible (that is, equals $\alpha n + s(n)$), which makes the other as small as possible (that is, equal to $(1-\alpha)n$). Therefore,

$$M(s(n_1)) + M(s(n_2)) \le M(s(\alpha n + s(n))) + M(s((1-\alpha)n))$$
$$\le M(s((\alpha+\delta)n)) + M(s((1-\alpha)n))$$
$$= M(cn^\sigma(\alpha+\delta)^\sigma) + M(cn^\delta(1-\alpha)^\sigma)$$

(see (14.8) and (14.15)). Applying here (14.14) and then (14.8) and (14.13), we deduce that

$$M(s(n_1)) + M(s(n_2)) \le (M((\alpha+\delta)^\sigma) + M((1-\alpha)^\sigma))M(cn^\sigma)$$
$$\le ((\alpha+\delta)^{\bar{\omega}\sigma} + (1-\alpha)^{\bar{\omega}\sigma})M(s(n)),$$

where $\bar{\omega}\sigma > 1$. The positive δ can be assumed to be arbitrarily close to 0, so that

$$M(s(n_1)) + M(s(n_2)) \le \nu M(s(n)) \qquad (14.23)$$

for a constant $\nu < 1$.

Combining (14.20)–(14.23) implies that

$$P(n,m) \le (2 + \nu c_0 + 2\gamma c_1 + 4\gamma c_2) M(s(n))$$
$$+ (2 + \gamma c_1 + 4\gamma c_2) M(s(n)) m/s(n) \qquad (14.24)$$
$$+ (1 + \gamma c_2) M(s(n))(m/s(n))^2$$

for two constants $\gamma < 1, \nu < 1$. We will choose c_2 large enough, so that $1 + \gamma c_2 \le c_2$, then choose c_1 large enough, so that $2 + \gamma c_1 + 4\gamma c_2 \le c_1$; and finally, choose c_0 large enough, so that $2 + \nu c_0 + 2\gamma c_1 + 4\gamma c_2 \le c_0$. Then (14.24) will imply the claim and, consequently, Theorem 14.4. ∎

14.3
Path Algebra Computations in Graphs

In this section, we will follow Pan and Reif [71, 75] and will extend the nested dissection algorithm of Section 14.2 to several problems of practical and theoretical interest that can be reduced to path algebra computations. The applications of path algebras include problems in: vehicle routing, investment and stock control, dynamic programming with discrete states and discrete time, network optimization, artificial intelligence and pattern recognition, labyrinths and mathematical games, encoding and decoding of information (see Section 14.3.1 and the references (Backhouse and Carre [5], Carre [16], Carre [17], Gondran and Minoux [35], Lawler [51], Tarjan [92, 93], Zimmermann [98]), also containing further bibliography). In particular, the papers Backhouse and Carre [5], Carre [16] and the later book of Gondran and Minoux [35] include several general sequential algorithms for such problems based on matrix operations in *dioids* (*semirings*) (see our next sections). These algorithms, however, are not powerful enough in the case of sparse input graphs. In the special case of the shortest path problems, there exist even more effective sequential algorithms (Frederickson [23]), but they apply neither to the case of general path algebra problems over dioids nor to the case of parallel computations. We will describe a substantial improvement of these general algorithms in the important case where the input matrix A is associated with a fixed undirected planar or, more generally, $s(n)$-separatable graph (see Definitions 14.3 and 14.4 in Section 14.2). Our improvement relies on our extension of the generalized nested dissection parallel algorithm of Section 14.2 to path algebra problems. From Section 14.2 we only need the black box subroutine for the recursive factorization (14.12), which reduces this factorization to a sequence of matrix inversions, multiplications and subtractions and which we, somewhat surprisingly, extend to a similar factorization over dioids.

We will define the algorithms over dioids (semirings) and will estimate the computational cost in terms of dioid operations, which will replace the arithmetic operations over real numbers, assumed in Section 14.2. In the major specific applications to the classes of the problems of path existence, optimization and counting, an operation over a dioid is an addition, a multiplication, a comparison of two numbers, or a selection of minimum or maximum of a few numbers; the numbers involved in the computation by our algorithms are usually represented with about the same precision (that is, with about the same number of binary digits) as the input values. Here is a table of the resulting complexity estimates where we assume that $s(n) = O(\sqrt{n})$.

TABLE 14.1
Complexity of Path Problems in an n-vertex Graph

Path problem	Sequential time	Parallel time	Processors	Precomputed separators
Single	$O(n^{1.5})$	$O((\log n)\sqrt{n})$	$n/\log n$	No
Source	$O(n^{1.5})$	$O(I(n)\log^2 n)$	$n^{1.5}/(I(n)\log n)$	Yes
All	$O(n^2 \log n)$	$O((\log n)\sqrt{n})$	$n^{1.5}$	No
Pair	$O(n^2 \log n)$	$O(I(n)\log^2 n)$	$n^2/(I(n)\log n)$	Yes

(The estimates can be extended to the case where s(n) is an arbitrary function, such that $s(n) = O(n^\sigma), 1/2 \le \sigma < 1$.) In the table and in the following $I(n)$ denotes the parallel time for the summation of n elements of a dioid on a computer with n processors. On EREW PRAM, $I(n) = O(\log n)$, on a randomized CRCW, $I(n) = O(1)$.

All our parallel algorithms have polylogarithmic parallel time, that is, $O(I(n)\log^3 n)$ or $O(I(n)\log^2 n)$, and have rather small processor bounds (which places our algorithms in NC). In many cases all the known alternate methods give inferior results. In particular, we use only $O(I(n)\log^2 n)$ parallel steps and either $n^{1.5}/(I(n)\log n)$ processors in order to solve the single source shortest path problem or $n^2/(I(n)\log n)$ processors in order to solve the all pair shortest path problem, whereas the other known polylogarithmic time parallel algorithms for both all pair shortest path problem and single source shortest path problem in planar graphs required n^3 processors, the same number as for the path computation in general graphs. Furthermore, our parallel algorithms for the shortest path computations in planar graphs, combined with the results of Johnson [43] and Klein and Reif [50], lead to a parallel algorithm for the evaluation of a maximum flow and a minimum cut in G using $O(I(n)\log^3 n)$ steps and $n^{1.5}/(I(n)\log n)$ processors or, alternatively, $O(I(n)\log^2 n)$ steps, $(n/\log n)^2$ processors, to compare with the alternate bounds of $O(\log^2 n)$ steps, n^4 processors (Johnson [43]). To yield such an extension to computing maxflow-mincut in a planar undirected network, we need to use a simple extension of our path algebra results to the case of directed graphs. (Such an extension may have other applications to maxflow-mincut computation and may be itself of independent interest.)

Several further applications can be expected; for instance, we may apply our parallel shortest path algorithms to feasibility testing of a multicommodity flow in a planar network in the case where all the k source-sink pairs lie on the boundary of the outer face, compare Frederickson [23], Hassin [38], Matsumoto, Nishizeki and Saito [59]. In this case the feasibility test, based on our parallel shortest path computation and on the algorithm of Klein and Reif [45] for constructing the auxiliary dual graphs (both applied within the construction of Matsumoto, Nishizeki and Saito [59]), requires only $O(I(n)\log n)$ parallel steps and $\min\{kn^{1.5}/(I(n)\log n), n^2/(I(n)\log n)\}$ processors, where k is the number of commodities. This improves the previously known processor bounds by more than the factor of n.

Furthermore, in Pan and Reif [75], for a more special but important path algebra problem of computing the minimum cost paths in a (planar) graph, pipelining is used in a nontrivial way in order to rearrange our parallel algorithms and to accelerate the computation further by a factor of $\log s(n)$ (which means a factor of $\sigma \log n + C$ for $s(n) = Cn^\sigma$, $\sigma = 0.5$ for planar graphs); we simultaneously preserve the original processor bounds (defined up to within a constant factor). This result also leads to the respective acceleration of parallel computation for mincut and maxflow and for other related problems (see Gazit and Reif [30]).

In the next section we will introduce some preliminary technical definitions required for our parallel algorithms; in particular, we will define dioids and will state some path algebra problems; we will also estimate the computational cost of solving these problems in the case of general graphs. In section 14.3, we will consider these problems for undirected $s(n)$-separatable input graphs (with $s(n)$ satisfying (14.8) of Section 14.2) and will present our main results. Our parallel algorithms for the shortest path computations are extended to computing maxflow and mincut in Section 14.3.3.

It is convenient for us to assume that hereafter all vectors are row vectors.

14.3.1 Path Algebra Problems for General Graphs

In this section we will recall some auxiliary results and definitions for path algebra problems for general graphs.

A Model Problem and Its Extension

As a model example, consider the special case of the shortest path problem in a graph $G = G(A)$ with n vertices defined by an $n \times n$ matrix $A = [a_{ij}]$ of nonnegative arc lengths where $a_{ij} = \infty$ if there is no arc between the vertices i and j in G. (A is symmetric if G is undirected.) We seek the vector $\mathbf{x} = [x(i)]$ of distances $x(i)$ (that is, of the lengths of shortest paths) from vertex 1 to all vertices i in G. This is the *single source shortest path problem (SS)*. The distances satisfy the following system of equations:

$x(1) = 0, x(i) = \min_j(x(j) + a_{ji}), i = 2, \ldots, n$, or equivalently:

$$x(1) = \min(\min_j(x(j) + a_{j1}), 0),$$
$$x(i) = \min(\min_j(x(j) + a_{ji}), \infty), \quad i = 2, \ldots, n.$$

We substitute $\oplus$ for min (a noninvertible operation!) and $*$ for $+$ and rewrite this system as follows:

$$\begin{aligned} x(1) &= \oplus_j(x(j) * a_{j1} \oplus 0), \\ x(i) &= \oplus_j(x(j) * a_{ji} \oplus \infty), \end{aligned} \quad i = 2, \ldots, n,$$

or, in matrix notation, denoting $\mathbf{i}^{(1)} = [0, \infty, \ldots, \infty]$:

$$\mathbf{x} = \mathbf{x} * A \oplus \mathbf{i}^{(1)}. \tag{14.25}$$

Here and hereafter we always assume that the operation $*$ precedes $\oplus$.

Extension. Seeking the matrix $X = [x(i,j)]$ of distances between all pairs of vertices in G (this is *the all pair shortest path problem, AP*) and denoting $I = [\delta_{ii}], \delta_{ii} = 0, \delta_{ij} = \infty$ if $i \neq j$, we similarly arrive at the following matrix equation:

$$X = X * A \oplus I. \tag{14.26}$$

Clearly, an SS of 14.25 is a subproblem of an AP of 14.26. Conversely, restricting 14.26 to the hth row, we arrive at the SS of computing the distances from the vertex h to all the vertices in G (for $h = 1$ we again arrive at (14.25)), so that an AP can be reduced to n SSs.

Two Algorithms

Some known algorithms of linear algebra can be extended to solve the systems (14.25) and (14.26); this may turn them into known combinatorial algorithms for the SS and/or the AP. Here are two examples (Backhouse and Carre [5], Carre [16], Gondran and Minoux [35]):

ALGORITHM 14.2
Set $\mathbf{x}^{(0)} = \mathbf{i}^{(1)}$; compute $\mathbf{x}^{(k+1)} = \mathbf{x}^{(k)} * A \oplus \mathbf{i}^{(1)}, k = 0, 1, \ldots$ until $\mathbf{x}^{(k+1)} = \mathbf{x}^{(k)}$; then output the vector $\mathbf{x} = \mathbf{x}^{(k)}$ satisfying 14.25.

Algorithm 14.2 extends Jacobi's method of linear algebra and amounts to the algorithm of Bellman [6] for the SS.

ALGORITHM 14.3
a) Set $A^{[1]} = A$.
b) For $k = 0, 1, \ldots, n-1$, compute $a_{ij}^{[k+1]} = a_{ij}^{[k]} \oplus a_{ik}^{[k]} * a_{kj}^{[k]}, i, j = 1, \ldots, n$.
c) Output $X = A^{[n]} \oplus I$. (The matrix X satisfies (14.26).)

Algorithm 14.3 extends Jordan's algorithm of linear algebra and amounts to the algorithm of Floyd [22] for the AP (compare also Algorithm 14.5 in section 14.3.2).

Definition of Dioids (Semirings)

Several other problems of path computation can be also reduced to the solution of the linear systems (14.25) or (14.26) or to some similar matrix computations performed by means of additions and multiplications only. We need to recall a general concept, already implicitly used in our reduction of the SS to (14.25) and of the AP to (14.26).

DEFINITION 14.9
A dioid (also called a semiring, because it extends noncommutative rings to the case where subtractions may not be defined) is a set S with two operations, $\oplus$ and $*$, such that for any triple of elements $a, b, c \in S$ and for two special elements e (unity) and ϵ (zero) of S, the following equations hold:

$$a \oplus b = b \oplus a \in S, (a \oplus b) \oplus c = a \oplus (b \oplus c), a \oplus \epsilon = a,$$

$$a * b \in S, (a * b) * c = a * (b * c),$$

$$a * e = a, a * \epsilon = \epsilon * a = \epsilon,$$

$$a * (b \oplus c) = (a * b) \oplus (a * c), (b \oplus c) * a = (b * a) \oplus (c * a).$$

In the above reduction of the SS to (14.25) and of the AP to (14.26), we used the dioid where $S = R \cup \{\infty\}$, R being the set of real numbers, $\oplus = \min$, $* = +$, $e = 0$, $\epsilon = \infty$. (This dioid is also used for other optimization path problems, see 3 below.) Generalizing (14.25) and (14.26) to arbitrary dioids, we define

$$i^{(1)} = [e, \epsilon, \ldots, \epsilon], I = [\delta_{ij}], \delta_{ii} = e, \delta_{ij} = \epsilon \text{ if } i \neq j. \tag{14.27}$$

A List of Path Problems

Here is a list of some classes of path problems, which can be reduced to solving the systems (14.25) and (14.26) or to similar matrix operations in appropriate dioids:

1. existence (problems of graph connectivity);
2. enumeration (elementary paths, multicriteria problems, generation of regular languages);
3. optimization (paths of maximum capacity, paths with the minimum number of arcs, shortest paths, longest paths, paths of maximum reliability, reliability of a network);
4. counting (counting of paths, Markov chains).

In particular, the class (1) includes the more specific problems of
a) the existence of paths having k (or at most k) arcs between vertices i and j in a given (di)graph G (for a fixed k);
b) computing the transitive closure of G;
c) testing G for being strongly connected and for having circuits.

An appropriate dioid for problems of class (1) is the Boolean algebra, $S = \{0, 1\}$, $\oplus = \max$, $* = \min$, $\epsilon = 0$, $e = 1$; in the incidence matrix $A = [a_{ij}]$ of G, $a_{ij} = 1$ if and only if $\{i, j\}$ is an arc of G.

The subclass of shortest path problems in (3) includes SS, AP (also in the versions where the shortest paths are required to have exactly k or at most k arcs), and testing a graph for having circuits of negative lengths.

Class (4) includes counting the numbers of
a) distinct paths having k (or at most k) arcs between i and j in G;
b) all the distinct paths between i and j in G.

In the dioid for this class, S is the set of integers, $\oplus = +$, $* = *$ (that is, $\oplus$ and $*$ are the conventional addition and multiplication, respectively), $\epsilon = 0$, $e = 1$; $A = [a_{ij}]$, $a_{ij} = 1$ if and only if $\{i, j\}$ is an arc of G.

Path Computation as Computing the Quasi-Inverse Matrix

The solution of most of the path problems listed above can be reduced to the evaluation (over the dioids) of the entries of the matrix $A^{(k)}$ (for the all pair path problems) or of the vector $bA^{(k)}$ (for the single source path problems) for some positive k, usually for $k = n - 1$. Here

$$A^{(q+1)} = A^{(q)} \oplus A^{q+1}, q = 0, 1, \ldots,$$

$A^{(0)} = I$ (see (14.27)), A is an $n \times n$ input matrix, $b = i^{(h)}$ is a fixed coordinate vector of dimension n. Here and hereafter, we assume that all computations, in particular, computing matrix sums, products and powers, are performed over the dioid associated with a given path problem; we simplify the notation, by writing uU and UV (rather than $\mathbf{u} * U$ and $U * V$, respectively) in order to denote the product of a vector $\mathbf{u}$ by a matrix U and the product of matrices U and V over dioids; similarly, for matrix powers over dioids.

There exists the *quasi-inverse* matrix, defined as

$$A^* = \lim_{q \to \infty} A^{(q)}, \qquad (14.28)$$

for the incidence matrix A of each of the path problems listed in the previous section, except for the shortest path and multicriteria problems where there exist circuits of negative lengths in G and for the counting problems where there exists a circuit in G. In both latter cases the existence of such circuits is detected by means of computing A^k or $(I \oplus A)^k$ over the dioids. Hereafter we will only consider the most typical case, where there exists the quasi-inverse A^*, and where, moreover,

$$A^* = A^{(n-1)}, A^{(q)} = A^{(q-1)} \text{ for } q \geq n. \qquad (14.29)$$

Our estimates of this section for the cost of the evaluation of A^* and bA^* under (14.29) can be immediately extended to the case of the evaluation of A^q, $A^{(q)}$ and $bA^{(q)}$ for $q \neq n-1$.

The equations (14.28) and (14.29) imply that A^* is the incidence matrix of the transitive closure of the graph of A for several problems of connectivity, existence, and optimization. In Abdali and Saunders [1], the study of the quasi-inverse A^* relies on an interesting concept of the eliminant (see Exercise 14.19, part b).

Evaluation of the Quasi-Inverse Matrix

(14.29) implies that

$$A^* = I \oplus A \oplus A^2 \oplus \ldots \oplus A^{n-1},$$

so A^* can be computed as follows:

$$A^* = (I \oplus A)(I \oplus A^2)(I \oplus A^4) \ldots (I \oplus A^{2^k}), k = \lceil \log_2 n \rceil. \qquad (14.30)$$

The known fast matrix multiplication algorithms (see Pan [64, 65]) cannot be generally applied over dioids, so we compute A^* by applying the straightforward matrix multiplication algorithm and using a total of $(4nk - k + 1)n^2$

operations in the dioid (compare (14.1) and (14.4) of Section 14.2). For many dioids, including the dioid that we associated with the shortest path computations, the operation $\oplus$ is *idempotent*, that is, $a \oplus a = a$ for all $a \in S$. For such dioids

$$A^* = \oplus_{r=0}^{n} A^r = \oplus_{r=0}^{n} C(n,r) A^r = (I \oplus A)^n = (I \oplus A)^{2^k} \qquad (14.31)$$

(where $C(n,r) = r!/n!(n-r)!, k = \lceil \log_2 n \rceil$), so, the quasi-inverse A^* can be computed by means of repeated squaring of $I \oplus A$, using only $n^2(k-1)(2n-1) + n$ dioid operations. The resulting asymptotic estimates for the cost of both algorithms (14.30) and (14.31) are the same: $O(n^3 \log n)$ operations or $O(I(n) \log n)$ parallel steps and $\lceil n^3/I(n) \rceil$ processors. The algorithms (14.30) and (14.31) are not efficient if A is sparse, for the sparsity of A is not generally preserved during the computation by these algorithms.

We may, of course, compute bA^* via computing A^*; and again, we have the same problems where A is sparse. Alternatively, we may perform n successive postmultiplications of the vectors $b \sum_{r=0}^{k} A^r$ by the matrix A for $k = 0, 1, \ldots, n-1$ and $n-1$ vector additions; this way exploits the sparsity of A but requires a sequence of n multiplications of vectors by A, that is, $nI(n)$ parallel steps.

If we agree to use the order of n parallel steps but only $O(n^3)$ dioid operations, we may compute A^* by applying the following alternate algorithm, generalizing Jordan elimination (compare Backhouse and Carre [5], Carre [16], Gondran and Minoux [35], p. 110):

ALGORITHM 14.4
(Evaluation of A^*)
Set $A^{[0]} = A$, $B^{[0]} = I$ and recursively compute $A^{[k]} = M^{[k]} A^{[k-1]}$, $B^{[k]} = M^{[k]} B^{[k-1]}$ for $k = 1, 2, \ldots, n$. Here, $M^{[k]}$ is obtained from the matrix I of 14.27 by replacing the (k,k) entry of I by $(a_{kk}^{[k-1]})^*$ and by replacing other entries of the $k-th$ column of I by the entries of the vector $[a_{ik}^{[k-1]} * (a_{kk}^{[k-1]})^*]$, where $i, k = 1, \ldots, n$. Output $B^{[n]}$.

Algorithm 14.4 extends (to the dioids) the Jordan elimination scheme for computing $B^{-1} = (I - A)^{-1}$ via the solution of the matrix equation $BX = I$. This equation can be interpreted as n linear systems, each consisting of n equations in n unknowns, having the same coefficient matrix, $B = I - A$, and having one of the n unit coordinate vectors on the right side. We will use the following result:

LEMMA 14.8 *(Compare Remark 14.8 in the next section.)*
Let $B^{[n]}$ be the output matrix of Algorithm 14.4, where for all the auxiliary matrices $a_{kk}^{[k-1]}$, there exist the quasi-inverse matrices $(a_{kk}^{[k-1]})^*$. Then $B^{[n]} = A^*$.

In the Appendix 14.B, we prove Lemma 14.8 by using, in particular, the arguments of Backhouse and Carre [5], Carre [16], Gondran and Minoux [35, pp. 108–110], and the next simple lemma, which can be immediately verified (see (14.28)):

LEMMA 14.9
The matrix $X = A^*$ satisfies (14.26).

PROOF OF LEMMA 14.8
Next, we give a proof of Lemma 14.8 that is slightly longer than one in the Appendix 14.B, but is more direct and informative. In this proof we start with the customary Jordan elimination scheme over a field, say, of real or rational numbers. In such an elimination scheme, n successive premultiplications by n matrices $M^{[k]}$ (for $k = 1, \ldots, n$) of a special format (see below) reduce the $n \times n$ input matrix $G = G^{[0]}$ to the identity matrix, so that $G^{[k]} = [g_{ij}^{[k]}] = M^{[k]} G^{[k-1]}$, $k = 1, \ldots, n$, $M^{[n]} M^{[n-1]} \ldots M^{[1]} G = I$, $M^{[n]} M^{[n-1]} \ldots M^{[1]} = G^{-1}$. It can be immediately verified that the latter matrix identities hold if the matrices $M^{[k]}$ are defined as follows: $M^{[k]} = [m_{ij}^{[k]}]$ where i, j, k range from 1 to n; $m_{ij}^{[k]} = 0$ if $i \neq j$, $j \neq k$; $m_{jj}^{[k]} = 1$ unless j=k; $m_{ik}^{[k]} = -g_{ik}^{[k-1]}/g_{kk}^{[k-1]}$ unless $i = k$; $m_{kk}^{[k]} = 1/g_{kk}^{[k-1]}$. These recursive expressions for $M^{[k]}$ through $G = G^{[0]}$ involve both subtractions and divisions. Now, in addition to the above sequences of matrices $G^{[k]}, M^{[k]}$, $k = 0, 1, \ldots, n$, consider also the sequence $A^{[k]}, k = 0, 1, \ldots, n$ where $A^{[0]} = A = I - G$, $A^{[k]} = [a_{ij}^{[k]}] = M^{[k]} A^{[k-1]}$ for $k = 1, 2, \ldots, n$. We will exploit the following correlation between the entries $a_{ik}^{[h]}$ and $g_{ik}^{[h]}$ for hk (which do not generally hold for $h \geq k$) : $a_{ik}^{[h]} = -g_{ik}^{[h]}$ unless $i = k$ and $a_{kk}^{[h]} = 1 - g_{kk}^{[h]}$ for $h = 0, 1, 2, \ldots, k-1$ for all k, so $m_{ik}^{[k]} = a_{ik}^{[k-1]}/(1 - a_{kk}^{[k-1]})$ unless $i = k$, $m_{kk}^{[k]} = 1/(1 - a_{kk}^{[k-1]})$. Replacing $1/(1-a)$ by the formal power series $a^* = 1 + a + a^2 + \ldots$ for $a = a_{kk}^{[k-1]}$, $k = 1, \ldots, n$ (such a series converges to $1/(1-a)$ if $|a| < 1$), we arrive at Algorithm 14.4 for computing the matrices $M^{[1]}, M^{[2]}, \ldots, M^{[n]}$ over dioids and at the desired matrix identity

$$B^{[n]} = M^{[n]} M^{[n-1]} \ldots M^{[1]} = I \oplus A \oplus A^2 \oplus \ldots = A^*.$$

This identity involves only additions and multiplications, so, it holds over dioids. ∎

$A^{[n]} = B^{[n]}A$ by the definition of $A^{[n]}$ and $B^{[n]}$ in Algorithm 14.4; therefore, $A^* = A^{[n]} \oplus I$; so, we may dispense with the evaluation of $B^{[k]}$ and simplify Algorithm 14.4 as follows:

ALGORITHM 14.5
(Evaluation of A^*)
 a) Set $A^{[0]} = A$.
 b) For k from 1 to n
$$a_{kk}^{[k]} := (a_{kk}^{[k-1]})^*$$
$$a_{ij}^{[k]} := a_{ij}^{[k-1]} \oplus a_{ik}^{[k-1]} * a_{kk}^{[k]} * a_{kj}^{[k-1]} \text{ for all } i,j \text{ except for } i = j = k$$
 c) Output $A^* = A^{[n]} \oplus I$.

Algorithm 14.5 turns into Algorithm 14.3 for the dioids where $a_{kk}^{(k)} = e$, which is the case, in particular, for the dioids associated with the shortest path problems.

Algorithms 14.4 and 14.5 can be applied to symmetric and nonsymmetric matrices A; we will focus on the symmetric case, where the matrix A is associated with an undirected graph, $G(A)$.

14.3.2 Solving Path Problems for Graphs Given with Small Separators

The Algorithm

In this section, we will consider a path problem for an undirected $s(n)$-separatable graph $G = (V, E)$, where $s(n) = cn^\sigma$, $1/2 < \sigma < 1$, c is a positive constant (see Definitions 14.3 and 14.4 and our comments on precomputation of the separator tree in Section 14.2). Let $A = A(G)$ denote the associated matrix. Then the path computations take form of the computation of A^* and bA^*. In both cases we extend the generalized nested dissection algorithms of Lipton, Rose and Tarjan [54], for sequential computation and of Section 14.2 for parallel computation. In their original form, these algorithms require the inversion of some auxiliary matrices of smaller sizes. Dioid elements and matrices over dioids may have no inverses, but in our extension of the algorithm of Section 14.2, we compute quasi-inverses of the auxiliary matrices over dioids by applying either the algorithms (14.30), (14.31) or Algorithms 14.4 and 14.5 of the previous section.

14.3. Path Algebra Computations in Graphs

To compute A^* in parallel, we *recursively factorize* the matrix $A_0^* = (PAP^T)^*$ (this extends the recursive factorization from (14.12) of Section 14.2 for A_0^{-1}, the inverse of $A_0 = PAP^T$). Here, P denotes the permutation matrix obtained in the ordering stage of the generalized nested dissection algorithm of Section 14.2 applied to the input matrix A. Here and hereafter (as in Section 14.2), W^T denotes the transpose of a matrix W; O denotes the null matrices, filled with the zeros ϵ; and I denotes the identity matrices of (14.27) of appropriate sizes. Here is our extension of the recursive factorization (14.12) of Section 14.2, where $h = 0, 1, \ldots, d-1$, $d = O(\log n)$:

$$A_h = \begin{bmatrix} X_h & Y_h^T \\ Y_h & Z_h \end{bmatrix}, A_{h+1} = Z_h \oplus Y_h X_h^* Y_h^T, \tag{14.32}$$

$$A_h^* = \begin{bmatrix} I & X_h^* Y_h^T \\ O & I \end{bmatrix} \begin{bmatrix} X_h^* & O \\ O & A_{h+1}^* \end{bmatrix} \begin{bmatrix} I & O \\ Y_h X_h^* & I \end{bmatrix}. \tag{14.33}$$

Note that the factorization (14.12) is invalid over dioids due to the use of divisions and subtractions, but our extension is valid. Let us verify that (14.33) indeed defines A_h^*. Expand the right side of (14.33) deleting h and replacing $h+1$ by 1 in the subscripts, to simplify the notation,

$$W = \begin{bmatrix} X^* \oplus X^* Y^T A_1^* Y X^* & X^* Y^T A_1^* \\ A_1^* Y X^* & A_1^* \end{bmatrix}. \tag{14.34}$$

LEMMA 14.10 *(See the proof in the Appendix 14.B and compare Remark 14.8 below.)*
Let $A = \begin{bmatrix} X & Y^T \\ Y & Z \end{bmatrix}$, let there exist the quasi-inverses X^* and $A_1^* = (Z \oplus YX^*Y^T)^*$, and let W be defined by (14.34). Then $WA \oplus I = W$.

Similarly to the proof of Lemma 14.8 (see Appendix 14.B), we deduce from Lemma 14.10 that $W = A^*$ under (14.34) (provided that there exist the quasi-inverses A_1^*, X^*). This immediately substantiates the validity of the recursive factorization (14.32), (14.33). (Note that computing a^* for $a \in S$ may require more than one operation in a dioid unless (14.29) holds; on the other hand, we may compute a^* as $(e-a)^{-1}$ in the dioids that have inverse operations to $\oplus$ and $*$.)

REMARK 14.8
The proofs of Lemmas 14.8 and 14.10 and consequently of the validity of Algorithms 14.4 and 14.5 and of the recursive factorization (14.32), (14.33) do not require that (14.29) be assumed. It is sufficient to use

the definition (14.28) of a quasi-inverse and to assume the existence of all the quasi-inverses included in Algorithms 14.4 and 14.5 and in the recursive factorization (14.32), (14.33). Furthermore, we may modify Algorithm 14.4 (including also the row and column interchanges) in order to ensure the existence of the quasi-inverses of all pivot entries (we may do this unless the quasi-inverse A^* does not exist for a given input matrix A).

REMARK 14.9
The matrix factorization (14.33) can be computed by means of Algorithm 14.4. Indeed, rewrite (14.33) as follows:

$$A_h^* = \begin{bmatrix} I & X_h^* Y_h^T A_{h+1}^* \\ O & A_{h+1}^* \end{bmatrix} \begin{bmatrix} X_h^* & O \\ Y_h X_h^* & I \end{bmatrix}.$$

Let $n = 2$ and let A_h replace A in Algorithm 14.4. Then Algorithm 14.4 computes the latter factorization applied to the 2×2 block matrix A_h. (This application of Algorithm 14.4 is valid because matrix algebras constitute a special class of dioids.)

Computational Cost Estimates

Next, we will estimate the cost of computing $\mathbf{b}A^*$ and A^*. Due to our analysis in Section 14.2, the recursive factorization has been reduced to computing the quasi-inverses X_h^*, the products $Y_h X_h^*$ (which also gives $X_h^*, Y_h^T = (Y_h X_h^*)^T$), and the matrices $A_{h+1}^* = (Z_h \oplus Y_h X_h^* Y_h^T)^*$ for $h = 0, 1, \ldots, d-1$, $d = O(\log n)$. We will proceed similarly to Section 14.2. Then, for each auxiliary matrix C that denotes an $s \times s$ diagonal block of the matrices X_h where $s \leq s(\alpha^h n)$, $\alpha < 1$; $h = k, k-1, \ldots, 0$, $k = O(\log n)$, we need to compute the quasi-inverse C^*. We may extend the assumed property that $A^{(q+1)} = A^{(q)}$ for $q \geq n - 1$ to the equations $C^{(q+1)} = C^{(q)}$ for $q \geq s(\alpha^h n) - 1$. Indeed, A and C are associated with the path problems of the same kind, having only different sizes, n and $s(\alpha^h n)$, respectively.

For faster parallel evaluation of C^*, given C, we may apply the algorithms (14.30), (14.31) cited in the dense matrix case. Then such algorithms will use $O(I(s) \log s)$ parallel steps, $\lceil 2s^3/(I(s) \log s) \rceil$ processors where $s = s(\alpha^h n)$. Alternatively, we may compute C^* by applying Algorithm 14.5; this would involve $O(s^3)$ (sequential) dioid operations or $O(s)$ parallel steps, s^2 processors. In the latter case the computations are arranged similarly to the algorithm of Lipton, Rose and Tarjan [49], except that computing the quasi-inverses X_h^* replaces the Cholesky factorization of X_h for all h.

Other arithmetic operations used in the algorithm of Section 14.2 are additions and multiplications, which are replaced by the similar dioid operations. Combining the above estimates for the complexity of computing the quasi-inverses and matrix products with our analysis of the complexity of recursive factorization (14.12) of Section 14.2, we arrive at the favorable complexity bounds of $O(I(n)\log^2 n)$ parallel steps and $s^3(n)/(I(n)\log n)$ processors for computing the recursive factorization (14.32), (14.33), and of $O(I(n)\log n)$ parallel steps and $((|E|/\log n) + s^2(n))/I(n)$ processors for computing $\mathbf{b}A^*$ for any fixed vector $\mathbf{b}$, provided that the recursive factorization is already available. Here, $|E|$ denotes the number of edges of the graph associated with the matrix A, $|E| = O(n)$ for planar graphs. Therefore, $O(I(n)\log^2 n)$ steps suffice for both single source path problems (where the vector $\mathbf{b}$ is fixed, and only the row $\mathbf{b}A^*$, but not the whole matrix A^*, must be computed); and all pair path problem (where we compute A^*, say, by evaluating $\mathbf{b}A^*$ for all the n coordinate vectors $\mathbf{b}$). In the former case, $(|E| + s^3(n))/(I(n)\log n)$ processors suffice, and in the latter case, we need $n(|E|/\log n + s^2(n))/I(n)$ processors. In a slower parallel algorithm, we need $O(s(n))$ parallel steps and either $s(n)^2 + |E|/s(n)$ processors to solve the single source path problem or $(|E| + s(n)^2 \log n)n$ processors to solve the all pair path problem. By multiplying the latter parallel time and processor bounds together, we arrive at the sequential time bounds of $O(|E| + s^3(n))$ in the case of the single source path problem and of $O((|E| + s^2(n)\log n)n)$ in the case of the all pair path problem.

If $s(n) = O(\sqrt{n})$ and $|E| = O(n)$, as in the case of planar graphs, we arrive at the estimates shown in Table 14.1.

14.3.3 Improvement of Parallel Evaluation of a Minimum Cut and of a Maximum Flow in an Undirected Planar Network

The best sequential algorithms for computing a minimum cut and a maximum flow in an undirected planar network $N = (G, c)$, (where $G = (V, E)$ denotes a graph, c denotes a set of the edge capacities), run in $O(n \log n)$ time and exploit the reduction to the shortest path computations (see Frederickson [23], Hassin and Johnson [39], Reif [84]). Specifically, Reif [84], presented $O(n \log^2 n)$ time algorithm for computing a mincut, Hassin and Johnson [39] extended that algorithm to computing a maximum flow, and Frederickson [23] improved the time bound to $O(n \log n)$. The previous best parallel polylog time algorithms of Johnson [43] for these problems parallelize the sequential scheme and require an order of n^4 processors using polylogarithmic time. Combining our results for the SS in planar graphs with the results of Klein and Reif [50], we arrive at the bounds of $O(I(n)\log^3 n)$ parallel steps and $n^{1.5}/(I(n)\log n)$ or alternatively, $O(\log^3 n)$ and $(n/\log n)^2$. More precisely, we should use the extensions of our results for the SS in *planar digraphs*; such extensions for both SS and AS immediately follow if we replace the matrices Y_h^T for all h by general matrices W_h of the same sizes but defined independently of Y_h (see Remark 14.7).

Johnson [43] computes a mincut and the value $v_{\max}$ of a maxflow using the following stages (see Hassin and Johnson [39], Johnson [43], Klein and Reif [50], and Reif [84] for further details):

1. compute a planar embedding of N, at the estimated cost of $O(\log^2 n)$ steps, n^4 processors;
2. find the planar dual network D(N); step, processor bounds are $O(\log n), n^3$;
3. compute the $\mu - path$, that is, the shortest path in the dual between the two faces F_s and F_t that adjoin the source s and the sink t of the primal network; step, processor bounds are $O(I(n)\log n), n^3$;
4. compute the consistent clockwise orderings for the faces on the $\mu - path$; step, processor count is $O(\log n), n^2$;
5. compute the F-minimum cut-cycles in $D(N)$ for every dual vertex F on the $\mu - path$ (that is, compute the cut-cycles of the minimum length in $D(N)$ passing through the vertex F); the latter stage can be reduced to solving the AP in the dual network $D(N)$; the cost is $O(I(n)\log n)$ steps, n^3 processors;
6. finally, compute the minimum value of the F-minimum cut-cycles over all the dual vertices F on the $\mu - path$; this gives $v_{\max}$ and a mincut; the cost is $O(\log n)$ steps, n processors.

The algorithm of Klein and Reif [50] performs the computation in the substage 1) using $O(\log^2 n)$ steps, n processors; the computation also includes substages 2) and 4) performed using $O(\log n)$ steps, n processors. Applying our parallel algorithm for the SS in stage 3) and our parallel algorithm for the AP in stage 5), we perform the computations in these stages in $O(I(n)\log^2 n)$ steps using $n^{1.5}/(I(n)\log n)$ processors in stage 3) and $n^2/(I(n)\log n)$ in stage 5). Summarizing, we arrive at $O(I(n)\log^2 n)$ steps, $(n/\log n)^2$ processors algorithm for computing $v_{\max}$ and a mincut. Alternatively, we may use $O(I(n)\log^3 n)$ steps and $O(n^{1.5}/(I(n)\log n))$ processors in stage 5); this would dominate the overall complexity. To arrive at these bounds, we apply the algorithm of Reif [84], which performs stage 5) by successively solving SSs in the dual networks derived from D(N). This is performed in at most $\lceil \log_2 n \rceil$ substages; in substage r, up to 2^r SSs are solved in the derived networks, having at most $2|E|+2^r$ edges, $r = 0, 1, \ldots$. Here, E is the edge set of the original planar network, $|E| = O(n)$, $2^r \leq 2^{1+\log n} = 2n$, so the total number of edges in the derived network is $O(n)$ in each substage. Therefore, our algorithm for the SS enables us to perform the computations in each substage using $O(I(n) \log^2 n)$ steps, $n^{1.5}/(I(n) \log n)$ processors, so, $O(I(n) \log^3 n)$ steps, $n^{1.5}/(I(n) \log n)$ processors suffice in all substages of stage 5) and, consequently, for the entire computation of $v_{\max}$ and a mincut.

When a mincut (passing through a vertex F on the $\mu - path$) and the value $v_{\max}$ are known, we may immediately reduce the computation of a maxflow to a SS in the dual network, by following Hassin and Johnson [39]. Specifically, this is the SS of computing the shortest distances between F and all other vertices in the dual network N. Thus, such a final stage only requires $O(I(n) \log^2 n)$ steps, $n^{1.5}/(I(n) \log n)$ processors.

Acknowledgements

This chapter relies on the material of author's papers with J. Reif. The author thanks Prof. Donald B. Johnson for helpful comments, Dr. Kamal Abdali for the useful reprint of his paper (with B.D. Saunders), and also Joan Bentley and Sally Goodall for typing the manuscript.

14.4
Exercises

14.1 a) Reduce INVERT to solving n linear systems.
b) Extend Theorem 14.2 to the complexity estimates for INVERT for sparse matrices.
c) Let A be an $n \times n$ nonsingular triangular matrix. Solve the linear system (14.5) using $O(n^2)$ arithmetic operations. Then solve it using $O(\log^2 n)$ steps and $M(n)$ processors.
d) Deduce the same estimates $O(\log^2 n)$ and $M(n)$ for computing $A^i \mathbf{v}, i = 1, 2, \ldots, n$, given an $n \times n$ matrix A and a vector $\mathbf{v}$ (see Keller-Gehrig [49], Borodin and Munro [14], page 128).

14.2 Verify Fact 14.1, parts a) and b). Extend them to the class of "near planar" graphs, which can be represented on the plane with $O(1)$ crossovers, Stearns and Hunt [91], p. 21.

14.3 Verify the estimates of Example 14.1.

664 Chapter 14. Parallel Solution of Sparse Linear and Path Systems

14.4 Define the elimination graphs for Gaussian elimination with no ordering and with minimum degree ordering for:

  a) a tridiagonal linear system,
  b) the linear system of Example 14.1,
  c) the linear system with the graph $G = (V, E)$, $V = \{1, 2, 3, 4, 5\}$, $E = \{\{1,2\}, \{1,6\}, \{2,3\}, \{2,4\}, \{3,5\}, \{5,6\}\}$,
  d) the linear systems with the 2×2 and 3×3 grid graphs.

14.5 Define $\sqrt{n}$-separator families for 2- and 3-dimensional grids.

14.6 Verify the estimates of Gilbert and Tarjan [33] for the number of arithmetic operations and for the fill-in (see Section 14.2.3) in the case of grid graphs.

14.7 Use the factorization (14.12) in order to reduce INVERT for a $(2n) \times (2n)$ matrix A_h to INVERT for a pair of $n \times n$ matrices and to few $n \times n$ matrix multiplications and subtractions. Recursively apply this reduction to deduce that $O_A(INVERT) \leq O_A(MULT)$ provided that $O_A(MULT)$ exceeds n^ω for $\omega > 2$ and $O_A(Problem)$ denotes the number of arithmetic operations required to solve the Problem. Observe that this reduction only implies a linear parallel time bound for INVERT.

14.8 Premultiply and then postmultiply unit coordinate vectors and then some random vectors of appropriate sizes by matrices $\begin{bmatrix} 0 & 1 \\ 1 & 0 \end{bmatrix}$, $\begin{bmatrix} 0 & 1 & 0 \\ 0 & 0 & 1 \\ 1 & 0 & 0 \end{bmatrix}$ and then by some random permutation matrices (filled with zeros and ones and such that every row and every column have exactly one entry filled with ones). Observe how this interchanges the coordinates of the vectors.

14.9 Deduce Lemma 14.3 from the interlacing properties of the eigenvalues of a symmetric matrix.

14.10 Let $f(n) = 4f(n/2) + kn^2 + O(n)$, $g(n) = g(n/2) + kn^2 \log_2 n + O(n^2)$, $h(n) = 2h(n/2) + kn^2 \log_2 n + O(n^2)$. Deduce that $f(n) = kn^2 \log_2 n + O(n)$, $g(n) = (4/3)kn^2 \log_n n + O(n^2)$, $h(n) = 2kn^2 \log_2 n + O(n^2)$.

14.11 Extend the factorizations (14.6) and (14.9)–(14.12) to the case of nonsymmetric input matrices.

14.12 Extend Definitions 14.1 and 14.3 to the nonsymmetric matrices A and associated digraphs.

14.13 Extend Theorem 14.2 (to the nonsymmetric case) omitting the numerical stability statements and Lemma 14.3.

14.14 Verify that all $n \times n$ matrices over a dioid (semiring) form a dioid (semiring).

14.15 Deduce the equation (14.26) of Section 14.3 for the AP shortest path problem (similarly to the derivation of the equation (14.25)). Similarly deduce the equations (14.25) and (14.26) for other path problems listed after Definition 14.8.

14.16 Deduce (14.30) from (14.29).

14.17 Deduce (14.31) from the assumption of idempotency, $a \oplus a = a$.

14.18 Verify all the details of the proofs of Lemmas 14.8 and 14.10.

14.19 The *eliminant* $|A|$ of an $n \times n$ matrix $A = [a_{ij}]$, $i, j = 1, \ldots, n$, is explicitly defined for $n = 1$ and 2, $|A| = a_{11}$ for $n = 1$, $|A| = a_{22} + a_{21}a_{11}^* a_{12}$ provided that $a^* = aa^* + 1 = a^*a + 1$. For $n = 3$, $|A| = |B|$ where $B = [b_{ij}]$, $i, j = 1, \ldots, n-1$, $b_{ij} = \begin{vmatrix} a_{11} & a_{1,j+1} \\ a_{i+1,1} & a_{i+1,j+1} \end{vmatrix}$. $\tilde{A}_{ij}^k$, the k,i,j- *select* of A for $i, j = 1, \ldots, n$, $k = 0, 1, \ldots, n$, is said to be the eliminant of the $(k+1) \times (k+1)$ matrix

$$\begin{bmatrix} a_{11} & \cdots & a_{1k} & a_{1j} \\ \cdots & \cdots & \cdots & \cdots \\ a_{k1} & \cdots & a_{kk} & a_{kj} \\ a_{i1} & \cdots & a_{ik} & a_{ji} \end{bmatrix}.$$

Prove that

a) $\tilde{A}_{nn}^{n-1} = |A|$,

b) $|A| = |\tilde{A}_{ij}^1|$ where $i, j = 1, \ldots, n$;

c) $|A| = |B|$ if $B = [\tilde{A}_{r+i,r+j}^r]$, $i, j = 1, \ldots, n-r$;

d) $A^* = |B|$ where A^* is defined by (14.28) and (14.29), $B = [b_{ij}]$, $i, j = 1, \ldots, n$, $b_{ij} = \begin{bmatrix} A & \mathbf{u}_j \\ \mathbf{u}_i^T & 0 \end{bmatrix}$, and $\mathbf{u}_k$ denotes the k-th unit coordinate vector, $\mathbf{u}_k^T = [0, \ldots, 0, 1, \ldots, 0]$, with k-th coordinate equal to 1.

14.20 Try to devise an algorithm for computing A^* based on the latter result d) of Exercise 14.19 and also deduce the equations $A^* = AA^* \oplus I = A^*A \oplus I$ using this result d).

Notes and References

Atkinson [4] and Isaacson and Keller [42] are good general texts on numerical computing. On computing with general matrices, see Golub and van Loan [34], Wilkinson [95], Pan [64, 65, 66, 67], Bini and Pan [11], Pan and Reif [70, 77], Coppersmith and Winograd [18], Ben-Israel [7], Hotelling [40, 41], Ben-Israel and Cohen [8], Bodewig [12], Newman [61], Borodin and Munro [14], Aho, Hopcroft and Ullman [2], Csanky [19], Galil and Pan [26], Preparata and Sarwate [83], Schreiber [89], Young [97], Varga [94].

On sequential computations with sparse matrices, see the general texts George and Liu [31], Pissanetsky [82] and the papers George [32], Lipton, Rose and Tarjan [54], Liu [57], Schreiber [89], and Rose [88]. On parallel computations with sparse matrices, see Ortega and Voight [63], Calahan [15], Gannon [27], Leiserson et al. [53], Liu [57], Opsahl and Reif [62], Pan and Reif [70, 76], Hafsteinson [36]. Also, on the graph-theoretical computations related to our sparse matrix computations, see Miller, Teng, and Vavasis [58], Djidjev [20], Bhatt and Leiserson [10], Bhatt and Leighton [9], Lipton and Tarjan [55], Gazit and Miller [28, 29], Klein and Reif [50], Matzumoto et al. [59]. Opsahl and Reif [62] and Leiserson et al. [53] describe two implementations of the parallel nested dissection algorithm of Section 14.2.

There are numerous surveys treating path problems algebraically, by using dioids (semirings) (see, in particular, Yoeli [96], Robert and Ferland [87], Backhouse and Carre [5], Carre [17], Tarjan [92, 93], Gondran and Minoux [35]; also see Aho, Hopcroft and Ullman [2], Pierce [81], Lehmann [52]). Besides, some specialized applications of path algebras have been studied in McNaughton and Yamada [60], Kam and Ullman [47].

The alternate algorithms for path problems have been cited already (see Bellman [6], Floyd [22], Frederickson [23], Hassin [38], Hassin and Johnson [39], Johnson [43]).

There are a number of recent improvements to parallel nested dissection. Armon and Reif [3] reduce the space bounds of parallel nested dissection algorithms to linear, with no asymptotic increase in time or processor bounds. A *pipeline* is a sequence of parallel processors, with a linear stream of data passing through the processors, where each data item is processed in sequence by successive processors. Pan and Reif [75] speed up the $O(\log^2 n)$ time parallel nested dissection algorithm for all pair min path problems in undirected graphs to $O(\log n)$ time, with no asymptotic increase in processor bounds. They use a technique called *stream contraction*, which pipelines the logarithmic recursive levels of nested dissection. Han, Pan and Reif [37] generalize parallel nested dissection algorithms to all pair path problems in directed graphs.

Bibliography

[1] Abdali, S.K., and B.D. Saunders, Transitive Closure and Related Semiring Properties via Elimination, *Theoretical Computer Science* 40, 257-274 (1985).

[2] Aho, A.V., J.E. Hopcroft, and J.D. Ullman, *The Design and Analysis of Computer Algorithms*, Addison-Wesley (1976).

[3] Armon, D., and J.H. Reif, An Optimal Space and Efficient Parallel Nested Dissection Algorithm, *4th Annual ACM Symposium on Parallel Algorithms and Architectures*, San Diego, CA, July 1992.

[4] Atkinson, K.E., *An Introduction to Numerical Analysis*, Wiley, New York (1984).

[5] Backhouse, R.C., and B.A. Carre, Regular Algebra Applied to Pathfinding Problems, *J. Inst. Math. Applics* 15 (1975), 161–186.

[6] Bellman, R., On a Routing Problem, *Quart. Appl. Math.* 16 (1958), 87–90.

[7] Ben-Israel, A., Note on Iterative Method for Generalized Inversion of Matrices, *Math. Computation,* 20, 439-440 (1966).

[8] Ben-Israel, A., and D. Cohen, Iterative Computation of Generalized Inverses and Associated Projections, *SIAM J. on Numerical Analysis* 3, 410–419 (1966).

[9] Bhatt, S.N., and F.T. Leighton, A Framework for Solving VLSI Graph Layout Problems, *J. of Computer and Systems Sciences*, 28(2), 300–343 (1984).

[10] Bhatt, S.N., and C.E. Leiserson, How to Assemble Tree Machines, *Advances in Computing Research* 2, 95–114 (1984).

[11] Bini, D., and V. Pan, *Numerical and Algebraic Computations with Matrices and Polynomials*, Birkhauser, Boston (1992).

[12] Bodewig, E., *Matrix Calculus*, Second Edition, Interscience Publishers, Inc., New York; North-Holland Company, Amsterdam (1959).

[13] Borodin, A., J. von zur Gathen, and J. Hopcroft, Fast Parallel Matrix and GCD Computations, *Proc. 23rd Ann. IEEE FOCS*, 65–71 (1982) and *Information and Control*, 53(3), 241–256 (1982).

[14] Borodin, A., and I. Munro, *The Computational Complexity of Algebraic and Numeric Problems*, American Elsevier, New York (1975).

[15] Calahan, D.A., Parallel Solution of Sparse Simultaneous Linear Equations, *Proc. 11th Allerton Conf.*, 729–738 (1973).

[16] Carre, B.A., An Algebra for Network Routing Problems, *J. Inst. Math. Applics* 7, (1971), 273–294.

[17] Carre, B.A., *Graphs and Networks*, The Clarendon Press, Oxford University Press (1979).

[18] Coppersmith, D., and S. Winograd, Matrix Multiplication via Arithmetic Progressions, *Proc. 19th Ann. ACM Symp. on Theory of Computing*, 1–6 (1987), *J. of Symbolic Computation* 9 (3), 251–280 (1990).

[19] Csanky, L., Fast Parallel Matrix Inversion Algorithms, *SIAM J. on Computing* 5(4), 618–623 (1976).

[20] Djidjev, H.N., The Problem of Partitioning Planar Graphs, *SIAM J. Algebraic Discrete Methods*, 3, 229–240 (1982).

[21] Eppstein, D., and Z. Galil, Parallel Algorithmic Techniques for Combinatorial Computation, *Annual Review of Computer Science*, 3, 233–283 (1988).

[22] Floyd, R.N., Algorithm 97, Shortest Path, *Comm. ACM* 5 (1962), 345.

[23] Frederickson, G.N., Fast Algorithms for Shortest Paths in Planar Graphs, with Applications, *SIAM J. on Computing* 16, 6 (1987), 1004–1022.

[24] Galil, Z., and V. Pan, Improving Processor Bounds for Algebraic and Combinatorial Problems in RNC, *Proc. 26th Ann. IEEE Symp. on Foundation of Computer Sci.*, 490–495, Portland, Oregon (1985).

[25] Galil, Z., and V. Pan, Improved Processor Bounds for Combinatorial Problems in RNC, *Combinatorica*, 8,2, 189–200 (1988).

[26] Galil, Z., and V. Pan, Parallel Evaluation of the Determinant and of the Inverse of a Matrix, *Inf. Proc. Letters*, 30, 41–45 (1989).

[27] Gannon, D.A., Note on Pipelining a Mesh-Connected Multiprocessor for Finite Element Problems by Nested Dissection, *Proc. Int. Conf. Par. Proc.*, 197–204 (1980).

[28] Gazit, H., and G.L. Miller, An $O(\sqrt{n}\log(n))$ Optimal Parallel Algorithm for a Separator for Planar Graphs, Manuscript, Computer Science Dept., USC, Los Angeles (1986).

[29] Gazit, H., and G.L. Miller, A Parallel Algorithm for Finding a Separator in Planar Graphs, *Proc. 28th Ann. IEEE Symp. FOCS*, 238–248 (1987).

[30] Gazit, H., and J. Reif, A Randomized Parallel Algorithm for Planar Graph Isomorphism, *Proc. 2nd Ann. ACM Symp. on Parallel Algorithms and Architecture*, 210–219 (1990).

[31] George, A., and Liu, J.W.-H., *Computer Solution of Large Sparse Positive Definite Systems*, Englewood Cliffs, NJ, Prentice-Hall (1981).

[32] George, J.A., Nested Dissection of a Regular Finite Element Mesh, *SIAM J. on Numerical Analysis* 10(2), 345–367 (1973).

[33] Gilbert, J.R., and R.E. Tarjan, The Analysis of a Nested Dissection Algorithm,. *Numer. Math.*, 50, 377–404 (1987).

[34] Golub, G.H., and C.F. van Loan, *Matrix Computations*, The Johns Hopkins University Press, Baltimore (1983, 1989).

[35] Gondran, M., and M. Minoux, *Graphs and Algorithms*, Wiley-Interscience, New York (1984).

[36] Hafsteinson, H., *Parallel Sparse Cholesky Factorization*, Ph.D. Thesis, Computer Science Dept., Cornell University, Ithaca, NY (1988).

[37] Han, Y., V. Pan, and J.H. Reif, Efficient Parallel Algorithms for Computing All Pair Shortest Paths in Directed Graphs (with Y. Han and V. Pan). University of Kentucky Technical Report 204-92. *4th Annual ACM Symposium on Parallel Algorithms and Architectures*, San Diego, CA, July 1992.

[38] Hassin, R., On Multicommodity Flows in Planar Graphs, *Networks* 14, (1985), 225–235.

[39] Hassin, R., and D.B. Johnson, An $O(n \log^2 n)$ Algorithm for Maximum Flow in Undirected Planar Networks, *SIAM J. on Computing* 14, 3 (1985) 612–624.

[40] Hotelling, H., Some New Methods in Matrix Calculation, *Ann. Math. Statist.* 14, 1–34 (1943a).

[41] Hotelling, H., Further Points on Matrix Calculations and Simultaneous Equations, *Ann. Math. Statist.* 14, 440–441 (1943b).

[42] Isaacson, E., and H.B. Keller, *Analysis of Numerical Methods*, Wiley, New York (1966).

[43] Johnson, D.B., Parallel Algorithms for Minimum Cuts and Maximum Flows in Planar Networks, *J. ACM* 34, 4 (1987) 950–967.

[44] Kaltofen, E., Greatest Common Divisor of Polynomials Given by Straight-Line Program, *J. of ACM*, 35, 1, 231–264 (1988).

[45] Kaltofen, E., and V. Pan, Processor Efficient Parallel Solution of Linear Systems over an Abstract Field, *Proc. 3rd Annual ACM Symp. on Parallel Algorithms and Architecture*, 180–191 (1991).

[46] Kaltofen, E., and V. Pan, Processor Efficient Parallel Solution of Linear Systems II. The Positive Characteristic and Singular Cases, *Proc. 33rd Ann. IEEE Symp. FOCS* (1992).

[47] Kam, J. B., and J. D. Ullman, Global Data Flow Analysis and Iterative Algorithms, *J. ACM* 23 (1) (1976) 158–171.

[48] Karp, R.M., and V. Ramachandran, A Survey of Parallel Algorithms for Soared Memory Machines, *Handbook of Theoretical Computer Science*, 869–941 (1990), North Holland, Amsterdam.

[49] Keller-Gehrig, W., Fast Algorithms for Characteristic Polynomial, *Theoretical Computer Science*, 36, 309–317 (1985).

[50] Klein, P., and J. Reif, An Efficient Parallel Algorithm for Planarity, *Proc. 27th Ann. IEEE Symp. FOCS* (1986), 465–477, invited to appear in *JCSS*.

[51] Lawler, E.L., *Combinatorial Optimization: Networks and Matroids*, Holt, Rinehart and Winston, N.Y. (1976).

[52] Lehmann, D. J., Algebraic Structures for Transitive Closure, *Theoret. Comput. Sci.* 4 (1) (1977) 59–76.

[53] Leiserson, C.E., J.P. Mesirov, L. Nekludova, S. Omohundro, and J. Reif, Solving Sparse Systems of Linear Equations on the Connection Machine, *Ann. SIAM Conference*, Boston, MA (1986).

[54] Lipton, R.J., D. Rose, and R.E. Tarjan, Generalized Nested Dissection, *SIAM J. Numer. Analysis* 16, 2 (1979), 346–358.

[55] Lipton, R.J., and R.E. Tarjan, A Separator Theorem for Planar Graphs, *SIAM J. Applied Math.* 36, 2 (1979), 177–189.

[56] Liu, J.W.-H. The Solution of Mesh Equations on a Parallel Computer, Technical Report CS-78-19, Department of Computer Science, University of Waterloo (1978).

[57] Liu, J.W.-H., Compact Row Storage Scheme for Cholesky Factors Using Elimination Trees, *ACM TOMS* 12(2), 127–148 (1986).

[58] Miller, G.L., S.-H. Teng, and S.A. Vavasis, A Unified Geometric Approach to Graph Separators, *32nd Ann. IEEE Symp. on FOCS*, 538–547 (1991).

[59] Matsumoto, K., T. Nishizeki, and N. Saito, An Efficient Algorithm for Finding Multicommodity Flows in Planar Networks, *SIAM J. on Computing* 14, 2 (1985), 289–302.

[60] McNaughton, R., and H. Yamada, Regular Expressions and State Graphs for Automata, *IRE Trans. EC9* (1960) 39–47.

[61] Newman, M., Matrix Computation, *Survey of Numerical Analysis,* 222–255 (S. Todd, Ed.), McGraw Hill, New York (1982).

[62] Opsahl, T., and J. Reif, Solving Very Large Sparse Systems on Mesh-Connected Computers, Manuscript (1986).

[63] Ortega, J.M., and R.G. Voight, Solution of Partial Differential Equations on Vector and Parallel Computers, *SIAM Review* 27(2), 149–240 (1985).

[64] Pan, V., *How to Multiply Matrices Faster*, Lecture Notes in Computer Science, 179, Springer-Verlag, Berlin (1984).

[65] Pan, V., How Can We Speed Up Matrix Multiplication?, *SIAM Review*, 26(3), 393–415 (1984a).

[66] Pan, V., Complexity of Parallel Matrix Computations, *Theoretical Computer Science*, 54, 65–85 (1987).

[67] Pan, V., Linear Systems of Algebraic Equations, *Encyclopedia of Physical Sciences and Technology*, Marvin Yelles, editor, vol. 7, 304–329, Academic Press, New York (1987a).

[68] Pan, V., Parametrization of Newton's Iteration for Computations with Structured Matrices and Applications, Tech. Report CUCS-032-90, Columbia University, Computer Science Dept. (1990) and *Computers and Math. (with Applications), to appear in 1992.*

[69] Pan, V.Y., and F.P. Preparata, Supereffective Slow-down of Parallel Computations, *Proc. 4th Ann. ACM Symposium on Parallel Algorithms and Architecture*, 402–409 (1992).

[70] Pan, V. and J. Reif, Efficient Parallel Solution of Linear Systems, *Proceedings of the 17th Annual ACM Symposium on Theory of Computing*, Providence, RI, May 1985, ACM-SIGACT, pp. 143–152.

[71] Pan, V., and J. Reif, Extension of the Parallel Nested Dissection Algorithm to the Path Algebra Problems, Tech. Rep. 85-9, Comp. Sci. Dept., SUNY, Albany, NY (1985a).

[72] Pan, V., and J. Reif, Parallel Nested Dissection for Path Algebra Computations, *Operations Research Letters* 5(4), 177–184 (1986a).

[73] Pan, V., and J. Reif, Efficient Parallel Linear Programming, *Operations Research Letters*, 5(3), 127–135 (1986b).

[74] Pan, V., and J. Reif, Fast and Efficient Algorithms for Linear Programming and for the Linear Least Squares Problem, *Computers & Mathematics*, 12A(12), 1217–1227 (1986c).

[75] Pan, V., and J. Reif, Parallel Computations of the Minimum Cost Paths in Graphs by Stream Contraction, Tech. Rep. 87-10, Computer Science Dept., SUNY Albany, NY (1987) and *Information Processing Letters*, 40 (1991) 79–83.

[76] Pan, V., and J. Reif, Fast and Efficient Parallel Solution of Sparse Linear Systems, Tech. Report TR 88-18, Computer Science Department, SUNY Albany (1988) and *SIAM J. on Computing*, (to appear), 1992.

[77] Pan, V. and J. Reif, Fast and Efficient Parallel Solution of Dense Linear Systems. *Computers and Mathematics with Applications*, Vol. 17, no. 11, pages 1481–1491, 1989.

[78] Pan, V., and J. Reif, Fast and Efficient Solution of Path Algebra Problems, *J. Comp. and System Sciences*, 38, 494–510 (1989a).

[79] Pan, V., and R. Schreiber, Improved Newton Iteration for the Generalized Inverse of a Matrix, with Applications, *SIAM J. Sci. Stat. Comp.* 12, 5 (1991).

[80] Parlett, B.N., *The Symmetric Eigenvalue Problem*, Prentice-Hall (1980).

[81] Pierce, A. R., Bibliography on Algorithms for Shortest Paths, Shortest Spanning Tree and Related Circuit Routing Problems, 1956–1974, *Networks* 5 (1975) 129–149.

[82] Pissanetsky, S., *Sparse Matrix Technology*, Academic Press, New York (1984).

[83] Preparata, F.P., and D.V. Sarwate, Improved Parallel Processor Bound in Fast Matrix Inversion, *Information Processing Letters* 7(3), 148–149 (1978).

[84] Reif, J., Minimum s-t Cut of a Planar Undirected Network in $O(n \log^2(n))$ Time, *SIAM J. Comput.* 12, 1 (1983), 71–81.

[85] Reischuck, R., A Fast Probabilistic Parallel Sorting Algorithm, *Proc. 22nd Ann. IEEE FOCS*, 212–219 (1981).

[86] Rice, J.R., *Numerical Methods, Software and Analysis*, McGraw-Hill (1983).

[87] Robert, P., and J. Ferland, Generalisation de L'algorithme de Warshall, *RAIRO Inform. Theor.* 2 (1968) 71–85.

[88] Rose, D.J., Graph-theoretic study of the numerical solution of sparse positive definite systems of linear equations, *Graph Theory and Computing*, edited by R. Read, 183–217 (1972), New York, Academic Press.

[89] Schreiber, R., New Implementation of Sparse Gaussian Elimination, *ACM Trans. on Math. Software*, 8, 256–276 (1982).

[90] Schreiber, R., Computing Generalized Inverses and Eigenvalues of Symmetric Matrices Using Systolic Arrays, *Computing Methods in Applied Science and Engineering* (R. Glowinski and J.-L. Lions, eds.), 285–295 (Elsevier Science, North Holland) (1984).

[91] Stearns, R.E., and H.B. Hunt, III, Lower Indices and Easier Hard Problems, *Math. Systems Theory* 23, 209–225, (1990).

[92] Tarjan, R.E., A Unified Approach to Path Problems, *J.ACM* 28, 3 (1981a), 577–593.

[93] Tarjan, R.E., Fast Algorithms for Solving Path Problems, *J. ACM* 28, 3 (1981b), 594–614.

[94] Varga, R. S., *Matrix Iterative Analysis*, Prentice-Hall (1962).

[95] Wilkinson, J.H., *The Algebraic Eigenvalue Problem*, Clarendon Press, Oxford (1965).

[96] Yoeli, M., A Note on a Generalization of Boolean Matrix Theory, *Amer. Math. Monthly* 68 (1961) 552–557.

[97] Young, D.M., *Iterative Solution of Large Linear Systems*, Academic Press, New York (1971).

[98] Zimmerman, V., *Linear and Combinatorial Optimization in Ordered Algebraic Structures*, Annals of Discrete Math., 10, North-Holland, Amsterdam (1981).

Appendix 14.A
Bounding the Size of the Fill-in in the Generalized Nested Dissection Algorithm

We will prove the following lemma (compare Lipton, Rose, and Tarjan [54], Theorem 2):

LEMMA 14.11
The total size of the fill-in in all stages of computing the recursive $s(n)$-factorization (14.9)–(14.12) is

$$|\cup_h (E_h - E_{h-1})| = O(n + s(n)^2 \log n),$$

where E_h denotes the edge set of the graph $G(A_h)$.

PROOF
We will prove the lemma in the case where $cs^2(n)\log n \geq n$ for a constant c. The number of fill-in edges with endpoints in $R_{h,k}$ is at most $s(n)^2 + s(n)m$ (here and hereafter in this appendix we use the notation of Sections 14.2.5–14.2.8; all logarithms will be to the base 2).

Let $F(n,m)$ denote the number of all the fill-in edges of $\cup_{h'\leq h}(E_{h'} - E_{h'-1})$ with endpoints in $V_{h,k}$. Then $F(n,m) \leq n(n-1)/2$ if $n \leq n_0$ and, otherwise,

$$F(n,m) \leq s(n)^2/2 + s(n)m + F(n_1, m_1) + F(n_2, m_2). \tag{14.35}$$

Lemma 14.11 immediately follows from the next claim (consider the case where $m = 0$). ∎

CLAIM 14.2
For all $n \geq 1$ and for all m, $F(n,m) \leq C_1(m + s(n)^2)\log n + C_2 s(n)m$ for some positive constants C_1, C_2.

PROOF OF CLAIM 14.2
We will proceed by induction on n. Let Claim 14.2 hold for all smaller n. Then the induction hypothesis and (14.35) imply that

$$\begin{aligned}F(n,m) \leq\;& (s(n)^2/2) + ms(n)\\&+ C_1((m_1 + s(n_1)^2)\log n_1\\&+ (m_2 + s(n_2)^2)\log n_2)\\&+ C_2(m_1 s(n_1) + m_2 s(n_2)).\end{aligned} \tag{14.36}$$

676 Chapter 14. Parallel Solution of Sparse Linear and Path Systems

Recall (14.15) and (14.19) and deduce that

$$\sum_{i=1}^{2}(m_i + s(n_i)^2)\log n_i \le \sum_{i=1}^{2}(m_i + s(n_i)^2)\log\max(n_1, n_2)$$

$$\le (\sum_i m_i + \sum_i s(n_i)^2)\log(\beta n) \quad (14.37)$$

$$\le (m + s(n)^2 + 3s(n))\log(\beta n), \beta < 1.$$

Indeed, $m_1 + m_2 \le m + 2s(n)$ (see (14.19)), and $s(n_1)^2 + s(n_2)^2 \le c(n_1^{2\sigma} + n_2^{2\sigma})$ where $2\sigma \ge 1$ (see (14.13)). Therefore $s(n_1)^2 + s(n_2)^2 \le c((\alpha n + s(n))^{2\sigma} + ((1-\alpha)n)^{2\sigma})$. The latter value equals $c(n + s(n)) = s(n)^2 + s(n)$ if $2\sigma = 1$ and is less than $cn^{2\sigma} = s(n)^2$ if $2\sigma > 1$, $n \ge n_0$, and n_0 is large enough. Thus

$$s(n_1)^2 + s(n_2)^2 \le s(n)^2 + s(n),$$

and (14.37) follows. Furthermore

$$m_1 s(n_1) + m_2 s(n_2) \le \max s(n_1), s(n_2)(m_1 + m_2) \quad (14.38)$$
$$\le s(\alpha n + s(n))(m + 2s(n)).$$

Combining (14.36)–(14.38) we obtain that

$$F(n, m) \le s(n)^2/2 + s(n)m + C_1(m + s(n)^2 + 3s(n))\log(\beta n)$$
$$+ C_2 s(\alpha n + s(n))(m + 2s(n)).$$

Note that $s(\alpha n + s(n)) \le s(\alpha, n) + c_0 n^{2\sigma - 1} \le s(\alpha n) + c_0 = \alpha^\sigma s(n) + c_0$ for a constant c_0 (see (14.8)) and deduce that

$$F(n, m) \le C_1(m + s(n)^2)\log n + (C_2\alpha^\sigma + 1)s(n)m$$
$$+ (s(n)^2/2) + 3C_1 s(n)\log n + 2C_2\alpha^\sigma s(n)^2 \quad (14.39)$$
$$+ C_2 c_0(m + 2s(n)) - C_1(m + s(n)^2)\log(1/\beta).$$

Assume that n_0 is large enough, so that for all $n \ge n_0$, $3s(n)\log n \le (1/2)s(n)^2 \log(1/\beta)$; choose $C_2 \ge \frac{1}{1-\alpha^\sigma}$, so that $C_2\alpha^\sigma + 1 < C_2$. Similarly choose C_1 large enough, so that

$$(1/2)C_1 s(n)^2 \log(1/\beta) \ge s(n)^2((1/2) + 2C_2\alpha^\sigma) + 2C_2 c_0 s(n)$$

and

$$C_1 m \log(1/\beta) \ge C_2 c_0 m,$$

and deduce Claim 14.2 from the inequality (14.39). ∎

Appendix 14.B
Proof of Lemmas 14.8 and 14.10

PROOF OF LEMMA 14.8

It is immediately verified that $B^{[n]} = A^*$ is the unique solution A^* of (14.26) in the case of the special dioid where S is the set of real matrices, $\oplus = +, * = *, \epsilon = 0, e = 1,$

$$|a_{kk}^{[k-1]}| < 1 \text{ for all } k, \tag{14.40}$$

and, say

$$n \max_{i,j} |a_{ij}| < 1. \tag{14.41}$$

The latter inequality immediately implies that $A^* = \sum_{h=0}^{\infty} A^h$ converges to $(I-A)^{-1}$, whereas $|a_{kk}^{[k-1]}| < 1$ implies that $(a_{kk}^{[k-1]})^* = \sum_{h=0}^{\infty} (a_{kk}^{[k-1]})^h$ converges to $(1 - a_{kk}^{[k-1]})^{-1}$.

Let us consider again an arbitrary dioid $(S, \oplus, *)$ where v^* is defined as the formal power series, $\oplus_{h=0}^{\infty} v^h$, $v^0 = e$ if $v \in S$, $v^0 = I$ if v is a matrix with the entries from S. Then the entries of $B^{[n]}$ and A^* are multivariate power series in the entries of A. The numerical values of the two power series representing the (i,j)-entries of $B^{[n]}$ and A^* for an arbitrary pair i, j must coincide with each other on any real matrix A such that (14.40) and (14.41) hold. It follows that such two power series coincide with each other also as formal power series. ∎

PROOF OF LEMMA 14.10

Let $X = X_h, Y = Y_h, Z = Z_h$ and let (14.27), (14.32), (14.34) define $I, A = A_h, A_1 = A_{h+1}, W$. Then the matrix $WA \oplus I$ has the upper left block

$$I \oplus X^*X \oplus X^*Y^T A_1^* Y X^*X \oplus X^*Y^T A_1^* Y$$
$$= (I \oplus X^*Y^T A_1^* Y)(X^*X \oplus I)$$
$$= X^* \oplus X^*Y^T A_1^* Y X^*$$

since $X^*X \oplus I = X^*$; has the upper right block

$$X^*Y^T \oplus X^*Y^T A_1^* Y X^*Y^T \oplus X^*Y^T A_1^* Z$$
$$= X^*Y^T \oplus X^*Y^T A_1^* (YX^*Y^T \oplus Z)$$
$$= X^*Y^T \oplus X^*Y^T A_1^* A_1$$
$$= X^*Y^T (I \oplus A_1^* A_1)$$
$$= X^*Y^T A_1^* \text{ since } I \oplus A_1^* A_1 = A_1^*;$$

has the lower left block

$$A_1^* Y X^* X \oplus A_1^* Y = A_1^* Y (X^* X \oplus I) = A_1^* Y X^*,$$

and has the lower right block

$$I \oplus A_1^* Y X^* Y^T \oplus A_1^* Z = I \oplus A_1^* (Y X^* Y^T \oplus Z) = I \oplus A_1^* A_1 = A_1^*.$$

Compare all this with the blocks of the matrix W of (14.34). ∎

15

Parallel Resultant Computation

Doug Ierardi

Department of Computer Science
University of Southern California
Los Angeles, California 90089
ierardi@flash.usc.edu

Dexter Kozen

Department of Computer Science
Cornell University
Ithaca, New York 14853
kozen@cs.cornell.edu

15.1
Introduction

The subject of this chapter is the computation of *resultants*. A resultant is a purely algebraic criterion for determining when a finite collection of polynomial equations has a common solution, expressed in terms of the coefficients of these polynomials. The investigation of such criteria belongs historically to the branch of mathematics known as Elimination Theory, the goal of which was to solve systems of polynomial equations by successive elimination of variables. Fundamental aspects of this project were developed by Hermann, Hilbert, Kronecker, Lasker, Macaulay and Noether at the turn of this century, marking the beginning of a fusion of algebra and geometry which later found fuller expression in the development of algebraic geometry.

Much of the fundamental work in Elimination Theory was pursued at a time when constructive methods in mathematics prevailed. In fact, the ostensible goal of the theory—solving systems of polynomial equations—had such obvious practical significance that the efficiency of algorithms was already a concern. Macaulay expressed this point of view in his 1914 tract, *The Algebraic Theory of Modular Systems*, when he wrote that the current body of knowledge in Elimination Theory

> might be regarded as in some measure complete if it were admitted that a problem is finished with when its solution has been reduced to a finite number of feasible operations. If however the operations are too numerous or too involved to be carried out in practice the solution is only a theoretical one; and its importance then lies not in itself, but in the theorems with which it is associated and to which it leads. Such a theoretical solution must be regarded primarily as a preliminary and not the final stage in the consideration of the problem.

The study of algorithms in elimination theory has not yet reached its "final stage": provably optimal algorithms are still lacking. Nevertheless, contemporary ideas in algorithm design and complexity are continually being brought to bear, and the most efficient sequential and parallel algorithms for these problems have been discovered during the last decade.

The theory of resultants rests on the well-known Nullstellensatz of Hilbert, which relates the algebra of the ring of polynomials over indeterminates $x_1, \ldots, x_n$ with coefficients in an algebraically closed field k (denoted $k[x_1, \ldots, x_n]$) and the geometry of the space k^n of n-tuples of elements of k. For example, over the complex numbers $\mathbf{C}$, the Nullstellensatz asserts that m polynomials $f_1, \ldots, f_m$ with complex coefficients have no common solutions in $\mathbf{C}^n$ exactly when there are m additional polynomials $g_1, \ldots, g_m$ such that

$$f_1 g_1 + f_2 g_2 + \cdots + f_m g_m = 1.$$

The foundation of elimination theory lies in the fact that the existence or nonexistence of these g_i's can be determined solely by examining the coefficients of the f_i's, and that an algebraic criterion can be constructed uniformly once the degrees of these polynomials are specified. From this observation the notion of a *resultant* arose—a polynomial in the coefficients of the given polynomials which vanishes exactly when a common solution exists.

Stated in this way, it might seem that the resultant yields no more than a decision procedure for the existence of solutions; but the fact that it provides a *purely algebraic criterion* extends its usefulness significantly. Resultants can be used in constructing solutions to systems of equations, both symbolically and numerically (by approximation). They have been employed successfully in the design of efficient parallel and sequential algorithms in symbolic algebra, computational geometry, and computational number theory, and have found important practical applications in solid modeling and robotics.

15.1.1 Outline of This Chapter

This chapter presents parallel algorithms to compute the resultants of both univariate and multivariate polynomials.

We begin in Section 15.2 with a review of some of the mathematical background in commutative algebra that will be required, including the necessary facts regarding graded algebras, and affine and projective spaces over arbitrary fields. The Nullstellensatz of Hilbert is presented in both its strong and weak forms.

In Section 15.2, we give a detailed account of the construction of the resultant of a pair of univariate polynomials. The treatment is also extended to deal also with several polynomials in a single variable. In exploring properties of these calculations, the theory of *subresultants* is developed in detail, and an efficient parallel algorithm for the computation of *polynomial remainder sequences* is derived in a natural way. We discuss the applications of subresultants in parallel greatest common divisor (gcd) algorithms and in computing the extended Euclidean scheme. These algorithms have played a major role in the recent development of parallel methods in real geometry. For example, the algorithm of Ben-Or, Kozen and Reif [1], which gives an efficient parallel decision procedure for questions in the theory of real closed fields, employs a variety of parallel resultant-based techniques; the extension and correction by Renegar [23] makes essential use of multivariate resultants, presented in Section 15.3 below. Although a complete discussion of these applications is beyond the scope of this chapter, our presentation should be sufficient to enable the interested reader to pursue more advanced topics in the literature. References to such applications are provided at the end of this chapter.

In Section 15.3, the theory of *multivariate resultants* is developed as a natural extension of the univariate case. Here we treat both classical results on the projective (homogeneous) case, as well as more recent results on the affine (inhomogeneous) case. In this section our presentation is necessarily more abbreviated. Some statements whose proofs rely on deep algebraic or geometric arguments are not explored in detail. However, the general strategy for obtaining the desired algorithms from these results is explored fully.

Constructions involving multivariate resultants have played a large role in the development of efficient sequential algorithms during the past decade, but only within the last few years have special cases of these results contributed significantly to the improvement of parallel algorithm design. The *u-resultant* of a set of n polynomials in n variables is perhaps the most important tool here, and we present a simple parallel algorithm for its computation. We also discuss the computation of so-called *generalized characteristic polynomials* and demonstrate how they aid in adapting resultants for the homogeneous case to the inhomogeneous case. These techniques have been developed recently in the design of parallel algorithms for deciding questions in the theories of real closed and algebraically closed fields and for eliminating quantifiers in these theories. Although an exposition of such applications is beyond the scope of this chapter, we have tried to gather together the fundamental results of the modern and classical theories and present them in sufficient detail to enable the interested reader to pursue the more recent work in this area.

Many of the problems considered in this chapter are ultimately reduced to computations in linear algebra, such as the computation of determinants or characterisitic polynomials. A variety of efficient parallel algorithms are known for these problems and are discussed in detail elsewhere in this volume (Chapter 13) and in [16]. We also do not analyze the processor efficiency of these algorithms. These data may be found in Chapter 13 or in the references.

15.1.2 Mathematical Preliminaries

This section presents much of the necessary mathematical background from commutative algebra and ring theory necessary for the following sections. We assume some familiarity with linear algebra and parallel algorithms in linear algebra as presented in Chapter 13 and [16]. The material on graded rings and projective spaces is used later in Section 15.3 to develop the theory of multivariate resultants and is not necessary for the understanding of univariate resultants. Omitted proofs in this section can be found in any standard text on algebra or algebraic geometry, such as [25].

Unless otherwise noted, k will denote an algebraically closed field of arbitrary characteristic. In general, lower case Roman letters will denote variables and lower case Greek letters will denote elements of the field k. We write $\bar{x}$ for a sequence of elements or variables $x_1, \ldots, x_n$. If $\bar{x}$ is a sequence of n variables and $E = (e_1, \ldots, e_n)$ is a multi-index, we write $\bar{x}^E$ for the monomial $x_1^{e_1} x_2^{e_2} \cdots x_n^{e_n}$. If R is a ring, then $R[\bar{x}]$ denotes the ring of polynomials in the variables $\bar{x}$ with coefficients in R.

Polynomial Rings and Ideals

Let $R = k[x_1, \ldots, x_n]$. An *ideal* of R is a subset I of R closed under addition and closed under multiplication by elements of R. A *basis* for an ideal I is a set of polynomials B which generates I in the sense that every $f \in I$ can be written

$$f = g_1 f_1 + \cdots + g_m f_m$$

for some $f_1, \ldots, f_m \in B$ and $g_1, \ldots, g_m \in R$. When an ideal has a finite basis, we say that it is *finitely generated* and write $(f_1, \ldots, f_m)$ for the ideal generated by the polynomials $f_1, \ldots, f_m$.

THEOREM 15.1 *Hilbert Basis Theorem*
Every ideal $I \subseteq R$ is finitely generated. In other words, every I is of the form $(f_1, \ldots, f_m)$ for some polynomials $f_1, \ldots, f_m \in R$.

Define the *(total) degree* of a monomial $\prod_{i=1}^{n} x_i^{e_i} \in R$ to be $\sum_{i=1}^{n} e_i$. The degree of a polynomial $f \in R$ is the maximum degree of any term of f. We write R_e for the subset of all polynomials in R of degree at most $e \geq 0$. Each R_e is in fact a *vector space* over k of dimension $\binom{e+n}{n}$. We take as a basis for R_e the set of all monic[1] monomials of degree at most e:

$$\{x_1^{e_1} x_2^{e_2} \cdots x_n^{e_n} \mid e_1 + e_2 + \cdots + e_n \leq e\}.$$

---
[1] I.e., with leading coefficient 1.

Geometric Background

In this section we assume that k is an algebraically closed field. For example, k may be $\mathbf{C}$, the field of complex numbers.

We denote by $\mathbf{A}_{\bar{x}}^n$ the space of n-tuples elements of k, called the n-dimensional affine space over k with coordinate functions $\bar{x} = x_1, \ldots, x_n$. We write $\mathbf{A}^n$ and omit the subscript when the coordinates are understood. For $f_1, \ldots, f_m \in R$, define

$$V(f_1, \ldots, f_m) = \{\bar{\xi} \in k^n \mid f_1(\bar{\xi}) = \cdots = f_m(\bar{\xi}) = 0\},$$

the set of common zeros of these polynomials in $\mathbf{A}^n$. A set of this form is called *algebraic*. A principal link between the geometry of algebraic sets in $\mathbf{A}^n$ and the ideal structure of the polynomial ring R is given by Hilbert's Nullstellensatz, or "theorem of the zeros". To state the Nullstellensatz in its so-called weak form, we let K be an arbitrary subfield of the algebraically closed field k and consider the polynomial ring $R' = K[x_1, \ldots, x_n] \subseteq R$. Let us begin with the case $n = 1$, where a proof of the weak form of Hilbert's theorem is more familiar. We know that any pair of univariate polynomials f_1 and f_2 have a common solution in k exactly when they have a nontrivial greatest common divisor (gcd), i.e., a gcd which is a non-constant polynomial. Since $K[x_1]$ is a Euclidean ring, the Euclidean Algorithm for computing gcds works here and implies that there exist additional polynomials $g_1, g_2 \in R'$ such that

$$g_1 f_1 + g_2 f_2 = \gcd(f_1, f_2).$$

So a necessary and sufficient condition for the existence of a common zero of f_1 and f_2 is that there are no polynomials g_1 and $g_2 \in R'$ such that $g_1 f_1 + g_2 f_2 = 1$. The Euclidean algorithm itself provides a sequential decision procedure in this case.

The Nullstellensatz provides an analogue of this result for the case of multivariate polynomials, in which the polynomial ring R is not Euclidean and the existence of a nontrivial gcd is not a necessary condition for the existence of common zeros.

THEOREM 15.2 *Hilbert's Nullstellensatz (weak form)*
Let $n \geq 1$ and let $R = K[x_1, \ldots, x_n]$. For $f_1, \ldots, f_m \in R$, there exist elements $\xi_1, \ldots, \xi_n \in k$ algebraic over K such that

$$f_1(\xi_1, \ldots, \xi_n) = \cdots = f_m(\xi_1, \ldots, \xi_n) = 0$$

if and only if $(f_1, \ldots, f_m) \neq R$.

15.1. Introduction

From the definition of ideals we know that $I = R$ exactly when $1 \in I$, so as an immediate corollary of the Nullstellensatz we obtain the following criterion for deciding when a set of polynomials in R has no common zero.

THEOREM 15.3
Let $f_1, \ldots, f_m \in R$. Then

$$V(f_1, \ldots, f_m) = \emptyset \iff 1 \in (f_1, \ldots, f_m).$$

Now if $f_1, \ldots, f_m$ are polynomials in R, then we know that they have no common algebraic zeros exactly when $1 \in (f_1, \ldots, f_m)$, or in other words when there exist additional polynomials $g_1, \ldots, g_m \in R$ such that

$$g_1 f_1 + g_2 f_2 + \cdots + g_m f_m = 1.$$

The strong form of the Nullstellensatz tells us more about the relation between the ideal I and its zero set $V(I)$.

THEOREM 15.4 *Hilbert's Nullstellensatz (strong form)*
Let $I \subseteq R$ be an ideal and let $f \in R$. Then f vanishes on every point in $V(I)$ if and only if $f^m \in I$ for some $m \geq 1$.

This theorem will be useful in defining a homogeneous analogue of the Nullstellensatz in the next section.

Homogeneous Polynomials

The following facts are used only in the development of multivariate resultants in Section 15.3 below. Our presentation parallels that of the previous sections, defining graded rings of homogeneous polynomials and a homogeneous version of the Nullstellensatz.

Graded Algebras

A *graded ring* is a ring S together with a collection $\{S_e \mid e \geq 0\}$ of subgroups of the additive group of S such that

- $S = \bigoplus_{e=0}^{\infty} S_e$, and[2]

- $S_d S_e \subseteq S_{d+e}$ for all $d, e \geq 0$.

---

[2] The direct sum ($\bigoplus$) signifies that every element of S can be written uniquely as a sum of the form $\sum_{i=0}^{\infty} f_i$ where each $f_i \in S_i$ and all but a finite number of these f_i's are zero.

An element $f \in S$ is called *homogeneous* if $f \in S_e$ for some $e \geq 0$. For $I \subseteq S$ an ideal, define $I_e = I \cap S_e$. The ideal I is called a *homogeneous ideal* if it is generated by its homogeneous elements,

$$I = \bigoplus_{d=0}^{\infty} I_d,$$

or equivalently, if I has a basis of homogeneous elements [19, Section 5.13].

A typical example of a graded ring—and the one on which we focus in Section 15.3—is the polynomial ring $S = R[x_0, \ldots, x_n]$, which can be graded in the following way. A polynomial $f \in R[x_0, \ldots, x_n]$ is said to be *homogeneous of degree e* if every term of f has total degree e and *homogeneous* if it is homogeneous of some degree. We define S_e to be the collection of all polynomials in S which are homogeneous of degree e. Since any polynomial $f \in S$ of degree e can be written uniquely as a sum

$$f = f_0 + f_1 + f_2 + \cdots + f_e$$

where each f_i is homogeneous of degree i, it follows that S is a graded ring. When R is a field, it is a simple exercise to show that each subgroup S_e is in fact a *vector space* over R. In this case, we take the set M_e of all monic monomials of degree e,

$$M_e = \{x_0^{e_0} x_1^{e_1} \cdots x_n^{e_n} \mid e_0 + e_1 + \cdots + e_n = e\},$$

as a basis for the R-vector space S_e. We write $n_e = \binom{e+n}{n} = |M_e|$.

Projective Spaces

We define the n-dimensional projective space $\mathbf{P}_x^n$ to be the set of $\doteq$-equivalence classes of points in $\mathbf{A}_x^{n+1} - \{(0, \ldots, 0)\}$, where

$$(\xi_0, \ldots, \xi_n) \doteq (\eta_0, \ldots, \eta_n)$$

if there exists a nonzero $\kappa \in k$ such that $\xi_i = \kappa \eta_i$, $0 \leq i \leq n$. Note that each point in $\mathbf{P}^n$ is just a line through (but excluding) the origin in $\mathbf{A}^{n+1}$. To more easily distinguish between points in affine and projective spaces, we write $(\xi_0 : \xi_1 : \cdots : \xi_n)$ for the $\sim$-equivalence class of affine points $\{(\kappa \xi_0, \ldots, \kappa \xi_n) \mid \kappa \in k \text{ and } \kappa \neq 0\}$.

Let $S = k[x_0, \ldots, x_n]$ be graded as in Section 15.1.2. Then $f \in S$ is homogeneous of degree e exactly when

$$f(\kappa \xi_0, \ldots, \kappa \xi_n) = \kappa^e f(\xi_0, \ldots, \xi_n)$$

15.1. Introduction

for all $\xi_0, \ldots, \xi_n$ and $\kappa \in k$. This implies that the zero sets of homogeneous polynomials respect $\sim$-equivalence classes, and it is meaningful to speak of the points in projective space that are the zeros of a homogeneous polynomial. Note that the affine point $(0, \ldots, 0)$ is a zero of every homogeneous polynomial and consequently has no counterpart in projective space.

The affine space $\mathbf{A}^n$ is embedded in the projective space $\mathbf{P}^n$ under the *standard embedding*

$$(\xi_1, \ldots, \xi_n) \mapsto (1 : \xi_1 : \cdots : \xi_n).$$

The only points of $\mathbf{P}^n$ not in the image of this map are the points of the so-called *hyperplane at infinity*:

$$\{(0 : \xi_1 : \cdots : \xi_n) \mid \text{not all } \xi_i = 0, \ 1 \leq i \leq n\}.$$

For any homogeneous ideal $I \subseteq S$, we will define

$$V(I) = \{\bar{\xi} \in \mathbf{P}^n \mid f(\bar{\xi}) = 0 \text{ for all } f \in I\},$$

the zero set of I in $\mathbf{P}^n$. As in the affine case, the geometry of $\mathbf{P}^n$ is related to the ideal structure of the graded ring S by the Nullstellensatz, now in a homogeneous form.

THEOREM 15.5 *Homogeneous Nullstellensatz*
Let $I \subseteq S$ be a homogeneous ideal and let $f \in S_e$ for some $e > 0$. Then f vanishes on every point in $V(I)$ if and only if $f^m \in I$ for some $m > 0$.

As in the affine case, we can use this theorem to define necessary and sufficient conditions for the emptiness of $V(I)$ for any homogeneous ideal I of S. The previous criterion that $1 \in I$ is still sufficient but no longer necessary. The difference arises because there is no projective counterpart of the affine point $(0, \ldots, 0)$.

THEOREM 15.6
Let $f_1, \ldots, f_m$ be homogeneous polynomials in S and let $I = (f_1, \ldots, f_m)$. Then $V(I) = \emptyset$ if and only if $I_d = S_d$ for some $d \geq 1$.

In particular, since S_d is generated by the monomial basis M_d as a vector space over k, it suffices to show that all monomials of degree d are in I, i.e., that $M_d \subseteq I_d$.

15.2
Univariate Resultants

In this section we develop the classical *univariate* or *Sylvester resultant*, an algebraic condition on the coefficients of a pair of univariate polynomials that determines whether they have a common root.

15.2.1 The Sylvester Matrix and the Resultant of Two Univariate Polynomials

Let k be an algebraically closed field and x an indeterminate. Consider two univariate polynomials $f, g \in k[x]$ of degree $\deg f$ and $\deg g$, respectively:

$$f(x) = \sum_{i=0}^{\deg f} \alpha_i x^i$$

$$g(x) = \sum_{i=0}^{\deg g} \beta_i x^i \ .$$

Arrange the coefficients of f and g in staggered columns to form a square matrix Φ as in the following figure, with $\deg g$ columns of coefficients of f and $\deg f$ columns of coefficients of g. The figure illustrates the case $\deg f = 5$ and $\deg g = 4$.

$$\Phi = \begin{bmatrix} \alpha_5 & 0 & 0 & 0 & \beta_4 & 0 & 0 & 0 & 0 \\ \alpha_4 & \alpha_5 & 0 & 0 & \beta_3 & \beta_4 & 0 & 0 & 0 \\ \alpha_3 & \alpha_4 & \alpha_5 & 0 & \beta_2 & \beta_3 & \beta_4 & 0 & 0 \\ \alpha_2 & \alpha_3 & \alpha_4 & \alpha_5 & \beta_1 & \beta_2 & \beta_3 & \beta_4 & 0 \\ \alpha_1 & \alpha_2 & \alpha_3 & \alpha_4 & \beta_0 & \beta_1 & \beta_2 & \beta_3 & \beta_4 \\ \alpha_0 & \alpha_1 & \alpha_2 & \alpha_3 & 0 & \beta_0 & \beta_1 & \beta_2 & \beta_3 \\ 0 & \alpha_0 & \alpha_1 & \alpha_2 & 0 & 0 & \beta_0 & \beta_1 & \beta_2 \\ 0 & 0 & \alpha_0 & \alpha_1 & 0 & 0 & 0 & \beta_0 & \beta_1 \\ 0 & 0 & 0 & \alpha_0 & 0 & 0 & 0 & 0 & \beta_0 \end{bmatrix} \quad (15.1)$$

$$\underbrace{}_{\deg g} \underbrace{}_{\deg f}$$

DEFINITION 15.1

The matrix Φ is called the Sylvester *or* resultant matrix *of f and g, and its determinant $\det \Phi$ is called the* resultant *of f and g.*

THEOREM 15.7

The univariate polynomials f and g have a common root in k if and only if Φ is singular, i.e. if and only if the resultant of f and g vanishes.

15.2. Univariate Resultants

PROOF
Equivalently, we must show that Φ is singular iff f and g have a nontrivial gcd, or in other words iff the degree of the lcm ℓ of f and g is strictly less than $\deg fg = \deg f + \deg g$.

Let $k[x]_d$ denote the space of polynomials in $k[x]$ of degree at most d. This is a vector space of dimension $d+1$ over k with standard basis $1, x, \ldots, x^d$. The spaces $k[x]_{\deg g - 1} \times k[x]_{\deg f - 1}$ and $k[x]_{\deg f + \deg g - 1}$ are both vector spaces of dimension $\deg f + \deg g$, and under the standard basis the matrix Φ denotes the linear map

$$\varphi : k[x]_{\deg g - 1} \times k[x]_{\deg f - 1} \to k[x]_{\deg f + \deg g - 1}$$
$$: (s, t) \mapsto sf + tg \ .$$

If Φ is singular, then the kernel of φ is nontrivial. Thus there exist nonzero s, t with $\deg s < \deg g$ and $\deg t < \deg f$ such that $sf = -tg$. Then ℓ divides $sf = -tg$, so its degree is strictly less than $\deg f + \deg g$. Conversely, if $\deg \ell < \deg f + \deg g$, then $(\frac{\ell}{f}, -\frac{\ell}{g}) \in \ker \varphi$, thus Φ is singular. ∎

The resultant of two univariate polynomials can be computed in NC using Csanky's algorithm [10] in characteristic 0 or Berkowitz' [2] or Chistov's algorithm [9] in arbitrary characteristic; see Chapter 13.

DEFINITION 15.2
Consider the coefficients of f and g as indeterminates $\bar{a} = a_{\deg f}, \ldots, a_0$, $\bar{b} = b_{\deg g}, \ldots, b_0$. Then the determinant of Φ is a polynomial in $k[\bar{a}, \bar{b}]$ of degree $\deg f + \deg g$. This polynomial is called the resultant polynomial.

For any specialization $\bar{\alpha}, \bar{\beta}$ of the indeterminates $\bar{a}, \bar{b}$ with $\alpha_{\deg f} \neq 0$ and $\beta_{\deg g} \neq 0$ giving polynomials $f, g \in k[x]$, the value of the resultant polynomial on $\bar{\alpha}, \bar{\beta}$ is the resultant of f and g.

15.2.2 Subresultants, Polynomial Remainder Sequences, and the Extended Euclidean Scheme

An important application of resultants is in the calculation of the *polynomial remainder sequences* (PRS) that accrue from the execution of the Euclidean algorithm. Coefficients of elements of the PRS can be expressed as signed quotients of products of minors of the Sylvester matrix [4]. This holds as well for the elements of the *extended Euclidean scheme* (EES), of which the PRS forms a part. Thus all coefficients of elements of the PRS and EES can be computed in NC [3, 12]. In this section, we describe this algorithm and prove its correctness.

The basic fact on which the Euclidean algorithm rests is that for any pair of polynomials f and $g \neq 0$, there exist a unique quotient $q \in k[x]$ and remainder $r \in k[x]$ such that $\deg r < \deg g$ and $f = qg + r$. The Euclidean algorithm calculates the sequence $f_0 = f$, $f_1 = g$, $f_2, f_3, \ldots, f_n, f_{n+1} = 0$, where for $2 \leq m \leq n+1$, f_m is the remainder obtained by dividing f_{m-2} by f_{m-1}. The polynomial f_n is the last nonzero polynomial in the sequence and is the gcd of f and g. This sequence is known as the *polynomial remainder sequence* (PRS) of f and g.

DEFINITION 15.3
For $2 \leq m \leq n+1$, let $q_m \in k[x]$ be the quotient obtained in the division of f_{m-2} by f_{m-1}; thus

$$f_m = f_{m-2} - q_m f_{m-1}.$$

Consider the polynomials $s_0, s_1, \ldots, s_n, s_{n+1}$ and $t_0, t_1, \ldots, t_n, t_{n+1}$ defined by

$$\begin{aligned} s_0 &= 1 & t_0 &= 0 \\ s_1 &= 0 & t_1 &= 1 \\ s_m &= s_{m-2} - q_m s_{m-1} & t_m &= t_{m-2} - q_m t_{m-1} \end{aligned}$$

for $2 \leq m \leq n+1$. The collection of all these polynomials f_m, s_m, t_m, and q_m is known as the extended Euclidean scheme *of f and g*.

The significance of s_m and t_m is given in the following lemma.

LEMMA 15.1
Assume that $\deg g \leq \deg f$. For $2 \leq m \leq n+1$,

(i) $\deg s_m = \deg g - \deg f_{m-1}$

(ii) $\deg t_m = \deg f - \deg f_{m-1}$

(iii) $s_m f + t_m g = f_m$.

Moreover, s_m and t_m are the unique polynomials s and t such that

(a) $\deg s < \deg g - \deg f_m$

(b) $\deg t < \deg f - \deg f_m$

(c) $sf + tg$ and f_m have the same degree and leading coefficient.

15.2. Univariate Resultants

PROOF

Statements (i–iii) are easily proved by induction on m. These properties also imply that s_m and t_m satisfy (a–c). Thus it remains to show uniqueness.

We note that

$$\begin{aligned}
\deg g_m &= \deg f_{m-2} - \deg f_{m-1} \\
&\geq 1 \quad \text{for all } m \\
\deg s_m &= \deg(q_m s_{m-1}) \\
&> \deg s_{m-1} \\
&\geq \deg s_{m-2} \quad \text{for all } m > 2 \\
\deg t_m &= \deg(q_m t_{m-1}) \\
&> \deg t_{m-1} \\
&> \deg t_{m-2} \quad \text{for all } m > 2
\end{aligned}$$

Let s and t be any polynomials satisfying (a–c). (Under assumption (c), the statements (a) and (b) are equivalent, so we will only need to use one.) We have

$$\begin{aligned}
\deg s_m(sf + tg) &< \deg s_m + \deg f_{m-1} \text{ by (c)} \\
&= \deg g \text{ by (i)} ,
\end{aligned}$$

$$\begin{aligned}
\deg s(s_m f + t_m g) &= \deg s + \deg f_m \text{ by (iii)} \\
&< \deg g \text{ by (a)} .
\end{aligned}$$

Subtracting, we get

$$\begin{aligned}
\deg(s_m(sf + tg) - s(s_m f + t_m g)) &= \deg(s_m t - s t_m)g \\
&< \deg g ,
\end{aligned}$$

which is possible only if $s t_m = s_m t$. But an easy inductive argument shows that s_m and t_m are relatively prime—in fact

$$s_m t_{m-1} - t_m s_{m-1} = (-1)^m$$

—therefore there exists a polynomial u such that $s = u s_m$ and $t = u t_m$. Then

$$\begin{aligned}
sf + tg &= u s_m f + u t_m g \\
&= u f_m .
\end{aligned}$$

But by (c), it must be that $u = 1$, therefore $s = s_m$ and $t = t_m$. ∎

At this point we are ready to define subresultants. For $0 \leq d \leq \deg g - 1$, let Φ_d be the $(\deg f + \deg g - 2d) \times (\deg f + \deg g - 2d)$ submatrix of Φ obtained by deleting the last d columns of coefficients of f, the last d columns of coefficients of g, and the last $2d$ rows. The following figure illustrates Φ_d for $\deg f = 5$, $\deg g = 4$, and $d = 2$:

$$\Phi_d = \begin{bmatrix} \alpha_5 & 0 & \beta_4 & 0 & 0 \\ \alpha_4 & \alpha_5 & \beta_3 & \beta_4 & 0 \\ \alpha_3 & \alpha_4 & \beta_2 & \beta_3 & \beta_4 \\ \alpha_2 & \alpha_3 & \beta_1 & \beta_2 & \beta_3 \\ \alpha_1 & \alpha_2 & \beta_0 & \beta_1 & \beta_2 \end{bmatrix} \quad (15.2)$$

$$\underbrace{}_{\deg g - d} \underbrace{}_{\deg f - d}$$

Under the standard basis, Φ_d represents the linear map

$$\varphi_d : k[x]_{\deg g - d - 1} \times k[x]_{\deg f - d - 1} \to k[x]_{\deg f + \deg g - 2d - 1}$$
$$: (s, t) \mapsto \text{the quotient obtained in the division of } sf + tg \text{ by } x^d.$$

DEFINITION 15.4
The matrix Φ_d is called the dth subresultant matrix of f and g, and its determinant $\det \Phi_d$ is called the dth subresultant of f and g.

THEOREM 15.8
The matrix Φ_d is nonsingular if and only if $d = \deg f_m$ for some f_m in the PRS. In this case, the vector of coefficients of s_m and t_m forms the unique solution x of the nonsingular system

$$\Phi_d x = (0, \ldots, 0, a_m)^T, \quad (15.3)$$

where a_m is the leading coefficient of f_m.

PROOF
Note that $d < \deg g$. If $d \neq \deg f_r$ for any r, let m be the largest number such that $\deg f_{m-1} > d$. Then $2 \leq m \leq n+1$, and by Lemma 15.1,

$$\begin{align*} \deg s_m &= \deg g - \deg f_{m-1} < \deg g - d \\ \deg t_m &= \deg f - \deg f_{m-1} < \deg f - d \\ \deg(s_m f + t_m g) &= \deg f_m < d, \end{align*}$$

so $(s_m, t_m) \in \ker \varphi_d$, therefore Φ_d is singular.

Now suppose that $d = \deg f_m$. Then $\varphi_d(s_m, t_m) = a_m$, the leading coefficient of f_m, therefore the vector x of coefficients of s_m and t_m satisfies

(15.3). Moreover, by Lemma 15.1, s_m and t_m are unique, therefore Φ_d is nonsingular. ∎

This theorem gives rise to an *NC* algorithm for calculating all elements of the EES.

ALGORITHM 15.1
Extended Euclidean Scheme
Input: Given polynomials f and g.
Output: The extended Euclidean scheme for f and g.

1. For each $d < \deg g$, compute the dth subresultant $\det \Phi_d$. The d for which Φ_d is nonsingular are exactly the degrees of the f_m in the PRS.
2. For each $d = \deg f_m$, $m \geq 2$, solve the nonsingular system
$$\Phi_d x = (0,\ldots,0,1)^T . \qquad (15.4)$$
This gives the coefficients of $s'_m = \frac{s_m}{a_m}$ and $t'_m = \frac{t_m}{a_m}$. (We do not yet know a_m.)
3. Compute $f'_m = s'_m f + t'_m g$. This is the monic associate $\frac{f_m}{a_m}$ of f_m.
4. Divide f'_{m-2} by f'_{m-1} using Algorithm 15.2 below. (Alternatively, solve equation (15.4) using the dth subresultant matrix of f'_{m-2} and f'_{m-1}.) This gives a constant b_m and polynomial p_m such that
$$b_m f'_m = f'_{m-2} - p_m f'_{m-1} .$$
5. Compute
$$a_m = \begin{cases} a_0 b_2 b_4 b_6 \cdots b_m , & m \text{ even} \\ a_1 b_3 b_5 b_7 \cdots b_m , & m \text{ odd} \end{cases} \qquad (15.5)$$
$$q_m = \frac{a_{m-2}}{a_{m-1}} p_m \qquad (15.6)$$
$$f_m = a_m f'_m$$
$$s_m = a_m s'_m$$
$$t_m = a_m t'_m .$$

The computations in Step 5 are justified by the following argument. In Step 4, we computed b_m and p_m such that
$$\frac{b_m}{a_m} f_m = \frac{f_{m-2}}{a_{m-2}} - p_m \frac{f_{m-1}}{a_{m-1}} ,$$

thus by the uniqueness of quotient and remainder,

$$\begin{aligned}\frac{b_m a_{m-2}}{a_m} f_m &= f_{m-2} - \frac{a_{m-2}}{a_{m-1}} p_m f_{m-1} \\ &= f_{m-2} - q_m f_{m-1} \\ &= f_m,\end{aligned}$$

whence follow (15.6) and the recurrence

$$a_m = b_m a_{m-2}$$

with solution (15.5).

The solution vector to (15.4) computed in Step 2 is the last column of Φ_d^{-1}, which by Cramer's rule is the last column of the adjoint of Φ_d divided by $\det \Phi_d$. This indicates that all the coefficients of s'_m and t'_m are signed quotients of minors of the Sylvester matrix.

Algorithm 15.1 can be implemented in NC using standard tools for linear algebra (see [11, 12]).

15.2.3 Polynomial Division with Remainder

Let f, g be polynomials, $\deg g \leq \deg f$, and let q and r be the quotient and remainder respectively obtained in the division of f by g. Algorithm 15.1 suggests an NC algorithm for polynomial division with remainder: compute the subresultants to find $d = \deg r$, then solve (15.4) with Φ_d.

However, this algorithm has two serious liabilities:

- It requires the computation of all the subresultants.

- It requires divisions in k.

The latter becomes a major problem when the coefficients of f and g are indeterminates. Algorithm 15.1 expresses the coefficients of q and r as quotients of polynomials in the coefficients of f and g. This is so even when the divisor g is monic, in which case the coefficients of q and r are polynomial functions of the coefficients of f and g rather than rational functions. In the computation of the multivariate resultant to be presented in Section 15.3, it will be essential to have a polynomial division algorithm for monic g that does not use any divisions in k, but only the ring operations $\cdot$ and $+$.

Here we give a resultant-style algorithm used by Canny [6] that alleviates these problems. The algorithm is based on the following theorem.

15.2. Univariate Resultants   695

THEOREM 15.9
Let $m = \deg f - \deg g + 2$, the dimension of $\Phi_{\deg g-1}$. If g is monic, then the coefficient of x^{m-i} in q is

$$(-1)^i \det \Phi_{\deg g-1}^{(i-1)}, \quad 2 \le i \le m,$$

where $\det \Phi_{\deg g-1}^{(i)}$ is the ith principal minor (determinant of the upper left $i \times i$ submatrix) of $\Phi_{\deg g-1}$.

PROOF
Let

$$f = \sum_{i=0}^{\deg f} \alpha_i x^i \qquad q = \sum_{i=0}^{\deg q} \gamma_i x^i$$

$$g = \sum_{i=0}^{\deg g} \beta_i x^i \qquad r = \sum_{i=0}^{\deg r} \delta_i x^i$$

Note that $\deg q = m - 2$. The equation $r = f - qg$ is expressed in the $(\deg f + 1) \times m$ linear system

$$\begin{bmatrix} \alpha_9 & 1 & 0 & 0 & 0 \\ \alpha_8 & \beta_5 & 1 & 0 & 0 \\ \alpha_7 & \beta_4 & \beta_5 & 1 & 0 \\ \alpha_6 & \beta_3 & \beta_4 & \beta_5 & 1 \\ \alpha_5 & \beta_2 & \beta_3 & \beta_4 & \beta_5 \\ \alpha_4 & \beta_1 & \beta_2 & \beta_3 & \beta_4 \\ \alpha_3 & \beta_0 & \beta_1 & \beta_2 & \beta_3 \\ \alpha_2 & 0 & \beta_0 & \beta_1 & \beta_2 \\ \alpha_1 & 0 & 0 & \beta_0 & \beta_1 \\ \alpha_0 & 0 & 0 & 0 & \beta_0 \end{bmatrix} \begin{bmatrix} 1 \\ -\gamma_3 \\ -\gamma_2 \\ -\gamma_1 \\ -\gamma_0 \end{bmatrix} = \begin{bmatrix} 0 \\ 0 \\ 0 \\ 0 \\ 0 \\ \delta_4 \\ \delta_3 \\ \delta_2 \\ \delta_1 \\ \delta_0 \end{bmatrix} \quad (15.7)$$

here illustrated for the case $\deg f = 9$, $\deg g = 6$, and $\deg r = 4$. The first m rows of this matrix comprise $\Phi_{\deg g-1}$.

Now consider the $m \times m$ system obtained by taking the first $m-1$ rows of (15.7) and last row $(1, 0, \ldots, 0)$:

$$\begin{bmatrix} \alpha_9 & 1 & 0 & 0 & 0 \\ \alpha_8 & \beta_5 & 1 & 0 & 0 \\ \alpha_7 & \beta_4 & \beta_5 & 1 & 0 \\ \alpha_6 & \beta_3 & \beta_4 & \beta_5 & 1 \\ 1 & 0 & 0 & 0 & 0 \end{bmatrix} \begin{bmatrix} 1 \\ -\gamma_3 \\ -\gamma_2 \\ -\gamma_1 \\ -\gamma_0 \end{bmatrix} = \begin{bmatrix} 0 \\ 0 \\ 0 \\ 0 \\ 1 \end{bmatrix} \quad (15.8)$$

and let A be the $m \times m$ matrix in (15.8). Certainly A is nonsingular, since its determinant is $(-1)^{m+1}$. The inverse of A is given by Cramer's rule: the i,j^{th} entry of A^{-1} is

$$(-1)^{i+j}\frac{\det A_{j,i}}{\det A} = (-1)^{i+j+m+1}\det A_{j,i},$$

where $A_{j,i}$ is the $(m-1) \times (m-1)$ submatrix obtained from A by dropping the jth row and i^{th} column. In particular, the last column of A^{-1}, which by (15.8) is the vector $(1, -\gamma_{m-2}, \ldots, -\gamma_0)^T$, contains

$$(-1)^{i+1}\det A_{m,i}, \quad 1 \le i \le m.$$

But note that for this particular matrix,

$$\det A_{m,i} = \det A^{(i-1)}$$
$$= \det \Phi^{(i-1)}_{\deg g - 1}, \quad 1 \le i \le m.$$

Thus for $2 \le i \le m$,

$$\gamma_{m-i} = -(-1)^{i+1}\det \Phi^{(i-1)}_{\deg g - 1}$$
$$= (-1)^i \det \Phi^{(i-1)}_{\deg g - 1}. \quad \blacksquare$$

Using Theorem 15.9, we can give the following simple division-free algorithm for polynomial division with remainder when the divisor is monic:

ALGORITHM 15.2
Polynomial Division with Remainder

Input: Polynomials f and g, $\deg f \ge \deg g$, g monic.
Output: Polynomials q and r such that $f = qg + r$.

1. Compute the principal minors of $\Phi_{\deg g - 1}$.
2. Set

$$\gamma_{m-i} = (-1)^i \det \Phi^{(i-1)}_{\deg g - 1}, \quad 2 \le i \le m,$$

where $m = \deg f - \deg g + 2$. Then γ_j is the coefficient of x^j in q.
3. Set $r = f - qg$.

If g is not monic, then divisions are inevitable. However, we can apply Algorithm 15.2 to the monic associate of g, then adjust q afterward by dividing by the leading coefficient of g.

All operations can be implemented in NC. The principal minors of $\Phi_{\deg g - 1}$ can be computed without division using Berkowitz's [2] or Chistov's [9] algorithm.

15.2.4 A Resultant System for Several Univariate Polynomials

The constructions of Section 15.2.1 and Section 15.2.2 can be modified to yield NC algorithms for testing whether a set of univariate polynomials has a common root and for computing their gcd. We reduce these problems to the case of two univariate polynomials over a larger field. In Section 15.2.6 below, we will show that in the presence of a source of randomness, these algorithms have implementations that are no less efficient than the algorithms of Section 15.2.1 and Section 15.2.2 for two polynomials.

Let $f_0, \ldots, f_n \in k[x]$. Let y be a new indeterminate, and consider the bivariate polynomial

$$f(x,y) = f_0(x) + f_1(x)y + f_2(x)y^2 + \cdots + f_{n-1}(x)y^{n-1} . \quad (15.9)$$

We regard f as a polynomial in the indeterminate x with coefficients in the transcendental extension $k(y)$ of k. Thus it makes sense to consider the gcd of f and f_n over $k(y)[x]$.

LEMMA 15.2
The gcd of f and f_n is the same as the gcd of $f_0, \ldots, f_n$. In other words, f, f_n and $f_0, \ldots, f_n$ generate the same principal ideal in $k(y)[x]$.

PROOF
Let g be the monic gcd of $f_0, \ldots, f_n$ in $k[x]$ and let h be the monic gcd of f and f_n in $k(y)[x]$. Certainly g divides h, since g divides f and f_n. To show that h divides g, it suffices to show that h divides $f_0, \ldots, f_n$. Since h divides f_n and h is monic, $h \in k[x]$. For $0 \leq i \leq n$, let q_i and r_i be the quotient and remainder, respectively, obtained in the division of f_i by h in $k[x]$. Then $f = qh + r$, where

$$q = \sum_{i=0}^{n-1} q_i y^i$$

$$r = \sum_{i=0}^{n-1} r_i y^i ,$$

and the degree of r as a polynomial in $k(y)[x]$ is less than $\deg h$. Since h divides f in $k(y)[x]$, r must be identically zero as a polynomial in $k(y)[x]$. Since y is transcendental, r is also identically zero as a polynomial in $k[x,y]$. Thus $r_i = 0$ and h divides f_i, $0 \leq i \leq n$. ∎

Now form the Sylvester matrix Φ of f and f_n. The entries of Φ are polynomials in $k[y]$ of degree at most $n - 1$, and the resultant $\det \Phi$ is a polynomial in $k[y]$ of degree at most $(n - 1) \cdot \deg f_n$.

THEOREM 15.10
The polynomials $f_0, \ldots, f_n$ have a common root in k if and only if $\det \Phi$ vanishes identically.

PROOF
The polynomials $f_0, \ldots, f_n$ have a common root in k iff they have nontrivial gcd. By Lemma 15.2, this occurs iff f and f_n have a nontrivial gcd in $k(y)[x]$, i.e., if f and f_n have a common root in the algebraic closure of $k(y)$. By Theorem 15.7, this occurs iff the resultant of f and f_n vanishes identically. ∎

This gives rise to the following *NC* algorithm:

ALGORITHM 15.3
Resultant of Several Polynomials
Input: Polynomials $f_0, \ldots, f_n \in k[x]$.
Output: The resultant of $f_0, \ldots, f_n$.

1. Let y be a new indeterminate. Form the polynomial

$$f(x,y) = \sum_{i=0}^{n-1} f_i y^i .$$

   (For efficiency, it makes sense to take f_n of minimum degree among $f_0, \ldots, f_n$.)
2. Form the Sylvester matrix Φ of f and f_n. The entries are polynomials in $k[y]$ of degree at most $n - 1$.
3. Calculate the resultant $\det \Phi$ of f and f_n using Berkowitz's [2] or Chistov's [9] algorithm. This is a polynomial in $k[y]$ of degree at most $(n - 1) \cdot \deg f_n$.
4. Check whether $\det \Phi$ vanishes identically. By Theorem 15.10, this occurs iff $f_0, \ldots, f_n$ have a common root.

As in Section 15.2.1, if the coefficients of $f_0, \ldots, f_n$ are indeterminates, then Algorithm 15.3 can be carried out symbolically. The entries of the Sylvester matrix of f and f_n are then polynomials in y and the indeterminate coefficients $\bar{a}$ of the f_i. The resultant is a polynomial $r(\bar{a}, y)$ of degree at most $(n-1) \cdot \deg f_n$ in y and $\deg f_n + \max_{i=0}^{n-1} \deg f_i$ in the $\bar{a}$. Considering $r(\bar{a}, y)$ as a polynomial in y with coefficients in $k[\bar{a}]$, Theorem 15.10 implies that all these coefficients vanish under a specialization $\bar{\alpha}$ of $\bar{a}$ iff the polynomials $f_0, \ldots, f_n$ resulting from the specialization $\bar{\alpha}$ have a common root in k. The coefficients of $r(\bar{a}, y)$ thus form a *resultant system* for $f_0, \ldots, f_n$.

It is possible to work in the polynomial ring $k[x, y, \bar{a}]$ explicitly and compute the symbolic resultant $r(\bar{a}, y)$ in NC using Berkowitz's [2] or Chistov's [9] algorithm. These algorithms produce a polylog-depth, polynomial-size circuit C for $r(\bar{a}, y)$ over $+$, $\cdot$, constants, and inputs $\bar{a}$ and y. For any specialization $\bar{\alpha}$ of $\bar{a}$, since $r(\bar{\alpha}, y)$ is of degree at most $(n-1) \cdot \deg f_n$, we can test in NC whether $r(\bar{\alpha}, y)$ vanishes identically by evaluating it at $(n-1) \cdot \deg f_n + 1$ sample elements of k using the circuit C.

15.2.5 The GCD of Several Univariate Polynomials

The construction of Section 15.2.4 can be extended to give a deterministic NC algorithm for computing the gcd of several polynomials. We will show in Section 15.2.6 below how to improve the efficiency in the presence of a source of randomness.

ALGORITHM 15.4
GCD of Several Polynomials
Input: Polynomials $f_0, \ldots, f_n \in k[x]$.
Output: $g \in k[x]$, the gcd of $f_0, \ldots, f_n$.

1. Let y be a new indeterminate, and form the polynomial

$$f(x, y) = \sum_{i=0}^{n-1} f_i y^i .$$

   (For efficiency, take f_n of minimum degree among $f_0, \ldots, f_n$.)
2. Form the subresultant matrices Φ_d of f and f_n. The entries of Φ_d are polynomials in y of degree at most $n - 1$.
3. Compute the subresultants over $k(y)[x]$ using Berkowitz's [2] or Chistov's [9] algorithm. The dth subresultant $\det \Phi_d$ is a polynomial in y of degree at most $(n-1) \cdot (\deg f_n - d)$.

4. Let d be the smallest number such that $\det \Phi_d$ does not vanish identically. This is the degree of the gcd of f_n and f.
5. Compute the monic gcd of f_n and f as in Algorithm 15.1 using polynomial arithmetic on the coefficients. By Lemma 15.2, this is also the gcd of $f_0, \ldots, f_n$.
6. Reduce the coefficients. As computed in Step 5, they are rational functions of y, i.e. quotients of polynomials in y, but they reduce to elements of k because the monic gcd is in $k[x]$.

15.2.6 Improving Efficiency with Randomness

If we have access to a source of randomness, then we can obtain significantly more efficient algorithms than those of Section 15.2.4 and Section 15.2.5 by using a randomly chosen element of k in place of the indeterminate y. This is in fact the method of choice in most implementations. This approach is based on the observation that a nonzero polynomial is not likely to vanish when evaluated on a random input chosen from a sufficiently large sample set.

This idea is made concrete in the following lemma, proven independently by Zippel [27] and Schwartz [24]:

LEMMA 15.3
Let $p(x_1, \ldots, x_n)$ be a nonzero polynomial of total degree d with coefficients in k, and let S be a finite subset of k. If p is evaluated on a random element $(s_1, \ldots, s_n) \in S^n$, then the probability that $p(s_1, \ldots, s_n) = 0$ is at most $\frac{d}{|S|}$.

This gives rise to the following probabilistic *NC* algorithm:

ALGORITHM 15.5
Resultant of Several Polynomials (Probabilistic Version)
Input: Polynomials $f_0, \ldots, f_n \in k[x]$.
Output: The resultant of $f_0, \ldots, f_n$.

1. Select a random element β uniformly from a large finite set $S \subseteq k$.
2. Form the polynomial

$$f_0(x) + f_1(x)\beta + f_2(x)\beta^2 + \cdots + f_{n-1}(x)\beta^{n-1} \ .$$

This is $f(x, \beta)$, where f is the polynomial (15.9).

3. Calculate the resultant r of $f(x, \beta)$ and $f_n(x)$. If $f_0, \ldots, f_n$ have a common root, then $r = 0$ with probability 1. If $f_0, \ldots, f_n$ do not

have a common root, then by Lemma 15.3, $r = 0$ with probability at most

$$\frac{(n-1) \cdot \deg f_n}{|S|} .$$

4. Reduce the probability of error by repeated trials.

This algorithm does not provide a way to check the accuracy of the output. This liability is corrected in the gcd algorithm below.

As shown in Lemma 15.2, if g is the gcd of $f_0, \ldots, f_n$ in $k[x]$, then g is also the gcd of f and f_n in $k(y)[x]$, where f is the polynomial (15.9). With high probability, g will also be the gcd of f_n and $f(x, \beta)$ for a random β chosen uniformly from a sufficiently large subset of k. This is because g always divides the gcd of f_n and $f(x, \beta)$, and by Theorem 15.8, g is not itself the gcd of f_n and $f(x, \beta)$ iff the dth subresultant of f_n and $f(x, \beta)$ vanishes, where $d = \deg g$. This is the dth subresultant of f_n and $f(x, y)$ evaluated at β, and again by Theorem 15.8, the dth subresultant of f_n and $f(x, y)$ does not vanish identically, thus Lemma 15.3 applies. This gives the following *RNC* algorithm:

ALGORITHM 15.6
GCD of Several Polynomials (Probabilistic Version)
Input: Polynomials $f_0, \ldots, f_n \in k[x]$ with gcd of degree d.
Output: The polynomial $g = gcd(f_0, \ldots, f_n)$.

1. Select a random element β uniformly from a large set $S \subseteq k$.
2. Form the polynomial

$$f_0(x) + f_1(x)\beta + f_2(x)\beta^2 + \cdots + f_{n-1}(x)\beta^{n-1} .$$

This is $f(x, \beta)$, where f is the polynomial (15.9).

3. Calculate the gcd h of $f(x, \beta)$ and f_n as in Algorithm 15.1. Then g divides h, and h does not divide g iff the dth subresultant of $f(x, \beta)$ and f_n vanishes. By Lemma 15.3, this happens with probability at most

$$p = \frac{(n-1) \cdot (\deg f_n - d)}{|S|} . \qquad (15.10)$$

4. Check whether h divides g by checking whether h divides f_i, $0 \leq i \leq n-1$, using Algorithm 15.2. If so, h is the desired gcd. If not, go back to Step 1 and repeat with a new random β.

Using the value p in (15.10) as a bound on the probability of failure in each trial, and assuming the trials are independent, the expected number of trials before successfully obtaining the gcd of $f_0, \ldots, f_n$ is at most $\frac{1}{1-p}$.

See von zur Gathen [12] for another approach, attributed therein to Steve Cook, which yields a deterministic NC^1 algorithm.

15.3
Multivariate Resultants

The resultant of univariate polynomials is a classical tool that has played a considerable role in modern algorithms in symbolic algebra and computational geometry. It provides an effectively computable algebraic criterion for deciding when two or more univariate polynomials have a common root. Quite early in this century, effective means were developed for generalizing the resultant to the case of multivariate polynomials. Here, however, there must be a somewhat different approach. Chevalley had noted that there can be no strictly algebraic criterion for the existence of common solutions to a system of multivariate polynomials in the sense that the projection of an algebraic set is not necessarily algebraic.

Because there is no purely algebraic criterion, classical attempts at algorithms for the multivariate case diverge. One direction saw the development of *resolvents* through iterated application of the classical univariate resultant to eliminate variables one at a time. In the work of Hermann, Kronecker, Macaulay and others, an algebraic criterion was found for the special case where the given polynomials are all homogeneous. Here one is dealing with the existence of common zeros in projective space. The solution, which is both elegant and reasonably efficient, parallels and generalizes the basic theory developed in Section 15.2 for the univariate case.

In this section we review some of the basic properties of multivariate resultants and their computation. We do not present detailed proofs, but instead state the properties of various algebraic objects and give constructions of such objects which have been found useful in the design of algebraic algorithms.

Below $S = k[x_0, \ldots, x_n]$ denotes the (graded) ring of polynomials in $n+1$ variables.

15.3.1 Resultant Systems

Classical elimination theory considered both necessary and sufficient conditions for homogeneous polynomials $f_1, \ldots, f_m$ to have no common projective zeros. Recall that the Homogeneous Nullstellensatz asserts that this occurs exactly when, for some d, every monomial $\bar{x}^E$ of degree d can be written

$$\bar{x}^E = g_1 f_1 + g_2 f_2 + \cdots + g_m f_m$$

for some polynomials $g_1, \ldots, g_m \in S$. An effective criterion for deciding whether the f_i's have a common projective zero can be derived by finding a bound on this degree d and on the degrees of the g_j's which must exist when there are no solutions.

An initial solution to this problem is provided by the next theorem. It is Lazard's [17] modern generalization of a classical result of Kronecker, proved using homological methods.

THEOREM 15.11 *Effective Homogeneous Nullstellensatz*
Let $m \geq n+1$ and $f_1, \ldots, f_m \in S = k[x_0, \ldots, x_n]$ be homogeneous polynomials generating the ideal I. Let $d = 1 + \sum_{i=1}^{n+1}(\deg f_i - 1)$. Then $V(f_1, \ldots, f_m) = \emptyset$ if and only if I contains every monomial of degree d.

When $m \leq n$ it is known that $V(f_1, \ldots, f_m)$ is never empty, a fact which follows from Krull's Hauptidealsatz and the Projective Dimension Theorem [13, Section 1.5]. When $m \geq n+1$, this degree bound is tight. For example, the polynomials $x_0^{d_0}, x_1^{d_1}, \ldots, x_n^{d_n}$ have no common projective zeros, although the ideal generated by them does not contain the monomial $x_0^{d_0-1} x_1^{d_0-1} \cdots x_n^{d_n-1}$ of degree $\sum_{i=0}^{n}(d_i - 1) = d - 1$. It does, however, contain every monomial of degree d.

We derive a parallel algorithm to verify this condition by reduction to a problem of linear algebra. Recall that the additive subgroup S_d of S is a vector space over k with basis M_d. By Theorem 15.11, to determine whether $V(f_0, \ldots, f_m) = \emptyset$, it is both necessary and sufficient to show that every monomial $\bar{x}^E \in M_d$ is in the ideal generated by the f_i's. Using the following Lemma, we reduce this verification to a problem of linear algebra.

LEMMA 15.4
Let $f_1, \ldots, f_m$ be homogeneous polynomials and $I = (f_1, \ldots, f_m)$. Let $d_i = \deg f_i$, $1 \leq i \leq m$.

If $h \in I_d$, then there are homogeneous polynomials $g_1, \ldots, g_m$ with $g_i \in S_{d-d_i}$ such that $h = g_1 f_1 + g_2 f_2 + \cdots + g_m f_m$.

PROOF
The lemma is proved by noting that if $h \in I_d$, then there are polynomials $g'_1, \ldots, g'_m \in k[\bar{x}]$ such that $h = \sum_{i=1}^{m} g'_i f_i$. Observe that all terms of g'_i which are not of of degree $d - d_i$ give rise to terms that must be

cancelled in the summation. So if we take g_i to be the part of g'_i that is homogeneous of degree $d - d_i$, then $h = \sum_{i=1}^{m} g_i f_i$ as well. ∎

It is easy to see that, as a consequence of Lemma 15.4, the map φ which takes every vector of m homogeneous polynomials $g_1, \ldots, g_m$, $g_i \in S_{d-d_i}$, to the sum $\sum_{i=1}^{m} g_i f_i$,

$$\varphi : \prod_{i=1}^{m} S_{d-d_i} \rightarrow S_d \qquad (15.11)$$

$$: (g_1, \ldots, g_m) \mapsto \sum_{i=1}^{m} g_i f_i \, ,$$

is a k-linear map of vector spaces. From Lemma 15.4 it also follows that the image of φ is exactly I_d. We can now define the matrix Φ of the map φ with respect to the bases M_d and M_{d_i}, $1 \leq i \leq m$, analogous to the matrix Φ of Section 15.2.1. We index the rows of this matrix by the elements of M_d and the columns by pairs $(i, \bar{x}^B)$ where $1 \leq i \leq m$ and $\bar{x}^B \in M_{d-d_i}$. The column indices comprise a basis for $\prod_{i=1}^{m} S_{d-d_i}$ of size $\sum_{i=1}^{m} n_{d-d_i}$. The entry of Φ in row $\bar{x}^A$ and column $(i, \bar{x}^B)$ is just the coefficient of the term $\bar{x}^A$ in the polynomial $\bar{x}^B f_i(\bar{x})$. When $n = 1$ and $m = 2$, Φ gives the Sylvester matrix (15.1) of the two univariate polynomials $f_1(1, x_1)$ and $f_2(1, x_1)$ defined in Section 15.2.1.

The Effective Homogeneous Nullstellensatz (Theorem 15.11) now implies that $V(f_1, \ldots, f_m) = \emptyset$ if and only if φ is surjective. This happens exactly when the matrix Φ has full rank $|M_d| = n_d$. So one way to verify that the given system of polynomials has no solution is to find a nonzero $n_d \times n_d$ minor of Φ, which exists if and only if Φ has rank n_d.

The multivariate resultant allows us to eliminate variables from some collections of multivariate polynomials. Suppose that $f_1(\bar{x}, \bar{y}), \ldots, f_m(\bar{x}, \bar{y})$ are polynomials in two sets of variables $\bar{x} = x_0, \ldots, x_n$ and $\bar{y} = y_1, \ldots, y_{n'}$, and in addition that they are homogeneous as polynomials in the variables $\bar{x}$. Construct the matrix $\Phi(\bar{y})$ with respect to $\bar{x}$, so that the entries of $\Phi(\bar{y})$ are polynomials in $\bar{y}$. Then the matrix $\Phi(\bar{y})$ has the following property: for any point $\bar{\gamma} \in \mathbf{A}^{n'}$, the system $f_1(\bar{x}, \bar{\gamma}), \ldots, f_m(\bar{x}, \bar{\gamma})$ has a solution in the variables $\bar{x}$ if and only if $\Phi(\bar{\gamma})$ has rank strictly less than n_d [26, Section 19]. Thus

$$\{\bar{\gamma} \in \mathbf{A}^{n'} \mid \exists \bar{\xi} \in \mathbf{P}^n \; f_1(\bar{\xi}, \bar{\gamma}) = \cdots = f_m(\bar{\xi}, \bar{\gamma}) = 0\,\}$$
$$= \{\bar{\gamma} \in \mathbf{A}^{n'} \mid \text{rank } \Phi(\bar{\gamma}) < n^d\,\}$$
$$= \{\bar{\gamma} \in \mathbf{A}^{n'} \mid \text{all } n_d \times n_d \text{ minors of } \Phi(\bar{\gamma}) \text{ are zero}\,\}.$$

As in Section 15.2, instead of concentrating on a single set of polynomial equations, we can go one step further and regard the coefficients as parameters. This allows us to compute a general algebraic criterion expressed in terms of the indeterminate coefficients. One can in principle compute a *resultant system* consisting of a set of polynomials in the indeterminate coefficients, then simply evaluate them on the coefficients of any given system and immediately determine whether or not a solution exists. Unfortunately, the number of polynomials which arise and their degree make the computation unrealistically complex in all but the most trivial cases. Nevertheless, we continue to discuss the computation of resultant systems in these terms for several reasons. First, there are in fact some essential results from classical elimination theory that require this form. In addition, this form allows us emphasize that the algorithms which we present can be expressed solely in terms of the basic operations of the ring of coefficients, and hence commute with substitution. Finally, it permits us to express the complexity of the quantities that arise in this computation in a purely algebraic manner in terms of the complexity of each coefficient.

Let $f_1, \ldots, f_m$ be a set of homogeneous polynomials in the variables $x_0, \ldots, x_n$ with indeterminate coefficients among the variables $\bar{c}$. In other words, we are considering polynomials f_i of the form $f_i(\bar{c}, \bar{x}) = \sum_A c_{i,A} \bar{x}^A$ where A ranges over multi-indices of some fixed degree d_i, each $\bar{x}^A$ is a monomial of degree d_i, and each coefficient $c_{i,A}$ is a distinct indeterminate.

DEFINITION 15.5

A resultant system for the polynomials $f_1, \ldots, f_m$ is a collection of polynomials $g_1, \ldots, g_r \in k[\bar{c}]$ with the following property: for every specialization $\bar{c} \mapsto \bar{\gamma}$ of the coefficients to elements of k, the polynomials $f_1(\bar{\gamma}, \bar{x}), \ldots, f_m(\bar{\gamma}, \bar{x})$ have a common solution in $\mathbf{P}^n$ if and only if $g_i(\bar{\gamma}) = 0$, $1 \leq i \leq r$. In other words

$$V(g_1, \ldots, g_r) = \{\bar{\gamma} \mid \exists \bar{\xi} \in \mathbf{P}^n \ f_1(\bar{\gamma}, \bar{\xi}) = \cdots = f_m(\bar{\gamma}, \bar{\xi}) = 0\}.$$

If $r = 1$, this polynomial is called a resultant *for the system $f_1, \ldots, f_m$.*

Additional details can be found in the texts of Macaulay [18, Ch. 1] and van der Waerden [25, Ch. 7].

Under current technology, the direct calculation of such resultant systems by computing all minors of the corresponding matrix $\Phi(\bar{c})$ is infeasible because of the large number of polynomials and their high degree. To improve this situation somewhat, we will use the parallel algebraic matrix rank algorithm of Mulmuley [20]. We summarize the relevant aspects of the construction in the statement of the next lemma. A more complete treatment can be found in [11].

LEMMA 15.5
Mulmuley's Rank Algorithm Let A be an $m \times n$ matrix over an arbitrary field k, $m \geq n$, and let z, w be indeterminates. It is possible to compute a polynomial $p \in k[w, z]$ of degree at most $(2m - 1)n$ in w and $2n$ in z such that the rank of A is $\frac{2n-j}{2}$, where z^j is the highest power of z that divides p. In particular, A is of full rank if and only if $j = 0$, i.e., if and only if $p(w, 0)$ does not vanish identically. Moreover, the computation uses only the ring operations of the field k and can be implemented by an arithmetic circuit of size polynomial in $m+n$ and depth $O(\log^2(m+n))$.

Since Mulmuley's algorithm uses only the ring operations of k, it can also be performed on matrices containing indeterminate entries. As a consequence, specialization of these indeterminates to elements of k give the same result as first performing the substitution and then applying the algorithm.

As a result we get the following theorem which provides an effective criterion for determining when there exists a solution to a system of homogeneous polynomial equations.

THEOREM 15.12
Let $f_1, \ldots, f_m \in k[\bar{y}][\bar{x}]$ be polynomials that are homogeneous in the variables $\bar{x}$ of degree $d_1, \ldots, d_m$, respectively, with coefficients in $k[\bar{y}]$. A necessary and sufficient algebraic condition for the existence of a common zero $\bar{\xi} \in \mathbf{P}^n$ expressed in terms of the parameters $\bar{y}$ can be computed in parallel polynomial time with respect to the elementary operations of the ring $k[\bar{y}]$. In other words, we can compute a set of polynomials $g_1, \ldots, g_r \in k[\bar{y}]$ such that

$$g_1(\bar{\gamma}) = \cdots = g_r(\bar{\gamma}) = 0 \quad \Leftrightarrow \quad \exists \bar{\xi} \in \mathbf{P}^n \; f_1(\bar{\xi}, \bar{\gamma}) = \cdots = f_m(\bar{\xi}, \bar{\gamma}) = 0 \;.$$

PROOF
Suppose $f_1, \ldots, f_m$ are as described in the statement of the theorem. Let Φ be the matrix of the linear map φ of (15.11). Then Φ is an $n_d \times (\sum_{i=1}^m n_{d_i})$ matrix over the ring $k[\bar{y}]$. To compute a resultant system, we will apply the parallel matrix rank algorithm of Mulmuley to Φ. From this algorithm, we obtain a polynomial $q(\bar{y}, w) = \sum_{j=0}^e q_j(\bar{y}) w^j$ that for every substitution of field elements for the variables $\bar{y}$ is identically zero as a polynomial in w if and only if Φ does not have full rank n_d. As a polynomial in w, $q(\bar{y}, w)$ is identically zero just when all of its coefficients are zero. The collection of coefficients $\{ q_i(\bar{y}) \mid 0 \leq i \leq e \}$

thus comprises a resultant system for the given set of polynomials. Since the computation involves only the operations of the coefficient ring, the computation also commutes with specialization of coefficients. ∎

If each $d_i \leq d$, then the entire computation requires $O(d^{3n})$ processors and time $O((n \log d)^2)$. The number of polynomials in the system, and their degree, are at most exponential in n and $\max_{i=1}^{m} d_i$.

15.3.2 The Resultant of n Polynomials in n Variables

When the number of homogeneous polynomials equals the number of variables, there is also a single resultant polynomial. The presentation in this section follows along classical lines. For an excellent, complete development when $k = \mathbf{C}$ which uses only elementary arguments, see Renegar [23].

LEMMA 15.6
For $n+1$ homogeneous polynomials $f_0, \ldots, f_n$ in the $n+1$ variables $x_0, \ldots, x_n$ with indeterminate coefficients $\bar{c}$, there exists a single resultant polynomial $r(\bar{c})$, which can be effectively constructed from the matrix Φ.

Let $\deg f_i = d_i$, $0 \leq i \leq n$, and define $d = 1 + (\sum_{i=0}^{n} d_i - 1)$ and $n_d = |M_d|$. Recall that the matrix Φ has n_d rows, each indexed by a monomial in M_d. To construct the resultant polynomial, we first partition M_d into $n+1$ sets. Say that a monomial $\bar{x}^A \in M_d$ is *reduced in* x_i if $\bar{x}^A$ is not divisible by $x_i^{d_i}$. For each i, $0 \leq i \leq n$, define M_d^i to be the set of monomials in M_d that are divisible by $x_i^{d_i}$ and reduced in the variables $x_0, \ldots, x_{i-1}$. Then the sets $M_d^0, \ldots, M_d^n$ comprise a partition of M_d.

Let A be the square submatrix of Φ obtained by selecting the columns of Φ labeled by $(i, \bar{x}^E)$ for $0 \leq i \leq n$ and $\bar{x}^E \in M_d^i$. It can be shown that, as a polynomial in the indeterminate coefficients $\bar{c}$, $\det A$ is divisible by the resultant r of the f_i's. Moreover the quotient of the determinant of A by the resultant r is itself the determinant of a submatrix B of A, obtained by

1. eliminating those columns in A corresponding to indices $(i, \bar{x}^E)$ where the monomial $\bar{x}^E$ is reduced in n of the variables $\bar{x}$; and

2. for each such column $(i, \bar{x}^E)$ removed in Step 1, eliminating the row containing the coefficient of $x_i^{d_i}$ in this column.

These facts are proven by Macaulay in [18]. It follows that the desired resultant can be computed as a polynomial in the indeterminates $\bar{c}$ by constructing the matrices A and B from Φ and finding the quotient $r = \det A / \det B$. When constructed in this manner, the submatrix B depends only on the coefficients of the polynomials $f_1\mid_{x_0=0},\ldots,f_n\mid_{x_0=0}$, and for each i, $r(\bar{c})$ is homogeneous of degree $\prod_{j\neq i} d_j$ in the coefficients of f_i. In [18, Ch. 1], Macaulay also shows that r is irreducible as a polynomial in $k[\bar{c}]$. See also [25, Ch. 7] for further details.

Calculation of the resultant as presented above requires a computation in which all of the coefficients are indeterminates. In most cases, this computation is prohibitively expensive, since there are $\sum_{i=0}^n n_{d_i}$ such coefficients. Most often we do not need to know r as a polynomial in the $\bar{c}$'s, but are interested only in the image of r under some substitution for these coefficients. For example, let us assume that we wish to compute the resultant of the homogeneous polynomials $f_0,\ldots,f_n \in k[x_0,\ldots,x_n]$, where all of the coefficients are constants in the field k. Note that all of the above calculations that require only ring operations commute with substitution for the indeterminates [6, Ch. 3], since substitution determines a ring homomorphism. The determinant, for example, is defined and computed using only ring operations; so we can either compute $\det A$ as a polynomial in $k[\bar{c}]$ and then specialize the variables $\bar{c}$ to the coefficients of the f_i's, or specialize the coefficients and then compute the determinant [2]. In both cases the result will be the same.

The only other operation required in the construction is the division of these determinants. Unfortunately this operation does not commute with specialization. For example, where all coefficients are elements of the field k as above, it is possible that $\det A = \det B = 0$ under the given substitution, while the resultant itself is nonzero. To overcome this obstacle, we modify the construction as follows. Observe that the coefficients c_i of the terms $x_i^{d_i}$ in f_i lie only on the diagonals of the matrices A and B, and that these comprise all diagonal entries. Instead of computing the determinants of A and B directly, we compute the *characteristic polynomials* of these matrices, i.e., the determinants

$$\begin{aligned} a(\bar{c},t) &= \det(tI - A) \\ b(\bar{c},t) &= \det(tI - B) \end{aligned}$$

where t is a new indeterminate. These determinants are obtained from the determinants of A and B by substituting $t-c_i$ for the coefficient c_i of the term $c_i x_i^{d_i}$ in each f_i and negating all other coefficients. It follows that b divides a, and the quotient $r(\bar{c},t)$ is just the resultant of the new system

$$tx_0^{d_0} - f_0(\bar{c},\bar{x}),\quad \ldots,\quad tx_n^{d_n} - f_n(\bar{c},\bar{x})\ . \tag{15.12}$$

Moreover, it is easy to see that neither $a(\bar{\gamma}, t)$ nor $b(\bar{\gamma}, t)$ vanishes identically for any $\bar{\gamma} \in k^n$. This implies that

$$b(\bar{\gamma}, t) \cdot r(\bar{\gamma}, t) = a(\bar{\gamma}, t) ,$$

so $r(\bar{\gamma}, t)$ is the resultant of the system (15.12) under the specified substitution, and $r(\bar{\gamma}, 0)$ the resultant of the original system $f_0, \ldots, f_n$. (It is easy to see that negation of the coefficients does not change the resultant.) Hence we can compute the resultant of $n+1$ homogeneous polynomials in $n+1$ variables with coefficients in a field k by computing the constant term of the quotient of the characteristic polynomials of A and B.

As noted in Section 15.2.3, computation of the quotient of two univariate polynomials such as $a(\bar{\gamma}, t)$ and $b(\bar{\gamma}, t)$ can be performed in parallel, using only the ring operations of the coefficient field when the divisor is monic. We claim further that the quotient of the characteristic polynomials $a(\bar{c}, t)$ and $b(\bar{c}, t)$ *as polynomials in* t is in fact the resultant $r(\bar{c}, t)$ of the polynomials in (15.12). For suppose this were not the case, and let r' be the quotient of a and b as polynomials in t. It must be that the actual resultant r divides r'. But if $r \neq r'$, then r' must have an additional factor $p(\bar{c})$ such that $r'(\bar{c}, t) = r(\bar{c}, t)p(\bar{c})$. So

$$a(\bar{c}, t) = b(\bar{c}, t)r(\bar{c}, t)p(\bar{c}) ,$$

and since

$$a(\bar{c}, t) = t^d + \sum_{i=0}^{d-1} a_i(\bar{c}) t^i$$

it is clear that $p = 1$.

Hence $r(\bar{c}, t)$ must be the resultant of (15.12), and the constant term $r(\bar{c}, 0)$ of this polynomial is always the resultant of the original system $f_0, \ldots, f_n$. Since the constant term of this quotient can be computed using only ring operations, the entire computation described above now commutes with specialization of the indeterminate coefficients. In other words, we can first substitute elements of any division ring for the coefficients of the f_i's and then perform the indicated computation. The result is guaranteed to be the same as would be obtained by constructing the actual resultant polynomial and then performing the same substitution.

This discussion suggests the following parallel algorithm for computing resultants. Let $f_0, \ldots, f_n$ be polynomials in the variables $x_0, \ldots, x_n$ with coefficients in a ring R'. Construct the matrices A and B and compute their characteristic polynomials $a, b \in R'[t]$, as discussed in [11]. For example, R' may be the polynomial ring $k[\bar{y}]$. Write

$$a(t) = t^{n_d} + a_{n_d-1}t^{n_d-1} + \cdots + a_1 t + a_0$$
$$b(t) = t^e + b_{e-1}t^{e-1} + \cdots + b_1 t + b_0$$

where $e = n_d - \prod_{i=0}^{n} d_i$, and compute the division a/b using Algorithm 15.2. Since b is monic, only the ring operations are needed. This gives the desired resultant.

See also [6, Section 3.1.3] for another method of computing these classical resultants.

THEOREM 15.13 *Resultant of n polynomials in n variables*
The resultant of a set of $n+1$ homogeneous polynomials $f_0, \ldots, f_n$ in $n+1$ variables $x_0, \ldots, x_n$ with coefficients in a ring $R' = K[\bar{y}]$ can be computed in parallel polynomial time relative to the elementary operations of the ring R'. Moreover, since the computation uses only ring operations on the coefficients, the calculation commutes with substitution.

The computation of the characteristic polynomials a and b require operations on matrices of size n_d. Using Chistov's algorithm (see Chapter 13), this can be done in parallel time $O(\log^2 n_d)$ in the elementary operations of the coefficient ring R'. The quotient of these two polynomials requires computing the determinant of a matrix of size $n_d - e = \prod_{i=0}^{n} d_i$, and so can be executed in parallel time $O(\log^2 \prod_{i=0}^{n} d_i) = O((\sum_{i=0}^{n} \log d_i)^2)$, again measured in terms of the elementary operations of the ring of coefficients R'. We can conservatively bound the size of elements in R' that arise in this computation by noting that the coefficients of a are polynomials of degree $\prod_{j \neq i} \deg d_i$ in the coefficients of each f_i. For example, assume that $f_0, \ldots, f_n$ are integral polynomials with maximum degree d, and let c be a bound on the number of bits necessary to express any one coefficient. Then the coefficients of a require fewer than $(n+1)d^n c$ bits, and those of the resultant no more than $(n+1)d^{2n}c$ bits. On the other hand, if the coefficients of the f_i's are polynomials in the ring $k[y]$ and have maximum degree d in $\bar{x}$ and e in y, then the coefficients of a are polynomials of degree no more than $(n+1)d^n e$ in y, and those of the resultant of degree at most $(n+1)d^{2n}e$ in y.

15.3.3 The u-Resultant

The u-resultant is a classical tool for solving systems of homogeneous equations which have only a finite number of projective solutions. Its use has

15.3. Multivariate Resultants

been revived by computational applications, as in the zero-finding algorithm of Lazard [17], the approximation algorithm of Renegar [21], and the work of Canny [6] in theoretical robotics and of Renegar [22, 23] in real algebraic geometry.

Behind the construction of u-resultants lies the following idea. Suppose that we have n homogeneous polynomial equations $f_1, \ldots, f_n$ in the $n + 1$ variables $x_0, \ldots, x_n$, with only a finite number of projective solutions $\bar{\xi}^{(1)}, \ldots, \bar{\xi}^{(s)} \in \mathbf{P}^n$. Then for almost every additional polynomial f_0 which we might add to this system, the enlarged system will have no common zeros. This is true even if the degree of f_0 is constrained to be 1.

More concretely, let $u_0, \ldots, u_n$ be new indeterminates and let f_0 be the linear form $\bar{u} \cdot \bar{x} = \sum_{i=0}^{n} u_i x_i$. We show now that for most assignments of values $\bar{v} \in \mathbf{P}^n$ to $\bar{u}$, the system $\bar{v} \cdot \bar{x}, f_1, \ldots, f_n$ has no common zeros. Construct the resultant $r(\bar{u})$ of these $n + 1$ polynomials as a polynomial in the variables $\bar{u}$. By the characterization of the resultant given in Section 15.3.2, we know that r is homogeneous in $\bar{u}$, and that for any point $\bar{v} \in \mathbf{P}^n$, $r(\bar{v}) = 0$ if and only if $f_1, \ldots, f_n$ and $\bar{v} \cdot \bar{x}$ have a common solution. Equivalently, $r(\bar{v}) = 0$ if and only if $\bar{v} \cdot \bar{\xi}^{(j)} = 0$ for some j. This means that

$$V(r) = V\left(\prod_{j=1}^{s} \bar{\xi}^{(j)} \cdot \bar{u}\right).$$

The Homogeneous Nullstellensatz now says that each of r and $\prod_{j=1}^{s} \bar{\xi}^{(j)} \cdot \bar{u}$ divides some power of the other, therefore r factors as a product of linear forms, each of the form $\bar{\xi}^{(j)} \cdot \bar{u}$ for some zero $\xi^{(j)}$ of the f_i's. Hence all of the zeros of the f_i's can be recovered by computing this resultant r and factoring it over $k[\bar{u}]$. The coefficients of these factors are the coordinates of the common zeros.

DEFINITION 15.6
Let $f_1, \ldots, f_n \in k[x_0, \ldots, x_n]$ be homogeneous polynomials generating a zero-dimensional ideal, and let $\bar{u} = u_0, \ldots, u_n$ be a set of new indeterminates. Then the resultant of the f_i's and the polynomial $\bar{u} \cdot \bar{x}$ with respect to the variables $\bar{x}$ is called the u-resultant of the f_i's.

The classical theorem on the u-resultant asserts that, when the set $V = V(f_1, \ldots, f_n)$ is finite, the points $\xi \in V$ and their multiplicities can be recovered from a factorization of the polynomial $r(\bar{u})$.

LEMMA 15.7 *The u-Resultant*
Let $f_1, \ldots, f_n \in k[x_0, \ldots, x_n]$ be homogeneous polynomials, and assume that $V = V(f_1, \ldots, f_n)$ is finite, where each $\bar{\xi} \in V$ has multiplicity

$\mu(\bar{\xi})$. Then the u-resultant $r(\bar{u})$ is a homogeneous polynomial of degree $\prod_{i=1}^{n} \deg f_i$, and[3]

$$r(\bar{u}) \doteq \prod_{\bar{\xi} \in V} \left(\sum_{i=0}^{n} \xi_i u_i \right)^{\mu(\bar{\xi})}. \qquad (15.13)$$

Note that, since the $\bar{\xi}$'s are points in projective space, this polynomial is unique only up to a nonzero constant factor.

The argument sketched above can be extended to show that $r(\bar{u}) \not\equiv 0$ if and only if $V(f_1, \ldots, f_n)$ is finite. From Lemma 15.6 it follows that this u-resultant is a quotient of determinants $a(\bar{u})$ and $b(\bar{u})$. By the same lemma, we can construct these polynomials so that b is independent of the coefficients of one of the given polynomials. In particular, we can construct a and b from the f_i's and the polynomial $\bar{u} \cdot \bar{x}$ so that b is independent of the variables $\bar{u}$. Thus when $f_1, \ldots, f_n \in k[\bar{x}]$, b is a constant in k. Whenever $b \neq 0$, $a \not\equiv 0$ and $a(\bar{u}) \doteq r(\bar{u})$, so that a is itself a u-resultant polynomial. In this case, the u-resultant can thus be constructed using a single determinant computation over $k[\bar{u}]$.

15.3.4 Generalized Characteristic Polynomials

The algorithms developed in Section 15.3.2 and Section 15.3.3 for computing resultants and u-resultants took advantage of the fact that the quotient of the characterisitic polynomials of the matrices A and B has two important features: it never vanishes identically, and it is the resultant (u-resultant) of a perturbation of the original system of equations by a new indeterminate. This observation was used to produce an efficient algebraic algorithm for computing the resultant of n polynomials in n variables directly from their coefficients, using only the operations of the coefficient ring. This quotient was given the name *generalized characteristic polynomial* by Canny [7] because it specializes in the case of linear equations to the characteristic polynomial of the matrix of the linear system.

Recently such polynomials have found wide use in extending resultants and u-resultants to the case of affine (inhomogeneous) sets. Let us begin by considering how the algorithms of Section 15.3.3 might be adapted to handle inhomogeneous polynomials. Recall the standard embedding of n-dimensional

---

[3] We write $f \doteq g$ to signify that f and g are equal up to a nonzero constant factor; i.e. there is a $\kappa \in k, \kappa \neq 0$ such that $f = \kappa g$.

affine space into n-dimensional projective space:

$$(\alpha_1, \ldots, \alpha_n) \mapsto (1 : \alpha_1 : \cdots : \alpha_n).$$

We can exploit this correspondence in the following manner. Let $f \in R$ be a (possibly inhomogeneous) polynomial in the n variables $x_1, \ldots, x_n$ and let x_0 be a new variable. We define the *homogenization* of f, written f^h, to be the polynomial

$$f^h(x_0, \ldots, x_n) = x_0^{\deg f} f(x_1/x_0, \ldots, x_n/x_0),$$

where $\deg f$ is the total degree of f. Operationally, this means that we multiply each term of f by a sufficiently large power of the new variable x_0 to bring it up to degree $\deg f$.

Although homogenization gives an operational way of obtaining a related homogeneous polynomial from an inhomogeneous one, we still have not shown that there is a relation between the roots of these polynomials which can be exploited in resultant-based algorithms. On the one hand, the reader can easily verify that $(\alpha_1, \ldots, \alpha_n) \in \mathbf{A}^n$ is a zero of f exactly when $(1 : \alpha_1 : \cdots : \alpha_n)$ is a zero of f^h. However, the polynomial f^h may have zeros which do not correspond to zeros of the original polynomial when we set $x_0 = 0$. Such solutions lie on the hyperplane at infinity and are called *improper* or *infinite*.

Let us consider the case of the u-resultant. We know that a u-resultant polynomial can be computed for n homogeneous polynomials in S whenever they have only finitely many projective zeros. Now suppose we are given n possibly inhomogeneous polynomials $f_1, \ldots, f_n \in R$ and wish to compute their u-resultant. We homogenize them and compute the u-resultant of $f_1^h, \ldots, f_n^h$. This u-resultant should be a constant multiple of the polynomial

$$\left(\prod_{\bar{\xi}} (u_0 + \sum_{i=1}^n \xi_i u_i) \right) \cdot \left(\prod_{\bar{\xi}'} \bar{\xi}' \cdot \bar{u} \right),$$

where $\bar{\xi}$ ranges over all common zeros of $f_1, \ldots, f_n$ and $\bar{\xi}'$ ranges over all improper solutions of $f_1^h, \ldots, f_n^h$, and from it we can still recover the common zeros of the original system. But this is only possible when the number of improper solutions is finite. For if there are infinitely many improper projective solutions, the u-resultant polynomial vanishes identically and no information about the proper solutions will be available.

What we would like is a guaranteed method of obtaining all proper solutions to the original system—perhaps together with a finite number of the improper solutions—in the manner that the u-resultant provides for the strictly homogeneous case. This is given by the following lemma, proved by Canny [7] for the case $k = \mathbf{C}$ and by Ierardi [14, Ch. 4] for fields of arbitrary characteristic. We state the lemma only in its simplest form.

LEMMA 15.8
Let $f_1, \ldots, f_n$ be inhomogeneous polynomials in n variables with only a finite number of common zeros. Then the u-resultant of the system

$$f_1^h(\bar{x}) - tx_1^{\deg f_1}, \quad \ldots, \quad f_n^h(\bar{x}) - tx_n^{\deg f_n}$$

with respect to the variables $x_0, \ldots, x_n$ is a polynomial

$$r(\bar{u}, t) \doteq \sum_{i=c}^{d} r_i(\bar{u}) t^i$$

in which the least nonzero coefficient $r_c(\bar{u})$ factors as

$$\left(\prod_{\bar{\xi}} (x_0 + \sum_{i=1}^{n} \xi_i u_i)^{\mu_\xi} \right) \cdot \left(\prod_{\bar{\xi}'} \bar{\xi}' \cdot \bar{u} \right),$$

where $\bar{\xi}$ ranges over all points in $V(f_1, \ldots, f_n)$. Here $d = \prod_{i=1}^n \deg f_i$ and the points $\bar{\xi}'$ correspond to certain points in the algebraic set which lie on the hyperplane at infinity.

15.3.5 Applications of u-Resultants

Many of the applications of resultants and u-resultants arise from the possibility of recovering the coordinates of points defined as the zero set of a number of multivariate polynomial equations. To a large extent, the efficiency and elegance of these parallel algorithms stems from their ability to recover information about these points *symbolically*, without resorting to numerical approximation.

For example, suppose that a u-resultant polynomial $r(\bar{u})$ has been constructed by one of the methods outlined in Section 15.3.4, and we wish to compute the ith coordinate of the affine points represented by this form. One way of obtaining this information is by choosing new indeterminates y and t and substituting into $r(\bar{u})$ the following values for the variables $\bar{u}$:

$$u_j \mapsto \begin{cases} y, & \text{if } j = 0, \\ t^i - 1, & \text{if } j = i, \\ t^j, & \text{otherwise.} \end{cases}$$

15.3. Multivariate Resultants

A simple calculation shows that if the original polynomial r factored as

$$r(\bar{u}) = \prod_{\bar{\xi}} \bar{\xi} \cdot \bar{u},$$

then after the substitution we obtain a polynomial r' which factors as

$$r'(y,t) = \prod_{\bar{\xi}} \left(y + (\sum_{j=0}^{n} \frac{\xi_j}{\xi_0} t^j) - \frac{\xi_i}{\xi_0} \right)$$

and the least nonzero coefficient of r' as a polynomial in t is a polynomial r_i,

$$r_i(y) = \prod_{\xi_0 \neq 0} (y - \frac{\xi_i}{\xi_0}),$$

the roots of which are just the ith coordinates of the affine solutions of the original system.

Using subresultant techniques and the notion of primitive elements, this method can be extended to provide a tool which is even more useful. As in the sketch above, the following result relies on the fact that the factors of the polynomial in which we are interested are all linear. Together with the construction of generalized characteristic polynomials presented in Section 15.3.4, it permits us to reduce the problem of finding all of the zeros of n polynomials in n variables to a univariate problem, provided that there are only a finite number of solutions altogether. More importantly, it provides a way of representing these solutions symbolically, as in the following theorem proven independently by Canny and Renegar.

THEOREM 15.14
Let $f_1, \ldots, f_n$ be polynomials with only a finite number of common zeros. We can compute a polynomial $q(t)$ and rational functions $r_1(t), \ldots, r_n(t)$ such that the points $(r_1(\theta), \ldots, r_n(\theta))$ include all of these common zeros as θ ranges over the roots of q.

The complete construction and its proof are presented in [22] and [8], where applications to problems in real geometry are also discussed. The constructions have already proven useful in the following context: when we have found a finite set of points defined by a set of multivariate equations, this algorithm allows us to reduce the multivariate problem to a univariate problem involving just those points.

15.3.6 Extensions to the Affine Case

In the affine case—the case of inhomogeneous polynomials—there cannot be a purely algebraic criterion for the existence of a common solution to sets of polynomial equations. However, recent results of Kollàr [15] and Galligo, Heintz and Morgenstern [5] do yield a parallel polynomial time algebraic algorithm for deciding this question. The algorithm again depends on obtaining good degree bounds for the Nullstellensatz.

According to the Nullstellensatz, whenever $(f_1, \ldots, f_m) \subseteq R$,

$$V(f_1, \ldots, f_m) = \emptyset \Leftrightarrow 1 \in (f_1, \ldots, f_m)$$

$$\Leftrightarrow \exists g_1, \ldots, g_m \in R \ \sum_{i=1}^{m} g_i f_i = 1 \ . \quad (15.14)$$

The discussion of the previous sections suggest that if we can find a degree bound for the polynomials g_i in (15.14), then we can reduce the problem of determining whether 1 is an element of this ideal to the problem of determining the rank of an appropriate matrix formed from the coefficients of the given polynomials. By reducing the problem to the problem of determining the rank of a matrix, one can then apply Mulmuley's algorithm to give a parallel polynomial-time solution to the problems of deciding the solvability of a set of polynomial equations, and even the problem of quantifier elimination in algebraically closed fields. The essential facts are stated in the following theorem of Kollàr [15].

THEOREM 15.15
Let $f_1, \ldots, f_m \in R$ be polynomials of degree at most d. If $V(f_1, \ldots, f_m) = \emptyset$, then there exist polynomials $g_1, \ldots, g_m$ satisfying (15.14) with $\deg f_i + \deg g_i \leq d^n$ for $1 \leq i \leq m$.

This bound now leads to the following algorithm for determining when there exist solutions to a system of polynomial equations.

For each $e \geq 0$, let R_e denote the additive subgroup of R consisting of all polynomials of degree at most e. Note that R_e is a k-vector space of dimension n_e, and as a basis we can take all monic monomials of degree at most e. Let Ψ be the matrix of the linear map

$$\psi \ : \ \prod_{i=1}^{m} R_{d^n - \deg f_i} \ \to \ R_{d^n}$$

$$: \ (g_1, \ldots, g_m) \ \mapsto \ \sum_{i=1}^{m} g_i f_i$$

with respect to this basis, and let $\hat{1}$ denote the vector corresponding to the multiplicative identity $1 \in R_{d^n}$. Now we know that the polynomials f_i have a solution if and only if $\hat{1}$ is in the linear span of the columns of Ψ. If it is, then the rank of Ψ is the same as the rank of the matrix $(\Psi, \hat{1})$, obtained by adding $\hat{1}$ as a new column of Ψ. If not, then the ranks differ. Invoking Mulmuley's NC algorithm for determining the rank of a matrix, we thus obtain an efficient parallel algorithm for this problem.

Acknowledgements

Doug Ierardi was supported in part by NSF grant CCR-8901061 and in part by the NSF Center for Discrete Mathematics and Theoretical Computer Science (DIMACS). Dexter Kozen was supported by NSF grant CCR-8901061, the John Simon Guggenheim Foundation, and the U.S. Army Research Office through the ACSyAM branch of the Mathematical Sciences Institute of Cornell University under contract DAAL03-91-C-0027.

15.4
Exercises

15.1 Let D be a square matrix with entries in a graded ring $S = \bigoplus_{d=0}^{\infty} S_d$. Suppose that every entry of D is nonzero and homogeneous, and every 2×2 minor of D is homogeneous. Prove that $\det D$ is homogeneous.

15.2 Let A be a set of distinct indeterminates, $|A| = d$, and let x be another indeterminate. Show that the coefficient of x^{d-i} in the polynomial $\prod_{a \in A}(x - a)$ is a homogeneous polynomial in $k[A]$ of degree i. (This coefficient is called the ith *elementary symmetric polynomial* in A.)

15.3 Let f, g be multivariate polynomials such that

- all irreducible factors of f are distinct (i.e., f is squarefree);
- f and g are homogeneous of the same degree; and
- $V(f) = V(g)$.

Show that $f \doteq g$.

15.4 Let A be the multiset of roots of a monic univariate polynomial f; thus $f = \prod_{\alpha \in A}(x - \alpha)$. Similarly, let B be the multiset of roots of a monic univariate polynomial g. Prove that the resultant of f and g is

$$\prod_{\substack{\alpha \in A \\ \beta \in B}} (\beta - \alpha) .$$

15.5 Show that the multivariate resultant of two homogeneous polynomials in two variables is essentially the same as the univariate resultant of two polynomials. Explain why this is so.

15.6 Let f be a univariate polynomial with rational coefficients. The resultant of f and its formal derivative f' is called the *discriminant* of f.

  a) Calculate the discriminant of the quadratic $ax^2 + bx + c$.

  b) Prove that f has a multiple root if and only if its discriminant vanishes.

15.7 Let s_m and t_m be as in Definition 15.3, and let p and q be arbitrary polynomials. Prove that for $0 \leq m \leq n$,

$$gcd(p, q) \;=\; gcd(s_m p + t_m q, s_{m+1} p + t_{m+1} q) .$$

15.8 Let $f_0, f_1, \ldots, f_n, f_{n+1}$ be the PRS of f_0 and f_1, and let s_m and t_m be as in Definition 15.3. Prove that for $0 \leq m \leq n$,

$$f_m \;=\; gcd(f_0 + (-1)^m t_m f_{m+1}, f_1 - (-1)^m s_m f_{m+1}) .$$

Bibliography

[1] M. Ben-Or, D. Kozen, and J. Reif. The complexity of elementary algebra and geometry. *JCSS*, 32(2):251–264, 1986.

[2] S.J. Berkowitz. On computing the determinant in small parallel time using a small number of processors. *Information Processing Letters*, 18:147–150, 1984.

[3] A. Borodin, J. von zur Gathen, and J. Hopcroft. Fast parallel matrix and gcd computations. *Information and Control*, 52(3):241–256, 1982.

[4] W. Brown and J. F. Traub. On Euclid's algorithm and the theory of subresultants. *J. ACM*, 18:505–514, 1971.

[5] L. Caniglia, A. Galligo, and J. Heintz. Some new effectivity bounds in computational geometry. Preprint.

[6] J. Canny. *The Complexity of Robot Motion Planning*. PhD thesis, MIT, 1987.

[7] J. Canny. Generalized characteristic polynomials. Preprint, 1988.

[8] J. Canny. Some algebraic and geometric problems in PSPACE. In *Proc. 20th ACM Symp. Theory of Computing*, pages 460–467. Assoc. Comput. Mach., May 1988.

[9] A. L. Chistov. Fast parallel calculation of the rank of matrices over a field of arbitrary characteristic. In *Proc. Conf. Foundations of Computation Theory, Lecture Notes in Computer Science 199*, pages 63–69. Springer-Verlag, 1985.

[10] L. Csanky. Fast parallel matrix inversion algorithms. *SIAM J. Comput.*, 5(4):618–623, 1976.

[11] J. von zur Gathen. Parallel linear algebra. *Chapter 13, this volume.*

[12] J. von zur Gathen. Parallel algorithms for algebraic problems. *SIAM J. Comput.*, 13(4):802–824, 1984.

[13] R. Hartshorne. *Algebraic Geometry*. Springer-Verlag, 1977.

[14] D. Ierardi. *The Complexity of Quantifier Elimination in the Theory of an Algebraically Closed Field*. PhD thesis, Cornell University, 1989.

[15] J. Kollàr. Sharp effective Nullstellensatz. Preprint.

[16] D. C. Kozen. *The Design and Analysis of Algorithms*. Springer-Verlag, 1991.

[17] D. Lazard. Résolution des systems d'équations algèbriques. *Theor. Comput. Sci.*, 15:77–110, 1981.

[18] F. S. Macaulay. *The Algebraic Theory of Modular Systems*. Cambridge Tracts in Mathematics and Mathematical Physics. Cambridge U. Press, 1916.

[19] H. Matsumura. *Commutative Ring Theory*. Cambridge Studies in Advanced Mathematics 8. Cambridge U. Press, 1988.

[20] K. Mulmuley. A fast parallel algorithm to compute the rank of a matrix over an arbitrary field. *Combinatorica*, 7(1):101–104, 1987.

[21] J. Renegar. On the worst case arithmetic complexity of approximating zeros of systems of polynomials. Technical Report 748, School of Operations Research and Industrial Engineering, Cornell U., Ithaca, New York, 1987.

[22] J. Renegar. A faster PSPACE algorithm for deciding the existential theory of the reals. Technical Report 792, School of Operations Research and Industrial Engineering, Cornell U., 1988.

[23] J. Renegar. On the computational complexity and geometry of the first-order theory of the reals, Parts I, II, III. Technical report, School of Operations Research and Industrial Engineering, Cornell U., Ithaca, New York, 1989.

[24] J. T. Schwartz. Fast probabilistic algorithms for verification of polynomial identities. *J. ACM*, 27(4):701–717, October 1980.

[25] B. L. van der Waerden. *Modern Algebra*, volume 2. F. Ungar Publishing Co., third edition, 1950.

[26] B. L. van der Waerden. *Modern Algebra*, volume 2. F. Ungar Publishing Co., fifth edition, 1970.

[27] R.E. Zippel. Probabilistic algorithms for sparse polynomials. In Ng, editor, *Proc. EUROSAM 79, Lecture Notes in Computer Science 72*, pages 216–226. Springer-Verlag, 1979.

Part VI

Extensions of Parallel Tree Contraction to Algebraic and Logical Problems

16

Dynamic Parallel Evaluation of Computation DAGs

Erich Kaltofen

Department of Computer Science
Rensselaer Polytechnic Institute
Troy, NY 12180-3590
kaltofen@cs.rpi.edu

16.1
Statement of Problem

One generic parallel evaluation scheme for algebraic objects, that of evaluating algebraic computation trees or formulas, is presented in Chapter 3 of this book. However, there are basic algebraic functions for which the tree model of computation seems not sufficient to allow an efficient—even sequential—decision-free algebraic computation. The formula model essentially restricts the use of an intermediate result to a single place, because the parse tree nodes have fan-out 1. If an intermediate result participates in the computation of several further nodes, in the tree model it must be recomputed anew for each of these nodes. It is a small formal change to allow node values to propagate to more than one node at a deeper level of the computation. Thus we obtain the *algebraic circuit model*, which is equivalent to the *straight-line program model*. Figure 16.1 exhibits a small example of such an algebraic computation. Note that the node v_1 propagates its value into the nodes v_3 and v_4. This distinguishes the algebraic circuit from an *expression tree (formula)*, where each node would be restricted to fan-out 1. We shall formally define this model in Section 16.1.2. Before discussing the parallelization results that are known in this context, we present algebraic circuits for several well-known algebraic objects. Since the size of the circuit corresponds to the length of the program that computes the object, it is important to keep the size as small as possible. For sake of the discussion we take the parallel time of an algebraic circuit to be the depth of that circuit, that is, the longest path in the corresponding directed acyclic graph (*DAG*).

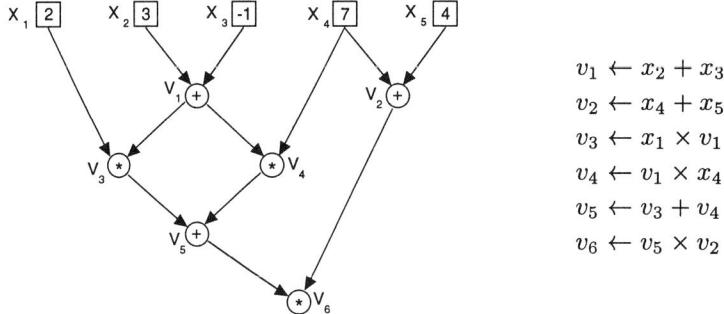

FIGURE 16.1
Sample algebraic circuit and corresponding straight-line program.

16.1.1 Algebraic Circuit Examples
Symmetric Functions

The n elementary symmetric functions σ_i, $1 \leq i \leq n$ in n indeterminates $x_1, \ldots, x_n$ can be defined by the coefficients of a polynomial of degree n whose roots are the (negated) indeterminates as follows:

$$\left.\begin{array}{l}(z+x_1)(z+x_2)\cdots(z+x_n) = z^n + \sigma_1(x_1,\ldots,x_n)z^{n-1} \\ \qquad + \sigma_2(x_1,\ldots,x_n)z^{n-2} + \cdots + \sigma_n(x_1,\ldots,x_n)\end{array}\right\} \quad (16.1)$$

Another way of computing σ_i is as the sum of the products of all combinations of i of the n indeterminates:

$$\sigma_i(x_1,\ldots,x_n) = \sum_{1 \leq j_1 < j_2 < \cdots < j_i \leq n} x_{j_1} \cdots x_{j_i} \qquad (16.2)$$

One important property is that these functions generate the algebra of symmetric polynomials in $x_1, \ldots, x_n$, where a symmetric polynomial f is one that remains invariant under permutation of the subscripts of the indeterminates, i.e., $f(x_1,\ldots,x_n) = f(x_{\tau(1)},\ldots,x_{\tau(n)})$ for all permutations τ on the n subscripts $1,\ldots,n$. The task is to compute the σ_i's from the x_j's. Clearly, by multiplying out the linear terms in (16.1) we can determine all σ_i's in $O(n^2)$ additions and multiplications. Let us assume here that the values for the x_j's come from a ring. The computation will be by an algebraic circuit, and the parallel time will be $O(\log(n)^2)$ if we perform the polynomial multiplications in a tree-like fashion.

The question arises how to compute the middle $\sigma_{\lfloor n/2 \rfloor}$ by an algebraic computation tree with polynomially (in n) many nodes. Clearly, the formula (16.2) for $i = \lfloor n/2 \rfloor$ contains exponentially many terms under the sum, approximately $2^n/\sqrt{\pi n/2}$. Our question is actually equivalent to the question of computing $\sigma_{\lfloor n/2 \rfloor}$ by an algebraic circuit of parallel time $O(\log n)$. This is, in fact, possible by using an evaluation/interpolation approach (see Exercise 16.3), but that method leads to a fairly large polynomial-sized formula for $\sigma_{\lfloor n/2 \rfloor}$. In general, the algebraic circuit model appears superior to the expression tree model. It is hard to imagine how one would devise a computation tree for $\sigma_{\lfloor n/2 \rfloor}$ with $O(n \log(n)^2 \log(\log n))$ nodes, which is the node count of the asymptotically smallest known algebraic circuit that computes this function (see Aho, et al. [3]).

Determinants

The basic invariant for linear system solving is the determinant of an $n \times n$ matrix. Treating the entries as indeterminates $x_{j,k}$, we can define it as a polynomial in these n^2 variables, namely,

$$\Delta_n = \text{Det}\left([x_{j,k}]_{1 \leq j,k \leq n}\right)$$
$$= \sum_{\tau \text{ a permutation on } 1,\ldots,n} \text{sign}(\tau)\, x_{1,\tau(1)} \cdots x_{n,\tau(n)}, \quad (16.3)$$

where $\text{sign}(\tau)$ is $+1$ or -1, depending whether τ is generated by an even or odd number of transpositions (i.e., permutations that switch two indices but leave the rest the same), respectively. Considering this sum of $n!$ terms alone, it is not obvious at all how to construct an algebraic circuit of polynomial size in n that computes this determinant. The solution hinges on the Gaussian elimination algorithm for solving linear systems. Consider the matrix $A^{(0)}$ of elements $a_{j,k}^{(0)} = x_{j,k}$. If we triangularize this matrix by elementary row operations, the determinant does not change. From (16.3) we conclude that the determinant of the resulting triangular matrix is the product of the elements on the diagonal. Let us assume for a moment that the elements come from a field. We have

$$\begin{aligned}&\textbf{for } i \leftarrow 1,\ldots,n-1 \textbf{ do}\\&\quad \textbf{for } j \leftarrow i+1,\ldots,n \textbf{ do}\\&\quad\quad \textbf{for } k \leftarrow i+1,\ldots,n \textbf{ do}\\&\quad\quad\quad a_{j,k}^{(i)} \leftarrow a_{j,k}^{(i-1)} - \frac{a_{j,i}^{(i-1)} a_{i,k}^{(i-1)}}{a_{i,i}^{(i-1)}}\\&\Delta_n \leftarrow a_{1,1}^{(0)} a_{2,2}^{(1)} \cdots a_{n,n}^{(n-1)}\end{aligned} \quad (16.4)$$

where the $a_{j,k}^{(i)}$ denote distinct circuit nodes.

Clearly, the obtained algebraic circuit has $O(n^3)$ nodes and requires $\Theta(n)$ parallel time. Since it contains divisions, it does not allow the evaluation of the determinant for any matrix of field elements. At this point, numerical procedures introduce decision statements that select proper nonzero elements before dividing. In general, the introduction of control of flow elements into a model of computation makes it difficult to parallelize sequential algorithms in that model. Furthermore, it is desirable to obtain division-free circuits for polynomial functions. We shall mention in Section 16.3.1 a method by Strassen that generically allows the removal of divisions in such a situation. For the determinant example here we shall now give a somewhat simpler solution.

First we observe that for the variables in the above program we have

$$a_{j,k}^{(i)} = \frac{\text{Det}(A^{(0)}(1,\ldots,i,j;\, 1,\ldots,i,k))}{\text{Det}(A^{(0)}(1,\ldots,i;\, 1,\ldots,i))}, \quad (16.5)$$

where $A^{(0)}(j_1, j_2, \ldots; k_1, k_2, \ldots)$ denotes the submatrix obtained from $A^{(0)}$ (the original input) by selecting the rows $j_1, j_2, \ldots$ and the columns $k_1, k_2, \ldots$ only. This fact is easy to prove; see for example Gantmacher [13], Chapter 3. If the entries in the input matrix come from an integral domain[1] D such as the integers or univariate polynomials over a field, we can simulate the arithmetic in the field of quotients by recording the numerator and denominator of a rational element separately. Defining the numerators and denominators of the right-hand side of (16.5),

$$\mathrm{Num}(a_{j,k}^{(i)}) = \mathrm{Det}(A^{(0)}(1, \ldots, i, j; 1, \ldots, i, k)), \mathrm{Den}(a_{j,k}^{(i)})$$
$$= \mathrm{Det}(A^{(0)}(1, \ldots, i; 1, \ldots, i))$$

we get

$$\frac{\mathrm{Num}(a_{j,k}^{(i)})}{\mathrm{Den}(a_{j,k}^{(i)})} = a_{j,k}^{(i)} = \frac{\mathrm{Num}(a_{j,k}^{(i-1)})\,\mathrm{Num}(a_{i,i}^{(i-1)}) - \mathrm{Num}(a_{j,i}^{(i-1)})\,\mathrm{Num}(a_{i,k}^{(i-1)})}{\mathrm{Den}(a_{i,i}^{(i-1)})\,\mathrm{Num}(a_{i,i}^{(i-1)})}$$

We conclude that $\mathrm{Den}(a_{i,i}^{(i-1)})$ can be divided out from the numerator of the right-hand-side expression. This observation leads to an algorithm for computing the determinant that uses subtractions, multiplications, and divisions by known factors. The following assumes that $B^{(0)}$ is initialized to the integral entries of the input matrix, and that $b_{0,0}^{(-1)} = 1$. Here the variables $b_{j,k}^{(i)}$ correspond to $\mathrm{Num}(a_{j,k}^{(i)})$.

> for $i \leftarrow 1, \ldots, n-1$ do
>   for $j \leftarrow i+1, \ldots, n$ do
>   for $k \leftarrow i+1, \ldots, n$ do
> $b_{j,k}^{(i)} \leftarrow \left(b_{j,k}^{(i-1)} b_{i,i}^{(i-1)} - b_{j,i}^{(i-1)} b_{i,k}^{(i-1)}\right) / b_{i-1,i-1}^{(i-2)}$
> $\Delta_n \leftarrow b_{n,n}^{(n-1)}$

We still have the problem of how to guarantee that the now exact division is not by zero. However, zero division can be avoided by considering a "characteristic" matrix $B^{(0)} = zI_n + A$ in place of A. Here z is a new indeterminate, and I_n is the $n \times n$ identity matrix. Essentially, we have switched from the integral domain D of entries to the domain D$[z]$ of entries. The

---
[1] i.e., a commutative ring D with the property that
$$\forall b, p, q \in \mathsf{D}: bp = bq \text{ and } b \neq 0 \implies p = q,$$
that is, if an element $a = bq$ is divisible by $b \neq 0$, then there exists a unique quotient q.

exact divisions necessary in the second version of the algorithm now become polynomial divisions by the polynomials

$$b_{i,i}^{(i-1)} = \text{Det}(z\, I_i + A(1,\ldots,i;\, 1,\ldots,i)) = z^i + \text{lower order terms}$$

We can now record the coefficients of z of the intermediate polynomial numerators explicitly (and drop the usage of z altogether from the computation). The exact divisions will be computations of polynomial quotients using polynomial divisors whose lead term is always 1. Since in the polynomial division process (Knuth [26], Section 4.6.1, Algorithm D) the only divisions are by the lead term of the divisor, the operations on the arising coefficient vector elements will not necessitate a division by a domain element. Hence these polynomial divisions can be carried out no matter what the entries of the original matrix A are. By (16.5) it is also easy to see that the $b_{j,k}^{(i)}$'s have degree in z no more than $i+1$, so the polynomial arithmetic for each individual assignment can be accomplished in $O(n^2)$ additions, subtractions, and multiplications in the domain D itself. The determinant of A finally is the constant coefficient of $b_{n,n}^{(n-1)}$. In summary, we have described a division-free algebraic circuit with $O(n^5)$ nodes that can find the determinant of a matrix over an integral domain, in fact over any ring, in parallel time of order no more than $O(n^2)$.

Optimal Matrix Chain Multiplication

In the previous two examples, the arithmetic was carried out over a ring. However, for certain optimization problems it is sometimes necessary to work over an even weaker algebraic structure, for example that of a commutative semiring. In a commutative semiring we have an associative and commutative addition $\oplus$ and multiplication $\otimes$, such that the multiplication distributes over addition, i.e., $a \otimes (b \oplus c) = (a \otimes b) \oplus (a \otimes c)$ for all semiring elements a, b, and c. The dynamic programming algorithm presented here supplies an example for an algebraic circuit over the semiring of nonnegative integers with the formal addition $a \oplus b = \min\{a,b\}$ and the formal multiplication $a \otimes b = a + b$. The problem is the following: We are given m matrices $A_1,\ldots,A_m$ with entries from a ring, such that A_i is of dimension $n_i \times n_{i+1}$ for all $1 \leq i \leq m$. We want to compute the product $A_1 A_2 \cdots A_m$ using standard matrix multiplication, but we wish to pick the order of multiplications in such a way that the total number of ring multiplications is minimized. For example, if $m=4$, $n_1 = n_4 = 1$, and $n_2 = n_3$, then an optimal order is $(A_1 A_2)(A_3 A_4)$ costing $O(n_2^2)$ operations, while the order $A_1(A_2 A_3) A_4$ requires the full $n_2 \times n_2$ matrix product $A_2 A_3$. An optimal order for the general problem is found by dynamic programming as follows (see also Aho, et al. [3], Section 1, for a more detailed discussion of this technique): Let $c_{i,j}$ denote the optimal number of operations to compute $A_i A_{i+1} \cdots A_j$ for $1 \leq i \leq j \leq m$ with the initial values $c_{i,i} = 0$. For all $1 \leq i < j \leq m$ we have

$$c_{i,j} = \min\{c_{i,k-1} + c_{k,j} + n_i n_k n_j \mid \text{for all } k \text{ with } i < k \leq j\} \qquad (16.6)$$

the interpretation being that we try all possible splits $(A_i \cdots A_{k-1})(A_k \cdots A_j)$. One can now evaluate $c_{1,m}$ by employing the recursive definition (16.6) together with *memoization* (see Abelson and Sussman [1] for the usage of this notion); that is, one computes an individual $c_{i,j}$ only once and looks up its value if it is needed again. Usually, the resulting algebraic circuit is presented bottom-up:

$$\text{for } l \leftarrow 1, \ldots, m-1 \text{ do}$$
$$\quad \text{for } i \leftarrow 1, \ldots, m-l \text{ do}$$
$$\qquad c_{i,i+l} \leftarrow \bigoplus_{k=i+1}^{i+l} c_{i,k-1} \otimes c_{k,i+l} \otimes (n_i n_k n_{i+l})$$

Notice that the integer $n_i n_k n_j$ corresponds to a circuit node containing a constant, so the algebraic circuit is actually over the semiring $(\mathbb{Z}_{\geq 0}, \oplus, \otimes)$. In fact, it is memoization that reduces the number of trial combinations from exponentially many to $O(n^3)$, corresponding exactly to reducing the exponential sized computation tree to a polynomial sized computation DAG. The bottom-up realization of this technique is referred to as *dynamic programming*.

16.1.2 Definitions and Main Results

The three examples given above all exhibit algebraic circuits for important multivariate polynomials. Circuits for the determinant and the optimal matrix product seemingly require fan-out higher than 1 in order to have only polynomially many nodes. Only the circuit for the symmetric functions has parallel time $O(\log(n)^2)$. The goal of this chapter is to give a transformation for any algebraic computation DAG with a certain (necessary) degree restriction that results in a DAG computing the same functions, but in log-squared depth. We now define the model of an algebraic circuit and its formal degree precisely, and state the main result.

DEFINITION 16.1
An algebraic circuit C over a semiring $\mathbb{R}$ is an edge-weighted directed acyclic graph with node set V and (directed) edges E satisfying the following conditions:

- Each edge has either an element from $\mathbb{R}$ as its weight, or the special symbol "unit-weight." For a (directed) edge $(v, w) \in E$, where v and w are nodes, we denote its weight by $W(v, w)$.
- Each node is labeled as one of three types: a leaf, an addition node, or a multiplication node.
- Any leaf has in-degree (fan-in) 0, i.e., there is no edge leading into the corresponding node, any addition node has in-degree no less than 1, and any multiplication node has in-degree exactly 2.
- Every leaf v is also assigned an element in $\mathbb{R}$, denoted by $v)$.

  A formula or expression tree is an algebraic circuit in which each node has at most one edge directed out of it.

This definition is more general than what we have used so far in two ways: first, there are weights on the edges with the interpretation that a semiring value is multiplied with the weight on an edge before being fed to the target node along that edge. These edge-weights are merely a conceptual tool for the transformations presented later. Clearly, one can realize the multiplication by a weight as an additional multiplication node. The reason we also need the special weight "unit-weight" is that in $\mathbb{R}$ we are not guaranteed to have a multiplicative unit. The second difference is that we allow arbitrarily many edges leading into an addition node. Such "super-nodes" must be ultimately realized by a balanced binary tree of addition nodes of fan-in 2. Otherwise, one could compute an n-dimensional determinant in parallel-time $O(\log n)$ by employing a tree for (16.3) with exponentially many nodes, which is clearly unrealistic.

The set of *children* of a node v in the circuit C are all those nodes $u \in V$ such that there is an edge $(u, v) \in E$ from u to v. With this notion we can recursively define the value of each node:

DEFINITION 16.2

The value of a leaf w is the associated w). Now let v be a non-leaf node of the circuit C, and let U be the set of all its children. We define $v)$ in terms of the $u)$ and the edge weights $W(u, v)$ for $u \in U$. Assume first that v is an addition node. Then we define

$$v) = \sum_{u \in U} propagate(u, v)$$

where

$$propagate(u, v) = \begin{cases} u) & \text{if } W(u, v) = \text{``unit-weight''} \\ u) \times W(u, v) & \text{if } W(u, v) \text{ is a semiring element} \end{cases}$$

16.1. Statement of Problem

If v is a multiplication node we have

$$v) = \prod_{u \in U} propagate(u, v)$$

Notice that in this case we have precisely two factors under the product.

There is an almost one-to-one correspondence between algebraic circuits and straight-line programs (refer back to Figure 16.1). We shall not give a precise definition for the latter (see, e.g., Strassen [41]), but we wish to point to some minor differences. In a straight-line program there exists a (somewhat artificial) total order of the intermediate variables that is not apparent in an algebraic circuit. Furthermore, in a straight-line program we are allowed to issue assignments of the form $v \leftarrow u \times u$, which are somewhat excluded in a circuit (there cannot be more than one edge from u to v). However, we clearly can simulate such an assignment by the pair of assignments $w \leftarrow u$; $v \leftarrow u \times w$, where w can be chosen to be an addition node with fan-in one in the corresponding algebraic circuit. This trick will be needed anyway in our results, since the transformation algorithm will be restricted to circuits that do not have any edges from a multiplication node to another multiplication node.

The *size* of an algebraic circuit is simply the number of its nodes, excluding the leaf nodes. The *depth* of a circuit is the longest path in its defining DAG. We observe that for an algebraic circuit of fan-in not higher than 2, the size corresponds to the number of arithmetic operations in a straight-line program that computes the node values, and that the depth corresponds to the parallel time. The goal of our parallelizing transformations is to construct a new algebraic circuit that computes all values of a given circuit but that has much smaller depth. For instance, we will construct a fan-in bounded circuit for the determinant of an $n \times n$ matrix that has depth $O(\log(n)^2)$. However, it is easily shown that not all algebraic circuits can be parallelized with such a dramatic gain in depth. Consider the straight-line program

$$v_1 \leftarrow x \times x; v_2 \leftarrow v_1 \times v_1; \ldots; v_n \leftarrow v_{n-1} \times v_{n-1};$$

the variable v_n computes the value x^{2^n}. Now the optimal algebraic circuit with respect to depth that computes the same function x^{2^n} must have depth $\Omega(n)$, for we can partition the nodes of that circuit into levels, where the ith level contains the nodes for which the maximum length path originating in any leaf node is of length exactly i. It follows by induction on i that the degrees of the values computed by the nodes on level i as polynomials in indeterminates replacing the leaf values is not higher than 2^i. Thus any algebraic circuit computing x^{2^n} must have depth at least n. Note that this argument is properly valid only for semirings in which the function x^{2^n} cannot be realized by a lower degree polynomial function in x. It is thus certainly true for any infinite integral domain. With this simple lower bound in mind, we will restrict our transformation to circuits that compute functions of polynomially bounded degree.

We now define the *formal degree* of a node in an algebraic circuit:

DEFINITION 16.3

The formal degree of a leaf node in an algebraic circuit is defined as the integer 1. The formal degree of an interior node v is defined recursively from the formal degrees of its children:

$$\text{degree}(v) = \begin{cases} \max\{\text{degree}(u) \mid u \text{ a child of } v\} & \text{if } v \text{ is an addition node} \\ \sum_{u \text{ a child of } v} \text{degree}(u) & \text{if } v \text{ is a multiplication node} \end{cases}$$

Finally, the formal degree of the entire circuit is the maximum of the formal degree of any node.

Essentially, the degree of a node is the value that the node has if we replace all leaf values in the circuit by 1, and convert all additions to taking the maximum and all multiplications to integer addition. As in the matrix chain product example above, we still have a circuit over a semiring at hand. It should be realized that the formal degree of a node might be higher than the degree of its algebraic value as a polynomial in the leaf values; this is because leading terms of the children of an addition node can cancel one another: If u_1 has value x^2 and u_2 has value $x - x^2$, then $v \leftarrow u_1 + u_2$ has formal degree 2, but actually its value is x. Note that over an arbitrary ring such "loss of leading terms" (to paraphrase the loss of significant digits in numerical computing) can also occur for multiplication nodes.

We now can formulate the main result of this chapter. Given is the description of an algebraic circuit in which no addition node has fan-in higher than 2. All leaf values and edge weights are represented as distinct symbols. The circuit has size n and formal degree d. We shall use the function $M(n)$ that denotes the asymptotic circuit size for $n \times n$ matrix multiplication over the semiring that is simultaneously of depth $O(\log n)$. The standard algorithm gives $M(n) = O(n^3)$ for any semiring, and the best fast method over a ring has $M(n) = O(n^{2.3755})$ (Coppersmith and Winograd [10]). The result is the following: One can compute, on a parallel random access machine with $M(n)$ processors running in time $O(\log(n) \log(dn))$ the description of a circuit of

$$\text{depth} = O(\log(n) \log(dn)) \text{ and size} = O(M(n) \log(dn)),$$

a subset of whose nodes attain all values of the input circuit. A similar result can be obtained for other models of parallel algebraic complexity, e.g., the parallel algebraic PRAM or arithmetic Boolean circuits discussed in Chapter 13. However, since we do not formally introduce these models, we have confined our discussion to the standard PRAM and circuit models.

16.1.3 Exercises

EXERCISE 16.1

Determine the formal degree of the algebraic circuit over the semiring $(\mathbb{Z}_{\geq 0}, \min, +)$ for computing the optimal matrix chain product, which is discussed in Section 16.1.1.

EXERCISE 16.2

Show that the formal degree of the division-free algebraic circuit for computing the determinant of an $n \times n$ matrix over a ring described in Section 16.1.1 is of order $O(n)$. [Hint: The exact division circuit over $D[z]$ computes polynomials that are homogeneous in $x_{j,k}$, the entries of A, and in z, the new indeterminate. Hence the coefficients of z^m have lower degree as polynomials in $x_{j,k}$ if m is larger.]

EXERCISE 16.3*[2]

Construct a circuit of depth $O(\log n)$ that computes $\sigma_{\lfloor n/2 \rfloor}(x_1, \ldots, x_n)$ over any ring. [Hint (see Eberly [12]): If the ring is the complex numbers, and ω is a primitive 2^k-th root of unity, $2^{k-1} < n+1 \leq 2^k$, one can interpolate the polynomial $(z + x_1) \cdots (z + x_n)$ at the points $1, \omega, \omega^2, \ldots, \omega^{2^k-1}$ in $O(\log n)$ time by the discrete fast Fourier transform.]

EXERCISE 16.4

Given is an algebraic circuit over a semiring with n nodes and with depth $O(\log n)$. Show that there exists a formula with $n^{O(1)}$ nodes that computes the values of each node of the circuit.

16.1.4 History of Solutions

The theory on parallelizing transformations of algebraic programs starts with Brent's [8] seminal paper on parallelizing the parse tree of an expression. A polynomial-sized, polylogarithmic depth circuit for the determinant and inverse of an $n \times n$ matrix over a field of characteristic 0 goes back to Csanky [11]. Hyafil [16] obtained polylogarithmic depth circuits for the general problem, but his construction needed superpolynomially many nodes.

---
[2]Starred exercises are difficult, and may contain/constitute publishable research results.

Valiant and Skyum [46] made the construction also work simultaneously in polynomial size (see also Valiant et al. [45]). With the recasting of Brent's problem as a dynamic question, i.e., by requiring the transformation itself to be a parallel algorithm, the techniques discussed in this chapter were introduced. Miller and Reif's [35] tree contraction process of 1985 solved the dynamic formula evaluation problem, and the dynamic circuit evaluation algorithm followed (Miller, Ramachandran, and Kaltofen [34]). Both algorithms have found their way into textbooks (Gibbons and Rytter [14], Leighton [31]). Follow-up papers containing additional results with respect to the circuit compression algorithm are Miller and Teng [36] and Mayr [33].

Circuits or straight-line programs have been used for a long time as a structured model of algebraic computation. We refer to Strassen [44] for an excellent survey of the known results and open problems, as well as to Karp and Ramachandran's [25] excellent survey on parallel algorithms in general. The problem of dealing with divisions will be discussed in detail in Section 16.3.1, and we delay the citation of work to that subsection. We will also introduce other related results in Section 16.3. The application of circuit compression to dynamic programming goes back to Valiant et al. [45], where the context-free language-recognition problem is solved in polylogarithmic time using that technique. We consider the application of this algorithm to the optimal matrix chain product circuit presented here as part of the folklore of that method.

16.2
Algebraic Circuit Compression

We now describe the algorithm by Miller et al. [34] and present its complexity analysis. This algorithm was developed by attempting to combine the ideas of Miller and Reif [35] for tree compression (see also Chapter 3 in this book) with the ideas by Valiant et al. [45] for straight-line code evaluation. However, in its final form it appears quite distinctive and even introduces a graph transformation (called the *shunt operation*) that has found its way back into the simple and elegant solution to tree compression by Abrahamson et al. [2] and by Kosaraju and Delcher [28].

16.2.1 The Compression Algorithm

We assume that the input circuit has no edges from multiplication nodes to multiplication nodes. As pointed out in Section 16.1.2, if this is not the case, then we add unary addition nodes along those edges that connect multiplication nodes. The algorithm will apply transformations to the DAG. These

16.2. Algebraic Circuit Compression

transformations can affect the circuit in the following ways: They may add a new edge with a certain weight between nodes previously unconnected, they may remove an edge, they may change an addition or multiplication node into a leaf node with a certain value, or they may change the weight on an existing edge. There are three types of transformations, the so-called *Matrix-Multiply* transformation, the *Rakes*, and the *Shunts*. A combination of these three transformations applied in that order is called a *Phase*. The algorithm repeats Phases until every interior node is turned into a leaf. The crucial theorem in its analysis is that this takes no more than $O(\log(nd))$ Phases, where n is the number of nodes and d is the formal degree of the input circuit. Let us now define the individual transformations. For this it is useful to label the circuit nodes in order as $v_1, \ldots, v_n$.

PROCEDURE 16.1
Matrix-Multiply
Consider the following $n \times n$ matrices derived from the edge-weights $W(v_i, v_j)$ on the nodes $v_1, \ldots, v_n$ of the circuit.

$$W[\![+,+]\!]_{i,j} = \begin{cases} W(v_i, v_j) & \text{if } v_i \text{ and } v_j \text{ are addition nodes} \\ 0 & \text{otherwise} \end{cases}$$

$$W[\![\,.\,,+]\!]_{i,j} = \begin{cases} W(v_i, v_j) & \text{if } v_j \text{ is an addition node} \\ 0 & \text{otherwise} \end{cases}$$

$$W[\![\,.\,,\,.\,]\!]_{i,j} = \begin{cases} W(v_i, v_j) & \text{if not both } v_i \text{ and } v_j \text{ are addition nodes} \\ 0 & \text{otherwise} \end{cases}$$

For simplicity, we shall assume that the semiring associated with the circuit has an additive zero, denoted by 0, and furthermore, that it also has a multiplicative unit, denoted by 1. We thus need no symbolic edge-weights "unit-weight" (see Exercise 16.5 below for removing these restrictions). Now, we compute the matrix product and sum

$$W' \leftarrow W[\![\,.\,,+]\!] \cdot W[\![+,+]\!] + W[\![\,.\,,\,.\,]\!] \qquad (16.7)$$

and replace all edge-weights $W(v_i, v_j)$ by $W'_{i,j}$. Notice, that this action not only replaces weights on edges, but it may introduce a new edge (in case $W'_{i,j} \neq 0$ but (v_i, v_j) is not an edge in the original circuit) or remove an old edge (in case $W'_{i,j} = 0$ but (v_i, v_j) is an edge). Furthermore, in case all edges into an addition node are removed, that node is converted into a leaf of value 0. The mechanics of this procedure are illustrated in Figure 16.2.

736 Chapter 16. Dynamic Parallel Evaluation of Computation DAGs

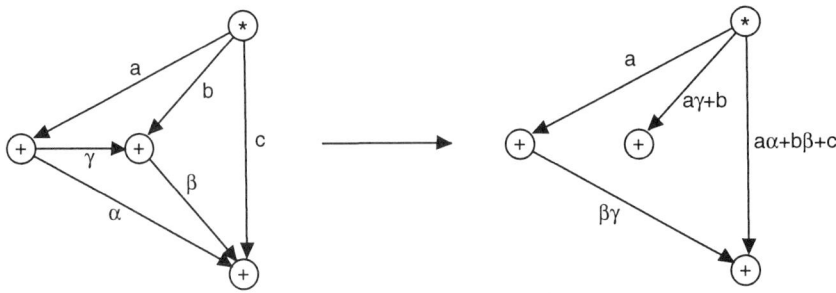

$$\begin{pmatrix} 0 & a & b & c \\ 0 & 0 & \gamma & \alpha \\ 0 & 0 & 0 & \beta \\ 0 & 0 & 0 & 0 \end{pmatrix} \begin{pmatrix} 0 & 0 & 0 & 0 \\ 0 & 0 & \gamma & \alpha \\ 0 & 0 & 0 & \beta \\ 0 & 0 & 0 & 0 \end{pmatrix} + \begin{pmatrix} 0 & a & b & c \\ 0 & 0 & 0 & 0 \\ 0 & 0 & 0 & 0 \\ 0 & 0 & 0 & 0 \end{pmatrix} = \begin{pmatrix} 0 & a & a\gamma + b & a\alpha + b\beta + c \\ 0 & 0 & 0 & \beta\gamma \\ 0 & 0 & 0 & 0 \\ 0 & 0 & 0 & 0 \end{pmatrix}.$$

FIGURE 16.2
Matrix-Multiply example.

Essentially, the matrix multiplication in (16.7) collects the combined weights from any node to a second level addition node through intermediate addition nodes. We add to that weight the weight that already is on the edge between those nodes. However, if those two nodes are addition nodes, no weight must be added, since that weight will be moved upward to a connection from an earlier node to the target node, explaining the definition of $W[\![\,.\,,\,.\,]\!]$. Note that this procedure does not alter the fan-in into a multiplication node, and the formal Definition 16.1 for an algebraic circuit remains valid for the resulting circuit.

PROCEDURE 16.2
Rake
Whenever all children of an interior node have become leaves, we can compute the value of that interior node. This is done as follows:

> **for all** addition nodes v all of whose children are leaves **do**
>   { set v to a leaf node
>   value$(v) \leftarrow \sum_{u \text{ a child of } v} W(u,v) \cdot u$
>   remove the edges (u,v) for all children u of v }

16.2. Algebraic Circuit Compression

for all multiplication nodes v all of whose children are leaves **do**
  { set v to a leaf node
    value$(v) \leftarrow \prod_{u \text{ a child of } v} W(u,v) \cdot u$
    remove the edges (u,v) for all children u of v }

The procedure essentially removes leaves from active participation in the computation, hence the name *rake*.

PROCEDURE 16.3
Shunt
This procedure processes those multiplication nodes one of whose children is a leaf. Then one can pass those leaf values together with the edge weights through to the edge between the nonleaf child and the parents of the multiplication node under consideration. Figure 16.3 exhibits this action on a small circuit, while the following code defines this procedure formally:

for all multiplication nodes v_j that have exactly one leaf child v_l and
         one child v_i that is an interior node **do**
for all parents v_k of v_j **do**
    $P_{i,j,k} \leftarrow W(v_i, v_j) \cdot W(v_j, v_k) \cdot W(v_l, v_j) \cdot \text{value}(v_l)$
for all pairs (i, k) considered previously **do**
    { $W'_{i,k} \leftarrow \sum_{\text{all considered intermediate } j} P_{i,j,k}$
    $W(v_i, v_k) \leftarrow W(v_i, v_k) + W'_{i,k}$
    set the weight of the edge (v_i, v_k) to $W(v_i, v_k)$, unless
    $W(v_i, v_k) = 0$, in which case the edge is removed}
for all (j, k) considered previously **do**
    remove the edge (v_j, v_k)

The **procedure** *Phase* is the consecutive execution of these three procedures in the order with which we presented them. The **algorithm** *Circuit Compression* now performs a sequence of phases until every node in the circuit has been converted into a leaf.

We first observe that the value of each node in the input circuit (see Definition 16.2) is not changed by any of our circuit transformation procedures. The formal proof is done by induction on the nodes $v_1, \ldots, v_n$ going from lowest level (the leaves) to the highest level. It is in this argument that we appeal to the axioms of a commutative semiring. Especially, the commutativity of multiplication plays a significant role; in fact, it is an open problem how

738 Chapter 16. Dynamic Parallel Evaluation of Computation DAGs

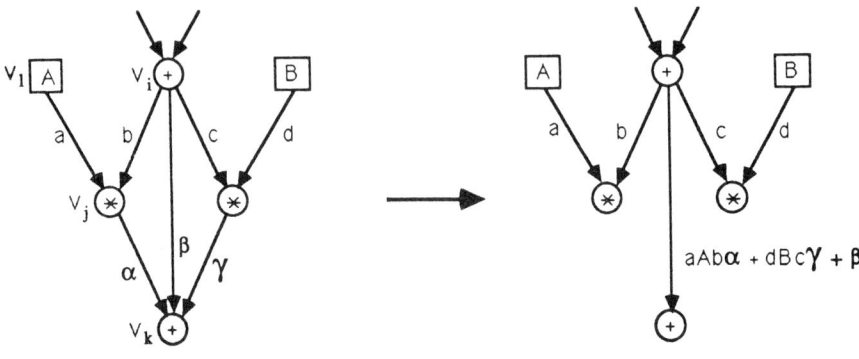

FIGURE 16.3
Shunt example.

to do without commutativity while remaining equally efficient (see the conclusion). Second, we note that the algorithm eventually obtains the value of every node. In fact, it will complete this quite quickly. We will prove next that after $O(\log(nd))$ phases all nodes are leaves, but it should be clear that the algorithm makes some progress towards completion in every one of the phases.

We have given a high level formulation of our algorithm as it would run on an algebraic random access machine (RAM) with the capability of performing semiring arithmetic. However, in the introduction we have emphasized the circuit transformation aspect of our result. Hence, one has to visualize an algorithm that does not actually perform the semiring arithmetic stated in our procedures, but that hardwires the corresponding algebraic subcircuits together for the new compressed circuit.

16.2.2 Analysis of the Compression Algorithm

We will carry out the run-time analysis in two steps. First, we are going to show that each individual procedure in a phase can be carried out efficiently on a PRAM. Second, we will show that the algorithm needs no more than $O(\log(nd))$ phases to complete, where n is the number of nodes and d is the formal degree of the circuit.

Our first goal is quite straightforward. The reader may consider a PRAM that has the circuit by means of a vector of leaf values and the matrix of weights W symbolically represented in its shared memory. In carrying out the Matrix Multiplication procedure using $M(n)$ processors, it stores the necessary circuit fragments in a place of shared memory that builds the compressed new circuit. The time needed is $O(\log n)$, as is in the matrix multiplication algorithm used. The details are somewhat technically involved, and we will not expand on them here. Exercise 16.7 below deals with the precise argument, and a reference to the solution of a very similar problem can be found there. The Rakes can be encoded by no more than n balanced binary addition or multiplication trees of no more than n leaves, which can be done by an n^2-processor PRAM in $O(\log n)$ time. A similar processor count/time measure yields the Shunts: The key observation is that every $P_{i,j,k}$ corresponds to exactly one edge (v_j, v_k), because the multiplication node v_j has fan-in 2. Thus, there are $O(n^2)$ elements $P_{i,j,k}$ to be considered for the summations to obtain $W'_{i,k}$. One uses a PRAM sorting procedure (say the one by Cole [9], also in Chapter 10 in this book) with the pairs of subscripts (i, k) as the keys. Then one sums up all $P_{i,j,k}$ with the same key, which are now located in adjacent memory locations. In summary, we can carry out procedure Shunt on a PRAM with n^2 processors in $O(\log n)$ parallel time.

We now estimate the number of phases necessary to evaluate every node in the input circuit. In order to obtain a sufficiently precise count, we measure the progress made in each such phase. A first idea is to express this progress in terms of the formal degrees of the circuits before and after each phase. Clearly, the procedures Shunt and Rake on multiplication nodes affect the formal degree of some nodes. But the procedure Rake on addition nodes and, more importantly, the procedure Matrix-Multiply may not, especially along long chains of addition nodes. For this reason, we introduce an artificial new formal degree, which we will name the *height* and which is similar to the formal degree (see Definition 16.2), but which also grows along paths containing only addition nodes.

DEFINITION 16.4

The height of a leaf is defined as the integer 1. The height of an interior node v is defined recursively from the heights of all its children: if v is an addition node we define

$\text{height}(v) = \max\left\{a + \frac{1}{2}, m\right\}$ *where*

$$\begin{cases} a = \max\{\text{height}(u) \mid u \text{ an addition child of } v\} \\ m = \max\{\text{height}(u) \mid u \text{ a multiplication child or a leaf child of } v\} \end{cases}$$

and if v is a multiplication node we define

$$\text{height}(v) = \sum_{u \text{ a child of } v} \text{height}(u).$$

Finally, we define the height of the entire circuit as the maximum of the heights of any node.

The following two lemmas are the key to the analysis of the algorithm. The proofs for the lemmas, which are based on induction on the structure of the circuit, are given at the end of this subsection. The first lemma estimates the progress made in each phase with respect to the height measure.

LEMMA 16.1
Consider a node v in an algebraic circuit C. We apply a single Phase to the circuit, obtaining a compressed circuit C' with the node v' corresponding to v. Now assume that the node v' in the circuit C' is not a leaf node or a node with fan-out 0, a so-called output node. Then the heights of the node before and after the single Phase compression are related by

$$\text{height}(v') \leq \tfrac{1}{2} \text{height}(v)$$

The second lemma relates the height of a circuit to its formal degree:

LEMMA 16.2
Consider an algebraic circuit: Let d be its formal degree, h its height, and $E[\![+,+]\!]$ the subset of its edges that go from an addition to an addition node. Then

$$h \leq \tfrac{1}{2} d e + d \quad \text{where e is the cardinality of $E[\![+,+]\!]$}$$

From these lemmas we can easily deduce an upper bound for the running time of our algorithm.

THEOREM 16.1
The Circuit Compression algorithm completes in no more than

$$O(\log(n) + \log(d))$$

consecutive applications of the procedure Phase, where n is the number of nodes and d is the formal degree of the input circuit.

PROOF
The number e of addition-to-addition edges is $O(n^2)$, hence the height of the input circuit (and by its definition the height of any of its nodes) is by Lemma 16.2 of order $O(n^2 d)$. Now, by Lemma 16.1 each Phase reduces the height of an interior node that does not become an output node by a divisor of at least 2. Hence after $O(\log_2(n^2 d))$ Phases there can be only leaves and output nodes left in the circuit. An additional Phase will by virtue of the procedure Rake evaluate all output nodes. The number of Phases necessary in the Compression algorithm is therefore bounded by $c \log_2(n^2 d) + 1 = O(\log(nd))$, where c is constant. ∎

16.2. Algebraic Circuit Compression

There are several corollaries that immediately follow from this theorem and the processor count given at the beginning of this subsection. The first corollary shows that all polynomials of polynomial formal degree—even quasi-polynomial, as we will state it—can be computed in polylogarithmic depth.

COROLLARY 16.1
Given is a multivariate polynomial $f(x_1, \ldots, x_m)$ with coefficients from a commutative semiring. Suppose that f is the value of an n-node algebraic circuit with leaf values $x_1, \ldots, x_m$, whose formal degree is

$$n^{O(\log(n)^c)} \quad \text{for a constant } c \geq 0,$$

i.e., quasi-polynomial in n. Then f is the value of an algebraic circuit with leaf values $x_1, \ldots, x_m$ that has bounded fan-in and

$$n^{O(1)} \text{ many nodes and } O(\log(n)^{c+2}) \text{ depth.}$$

The second corollary addresses the problem of computing all values of an algebraic circuit in parallel. Since we have not formally introduced the notion of an algebraic PRAM here, we shall formulate the corollary for integers as our semiring.

COROLLARY 16.2
Consider an exclusive-read/exclusive-write PRAM that has the description of an n-node algebraic circuit over the integers stored in its shared memory. Let d be the formal degree of the circuit, and assume that the PRAM has $M(n)$ many processors. Than the values of all nodes of the circuit can be computed in $O(\log(n) \log(nd))$ arithmetic parallel PRAM steps, where arithmetic refers to the fact that we will count an integer arithmetic operation as a single step independently of how many bits the individual integer operands have.

Notice that this corollary shows how to find the formal degree of an algebraic circuit dynamically. One replaces the leaf values by the integer 1, addition nodes by the maximum function, and multiplication nodes by integer addition. The Circuit Compression algorithm executed on a PRAM will complete in $O(\log(n) \log(dn))$ time, where d is the wanted formal degree. Note that d appears in the run-time estimate, but nothing needs to be known for it (cf. Theorems 16.2 and 16.3 below).

We conclude this subsection with the proofs of the Lemmas 16.1 and 16.2. Both proofs will be by induction on the size of the circuit. We will consider the subcircuit C_v for a circuit C and one of its nodes v, which we define as the circuit of all those nodes and edges of C that participate in the determination of the value of v (see Definition 16.4). We will also use the notion of a *dominant child* of an addition node v. If among the children of v there is a multiplication or leaf node u with $\text{height}(u) = \text{height}(v)$, we call u a dominant child. If there is no such node, we call a child u with $\text{height}(v) = \text{height}(u) + \frac{1}{2}$ a dominant child. Essentially, a dominant child determines the height of v.

PROOF OF LEMMA 16.1

We consider the node v' in the circuit C'. The proof proceeds by induction on the depth of the subcircuit $C'_{v'}$. The basis case is where all children of v' are leaves. We argue this case first for v' being an addition node. We must show that the height of v is no less than 2, where v is the node corresponding to v' before the application of a Phase. Assume to the contrary that $\text{height}(v) \leq 3/2$. By the definition of height, the depth of C_v can be no more than 2 and cannot contain a multiplication node. Clearly the procedures Matrix-Multiply and Rake would convert v into a leaf node in C', in contradiction to our assumption that v' is not a leaf.

The second part of the basis case is when v' is a multiplication node. It suffices to show that both children of v have height at least 2. Suppose to the contrary that a child u has $\text{height}(u) < 2$. After raking addition nodes, this child will be a leaf. If the second child is also a leaf at that point, raking the multiplication nodes will turn v' into a leaf. Otherwise, the procedure Shunt will disconnect v' from the rest of C', thus making v' an output node. Both are in contradiction to our assumption that v' is not a leaf or an output node.

We now present the inductive argument. The easier case here is when v' is a multiplication node. If none of its children u'_1 and u'_2 are leaves, we have by the induction hypothesis

$$\begin{aligned}\text{height}(v') &= \text{height}(u'_1) + \text{height}(u'_2) \\ &\leq \tfrac{1}{2}\left(\text{height}(u_1) + \text{height}(u_2)\right) \\ &= \tfrac{1}{2}\text{height}(v).\end{aligned}$$

If one of its children u'_i is a leaf, we note that $\text{height}(u_i) \geq 2$, for otherwise the node v would have participated in a Shunt and been turned into an output node, as was argued in the basis case. Thus the same inequality holds, leaving only the case where v' is an addition node.

16.2. Algebraic Circuit Compression

Let u' be a dominant child of the addition node v'. If u' is a multiplication node, then

$$\text{height}(v') = \text{height}(u') \leq \tfrac{1}{2}\text{height}(u) \leq \tfrac{1}{2}\text{height}(v)$$

by the induction hypothesis and the fact that the height function does not decrease along paths. The node u' cannot be a leaf, because otherwise that case would be the basis case. Thus we are left with the possibility that u' is an addition node. It suffices to establish in this last case that

$$\text{height}(u) \leq \text{height}(v) - 1 \qquad (16.8)$$

for then we have by induction hypothesis

$$\text{height}(v') = \text{height}(u') + \tfrac{1}{2} \leq \tfrac{1}{2}\text{height}(u) + \tfrac{1}{2} \leq \tfrac{1}{2}\text{height}(v)$$

It remains to show (16.8). Note that both u and v are addition nodes. If there is a path from u to v containing an additional node, (16.8) follows from Definition 16.2. Finally, suppose that the *only* path from u to v in C is a single edge. This configuration is impossible, however, since the procedure Matrix-Multiply will remove that edge, and no edge (u', v') can be introduced because there is no other path from u to v. ∎

PROOF OF LEMMA 16.2
The proof is by induction on the number of nodes in C. Clearly, for a single node circuit, which must be a single leaf, the estimate is true with $e = 0$ and $h = d = 1$. Suppose now that the theorem is true for all circuits with less than n nodes, and let C be a circuit with n nodes. For every non-output node u we have by induction hypothesis

$$\text{height}(u) \leq \text{degree}(u)\left(\frac{e_u}{2} + 1\right) \leq d\left(\frac{e}{2} + 1\right), \qquad (16.9)$$

where e_u is the number of addition to addition edges in C_u, which satisfies $e_u \leq e$, because the circuit C_u is a subcircuit of C. It thus remains to prove the estimate for every output node v in C. First let v be a multiplication node, and let u_1 and u_2 be its children. Then by (16.9) we have

$$\begin{aligned}\text{height}(v) &= \text{height}(u_1) + \text{height}(u_2) \\ &\leq (\text{degree}(u_1) + \text{degree}(u_2))\left(\frac{e}{2} + 1\right) \\ &\leq d\left(\frac{e}{2} + 1\right)\end{aligned}$$

Second, assume that v is an addition node. If the dominant child is a multiplication or leaf child, height(v) will be equal to the height of that child, which by (16.9) already satisfies the estimate. So we are finally left with the situation that the dominant child u of v is an addition child. We then have

$$\begin{aligned}\text{height}(v) &= \text{height}(u) + \tfrac{1}{2} \\ &\leq \text{degree}(u)\left(\frac{e_u}{2}+1\right) + \tfrac{1}{2} \\ &\leq d\left(\frac{e-1}{2}+1\right) + \tfrac{1}{2}\end{aligned} \qquad (16.10)$$

because $d \geq \text{degree}(v) \geq \text{degree}(u)$ and $e_u + 1 \leq e$, the latter since the subcircuit C_u does not contain the edge (u,v). Expanding out the right-hand side of (16.10) and observing that $d \geq 1$ completes the proof. ∎

16.2.3 Exercises

EXERCISE 16.5
Modify the description of the Circuit Compression algorithm so that it is valid over a semiring without a multiplicative unit. [Hint: The problem is that the Matrix-Multiply procedure may generate the addition $1+1$ for the formal unit edge weight; introduce an additional integer weight on each edge with the interpretation that $2 \cdot a = a + a$ for any semiring element a.]

EXERCISE 16.6
Show how to accomplish circuit compression over a fixed finite non-commutative ring. [See Miller and Teng [36] for a solution.]

EXERCISE 16.7
This exercise relates the problem of building a description of circuits on a standard PRAM to evaluating circuits on an algebraic PRAM. Given is a $P(n)$-processor algebraic PRAM that evaluates an algebraic circuit in $O(T(n))$ steps without testing an element of the semiring for zero. Show that there exists a standard $P(n)$-processor PRAM that on the same input constructs in $O(T(n))$ time the formal description of a circuit of size $O(P(n)\,T(n))$ and depth $O(T(n))$ that computes all values of the input circuit. [The solution to the sequential counterpart of this problem can be found in Kaltofen [21], Theorem 4.1; the notion of an algebraic RAM is also defined there.]

EXERCISE 16.8
Consider an $n \times n$ matrix with polynomial entries of degree no more than n and coefficients from a ring. Construct a circuit of size $n^{O(1)}$ and depth $O(\log(n)^2)$ that computes the coefficients of the determinant polynomial of the matrix.

EXERCISE 16.9
Exhibit an n-node circuit of formal degree polynomially bounded in n, such that the Circuit Compression algorithm requires $\Omega(\log(n)^2)$ parallel steps.

EXERCISE 16.10*
Given are $n \times n$ matrices $A_1, A_2 \ldots$ and an n-dimensional column vector b over a ring. Consider the sequential circuit of size $O(n^2 m)$ for computing $A_1 \cdots A_m b$ by multiplication from right to left. Give a sharper than $O(M(n^2 m))$ estimate for the number of nodes in the circuit produced by the Circuit Compression algorithm. [Open problem: Is there a circuit of depth $O(\log(n) \log(m))$ and size $O(n^2 m)$ that computes that matrix chain product?]

EXERCISE 16.11*
Given is a non-singular $n \times n$ matrix over a field. Find a circuit for the Gram-Schmidt orthogonalization of that matrix that has $n^{O(1)}$ nodes and $O(\log(n)^2)$ depth Kozen [29].

The following exercises are taken from Leighton et al. [31], Problem Set 1.

EXERCISE 16.12
Show that for a ring, the Compression Algorithm can produce a parallel circuit of depth $O(\log(n) \log(d))$, where n is the number of nodes of the input circuit and d is the formal degree. [Hint: Eliminate all plus-plus edges in the input circuit by applying the procedure Matrix Multiply repeatedly before performing the full compression.]

EXERCISE 16.13*
Consider an algebraic circuit C with n nodes and formal degree d over the integers. Construct a Boolean circuit that evaluates C for l-bit integers, and that has $(dnl)^{O(1)}$ many gates and $O(\log(dnl) \log(dn))$ delay. [Hint: One can add N L-bit integers in $O(\log(LN))$ delay (Muller and Preparata [38]). The value of each node v, $v) \in \mathbb{Z}[x_1, \ldots, x_m]$, where the x_i are indeterminates for the leaf values. Since $\|v)\|_1 \leq 2^{d+n}$ (cf. Lemma 16.2), where $\|v)\|_1$ denotes the sum of the absolute values of the

integral coefficients of v) as a polynomial in $x_1, \ldots, x_m$, all node values are bounded absolutely by 2^{dl+d+n}. It therefore suffices to compute the node values modulo 2^L with $L = dl + d + n + 1$. Now the procedures Matrix-Multiply, Rake, and Shunt add up at most $O(n^2)$ integers with L bits, which by the above reference is doable in $O(\log(dln))$ delay. The remaining multiplications modulo 2^L are doable in $O(\log L)$ delay (Wallace [47]), hence each Phase can be done in delay $O(\log(dnl))$.] Note that this exercise shows that one can compute the determinant of an $n \times n$ matrix with the entries being n-bit integers on a Boolean circuit of depth $O(\log(n)^2)$; this specific result was proven in (Borodin et al. [7]).

16.3 Related Questions

In this section we survey results and problems that are related to the Circuit Compression algorithm. We shall not give complete proofs for the theorems stated here, but we confine ourselves to sketching the ideas and/or giving the reference to the literature.

16.3.1 Circuits with Division

In this subsection we only consider circuits over a field K. It is natural to permit division nodes in such circuits. Note that a division node has fan-in 2, and that the in-edge for the dividend is distinguished from the in-edge for the divisor. The Gaussian elimination circuit for computing the determinant described in Section 16.1.1 is an example of a circuit with divisions. There is the possibility of introducing a division by zero into such circuits. By definition, we exclude circuits that always divide by zero, no matter what input values the leaves assume. For example, the circuit corresponding to the program

$$v_1 \leftarrow x; \; v_2 \leftarrow v_1 - x; \; v_3 \leftarrow 1 \div v_2$$

is excluded from consideration, since v_2 always assumes the value 0. As values for leaves we consider elements from the algebraic closure of K.[3] Thus, any such circuit must not divide by zero when retaining the leaf values as indeterminates $x_1, \ldots, x_m$, i.e., when evaluating over the field of rational functions $K(x_1, \ldots, x_m)$. Note that for finite fields K the problem of deciding whether

---
[3] If K is the field of rational numbers, the algebraic closure of K is contained in the field of complex numbers.

a circuit divides by zero for all leaf values from K is co-$\mathcal{N}P$-complete (Ibarra and Moran [17]).

The first problem concerns circuits with divisions that are used to compute selected outputs $f_i(x_1,\ldots,x_m)$, $1 \leq i \leq k$, that are polynomials in $\mathsf{K}[x_1,\ldots,x_m]$. Again, the Gaussian elimination determinant circuit falls into this category. A fundamental result, due to Strassen [42], shows how to avoid the use of divisions for computing all f_i. We first state the theorem and then give an idea for its proof.

THEOREM 16.2
Let $f_i(x_1,\ldots,x_m) \in \mathsf{K}[x_1,\ldots,x_m]$, $n-k < i \leq n$, be k selected polynomial values of an n-node circuit with divisions and that has as leaf values the indeterminates $x_1,\ldots,x_m$. Suppose we are given a bound δ for the algebraic degrees of all f_i, and the values of all nodes for a specific input $x_1 \leftarrow a_1 \in \mathsf{K}, \ldots x_m \leftarrow a_m \in \mathsf{K}$. Note that the occurring divisors must not evaluate to zero at those input points $a_1,\ldots,a_m$. Then we can (on a PRAM) construct the description of a circuit without divisions that also computes all f_i, $n-k < i \leq n$, that has

$$O(n\,\delta \log(\delta) \log(\log \delta)) \ \text{nodes}$$

and that is of formal degree no more than δ.

The idea behind this construction is similar to our characteristic matrix trick for eliminating divisions from the Gaussian elimination circuit. For $1 \leq i \leq n$ let $f_i(x_1,\ldots,x_m) \in \mathsf{K}(x_1,\ldots,x_m)$ be the rational function values of all nodes. We consider the auxiliary functions

$$g_i(y_1,\ldots,y_m,z) = f_i(y_1 z + a_1,\ldots,y_m z + a_m). \qquad (16.11)$$

Since $g_i(0,\ldots,0,z) = f_i(a_1,\ldots,a_m) \in \mathsf{K}$, the coefficient $c_{i,0}$ of the constant term of g_i as an extended power series in z over $\mathsf{K}(y_1,\ldots,y_m)$,

$$g_i(y_1,\ldots,y_m,z) = \sum_{j=-l_i}^{+\infty} c_{i,j}(y_1,\ldots,y_m) z^j, \quad l_i \geq 0, \qquad (16.12)$$

is an element in K and is given as input. Furthermore, if the i-th node is the divisor of another node, then $c_{i,j} \neq 0$, for otherwise the circuit evaluated at $a_1,\ldots,a_m$ would divide by zero. We also remark that for $n-k < i \leq n$ the coefficients $c_{i,j}$ are equal 0 for all $j < 0$ or $j > \delta$, because the corresponding f_i are polynomials of degree no more than δ and the extended power series representation (16.12) is a canonical, thus unique, form.

It now follows easily by induction on the depth of the input circuit that no g_i has terms in z of negative order. That is, for all $j < 0$ we have $c_{i,j} = 0$; and it follows that for all $j > 0$ the $c_{i,j}(y_1,\ldots,y_m)$ are polynomials in $\mathsf{K}[y_1,\ldots,y_m]$. The division-free output circuit of the above theorem computes the $c_{i,j}(y_1,\ldots,y_m)$ for $1 \leq j \leq \delta$ by truncated power series arithmetic. Finally, the original output polynomials f_i, $n - k < i \leq n$, are computed as

$$f_i(x_1,\ldots,x_m) = f_i(a_1,\ldots,a_m) + \sum_{j=1}^{\delta} c_{i,j}(x_1 - a_1,\ldots,x_m - a_m),$$

which undoes the translations in (16.11) by setting $z = 1$ and $y_i = x_i - a_i$. The exact circuit transformation operations are described in Kaltofen [21], Section 7.

Strassen used this result to prove that divisions do not aid in the construction of asymptotically fast $n \times n$ matrix multiplication circuits. In that setting the f_i are the n^2 row-column inner products, whose algebraic degree is two. In light of the Circuit Compression algorithm, which excludes division nodes, the method has been scrutinized again. Take the Gaussian elimination circuit (16.4), for instance. A point at which the circuit avoids zero-division is the identity matrix I_n. We certainly know the values of all $a_{j,k}^{(i)}$ for that matrix. Thus, by combining Theorem 16.2 and Corollary 16.1 we have found a division-free circuit for the $n \times n$ determinant that has depth $O(\log(n)^2)$ and $n^{O(1)}$ many nodes.[4] Note that the method generates a circuit of formal degree no more than δ (cf. Exercise 16.2). This transformation also shows that if one is given a bound for the algebraic degree of a polynomial computed by a circuit of much higher formal degree, that circuit can be transformed into one whose formal degree is no more than the actual degree bound of the polynomial.

There are two obvious questions that one may ask at this point: How does one find, in general, a degree bound δ or an appropriate point set $a_1 \ldots, a_m$ and the corresponding node values? And, what if the circuit computes a rational function in $\mathsf{K}(x_1,\ldots,x_m)$ rather than a polynomial? The problem of finding a point that avoids zero-division in a circuit is intimately related to the problem of avoiding the zeros of a multivariate polynomial. A sequential Monte-Carlo randomized polynomial-time algorithm, based on a lemma by Schwartz [40] (see also Zippel [48]), is described in detail in

---

[4] n exponent "big-oh" of 1 indeed! Plugging in the upper bounds from both constructions we get $O(M(n^4 \log(n) \log(\log n)))$ many nodes in the circuit. This ring-independent construction is due to Borodin et al. [7].

Kaltofen [21], Lemma 4.2. We note that for finite fields K of small cardinality one can simulate evaluation of the input circuit at values from a sufficiently large algebraic extension by arithmetic in K itself (Kaltofen [20], Lemma 1) thus keeping the methods of Theorem 16.2 valid for circuits over any field K. The maximum degree of all output polynomials f_i can be determined by a sequential Monte-Carlo polynomial-time algorithm as well (see Kaltofen [21], Section 5, Kaltofen and Trager [24], Section 2, Footnote). However, all known solutions to the first question are sequential, as they require the evaluation of the input circuit at random leaf values.

The second question concerns the division free computation of the polynomial numerator and denominator of the rational function value of a circuit. It is assumed the the numerators and denominators are reduced, i.e., that they do not share a polynomial common factor. This problem was first posed by Strassen in [42] and was settled by the author of this chapter in 1985. Note that the solution is needed if the Circuit Compression algorithm were to be applied to circuits computing rational functions rather than polynomials. Using power series expansions in an auxiliary indeterminate, as we did above for polynomial outputs, and Padé approximations to such series, one can obtain the following theorem (see Kaltofen [21], Section 8, and Kaltofen and Trager [24], Section 3).

THEOREM 16.3
Let $f, g \in \mathsf{K}[x_1, \ldots, x_m]$ be two relatively prime polynomials over the infinite field K. Given is an n node algebraic circuit that computes from leaf values $x_1, \ldots, x_m$ the rational function f/g, and given is an upper bound δ for the degrees of f and g. Then there exists a circuit with

$$O(n\,\delta + \delta \log(\delta)^2 \log(\log \delta))$$

many nodes that computes the polynomials cf and g/c where c is a non-zero element of K.

As for the first set of problems, this construction of the circuit for f and g can again be realized by a sequential polynomial-time Monte-Carlo randomized algorithm, and remains valid for small finite fields but with a worse asymptotic node count. Moreover, one can also determine the degrees of f and g by a sequential Monte-Carlo polynomial-time algorithm (see the references to the theorem). Now, we can use the methods in Theorem 16.2 to remove the divisions in the circuit constructed in Theorem 16.3, and then use the Circuit Compression algorithm to obtain the following far-reaching corollary.

COROLLARY 16.3

Let $f, g \in \mathsf{K}[x_1, \ldots, x_m]$ be two relatively prime polynomials of degree no more than d with coefficients from the field K such that f/g is computed by an algebraic circuit with n nodes. Then f/g can also be computed by an algebraic circuit with $(nd)^{O(1)}$ many nodes and with depth $O(\log(nd)^2)$.

16.3.2 Reducing the Fan-Out

In Section 16.2.1 we constructed algebraic circuits of bounded fan-in and shallow depth. There are situations where one wishes to control the maximum fan-out as well. A simple idea is to replace a node of high fan-out by a binary tree of duplication nodes, say addition nodes with fan-in 1 and fan-out 2. As the fan-in is assumed to be bounded, it is easy to see that the number of nodes in the new circuit stays within a constant factor of the original number of nodes. Choosing balanced binary trees might, however, increase the depth of the new circuit asymptotically by a log factor of the maximum original fan-out. Hoover et al. [15] select nonbalanced binary trees for the duplication process in such a way that a distance measure from the source node to all possibly reachable output nodes is optimized (see Figure 16.4). They can then prove the following theorem.

THEOREM 16.4

Given is an algebraic circuit with n nodes and of depth t such that no node has fan-in higher than 2. Then there is an algebraic circuit with $O(n)$ nodes and depth $O(t)$ where no node has fan-out higher than 2, and a subset of whose nodes determines the values of all nodes of the input circuit.

16.3.3 Partial Derivatives

The circuit transformation of the previous subsection preserved size and depth within a constant factor. We now give an algebraic transformation that also shares this property. Again, let us consider circuits with divisions over a field K. By labeling the input nodes with the indeterminates $x_1, \ldots, x_n$, a selected output node has a rational function $f(x_1, \ldots, x_m) \in \mathsf{K}(x_1, \ldots, x_m)$ as its value. At issue is to construct a circuit that has all the first order partial derivatives

$$\frac{\partial f}{\partial x_1}, \ldots, \frac{\partial f}{\partial x_m} \qquad (16.13)$$

as a subset of its values. One certainly can compute the value of a single derivative D (defined on $\mathsf{K}(x_1, \ldots, x_m)$) along with the value of each node

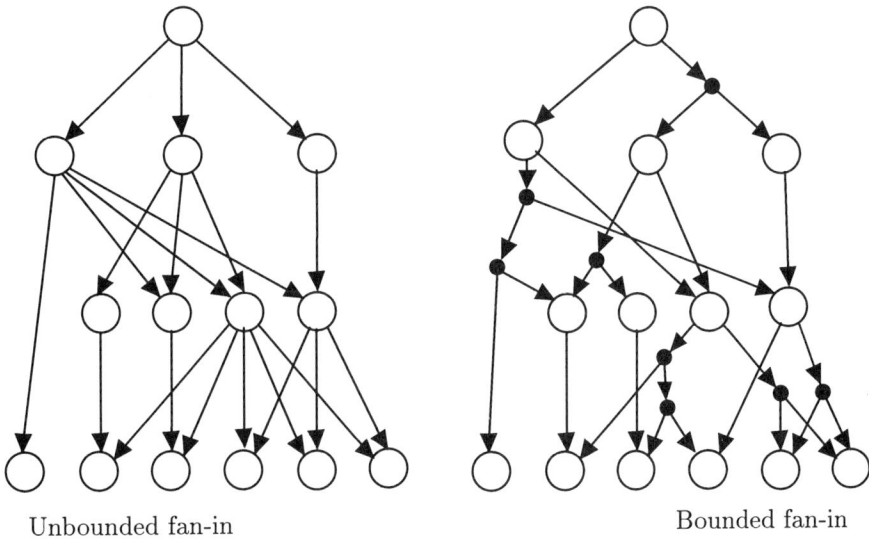

Unbounded fan-in   Bounded fan-in

FIGURE 16.4
Fan-out reduction.

by inductively applying the addition, product, and quotient rules for D. For example, for the assignment $v \leftarrow u_1 \div u_2$ we construct the circuit fragment for $v' \leftarrow (u_1' - v \times u_2')/u_2$, where the nodes u_1', u_2', and v' attain the values $D(u))$, $D(v))$, and $D(v))$, respectively. The resulting circuit is of size no more than 4 times the input circuit that also computes $D(f)$. It is Baur and Strassen's [5] accomplishment to establish that all partial derivatives (16.13) of f can be computed in essentially the same number of operations. Using their construction in conjunction with Theorem 16.4, we can actually show that the depth is preserved as well (Kaltofen and Singer [23]).

THEOREM 16.5
Given an n node algebraic circuit of depth t over the field K that computes in one of its output nodes the rational function $f \in \mathsf{K}(x_1, \ldots, x_m)$, where $x_1, \ldots, x_m$ are the leaf values. Then there is a circuit with no more than $4n$ nodes and with depth $O(t)$ that computes in a subset of its output nodes f and all first order partial derivatives $\partial f/\partial x_i$, $1 \leq i \leq m$.

Baur and Strassen's original motivation is still useful in the setting of parallel algebraic complexity. Consider a circuit that computes the determinant

$\Delta(x_{1,1}, \ldots, x_{n,n})$ of an $n \times n$ matrix $A = [x_{i,j}]_{1 \le i,j \le n}$ with leaf values $x_{i,j}$. Then by Theorem 16.5, the inverse of A can be computed by a circuit of the same asymptotic size and depth, namely

$$A^{-1} = \left[\frac{(-1)^{i+j}}{\Delta} \frac{\partial \Delta}{\partial x_{j,i}} \right]_{1 \le i,j \le n}$$

In fact, one can also establish for Theorem 16.5 that the circuit for the partial derivatives avoids any zero-division for all those inputs for which the input circuit for f does not divide by zero.

16.3.4 Exercises

EXERCISE 16.14*
Show that the construction by Hoover et al. [15] for Theorem 16.4 can be carried out on a PRAM in $O(\log(n)^2)$ time [Hint: See Atallah et al. [4] for a parallel construction of the optimal duplication trees.]

EXERCISE 16.15
Develop a theorem similar to Theorem 16.5 for circuits that allow fan-in 1 nodes of the form $v \leftarrow \sqrt{u}$, or $v \leftarrow \exp(u)$, or $v \leftarrow \log(u)$ (Morgenstern [37]).

EXERCISE 16.16*
Show that if a circuit over the complex numbers of size n with divisions and nodes of the form $v \leftarrow \sqrt{u}$ computes a polynomial f in the leaf values and that has degree d, then there exists a circuit of size $(dn)^{O(1)}$ without divisions and square roots that also computes f (Lickteig [32]). An example of such a circuit is the method of computing the determinant of an $n \times n$ real matrix A by orthogonalizing the matrix to U, computing the squares of the lengths of the basis vectors as $D = UU^T$, and then finding the determinant as $\text{Det}(A) = \sqrt{\text{Det}(D)}$. Note that the program is valid in an open neighborhood of the identity matrix in Euclidean n^2-space.

16.4 Recent Progress

Within the last two years, several problems related to the topic discussed here have been solved. The first is the question on the use of commutativity of the multiplicative operation in the circuit compression algorithm. It turns out that in noncommutative rings, it is not possible to achieve polylogarithmic speedup. Nisan [39] has shown that any circuit C_n that evaluates the function

$$f_n(x_1, x_2) = \sum_{\chi:\{1,\ldots,n\}\to\{1,2\}} x_{\chi(1)} x_{\chi(2)} \cdots x_{\chi(n)} x_{\chi(n)} \cdots x_{\chi(2)} x_{\chi(1)}$$
$$= x_1 f_{n-1}(x_1, x_2) x_1 + x_2 f_{n-1}(x_1, x_2) x_2$$

of formal degree $2n$ has depth $\Omega(n)$. Kosaraju [27] gives a similar family of functions over the semiring $\mathcal{L}$ of languages over $\{\mathsf{a},\mathsf{b},\mathsf{c}\}$ with $\oplus$ to be language union and $\otimes$ to be language concatenation, i.e.,

$$L_1 \oplus L_2 := \{w \mid w \in L_1 \text{ or } w \in L_2\}, \quad L_1 \otimes L_2 := \{w \mid w = uv \text{ with } u \in L_1$$
$$\text{and } v \in L_2\}$$

Any circuit computing the function

$$f_n(x) = \{\mathsf{a}\} \otimes f_{n-1}(x) \otimes \{\mathsf{a}\} \oplus \{\mathsf{b}\} \otimes f_{n-1}(x) \otimes \{\mathsf{b}\}, \quad f_0(x) = x$$

over $\mathcal{L}$ must have depth no less than n.

A second problem that has been solved in part is the question of constructing $\log(n)^{O(1)}$ deep circuits with $O(M(n))$ many arithmetic nodes that can compute the determinant of an $n \times n$ matrix; by Theorem 16.4 one then also can compute matrix inverses and solve linear systems. V. Pan and the author [22] prove that there exists a randomized algebraic circuit with n^2 inputs, $O(n)$ nodes that denote random (input) elements, and of

$$O(n^\omega \log(n)) \text{ size and } O(\log(n)^2) \text{ depth,}$$

with the following property. If the inputs are the entries of nonsingular matrix $A \in \mathsf{K}^{n \times n}$, where K is a field of characteristic zero or greater than n, and if the random nodes uniformly select field elements in $S \subset \mathsf{K}$, then with probability no less than $1 - 3n^2/\text{card}(S)$, the circuit outputs the determinant of A. On the other hand, if the random choices are unlucky or if the input matrix is singular, the circuit divides by zero. On nonsingular inputs zero-divisions occur with probability no more than $3n^2/\text{card}(S)$. A similar construction can certify the matrix to be singular.

A major set of open problems is concerned with the smallest depth possible to compute selected polynomials. For instance, is it possible to compute the power A^n of an $n \times n$ matrix A by a circuit of size $n^{O(1)}$ and depth $o(\log(n)^2)$, i.e., asymptotically better than $\log(n)^2$? Or, can one establish a lower-bound $\Omega(t(n))$, where $\lim_{n\to\infty} \log(n)/t(n) = 0$, for the minimum depth of any polynomial sized circuit that computes A^n? In fact, we know not a single such family of functions over a ring of polynomially bounded algebraic degree that would not be computable by formulas of polynomial size, while being computable by polynomial sized circuits. Some related lower-bound references are: Strassen [42], Jerrum and Snir [18], Kalorkoti [19]; see also Chapter 13 by von zur Gathen for $\log(n)$-reductions of related problems such as the determinant to matrix power.

Acknowledgement

I wish to express my gratitude to Jeff Ullman and Vijaya Ramachandran for their detailed remarks about this paper, to an anonymous referee and to several of my students in my Design and Analysis of Algorithms Course in the Fall 1990 for spotting misprints, and to my wife Hoang for preparing the figures in MacDraw.

Bibliography

[1] Abelson, H. and Sussman, G. J., *Structure and Interpretation of Computer Program*; MIT Press, Cambridge, MA, 1985.

[2] Abrahamson, K., Dadoun, N., Kirkpatrick, D. G., and Przytycka, T., "A simple parallel tree contraction algorithm," *J. Algorithms* **10**, pp. 287–302 (1989).

[3] Aho, A., Hopcroft, J., and Ullman, J., *The Design and Analysis of Algorithms*; Addison and Wesley, Reading, MA, 1974.

[4] Atallah, M. J., Kosaraju, S. R., Larmore, L. L., Miller, G. L., and Teng, S.-H., "Constructing trees in parallel," *Proc. 1989 ACM Symp. Parallel Algorithms and Architectures*, pp. 421–431 (1989).

[5] Baur, W. and Strassen, V., "The complexity of partial derivatives," *Theoretical Comp. Sci.* **22**, pp. 317–330 (1983).

[6] Borodin, A., Cook, S. A., and Pippenger, N., "Parallel computations for well-endowed rings and space-bounded probabilistic machines," *Information Control* **58**/1–3, pp. 113–136 (1983).

[7] Borodin, A., von zur Gathen, J., and Hopcroft, J. E., "Fast parallel matrix and GCD computations," *Inf. Control* **52**, pp. 241–256 (1982).

[8] Brent, R. P., "The parallel evaluation of general arithmetic expressions," *J. ACM* **21**, pp. 201–208 (1974).

[9] Cole, R., "Parallel merge sort," *SIAM J. Comput.* **17**/4, pp. 770–785 (1988).

[10] Coppersmith, D. and Winograd, S., "Matrix multiplication via arithmetic progressions," *J. Symbolic Comput.* **9**/3, pp. 251–280 (1990).

[11] Csanky, L., "Fast parallel matrix inversion algorithms," *SIAM J. Comput.* **5**/4, pp. 618–623 (1976).

[12] Eberly, W., "Very fast parallel polynomial arithmetic," *SIAM J. Comput.* **18**/5, pp. 955–976 (1989).

[13] Gantmacher, F. R., *The Theory of Matrices, Vol. 1*; Chelsea Publ. Co., New York, N. Y., 1960.

[14] Gibbons, A. and Rytter, W., *Efficient Parallel Algorithms*; Cambridge Univ. Press, Cambridge, 1988.

[15] Hoover, H. J., Klawe, M. M., and Pippenger, N. J., "Bounding fan-out in logical networks," *J. ACM* **31**/1, pp. 13–18 (1984).

[16] Hyafil, L., "On the parallel evaluation of multivariate polynomials," *SIAM J. Comp.* **8**, pp. 120–123 (1979).

[17] Ibarra, O. H., and Moran, S., "Probabilistic algorithms for deciding equivalence of straight-line programs," *J. ACM* **30**, pp. 217–228 (1983).

[18] Jerrum, M. and Snir, M., "Some exact complexity results for the straight-line computations over semi-rings," *J. ACM* **29**/3, pp. 874–897 (1982).

[19] Kalorkoti, K. A., "A lower bound for the formula size of rational functions," *SIAM J. Comp.* **14**, pp. 678–687 (1985).

[20] Kaltofen, E., "Single-factor Hensel lifting and its application to the straight-line complexity of certain polynomials," *Proc. 19th Annual ACM Symp. Theory Comp.*, pp. 443–452 (1987).

[21] Kaltofen, E., "Greatest common divisors of polynomials given by straight-line programs," *J. ACM* **35**/1, pp. 231–264 (1988).

[22] Kaltofen, E., and Pan, V., "Processor efficient parallel solution of linear systems over an abstract field," in *Proc. 3rd Ann. ACM Symp. Parallel Algor. Architecture*; ACM Press, pp. 180–191, 1991.

[23] Kaltofen, E. and Singer, M. F., "Size efficient parallel algebraic circuits for partial derivatives," *Tech. Report* **90-32**, Dept. Comput. Sci., Rensselaer Polytechnic Inst., Troy, N.Y., October 1990.

[24] Kaltofen, E. and Trager, B., "Computing with polynomials given by black boxes for their evaluations: Greatest common divisors, factorization, separation of numerators and denominators," *J. Symbolic Comput.* **9**/3, pp. 301–320 (1990).

[25] Karp, R. M. and Ramachandran, V., "Parallel algorithms for shared-memory machines," in *Handbook of Theoretical Computer Science, Algorithms and Complexity (Volume A)*, edited by J. van Leeuwen; Elsevier Science Publ., Amsterdam, pp. 869–941, 1990.

[26] Knuth, D. E., *The Art of Programming, Vol. 2, Semi-Numerical Algorithms, Ed. 2*; Addison Wesley, Reading, MA, 1981.

[27] Kosaraju, S. R., "On the parallel evaluation of classes of circuits," in *Foundations of Software Technology and Theoretical Computer Science*, Springer Lect. Notes Comput. Sci. **472**, edited by Nori, K. V. and Veni Madhavan, C. E.; pp. 232–237, 1990.

[28] Kosaraju, S. R. and Delcher, A. L., "Optimal parallel evaluation of tree-structured computations by raking," *Proc. AWOC 88, Springer Lec. Notes Comp. Sci.* **319**, pp. 101–110 (1988).

[29] Kozen, D., "Fast parallel orthogonalization," *SIGACT News* **18**/2, 47 (1987).

[30] Leighton, F. T., *Introduction to Parallel Algorithms and Architectures: Arrays, Trees & Hypercubes*; Morgan Kaufmann Publ., San Mateo, California, 1991.

[31] Leighton, T., Leiserson, C. E., Maggs, B., Plotkin, S., and Wein, J., *Advanced parallel and VLSI computation*; Research Seminar Series **RSS 1**; Lab. Comp. Sci., MIT, March 1988.

[32] Lickteig, T. M., "On semialgebraic decision complexity," *Tech. Report* **TR-90-052**, Internat. Computer Sci. Inst., Berkeley, California, September 1990. Habilitationsschrift.

[33] Mayr, E. W., "The dynamic tree expression problem," in *Concurrent Computations Algorithms, Architecture, and Technology*, edited by Tewksbury, S. K., Dickinson, B. W., and Schwartz, S. C.; Plenum Press, New York, pp. 157–179, 1987.

[34] Miller, G. L., Ramachandran, V., and Kaltofen, E., "Efficient parallel evaluation of straight-line code and arithmetic circuits," *SIAM J. Comput.* **17**/4, pp. 687–695 (1988).

[35] Miller, G. L. and Reif, J. H., "Parallel tree contraction Part 1: Fundamentals," in *Randomness in Computation*, Advances in Computing Research **5**, edited by S. Micali; JAI Press Inc., Greenwich, CT., pp. 47–72, 1989.

[36] Miller, G. L. and Teng, S.-H., "Dynamic complexity of computational circuits," *Proc. 19th Annual ACM Symp. Theory Comp.*, pp. 254–263 (1987).

[37] Morgenstern, J., "How to compute fast a function and all its derivations," *SIGACT News* **16**/4, pp. 60–62 (1985).

[38] Muller, D. E. and Preparata, F. P., "Bounds to complexities of networks for sorting and switching," *J. ACM* **22**/2, pp. 195–201 (1975).

[39] Nisan, N., "Lower bounds for non-commutative computation," in *Proc. 23rd Ann. ACM Symp. Theory Comput.*; ACM Press, pp. 410–418, 1991.

[40] Schwartz, J. T., "Fast probabilistic algorithms for verification of polynomial identities," *J. ACM* **27**, pp. 701–717 (1980).

[41] Strassen, V., "Berechnung und Programm I," *Acta Inf.* **1**, pp. 320–335 (1972). In German.

[42] Strassen, V., "Vermeidung von Divisionen," *J. reine u. angew. Math.* **264**, pp. 182–202 (1973). In German.

[43] Strassen, V., "Die Berechnungskomplexität von elementarsymmetrischen Funktionen und von Interpolationskoeffizienten," *Numer. Math.* **20**, pp. 238–251 (1973). In German.

[44] Strassen, V., "Algebraische Berechnungskomplexität," in *Anniversary of Oberwolfach 1984*, Perspectives in Mathematics; Birkhäuser Verlag, Basel, pp. 509–550, 1984.

[45] Valiant, L., Skyum, S., Berkowitz, S., and Rackoff, C., "Fast parallel computation of polynomials using few processors," *SIAM J. Comp.* **12**, pp. 641–644 (1983).

[46] Valiant, L. G. and Skyum, S., "Fast parallel computation of polynomials using few processors," *Proc. ICALP 81, Springer Lect. Notes Comput. Sci.* **118**, pp. 132–139 (1981).

[47] Wallace, C. S., "A suggestion for a fast multiplier," *IEEE Trans. Electronic Comput.* **EC-13**, pp. 14–17 (1964).

[48] Zippel, R. E., "Probabilistic algorithms for sparse polynomials," *Proc. EUROSAM '79, Springer Lec. Notes Comp. Sci.* **72**, pp. 216–226 (1979).

17

The Parallel Complexity of Logical Inference

Jeffrey D. Ullman

Department of Computer Science
Stanford University
Stanford, CA 94305
ullman@cs.stanford.edu

17.1
Logical Rules

Our model of logical inference is database-oriented; it generalizes the parallel execution of queries written in a relational database language such as SQL. The logic with which we deal is "datalog," that is, first-order "if-then" rules with no function symbols permitted in arguments. Formally, we shall write rules in the form

$$p(X_1, \ldots, X_n) \leftarrow S_1, \ldots, S_m$$

where $p(X_1, \ldots, X_n)$ is the *head* of the rule, and $S_1, \ldots, S_m$ are the *subgoals*, which collectively form the *body* of the rule. Each subgoal, like the head, has a form in which a predicate symbol (e.g., p in the head) is applied to a list of arguments ($X_1, \ldots, X_n$ in the head). In general, an argument can be either a variable or a constant. Conventionally, as in Prolog, capital letters will be used for variables, and lower-case letters for constants and predicates. We assume that any variable appearing in the head also appears in the body, but variables appearing in the body may not appear in the head.

EXAMPLE 17.1
The following rules

(1) cousin(X,Y) <- sibling(X,Y)
(2) cousin(X,Y) <- parent(X,Xp),parent(Y,Yp),cousin(Xp,Yp)

define generalized "cousins," where siblings are regarded as "0th cousins." Intuitively, we want $cousin(a, b)$ to be true if a and b have (not necessarily proper) ancestors c and d, respectively, such that c and d are siblings, and the number of generations from a to c is the same as the number of generations from b to d. Zero generations is a possibility, in which case $a = c$ and $b = d$.

For example, in rule (2), the head is $cousin(X, Y)$, and the subgoals are $parent(X, Xp), parent(Y, Yp)$, and $cousin(Xp, Yp)$. These three subgoals form the body. Rule (1) says that siblings are cousins, and rule (2) says that if X has a parent Xp, and Y has a parent Yp, and Xp and Yp are cousins, then X and Y are also cousins.

17.2
The Fixpoint Meaning of Logical Rules

Formally, the meaning of logical rules is defined by a fixpoint operation.

17.2. The Fixpoint Meaning of Logical Rules

To understand the meaning of rules, we require, for simplicity only, that predicate symbols are divided into two classes:

1. EDB (*extensional database*) predicates, which are defined by storing the tuples (lists of argument values) for which the predicate is true, and
2. IDB (*intensional database*) predicates, where the set of tuples for which the predicate is true is defined by the rules.

Only IDB predicates can appear in the head of a rule, although either kind can appear in the body. In Example 17.1, we suppose *parent* and *sibling* are EDB predicates; that is, there is a list of pairs (a, b) such that b is a parent of a, and similarly for *sibling*.

To compute the set of tuples for which each IDB predicate is true, we start by assuming no such predicate is true for any tuple. With this "estimate" of the IDB predicates, we substitute constants for the variables of the rules in all possible ways, causing each subgoal to become essentially a tuple of the relation for its predicate. If all the subgoals of the body become true, then we have inferred that the head, with the same substitution of constants for variables, is also true. That gives us a tuple for which the IDB predicate at the head is true. Thus, on the second round, we have some tuples for the IDB predicates, which may yield still more tuples, and so on.

However, this process cannot go on indefinitely. Because we require that a variable in the head also appear in the body, every tuple we infer must consist only of symbols that appear either in the database (i.e., in some tuple for an EDB predicate), or in the rules themselves. Assuming the database and rules are finite, we have a finite upper bound on the number of possible tuples for the various IDB predicates. Since the iterative discovery of new IDB tuples is monotone (a tuple discovered on one round will continue to be discovered on subsequent rounds), eventually we reach a point at which no new tuples are discovered on a round. Subsequent rounds will then repeat the same calculation, and we never discover any new tuples. We have thus converged to the least fixpoint of the rules and database. It is not hard to show that this fixpoint is exactly the set of IDB facts that may be inferred using the rules.

EXAMPLE 17.2

Suppose we have the rules from Example 17.1, with the database suggested by Fig. 17.1. There, horizontal lines represent the sibling relation, and vertical lines run from the parent at the upper end to the child at

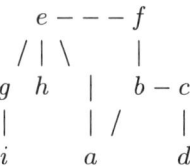

FIGURE 17.1
Example database

the lower end. On the first round, rule (2), which has an IDB subgoal, cannot infer anything, because whatever constants we substitute for the variables, $cousin(Xp, Yp)$ cannot be made true. The reason is that our initial estimate of *cousin* is that it is true for no tuple. However, rule (1) can be made to have a true body if we substitute $X = e$ and $Y = f$, because $sibling(e, f)$ is true. That substitution lets us infer the tuple $cousin(e, f)$. Similarly, substituting $X = b$ and $Y = c$ allows us to infer $cousin(b, c)$.

On the second round, we have these two *cousin* facts to make the third subgoal of rule (2) true. For example, the substitution $X = a$, $Xp = e$, $Y = b$, and $Yp = f$ makes the body of rule (2) have subgoals $parent(a, e)$, $parent(b, f)$, and $cousin(e, f)$, all of which are known to be true. Thus, we can infer the tuple $cousin(a, b)$. Similarly, on the second round, we can infer $cousin(g, b)$, $cousin(h, b)$, and $cousin(a, d)$.

We now have six *cousin* tuples, and these can all be used on the third round as ways to make the $cousin(Xp, Yp)$ subgoal true in rule (2). There is only one new tuple to infer on the third round. If we substitute $X = i$, $Xp = g$, $Y = a$, and $Yp = b$, we infer $cousin(i, a)$, using rule (2). The fourth round yields no new tuples, so the seven tuples inferred for *cousin*, namely the pairs

$$\{(e, f),\ (b, c),\ (a, b),\ (g, b),\ (h, b),\ (a, d),\ (i, a)\}$$

form the "meaning" of the IDB predicate *cousin* for this database.

17.3
Complexity of Logical Inference

For each *datalog program*, that is, collection of rules, we define the *inference problem* as follows. Given a database, i.e., finite sets of tuples for the

EDB predicates, and given a particular tuple t for some IDB predicate p, determine whether $p(t)$ follows from the database. Put another way, we want to determine whether the least fixpoint of the rules and given database includes t in the set of true tuples for p. It is important to understand that the inference problem treats the rules as fixed, and only the database is allowed to vary. Thus, for any given datalog program, we know there are constants limiting the number of predicates, the number of arguments in a predicate, and any other property that depends only on the program, and not on its data. Thus, it is not hard to prove the following.

THEOREM 17.1
For every datalog program, the inference problem is in $\mathcal{P}$.

PROOF
If the database is of size n, then there are no more than n constants appearing as components. If k is the size of the program, then there are no more than k constants in the rules, no more than k predicates, and no more than k arguments for any predicate. The number of IDB facts that can ever be derived is thus no greater than k (the number of predicates) times $(n+k)^k$ (the number of tuples of length k), with each component chosen from the constants that appear either in the database or the rules. It is not hard to find an algorithm that makes all inferences in no more time than the square of this number of possible facts. Thus, $O\bigl(k^2(n+k)^{2k}\bigr)$ is an upper bound on the time for inference. As k is a constant, the time is polynomial. ∎

Note that different datalog programs have different exponents to their polynomial. There is unlikely to be a uniform exponent that works independent of the program, since if the program is part of the input, the problem of inference is PSPACE-complete [8].

For some datalog programs, the inference problem is quite easy, as we shall see. However, there are others for which the inference problem is as hard as can be. Specifically,

THEOREM 17.2
There is a datalog program whose inference problem is $\mathcal{P}$-complete

We shall see a number of examples further on.

Since membership in the class $\mathcal{N}\mathcal{C}$ is a hopeful sign that the problem can be parallelized, and $\mathcal{P}$-completeness is a sign that it cannot, the natural question to ask is whether a given datalog program has a $\mathcal{P}$-complete or $\mathcal{N}\mathcal{C}$ inference problem.

17.4
Relationship Between Logical Rules and Alternating Turing Machines

The inference problem is closely related to the problem of acceptance by alternating Turing machines (ATM's). These are abstract devices that were studied to help understand the complexity of solving problems that involve the alternation of quantifiers "for all" and "there exists." ATM's are Turing machines that can make both nondeterministic or "existential" branches, where acceptance occurs if either branch leads to acceptance, and "universal" branches, where acceptance occurs only when both branches lead to acceptance. As for ordinary Turing machines, an ATM accepts immediately when it enters one of its states designated "accepting."

Intuitively, the existence of several rules for a predicate is equivalent to an existential branch, while the existence of several subgoals in one rule is equivalent to a universal branch. That is, a predicate can be satisfied if any of its rules are satisfied, but a rule is satisfied only if all its subgoals are satisfied. The relationship can be seen formally in the next two theorems.

THEOREM 17.3

For any datalog program P, the inference problem can be solved by a logspace-bounded alternating Turing machine.

PROOF

We suppose that the goal and the EDB are encoded on the input of the ATM so that each symbol in the database has a binary code of length $O(\log n)$, where n is the total length of the EDB and goal. The rules of P are "built in" to the ATM. At certain times, the ATM will have a goal on its tape, and it must reach acceptance if this goal is true. Note that any goal can be written down in $O(\log n)$ space, since the number of arguments of any predicate is fixed beforehand, and each argument can be written in $O(\log n)$ space.

To determine the truth of a goal with an IDB predicate, the ATM guesses (i.e., makes an existential branch) a rule and an instantiation for the rule, that is, an assignment of constants to each of the variables of the rule. Again, since the rules of P are fixed in advance, an instantiation can be written in $O(\log n)$ space. Now, the ATM uses a universal branch to check that each of the subgoals of the rule is true.

To determine whether an EDB goal is true, the ATM simply examines the EDB on its input tape. Thus, if we start the ATM with the query

17.4. Relationship Between Logical Rules and Alternating Turing Machines

goal on its tape, it will accept if and only if the query follows from the EDB and the rules. ∎

Note that it is easy to transform instances of the inference problem for P into ATM inputs using logarithmic scratch space. Thus, Theorem 17.3 shows that inference for datalog programs logspace-reduces to logspace-ATM acceptance.

THEOREM 17.4

For any logspace ATM A, there is a datalog program P and a logspace transformation from ATM inputs to EDB's and queries, such that A accepts an input if and only if the corresponding instance of the inference problem for P has answer "true."

PROOF

The constants of the database are coded instantaneous descriptions (ID's) of A. For an input of length n, the ID's of A require $O(\log n)$ bits to encode the input head position, the scratch tape contents, and the state. The datalog program P will use an EDB predicate $move(I, J)$, which means that

1. ID I does not have a universal-branch state, and
2. One possible move of A from ID I takes A into ID J.

For universal branching states, we shall assume without loss of generality that there are exactly two choices of move. Thus, we can construct an EDB predicate $ubranch(I, J, K)$, which says that A, in ID I, branches universally, and the two next ID's are J and K. Lastly, there is an EDB predicate $final(I)$ that means the state in ID I is a final state.

P will also have IDB predicate $accept$, and the rules of P are

```
accept(I) <- final(I).
accept(I) <- move(I,J) & accept(J).
accept(I) <- ubranch(I,J,K) & accept(J) & accept(K).
```

The goal is $accept(I_0)$, where I_0 is the initial ID. We leave to the reader the details of the logspace transformation that takes an input of length n into the appropriate EDB. ∎

17.5
The Polynomial Fringe Property

If the goal follows from the database, then we can find a *proof tree* for the goal, which is a tree with the following properties.

1. Each node is labeled by a predicate with constant arguments (called a *ground atom*).
2. The label of each leaf is a fact in the EDB.
3. The root is labeled by the goal.
4. For each interior node N, there is a rule and a substitution of constants for variables in the rule, such that the head of the rule becomes the label of N, and the subgoals become the labels of N's children, from left to right.

EXAMPLE 17.3
Suppose we use the "cousin" rules of Example 17.1, with the EDB given in Fig. 17.1. Then a proof tree for goal $cousin(i, a)$ is shown in Fig. 17.2.

Suppose datalog program P has the property that there is a polynomial $p(n)$ such that whenever we are given a size-n instance of the inference problem for P that has answer "true," there is a proof tree for the goal that has at most $p(n)$ leaves. Then we say that P has the *polynomial fringe property*. There are datalog programs that do not have the polynomial fringe property, although every datalog program has an "exponential fringe property," where the number of leaves is allowed to grow exponentially with the size of the database.

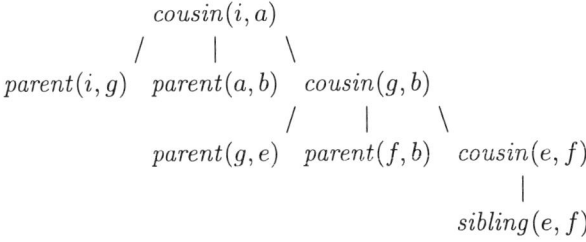

FIGURE 17.2
Proof tree

EXAMPLE 17.4
Consider the datalog program

```
p(X,Y) <- a(X,Y).
p(X,Y) <- b(X,U) & p(U,V) & p(V,W) & c(W,Y).
```

and let the EDB consist of the following facts about integers from 1 to n.

1. $a(n,n)$.
2. $b(i, i+1)$ for $i = 1, 2, \ldots, n-1$.
3. $c(i, i-1)$ for $i = 2, 3, \ldots, n$.

Then the proof tree for $p(i,i)$ uses two copies of the proof tree for $p(i+1, i+1)$. Hence, $p(1,1)$ is true, but the smallest proof tree for that fact has $3 \times 2^{n-1} - 2$ leaves.

However, there are a number of families of datalog programs that do guarantee the polynomial fringe property. For example, a datalog program is defined to be *linear* if each rule has at most one subgoal with an EDB predicate. For instance, the cousin rules of Example 17.1 are linear. Rule (1) has no EDB subgoal, and rule (2) has only one.

THEOREM 17.5
Every linear datalog program P has the polynomial fringe property.

PROOF
A proof tree must look like Fig. 17.3, where there is one path labeled by ground atoms with IDB predicates, and all nodes not on the path are leaves. Let n be the size of the database, and let k be the size of P. As we argued in Theorem 17.1, there are no more than $k(n+k)^k$ ground atoms. Thus, if the path were longer than $k(n+k)^k$, there would have to be a repeated label on the path, and we could find a shorter proof tree for the same goal.

Now, assume the path is no longer than $k(n+k)^k$. As there can be no more than k subgoals in any rule, the number of leaves in the proof tree cannot exceed k times the length of the path, that is, $k^2(n+k)^k$. Since k is a constant independent of the database size n, we conclude that there is a polynomial bound on the number of leaves in the smallest proof tree for any goal. ∎

The following theorem generalizes Theorem 17.5; we omit the proof, which is similar. Define a datalog program to be *metalinear* if there is a way

FIGURE 17.3
Proof tree for a linear program

to assign ranks to the predicates so that in each rule, there are no subgoals whose ranks exceed the rank of the head, and there is at most one subgoal of rank equal to that of the head. For instance, a linear program is a metalinear program in which all EDB predicates can be given rank 0 and all IDB predicates get rank 1.

THEOREM 17.6
Every metalinear program has the polynomial fringe property.

17.6
$\mathcal{NC}$ and the Polynomial Fringe Property

At first glance, the polynomial fringe property does not seem to help find proof trees efficiently in parallel. Even a linear datalog program can have smallest proof trees with paths of length polynomial in the database size. If we try to construct proof trees either top-down (working from the goal to the leaves), or bottom-up (working from the leaves to the goal), and we can only cover one level in one time unit, then polynomial time is required even if we work on many possibilities for proof trees in parallel. However, it turns out that it is sufficient to both work bottom-up, that is, to infer the head of an instantiated rule when all its subgoals have been inferred, and to prove a limited family of "lemmas," of the form that the truth of one ground atom (which we do not yet know) implies the truth of another ground atom. The polynomial fringe property is sufficient to guarantee that all this work can be carried out with a polynomial number of processors.

THEOREM 17.7
Every datalog program with the polynomial fringe property has its inference problem in $\mathcal{NC}$, and in fact, the inference problem can be solved in $O(\log n)$ parallel time on a CRCW PRAM.

PROOF
Let n be the size of the given EDB. Intuitively, what we do is run the Miller-Reif path compression algorithm [4] on every hypothetical proof tree for the query goal. Specifically, in shared memory, we have a bit for each ground atom; note the number of these ground atoms is polynomial in n, once the datalog program is fixed. For each instantiation of a rule, we assign a processor whose job it is to check whether the bits for all the subgoals have been set to 1, or if the bits for all but one of the subgoals have been set to 1.

Also, for each pair of ground atoms (G_1, G_2), we have a bit of shared memory, indicating whether the "lemma" $G_2 \to G_1$ has been proven. For each triple of ground atoms (G_1, G_2, G_3), we assign one processor to check whether we can prove the lemma $G_3 \to G_1$ because both $G_3 \to G_2$ and $G_2 \to G_1$ have been proved.

The algorithm proceeds in rounds. First, all the processors for the instantiated rules check their subgoals. If all have associated value 1, then the bit for the head H is set to 1. If all but one subgoal, say G, has value 1, then the bit for the lemma $G \to H$ is set to 1. This stage corresponds to "sweeping leaves."

Second, the processors for the triples of ground atoms (G_1, G_2, G_3) check whether both (G_1, G_2) and (G_2, G_3) have their bits set to 1, and set the bit for (G_1, G_3), if so. This stage corresponds to path compression.

Each round takes $O(1)$ parallel time. The proof that $O(\log n)$ rounds are needed to infer any true fact is exactly the same as for the theorem of Miller and Reif [4]. Thus, we can check whether the bit for the goal ground atom has been set after this time, and correctly tell whether it follows from the database. ∎

17.7 Chain Programs

Many common datalog programs have a special property that makes it relatively easy to tell whether their inference problem is in $\mathcal{NC}$ or is $\mathcal{P}$-complete. We shall define the property only for binary predicates, although the generalization to predicates with more arguments is straightforward (but not particularly useful). Say that a rule is a *chain rule* if it is of the form

$$p(X_0, X_n) \leftarrow q_1(X_0, X_1) \& q_2(X_1, X_2) \& \cdots \& q_n(X_{n-1}, X_n).$$

where the X's are distinct variables. Notice how the pattern of variables forms a chain, from the first argument of the head, through the subgoals in order, and finally back to the second argument of the head. As a special case, we allow a rule

$$p(X_0, X_0).$$

to be considered a chain rule; its meaning is that p is always true when its two arguments are equal.

A datalog program is a *chain program* if all its rules are chain rules.

EXAMPLE 17.5

The "cousin" rules of Example 17.1 form a chain program. However, in the second rule, we must reorder the subgoals and rename one use of the EDB predicate *parent*, in order to see the chain, as

```
cousin(X,Y) <- parent(X,Xp) & cousin(Xp,Yp) & child(Yp,Y)
```

Here, $child(A, B)$ is intended to be the same as $parent(B, A)$. Then $X_0, \ldots, X_3$ are X, Xp, Yp, and Y, respectively.

For another example, consider the transitive closure rules

```
path(X,Y) <- arc(X,Y).
path(X,Y) <- path(X,Z) & path(Z,Y).
```

These rules are also chain rules. In the second rule, $X_0, \ldots, X_2$ are X, Z, and Y. Finally, the rules of Example 17.4 are likewise chain rules.

Chain programs are special in that, once we are told the program has this property, we can effectively ignore the variables; up to renaming, we can always fill them in. Moreover, when we build proof trees for a chain program, the chain property holds for the proof trees as well, in the following sense. If we list the leaves of a proof tree in left-to-right order, then the list must be of the form $q_1(a_0, a_1), q_2(a_1, a_2), \ldots, q_k(a_{k-1}, a_k)$, and the root will be of the form $p(a_0, a_k)$. That is, the root and the leaves look like the instantiation of a giant chain rule.

As a consequence, there is a close relationship between proof trees for a chain program and parse trees of a corresponding context-free grammar (CFG). This CFG is obtained from the chain program by

1. Treat each IDB predicate as a nonterminal.
2. Treat each EDB predicate as a terminal.
3. If there is a rule with head predicate p and subgoals with predicates $q_1, \ldots, q_k$, in that order (note that the chain rule is now completely determined, up to renaming of variables), then we have production $p \to q_1, \ldots, q_k$. If $k = 0$, that is, the rule is just $p(X_0, X_0)$, then the production is $p \to \epsilon$ (ϵ stands for the empty string).

EXAMPLE 17.6
For the version of "cousin" given in Example 17.5, we have the following grammar.

$$C \to s$$
$$C \to pCc$$

Here, C is the nonterminal corresponding to *cousin*, and s, p, and c are the terminals for *sibling*, *parent*, and *child*, respectively. Notice that the language of this grammar, with C as the start symbol, is $\{p^i s c^i \mid i \geq 0\}$. That language expresses the way we prove cousin facts: go up i generations, find a sibling, and go down exactly i generations.

The productions for the chain program of Example 17.4 are

$$P \to bPPc \mid a$$

while

$$P \to PP \mid a$$

are the productions for the path program of Example 17.5.

17.8
The Polynomial Stack Theorem

There is a powerful tool from language theory that lets us show easily that certain chain programs have the polynomial fringe property, and that therefore, their inference problem is in $\mathcal{NC}$. We assume that the reader is familiar with the *pushdown automaton* (PDA), a device with a one-way input and a stack. The PDA is nondeterministic, and in one step, based on its current state, the symbol on top of its stack, and the current input symbol, has a finite number of choices for its

1. next state,
2. string with which to replace the top stack symbol, and
3. whether or not to advance the input.

It starts with a designated symbol on its stack, in the "start state," and it accepts whatever input it consumes before erasing its stack. It is well known that all and only the context-free languages are recognized by pushdown automata. See [3] for details and background.

With a given input, the PDA may make many sequences of moves, since it is nondeterministic. Possibly, many of these sequences lead to acceptance of the input. For any one accepting computation, there will be a sequence of stack contents. Let us divide the various strings that appear on the stack during this one accepting computation according to their length, and let $f(i)$ be the number of different stacks of length i. If there is a polynomial $p(i)$, such that for any input, and for any accepting computation of that input, $f(i)$ is guaranteed to be upper bounded by $p(i)$, then we say that the PDA has the *polynomial stack property*.

EXAMPLE 17.7
The definition is a bit subtle, and the following example should emphasize the subtlety. Consider the PDA that accepts all inputs in $(0+1)^*$ by copying its input onto the stack, and after reading each input symbol, nondeterministically deciding whether to read another symbol or erase its stack and accept. For any input $w = a_1 \cdots a_n$, there is one accepting computation, in which the sequence of stacks is $Z_0, Z_0 a_1, Z_0 a_1 a_2, \ldots, Z_0 a_1 \cdots a_n$, then the reverse of this sequence of steps, and finally, the empty stack. Here, Z_0 is the bottom-of-stack marker, found on the stack initially. For this sequence, $f(i) = 1$ for $i = 0, 1, \ldots, n+1$, and $f(i) = 0$ otherwise. The constant polynomial $p(i) = 1$ thus serves as an upper bound on the number of stacks of length i, regardless of the input or the particular accepting computation. We conclude that this PDA has the polynomial stack property.

We should not be deceived by the observation that, if we consider all inputs of length n, then there are 2^i different stacks of length $i+1$, for $i = 1, 2, \ldots, n$. It is not necessary to consider more than one computation at a time, when counting the number of stacks of any given length.

EXAMPLE 17.8

Consider the chain program

```
p(X,Y) <- a(X,Y).
p(X,Y) <- p(X,U) & b(U,V) & p(V,Y).
```

This is not a metalinear program, so Theorem 17.6 does not help us prove the polynomial fringe property. The CFG for this program is

$$P \to PbP \mid a$$

Consider the obvious PDA for recognizing L, the language of this grammar, top down. It begins with P alone on its stack. If at some time, b is on top of the stack, then it checks that the input is b. If so, it advances the input and pops b off the stack. If not, it dies (in this series of nondeterministic moves). If there is P at the top of the stack, the PDA nondeterministically chooses to

1. Replace P by PbP on the top of the stack; that is, push b then P onto the stack.
2. Check that the input is a, and if so, pop P from the stack and advance the input.

It is not hard to check that this PDA simulates all possible derivations, starting from P, and thus accepts the correct language. Also, the stack is always an alternating sequence of P's and b's, beginning at the bottom with P, as an easy induction on the number of moves will show. Thus, the constant polynomial $p(i) = 1$ again bounds the number of stacks of length i during any one accepting computation. We conclude that this PDA has the polynomial stack property. We shall see that the polynomial stack property for a PDA accepting L is sufficient to conclude that the inference problem for the chain program is in $\mathcal{NC}$. It is easier to see that result for this special case by observing that L is also generated by the CFG

$$P \to abP \mid a$$

More to the point, the language L is $(ab)^*a$, and it is easy to prove that both grammars generate this language. However, the above grammar is linear, and it is easy to see that linear CFG's come from linear chain programs, and vice versa. Thus, Theorem 17.5 tells us the inference problem for the given chain program is in $\mathcal{NC}$.

THEOREM 17.8
Let P be a chain program, L the language of its corresponding CFG, and suppose that there is a PDA A with the polynomial stack property that accepts L. Then the inference problem for P is in $\mathcal{NC}$.

PROOF
Using the standard construction, we convert PDA A into a CFG G, that generates L. We then convert G into a chain program Q. Since the CFG's from programs P and Q each generate the same context-free language, it is not hard to show that the sets of goals provable by either program, from a given database, must be the same. Thus, in order to solve the inference problem for P, it suffices to solve the inference problem for Q. If we can show that Q has the polynomial fringe property, then we are done, by Theorem 17.7.

We shall not give all the details here, since they depend on understanding the construction of a CFG from a PDA as given in [3]. However, the following observations should help convince the reader who is familiar with that construction. First, we note that the argument of Theorem 17.5 about how long paths in a proof tree could get really applies not only to linear programs, but to all datalog programs, since whenever we find a path longer than the number of possible ground atoms, there must be a repeat on the path, and we can find a shorter proof tree. We conclude that for any datalog program whatsoever, there is a polynomial $p(n)$ such that, when the EDB is of size n, the depth of a proof tree may be assumed bounded above by $p(n)$.

The part of the proof we shall omit is the claim that parse trees in the grammar G constructed from A have paths from the root to interior nodes that correspond to stacks of A. In particular, if we have a parse tree with yield w, then the number of paths from the root, having length i, is bounded polynomially by the number of different stacks of length i that A attains, during the accepting computation of w that the parse tree mimics. As A is assumed to have the polynomial stack property, there is a polynomial $q(m)$ such that parse trees in G have no more than $q(i)$ nodes of depth i, for any i.

To put the pieces together, consider a given EDB of size n and a goal $r(a,b)$ that is inferred from this EDB according to P, and therefore according to Q. Then there is a proof tree with root $r(a,b)$ and depth no more than $p(n)$. By removing the arguments, we can convert this proof tree into a parse tree of G. The number of nodes of depth i in this tree is bounded by $q(i)$. Thus, the total number of nodes in the tree, and therefore the number of leaves, is bounded above by $\sum_{i=1}^{p(n)} q(i)$, which is $O\big(p(n)q(p(n))\big)$, surely a polynomial. Thus, Q has the polynomial fringe property, and its inference problem is in $\mathcal{NC}$. Since Q's inference problem is in $\mathcal{NC}$, the inference problem for P, which is the same, is in $\mathcal{NC}$. ∎

EXAMPLE 17.9

The chain program of Example 17.8, whose grammar is $P \to PbP \mid a$, was seen there to have the polynomial stack property. We now have another proof that the inference problem for this program is in $\mathcal{NC}$.

Consider the program

```
p(X,Y) <- p(X,Z) & q(Z,Y).
p(X,Y) <- a(X,Y).

q(X,Y) <- q(X,Z) & b(Z,Y).
q(X,Y) <- c(X,Y).
```

The obvious PDA for recognizing the language of the underlying grammar,

$$P \to PQ \mid a$$
$$Q \to Qb \mid c$$

guesses how many times to replace P on top of the stack by PQ, and then, after consuming an a from the input and popping P, guesses how many times to replace each Q by Qb. Thus, there are $n+1$ stacks of length n, all the strings of length n in the regular expression $PQ^* + Qb^*Q^*$. This language too has the polynomial stack property, and we conclude that the inference problem for the above chain program is in $\mathcal{NC}$. In fact, we can similarly prove that every metalinear chain program has the polynomial stack property, but of course we already knew that these programs had an inference problem in $\mathcal{NC}$.

17.9
Simple Chain Programs

For the special case of chain programs with a single *basis rule* of the form

$$p(X,Y) \leftarrow p_0(X,Y).$$

and a single recursive chain rule not involving the predicate p_0 (called a *simple chain program*), we can characterize exactly when the inference problem is in $\mathcal{NC}$. In fact, there is a remarkable fact, due to [1], that there is no middle ground among simple chain programs; either the inference problem is in $\mathcal{NC}$, or it is $\mathcal{P}$-complete. This result probably represents the most complex class of problem instances of any type, where a sharp division between easy and hard cases has been proved.

THEOREM 17.9
A simple chain program has an inference problem in $\mathcal{NC}$ if and only if the recursive rule corresponds to one of the following seven forms of productions or their reversals, where w and x can stand for any string of 0 or more terminal symbols, and i and j are any nonnegative integers.

1. $P \to wPx$.
2. $P \to w(xP)^j$.
3. $P \to Pw(xP)^j$.
4. $P \to wP(xP)^j$.
5. $P \to PwP(xP)^j$.
6. $P \to (Pw)^i PxP(wP)^j$.
7. $P \to PwP(xPwP)^j$.

PROOF
In one direction, one exhibits PDA's with the polynomial stack property accepting the corresponding languages. Showing the converse is a very difficult combinatorial argument, and we omit it. ∎

EXAMPLE 17.10
The recursive rule of Example 17.8 corresponds to production $P \to PbP$. The right side of this rule is of form (3), with $w = \epsilon$, $x = b$, and $j = 1$.

The simple chain program corresponding to the grammar

$$P \to a$$
$$P \to PbPcdPbPbP$$

is of form (6), with $w = b$, $x = cd$, $i = 1$, and $j = 2$. Note that a plays the role of p_0 here, and must not appear in the recursive rule.

The program of Example 17.4, whose corresponding grammar has recursive production $P \to bPPc$, is of none of these forms. It cannot be of form (1), because it has two occurrences of P. It is not of any other form, or its reverse, because we can easily check that none of forms (2)–(7) can both begin and end with a terminal.

17.10 Exercises

17.1 The following rules suppose that there are EDB predicates $isa(X, Y)$, meaning that an X is a Y (e.g., a cat is an animal), $mother(X, Y)$, meaning that Y is the mother of X, and $father(X, Y)$ with the analogous meaning. They define the IDB predicate $isa\text{-}star(X, Y)$, to mean intuitively that there is a chain of isa's from X to Y. However, the rules also allow inferences according to roles, i.e., if Garfield is a cat, then Garfield's mother is the mother of a cat.

```
isa-star(X,Y) ← isa(X,Y).
isa-star(X,Y) ← isa-star(X,Z) & isa-star(Z,Y).
isa-star(X,Y) ← isa-star(U,V) & mother(X,U) & mother(Y,V).
isa-star(X,Y) ← isa-star(U,V) & father(X,U) & father(Y,V).
```

Answer the following questions about these rules.

   a) Are these rules linear?
   b) Are the rules chain rules?
   c) Do these rules have the polynomial fringe property?
   d) Does the natural PDA for these rules have the polynomial stack property?
   e) Is inference for these rules $\mathcal{P}$-complete? Is it in $\mathcal{NC}$?

17.2 Suppose that the EDB predicates from Exercise 17.1 have the following facts:

$isa(a, b)$ $mother(c, d)$ $father(b, f)$
$isa(b, c)$ $mother(b, e)$ $father(h, g)$
               $mother(d, h)$
               $mother(g, i)$

   a) What are all the $isa\text{-}star$ facts we can infer?
   b) Draw a proof tree for the fact $isa\text{-}star(i, h)$.

17.3 For each of the following grammars, (*i*) convert to a chain program (*ii*) tell whether the program is a simple chain program (*iii*) tell whether inference for the program is in $\mathcal{NC}$ or is $\mathcal{P}$-complete.

   a) $P \to PP \mid aPb \mid c$.
   b) $P \to aQ \mid b$; $Q \to bP \mid a$.
   c) $P \to aaPbP \mid c$.
   d) $P \to PaPbPaPbPaP \mid c$.

Notes and References

Theorem 17.1, relating datalog and the class $\mathcal{P}$, is from Chandra and Harel [2], and Theorem 17.2 is from Vardi [8]. Theorems 17.3 and 17.4, relating logical inference to ATM's are from Shapiro [6]. Theorems 17.5 and 17.6, that (meta)linear logic programs have the polynomial fringe property are essentially "folklore," having been noticed by a number of different people.

Theorem 17.7, that the polynomial fringe property implies membership in $\mathcal{NC}$, was shown by Ruzzo [5], stated in terms of ATM's. The sketch of a proof we offer here is from Ullman and Van Gelder [7]. It has a better processor count than that of Ruzzo [5].

The polynomial stack property and Theorems 17.8 and 17.9 are from Afrati and Papadimitriou [1]. The $\mathcal{P}$-completeness of inference for the simple chain program of Example 17.4 was shown by Ullman and Van Gelder [7]; it also follows from Theorem 17.9.

Bibliography

[1] Afrati, F. and C. H. Papadimitriou, "The parallel complexity of simple chain queries," *Proc. Sixth ACM Symposium on Principles of Database Systems*, pp. 210–213.

[2] Chandra, A. K., and D. Harel, "Horn clause queries and generalizations," *J. Logic Programming* 4:1, pp. 1–15.

[3] Hopcroft, J. E. and J. D. Ullman, *Introduction to Automata Theory, Languages, and Computation*, Addison Wesley, Reading, Mass., 1979.

[4] Miller, G. L. and J. H. Reif, "Parallel tree contraction and its applications," *Proc. 26th Annual IEEE Symposium on Foundations of Computer Science*, pp. 478–489.

[5] Ruzzo, W. L., "Tree size bounded alternation," *J. Computer and System Sciences* **21**:2, pp. 218–235.

[6] Shapiro, U., "Alternation and computational complexity of logic programs," *J. Logic Programming* **1**:1, pp. 19–23.

[7] Ullman, J. D. and A. Van Gelder, "Parallel complexity of logical query programs," *Algorithmica* **3**:1, pp. 5–42.

[8] Vardi, M. Y., "Complexity of relational queries," *Proc. Fourteenth Annual ACM Symposium on the Theory of Computing*, pp. 137–145.

Part VII

Parallel Combinatorial Optimization

18

Parallel Graph Matching

Vijay V. Vazirani

Dept. of Computer Science & Engineering
Indian Institute of Technology
New Delhi, India
vazirani@cse.iitd.ernet.in

18.1 Introduction

Matching is a classical problem in the study of algorithms. Its importance lies in that it lends itself to efficient solutions; however, the solutions require non-trivial algorithmic ideas that have resulted in several conceptual breakthroughs in the past. Additionally, matching has intimate connections with several other fundamental problems, and is used as a subroutine for solving some of them. Studying this problem from the perspective of parallel computation has also been quite fruitful, as elaborated below.

Unlike sequential matching algorithms, which are combinatorial, the known parallel algorithms use randomization and tools from linear algebra. A key ingredient in this new approach is a theorem of Tutte [27] which states that a graph has a perfect matching iff a certain matrix of indeterminates, called the Tutte matrix, is non-singular. The first algorithmic use of this theorem was proposed by Lovasz [19]. Using the insight that polynomial identities can be efficiently tested by randomization [26, 30], Lovasz reduced the decision problem 'Does the given graph have a perfect matching?' to testing if a given integer matrix is non-singular. Borodin, von zur Gathen and Hopcroft [3] observed that since the latter problem is in NC [6], this yields an RNC algorithm for the former problem. Rabin and Vazirani [25] used this approach to give a simple sequential matching algorithm, ideas from which were useful in the parallel setting as well.

The search problem, i.e., actually finding a maximum matching in parallel, is much harder. The first RNC algorithm for this problem was given by Karp, Upfal and Wigderson [15]. Subsequently, Galil and Pan [9] gave a more processor-efficient implementation of this algorithm, and Karloff [14] gave a complementary Monte Carlo algorithm for bounding the size of maximum matching from above, thus yielding the Las Vegas extension. A more direct and somewhat faster matching algorithm was given by Mulmuley, Vazirani and Vazirani [22].

In this chapter, we will first present the NC algorithms of Israeli and Shiloach [13] for finding a maximal matching in general graphs, and of Lev, Pippenger and Valiant [18] for finding a perfect matching in regular bipartite graphs. Both these algorithms are combinatorial. We will then turn to the maximum matching problem. We will first present the simpler algorithm of [22], and then the algorithm of [15] with the processor-efficient extension [9].

An important open problem is to find a (deterministic) NC algorithm for finding a perfect matching. In this context, we will present the NC algorithm of Grigoriev and Karpinski [11] for the case when the graph has only polynomially many perfect matchings.

The study of parallel matching algorithms has led to some understanding of the relative difficulty of parallelizing the decision and search versions of problems [16, 22]. It has also yielded probabilistic tools, such as the Isolating Lemma [22], that have found applications outside the area of matchings. More recently, the new methodology of using randomization and linear algebra has been used for obtaining RNC algorithms for several fundamental matroid problems and their applications [23].

In this chapter, we have restricted attention to NC and RNC algorithms. It should be noted, however, that there are other parallel matching algorithms that achieve substantial speedup, though they do not run in polylog time; see for example [8, 10].

18.2
A Simple Parallel Maximal Matching Algorithm

The maximal matching algorithm of Israeli and Shiloach [13] works iteratively in phases. In each phase, a "large" matching is found in the current graph, and the matched vertices are deleted to obtain the graph for the next phase. The algorithm halts when the current graph is either empty or has only isolated vertices. Clearly, the union of the matchings found forms a maximal matching in the original graph.

In each phase, the algorithm successively prunes away edges from the current graph, say G', until every vertex has degree at most 3; the pruning is done in such a way that at this stage, the set of vertices of non-zero degree, say S, forms a vertex cover for G'. It then finds a matching M' among the left-over edges, such that $|M'| \geq \frac{|S|}{4}$. Let us first bound the number of phases executed.

LEMMA 18.1
The algorithm requires at most $O(\log n)$ phases.

PROOF
We will prove that the size of a minimum vertex cover in G' drops by a constant factor with each phase. Let C be such a cover. Clearly, $|S| \geq |C|$. The key observation is that C must include at least one end point of each edge in M'. Hence, the number of vertices of C that are matched off

by M' is $\geq |M'| \geq \frac{|S|}{4} \geq \frac{|C|}{4}$. Next observe that if we delete some vertices of G', then the left-over vertices of C must form a vertex cover for the remaining graph. Hence, the unmatched vertices of C must form a vertex cover for the graph passed on to the next phase (i.e., the graph obtained on deleting all vertices matched by M'). Therefore, this graph has an optimal vertex cover of size $\leq \frac{3}{4}|C|$. Hence, the size of an optimal vertex cover drops by a factor of 3/4 with each phase. The lemma follows. ∎

Next, we describe a phase of the algorithm. A phase consists of $O(\log n)$ iterations; in each iteration, edges are pruned using an Euler circuit algorithm as a subroutine [1]. Let H be the graph at the beginning of an iteration, and let $\delta(H)$ denote the maximum degree of a vertex in H. An iteration is executed as follows:

ALGORITHM 18.1
   Step 1   Find the integer k such that $2^k \leq \delta(H) \leq 2^{k+1} + 1$.

   Step 2   Find the set of vertices of H having degree in the range $[2^k, 2^{k+1} + 1]$; these are called *active vertices*.

   Step 3   Construct the graph H' consisting of all edges of H that are incident on at least one active vertex.

   Step 4   Make H' a graph of even degrees by throwing in a new vertex v^*, and an edge from v^* to any vertex of odd degree.

   Step 5   Find an Euler circuit in every connected component of H'.

   Step 6   Delete alternate edges of the circuit in each connected component, and then delete v^*.

It is easy to see that after the iteration, the degree of each active vertex is in the range $[2^{k-1}, 2^k + 1]$. So, an active vertex will be active in the next iteration as well. The process is continued until $1 \leq \delta(H) \leq 3$. Let S be the set of vertices of non-zero degree at this stage, and as before, let G' denote the graph at the beginning of the current phase. Notice that S contains all vertices that were declared active in the current phase. Let us show that S is a vertex cover for G': certainly S covers the edges remaining in H. On the other hand, if an edge is deleted, one of its end points must be active, and is therefore in S.

Consider the connected components of H at this stage. Let H_i be a component, and let S_i be its vertices, $S_i \subseteq S$. If H_i has $\geq |S_i|$ edges, two more iterations are executed on H_i, with the extra caution that in Step 6, of the two sets of alternate edges, we will delete the one containing fewer edges of H_i. This ensures that at the end we will have $\geq \lceil \frac{|S_i|}{4} \rceil$ edges remaining. Since each vertex has degree at most 1, these edges form a matching.

If H_i has $< |S_i|$ edges, it must have $|S_i| - 1$ edges, and it must be a tree. In this case, pick an edge incident at a leaf, match it off, and remove its two end points. Since the other end point has degree ≤ 3, we have deleted at most 3 edges in this process. Execute two more iterations, as in the previous case, on the remaining graph. Since this graph contains $\geq |S_i| - 4$ edges, we get a total of at least $\lceil \frac{|S_i|}{4} \rceil$ matched edges in H_i (on including the first matched edge).

Since Euler circuits can be found in $O(\log n)$ time using $O(m)$ processors on a CRCW-PRAM [1], the above algorithm takes $O(\log^3 n)$ time and $O(m)$ processors.

THEOREM 18.1 *Israeli and Shiloach [13]*
A maximal matching in general graphs can be found in $O(\log^3 n)$ time using $O(m)$ processors.

18.3
Finding a Perfect Matching in Regular Bipartite Graphs

In this section we will present the algorithm of Lev, Pippenger and Valiant [18] for edge colouring a bipartite graph $G = (U, V, E)$ with Δ colours, where Δ is the maximum degree of a vertex in G; if G is Δ-regular, then each colour class will be a perfect matching in G.

First consider the case that Δ is a power of 2, say $\Delta = 2^k$. First each vertex pairs off edges incident at it; one edge remains unpaired if an odd number of edges are incident at the vertex. This pairing relation partitions the edges into even length cycles and paths. Throw alternate edges of each cycle and path into two sets E_1 and E_2. The partition (E_1, E_2) can be easily obtained in $O(\log n)$ time by standard pointer doubling. Under each partition, the degree of each vertex is $\leq 2^{k-1}$. The first 2^{k-1} colours are used for colouring E_1, and the rest for colouring E_2; each partition is coloured by this same recursive procedure. Clearly, at the end a valid colouring is obtained, and the algorithm runs in $O(\log n \log \Delta)$ time on $O(m)$ processors.

Next, assume that Δ is arbitrary. First, each vertex assigns distinct colours to the edges incident at it. Thus each edge is assigned two colours—one from each of its endpoints. Say that an edge is *good* if the two colors are the same, and *bad* otherwise. The algorithm now iteratively recolours the edges, decreasing the number of bad edges by a constant fraction ($\frac{1}{6}$) in each iteration. Hence, at the end of $O(\log n)$ iterations, it obtains a valid edge colouring with Δ colours.

The recolouring strategy is as follows. The Δ colours are partitioned into four sets C_1, C_2, C_3 and C_4, of as equal sizes as possible. Let set S_{ij} consist of all edges whose two colours fall in $C_i \cup C_j$, for $1 \leq i \leq j \leq 4$. In general, S_{ij} contains good and bad edges. Among these six sets, pick the one having the most number of bad edges, say this is S_{12}. Clearly, S_{12} has at least $\frac{1}{6}$ of the bad edges. Let $C = C_1 \cup C_2$. If $|C|$ is not a power of two, throw some of the remaining colours into C till it reaches the next power of two; notice that there must be enough colours remaining to carry out this step. Let S consist of all edges whose two colours fall in C. Clearly the number of S edges incident at a vertex is $\leq |C|$. Now, recolour the edges in S with colours in C using previously-stated algorithm. After this, each vertex examines its bad edges and ensures that they are not coloured with colours in C; this is clearly possible because every vertex has Δ colours available to it. The algorithm stops when there are no bad edges left.

THEOREM 18.2 *Lev, Pippenger and Valiant [18]*
A perfect matching can be found in a Δ − regular bipartite graph in $O(\log^2 n \log \Delta)$ time using $O(m)$ processors.

18.4
Tutte's Theorem and its Generalizations

Tutte's Theorem [27] helps carry the graph-theoretic matching problem over to the realm of linear algebra, where a lot of powerful tools are available. The known RNC matching algorithms use this theorem and some of its generalizations. Ideas from the proof of Tutte's theorem will also be used in subsequent sections to derive other useful facts about certain matrices obtained from the Tutte matrix.

Notation: We will represent the $(i,j)^{th}$ element of matrix A by (lower case) a_{ij}, the minor obtained by removing the ith row and the jth column by A_{ij}, the determinant of A by $|A|$, and the adjoint of A by $adj(A)$.

18.4. Tutte's Theorem and its Generalizations

Before introducing the Tutte matrix, which handles general graph matching, let us consider the simpler case of bipartite graphs, which uses a result of Edmonds [7]. Let $G = (U, V, E)$ be a bipartite graph, with bipartition $|U| = |V| = n$. Let D be the $n \times n$ adjacency matrix of G, i.e.,

$$d_{ij} = \begin{cases} 1 & \text{if } (u_i, v_j) \in E \\ 0 & \text{otherwise} \end{cases}$$

Edmonds obtained an $n \times n$ matrix from D, say A, as follows: if $d_{ij} = 1$, replace it by indeterminate x_{ij}, and leave the 0 entries of D unchanged. All the facts that we will state for general graphs using the Tutte matrix can be proven more easily for bipartite graphs using Edmonds' matrix. For example, it is easy to see that:

1. $|A| \neq 0$ iff G has a perfect matching.
2. the size of a maximum matching in G is given by $rank(A)$.

We will prove facts for general graphs using the Tutte matrix; the reader is encouraged to work out the corresponding facts for bipartite graphs using Edmonds' matrix (see Exercise 18.3).

DEFINITION 18.1
Given a graph $G = (V, E)$, the adjacency matrix of G is an $n \times n$ symmetric matrix D such that $d_{ij} = 1$ if $(v_i, v_j) \in E$, and 0 otherwise. The Tutte matrix of G is an $n \times n$ skew-symmetric matrix A, obtained as follows from D: if $d_{ij} = d_{ji} = 1$, replace them by indeterminates x_{ij} and $-x_{ij}$, so that the entries above the diagonal are positive, and leave the 0 entries of D unchanged.

DEFINITION 18.2
Let A be an $n \times n$ matrix. For permutation σ on $1, \ldots n$ define

$$value(\sigma) = \prod_{i=1}^{n} a_{i\sigma(i)}.$$

We will say that σ is non-vanishing if $value(\sigma) \neq 0$.

THEOREM 18.3 Tutte [27]
Let $G = (V, E)$ be a graph and let A be its Tutte matrix. Then, $|A| \neq 0$ iff there is a perfect matching in G.

PROOF
Let us first show that if G contains a perfect matching, say M, then $|A| \neq 0$. Substitute the variables in A as follows to obtain matrix $\tilde{A}$: if $(v_i, v_j) \in M$, let $x_{ij} = 1$, and otherwise let $x_{ij} = 0$. It is easy to see that $|\tilde{A}| = (-1)^{\frac{n}{2}}$, and therefore $|A| \neq 0$.

Next suppose $|A| \neq 0$. Clearly, σ is non-vanishing iff $(v_i, v_{\sigma(i)}) \in E$, for $1 \leq i \leq n$.

By definition, $|A| = \sum_\sigma sign(\sigma) \times value(\sigma)$

where the sum is over all permutations σ on $1, \ldots n$, and $sign(\sigma)$ is $+1$ if σ is an even permutation and -1 otherwise.

Define the *trail* of a non-vanishing permutation σ to be the subgraph of G consisting of the edges $(v_i, v_{\sigma(i)})$, for $1 \leq i \leq n$. Clearly, each vertex will have two such edges incident at it (maybe with repetition), and the subgraph will in general consist of disjoint cycles and single edges traversed twice. Corresponding to each perfect matching M in G there is a non-vanishing permutation whose trail is M itself. For every permutation σ containing an odd cycle, there is a corresponding permutation σ' which differs from σ only in that this odd cycle is traversed in the reverse direction. Clearly, $value(\sigma) = -value(\sigma')$ and $sign(\sigma) = sign(\sigma')$. Therefore, the permutations containing odd cycles cancel each other out, and do not contribute to $|A|$. Since $|A| \neq 0$, there must be a permutation, say σ, whose trail contains no odd cycles. We can now extract a perfect matching from the trail of σ by picking alternate edges from every even cycle, together with all the edges traversed twice. ∎

We will use Tutte's Theorem for testing whether the given graph has a perfect matching. The main difficulty is that $|A|$ may have exponentially many terms, and so it is not efficient to compute it completely. The following lemma shows that if the variables are substituted from a polynomially large set of integers, then with high probability, the substituted matrix will be non-singular, provided $|A| \neq 0$. Of course, if $|A| = 0$ then the substituted matrix will be singular.

LEMMA 18.2 *Schwartz [26], Zippel [30]*
Let $p(x_1 \ldots x_n)$ be a polynomial in n indeterminates of degree $d \geq 1$ over $\mathbf{Z}$, and let S be a non-empty set of integers. Substitute for the

indeterminates randomly and independently from S, say $a_1 \ldots a_n$. Then

$$Pr[p(a_1, \ldots, a_n) = 0] \leq \frac{d}{|S|}.$$

PROPOSITION 18.1
There is an RNC algorithm for testing if a given graph has a perfect matching.

PROOF
Notice that $|A|$ is a polynomial of degree $2n$. So, by Lemma 18.2 taking $|S| = 4n$ will be sufficient to guarantee that the substituted matrix is non-singular with probability $\geq \frac{1}{2}$, provided $|A| \neq 0$. The probability of error can be reduced by repeated trials. The determinant of the substituted matrix can be computed efficiently using the algorithm of Pan [24]. This requires $O(\log^2 n)$ time and $M(n)$ processors, where $M(n)$ is the number of arithmetic operations required to multiply two $n \times n$ matrices. The current bound for $M(n)$ in $O(n^{2.396})$ [5]. ∎

The following generalization of Tutte's Theorem helps in computing the size of a maximum matching.

THEOREM 18.4   Lovasz [19, 20]
Let A be the Tutte matrix of graph G, and let m be the size of a maximum matching in G. Then, $rank(A) = 2m$.

Notation: Let A be an $n \times n$ matrix, and α and β be n-dimensional 0/1 vectors. By the *set of indices picked by* α, we mean the set of indices at which α has 1's. Then, $A_{\alpha\beta}$ will denote the submatrix of A obtained by choosing the rows of A corresponding to indices picked by α, and columns corresponding to indices picked by β. Similarly, G_α will denote the subgraph of G induced on vertices corresponding to indices picked by α. The number of 1's in α will be denoted by $|\alpha|$.

DEFINITION 18.3
Let M be a matching in G. Vector α is said to be the matching vector of *M if the indices picked by α correspond exactly to the vertices matched by M. The matching vector of a maximum matching in G is said to be a* maximum matching vector.

A simple proof of Lovasz's theorem is given in [25], using the following theorem of Frobenius. This proof is presented below.

THEOREM 18.5 Frobenius [17]

Let A be an $n \times n$ skew-symmetric matrix, and let α and β be n-dimensional 0/1 vectors such that $|\alpha| = |\beta| = rank(A)$. Then,

$$|A_{\alpha\alpha}||A_{\beta\beta}| = (-1)^{|\alpha|}|A_{\alpha\beta}|^2.$$

PROOF OF THEOREM 18.4

We will prove the theorem in two parts:

$$(i) \quad rank(A) \geq 2m.$$

Choose any matching of size m, and let α be its matching vector. Since $A_{\alpha\alpha}$ is the Tutte matrix of G_α, and G_α has a perfect matching, $|A_{\alpha\alpha}| \neq 0$. Hence, $rank(A) \geq |\alpha| = 2m$.

$$(ii) \quad rank(A) \leq 2m.$$

Suppose $rank(A) = k$. Let $A_{\alpha\beta}$ be a non-singular submatrix of A, with $|\alpha| = |\beta| = k$. We want to show that G has a matching of size at least $k/2$. Notice that if $A_{\alpha\beta}$ is *symmetrically located*, i.e., $\alpha = \beta$, we are done, since G_α has a perfect matching by Tutte's Theorem.

Since $|A_{\alpha\beta}| \neq 0$ and $|\alpha| = |\beta| = rank(A)$, by Frobenius' Theorem, $|A_{\alpha\alpha}| \neq 0$. Since this is a symmetrically located submatrix of A, by the previous remark, G has a matching of size $k/2$ (notice that k must be even, since a skew-symmetric matrix of odd dimension is always singular). ∎

PROPOSITION 18.2

There is an RNC algorithm for computing the size of a maximum matching in a graph.

PROOF

As before, substitute for the variables in A randomly from a set of integers of size $4n$ to obtain $\tilde{A}$. Let B be a non-singular submatrix of A of dimension $rank(A)$. Since $|B|$ is a polynomial of degree at most $2n$, by Lemma 18.2, $|\tilde{B}| \neq 0$ with probability at least $1/2$. Since, every submatrix of A of larger dimension is singular, the corresponding submatrices of $\tilde{A}$ will also be singular. Therefore, with high probability, $rank(\tilde{A}) = rank(A)$. Since the rank of an integer matrix can be computed in RNC [3], the claim follows. ∎

Notation: We will represent a base of an $n \times n$ matrix A as an n-dimensional 0/1 vector, whose i^{th} entry is 1 if the ith row of A is in the base.

The following generalization helps in actually finding a maximum matching vector.

THEOREM 18.6 *Vazirani and Vazirani [28]*
Let A be the Tutte matrix of a graph $G = (V, E)$. An n-dimensional 0/1 vector α is a base of A iff it is a maximum matching vector.

PROOF
Let α be a maximum matching vector in G. Since G_α has a perfect matching, by Tutte's Theorem $|A_{\alpha\alpha}| \neq 0$. Thus the rows of A indexed by α are linearly independent. Moreover, by Theorem 18.4, $rank(A) = |\alpha|$. Hence α is a base of A.

Conversely, let α be a base of A. Then, there is an n-dimensional vector β such that $|\alpha| = |\beta|$, and $|A_{\alpha\beta}| \neq 0$. Since $rank(A) = |\alpha| = |\beta|$, by Frobenius' Theorem $|A_{\alpha\alpha}| \neq 0$. Now, by Tutte's Theorem, G_α has a perfect matching, and by Theorem 18.4, this is a maximum matching in G. Hence α is a maximum matching vector in G. ∎

PROPOSITION 18.3
There is an RNC algorithm for finding the lexicographically first maximum matching vector.

PROOF
By Theorem 18.6, the lexicographically first base of A is also the lexicographically first maximum matching vector. The rest of the proof is similar to that of Proposition 18.2. The only difference is that we let B be a non-singular submatrix of the lexicographically first base of A. Then, it follows that with high probability, the lexicographically first base of $\tilde{A}$ is the same as that of A. Since the lexicographically first base of an integer matrix can be found in RNC [3], the claim follows. ∎

18.5
Finding a Maximum Matching in Parallel

In this section we will present the maximum matching algorithm of Mulmuley, Vazirani and Vazirani [22]; this algorithm requires $O(\log^2 n)$ time and $O(n^{3.5}m)$ processors. As stated in Section 18.4, the decision problem is easily seen to be in RNC. The main difficulty in solving the search problem can be viewed as follows: the given graph may have exponentially many maximum matchings; how do we coordinate the processors so they seek the same matching in parallel? The key ingredient in the algorithm is a probabilistic lemma called the *Isolating Lemma*. This lemma helps single out one maximum matching in the graph.

18.5.1   The Isolating Lemma

DEFINITION 18.4
A set system (S, F) consists of a finite set, $S = \{x_1, x_2, \ldots, x_n\}$, and a family F of subsets of S, i.e. $F = \{S_1, S_2, \ldots S_k\}, S_j \subseteq S$, for $1 \leq j \leq k$.

Let us assign a weight w_i to each element $x_i \in S$, and let us define the weight of the set S_j to be $\sum_{x_i \in S_j} w_i$.

LEMMA 18.3
Let (S,F) be a set system whose elements are assigned integer weights chosen uniformly and independently from $[1,2n]$. Then,

$$Pr[\text{There is a unique minimum weight set in } F] \geq \frac{1}{2}.$$

PROOF
Fix the weights of all elements except x_i. Define the *threshold* for element x_i to be the real number α_i such that if $w_i \leq \alpha_i$ then x_i is contained in some minimum weight subset, S_j, and if $w_i > \alpha_i$ then x_i is in no minimum weight subset.

Clearly, if $w_i < \alpha_i$ then x_i must be in *every* minimum weight subset. Thus ambiguity about element x_i occurs iff $w_i = \alpha_i$, since in this case there is a minimum weight subset that contains x_i and another which does not. In this case we shall say that the element x_i is *ambiguous*.

We now make the crucial observation that the threshold, α_i, was defined without reference to the weight, w_i, of x_i. It follows that the random variable α_i is *independent* of w_i. Since w_i is a uniformly distributed integer in $[1, 2n]$,

$$Pr[\text{Element } x_i \text{ is ambiguous, i.e. } w_i = \alpha_i] \leq \frac{1}{2n}.$$

Since S contains n elements,

$$Pr[\text{There exists an ambiguous element}] \leq \frac{1}{2n} \times n = \frac{1}{2}.$$

Thus, with probability at least $\frac{1}{2}$, no element is ambiguous. In this case each element is either in every minimum weight subset or in none. It follows that the minimum weight subset is unique. ∎

18.5.2 Parallel Algorithm for Perfect Matching

We will first present a parallel algorithm for finding a perfect matching, assuming that the given graph has one. We will view the edges in E and the set of perfect matchings in G as a set system. Let us assign random integer weights to the edges of the graph, chosen uniformly and independently from $[1, 2m]$, where $m = |E|$. Now, by the Isolating Lemma, the minimum weight perfect matching in G will be unique with probability at least $\frac{1}{2}$. Our parallel algorithm will pick out this perfect matching.

We will obtain an integer matrix B from the Tutte matrix by substituting for the indeterminates x_{ij} the integers $2^{w_{ij}}$, where w_{ij} is the weight assigned to the edge (v_i, v_j).

LEMMA 18.4
Let $G = (V, E)$ be a graph with weights assigned to its edges, and let B be the matrix described above. Suppose the minimum weight perfect matching in G is unique, and its weight is w. Then $|B| \neq 0$; moreover, the highest power of 2 which divides $|B|$ is 2^{2w}.

PROOF
Let M be the minimum weight perfect matching, and let w be its weight. The value of the permutation whose trail is M is $(-1)^{n/2} 2^{2w}$. We want to argue that the value of any other non-vanishing permutation which does not contain an odd cycle must be a higher power of 2. Certainly this is true if the trail of the permutation is a perfect matching. On the other hand, the edges of an even cycle can be partitioned into two matchings. Therefore, if the trail of a permutation contains even cycles, we can demonstrate two perfect matchings M_1 and M_2 whose union is the trail of α. Clearly, $|value(\alpha)| = 2^{w(M_1)+w(M_2)} > 2^{2w}$. ∎

Thus by evaluating $|B|$, we can determine the weight of the minimum weight perfect matching. The next lemma will enable us to obtain the matching itself.

LEMMA 18.5
Let M be the unique minimum weight perfect matching in G, and let w be its weight. The edge (v_i, v_j) belongs to M iff

$$\frac{|B_{ij}|2^{w_{ij}}}{2^{2w}} \text{ is odd.}$$

PROOF
First notice that

$$|B_{ij}|2^{w_{ij}} = \sum_{\alpha:\alpha(i)=j} sign(\alpha) \times value(\alpha).$$

Since n is even, if a non-vanishing permutation contains an odd-cycle in its trail, then it contains at least two such cycles. Now by the argument in Tutte's Theorem, such permutations cancel each other out and do not contribute to $|B_{ij}|2^{w_{ij}}$.

If $(v_i, v_j) \in M$, the permutation whose trail is M has value $\pm 2^{2w}$, and the value of every other permutation is a higher power of 2. On the other hand, if $(v_i, v_j) \notin M$, all permutations have values which are higher power of 2. The lemma follows. ∎

The algorithm to find M is now straightforward:

ALGORITHM 18.2
  Step 1 Compute $|B|$, and obtain w.

  Step 2 Compute $adj(B)$; its $(j,i)^{th}$ entry will be the minor $|B_{ij}|$.

  Step 3 For each edge (v_i, v_j) do in parallel:
  Compute $\frac{|B_{ij}|2^{w_{ij}}}{2^{2w}}$;
  If this quantity is odd, include (v_i, v_j) in the matching.
  end;

Notice that the only non-trivial computational effort involved is in evaluating the determinant and adjoint of B. We will use Pan's [24] randomized matrix-inversion algorithm, which computes $|B|$ and $adj(B)$ in order to compute B^{-1}. It requires $O(\log^2 n)$ time and $O(n^{3.5}m)$ processors for inverting an $n \times n$ matrix whose entries are m-bit integers.

THEOREM 18.7 *Mulmuley, Vazirani and Vazirani [22]*
There is an $O(\log^2 n)$ time parallel algorithm which find a perfect matching in the given graph, using $O(n^{3.5}m)$ processors.

Although the sequential version of this algorithm is less efficient than conventional matching algorithms, it has the advantage of being easy to program, especially if a subroutine for matrix inversion is available. The most efficient known sequential algorithms take $O(m\sqrt{n})$ steps. See [12] for bipartite graphs, and [21, 29] for general graphs.

18.5.3 Parallel Algorithms for Maximum Matching and Related Problems

a) We first address the problem of finding a *minimum weight perfect matching* in a graph $G = (V, E)$, given edge-weights $w(e)$ for each edge $e \in E$ in *unary*. First notice that if the weight of each edge is scaled up by a factor of mn, the minimum weight perfect matchings will be lighter than the rest by at least mn. We can now use the Isolating Lemma to isolate one of these minimum weight matchings: to edge $e \in E$ assign the weight $mnw(e) + r_e$, where r_e is chosen uniformly and independently from $[1, 2m]$. The proof of the Isolating Lemma works in this setting as well (see Exercise 18.4). As such this algorithm will require $0(n^{3.5}mW)$ processors, where W is the weight of the heaviest edge. Hence if the edge weights are in unary, this problem is in RNC^2. The parallel complexity of this problem for binary edge-weights is as yet unresolved.

b) The problem of finding a *maximum matching* in a graph can now be reduced to minimum weight perfect matching as follows: extend G into a complete graph by throwing in new edges. Assign weight 0 to each edge of G, and 1 to each of the new edges, and find a minimum weight perfect matching.

c) The *vertex-weighted matching* problem is the following:

Input: Graph $G = (V, E)$, and a positive weight for each vertex $v \in V$.

Problem: Find a matching in G whose vertex-weight is maximum. The vertex-weight of a matching is defined to be the sum of the weights of the vertices covered by the matching.

First notice that the desired matching will be a maximum matching. This is so because any non-maximum matching can be augmented into a maximum matching without unmatching any vertices in the process. Hence, the solution consists of finding the heaviest maximum matching vector, α, and a perfect matching in G_α. Sort the vertices of G by decreasing weight.

LEMMA 18.6
The lexicographically first maximum matching vector is the heaviest.

PROOF
Suppose the lexicographically first maximum matching vector is not the heaviest. Let L and H be maximum matchings which give the lexicographically first and the heaviest maximum matching vectors respectively; moreover, among all such matchings L and H, choose a pair that have the most number of edges in common. Let u be the first vertex in the sorted order where the two vectors differ. The vertex u will be matched in L but not in H. Consider the symmetric difference of L and H. This will have an even length alternating path from u to a vertex, say v. Clearly, u is at least as heavy as v. The symmetric difference of this path and H will yield a maximum matching that is at least as heavy as H, and has more edges in common with L. The contradiction proves the lemma. ∎

In Proposition 18.3, we presented an RNC algorithm for finding the lexicographically first maximum matching set.

18.6
The Las Vegas Extension

The maximum matching algorithm given in the previous section is Monte Carlo in the sense that it may sometimes give the wrong answer (i.e., either a matching that is not maximum, or a subset of edges that is not even a matching). We now give a complementary Monte Carlo parallel algorithm for upper bounding the size of maximum matching in the graph; this algorithm always outputs a number $\geq$ the size of maximum matching, and is correct with high probability. The two algorithms can be run simultaneously until the size of matching found by the first algorithm agrees with the number output by the second algorithm. This is a Las Vegas algorithm; it outputs the correct answer in *expected* running time $O(\log^2 n)$.

DEFINITION 18.5
For graph $G = (V, E)$ and $U \subseteq V$, let $\theta(V - U)$ denote the number of connected components having an odd number of vertices in the subgraph induced on $V - U$. Let $\nu(G)$ denote the size of a maximum matching in G. Notice that $\theta(V - U) - |U|$ is a lower bound on the number of vertices left unmatched by any maximum matching in G.

THEOREM 18.8 *Tutte and Berge*
In graph $G = (V, E)$,

$$\nu(G) = \min_{U \subseteq V} \frac{|V| + |U| - \theta(V - U)}{2}.$$

A set $V' \subset V$ for which the minimum is achieved is called a Tutte set.

By this theorem, it is sufficient to give a Monte Carlo algorithm for finding a Tutte set. Below we give an algorithm due to Lovasz (appearing in [14]). This algorithm is based on the Gallai-Edmonds structure theorem. We refer the reader to [14] or [20] for details of this theorem.

DEFINITION 18.6
Say that vertex $v \in V$ is critical if $\nu(G(V - \{v\})) = \nu(G) - 1$. For $S \subseteq V$, let $N(S)$ denote the neighborhood of S, i.e.

$$N(S) = \{v \in V - S | v \text{ is adjacent to a vertex in } S\}.$$

Let $D = \{v \in V \mid v \text{ is not critical}\}$ and $A = N(D)$. Then, by the Gallai-Edmonds structure theorem, A is a Tutte set. The set A can be found by $n + 1$ calls to a Monte Carlo algorithm for testing the size of maximum matching in a graph (see Proposition 18.2); the calls are made once with G as input, and once with $G(V - \{v\})$ as input, for each vertex v. We may assume that the Monte Carlo algorithm has error probability less than $\frac{1}{2(n+1)}$. Then, with probability at least $\frac{1}{2}$, the set found will be a Tutte set.

THEOREM 18.9 *Karloff [14]*
There is an $O(\log^2 n)$ time $O(n^2 M(n))$ processor Monte Carlo parallel algorithm for upper bounding the size of a maximum matching in a graph.

18.7
Improving the Processor Efficiency

In this section we will present the matching algorithm of Karp, Upfal and Wigderson [15], together with the Galil-Pan extension [9]. This achieves a processor bound of $O(n^{3.5})$, with the slightly higher time bound of $O(\log^3 n)$.

The algorithm requires $O(\log n)$ iterations; in each iteration it prunes out a constant fraction of the edges, so that in the end it is left with a perfect matching. The key idea lies in pruning out edges in such a way that after each iteration, the remaining graph still has a perfect matching. For this purpose, we need the following definitions:

DEFINITION 18.7
For $S \subseteq E$, define $rank(S)$ to be the size of the largest intersection of S with some perfect matching M in G, i.e.

$$rank(S) = \max_{M} |M \cap S|$$

Define $redundant(S)$ to be edges in $E - S$ that do not increase the rank of S, i.e.,

$$redundant(S) = \{e \in E - S \mid rank(S) = rank(S \cup \{e\})\}.$$

The algorithm removes edges in $redundant(S)$, corresponding to a well-chosen set S. By the next lemma, the remaining graph still has a perfect matching.

LEMMA 18.7
Let S be an arbitrary subset of E, and let G' be the graph obtained by removing the edges in $redundant(S)$ from G. Then G' has a perfect matching.

PROOF
Let M be a perfect matching in G such that $|M \cap S| = rank(S)$. Clearly, each edge of $M \cap S$ will be retained in G'. The lemma follows on observing that each edge of $M - S$ is non-redundant, and so will also be retained in G'. ∎

The next task is to choose a set S so that $redundant(S)$ is large. Assume that the graph has at least $\frac{3}{4}n$ edges; if not, the graph can be handled quite simply as shown later. The next lemma shows that if set S is picked by picking a random integer i in $[1 \ldots m]$, and picking a random subset of E of cardinality i, then $redundant(S)$ will be large, with high probability. The intuition is as follows: think of S as being built up one step at a time, by picking edges at random. In any run, the rank increases on only $\frac{n}{2}$ steps. Therefore, on most steps, the probability that a random edge will increase the rank is small, i.e., many edges are redundant.

LEMMA 18.8
Suppose $m \geq \frac{3}{4}n$. Pick a set $S \subseteq E$ by picking a random number i in $[1 \ldots m]$, and picking a random subset of E of cardinality i. Then

$$Pr\left[|redundant(S)| \geq \frac{m}{72}\right] \geq \frac{1}{72}.$$

18.7. Improving the Processor Efficiency

PROOF
Let T_i be a random subset of E of cardinality i, $0 \leq i < m$, and let e be a random edge in $E - T_i$. Denote by p_i the probability that e is not redundant $w \cdot r \cdot t \cdot T_i$, i.e.,

$$Pr_{T_i, e}\left[rank(T_i \cup \{e\}) = rank(T_i) + 1\right] = p_i.$$

Then,

$$E_{T_i}\left[|redundant(T_i)|\right] = (1 - p_i)(m - i).$$

Also, since T_i may be thought of as being built up by adding i edges at random, starting with the empty set,

$$E\left[rank(T_i)\right] = \sum_{j=0}^{i-1} p_j.$$

Since $rank(E) = \frac{n}{2}$, we get

$$E\left[rank(E)\right] = \sum_{j=0}^{m-1} p_j = \frac{n}{2}.$$

Finally, let us lower bound the expected size of $|redundant(S)|$.

$$\begin{aligned}
E(|redundant(S)|) &= \frac{1}{m}\sum_{i=1}^{m}(1 - p_i)(m - i) \\
&\geq \frac{1}{m}\sum_{i=1}^{\frac{5}{6}m}(1 - p_i)\frac{1}{6}m \\
&\geq \frac{1}{6}\left[\frac{5m}{6} - \sum_{i=0}^{m-1} p_i\right] = \frac{1}{6}\left[\frac{5}{6}m - \frac{n}{2}\right] \\
&\geq \frac{m}{36}.
\end{aligned}$$

Now,

$$\frac{m}{36} \leq E(|redundant(S)|) = \sum_{i=1}^{m} i\left(Pr\left[|redundant(S)| = i\right]\right)$$

$$\leq \frac{m}{72} Pr\left[|redundant(S)| < \frac{m}{72}\right] + m Pr\left[|redundant(S)| \geq \frac{m}{72}\right]$$

Therefore, $Pr\left[|redundant(S)| \geq \frac{m}{72}\right] \geq \frac{1}{72}$. ∎

Next, let us clear up the case that $m < \frac{3}{4}n$. In this case at least $\frac{n}{2}$ vertices have only one edge incident at them. Edges incident at these vertices are included in the matching, and their endpoints are deleted. Hence, in this case also a constant fraction of the edges are deleted. Consequently, the algorithm requires $O(\log n)$ iterations. An outline of the algorithm follows:

ALGORITHM 18.3

> while $m > 0$ do
>   if $m < \frac{3}{4}n$ then
>   accumulate edges incident at vertices of degree 1 in the matching, and remove the matched vertices from the graph;
>   else   pick a random integer i in $[1 \ldots m]$;
>   pick a random set $S \subseteq E$ of cardinality i;
>   remove edges in $redundant(S)$ from the graph
> end;

The remaining task is to find $redundant(S)$ efficiently. For this, we shall modify the Tutte matrix as follows:

LEMMA 18.9
Let $T \subseteq E$. Define an $n \times n$ matrix $A(T, y)$ as follows:

$$a_{ij}(T,y) = \begin{cases} y\, x_{ij} & \text{if } (i,j) \in T \text{ and } i < j \\ -y\, x_{ij} & \text{if } (i,j) \in T \text{ and } i > j \\ x_{ij} & \text{if } (i,j) \in E - T \text{ and } i < j \\ -x_{ij} & \text{if } (i,j) \in E - T \text{ and } i > j \\ 0 & \text{if } (i,j) \notin E. \end{cases}$$

Then $deg_y(|A(T,y)|) = 2\, rank(T)$.

PROOF
Let M be a perfect matching in G such that $|M \cap T| = rank(T)$. Let σ be the permutation whose trail is M, i.e., $\sigma(i) = j$ for each edge $(i,j) \in M$. Then, $deg_y(value(\sigma)) = 2\, rank(T)$. Notice that $value(\sigma)$ will not be cancelled by any term in $|A(T,y)|$. Therefore, $deg_y(|A(T,y)|) \geq 2\, rank(T)$.

To prove the other direction, first notice that as in the proof of Tutte's theorem, permutations having an odd cycle in their trail will cancel each other out. Finally, consider a permutation σ whose trail has only even cycles and edges traversed twice. By appropriately choosing alternate edges from the even cycles, we can obtain a permutation σ' whose trail is a perfect matching and $deg_y(value(\sigma')) \geq deg_y(value(\sigma))$. Clearly, for any permutation σ whose trail is a perfect matching, $deg_y(value(\sigma')) \leq 2rank(T)$. Therefore, $deg_y(|A(T,y)|) \leq 2rank(T)$. ∎

As a consequence, $rank(S)$ can be computed by substituting the x_{ij}'s randomly and computing $|A|$, which will be a polynomial in y, say $p(y)$. As before, by Lemma 18.2, $deg_y(p(y)) = deg_y(|A(t,y)|)$, with high probability. The following result shows that $|A(T,y)|$ can be computed in RNC.

THEOREM 18.10   *Borodin, Cook and Pippenger [2]*
The determinant of an $n \times n$ matrix, whose entries are polynomials with integer coefficients, over a constant number of variables, having degree at most n, can be computed in $O(\log^2 n)$ time using $O(n^{4.5})$ processors.

For computing $redundant(S)$, the algorithm computes $rank(S \cup e)$ for each edge $e \in E - S$.

THEOREM 18.11   *Karp, Upfal and Wigderson [15]*
There is an $O(\log^3 n)$ time $O(n^5 M(n))$ processor randomized parallel algorithm for finding a perfect matching in a graph.

Next we present the Galil-Pan extension which reduces the processor bound considerably, to $O(nM(n))$. The saving in processors comes by operating on integers rather than polynomials (i.e., substituting for indeterminate y also), and by computing the adjoint of the substituted matrix. The adjoint has all n^2 minors, which give $rank(S \cup e)$ for each edge $e \in E - S$. On the other hand, computing the adjoint is no more costly than computing the determinant!

Consider the bipartite case first. For graph $G(U,V,E), |U| = |V| = n$, and $T \subseteq E$, define an $n \times n$ matrix $B(T,y)$ as follows:

$$b_{ij}(T,y) = \begin{cases} yx_{ij} & \text{if } (i,j) \in T \\ x_{ij} & \text{if } (i,j) \in E - T \\ 0 & \text{if } (i,j) \notin E. \end{cases}$$

It is easy to see that $deg_y(|B(T,y)|) = rank(T)$.

DEFINITION 18.8
For prime p and integer g, define the p-adic order of g to be the largest integer k such that p^k divides g. This is denoted by $|g|_p$. Let x denote a set of variables $\{x_{ij}\}$. The p-adic order of a non-zero polynomial $g(x)$ is

defined as the least p-adic order of the coefficients of $g(x)$, and is denoted by $|g(x)|_p$.

Define an $n \times n$ matrix $C(T, p)$, where p is a prime, as follows:

$$c_{ij}(T,p) = \begin{cases} x_{ij} & \text{if } (i,j) \in T \\ px_{ij} & \text{if } (i,j) \in E - T \\ 0 & \text{if } (i,j) \notin E \end{cases}$$

By relating $C(T, p)$ to $B(T, y)$, it is easy to see that $n - |C(T,p)|_p = deg_y(|B(T,y)|) = rank(T)$. Furthermore, since $C(S \cup e, p)$ differs from $C(S, p)$ only in one entry, by expanding the first determinant by row i and collecting appropriate terms, we get

$$|C(S \cup e, p)| = |C(S, p)| + (1 - p)x_{ij}(adj\, C(S, p))_{ij} \quad \ldots (i)$$

where $e = (i, j)$. The idea now is to substitute for the variables in $C(S, p)$ randomly, pick a prime p, and compute the determinant and adjoint of the resulting matrix. This will yield, with high probability, $|C(S,p)|_p$ and $|C(S \cup e, p)|_p$ for each edge $e \in E - S$. The following fact will be used for proving that the p-adic order of the determinant of the substituted matrix is the same as $|C(S,p)|_p$, with high probability.

Fact: Let $g(x)$ be a polynomial with $|g(x)|_p = s$, and let $g_o(x)$ denote the sum of all monomials of $g(x)$ having p-adic order s. Let $x^o = \{x_{ij}^o\}$ denote an integer substitution for the variables x_{ij}. If p^{s+1} does not divide $g_o(x^o)$ then $|g(x^o)|_p = |g(x)|_p$.

The algorithm is as follows:

ALGORITHM 18.4

Step 1 Choose a random prime p, $n \leq p \leq n^r$, where $r > 4$.

Step 2 Choose x_{ij}'s randomly from $[1 \ldots n^r]$. Let $\tilde{C}$ denote the substituted matrix.

Step 3 Compute $|\tilde{C}|$ and $adj(\tilde{C})$.

Step 4 Use (i) to compute $|\tilde{C}(S \cup e, p)|$ for each $e \in E - S$.

Step 5 Compute the p-adic orders of $|\tilde{C}|$ and $|\tilde{C}(S \cup e, p)|$ for each $e \in E - S$, thereby obtaining $rank(S)$ and $rank(S \cup e)$.

The above procedure outputs the ranks of all sets S and $S \cup e$, for each edge e correctly with probability at least $1 - O(1/n^{r-4})$. For a detailed argument, see Lemma 2·3 in [9]. The main idea is to upper-bound the determinants using Haddamard's inequality; the random choice of p and the above-stated Fact then give the result.

Finally, let us consider the case of general graphs. For this we will use Lemma 18.9 and:

LEMMA 18.10
Let $T \subseteq E$ and $rank(T) = k$, and let $e \in E - T, e = (i,j)$. Define an $n \times n$ matrix $A^+(T,e,y)$ which differs from $A(T,y)$ only in two entries (assume $i < j$):

$$a_{ij}^+(T,e,y) = yx_{ij}$$
$$a_{ji}^+(T,e,y) = -x_{ij}$$

Then $deg_y(|A^+(T,e,y)|)$ is $2k+1$ if $rank(T \cup e) > rank(T)$, and $2k$ otherwise.

PROOF
The proof is similar to that of Lemma 18.9. One new idea is needed: notice that any permutation, σ, containing an odd cycle, contains at least two such cycles (since n is even). At least one of these odd cycles does not contain edge e. By reversing this odd cycle we obtain another permutation σ' that will cancel σ. Thus once again all permutations having odd cycles cancel each other out. ∎

By Lemmas 18.9 and 18.10,

$$deg_y(|A^+(S,e,y)|) - deg_y(|A(S,y)|) = rank(S \cup e) - rank(S).$$

Since $A^+(S,e,y)$ and $A(S,y)$ differ only in one entry, using (i), $|A^+(S,e,y)|$ can be computed from $|A(S,y)|$ and $adj(A(S,y))$. The rest of the algorithm is similar to Algorithm 18.4.

THEOREM 18.12 *Galil and Pan [9]*
There is an $O(\log^3 n)$ time $O(nM(n))$ processor randomized parallel algorithm for finding a perfect matching in a graph.

18.8
Is Matching in NC?

The title states an outstanding open problem in the area of parallel computation (the author believes that the answer lies in the affirmative). In this section we will present the NC algorithm of Grigoriev and Karpinski [11] for finding a perfect matching in case the graph has only a polynomially bounded number of perfect matchings. We will restrict attention to the bipartite case, and will leave the general graph case as an exercise (see Exercise 18.6).

Let us first consider the special case that the graph has a unique perfect matching. Rabin and Vazirani [25] observed that in this case, the perfect matching can be found in NC. Let A be the adjacency matrix of the bipartite graph, and let M be the unique perfect matching in it. Since only the permutation corresponding to M will be non-vanishing, $|A| = \pm 1$. Furthermore, notice that edge $(u_i, v_j) \in M$ iff $a_{ij} = 1$ (i.e., $(u_i, v_j) \in E$) and $|A_{ij}| = \pm 1$ (i.e., the graph obtained by removing u_i and v_j has a unique perfect matching). Hence, the perfect matching M can be obtained in NC by inverting A, since the inverse has all n^2 minors.

Suppose bipartite graph $G = (U, V, E)$ has at most n^k perfect matchings for some constant k. Let A be the $n \times n$ adjacency matrix of G. Find $|E|$ distinct primes $p_1 \ldots p_{|E|}$, and replace each 1 in A by a distinct prime to obtain matrix Q. Now obtain matrices B_l, $0 \leq l < n^k$ as follows: the $(i,j)^{th}$ entry of B_l is $(q_{ij})^l$. Notice that $B_0 = A$ and $B_1 = Q$. In parallel, compute $|B_l|$, $0 \leq l < n^k$.

LEMMA 18.11
G has a perfect matching iff $|B_l| \neq 0$ for some l, $0 \leq l < n^k$.

PROOF
Clearly, if G has no perfect matchings, then $|B_l| = 0$ for $0 \leq l < n^k$. Next suppose G has r perfect matchings $M_1 \ldots M_r$, $r \leq n^k$. For each matching M_i define $weight(M_i)$ to be the product of the n primes corresponding to the n edges of M_i. Let C be an $r \times r$ matrix whose $(i,j)^{th}$ entry is $(weight(M_i))^{j-1}$. Since $weight(M_i)$ is distinct for each matching, C is a Vandermonde matrix and is non-singular. Let $sign(M_i)$ be $+1$ if the permutation corresponding to M_i is even, and -1 otherwise, and let v be an r-dimensional column vector whose i^{th} entry is $sign(M_i)$. Notice that the ith entry of Cv is $|B_{i-1}|$. Since C is non-singular, and v is a non-zero vector, $Cv \neq \mathbf{0}$. Therefore $|B_l| \neq 0$ for some l, $0 \neq l < r$. ∎

By the above-stated lemma, there is an NC algorithm for testing whether G has a perfect matching. The idea for finding one perfect matching in polylog time is to first find a set S of at most $k \log n$ edges such that there is only one perfect matching containing all these edges. Once such a set S of edges is obtained, the end points of these edges are removed from G, and the unique perfect matching in the remaining graph is found as described earlier.

The set S is found as follows. If G has more than one perfect matching, we can obtain (by invoking the test for perfect matching) two edges (u_i, v_{j_1}) and (u_i, v_{j_2}) which belong to perfect matchings. Let G_1 and G_2 be the graphs obtained by deleting the end points of (u_i, v_{j_1}) and (u_i, v_{j_2}) respectively. Repeat this process recursively on G_1 and G_2 separately. Notice that the sum of the number of perfect matchings in G_1 and G_2 is at most the number of perfect matchings in G. Therefore, one of these two graphs has at most half the number of perfect matchings in G. Hence, after at most $k \log n$ steps we will obtain a graph having a unique perfect matching.

THEOREM 18.13 *Grigoriev and Karpinski [11]*
For any fixed integer k, there is an NC algorithm for finding a perfect matching in graphs having $O(n^k)$ perfect matchings.

Cheriyan and Reif [4] give output sensitive (where the complexity depends on the output) algorithms for various classes of matchings, including bipartite graphs where one of the vertex partitions is substantially smaller than the other, and also computing the maximum vertex weighted matchings.

Acknowledgement

I wish to thank Richard Karp and Eva Tardos for valuable discussions.

18.9
Exercises

18.1 For each of the following, give an NC reduction from the first problem to the second, i.e., show that if the second problem is in NC then so is the first:
   a) Computing the size of a maximum matching; testing if a graph has a perfect matching.
   b) Finding a maximum matching; finding a perfect matching if the graph has one, else outputting "no".
   c) Finding the lexicographically first maximal matching; finding the lexicographically first perfect matching.
   d) Finding a maximal matching; finding a maximal independent set, i.e., a maximal set of vertices no two of which are connected by an edge.

808 Chapter 18. Parallel Graph Matching

e) Finding a maximum s-t flow in a directed graph whose edge capacities are (i) all unity, and (ii) given in unary; finding a maximum matching in a bipartite graph.

18.2 A is an $n \times n$ matrix of rank r, whose entries are polynomials of degree d in k variables over the integers, for fixed k and d. The variables are substituted at random from a set S of integers, to obtain the substituted matrix $\tilde{A}$, whose entries are integers. How large should S be to ensure that with probability at least 1/2, each base of A is also a base of $\tilde{A}$, where as before, a base of A is a set of r linearly independent rows of A.

18.3 For each theorem proved in Section 18.4, prove the corresponding version for bipartite graphs, using Edmonds' matrix. Also, adapt the parallel maximum matching algorithms to bipartite graphs using Edmonds' matrix, and prove their correctness.

18.4 Prove that the Isolating Lemma holds even with the following modifications:

a) Each element x_i has an initial weight a_i, and the assigned random weight w_i is added to the initial weight to obtain the weight of x_i. (This modification is used in Section 18.5.3 for finding a minimum weight perfect matching.)

b) The weight of a set is defined to be the product of the weights of elements in it, i.e.,
$$wt(S_j) = \Pi_{x_i \in S_j} w_i$$
Are there other natural definitions of $wt(S_j)$ for which the Isolating Lemma holds?

18.5 Given a graph G with a subset of edges coloured red, and an integer k, the *exact matching* problem asks for a perfect matching containing exactly k red edges. Give RNC algorithms for the decision and search versions of this problem. Interestingly enough, this problem is not yet known to be in P. (Hint: use Lemma 18.2 and the Isolating Lemma respectively.)

18.6 Extend the algorithms presented in Section 18.8 to general graphs, using the Tutte matrix. For the case of unique perfect matching, first prove the following fact using Frobenius' Theorem: Let A be an $n \times n$ skew-symmetric integer matrix, where n is an even number. Then, for $1 \leq i, j \leq n$, if $|A_{ij}| \neq 0$ then $|A_{ii,jj}| \neq 0$, where $A_{ii,jj}$ is the submatrix of A obtained on deleting the ith and the jth rows and columns.

18.7 Let A and B be two $n \times m$ integer matrices of rank n. Give an RNC algorithm for picking a set of n columns from A, and the same numbered columns from B so that each set is of full rank. This is a special case of the matroid intersection problem for linearly representable matroids. (Hint: use the Isolating Lemma, and the Binet-Cauchy Theorem which states that

$$|AB^T| = \sum_\sigma |A_\sigma||B_\sigma|,$$

where the sum ranges over all possible ways, σ, of picking n elements out of m.)

Bibliography

[1] B. Awerbuch, A. Israeli, and Y. Shiloach. Finding Euler circuits in logarithmic parallel time. In *Proc. 16th Annual ACM Symposium on Theory of Computing*, pages 249–257, 1984.

[2] A. Borodin, S.A. Cook, and N. Pippenger. Parallel computation for well-endowed rings and space bounded probabilistic machines. *Information and Control*, 58:113–136, 1983.

[3] A. Borodin, J. von zur Gathen, and J. Hopcroft. Fast parallel matrix and gcd computations. *Inform. and Control*, 53:241–256, 1982.

[4] J. Cheriyan and J. Reif, Parallel and Output Sensitive Algorithms for Combinatorial and Linear Algebra Problems, manuscript, 1992.

[5] D. Coppersmith and S. Winograd. Matrix multiplication via arithematic progressions. In *Proc. 19th Ann. ACM Symp. on Theory of Computing*, pages 1–6, 1987.

[6] L. Csanky. Fast parallel matrix inversion algorithms. *SIAM J. Computing*, 5:618–23, 1976.

[7] J. Edmonds. Systems of distinct representatives and linear algebra. *J. Res. Nat. Bureau of Standards*, 71B, 4:241–245, 1967.

[8] H.N. Gabow and R.E. Tarjan. Almost-optimal speed-ups of algorithms for bipartite matching and related problems. In *Proc. 20th Ann. ACM Symp. on Theory of Computing*, pages 514–527, 1988.

[9] Z. Galil and V. Pan, *Improved Processor Bounds for Combinatorial Problems in RNC*, Combinatorica **8** (1988), 189–200.

[10] S.A. Plotkin Goldberg, A.V. and P.M. Vaidya. Sublinear-time parallel algorithms for matching and related problems. In *Proc. 29th Annual IEEE Symposium on Foundations of Computer Science*, pages 174–185, 1988.

[11] D. Grigoriev and M. Karpinski. The matching problem for bipartite graphs with polynomially bounded permanents is in *NC*. In *Proc. 28th Annual IEEE Symposium on Foundations of Computer Science*, pages 166–172, 1987.

[12] J. Hopcroft and R.M. Karp. An $n^{5/2}$ algorithm for maximum matching in bipartite graphs. *SIAM J. Comput.*, 2:225–231, 1973.

[13] A. Israeli and Y. Shiloach. An improved parallel algorithm for maximal matching. *Inf. Proc. Letters*, 22(2):57–60, 1986.

[14] H. Karloff. A randomized parallel algorithm for the odd set cover problem. *Combinatorica*, 6:387–391, 1986.

[15] M. Karp, E. Upfal, and A. Wigderson. Finding a maximum matching is in Random *NC*. *Combinatorica*, 6:35–48, 1986.

[16] M. Karp, E. Upfal, and A. Wigderson. The complexity of parallel search. *J. Comput. and System Sci.*, 36:225–253, 1988.

[17] G. Kowalewski. Einfuhrung in die determinate theorie, 1909.

[18] A. Lev, N. Pippenger, and L.G. Valiant. A fast parallel algorithm for routing in permutation networks. *IEEE Trans. on Computers*, C30(2):93–100, 1981.

[19] L. Lovasz. On determinants, matchings and random algorithms. In L. Budach, editor, *Fundamentals of Computing Theory*. Akademia-Verlag, Berlin, 1979.

[20] L. Lovasz and M. Plummer. *Matching Theory*. Academic Press, Budapest, Hungary, 1986.

[21] S. Micali and V.V. Vazirani. An $O(\sqrt{|V|}|E|)$ algorithm for finding maximum matching in general graphs. *Twenty First Annual IEEE Symp. on the Foundations of Computer Science*, pages 17–27, 1980.

[22] K. Mulmuley, U. Vazirani, and V. Vazirani. Matching is as easy as matrix inversion. *Combinatorica*, 7:105–113, 1987.

[23] H. Narayanan, H. Saran, and V.V. Vazirani. Randomized parallel algorithms for matroid union and intersection, with applications to arborescences and edge-disjoint spanning trees. In *Proc. Fourth Annual ACM-SIAM Symposium on Discrete Algorithms*, pages 357–366, 1992.

[24] V. Pan. Fast and efficient algorithms for the exact inversion of integer matrices. In *Fifth Annual Foundations of Software Technology and Theoretical Computer Science Conference*, pages 504–521, 1985.

[25] M.O. Rabin and V.V. Vazirani. Maximum matching in general graphs through randomization. *J. Algorithms*, 10:557–567, 1989.

[26] J.T. Schwartz. Fast probabilistic algorithms for verification of polynomial identities. *J. of ACM*, 27(4):701–717, 1980.

[27] W. T. Tutte. The factorization of linear graphs. *J. London Math Society*, 22:107–111, 1947.

[28] U.V. Vazirani and V.V. Vazirani. The two-processor scheduling problem is in Random NC. *SIAM J. Computing*, 18(6):1140–1148, 1989.

[29] V.V. Vazirani. A theory of alternating paths and blossoms for proving correctness of the $O(\sqrt{|V|}|E|)$ general graph matching algorithm. To appear in *Combinatorica*.

[30] R.E. Zippel. Probabilistic algorithms for sparse polynomials. In *Proc. EUROSAM '79, Springer Lecture Notes in Computer Science 72*, pages 216–226, 1979.

19

Parallel Algorithms for Network Flow Problems

Andrew V. Goldberg

Department of Computer Science
Stanford University
Stanford, CA 94305
goldberg@cs.stanford.edu

19.1 Introduction

In this chapter we discuss parallel algorithms for the maximum flow and the minimum-cost flow problems. These problems have many applications and have been studied for decades. The problems are P-complete [19], so it is unlikely that they can be solved in NC. As we shall see, however, these problems can be solved substantially faster on parallel machines.

This speedup is achieved by implementing sequential flow algorithms using parallel primitives like sorting and parallel prefix computations. However, not all flow algorithms have parallel implementations. In fact, parallelism was one of the main motivations for the development of the recent network flow methods.

In the following discussion, n and m denote the number of vertices and arcs in the input network. Also, U denotes the biggest arc capacity and C denotes the biggest arc cost. When U (or C) appear in the bound, the capacities (costs) are assumed to be integral. For more complete definitions, see the following section.

The first parallel algorithm for the maximum flow problem is due to Shiloach and Vishkin [31]. This algorithm is based on the blocking flow algorithm of Karzanov [22] and runs in $O(n^2 \log n)$ time using n processors and $O(n^2)$ memory. Goldberg [9] gave an algorithm that runs in the same time and processor bounds but uses $O(m)$ memory. This algorithm is the first of the *push-relabel* maximum flow algorithms. The *push-relabel* method was developed by Goldberg and Tarjan as a generalization of it.

The *push-relabel* method is the basis for the best currently known parallel algorithms for the maximum flow problem. In this chapter we describe an implementation of this method that runs in $O(n^2 \log n)$ time using $\sqrt{m}$ processors and $O(m)$ memory (or slightly faster using slightly more memory). A parallel implementation of a scaling version of the *push-relabel* method, due to Ahuja and Orlin [2], runs in $O(n^2 \log(U) \log(n))$ time using m/n processors and $O(m)$ memory.

The minimum-cost flow problem is a generalization of the maximum flow problem, and most sequential algorithms for the latter problem generalize to the former. However, this is not the case with the parallel algorithms [10, 17].

The most promising parallel algorithms currently known are based on the blocking flow method first used by Dinic [6] and fully developed by Karzanov [22]. The maximum flow problem can be solved in $O(n)$ blocking flow computations on layered networks [6, 22]. Goldberg and Tarjan have shown that the minimum-cost flow problem can be solved in $O(n \log(nC))$ blocking flow computations on an acyclic network. Thus a parallel algorithm for finding a blocking flow in an acyclic network implies a parallel algorithm for the minimum-cost flow problem. The first parallel algorithm for the blocking flow problem in acyclic graphs is due to Goldberg and Tarjan [17]; this algorithm runs in $O(n^2 \log(nC) \log n)$ time and uses $O(m)$ processors and $O(nm)$ memory. Vishkin [36] shows that the resource bounds for the Shiloach-Vishkin algorithm (mentioned above) for the acyclic case are only by a constant factor worse than those for the layered network case; these bounds are $O(n^2 \log n)$ time, $O(n)$ processors, and $O(n^2)$ space. This algorithm is probably the most practical parallel algorithm for the problem known today.

Given a sufficient number of processors (as many as needed to do matrix multiplication in $O(\log^2 n)$ time), the minimum-cost flow problem can be solved faster. With this many processors, the interior-point linear programming algorithms of Renegar [30] and Ye [37], applied to the minimum-cost flow problem, run in $O(\sqrt{m} \log(CUn) \log^2 n)$ time, and Orlin's minimum-cost flow algorithm [27] runs in $O(m \log^3 n)$ time. The high processor requirement of these algorithms limits their practicality.

Using parallelism is one way to improve efficiency of network flow algorithms. Using sophisticated data structures is another way. For example, Sleator and Tarjan [32, 33] introduced the dynamic tree data structure and used it to obtain an $O(nm \log n)$ algorithm for the maximum flow problem. This bound has been slightly improved in [3, 16]. The ideas leading to these improved bounds do not seem to carry over to the context of parallel computation.

For more information on network flow algorithms, see, e.g., [1, 14, 24, 28, 34].

This chapter is limited to network flow problems with arbitrary capacities. Better algorithms exist for the special cases of zero-one capacities and bipartite matching. For details, see [8, 12, 13, 21, 25].

The chapter is organized as follows. Section 19.2 introduces definitions and background. Section 19.3 describes the push-relabel method for the maximum flow problem and its parallel implementations. Section 19.4 describes the Goldberg-Tarjan algorithm for finding blocking flows in acyclic networks, which is the basis for a parallel algorithm for the minimum-cost flow problem.

19.2
Background

We use the following definitions. Let $G = (V, E)$ be a directed graph with vertex set V of size n and arc set E of size m. For ease in stating time bounds, we assume that $m > n$ and therefore $\log(m/n) > 0$. Define $E^{-1} = \{(w,v)|(v,w) \in E\}$ and $E^+ = E \bigcup E^{-1}$. For any vertex w we denote by $E(w)$ the set of vertices *adjacent out from* w, $E(w) = \{x|(w,x) \in E\}$, and by $E^{-1}(w)$ the set of vertices *adjacent into* w, $E^{-1}(w) = \{v|(v,w) \in E\}$. Graph G is a *network* if it has two distinguished vertices, a *source* s and a *sink* t, and a nonnegative real-valued capacity $u(v, w)$ on every arc (v, w). A *preflow* on a network is a nonnegative real-valued function f on the arcs such that $f(v,w) \leq u(v,w)$ for every arc (v,w) and $\sum_{v \in E^{-1}(w)} f(v,w) \geq \sum_{x \in E(w)} f(w,x)$ for every vertex $w \neq s$. The quantity $e_f(w) = \sum_{v \in E^{-1}(w)} f(v,w) - \sum_{x \in E(w)} f(w,x)$ is called the *excess* at vertex w. A preflow f is a *flow* if $e_f(w) = 0$ for every vertex $w \notin \{s,t\}$. A *cost function* $c : E \to R$ assigns a cost to arcs of the network. We assume that costs are integers in the range $[-C, \ldots, C]$.

The residual graph is induced by the set of residual arcs $E_f = E_1 \bigcup E_2$, where

$$E_1 = \{(v,w)|(v,w) \in E \text{ and } f(v,w) < u(v,w)\}$$

and

$$E_2 = \{(w,v)|(w,v) \in E^{-1} \text{ and } f(v,w) > 0\}.$$

The residual capacity function is defined on E_f as follows: for arcs $(v,w) \in E_1$, $u_f(v,w) = u(v,w) - f(v,w)$; for arcs $(w,v) \in E_2$, $u_f(w,v) = f(v,w)$. Arc (v,w) is *saturated* if $u_f(v,w) = 0$. A preflow is *blocking* if every path in G from s to t contains at least one saturated arc, i.e., there is no path of residual arcs from s to t.

A *value* of a flow f is the excess of the sink $e_f(t)$. The *maximum flow* problem is to find a flow of the biggest value. The *minimum-cost flow* problem is to find a maximum flow of minimum cost. The *blocking flow* problem is to find a blocking flow.

Given a flow f, we define an *augmenting path* to be a source-to-sink path in the residual graph. The following theorem, due to Ford and Fulkerson, gives an optimality criterion for maximum flows.

THEOREM 19.1 [7]
A flow is optimal if and only if its residual graph contains no augmenting path.

Suppose the graph G is acyclic. We say that G is *layered* if each vertex v can be assigned an integer *layer* $L(v)$ such that $L(w) = L(v) + 1$ for every arc (v, w). The blocking flow problem in a layered and acyclic graph are of a special interest because of the following theorems.

THEOREM 19.2 [22]
The maximum flow problem can be reduced to $O(n)$ computations of blocking flows in layered networks.

The above theorem holds because the distance from the source to the sink in the residual graph increases after each blocking flow computation. Scaling techniques allow a generalization of this result to the minimum-cost flow case as follows.

THEOREM 19.3 [18]
The minimum-cost flow problem can be reduced to $O(n \log(nC))$ computations of blocking flows in acyclic networks.

To get better running time bounds, we also assume, without loss of generality, that the maximum degree of a vertex in the input graph is at most $\Delta = 2m/n$. To justify this assumption, consider an arbitrary graph and replace each vertex v with degree $deg(v) > \Delta$ by $k = \lceil deg(v)/\Delta \rceil$ vertices $v_1, \ldots, v_k$ connected in a ring with arcs of a very high capacity (e.g., the sum of all original capacities). Distribute arcs of v along $v_1, \ldots, v_k$ so that the degree of the new vertices is bounded by Δ. If the original graph has n vertices and m arcs, the transformed graph has at most $2n$ vertices and $m + n$ arcs.

19.3
The Maximum Flow Problem

In this section we describe a method for solving the maximum flow problem based on the basic operations push and relabel. This method was introduced by Goldberg [9] and fully developed by Goldberg and Tarjan [16]. We describe the general method first, and then we describe its parallel implementations developed in [10, 11, 31].

19.3.1 A Generic Algorithm

To describe the generic *push-relabel* algorithm, we need the following definition. For a given preflow f, a *distance labeling* is a function d from the vertices to the non-negative integers such that $d(t) = 0$, $d(s) = n$, and $d(v) \leq d(w) + 1$ for all residual arcs (v, w). To see the intuition behind this definition, define a *distance graph* G_f^* as follows. Add an arc (s,t) to G_f. Define the length of all residual arcs to be equal to one and the length of the arc (s,t) to be n. Then d is a "locally consistent" estimate on the distance to the sink in the distance graph. (In fact, it is easy to show that d is a lower bound on the distance to the sink.) We denote by $d_{G_f^*}(v,w)$ the distance from vertex v to vertex w in the distance graph.

The generic algorithm maintains a preflow f and a distance labeling d for f, and updates f and d using *push* and *relabel* operations. To describe these operations, we need the following definitions. We say that a vertex v is *active* if $v \notin \{s,t\}$ and $e_f(v) > 0$. Note that a preflow f is a flow if and only if there are no active vertices. An arc (v,w) is *admissible* if $(v,w) \in E_f$ and $d(v) = d(w) + 1$.

The algorithm begins with the preflow f that is equal to the arc capacity on each arc leaving the source and zero on all arcs not incident to the source, and with some initial labeling d. The algorithm then repetitively performs, in any order, the *update operations, push* and *relabel*, described in Figure 19.2. When there are no active vertices, the algorithm terminates. A summary of the algorithm appears in Figure 19.1.

procedure *generic* (V, E, u);
  [initialization]
  $\forall (v,w) \in E$ **do begin**
    $f(v,w) \leftarrow 0$;
    if $v = s$ **then** $f(s,w) \leftarrow u(s,w)$;
  end;
  $\forall w \in V$ **do begin**
    $e_f(w) \leftarrow \sum_{(v,w) \in E^{-1}} f(v,w)$;
    if $w = s$ **then** $d(w) = n$ **else** $d(w) = 0$;
  end;
  [loop]
  while $\exists$ an active vertex **do**
    select an update operation and apply it;
  return(f);
end.

FIGURE 19.1
The generic maximum flow algorithm.

$push(v, w)$.
Applicability: v is active **and** (v, w) is admissible.
Action: send $\delta \in (0, \min(e_f(v), u_f(v, w))]$ units of flow from v to w.

$relabel(v)$.
Applicability: either s or t is reachable from v in G_f **and**
$\forall w \in V \; u_f(v, w) = 0$ or $d(w) \geq d(v)$.
Action: replace $d(v)$ by $\min_{(v,w) \in E_f} \{d(w)\} + 1$.

FIGURE 19.2
The update operations. The *push* operation updates the preflow, and the *relabel* operation updates the distance labeling.

The update operations modify the preflow f and the labeling d. A *push* from v to w increases $f(v, w)$ (or decreases $f(w, v)$) and increases $e_f(w)$ by up to $\delta = \min\{e_f(v), u_f(v, w)\}$, and decreases $e_f(v)$ by the same amount. The push is *saturating* if $u_f(v, w) = 0$ after the push and *nonsaturating* otherwise. A *relabeling* of v sets the label of v equal to the largest value allowed by the valid labeling constraints.

There is one part of the algorithm we have not yet specified: the choice of an initial labeling d. The simplest choice is $d(s) = n$ and $d(v) = 0$ for $v \in V - \{s\}$. A more accurate choice (indeed, the most accurate possible choice) is $d(v) = d_{G_f^*}(v, t)$ for $v \in V$, where f is the initial preflow. The latter labeling can be computed in $O(m)$ time using backwards breadth-first searches from the sink and from the source in the residual graph. The resource bounds we shall derive for the algorithm are correct for any valid initial labeling. For simplicity we assume that the algorithm starts with the simple labeling. In practice, it is preferable to start with the most accurate values of the distance labels, and to update the distance labels periodically by using backward breadth-first search.

Next we turn our attention to the correctness and termination of the algorithm. Our proof of correctness is based on Theorem 19.1. The following lemma is important in the analysis of the algorithm.

LEMMA 19.1
If f is a preflow and v is a vertex with positive excess, then the source s is reachable from v in the residual graph G_f.

Using this lemma and induction on the number of update operations, it can be shown that one of the two update operations must be applicable to an active vertex, and that the operations maintain a valid distance labeling and preflow.

THEOREM 19.4 [16]
Suppose that the algorithm terminates. Then the preflow f is a maximum flow.

PROOF
When the algorithm terminates, all vertices in $V - \{s,t\}$ must have zero excess, because there are no active vertices. Therefore f must be a flow. We show that if f is a preflow and d is a valid labeling for f, then the sink t is not reachable from the source s in the residual graph G_f. Then Theorem 19.1 implies that the algorithm terminates with a maximum flow.

Assume by way of contradiction that there is an augmenting path $s = v_0, v_1, \ldots, v_l = t$. Then $l < n$ and $(v_i, v_{i+1}) \in E_f$ for $0 \le i < l$. Since d is a valid labeling, we have $d(v_i) \le d(v_{i+1}) + 1$ for $0 \le i < l$. Therefore, we have $d(s) \le d(t) + l < n$, since $d(t) = 0$, which contradicts $d(s) = n$. ∎

The key to the running time analysis of the algorithm is the following lemma, which shows that distance labels cannot increase too much.

LEMMA 19.2
At any time during the execution of the algorithm, for any vertex $v \in V$, $d(v) \le 2n - 1$.

PROOF
The lemma is trivial for $v = s$ and $v = t$. Suppose $v \in V - \{s,t\}$. Since the algorithm changes vertex labels only by means of the *relabeling* operation, it is enough to prove the lemma for a vertex v such that s or t is reachable from v in G_f. Thus there is a simple path from v to s or t in G_f. Let $v = v_0, v_1, \ldots, v_l$ be such a path. The length l of the path is at most $n - 1$. Since d is a valid labeling and $(v_i, v_{i+1}) \in E_f$, we have $d(v_i) \le d(v_{i+1}) + 1$. Therefore, since $d(v_l)$ is either n or 0, we have $d(v) = d(v_0) \le d(v_l) + l \le n + (n-1) = 2n - 1$. ∎

Lemma 19.2 limits the number of relabeling operations, and allows us to amortize the work done by the algorithm over increases in vertex labels. The next two lemmas bound the number of relabelings and the number of saturating pushes.

LEMMA 19.3
The number of relabeling operations is at most $2n - 1$ per vertex and at most $(2n - 1)(n - 2) < 2n^2$ overall.

PROOF
Relabeling operations apply only to vertices $v \in V - \{s, t\}$. A relabeling of v increases $d(v)$. The lemma follows immediately from Lemma 19.2. ∎

LEMMA 19.4
The number of saturating pushes is at most nm.

PROOF
For an arc $(v, w) \in E$, consider the saturating pushes from v to w. After one such push, $u_f(v, w) = 0$, and another such push cannot occur until $d(w)$ increases by at least 2, a push from w to v occurs, and $d(v)$ increases by at least 2. If we charge each saturating push from v to w except the first to the preceding label increase of v, we obtain an upper bound of n on the number of such pushes. ∎

The most interesting part of the analysis is obtaining a bound on the number of nonsaturating pushes. For this we use amortized analysis and in particular the *potential function* technique (see, e.g., [35]).

LEMMA 19.5
The number of nonsaturating pushing operations is at most $2n^2m$.

PROOF
We define the *potential* Φ of the current preflow f and labeling d by the formula $\Phi = \sum_{\{v | v \text{ is active}\}} d(v)$. We have $0 \leq \Phi \leq 2n^2$ by Lemma 19.2. Each nonsaturating push, say from a vertex v to a vertex w, decreases Φ by at least one, since $d(w) = d(v) - 1$ and the push makes v inactive. It follows that the total number of nonsaturating pushes over the entire algorithm is at most the sum of the increases in Φ during the course of the algorithm, since $\Phi = 0$ both at the beginning and at the end of the computation. Increasing the label of a vertex v by an amount k increases Φ by k. The total of such increases over the algorithm is at most $2n^2$. A saturating push can increase Φ by at most $2n - 2$. The total of such increases over the entire algorithm is at most $(2n - 2)nm$. Summing gives a bound of at most $2n^2 + (2n - 2)nm \leq 2n^2m$ on the number of nonsaturating pushes. ∎

THEOREM 19.5 [16]
The generic algorithm terminates after $O(n^2m)$ update operations.

PROOF
Immediate from Lemmas 19.3, 19.4, and 19.5. ■

The running time of the generic algorithm depends upon the order in which update operations are applied and on implementation details. In the next sections we explore these issues. In Section 19.3.2 we describe an ordering of the operations that reduces the sequential and parallel running time. Then we describe two implementations of the parallel algorithm, a simple implementation and a more sophisticated one that requires significantly fewer processors.

19.3.2 Efficient Orderings of Basic Operations

In this section we describe an ordering of the operations that leads to an $O(n^2 \sqrt{m})$-time sequential algorithm. As we shall see later, this algorithm has a substantial degree of parallelism. Since parallel algorithms are the main topic of this book, we omit low-level details of the sequential algorithm in our description. These details can be found in [14, 16].

Our first step toward an efficient implementation is a way of combining the update operations locally. The *discharge* operation, described in Figure 19.3, accomplishes this. The *discharge* operation is applicable to an active vertex v. Discharge iteratively reduces the excess at v by pushing it through admissible arcs going out of v if such arcs exist; otherwise, *discharge* relabels v. The operation stops when the excess at v is reduced to zero or v is relabeled. Note that *discharge* relabels v only when the relabel operation applies.

The second step to an efficient ordering of basic operation is to restrict the order in which active vertices are processed. Two natural orders were suggested in [15, 16]. One, the *FIFO algorithm*, is to maintain the set of active vertices as a first-in, first-out queue. The corresponding parallel algorithm processes all active vertices at once. Another, the *maximum distance discharge*

$discharge(v)$.
Applicability: v is active.
Action: while $e_f(v) > 0$ **and** v is not relabeled **do**
        if $\exists$ an admissible arc (v, w)
           then $push(v, w)$
           else $relabel(v)$;

FIGURE 19.3
The discharge operation.

(MDD) algorithm, is to discharge a vertex with the largest label at every step. The corresponding parallel algorithm processes all such vertices at once. The FIFO algorithm runs in $O(n^3)$ time [15, 16] and the MDD algorithm runs in $O(n^2\sqrt{m})$ time [5]. We shall discuss the latter algorithm because it is more efficient.

The sequential implementation of the MDD algorithm maintains an array of sets B_i, $0 \leq i \leq 2n - 1$, and an index b into the array. Set B_i consists of all active vertices with label i, represented as a doubly-linked list, so that insertion and deletion take $O(1)$ time. The index b is the largest label of an active vertex. During the initialization, when the arcs going out of the source are saturated, the resulting active vertices are placed in B_0, and b is set to 0. At each iteration, the algorithm removes a vertex from B_b, processes it using the *discharge* operation, and updates b. The algorithm terminates when b becomes negative, i.e., when there are no active vertices. This processing of vertices, which implements the *while* loop of the generic algorithm, is described in Figure 19.4.

To understand why the *process-vertex* procedure correctly maintains b, note that $discharge(v)$ either relabels v or gets rid of all excess at v, but not both. In the former case, v is the active vertex with the largest distance label, so b must be increased to $d(v)$. In the latter case, the excess at v has been moved to vertices with distance labels of $b - 1$, so if B_b is empty, then b must be decreased by one. The total time spent updating b during the course of the algorithm is $O(n^2)$.

procedure *process-vertex*;
    remove a vertex v from B_b;
    old-label $\leftarrow d(v)$;
    $discharge(v)$;
    add each vertex w made active by the discharge to $B_{d(w)}$;
    if $d(v) \neq$ *old-label* **then begin**
        $b \leftarrow d(v)$;
        add v to B_b;
    end
    else if $B_b = \emptyset$ **then** $b \leftarrow b - 1$;
end.

FIGURE 19.4
The *process-vertex* procedure.

The bottleneck of the MDD algorithm is the nonsaturating pushes. We shall obtain an $O(n^3)$ bound on the number of such pushes by dividing the computation into *phases*. A phase consists of a maximal interval of time during which b remains constant.

LEMMA 19.6
The number of phases during the running of either the MDD algorithm is at most $4n^2$.

PROOF
We use the value of b as a potential function. Initially $b = 0$. The algorithm terminates as soon as b becomes negative. A phase that does no relabeling decreases b by one. The only increases in b are due to label increases; an increase of a label by k can cause b to increase by up to k. By Lemma 19.2, the sum of the increases in b over the computation is at most $2n^2$, and thus the number of phases that do no relabeling is at most $2n^2$. ∎

THEOREM 19.6 [16]
The MDD algorithm runs in $O(n^3)$ time.

PROOF
There is at most one nonsaturating push per vertex per phase. Thus by Lemma 19.6 the total number of nonsaturating pushes is $O(n^3)$. The number of saturating pushes is $O(nm)$ by Lemma 19.4. Note that the work done by *relabel(v)* is proportional to the degree of v. Since each vertex is relabeled $O(n)$ times, the total amount of work involved in relabeling is $O(nm)$. ∎

Cheriyan and Maheshwari [5], by means of an elegant balancing argument, were able to improve the bound on the number of nonsaturating pushes in the MDD algorithm to $O(n^2 \sqrt{m})$, obtaining the following result:

THEOREM 19.7 [5]
The MDD algorithm runs in $O(n^2 \sqrt{m})$ time.

Lemma 19.6 suggests a parallel version of the algorithm, where all largest-label active vertices are processed in parallel. The running time of the resulting algorithm is $O(n^2)$ times the time needed for the parallel processing of the vertices. Sections 19.3.3 and 19.3.4 describe such implementations.

19.3.3 A Simple Parallel Implementation

In this section we describe a parallel implementation of the MDD algorithm that uses $O(m)$ processors and runs in $O(n^2 \log(m/n))$ time. The disadvantage of this implementation is its processor-time product of $O(n^2 m \log(m/n))$ (compared with the $O(n^2 \sqrt{m})$ running time of the underlying sequential algorithm). However, this implementation is very simple and maps well onto high-grain SIMD architectures.

The simple implementation uses the following representation of the graph. There is a processor associated with every vertex and every arc of the network. Variables for arcs going out of a vertex v are allocated to facilitate parallel prefix computations. Note that in the parallel implementation it is not necessary to maintain the sets B_i explicitly. Given an index j, each vertex processor can in constant time determine if the corresponding vertex belongs to B_j.

The parallel implementation works in pulses, each of which consists of two stages. In the first stage, the active vertices in B_b push the flow out. In the second stage the vertices that still have positive excess are relabeled.

The first stage works as follows. For every active vertex v in B_b, the excess $e_f(v)$ is distributed among the admissible arcs going out of v, giving preference to the arcs at the beginning of the edge list. For each arc (v, w), the corresponding processor sets $u'(v, w)$ to $u_f(v, w)$ if (v, w) is admissible and to 0 otherwise. Then the parallel prefix *sum* operation if performed on $u'(v, w)$ variables, and the result is stored in $u''(v, w)$ variables. For each arc (v, w), $u''(v, w) - u'(v, w)$ is the total residual capacity of the admissible arcs that precede (v, w) on the edge list. Therefore, for every admissible arc (v, w), $f(v, w)$ is increased by

$$\delta(v,w) = \begin{cases} u_f(v,w) & \text{if } u''(v,w) \leq e_f(v), \\ 0 & \text{if } u''(v,w) - u_f(v,w) \geq e_f(v), \\ e_f(v) - (u''(v,w) - u_f(v,w)) & \text{otherwise.} \end{cases}$$

For nonadmissible arcs, $\delta(v, w) = 0$. Then $e_f(v)$ is updated using the value of u'' at the last arc on the edge list for v. The active vertices that participated in discharges but did not succeed in reducing their excess to zero are marked for relabeling.

To add the flow pushed into vertices, we copy the values $\delta(v, w)$ to $\delta'(w, v)$, use parallel a prefix *sum* operation in parallel on all the edge lists to compute the total amount of flow pushed into each vertex, and then update the excesses at each vertex.

It is easy to see that the first stage runs in $O(\log n)$ time. In fact, since we have assumed that the maximum vertex degree is $O(m/n)$, the first stage runs in $O(\log(m/n))$ time.

The second stage relabels the marked vertices and computes the new value for b. The *relabel(v)* operation is implemented using parallel prefix computations as follows. For each arc (v, w), the corresponding processor sets $d'(v, w)$ to $d(v)$ if (v, w) is admissible and to $2n$ otherwise. Then the parallel prefix *min* computation is done on $d'(v, w)$ variables to compute the smallest distance label d' of a residual neighbor of v. (Note that since $e_f(v) > 0$, for at least one arc, $d'(v, w) < 2n$). Then $d(v) \leftarrow d'(v) + 1$. The index b is decreased by one if no relabeling occurred, and set to the maximum new label otherwise. The second stage takes $O(\log(m/n))$ time, the same as the first stage.

By an argument similar to that of Lemma 19.6, the algorithm terminates in $O(n^2)$ pulses and therefore in $O(n^2 \log(m/n))$ time. It uses $O(m)$ processors and $O(m)$ memory.

Note that segmented parallel prefix operations can be used in the above implementation. The resulting implementation works on a SIMD machine. Since many architectures allow a very efficient implementation of segmented prefix operations (with running time comparable to a routing cycle), this implementation will be fast given that the machine has enough processors (with respect to the input graph size). Experimental results obtained by the author on the Connection Machine [20] are very encouraging. See [10] for details.

19.3.4 Processor-Efficient Parallel Implementations

The implementation of the previous section uses $O(m)$ processors and runs in $O(n^2 \log(m/n))$ time, but the total number of operations of the underlying sequential method is $O(n^2 \sqrt{m})$. Therefore most processors are idle most of the time. In this section we show that a careful scheduling allows a more efficient use of processors. The resulting implementations are similar to that of the Shiloach-Vishkin algorithm [31]. We show two slightly different implementations. Both use $p = O(\sqrt{m})$ processors. The first implementation runs in $O(n^2 \log(\Delta + p) \frac{\sqrt{m}}{p})$ time and uses $O(m + n \log n)$ space. The second implementation runs in $O(n^2 \log n \frac{\sqrt{m}}{p})$ time and uses $O(m)$ space. The time-processor product of this parallel implementation is within a logarithmic factor of the number of operations of the underlying sequential method. Because of the scheduling, this implementation is more complex than the one described above. It is more suitable for machines with a relatively small number of powerful processors.

The implementations maintain the sets B_i of active vertices with the distance label i, for $0 \leq i \leq 2n - 1$. The index b is maintained as in the sequential implementation. The sets are maintained so that in $O(\log p)$ time, several processors can add a vertex each to the sets, and several elements of B_b can be assigned to different processors and removed from the set.

A straightforward way to implement these operations is to use an array of length n for each set. The elements of B_i occupy the first $|B_i|$ locations of the corresponding array. The processors that want to add elements to $|B_i|$ are enumerated and add their elements to the array position determined by their rank and $|B_i|$; after this is done, $|B_i|$ is updated. To assign elements of B_b to a set of processors of size $k \leq B_b$, the processors are ranked and assigned elements starting from the end of the corresponding array. Then $|B_b|$ is decreased by k. This straightforward implementation meets the desired time bound but uses $O(n^2)$ space.

To reduce the space requirement, we take advantage of the fact that the total number of elements in all sets B_i is at most n. We discuss two parallel data structures that can be used to maintain sets B_i in a space-efficient way. The two implementations mentioned above differ only in the choice of the data structure to maintain the sets.

One such data structure, used by our first implementation, is a *dynamic array*. A dynamic array consists of an ordered list of segments. If the dynamic array contains k elements, the number of segments in the list is $\lceil \log k \rceil + 1$. Each segment is an array; the length of the first two segments is 1, and for $j > 2$, the length of the jth segment is 2^{j-1}. Note that more than half of the space allocated to the nonempty segments is actually used and all segments are of size $O(n)$.

Using this observation, it is easy to see that the total space required is $O(n \log n)$. Let 2^r be the smallest power of two greater or equal to n. We allocate 2^{r-j} segments of length 2^j for every $j \in [0 \ldots r]$. Each of the allocated segments is used to represent at most one set B_i, but we never need more than the allocated segments. Suppose all segments of length 2^j are used for maintaining the sets B_i. Note that in our implementation, the sets are disjoint, so the total number of elements in these sets is at most n. If we use all segments of length 2^j to maintain B_i's, then the total number of elements in the maintained sets is at least $2^j \times 2^{r-j} \geq n$, so we do not need more than the allocated number of segments of each length. Since $r = O(\log n)$ and for each j, the total size of the segments of length 2^j is $O(n)$, the total $O(n \log n)$ space is required.

We can implement the lists of segments by arrays of pointers to the segments. These arrays take $O(n \log n)$ space. The time required for l processors to add elements to the sets B_i or to remove l elements from B_b is $O(\log l)$ (using sorting or ranking of processors operating on the same set).

Our second implementation uses the *parallel cardinality stack* data structure [23], which was inspired by the parallel 2-3 trees [29]. This data structure allows addition (deletion) of l elements by l processors to (from) a set of k elements in $O(\log k + \log l)$ time and $O(k+l)$ space. In our application $k \leq n$ and $l \leq p = O(n)$, so the bounds can be rewritten as $O(\log n)$ time and $O(n)$ space.

The implementations of the *MDD* algorithm work in iterations, each of which implements a pass of the sequential algorithm. Each iteration is divided into three phases. During the first phase flow is pushed out of the active vertices in B_b. During the second phase this flow is collected at the destination vertices. The last phase relabels the appropriate vertices.

For the purpose of scheduling, one has to keep track of the number of processors needed to perform a *relabel* or a *discharge* operation. In the sequential algorithm, the number of steps required to relabel a vertex v is linear in the degree of v, so in the parallel implementation we assign for this operation the number of processors equal to the degree of v. Discharging a vertex v requires the number of processors equal to the number of pushes performed during the discharge. We maintain a data structure at each vertex that allows fast computation of this number by a single processor. This data structure is also used for pushing the flow.

The data structure we use is a variant of the *partial sum tree* data structure [31]. A partial sum tree is a balanced binary tree with leaves corresponding to edges adjacent to a vertex. Each vertex v has two trees associated with it, the *out-tree(v)* which is used to push flow out of the vertex, and the *in-tree(v)* which is used to collect flow pushed into the vertex.

The *out-trees* are used in the first phase. Leaves of the *out-tree(v)* correspond to the arcs (v, w). Each node x of the tree has two labels, $a(x)$ and $b(x)$. The label values are defined as follows. If x is a leaf corresponding to the arc (v, w), then

$$a(x) = \begin{cases} u_f(v, w) & \text{if } (v, w) \text{ is admissible} \\ 0 & \text{otherwise} \end{cases}$$

and

$$b(x) = \begin{cases} 1 & \text{if } (v, w) \text{ is admissible} \\ 0 & \text{otherwise.} \end{cases}$$

If x is not a leaf, then $a(x)$ and $b(x)$ are equal to the sums of the corresponding values of the children of x. In other words, $a(x)$ is equal to the sum of residual capacities of the admissible arcs corresponding to the leaves of the subtree rooted at x, and $b(x)$ is the number of such admissible arcs.

Suppose v is an active vertex to be discharged. First a processor is assigned to v to determine the number of pushes $p(v)$ that will be made from v. Using a and b values of *out-tree(v)* and the value of $e_f(v)$, this can be done in $O(\log \Delta)$ time.

To push the flow out of v, we assign $p(v)$ processors to v. Then we associate each processor with the arc it will push the flow through. To do this, we rank the processors assigned to v, which takes $O(\log p(v))$ time. Then each processor goes down the tree starting from the root and picking at each step the left or the right child of its current node x depending on the processor's rank and on the b values of the children of x. At the end, the ith processor will be at the leaf corresponding to the ith admissible arc of v. Note that this process requires concurrent read. Then each of the processors computes the amount $\delta(v,w)$ to be pushed along the arc corresponding to the processor. For all but the last processor assigned to v, this amount is equal to $u_f(v,w)$, since the corresponding pushes are saturating. For the last processor, this amount is equal to $u_f(v,w)$ if $a(root(\mathit{out\text{-}tree}(v))) \leq e_f(v)$ and to $u_f(v,w) - a(root(\mathit{out\text{-}tree}(v))) + e_f(v)$ otherwise. The processors update the flow function on the corresponding arcs and the last processor updates $e_f(v)$.

Next the *out-trees* are updated from the bottom level of the tree up. Initially each processor updates the a and b labels at the leaf assigned to it. Then the processor decides if it will stop updating or not. The processor stops the update only if it is currently at the root of the tree or if it is at a tree node x which is a right son of its father y, and the left son of y has just been updated, i.e., also has a processor working on it. The process is repeated until the root of the tree is reached and updated.

In the second stage, the flow pushed into vertices w is collected using the in-trees. Leaves of *in-tree(w)* correspond to arcs entering w. A processor that was assigned to the arc (v,w) when pushing flow out of v is assigned to the same arc when processing the flow pushed into w. Every node x of *in-tree(w)* has a variable $a'(x)$ associated with it. If x is a leaf corresponding to an arc (v,w), then $a'(x)$ is set to the amount equal to that just pushed along the arc. If x is not a leaf, then $a'(x)$ is equal to the sum of the values of the the corresponding variables of its children. The values of the a' variables are propagated going from the leaves to the root in the same way as the values of a variables of the out-trees. The update takes $O(\log \Delta)$ time. After the update,

$e_f(w)$ is increased by $a'(\textit{in-tree}(\textit{root}(v)))$. Then the values of a' variables are reinitialized to 0 by making another leaves-to-root pass.

Relabelings are implemented by maintaining an array of vertices to be relabeled. (Note that this array has at most n items). Vertices that were unable to get rid of their excesses during a discharge are added to the end of the array using ranking of the processors that want to add the vertices to the array. During the relabeling stage, processors are assigned to the last p elements of the array or to every element of the array if there are less than p elements in it. Using parallel prefix computations on the portion of the array for which the processors have been assigned, vertices needing relabeling are assigned the number of processors equal to their degrees, and the relabeling is performed by doing a parallel prefix computation on edge list of every vertex that needs to be relabeled.

The analysis is based on the following theorem of Brent.

THEOREM 19.8 [4]
Any synchronized parallel algorithm of depth d that consists of a total of x elementary operations can be implemented by p processors within a depth of $\lceil x/p \rceil + d$.

Let *macro-operations* be any standard unit-time PRAM operations plus sequences of operations performed by individual processors while working on an *in-tree* or an *out-tree* or performing an operation of ranking, sorting, or computing parallel prefix. Note that the macro-operations used by the algorithm take $O(\log(\Delta + p))$ time. By Lemma 19.6, the depth of the algorithm is $O(n^2)$ macro-operations. The total number of macro-operations used by the algorithm is $O(n^2\sqrt{m})$. Applying Theorem 19.8 for $p = O(\sqrt{m})$ and using the fact that each macro-operation takes $O(\log(\Delta + p))$ time, we get the following result.

THEOREM 19.9
The parallel implementations of the MDD algorithm run in $O(n^2 \log(2m/n + p)(\sqrt{m}/p))$ time using $p = O(\sqrt{m})$ processors and $O(m + n \log n)$ memory, or in $O(n^2 \log n(\sqrt{m}/p))$ time using $p = O(\sqrt{m})$ processors and $O(m + n)$ memory.

19.4
The Minimum-Cost Flow Problem

Theorem 19.3 shows that a parallel algorithm for finding blocking flows in acyclic networks yields a parallel algorithm for the minimum-cost circulation problem. In this section we describe a parallel blocking flow algorithm due to Goldberg and Tarjan [17]. The algorithm runs in $O(n \log n)$ time using m processors and $O(nm)$ memory. This algorithm uses more memory and processors than the Shiloach-Vishkin algorithm [31], as recently shown by Vishkin [36]. (The Shiloach-Vishkin algorithm runs in $O(n \log n)$ time using n processors and $O(n^2)$ memory.) However, the algorithm we describe is somewhat simpler. Furthermore, the algorithm has a distinctly parallel flavor: it treats the flow as a collection of "atoms" which, in parallel, try to get from the source to the sink, and return back to the source along their path if fail to reach the sink.

The above mentioned results, combined with Theorem 19.3, imply that the minimum-cost flow problem can be solved in $O(n^2 \log n \log(nC))$ time using n processors and $O(n^2)$ memory.

19.4.1 The Atomic Method

The general method behind the algorithm is the same as that used by Karzanov [22]. The algorithm begins with a blocking preflow and moves flow excess through the network while maintaining a blocking preflow, until eventually this flow movement produces a blocking flow. The algorithm maintains a partition of the vertices into two states: *blocked* and *unblocked*. We call an arc (v, w) *admissible* if it is residual and w is unblocked. The algorithm blocks a vertex v when it discovers that none of the arcs leaving v is admissible; once v is blocked, every path from v to t contains a saturated arc. Excess on blocked vertices is returned from whence it came, by decreasing the flow on appropriate incoming arcs.

To keep track of the detailed flow movements, the algorithm maintains a partition of the flow excess into *atoms*. Consider a time during an execution of the algorithm. An atom is a maximal quantity of excess that has moved in exactly the same way so far. An atom a at a vertex v consists of an amount of excess denoted by $size(a)$; the vertex v is denoted by $position(a)$. An atom located at a vertex other than s or t is called *active*.

Associated with an atom a at a vertex v is a path of arcs in E^+ from s to v that the atom followed in arriving at v. This path is denoted by $trace(a)$. Also associated with a is a simple path from s to v, denoted by $path(a)$, of arcs in E through which the atom moved forward but not backward in the course of reaching v from s. The relationship between $trace(a)$ and $path(a)$ is that $path(a)$ contains each arc (v, w) such that (v, w) but not (w, v) is on $trace(a)$. The algorithm maintains the atom paths but not the atom traces, which are needed for analysis only. The intuition behind the algorithm is that each atom does a depth-first search from s in an attempt to reach t. The graph being searched changes dynamically as arcs become saturated and vertices become blocked.

During initialization, the algorithm saturates every arc (s, v) leaving the source, creating at each neighbor v of s an atom of size $u(s, v)$ and trace (s, v). At each iteration, the algorithm selects an active atom a and processes it as described in Figure 19.5. Let $w = position(a)$. If w is not blocked, the algorithm tries to move a forward along an arc with positive residual capacity. If no such arc exists, w becomes blocked. If there is such an arc, the algorithm picks one, say (w, x). If $size(a) > u_f(w, x)$, atom a is split into two parts. One part, of size equal to $size(a) - u_f(w, x)$, gets a new name a'. The other

procedure *Process-Atom(a)*.
begin
    $w \leftarrow position(a)$;
    if w is unblocked **then**
        if $\exists\, (w, x) : u_f(w, x) > 0$ and x is unblocked **then begin**
            if $size(a) > u_f(w, x)$ **then begin**
                [split a]
                create a new atom a';
                $path(a') \leftarrow path(a)$;
                $size(a') \leftarrow size(a) - u_f(w, x)$;
                $size(a) \leftarrow u_f(w, x)$;
            end;
            $position(a) \leftarrow x$;
            append (w, x) to $path(a)$;
            $u_f(w, x) \leftarrow u_f(w, x) - size(a)$;
        end
        else mark w as blocked;
    if w is blocked **then begin**
        $(v, w) \leftarrow$ last arc on $path(a)$;
        $position(a) \leftarrow v$;
        delete (v, w) from $path(a)$;
        $u_f(v, w) \leftarrow u_f(v, w) + size(a)$;
    end;
end.

FIGURE 19.5
The Process-Atom procedure. Note that the flow is maintained implicitly as a difference between u and u_f.

part, of size equal to $u_f(w,x)$, retains the name a. Atom a' remains at vertex w to be processed later; atom a moves to vertex x. Finally, if atom a has not moved (i.e., vertex w is blocked), atom a is returned to the vertex, say v, from which it first reached w.

Note that an atom can move in two ways: forward from w to x or backward from w to v. In the former case, w is unblocked and $f(w,x)$ increases. In the latter case, w is blocked and $f(v,w)$ decreases. An atom can move backward from w to v only if at a previous time it moved forward from v to w. Thus the flow through an arc never becomes negative. During the course of the algorithm, for any arc (w,x), the flow on (w,x) first increases, until x becomes blocked, after which the flow decreases.

Note that we have not specified the way in which we select an atom to be processed next. In the parallel implementation of the algorithm, all active atoms, and atoms arising from them by iterated splitting, are processed concurrently. In the sequential implementation, any constant-time selection rule leads to an $O(nm)$ time bound. For example, one can maintain the set of active atoms as a queue or a stack. Alternatively, at each vertex one can maintain a list of the atoms located at the vertex, and keep a queue or a stack of vertices with nonempty lists of atoms.

We begin our analysis of the algorithm by bounding the number of atoms.

LEMMA 19.7
The total number of atoms created during an execution of the atomic algorithm is at most m.

PROOF
We claim that each increase in the number of atoms corresponds to an arc saturation. Atoms created during initialization are charged to the saturation of the corresponding arcs. An atom created by splitting in procedure *Process-Atom* is charged to the saturation of the arc (w,x) in the same execution of the procedure. Thus the claim is true. Since each arc becomes saturated only once, the lemma is true. ∎

The next lemma gives the key property of the algorithm. Intuitively, the lemma holds because the trace of an atom is a partial traversal of a tree rooted at s.

LEMMA 19.8
Consider an atom a at some time during execution of the algorithm. Then the length of the trace of a is at most $2n - 3$.

PROOF

An atom a only moves backward from a vertex $w \notin \{s,t\}$ once w is blocked. Just after a moves backward from w, w is not on $path(a)$, and a never visits w again. It follows that, for each vertex $w \neq t$, $E \cap trace(a)$ contains at most one arc of the form (v,w); and, for each vertex $w \notin \{s,t\}$, $E^{-1} \cap trace(a)$ contains at most one arc of the form (w,v). This gives a bound of $2n - 3$ on the length of $trace(a)$. ∎

We define *phases* of the algorithm as follows. Initialization is phase 1. Phase i for $i > 1$ begins at the end of phase $i - 1$ and ends as soon as every atom that existed at the end of phase $i-1$ and every atom created by splitting since the end of phase $i - 1$ have moved at least one step. Since every atom moves (either forward of backward) at least once during each phase, we have the following lemma, which is crucial for the analysis of parallel versions of the atomic method.

LEMMA 19.9

The number of phases during an execution of the algorithm is at most $2n - 3$.

To obtain an efficient implementation of the algorithm, we maintain the path of each atom as a stack of arcs. When an atom moves forward along an arc, the arc is pushed on top of the stack. To move an atom backward, we move it to the tail vertex of the top-of-stack arc and pop the stack.

Using stacks allows the algorithm to move atoms forward and backward in constant time. Splitting an atom, however, requires copying a stack. For ordinary stacks, this requires linear time. A very simple implementation of *persistent stacks* [26] allows the copy operation, as well as the push and pop operations, to be done in constant time. In combination with Lemmas 19.7 and 19.8, this fact gives the following result.

THEOREM 19.10

The atomic algorithm, implemented using persistent stacks, runs in $O(nm)$ time and $O(nm)$ space.

19.4.2 A Parallel Implementation

Now we describe a parallel implementation of the atomic method. The parallel implementation works in pulses; at each pulse, all atoms, including those that arise by splitting, move either forward or backward or both. Thus each pulse completes at least one phase, where a phase is as defined in Section 19.4.1.

19.4. The Minimum-Cost Flow Problem

The parallel implementation consists of the following four steps:

Step 1 (initialize).
For each arc (s, v), set $f(s, v) = u(s, v)$. Create an atom at v of size $f(s, v)$ with stack containing only (s, v). For each arc (v, w) with $v \neq s$, set $f(v, w) = 0$. Block vertex s and unblock all other vertices.

Step 2 (push flow forward).
For each unblocked vertex $w \notin \{s, t\}$, in parallel, do the following.
Arbitrarily order the atoms at w, say $a_1, a_2, \ldots, a_k$, and the admissible arcs (w, x), say $(w, x_1), (w, x_2), \ldots, (w, x_l)$. For $1 \leq j \leq k$, compute a *cumulative size* $S(j) = \sum_{i=1}^{j} size(a_i)$. For each $1 \leq j \leq l$, compute a *cumulative residual capacity* $R(j) = \sum_{i=1}^{j} u_f(w, x_i)$. Assign the atoms a_i to the admissible arcs (w, x_j) as follows:

1. If $S(i-1) \geq R(j-1)$ and $S(i) \leq R(j)$, assign all of atom a_i to (w, x_j).
2. If $S(i-1) \geq R(j-1)$ and $S(i) > R(j)$, assign an amount $\max\{0, R(j) - S(i-1)\}$ of atom a_i to (w, x_j).
3. If $S(i-1) < R(j-1)$ and $S(i) > R(j)$, assign an amount $u_f(w, x_j)$ of atom a_i to (w, x_j).
4. If $S(i-1) < R(j-1)$ and $S(i) < R(j)$, assign an amount $\max\{0, S(i) - R(j-1)\}$ of atom a_i to (w, x_j).

(This assignment associates with each admissible arc a total amount of excess less than or equal to its residual capacity. At most one such arc receives an amount that is positive but less than its residual capacity. The total amount assigned to admissible arcs equals the minimum of the excess at w and the sum of the residual capacities of the admissible arcs (w, x_j).)

Split any atom assigned to more than one arc into two or more atoms, one per assigned arc, each of size equal to the amount of the original atom assigned to the arc. Each of the new atoms inherits the assignment of the corresponding amount of the old atom, as well as a copy of the stack of the old atom.

For each admissible arc (w, x_j), increase $f(w, x_j)$ by the sum of the sizes of the atoms assigned to (w, x_j), and move each such atom to x_j, pushing

(w, x_j) onto its stack. If all arcs (w, x_j) are now saturated, mark w to be blocked. (Do not block w yet.)

Step 3 (block vertices).
Block every vertex marked to be blocked in Step 2.

Step 4 (return flow).
For each blocked vertex $w \notin \{s, t\}$, in parallel, do the following:

For each atom a at w, let (v_a, w) be the top arc on $stack(a)$. Pop $stack(a)$. Decrease $f(v_a, w)$ by $size(a)$ and move a to v.

Step 5 (loop).
If every atom is at s or t, stop. Otherwise, go to Step 2.

The implementation details are similar to those for the maximum flow algorithms, and we leave them to the reader.

The running time of the entire algorithm is then $O(n \log n)$ by Lemma 19.9. The space required is dominated by the space for the paths of atoms, which is $O(nm)$.

Acknowledgements

I would like to thank Robert Kennedy, Serge Plotkin, and Bob Tarjan for helpful discussions and suggestions. This work was supported in part by NSF Presidential Young Investigator Grant CCR-8858097, a grant from 3M Corporation, a grant from Mitsubishi Corporation, and ONR Contract N00014–88–K–0166.

Bibliography

[1] G. M. Adel'son-Vel'ski, E. A. Dinits, and A. V. Karzanov. *Flow Algorithms*. Nauka, Moscow, 1975. In Russian.

[2] R. K. Ahuja and J. B. Orlin. A Fast and Simple Algorithm for the Maximum Flow Problem. Sloan Working Paper 1905-87, Sloan School of Management, M.I.T., 1987.

[3] R. K. Ahuja, J. B. Orlin, and R. E. Tarjan. Improved Time Bounds for the Maximum Flow Problem. Technical Report CS-TR-118-87, Department

of Computer Science, Princeton University, 1987. (SIAM J. Comput., to appear).

[4] R. P. Brent. The Parallel Evaluation of General Arithmetic Expressions. *J. Assoc. Comput. Mach.*, 21:201–206, 1974.

[5] J. Cheriyan and S. N. Maheshwari. Analysis of Preflow Push Algorithms for Maximum Network Flow. *SIAM J. Comput.*, 18:1057–1086, 1989.

[6] E. A. Dinic. Algorithm for Solution of a Problem of Maximum Flow in Networks with Power Estimation. *Soviet Math. Dokl.*, 11:1277–1280, 1970.

[7] L. R. Ford, Jr. and D. R. Fulkerson. *Flows in Networks*. Princeton Univ. Press, Princeton, NJ, 1962.

[8] H. N. Gabow and R. E. Tarjan. Almost-Optimal Speed-ups of Algorithms for Matching and Related Problems. In *Proc. 20th Annual ACM Symposium on Theory of Computing*, pages 514–527, 1988.

[9] A. V. Goldberg. A New Max-Flow Algorithm. Technical Report MIT/LCS/TM-291, Laboratory for Computer Science, M.I.T., 1985.

[10] A. V. Goldberg. *Efficient Graph Algorithms for Sequential and Parallel Computers*. PhD thesis, M.I.T., January 1987. (Also available as Technical Report TR-374, Lab. for Computer Science, M.I.T., 1987).

[11] A. V. Goldberg. Processor-Efficient Implementation of a Maximum Flow Algorithm. *Information Processing Let.*, pages 179–185, 1991.

[12] A. V. Goldberg, S. A. Plotkin, D. Shmoys, and É. Tardos. Interior-Point Methods in Parallel Computation. Technical Report STAN-CS-89-1259, Stanford University, 1989.

[13] A. V. Goldberg, S. A. Plotkin, and P. M. Vaidya. Sublinear-Time Parallel Algorithms for Matching and Related Problems. Technical Report STAN-CS-88-1211, Stanford University, 1988.

[14] A. V. Goldberg, É. Tardos, and R. E. Tarjan. Network Flow Algorithms. In B. Korte, L. Lovász, H. J. Prömel, and A. Schrijver, editors, *Flows, Paths, and VLSI Layout*, pages 101–164. Springer Verlag, 1990.

[15] A. V. Goldberg and R. E. Tarjan. A New Approach to the Maximum Flow Problem. In *Proc. 18th Annual ACM Symposium on Theory of Computing*, pages 136–146, 1986.

[16] A. V. Goldberg and R. E. Tarjan. A New Approach to the Maximum Flow Problem. *J. Assoc. Comput. Mach.*, 35:921–940, 1988. A preliminary version appeared in *Proc. 18th ACM Symp. on Theory of Comp.*, 136-146, 1986.

[17] A. V. Goldberg and R. E. Tarjan. A Parallel Algorithm for Finding a Blocking Flow in an Acyclic Network. *Information Processing Let.*, 31:265–272, 1989.

[18] A. V. Goldberg and R. E. Tarjan. Finding Minimum-Cost Circulations by Successive Approximation. *Math. of Oper. Res.*, 15:430–466, 1990. A preliminary version appeared in *Proc. 19th ACM Symp. on Theory of Comp.*, 7-18, 1987.

[19] L. M. Goldschlager, R. A. Shaw, and J. Staples. The Maximum Flow Problem is Log Space Complete for P. *Theoretical Computer Sci.*, 21:105–111, 1982.

[20] W. D. Hillis. *The Connection Machine*. MIT Press, 1985.

[21] R. M. Karp, E. Upfal, and A. Wigderson. Constructing a Maximum Matching is in Random NC. *Combinatorica*, 6:35–48, 1986.

[22] A. V. Karzanov. Determining the Maximal Flow in a Network by the Method of Preflows. *Soviet Math. Dok.*, 15:434–437, 1974.

[23] R. Kennedy. Parallel Cardinality Stacks and an Application. *Information Processing Let.*, to appear.

[24] E. L. Lawler. *Combinatorial Optimization: Networks and Matroids*. Holt, Reinhart, and Winston, New York, NY., 1976.

[25] K. Mulmuley, U. V. Vazirani, and V. V. Vazirani. Matching is as Easy as Matrix Inversion. *Combinatorica*, pages 105–131, 1987.

[26] E. W. Myers. An Applicative Random-Access Stack. *Information Processing Let.*, 17:241–248, 1983.

[27] J. B. Orlin. A Faster Strongly Polynomial Minimum Cost Flow Algorithm. In *Proc. 20th Annual ACM Symposium on Theory of Computing*, pages 377–387, 1988.

[28] C. H. Papadimitriou and K. Steiglitz. *Combinatorial Optimization: Algorithms and Complexity*. Prentice-Hall, Englewood Cliffs, NJ, 1982.

[29] W. Paul, U. Vishkin, and H. Wagener. Parallel Dictionary. In *Proc. 10th International Colloquium on Automata, Languages and Programming*, pages 597–609. Springer-Verlag, 1983.

[30] J. Renegar. A Polynomial Time Algorithm, Based on Newton's Method, for Linear Programming. *Math. Prog.*, 40:59–94, 1988.

[31] Y. Shiloach and U. Vishkin. An $O(n^2 \log n)$ Parallel Max-Flow Algorithm. *J. Algorithms*, 3:128–146, 1982.

[32] D. D. Sleator. An $O(nm \log n)$ Algorithm for Maximum Network Flow. Technical Report STAN-CS-80-831, Computer Science Department, Stanford University, Stanford, CA, 1980.

[33] D. D. Sleator and R. E. Tarjan. A Data Structure for Dynamic Trees. *J. Comput. System Sci.*, 26:362–391, 1983.

[34] R. E. Tarjan. *Data Structures and Network Algorithms*. Society for Industrial and Applied Mathematics, Philadelphia, PA, 1983.

[35] R. E. Tarjan. Amortized Computational Complexity. *SIAM J. Alg. Disc. Math.*, 6:306–318, 1985.

[36] U. Vishkin. A Parallel Blocking Flow Algorithm for Acyclic Networks. *J. Algorithms*, Vol. 13:489–501, 1992.

[37] Y. Ye. An $O(n^3 L)$ Potential Reduction Algorithm for Linear Programming. Unpublished manuscript, The University of Iowa, 1989.

Part VIII

Inherent Limitations of Parallel Computations

20

The Complexity of Computation on the Parallel Random Access Machine

Faith E. Fich

Department of Computer Science
University of Toronto
Toronto, Ontario, CANADA
fich@cs.toronto.edu

20.1 Introduction

20.1.1 The PRAM Model

The PRAM is a synchronous model of parallel computation in which processors communicate via a shared memory. It consists of m shared memory cells $M_1, \ldots, M_m$ and p processors $P_1, \ldots, P_p$. Each processor is a random access machine (RAM) with a private local memory. During every step of a computation, a processor may read from one shared memory cell, perform a local operation, and write to one shared memory cell. Different processors may execute different operations during a step and may make reference to their own index. Reads, local operations, and writes are viewed as occurring during three separate phases. This simplifies analysis and only changes the running time of an algorithm by a constant factor.

An input x consists of n values $x_1, \ldots, x_n$. At the beginning of the computation, these values are located in shared memory cells $M_1, \ldots, M_n$, respectively, provided $n \leq m$. If there are $n' \leq m$ output values, they appear in the first n' shared memory cells at the end of the computation. When the number of shared memory cells is too small, the input and/or output values can be distributed approximately equally among the processors. Another possibility is to have a separate, read-only memory containing the input values [88, 57, 29]. In this case, each processor is also allowed to read from one of the n read-only memory cells during each computation step.

EXERCISE 20.1

Explain why it doesn't matter whether the input is initially located in the shared memory, the processors' local memories, or a separate read-only memory, provided the number of shared memory cells is large enough to contain the input (i.e., $m \geq n$).

If the number m of shared memory cells is small, then where the input is located is more important. For example, determining whether the input contains two consecutive variables with value 1 requires $\Omega(n/m)$ steps when the input is distributed among the processors' local memories (Exercise 20.45). However, when the input is in a separate read-only memory, this computation can be done significantly faster (Exercise 20.8).

The *word size* of a PRAM is the maximum number of bits that can be contained in each cell of shared memory. Limiting the word size can significantly affect the amount of time required to solve certain problems. Different organizations of memory, such as a small number of shared memory cells with large word size or a large number with small word size, can be better for different problems [6, 7].

Limiting the amount of shared memory in a PRAM corresponds to restricting the amount of information that can be communicated between processors in one step. For example, a set of processors connected to one another by a single bus may be viewed as a PRAM with one shared memory cell.

THEOREM 20.1
Any problem that can be solved by a PRAM using p processors in t steps can also be solved using $p' \leq p$ processors in $O(tp/p')$ steps.

PROOF
The original p processors are partitioned into p' groups of size at most $\lceil p/p' \rceil$. Each of the p' processors in the simulating machine is associated with one of these groups. To simulate one step of the original computation, each of the p' processors sequentially simulates the read and local computation phases performed by all the processors in its group and then sequentially simulates their write phases. ∎

EXERCISE 20.2
Prove that any problem that can be solved by a PRAM using m shared memory cells and p processors in t steps can also be solved using $m' \leq m$ shared memory cells and $\max\{p, m'\}$ processors in $O(tm/m')$ steps.

EXERCISE 20.3 [42]
A *forking PRAM* is a PRAM in which a new processor is created when an existing processor executes a *fork operation*. In addition to creating the new processor, this operation specifies the task that the new processor is to perform (starting at the next time step). Initially there is one processor. Prove that if a forking PRAM algorithm uses t steps and p processors, then this algorithm can be implemented on a PRAM using $O(t \log p)$ steps and $O(p/\log p)$ processors. Explain explicitly how work is allocated to the PRAM processors.

One consequence of Theorem 20.1 is that the processor-time product $p \cdot t$ of any algorithm solving a problem is at least a constant factor times the sequential time complexity of the problem. Thus, a lower bound on the sequential time complexity gives a lower bound on the number of processors needed for a PRAM to solve the problem in a given amount of time, as well as a lower bound on the amount of time needed for a PRAM to solve the problem using a given number of processors.

A PRAM algorithm is *optimal* if it is not possible to simultaneously improve both the time and the number of processors by more than a constant factor. If a PRAM algorithm solves a problem in t steps using p processors, where $p \cdot t$ is within a constant factor of the sequential time complexity of the problem, then it is called *efficient* [53]. All efficient algorithms are optimal, but not all optimal algorithms are efficient. For many problems, it is interesting and important to find the fastest possible efficient algorithm.

Some people call a parallel algorithm optimal if its processor-time product is within a constant factor of the running time of the fastest known sequential algorithm solving the same problem. Unless there is a matching sequential lower bound for the problem, difficulties may arise with this definition. Specifically, if a faster sequential algorithm is found, suddenly, the parallel algorithm is no longer optimal. Furthermore, this definition of optimality may preclude calling an algorithm optimal even when it is provably impossible to simultaneously improve both the time and the number of processors. For example, PARITY has linear sequential time complexity, but an exponential number of processors are required by any PRAM that computes it in constant time (Theorem 20.29).

The PRAM is a natural generalization of the unit cost RAM [20, 3], a commonly used model of sequential computation. Numerous parallel algorithms have been designed for the PRAM because it is a simple, precise model in which parallel algorithms can be expressed using a high level language. The PRAM often corresponds to the programmer's view of parallel computation, ignoring lower level architectural details such as memory organization, routing, memory contention, and synchronization. In essence, PRAM programmers do not have to be concerned with how communication is accomplished, but only what is to be communicated where. PRAMs can be simulated by more realistic models [59, 86, 67, 2, 50] and some real parallel computers [38, 72, 74]. Thus, programs written for the PRAM can be compiled into programs for these machines. Although some of the assumptions (such as constant time access to shared memory) are not generally valid, the performance of algorithms on the PRAM can be a good predictor of their relative performance on real machines, especially as problem sizes get large. Furthermore, because the PRAM is a powerful model, the lower bounds obtained for it are automatically applicable to a wide variety of less powerful, more realistic machines.

20.2
Restrictions on Access to Shared Memory

An important parameter of the PRAM model is the extent to which concurrent access to shared memory cells is allowed. If at most one processor can read from each memory cell at a particular step and at most one processor can write to each memory cell at a particular step, then the PRAM is called *exclusive read exclusive write* (EREW) [56]. In a *concurrent read exclusive write* (CREW) PRAM [36], any number of processors can simultaneously read from the same memory cell, but at most one processor can write to each memory cell at a given step. Many algorithms designed for CREW and EREW PRAMs avoid write conflicts by having each processor own one cell to which all its writes are performed. These restricted models are called the *concurrent read owner write* (CROW) PRAM [24] and the *exclusive read owner write* (EROW) PRAM [37], respectively.

When concurrent writes are allowed, it is necessary to specify how conflicts are resolved. Three of the most frequently used *concurrent read concurrent write* (CRCW) PRAMs are considered here. A number of others are discussed in the literature. (For example, see [33, 43, 27, 51, 54].) The COMMON model [55] requires that all processors simultaneously writing to the same memory cell write a common value. On the ARBITRARY model [87], an arbitrary one of the values written to a memory cell at a given step will appear in the cell. Any algorithm for this model must work regardless of which value is chosen (say, by an adversary) to resolve each write conflict that arises. In the PRIORITY model [41], processors are assigned fixed, distinct priorities. The processor of highest priority among those that simultaneously write to a memory cell succeeds. Without loss of generality, we assume that lower indexed processors have higher priority.

Any algorithm that runs on ARBITRARY will run unchanged on PRIORITY. Thus PRIORITY is at least as powerful as ARBITRARY. Similarly, ARBITRARY is at least as powerful as COMMON, COMMON is at least as powerful as the CREW PRAM, and the CREW PRAM is at least as powerful as the EREW PRAM. Diagrammatically,

PRIORITY $\geq$ ARBITRARY $\geq$ COMMON
$\geq$ CREW PRAM $\geq$ EREW PRAM.

It is important to understand the relationships between these models. Specifically, are some of these models strictly more powerful than others? Can additional resources compensate for restricted access to shared memory?

Tradeoffs can help a computer architect choose among a number of design alternatives or help a programmer compare algorithms written on different models. A step by step simulation of a more powerful model by a less powerful model is particularly useful because it allows the straightforward transformation of algorithms designed for the former into algorithms that can run on the latter.

Table 20.1 summarizes some of the known relationships between the PRAM models discussed above. The remainder of this section discusses these and related results and the conditions under which they apply. In particular, all the simulations presented here are deterministic. (Randomized simulations have also been studied and, in certain cases, they are significantly faster. See [39, 45, 28].) Some of the lower bounds appear in Section 20.4.

The first simulation is quite simple. Moreover, it contains ideas that are useful for other simulations in this section.

TABLE 20.1
Relationships Between PRAM Models
Bounds are given for the amount of time for the simulating model, using the indicated number of processors, to simulate one step of the original model with p processors. For the last three rows, the input is initially distributed among the processors' local memories and both the original and simulating models have m shared memory cells.

Original Model	Simulating Model	Number of Processors	Amount of Time	Reference
PRIORITY ARBITRARY COMMON	EREW PRAM CREW PRAM	$\geq p$	$\Theta(\log p)$	Thm 20.2 Thm 20.3 Thm 20.20
PRIORITY ARBITRARY	COMMON	kp	$\Theta\left(\frac{\log p}{k(\log\log p - \log p)}\right)$	Thm 20.7 Ex 20.11
PRIORITY	ARBITRARY	kp	$O\left(\frac{\log\log p}{\log(k+1)}\right)$	Ex 20.13
CREW PRAM	EREW PRAM	$\geq p$	$\Omega(\sqrt{\log p}/\log\log p)$	Ex 20.41
PRIORITY ARBITRARY COMMON	EREW PRAM CREW PRAM	$\geq p$	$O(p/m + \log m)$ $\Omega(p/m)$	Ex 20.6 Ex 20.46
PRIORITY ARBITRARY	COMMON	p	$\Omega(\log(p/m))$	Ex 20.44
PRIORITY	ARBITRARY	p	$\Omega(\log(p/m))$	Thm 20.35

20.2. Restrictions on Access to Shared Memory

THEOREM 20.2 *[33]*
One step of PRIORITY with p processors and m shared memory cells can be simulated by an EREW PRAM in $O(\log p)$ steps with p processors and mp shared memory cells.

PROOF
Each PRIORITY processor is simulated by a corresponding EREW PRAM processor. Each original shared memory cell in PRIORITY will be associated with p shared memory cells in the EREW PRAM. One of these will contain the same sequence of values as the original cell. The other $p-1$ cells are assumed to be initialized to 0 and will be used to resolve conflicting accesses to the cell.

To simulate the write phase, each processor must know whether it is the leftmost processor (i.e., the processor of lowest index) wishing to write into some cell and, if so, it can perform the write. This leads to the definition of the following problem.

> LEFTMOST WRITERS
> Each processor P_i has a value in the range $\{0,\ldots,m\}$ in its local memory, denoting the shared memory cell that P_i wants to access. The value 0 indicates that P_i does not want to access any cell. Each processor P_i must determine whether it is the leftmost processor among those with the same nonzero value.

One way to solve this problem is to solve m simultaneous instances of the following problem, one for each of the original shared memory cells.

> LEFTMOST PRISONER
> Each processor has a bit, known only to itself, which is 1 if the processor wants to access the given memory cell. Each processor with value 1 must determine whether it is the leftmost processor with value 1. Only processors with value 1 can participate in the computation.

The name of this problem comes from imagining the processors to be prisoners who cooperate with one another to determine the lowest numbered prison cell that contains a prisoner. Note that all m instances can be solved simultaneously because every processor participates in at most one computation.

The additional $p-1$ memory cells associated with each original memory cell are used to solve its LEFTMOST PRISONER problem. They are viewed as representing the internal nodes of a binary tree with p leaves (one corresponding to each processor) and depth $\lceil \log_2 p \rceil$.

All processors with value 1 are considered to be initially located at the corresponding leaves of the tree. The internal nodes of the tree are processed one level at a time, starting from the bottom. Each processor located at the right child of a node in the current level reads the value stored at the left child. When this value is 0, there is no processor located at the left child. In this case, the processor proceeds to the node and writes the value 1 there. Simultaneously, each processor located at the left child of a node in the current level proceeds to the node and writes the value 1 there. The processor that reaches the root of the tree is the leftmost processor that wants to access the given shared memory cell. Hence, it can perform the write to the cell. (An example appears in Figure 20.1.) Finally, processors retrace their steps back down the tree, erasing the values they have written at the nodes.

To simulate the read phase, it is sufficient to choose one processor to read each desired memory cell and distribute its contents to all interested processors. This can also be accomplished by performing the LEFTMOST PRISONER algorithm and having the processor that reaches the root read the given shared memory cell. As the processors go back down the tree, this value is written at each internal node that a processor reaches and is then erased. A processor waiting at a node can read the value from its parent at the appropriate time. ∎

A simplified version of this algorithm can be performed on COMMON without using more shared memory than the original machine.

EXERCISE 20.4   [33]
Prove that one step of PRIORITY with p processors and m shared memory cells can be simulated by COMMON in $O(\log p)$ steps with p processors and m shared memory cells.

The following simulation is more complicated than the simulation in Theorem 20.2, but uses fewer shared memory cells. When $m \in \Omega(p)$, Exercise 20.2 and Theorem 20.1 can be applied to this result to show that no additional processors or shared memory cells are needed for simulating PRIORITY by an EREW PRAM in $O(\log p)$ steps.

20.2. Restrictions on Access to Shared Memory 851

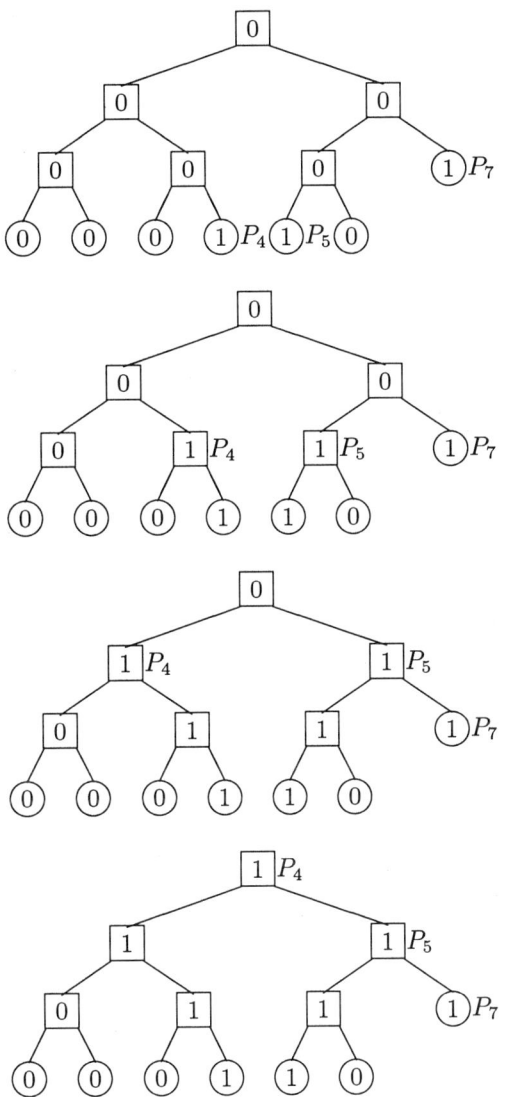

FIGURE 20.1
An example illustrating the processors going up the tree during an execution of the LEFTMOST PRISONER algorithm described in the proof of Theorem 20.2. Here processors P_4, P_5, and P_7 want to access the shared memory cell and processors P_1, P_2, P_3 and P_6 do not.

THEOREM 20.3 *[25, 87, 62]*

One step of PRIORITY with p processors and m shared memory cells can be simulated by an EREW PRAM in $O(\log p)$ steps with p processors and $m + p$ shared memory cells.

PROOF

As in the proof of Theorem 20.2, each original shared memory cell has a corresponding shared memory cell in the EREW PRAM and the local operations of each original processor are performed by a corresponding processor in the EREW PRAM. An auxiliary array A of size p is also used.

At a given read or write phase in the PRIORITY algorithm, if processor P_i wants to access cell M_j, it writes the pair (j,i) to memory cell A_i. If P_i doesn't want to access any shared memory cell, it writes $(0,i)$ to A_i. Notice that each of the indices $1, \ldots, p$ occurs as the second component of exactly one element of A. For example, if P_1 wants to access cell M_2, P_2 wants to access cell M_4, P_3 wants to access cell M_2, P_4 wants to access cell M_1, P_5 wants to access cell M_4, P_6 wants to access cell M_2, and P_7 does not want to access any cell, the array A would have the following contents.

(2,1)	(4,2)	(2,3)	(1,4)	(4,5)	(2,6)	(0,7)

The array A is then sorted into lexicographic order. This takes $O(\log p)$ steps on an EREW PRAM with p processors. (See Chapter 10.)

(0,7)	(1,4)	(2,1)	(2,3)	(2,6)	(4,2)	(4,5)

Next, each processor P_i appends a bit to the contents of cell A_i. This bit is 0 when the first component of A_i is either 0 or the same as the first component of A_{i-1}; otherwise this bit is 1. Observe that processor P_i is the highest priority processor wishing to access cell M_j if and only if the triple $(j,i,1)$ appears in the array A.

(0,7,0)	(1,4,1)	(2,1,1)	(2,3,0)	(2,6,0)	(4,2,1)	(4,5,0)

In the example, P_4 is the highest priority processor that wants to access M_1, P_1 is the highest priority processor that wants to access M_2, and P_2 is the highest priority processor that wants to access M_4.

From this point, it is easy to finish the simulation of a write phase. First, each processor P_k reads the triple (j,i,b) in A_k and writes this triple into A_i. Then processor P_i reads the triple (j,i,b) from A_i. If the bit b has value 1, then P_i is the highest priority processor that wants to write to M_j, so it can perform its write.

| (2,1,1) | (4,2,1) | (2,3,0) | (1,4,1) | (4,5,0) | (2,6,0) | (0,7,0) |

In this case, P_1 writes to cell M_2, P_2 writes to cell M_4, and P_4 writes to cell M_1.

Once the appropriate bit has been appended to the contents of each cell of A, the simulation of a read phase can be completed as follows. Each processor P_k reads the triple (j,i,b) in A_k. If the bit b has value 1, then P_k reads shared memory cell M_j and overwrites the third component of A_k with the value v_j it read. These values are then duplicated as appropriate. Specifically, for $l = \lceil \log_2 p \rceil, \ldots, 1$, processor P_k overwrites the third component of $A_{k+2^{l-1}}$ with the third component of A_k, provided they have the same first components. Here are the successive contents of the array in the example.

| (0,7,0) | (1,4,v_1) | (2,1,v_2) | (2,3,0) | (2,6,0) | (4,2,v_4) | (4,5,0) |

| (0,7,0) | (1,4,v_1) | (2,1,v_2) | (2,3,v_2) | (2,6,0) | (4,2,v_4) | (4,5,v_4) |

| (0,7,0) | (1,4,v_1) | (2,1,v_2) | (2,3,v_2) | (2,6,v_2) | (4,2,v_4) | (4,5,v_4) |

Finally, each processor P_k reads the triple (j,i,b) in A_k and writes it into A_i.

| (2,1,v_2) | (4,2,v_4) | (2,3,v_2) | (1,4,v_1) | (4,5,v_4) | (2,6,v_2) | (0,7,0) |

Now processor P_i can read the value it needs from A_i. ∎

Concurrent write PRAMs with p processors cannot, in general, be simulated by exclusive write PRAMs without an $\Omega(\log p)$ factor increase in time, even with an infinite number of processors and shared memory cells. This is because the OR of n Boolean variables can be computed in one step on COMMON with n processors and one shared memory cell, but requires $\Omega(\log n)$ steps on a CREW PRAM with an infinite number of processors and shared memory cells. (See Theorem 20.20.)

EXERCISE 20.5
Prove that the OR of n Boolean variables can be computed in $O(\log n)$ steps on an EROW PRAM with $n/\log_2 n$ processors and $n/\log_2 n$ shared memory cells.

When the amount of shared memory m is small, significantly more time is needed to compute the OR of n input bits on a CREW PRAM, even with an infinite number of processors. If the input bits are initially located in the processors' local memories, then $\Omega(n/m)$ steps are required (see Exercise 20.46), whereas, if they are located in a separate read-only shared memory, $\Omega\left(\sqrt{n/m}\right)$ steps are required [4, 88].

EXERCISE 20.6
Prove that one step of PRIORITY with p processors and m shared memory cells can be simulated by an EREW PRAM in $O(p/m + \log m)$ steps with p processors and m shared memory cells.

EXERCISE 20.7 [88]
Prove that the OR of n Boolean variables located in a separate read-only shared memory can be computed in $O\left(\sqrt{n/m} + \log m\right)$ steps on an EREW PRAM with $O(\sqrt{nm})$ processors and m shared memory cells.

EXERCISE 20.8
Consider the problem of determining whether an input consisting of n bits located in a separate read-only shared memory contains two consecutive variables with value 1. Prove that an EREW PRAM with $O(\sqrt{nm})$ processors and m shared memory cells can solve this problem in $O(\sqrt{n/m} + \log m)$ steps and that COMMON with n processors and 1 shared memory cell can solve it in $O(1)$ steps.

Both COMMON and ARBITRARY can simulate PRIORITY with only a constant factor increase in time, by increasing the number of processors and the amount of shared memory.

THEOREM 20.4 [55]
One step of PRIORITY with p processors and m shared memory cells can be simulated by COMMON in $O(1)$ steps with $\binom{p}{2}$ processors and $m + p$ shared memory cells.

PROOF
Processors $P_1, \ldots, P_p$ are responsible for simulating the original processors. Simulation of the read and compute phases is trivial, since the two

models do not differ in this respect. To simulate the write phase, each processor P_i writes the index of the location to which it wishes to write into the auxiliary shared memory cell A_i. As in the proof of Theorem 20.3, the value 0 denotes the fact that P_i does not wish to write during this step. For example, the array A might contain the following values.

$$\boxed{2\,|\,4\,|\,2\,|\,1\,|\,4\,|\,2\,|\,0}$$

Each of the $\binom{p}{2}$ processors then reads a different pair of memory cells in the array. If A_i and A_j contain the same value and $i < j$, the processor that read these two cells writes 0 into A_j. In the example, this results in the following array.

$$\boxed{2\,|\,4\,|\,0\,|\,1\,|\,0\,|\,0\,|\,0}$$

Finally, for $1 \leq i \leq p$, memory cell A_i is read by processor P_i. If A_i contains a nonzero value j, then processor P_i is the lowest indexed processor that wants to write to location M_j and, thus, it can perform the write. ∎

The number of processors used to achieve a constant time simulation can be reduced if more shared memory is available.

EXERCISE 20.9 [33]
Prove that one step of PRIORITY with p processors and m shared memory cells can be simulated by COMMON in $O(1/\varepsilon)$ steps with $\min\{p + mp^\varepsilon, p^{1+\varepsilon}\}$ processors and $mp^{\varepsilon/2}$ shared memory cells, for any $\varepsilon \in O(1)$.

EXERCISE 20.10 [17]
A *semi-oblivious PRAM algorithm* is a PRAM algorithm such that, for each processor and at each step of the computation, there is at most one shared memory cell to which it writes. However, whether a particular processor decides to write at a particular step may depend on the state it is in (and hence on the input values). Prove that any step of a semi-oblivious PRIORITY algorithm that uses p processors and m shared memory cells can be simulated in constant time by a semi-oblivious COMMON algorithm using p processors and $m+p$ shared memory cells.

THEOREM 20.5 *[15]*
One step of PRIORITY with p processors and m shared memory cells can be simulated by COMMON in $O(1)$ steps with $p\log_2 p$ processors and $mp + p$ shared memory cells.

PROOF

As in the proof of Theorem 20.2, for each shared memory cell M_j, the LEFTMOST PRISONER problem is solved using a binary tree with p leaves. However, instead of having processors proceed up the tree one level at a time, the bit values at all the internal nodes are determined simultaneously.

For every original processor P_i, there are $\log_2 p$ processors in the simulating machine. If processor P_i wants to write to memory cell M_j, these processors simultaneously write the value 1 to each ancestor of the ith leaf that has the ith leaf in its left subtree.

Note that, after this step, an internal node has value 1 if and only if there is a leaf in its left subtree corresponding to an original processor that wants to write to M_j. Thus, processor P_i is the leftmost processor that wants to write to memory cell M_j if and only if every ancestor of the ith leaf that has the ith leaf in its right subtree contains the value 0. This can be determined in constant time by computing the OR of these values using one additional memory cell associated with processor P_i.

Consider the example in Figure 20.2. Since P_3 is in its parent's and great grandparent's left subtree and in its grandparent's right subtree, the OR that is computed has value 0. However, P_4 is in its parent's and grandparent's right subtrees and P_5 is in its great grandparent's right subtree, so the ORs that are computed for both these processors have value 1. ∎

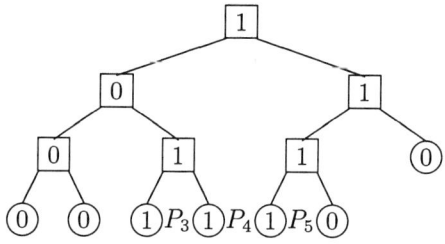

FIGURE 20.2
The result of the first step of the LEFTMOST PRISONER algorithm in the proof of Theorem 20.5. In this example, $p = 7$ and processors P_3, P_4, and P_5 wish to write to the shared memory cell.

Even without increasing the number of processors, it is still possible to simulate PRIORITY faster on COMMON than on the EREW PRAM.

THEOREM 20.6 *[34]*
One step of PRIORITY with p processors and m shared memory cells can be simulated by COMMON in $O(\log p/\log\log p)$ steps with p processors and $O(mp)$ shared memory cells.

PROOF
Here the LEFTMOST PRISONER problem is solved using a completely different approach. For $p = 2$, the problem can be solved in one write phase and one read phase using one shared memory cell that is initialized to 0. Specifically, if P_1 has value 1, it writes 1 into the cell. In this case, it is the leftmost processor with value 1. If P_2 has value 1, it reads the cell. It is the leftmost processor with value 1 if and only if it reads the value 0.

Given an algorithm that solves the problem for $p = t!$ in $t - 1$ write and read phases using m_{t-1} shared memory cells, we can construct an algorithm that solves the problem for $p = (t + 1)!$ in t write and read phases using $m_t = (t + 1)m_{t-1} + 1$ shared memory cells, as follows. The problem is divided into $t + 1$ subproblems of size $t!$. Group 1 consists of the first $t!$ processors, group 2 consists of the next $t!$ processors, etc. Each subproblem is solved independently, using m_{t-1} shared memory cells and $t - 1$ of the write and read phases. One shared memory cell is used for the groups to interact with one another, so that the group containing the leftmost processor with value 1 can be determined. This cell initially contains the value 0. In the ith write phase, all processors (with value 1) in group i write 1 to the cell and in the next read phase all processors (with value 1) in group $i + 1$ read the cell. Notice that after the ith write phase, the cell contains the value 0 if and only if no processors in the first i groups have the value 1. Thus if there are any processors with value 1 in group $i + 1$, they will learn whether their group contains the leftmost processor with value 1.

The number of shared memory cells used to solve each leftmost prisoner problem satisfies the recurrence

$$\begin{aligned} m_1 &= 1 \\ m_t &= (t+1)m_{t-1} + 1, \text{ for } t > 1. \end{aligned}$$

This implies $m_t \in O((t + 1)!) = O(p)$. Therefore the total number of shared memory cells used in the simulation is $O(mp)$. ∎

The ideas in the proofs of the two preceding theorems can be combined to get increasingly faster simulations of PRIORITY by COMMON as the number of processors in the simulating machine increases.

EXERCISE 20.11 [66, 10]
Prove that one step of PRIORITY with p processors and m shared memory cells can be simulated by COMMON in t steps with kp processors, for $1 \leq k \leq \log_2 p$, provided $t \log_2 t \in O(\frac{\log p}{k})$ i.e. $t \in O(\frac{\log p}{k(\log \log p - \log k)})$. How many shared memory cells do you use?

A matching $\Omega(\frac{\log p}{k(\log \log p - \log k)})$ lower bound for solving the LEFTMOST PRISONER problem on COMMON has been achieved [66]. This does not imply that the simulation of PRIORITY by COMMON in Exercise 20.11 is optimal; it only implies that a better simulation cannot be achieved by this approach. However, the algorithm is known to be optimal for simulating either PRIORITY or ARBITRARY with *sufficiently large* amounts of shared memory. Consider the following problem.

ELEMENT DISTINCTNESS
Given n input values $x_1, \ldots, x_n \in \{1, \ldots, r\}$, determine whether they are pairwise distinct (i.e., whether $x_i \neq x_j$ for all $i \neq j$).

ELEMENT DISTINCTNESS is easy to solve on ARBITRARY (and, hence, PRIORITY), given sufficient shared memory.

EXERCISE 20.12
Prove that ELEMENT DISTINCTNESS can be solved in $O(1)$ steps on ARBITRARY using n processors and r shared memory cells.

An algorithm for solving ELEMENT DISTINCTNESS on COMMON can be obtained by simulating the algorithm in Exercise 20.12, using the simulation in Exercise 20.11. If the size, r, of the input domain is sufficiently large, then the resulting algorithm is optimal.

THEOREM 20.7 *[10, 26]*
COMMON with kn processors and an infinite amount of shared memory requires $\Omega(\frac{\log n}{k(\log \log n - \log k)})$ steps to solve ELEMENT DISTINCTNESS, provided the size of the input domain is sufficiently large.

This result implies that, for r sufficiently large, COMMON with kn processors and infinite shared memory requires $\Omega(\frac{\log n}{k(\log \log n - \log k)})$ times as many steps to solve ELEMENT DISTINCTNESS as PRIORITY or ARBITRARY with n processors and r shared memory cells. Thus, the simulation mentioned in Exercise 20.11 is optimal whenever the machine being simulated has a sufficiently large shared memory. However, when the amount of shared memory is smaller (for example, only exponential in the number of processors), the optimality of this simulation is unknown. The difficulty with using ELEMENT DISTINCTNESS to separate COMMON from ARBITRARY and PRIORITY in this case is that, if the domain size r is small, there are not good lower bounds known for ELEMENT DISTINCTNESS on COMMON and, if r is large as compared to the number of shared memory cells, there are not good upper bounds known for ELEMENT DISTINCTNESS on ARBITRARY or PRIORITY.

ARBITRARY can simulate PRIORITY substantially faster than COMMON can.

THEOREM 20.8 *[15]*
One step of PRIORITY with p processors and m shared memory cells can be simulated by ARBITRARY in $O(\log \log p)$ steps using p processors and $m(p-1)$ shared memory cells.

PROOF
This algorithm also works by solving the LEFTMOST PRISONER problem. The processors are divided into $\sqrt{p}$ contiguous groups of size $\sqrt{p}$.

Each group chooses a representative processor with value 1, if there is one. This is done as follows, using one cell for each group. The cell is initialized to 0. Then all processors that are in the group and have value 1 attempt to write their indices into the cell. The processor whose index appears in the cell is chosen to be the representative for the group. All processors (with value 1) in the group can then read the cell to find out which processor is the representative.

For example, if there are nine processors, of which P_5, P_6, P_7, and P_9 have value 1, then the memory cells associated with the three groups $\{1, 2, 3\}$, $\{4, 5, 6\}$, and $\{7, 8, 9\}$ could contain the values 0, 6, and 7, respectively.

Determining the leftmost group that contains a processor with value 1 is a subproblem of size $\sqrt{p}$ that is solved recursively by the chosen representatives. Those processors (with value 1) that were not chosen as representatives are used to recursively determine the leftmost processors with value 1 in their own groups. Each of these subproblems has size less than $\sqrt{p}$.

The total time to solve a LEFTMOST PRISONER problem of size p satisfies the recurrence $t(p) = t(\sqrt{p}) + O(1)$. This implies $t(p) \in O(\log \log p)$.

It can also be shown inductively that at most $p-1$ memory cells are used to solve a LEFTMOST PRISONER problem of size p. ∎

EXERCISE 20.13 [66]
Prove that one step of PRIORITY with p processors and m shared memory cells can be simulated by ARBITRARY in $O(\frac{\log\log p}{\log(k+1)})$ steps with kp processors, for $1 \leq k \leq \log_2 p$. How many shared memory cells do you use?

In particular, the result in Exercise 20.13 implies that one step of PRIORITY with p processors can be simulated by ARBITRARY in $O(1/\varepsilon)$ steps using $O(p(\log p)^\varepsilon)$ processors, for $0 < \varepsilon \leq 1$. There is a matching lower bound of $\Omega(\frac{\log\log p}{\log(k+1)})$ steps on ARBITRARY for solving a generalized version of the LEFTMOST PRISONER problem in which each input bit is known to a group of k processors [66]. It is open whether a better simulation of PRIORITY by ARBITRARY can be achieved in some other way.

When the amount of shared memory is small and the input is initially distributed among the processors' local memories, $o(\log p)$ time simulations of CRCW PRAMs with stronger write conflict resolution mechanisms by CRCW PRAMs with weaker ones are not possible without increasing the number of processors or shared memory cells. Specifically, $\Omega(\log(p/m))$ steps are required to simulate either one step of ARBITRARY by COMMON or one step of PRIORITY by ARBITRARY, when the models have p processors and m shared memory cells. (See Exercise 20.44 and Theorem 20.35.) If $m \in O(p^{1-\varepsilon})$ for some constant $\varepsilon > 0$, this implies that $\Omega(\log p)$ steps are required.

It is more difficult to obtain lower bounds when the input is located in a separate read-only shared memory. In this case, $\Omega(\log\log(p/m))$ steps are known to be required to simulate ARBITRARY with p processors and m shared memory cells by COMMON with p processors and m shared memory cells and $\Omega(\log(p/m)/\log\log(p/m))$ steps are known to be required to simulate PRIORITY with p processors and m shared memory cells by ARBITRARY with m shared memory cells and any number of processors [29].

20.3
Relationships Between PRAMs and Other Models

The PRAM is closely related to other models of computation. This section considers relationships between PRAMs and four theoretical models: Boolean circuits, unbounded fan-in Boolean circuits, Turing machines, and

20.3. Relationships Between PRAMs and Other Models

decision trees. These relationships are important for obtaining a better understanding of parallel computation and because good upper and lower bounds on PRAMs can often be obtained from corresponding bounds on these models.

A *Boolean circuit* is a directed acyclic graph whose nodes are either *input nodes*, which have fan-in 0, or *gates*, which have fan-in at most 2. The input nodes are labelled by distinct variables. Each gate is labelled by a Boolean function whose arity is equal to the fan-in of the gate. Some nodes are also designated as *output nodes*. The *depth* of a Boolean circuit is the length of the longest directed path from an input node to an output node and its *size* is the number of gates it contains.

A CROW PRAM can easily simulate a Boolean circuit using one processor for each gate. The processor reads the values it needs as they become available, performs the gate computation, and writes the result in its shared memory cell. If the fan-out of every node in the circuit is bounded by a constant, then the simulation can be done using only exclusive reads. However, any Boolean circuit can be converted to a Boolean circuit with nodes of fan-out at most 2 and only constant factor increases in size and depth [49]. Therefore concurrent reads are not necessary.

THEOREM 20.9
Any Boolean circuit of depth d and size s can be can be simulated by an EROW PRAM in $O(d)$ steps using s processors and s shared memory cells of word size 1.

An *unbounded fan-in Boolean circuit* is a directed acyclic graph composed of input nodes with fan-in 0 (each labelled by a distinct variable), NOT gates with fan-in 1, and unbounded fan-in AND and OR gates. Its *depth* is the length of the longest directed path from an input node to an output node and its *size* is the number of edges in the graph. Since COMMON can compute the AND or OR of n bits in one step using n processors and one shared memory cell, it can easily simulate any unbounded fan-in Boolean circuit.

THEOREM 20.10 *[79]*
Any unbounded fan-in Boolean circuit of depth d and size s can be simulated by COMMON in d steps using s processors and s shared memory cells of word size 1.

Any Boolean function $f : \{0,1\}^n \to \{0,1\}$ can be expressed by a formula of size $O(n2^n)$ in disjunctive or conjunctive normal form and thus can be computed by COMMON in two steps using exponentially many processors and cells. More generally, COMMON can use *table look-up* to compute any function of n Boolean variables in two steps with $n2^n$ processors and 2^n shared memory cells. The idea is to allocate n processors to each of the 2^n possible inputs. Each group checks if it corresponds to the true input by reading the input and performing an AND. One processor from the unique group whose AND is 1 writes the answer. Thus it is impossible to get nontrivial lower bounds on time for CRCW PRAMs without restricting the number of processors or shared memory cells.

EXERCISE 20.14 [21]
Prove that any Boolean function computed by a Boolean circuit of depth $O(\log \log n)$ can be computed by an unbounded fan-in circuit of depth two and $n^{O(1)}$ size.

The complexity classes NC^k and AC^k consist of those Boolean functions that can be computed in $O((\log n)^k)$ depth and $n^{O(1)}$ size by Boolean circuits and unbounded fan-in Boolean circuits, respectively. Theorem 20.9 implies that any problem in NC^k can be solved by an EROW PRAM in $O((\log n)^k)$ steps using $n^{O(1)}$ processors. Similarly, by Theorem 20.10, any problem in AC^k can be solved by COMMON in $O((\log n)^k)$ steps using $n^{O(1)}$ processors. It follows from Exercise 20.14 that any problem in NC^k can be computed by an unbounded fan-in circuit of depth $O((\log n)^k/\log \log n)$ and size $n^{O(1)}$. Combined with Theorem 20.10, this result shows that any problem in NC^k can be solved by COMMON in $O((\log n)^k/\log \log n)$ steps using $n^{O(1)}$ processors and $n^{O(1)}$ memory cells. Thus the upper bound on the time for solving problems in NC^k can be improved by allowing concurrent writes.

Next, we consider some sequential complexity classes. Specifically, let $DLOG$ denote the class of problems that can be solved by deterministic Turing machines using logarithmic space. Let $NLOG$ be the analogous class for nondeterministic Turing machines. Theorems 20.11 and 20.12 will show that all problems in the classes $DLOG$ and $NLOG$ can be solved in logarithmic time by the EROW PRAM and COMMON, respectively, using a polynomial number of processors.

The *configuration graph* of a Turing machine on a given input has one node for each possible combination of state, head positions, work tape contents, and time. Directed edges correspond to transitions. Note that the configuration graph is acyclic, because each transition takes one unit of time. Furthermore, since a Turing machine has a constant size alphabet, every node has constant fan-in and fan-out.

20.3. Relationships Between PRAMs and Other Models

If a Turing machine uses $O(\log n)$ space, then it uses $n^{O(1)}$ time and its configuration graph has $n^{O(1)}$ nodes. Moreover, the configuration graph can easily be constructed by an EROW PRAM in $O(\log n)$ steps using $n^{O(1)}$ processors. Each node in the configuration graph of a deterministic Turing machine has fan-out at most one. In this case, parallel tree contraction (see Chapter 3) can be used to determine whether there is a path from the initial configuration to an accepting configuration.

THEOREM 20.11 *[36, 52]*
Any problem in DLOG can be solved by an EROW PRAM in $O(\log n)$ steps using $n^{O(1)}$ processors.

For nondeterministic Turing machines, it suffices to compute the transitive closure of the configuration graph. This can be done by an unbounded fan-in Boolean circuit with $O(\log n)$ depth and $n^{O(1)}$ size [11] and, hence, from Theorem 20.10, by COMMON in $O(\log n)$ steps using $n^{O(1)}$ processors.

THEOREM 20.12
Any problem in NLOG can be solved by COMMON in $O(\log n)$ steps using $n^{O(1)}$ processors.

For any problem in *NP*, a superlogarithmic lower bound on the EREW PRAM with a polynomial number of processors would imply that *DLOG* is strictly contained in *NP*. Similarly, $\omega(\log n/\log \log n)$ lower bounds on COMMON with a polynomial number of processors for problems in *NP* would imply that NC^1 is strictly contained in *NP*. Since it is currently unknown whether the inclusions $NC^1 \subseteq DLOG \subseteq NP$ are strict, such lower bounds could be difficult to obtain.

Unbounded fan-in Boolean circuits can also simulate PRAMs. A *restricted instruction set PRAM* is a PRAM in which only addition, subtraction, comparison, bitwise Boolean operations, read, write, and indirect addressing are allowed. Multiplication and division of small numbers can also be permitted without changing the results.

THEOREM 20.13 *[79]*
A restricted instruction set PRAM with inputs of word size μ using t steps and p processors can be simulated by an unbounded fan-in circuit of depth $O(t)$ and size $(pt\mu)^{O(1)}$.

In particular, any Boolean function that can be computed by a restricted instruction set PRAM in $O((\log n)^k)$ steps using $n^{O(1)}$ processors is in AC^k.

$$AC^0 \subseteq NC^1 \subseteq DLOG \subseteq NLOG \subseteq AC^1$$
$$\| \quad\quad\quad \cap \quad\quad\quad\quad\quad\quad\quad\quad\quad\quad \|$$
$$\text{CRCW}(1) \quad \text{CRCW}\left(\tfrac{\log n}{\log \log n}\right) \quad\quad\quad\quad \text{CRCW}(\log n)$$

$$\subseteq \cdots \subseteq \quad NC^k \quad \subseteq \quad AC^k \quad \subseteq \cdots \subseteq \quad P$$
$$\cap \quad\quad\quad \|$$
$$\text{CRCW}\left(\tfrac{(\log n)^k}{\log \log n}\right) \quad \text{CRCW}((\log n)^k).$$

FIGURE 20.3
Some Relationships Between Complexity Classes.

The relationships described above among Boolean circuits, unbounded fan-in Boolean circuits, Turing machines, and restricted instruction set PRAMs are summarized in Figure 20.3. CRCW(t) denotes those Boolean functions that can be computed by COMMON, ARBITRARY, or PRIORITY in $O(t)$ steps using $n^{O(1)}$ processors with restricted instruction set. Similarly, EROW(t) denotes those Boolean functions that can be computed by a restricted instruction set EROW PRAM in $O(t)$ steps.

PRAMs that allow a limited number of other instructions such as arbitrary multiplication, division, and shifts have also been studied [85, 84].

An easy counting argument [71] shows that almost all Boolean functions require exponential size unbounded fan-in circuits. Combining this result with Theorem 20.13 gives an analogous result for restricted instruction set PRAMs.

COROLLARY 20.1
Almost all Boolean functions require an exponential number of processors to be computed by a restricted instruction set PRAM in polynomial time.

Lower bounds for specific functions on unbounded fan-in Boolean circuits can be translated into lower bounds on restricted instruction set PRAMs. For example, consider the following lower bound for PARITY.

THEOREM 20.14 *[46]*
Any unbounded fan-in circuit of depth k that computes PARITY has size $2^{\Omega\left(n^{1/(k-1)}\right)}$.

In particular, any unbounded fan-in circuit with $n^{O(1)}$ size computing the PARITY of n input bits (i.e., $x_1 \oplus \cdots \oplus x_n$) has depth $\Omega(\log n/\log\log n)$. Together with Theorem 20.13, this implies that any restricted instruction set

PRAM with $n^{O(1)}$ processors requires $\Omega(\log n/\log\log n)$ steps to compute PARITY. Since PARITY $\in NC^1$, it follows that this lower bound is tight to within a constant factor.

Lower bounds for computing specific functions, such as integer addition, have also been directly obtained on restricted instruction set CRCW PRAMs [58, 68, 22]. These proofs and the lower bounds obtained using Theorem 20.13 depend in an essential way on the restricted instruction set.

In contrast, an *abstract* or *ideal* PRAM, places no restriction on the instruction set. At each step, a processor can compute any function of its local information. Furthermore, shared memory cells are allowed to contain arbitrarily large values. Although this model is unrealistic, lower bounds in this model have great generality since they do not depend on any assumptions about the instruction set. These lower bounds are actually lower bounds on the amount of communication, as opposed to the amount of computation, necessary to solve a problem. They give insight into the nature of communication between processors and may point out where bottlenecks in communication arise. Abstract PRAMs also approximate the situation where communication to and from shared memory is much more expensive than local operations; for example, where each processor is located on a separate chip and access to shared memory is through a combining network.

Not surprisingly, abstract PRAMs can be much more powerful than restricted instruction set PRAMs.

THEOREM 20.15
Any function of n variables can be computed by an abstract EROW PRAM in $O(\log n)$ steps using $n/\log_2 n$ processors and $n/2\log_2 n$ shared memory cells.

PROOF
Each processor begins by reading $\log_2 n$ input values and combining them into one large value. The information known by all the processors is then combined in a binary-tree-like fashion, as follows. In each round, the remaining processors are grouped into pairs. In each pair, one processor communicates the information it knows about the input to the other processor and then leaves the computation. After $\lceil \log_2 n \rceil$ rounds, one processor knows all n input values. Finally, this processor computes the answer in a single additional step. ∎

In particular, superlogarithmic lower bounds for time cannot be obtained on an abstract PRAM without severely restricting the numbers of processors or shared memory cells.

EXERCISE 20.15 [4, 5]
Prove that any function $f : \{0, 1\}^n \to R$ can be computed by abstract COMMON in $\log_2 n - \log_2 \log_2(p/n) + O(1)$ steps using $p \geq 2n$ processors and $p/\log_2(p/n)$ shared memory cells.

Processors in an abstract PRAM can read and write arbitrarily large values. However, to handle large input and output values, unbounded fan-in Boolean circuits need large numbers of gates. Except for this technicality, unbounded fan-in Boolean circuits can simulate abstract PRAMs.

THEOREM 20.16 [57]
If an (abstract CRCW) PRAM computes a Boolean function in t steps using p processors, then the function can be computed by an unbounded fan-in Boolean circuit of depth $O(t)$ and size $p^{2^{t+O(1)}}$.

This result was proved using Kolmogorov complexity. Specifically, given a PRAM, the number of bits needed to describe the states of processors and the addresses and contents of accessed shared memory cells does not grow too quickly with time. In constant depth, an unbounded fan-in Boolean circuit can compute these descriptions (rather than the actual states, addresses, and contents) at any given step from the descriptions at the previous step.

COROLLARY 20.2 [8, 4]
Almost all Boolean functions of n variables require $\log_2 n - \log_2 \log_2 p + \Omega(1)$ steps to be computed by an (abstract CRCW) PRAM with p processors.

Together, Theorems 20.14 and 20.16 imply that a PRAM with $n^{O(1)}$ processors requires $\Omega(\sqrt{\log n})$ steps to compute the PARITY of n bits. Although this lower bound is not tight, the proof is much easier than the direct proof of Theorem 20.29.

For reasonable bounds on the word size, improved versions of Theorem 20.16 and Corollary 20.2 can be obtained.

THEOREM 20.17 [6, 7]
If an (abstract CRCW) PRAM with word size μ computes a Boolean function using t steps and p processors, then the function can be computed by an unbounded fan-in circuit of depth $O(t)$ and size $p^{O(1)} 2^{O(t\mu)}$.

COROLLARY 20.3
Almost all Boolean functions of n variables require $\frac{n - \log p}{\mu}$ steps to be computed by an (abstract CRCW) PRAM with word size μ using p processors.

20.3. Relationships Between PRAMs and Other Models

From Theorems 20.14 and 20.17, it also follows that an (abstract CRCW) PRAM with $(\log n)^{O(1)}$ word size and $n^{(\log n)^{O(1)}}$ processors requires $\Omega(\frac{\log n}{\log \log n})$ steps to compute the PARITY of n input bits. Since PARITY is in NC^1, this lower bound on time is tight to within a constant factor.

The *decision tree* is a widely used model for obtaining sequential lower bounds. An algorithm consists of a tree in which every internal node is labelled by an input variable and every leaf is labelled by a possible answer. An internal node has one child corresponding to each possible value of the input variable that labels it. The execution of a decision tree algorithm begins at its root. When an internal node is visited, the value of the input variable labelling the node determines the appropriate child to visit next. The output is the label of the leaf that is reached. The *decision tree complexity* of a problem is the minimum height of any decision tree that solves the problem (expressed as a function of the number of input variables). Like the abstract PRAM, the decision tree ignores the computation used to solve a problem. Rather, it focuses attention on how much of the input must be examined to determine the answer. There is a very close correspondence between the decision tree model and the CROW PRAM.

THEOREM 20.18 *[65]*
Any function that can be computed by a CROW PRAM in t steps can be computed by a decision tree of height 2^t.

PROOF
Consider any CROW PRAM that computes a function in t steps using p processors. For $1 \leq i \leq p$ and $1 \leq t' \leq t$, let $S(i, t')$ denote an ordered pair consisting of the state of processor P_i and the contents of its corresponding memory cell M_i immediately after step t'. We will inductively define decision trees $T(i, t')$ that compute $S(i, t')$. The desired decision tree can then be obtained from $T(1, t)$ by erasing the first component of the ordered pair labelling each leaf.

Let q_i be the initial state of processor P_i. If M_i initially contains the ith input value, then $T(i, 0)$ is a tree of height 1. Its root is labelled x_i and there is a leaf labelled (q_i, v) for each possible value v of x_i. Otherwise, M_i initially contains the value 0 and $T(i, 0)$ consists of a single node labelled $(q_i, 0)$.

The decision tree $T(i, t' + 1)$ is created by modifying the decision tree $T(i, t')$. Each leaf labelled (q, v) in $T(i, t')$ indicates that processor P_i is in state q and memory cell M_i contains v at the end of step t' on those inputs that lead to the leaf. Suppose that, in this state, P_i reads memory cell $M_{i'}$ during step $t'+1$. Then each leaf of $T(i,t')$ labelled (q,v) is replaced by the decision tree $T(i',t')$ and the labels of the leaves are changed appropriately. Specifically, the label (q',v') of a leaf in this subtree is changed to (q'',v'') if P_i goes into state q'' and writes the value v'' as a result of reading the value v' from $M_{i'}$.

It is easy to verify that the height of $T(i,t')$ is at most $2^{t'}$. ∎

THEOREM 20.19 *[65]*
If a function can be computed by a decision tree of height h, then it can be computed by a CROW PRAM in $1 + \lceil \log_2 h \rceil$ steps.

PROOF
Consider any decision tree of height h. Associate one CROW PRAM processor with each node in the decision tree. Let processor P_1 be associated with the root. To begin the computation, each processor associated with an internal node reads the value of the input variable labelling its node and writes a pointer to the processor associated with the corresponding child. Then, using pointer jumping, the path in the decision tree from the root to a leaf can be determined in $\lceil \log_2 h \rceil$ more steps. ∎

From Theorem 20.18, a lower bound of h on a decision tree implies a lower bound of $\lceil \log_2 h \rceil$ on a CROW PRAM. Thus the logarithm of the decision tree complexity of a problem almost exactly characterizes the time it takes to solve the problem on a CROW PRAM.

EXERCISE 20.16 [76]
Prove that a CROW PRAM requires $\lceil \log_2 n \rceil$ steps to compute the OR of n input bits, even if it is known that all possible inputs contain at most one bit with value 1.

EXERCISE 20.17 [32]
Prove that a CROW PRAM requires $\lceil \log_2(n-1) \rceil$ steps to reverse the direction of the links in a singly linked list of length n.

20.4 Lower Bound Techniques

Many of the lower bound proofs created specifically for (abstract) PRAMs look at how processors and memory cells accumulate knowledge about the input as computation proceeds and what kinds of knowledge can be accumulated. They also find ways of measuring that knowledge and then showing that it cannot increase too quickly.

One of the facts that makes proving PRAM lower bounds difficult is that processors can communicate information by deciding not to write to a particular shared memory cell. This is nicely illustrated by the following example in which $x_1 \vee x_2 \vee x_3 \vee x_4 \vee x_5$ is computed in 2 steps on an EREW PRAM. Note that, by Exercise 20.16, a CROW PRAM requires 3 steps for this computation.

Initially, memory cells M_1, M_2, M_3, M_4, and M_5 contain the input bits x_1, x_2, x_3, x_4, and x_5, respectively. During step 1,

processor P_1 reads M_2 and, if $x_2 = 1$, writes 1 into M_1,
processor P_3 reads M_4 and, if $x_4 = 1$, writes 1 into M_3, and
processor P_5 reads M_5,

resulting in the following memory contents.

$x_1 \vee x_2$	x_2	$x_3 \vee x_4$	x_4	x_5

At step 2, processor P_5 reads M_3. If $x_5 \vee (x_3 \vee x_4) = 1$, then P_5 writes 1 into M_1. At this point, memory cell M_1 contains the value $x_1 \vee x_2 \vee x_3 \vee x_4 \vee x_5$, as desired. Note that processor P_5 communicates the fact that $x_3 \vee x_4 \vee x_5 = 0$ by not writing into memory cell M_1.

This idea can be generalized to show that the OR of n bits can be computed almost 1.4 times faster on an EREW PRAM than on a CROW PRAM. The tth Fibonacci number, F_t, is defined by the recurrence

$$\begin{aligned} F_0 &= 0, \\ F_1 &= 1, \text{ and} \\ F_t &= F_{t-1} + F_{t-2}, \text{ for } t \geq 2 \end{aligned}$$

and satisfies the inequalities $((1+\sqrt{5})/2)^{t-2} < F_t \leq ((1+\sqrt{5})/2)^{t-1}$ for $t > 0$.

EXERCISE 20.18 [16]
Prove that an EREW PRAM can compute the OR of n input bits using n processors and shared memory cells in t steps, provided $F_{2t+1} \geq n$.

In the example above, there is at most one processor that writes into a particular memory cell at a particular time step on any input. This is not the case for all EREW and CREW PRAM algorithms. Depending on the input, different processors may write into a particular cell at a given step. However, the conditions under which the different processors perform these writes are mutually exclusive. The fact that none of these conditions are satisfied by the input is conveyed when no write occurs. Under some circumstances, this could be a lot of information. For example, if the input is known to contain at most one bit with value 1, then the OR of the input bits can be computed in one step on an EREW PRAM. This is accomplished by having each processor read an input bit and, if it reads the value 1, write 1 to memory cell M_1. In contrast, by Exercise 20.16, a CROW PRAM requires $\lceil \log_2 n \rceil$ steps to perform this computation.

20.4.1 Sensitivity, Block Sensitivity, and Degree

The time needed to compute total n-ary functions on a CREW PRAM can be characterized, to within a small constant factor, in terms of certain simple properties. In particular, Theorem 20.26, shows that the the upper bound in Exercise 20.18 is tight, even on a CREW PRAM.

A function f with domain $D \subseteq D_1 \times \cdots \times D_n$ and range R is *sensitive to the set of coordinates* $S \subseteq \{1,\ldots,n\}$ on input $x \in D$ if there exists an input $y \in D$ such that $f(x) \neq f(y)$ and $x_j = y_j$ for all $j \notin S$. The *sensitivity* or *critical complexity* of f is

$$\max_{x \in D} \#\{i | f \text{ is sensitive to } \{i\} \text{ on input } x\}.$$

For example, the OR and PARITY of n Boolean variables have sensitivity n. This is because OR is sensitive to every coordinate on the all 0 input and PARITY is sensitive to every coordinate on every input. The function that computes the MAXIMUM of any n input values also has sensitivity n. An example of a function with sensitivity $\lfloor n/2 \rfloor + 2$ is the Boolean function of n variables that has value 1 when exactly $\lfloor n/2 \rfloor$ or $\lfloor n/2 \rfloor + 1$ of the input bits have value 1. (Complementing a bit with value 1 changes the value of this function only for inputs with $\lfloor n/2 \rfloor$ or $\lfloor n/2 \rfloor + 2$ ones. Complementing a bit with value 0 changes the function value only for inputs with $\lfloor n/2 \rfloor - 1$ or $\lfloor n/2 \rfloor + 1$ ones.)

The sensitivity of a function can be used to obtain a lower bound on the amount of time necessary to compute the function on a CREW PRAM. This is done by showing that the number of coordinates affecting the state of a processor or the contents of a memory cell cannot grow too quickly as a computation proceeds. Formally, coordinate i *affects* processor P (or memory cell M) on input $x \in D_1 \times \cdots \times D_n$ at time t if the state of P (or contents of M) immediately after step t is different on the inputs x and y, for some input $y \in D_1 \times \cdots \times D_n$ that is the same as x except in coordinate i. On every input, the coordinates to which the function is sensitive must all affect the answer cell at the end of the computation. This leads to a lower bound on time.

20.4. Lower Bound Techniques

THEOREM 20.20 *[16]*
A CREW PRAM requires at least $\log_b(sensitivity(f))$ steps to compute a function $f : D_1 \times \cdots \times D_n \to R$, where $b = (5 + \sqrt{21})/2$.

PROOF
Let $s(t)$ and $c(t)$ denote the maximum number of coordinates affecting any processor or shared memory cell, respectively, on any input at time t. Then

$$s(0) = 0 \text{ and } c(0) = 1.$$

Suppose processor P reads from memory cell M on input x during step t. Consider any input y that differs from x only in coordinate i and causes P to have a different state immediately after step t. If, on input y, P is in the same state immediately after step $t-1$ as it is on x, then it reads from cell M during step t. In this case, M must contain different values on x and y immediately after step $t-1$; otherwise processor P would be in the same state immediately after step t on both these inputs. Thus, any coordinate i affecting P on input x at time t either affects P on input x at time $t-1$ or affects M on input x at time $t-1$. Hence

$$s(t) \leq s(t-1) + c(t-1).$$

Now consider any shared memory cell M on any input x. If some processor P writes to M on input x during step t, then any coordinate that does not affect P on input x at time t cannot affect M on input x at time t. By definition, there are at most $s(t)$ coordinates that affect P on input x at time t. Hence, in this case, there are at most $s(t)$ coordinates that affect M on input x at time t.

The other case is when no processor writes to M on input x during step t. Let i be any coordinate that affects M on input x at time t, but does not affect M on input x at time $t-1$. Then there is an input $y^{(i)}$ that differs from x only in coordinate i and a processor $P^{(i)}$ that writes to M on input $y^{(i)}$ during step t.

Consider any other coordinate j that affects M on input x at time t, but does not affect M on input x at time $t-1$. If $P^{(i)} \neq P^{(j)}$, then either coordinate j affects $P^{(i)}$ on input $y^{(i)}$ or coordinate i affects $P^{(j)}$ on input $y^{(j)}$ at time t. Otherwise, during step t, processors $P^{(i)}$ and $P^{(j)}$

would both write to M on input z, where

$$z_k = \begin{cases} y_i^{(i)} & \text{if } k = i \\ y_j^{(j)} & \text{if } k = j \\ x_k & \text{otherwise.} \end{cases}$$

Let v be the number of coordinates that affect M on input x at time t, but not at time $t - 1$. Consider a graph whose vertices are these v coordinates. In this graph there is an edge from i to j if and only if $P^{(i)} \neq P^{(j)}$ and coordinate j affects $P^{(i)}$ on input $y^{(i)}$ at time t. Since no processor is affected by more than $s(t)$ coordinates at time t on a given input, this graph contains at most $v \cdot s(t)$ edges.

For any processor P, if $P = P^{(i)}$, then coordinate i affects P on input x at time t. Thus at most $s(t)$ of the v coordinates are associated with the same processor. This implies that there are at least $v \cdot (v - s(t))$ ordered pairs (i, j) such that $P^{(i)} \neq P^{(j)}$. At least half of these must be edges in the graph. Therefore $v \cdot s(t) \geq v \cdot (v - s(t))/2$, so $v \leq 3s(t)$. At time t on input x, M is affected by the at most $c(t-1)$ coordinates that affect M at time $t - 1$, and by the v coordinates that affect M at time t, but not at time $t - 1$. Hence

$$c(t) \leq c(t-1) + 3s(t).$$

From the solution of the resulting recurrence, it follows that $s(t), c(t) < b^t$ so at least $\log_b(sensitivity(f))$ steps are required to compute f. ∎

EXERCISE 20.19 [16]
Show that the OR of n input bits can be computed in $k + O(\log k)$ steps on a CREW PRAM if the input is known to contain at most k bits with value 1.

EXERCISE 20.20
Where does the proof of Theorem 20.20 break down for computing the OR of n input bits, when the input is known to contain at most one bit with value 1?

EXERCISE 20.21
Prove that a CREW PRAM requires $\Omega(\min\{k, \log n\})$ steps to compute the OR of n input bits, when the input is known to contain at most k bits with value 1.

EXERCISE 20.22 [64]
Consider a CREW PRAM in which processors never forget information and suppose that no processor writes to shared memory cell M on

input x during step t. For any processor P, let $E(P)$ be the set of coordinates i for which there is an input $y^{(i)}$ that differs from x only in coordinate i and that causes processor P to write to M during step t. Prove that for all $i, j \in E(P)$, either i affects P on input $y^{(j)}$ at time t, or j affects P on input $y^{(i)}$ at time t, or both i and j affect P on input x at time $t-1$.

EXERCISE 20.23 [64]
Prove that a CREW PRAM requires at least $\log_4(sensitivity(f))$ steps to compute a function $f : D_1 \times \cdots \times D_n \to R$. (Hint: use Exercise 20.22.)

EXERCISE 20.24 [4]
Prove that no coordinate affects more than $(2+\sqrt{3})^t$ processors or memory cells after t steps of an EREW PRAM computing a Boolean function.

The function f *depends on coordinate* i if f is sensitive to $\{i\}$ on some input $x \in D_1 \times \cdots \times D_n$. In other words, there are two inputs x and y, differing only on coordinate i, such that $f(x) \neq f(y)$.

THEOREM 20.21 [73]
Every function $f : \{0,1\}^n \to \{0,1\}$ that depends on k of its coordinates has sensitivity $\Omega(\log k)$.

COROLLARY 20.4 [73]
A CREW PRAM requires $\Omega(\log \log k)$ steps to compute any function $f : \{0,1\}^n \to \{0,1\}$ that depends on k of its coordinates.

EXERCISE 20.25
Consider the Boolean addressing function $f : \{0,1\}^n \to \{0,1\}$ defined by
$$f(x_1, \ldots, x_r, y_0, \ldots, y_{2^r-1}) = y_j,$$
where $j = \sum_{i=1}^{r} x_i 2^{r-i}$ and $n = r + 2^r$. It uses the binary number formed by concatenating the first r bits as an index to select one of the remaining bits. Prove that this Boolean function has sensitivity $r+1$ and depends on all n of its coordinates. How quickly can you compute this function on a CREW PRAM?

EXERCISE 20.26
Show that the addressing function $f : \{0, \ldots, n-1\} \times \{0,1\}^n \to \{0,1\}$ defined by
$$f(i, y_0, \ldots, y_{n-1}) = y_i$$

can be computed by a PRAM in constant time using only one processor. Explain the difference between this result and Exercise 20.25.

A very useful generalization of sensitivity is *block sensitivity* [61]. For any function f with domain $D \subseteq D_1 \times \cdots \times D_n$, it is defined to be

$$\max_{x \in D} \max\{k \mid f \text{ is sensitive to } k \text{ disjoint subsets of coordinates on input } x\}.$$

By definition, the sensitivity of a function f is bounded above by its block sensitivity. However, there are functions whose sensitivity is less than their block sensitivity. For example, the function $f : \{0,1\}^n \to \{0,1\}$ such that

$$f(x) = \begin{cases} 1 & \text{if } x \text{ contains exactly } \lfloor n/2 \rfloor \text{ or } \lfloor n/2 \rfloor + 1 \text{ ones}, \\ 0 & \text{otherwise} \end{cases}$$

has sensitivity $\lfloor n/2 \rfloor + 2$, but has block sensitivity $\lfloor 3n/4 \rfloor$ [61, 89]. To see that the block sensitivity of this function is at least $\lfloor 3n/4 \rfloor$, consider any input with exactly $\lfloor n/2 \rfloor$ ones. On this input, the function is sensitive to any set consisting of one component with value 1 or two components with value 0. It is an open question whether the sensitivity of every function is always at least some polynomial function of its block sensitivity or whether it can be exponentially smaller.

EXERCISE 20.27 [70]
Consider the Boolean function of n variables that has value 1 if and only if there is a sequence of $\sqrt{n}$ consecutive coordinates the first two of which have value 1 and the remaining $\sqrt{n} - 2$ of which have value 0. What are the sensitivity and block sensitivity of this function?

The lower bounds in Theorem 20.20 and Exercise 20.23 can be extended to block sensitivity.

THEOREM 20.22 [61]
A CREW PRAM requires at least $\log_4(\text{block sensitivity}(f))$ steps to compute a function $f : D_1 \times \cdots \times D_n \to R$.

PROOF
Suppose the block sensitivity of f is k and, on input x, f is sensitive to the disjoint sets of coordinates $S_1, \ldots, S_k$. For $j = 1, \ldots, k$, let $y^{(j)} \in D_1 \times \cdots \times D_n$ be an input such that $f(y^{(j)}) \neq f(x)$ and $y_i^{(j)} = x_i$ for all $i \notin S_j$. Given $(z_1, \ldots, z_k) \in \{0,1\}^k$, define $g(z_1, \ldots, z_k)$ to be the value of the function f on the input $w = (w_1, \ldots, w_n)$, constructed as follows. If $z_j = 1$, let w agree with $y^{(j)}$ on all coordinates in the set S_j. If $z_j = 0$,

let w agree with x on all coordinates in the set S_j. On those coordinates not in any of the sets $S_1, \ldots, S_k$, let w agree with x. In other words,

$$w_i = \begin{cases} y_i^{(j)} & \text{if } i \in S_j \text{ and } z_j = 1 \\ x_i & \text{otherwise.} \end{cases}$$

The function $g : \{0,1\}^k \to R$ has sensitivity k and can be computed by a CREW PRAM at least as quickly as the function f. It follows from Exercise 20.23 that any CREW PRAM requires at least $\log_4(k)$ steps to compute f. ∎

A set of coordinates C is a *certificate* for a function f on an input x if $f(x) = f(y)$ for every input y that agrees with x on all coordinates in C. In other words, knowing the values x_i for all $i \in C$ determines the value of $f(x)$. For example, a set of coordinates corresponding to the variables in a minterm or maxterm of a Boolean function is a certificate. The *certificate complexity* or *nondeterministic decision tree complexity* of a function f with domain $D \subseteq D_1 \times \cdots \times D_n$ is

$$\max_{x \in D} \min\{\#C | C \text{ is a certificate for } f \text{ on input } x\}.$$

The block sensitivity, certificate complexity, and decision tree complexity of a function $f : D_1 \times \cdots \times D_n \to R$ are always closely related.

EXERCISE 20.28
Prove that, for any function $f : D_1 \times \cdots \times D_n \to R$, block sensitivity$(f)$ $\leq$ certificate complexity(f) $\leq$ decision tree complexity(f).

THEOREM 20.23 *[61]*
For any function $f : D_1 \times \cdots \times D_n \to R$, certificate complexity$(f) \leq$ (block sensitivity$(f))^2$.

PROOF
Suppose the block sensitivity of f is k. For any input x, let $\mathcal{B}$ be a maximal collection of disjoint sets of coordinates such that each $S \in \mathcal{B}$ is a minimal set of coordinates to which f is sensitive on input x. Note that $|\mathcal{B}| \leq k$. Let $C = \cup\{S | S \in \mathcal{B}\}$ be the coordinates that occur in the sets in $\mathcal{B}$.

C is a certificate for f on input x. To see why, suppose there is an input y that agrees with x on all coordinates in C, but $f(x) \neq f(y)$. Let C' be the set of coordinates on which y differs from x. Then f is sensitive to C' on input x. Let C'' be a minimal subset of C' to which f is sensitive on input x. Since C and C'' are disjoint, C'' is disjoint from each $S \in \mathcal{B}$. But then $\mathcal{B} \cup \{C''\}$ is a collection of disjoint minimal sets of coordinates to which f is sensitive on input x, contradicting the maximality of $\mathcal{B}$.

To complete the proof, it suffices to show that $|S| \leq k$ for each $S \in \mathcal{B}$, because this implies that $|C| \leq k^2$. Let $S \in \mathcal{B}$ and let y be an input that differs from x only on coordinates in S and such that $f(y) \neq f(x)$. For each $i \in S$, consider the input that agrees with x on coordinate i, agrees with y on all coordinates in $S - \{i\}$, and agrees with both x and y on all other coordinates. By the minimality of S, f has the same value on this input as it does on input x and, hence, has a different value than on input y. Thus $\{i\}$ is a set of coordinates to which f is sensitive on input y. Clearly $\{\{i\} | i \in S\}$ is a disjoint collection of sets. Since the block sensitivity of f is k, it follows that $|S| \leq k$. ∎

THEOREM 20.24 *[9, 48, 83]*
For any function $f : D_1 \times \cdots \times D_n \to R$, decision tree complexity(f) $\leq$ (certificate complexity(f))^2.

PROOF
Let $f : D_1 \times \cdots \times D_n \to R$, let $r \in R$, and let $0 \leq l \leq k$. Suppose f has a certificate of size at most k on every input x such that $f(x) = r$ and suppose f has a certificate of size at most l on every input y such that $f(y) \neq r$. We prove by induction on $|R|$ and l that *decision tree complexity(f)* $\leq kl$.

If $|R| = 1$, then f is a constant function and *decision tree complexity(f)* $= 0 \leq kl$. Therefore, assume $|R| > 1$.

Suppose there is no input in its domain for which f has value r. Then consider the function f' that is identical to f, but without r in its codomain. Since f' has a certificate of size at most l on every input, it follows from the induction hypothesis that *decision tree complexity(f)* $=$ *decision tree complexity(f')* $\leq l^2 \leq kl$.

The remaining case is when there exists an input z such that $f(z) = r$. Let C be a certificate of size at most k for f on input z. For each possible assignment ρ of values to the variables with index in C, consider the restriction $f|_\rho$ of f. Since f has a certificate of size at most k on every input x such that $f(x) = r$, $f|_\rho$ has a certificate of size at most k on every input x' such that $f|_\rho(x') = r$.

On each input y such that $f(y) \neq r$, f has a certificate C' of size at most l. Furthermore, C' intersects C; otherwise it would be possible to construct an input w consistent with z at all coordinates in C and consistent with y at all coordinates in C', which implies that $f(w) = f(z) = r$ and $f(w) = f(y) \neq r$. Thus $f|_\rho$ has a certificate of size at most $l - 1$ on every input y' such that $f|_\rho(y') \neq r$. By the induction hypothesis, $f|_\rho$ has a decision tree of height at most $k(l - 1)$.

Construct a decision tree whose first $|C|$ levels are labelled by the variables with index in C. Each possible assignment ρ of values to these variables leads to one node at depth $|C|$. At that node, root a decision tree for $f|_\rho$ of height at most $k(l - 1)$. The resulting decision tree computes f and has height at most $|C| + k(l - 1) \leq kl$. ∎

EXERCISE 20.29 [13]
Prove that, for any monotone Boolean function f, sensitivity(f) = certificate complexity(f).

From Theorems 20.24, 20.23, and 20.22, if $f : D_1 \times \cdots \times D_n \to R$, then *decision tree complexity* $(f) \leq$ (*certificate complexity* $(f))^2$, *certificate complexity* $(f) \leq$ (*block sensitivity* $(f))^2$, and $\log_4($*block sensitivity* $(f))$ is a lower bound on the time for a CREW PRAM to compute f. The time to compute f on a CREW PRAM is bounded above by the time to compute f on a CROW PRAM, which, by Theorem 20.19, is bounded above by $1 + \lceil \log_2($*decision tree complexity*$(f)) \rceil$. Thus the time to compute f on a CREW PRAM can be characterized, to within a small constant factor, by $\log_2($*decision tree complexity* $(f))$, $\log_2($*certificate complexity* $(f))$, and $\log_2($*block sensitivity* $(f))$. Moreover, given a CREW PRAM algorithm computing f, it is possible to construct a CROW PRAM algorithm to compute f to within a constant factor as quickly. However, the new algorithm is not necessarily a step by step simulation of the original algorithm and it might use exponentially more processors. It is an open question whether such a large blowup in the number of processors is necessary.

The proofs of many of these results rely on the fact that the domain of the function f is *complete*, i.e., if there are n input variables, then the domain of f can be expressed as the direct product of n sets of values. In other words, the value of any input variable is not constrained by the values of the other input variables. The results are not necessarily true when the domain of f is not complete.

For example, the OR of n bits when it is known that at most one bit is 1 can be computed in one step on a CREW PRAM, but the sensitivity of this function is n and it requires $\log_2 n$ steps on a CROW PRAM. (See

Exercises 20.19 and 20.16.) Similarly, reversing a singly linked list or a disjoint collection of circular singly linked lists of length n has decision tree complexity $n-1$ (Exercise 20.17), although this can be done in one step on a CREW PRAM [32]. In contrast, when reverse pointers are also present, the situation is quite different, as the following theorem shows.

THEOREM 20.25 [32]
A CROW PRAM can compute any function of a disjoint collection of circular doubly linked lists to within a constant factor as fast as a CREW PRAM.

Additional work needs to be done to understand what problems can be solved more quickly by a CREW PRAM than by a CROW PRAM.

For certain Boolean functions, lower bounds that exactly match the upper bounds have been obtained by considering yet another property. Let $f : \{0,1\}^n \to \{0,1\}$ be a Boolean function. Then there is a unique multilinear polynomial

$$P_f(x) = \sum_{I \subseteq \{1,\ldots,n\}} a_I \prod_{i \in I} x_i$$

with integer coefficients that represents f in the sense that $P_f(x) = f(x)$ whenever $x \in \{0,1\}^n$. Moreover, the coefficients a_I have absolute value at most 2^{n-1} [75]. The *degree* of the Boolean function f is defined to be the degree of the polynomial P_f. For example, if $f(x_1,\ldots,x_n) = x_1 \wedge \cdots \wedge x_n$ and $g(x_1,\ldots,x_n) = x_1 \vee \cdots \vee x_n$, then $P_f(x_1,\ldots,x_n) = x_1 \cdots x_n$ and $P_g(x_1,\ldots,x_n) = 1 - (1-x_1)\cdots(1-x_n)$. Both these Boolean functions have degree n.

EXERCISE 20.30 [23]
Prove the following facts about the degrees of Boolean functions.

1. degree($\neg f$) = degree(f).
2. degree($f \wedge g$) $\leq$ degree(f) + degree(g).
3. degree($f \vee g$) $\leq$ degree(f) + degree(g).
4. If $f \wedge g \equiv 0$, then degree($f \vee g$) $\leq$ max{degree(f), degree(g)}.

Since the OR of n bits has degree n, the following result gives a lower bound on a CREW PRAM that exactly matches the upper bound for computing OR on an EREW PRAM (Exercise 20.18).

THEOREM 20.26 [23]
If a CREW PRAM computes a Boolean function $f : \{0,1\}^n \to \{0,1\}$ in t steps, then $F_{2t+1} \geq degree(f)$.

PROOF
Without loss of generality, we make the following two assumptions. First, a processor's state is merely (an encoding of) the sequence of values it has read at each step (so a processor never forgets information). Second, whenever a processor writes, it identifies itself and its current state (i.e., it communicates everything it knows).

For each processor, partition the set of inputs $\{0,1\}^n$ so that inputs are in the same block if and only if they cause the processor to be in the same state immediately after step t. Let $s(t)$ denote the maximum degree of the characteristic function of any block of one of these partitions. Similarly, let $c(t)$ denote the maximum degree of the characteristic function of any block of the partition of $\{0,1\}^n$ induced by a shared memory cell's contents immediately after step t. Then, using Exercise 20.30, it follows as in the proof of Theorem 20.20 that

$$\begin{aligned} s(0) &= 0, \\ c(0) &= 1, \text{ and} \\ s(t) &\leq s(t-1) + c(t-1), \text{ for } t > 0. \end{aligned}$$

Now consider the characteristic function of any block of the partition induced by the contents of a shared memory cell M immediately after step t. If, in state q, processor P_i writes into M during step t, then the characteristic function $g_{i,q} : \{0,1\}^n \to \{0,1\}$ of the set of inputs that cause P_i to be in state q at time t is also the characteristic function of the set of inputs for which M contains the value (i,q) immediately after step t. This function has degree at most $s(t)$, by definition.

Suppose, instead, that no processors write to M during step t and M contains the value v immediately after step t. Let g be the characteristic function identifying those inputs for which M contains the value v immediately after step $t-1$. Then the characteristic function g' of the corresponding block of the partition induced by M's contents at time t can be expressed as

$$g' = g \wedge \neg \bigvee_{i,q} g_{i,q}$$

where the OR is taken over all values of i and q such that, in state q, processor P_i writes to M during step t. Since concurrent writes to M

cannot occur, at most one of the functions $g_{i,q}$ has value 1 for a given input. From Exercise 20.30, it follows that $degree\left(\neg\bigvee_{i,q} g_{i,q}\right) \leq s(t)$. By definition, $degree(g) \leq c(t-1)$. Thus $degree(g') \leq c(t-1) + s(t)$ and, hence,

$$c(t) \leq c(t-1) + s(t) \text{ for } t > 0.$$

It is easy to prove by induction that $s(t) \leq F_{2t}$ and $c(t) \leq F_{2t+1}$. The theorem follows from the second inequality and the fact that $degree(f)$ is bounded above by the maximum degree of the characteristic functions of the blocks of the partition induced by the contents of the answer cell at the end of the computation. ∎

EXERCISE 20.31 [23]
Let $n = k^2$ and consider the function

$$f(x_1, \ldots, x_n) = (x_1 \wedge \cdots \wedge x_k) \vee \cdots \vee (x_{n-k+1} \wedge \cdots \wedge x_n).$$

Prove that the certificate complexity of f is k and its degree is n. What is its decision tree complexity?

EXERCISE 20.32
Prove that the degree of any Boolean function is always bounded above by its decision tree complexity.

THEOREM 20.27 [82, 63]
The block sensitivity of any Boolean function is bounded above by twice the square of its degree.

EXERCISE 20.33 [23, 63]

Let $g_1(x_1, x_2, x_3) = \begin{cases} 0 & \text{if } x_1 = x_2 = x_3 \\ 1 & \text{otherwise} \end{cases}$

and for $d > 1$, let

$$g_d(x_1, \ldots, x_{3^d}) = g_1(g_{d-1}(x_1, \ldots, x_{3^{d-1}}), \ldots, g_{d-1}(x_{2 \cdot 3^{d-1}+1}, \ldots, x_{3^d})).$$

The function g_d can be computed by a complete ternary tree of depth d, all of whose gates compute the function g_1. What is the sensitivity, block sensitivity, and degree of the function g_d?

For special classes of Boolean functions, better results can be shown.

EXERCISE 20.34 [23]
Prove that the sensitivity of a monotone Boolean function is bounded above by its degree.

EXERCISE 20.35 [13]
Prove that the sensitivity of any nonconstant, symmetric, Boolean function of n variables is at least $\lceil (n+1)/2 \rceil$.

THEOREM 20.28 *[23]*
Any nonconstant, symmetric, Boolean function of n variables has degree larger than $n/2$.

The time taken by a randomized CREW PRAM to compute a Boolean function is also related to the function's block sensitivity and degree [23]. These results imply that a randomized CREW PRAM cannot compute a Boolean function more than a constant factor faster than a deterministic CROW PRAM.

20.4.2 Simplifying the Algorithmic Structure by Restricting the Input

A number of lower bounds for the PRAM have been obtained using the following approach. Given an algorithm to solve a problem, find a restricted problem, either on a smaller number of input variables or a smaller input domain, on which the algorithm behaves in a considerably simpler manner. Then directly prove lower bounds for this class of simpler algorithms.

It is useful to consider the partitions of the set of inputs into blocks that are indistinguishable to a processor or a memory cell during the first t steps of the computation. As time increases, these partitions become more complicated. The proofs of the next results use the technique of random restrictions to show that these partitions do not become sufficiently complex too quickly. Essentially, after each step of the computation, the values of relatively few randomly chosen bits are fixed. This leaves only a slightly smaller instance of the problem. However, it can be shown that the resulting partitions of the set of inputs are likely to remain quite simple in structure.

THEOREM 20.29 *[8]*
If PRIORITY computes the PARITY of n input bits in t steps, then it uses $2^{\Omega(n^{1/t})}$ processors and $2^{\Omega((n/t!)^{1/t})}$ shared memory cells.

COROLLARY 20.5 *[8]*
PRIORITY with $n^{O(1)}$ processors or $n^{O(1)}$ shared memory cells requires $\Theta(\log n/\log\log n)$ steps to compute the PARITY of n input bits.

THEOREM 20.30 *[8]*
For all $t \in \frac{1}{3} \log n / \log \log n - \omega(\log n/(\log \log n)^2)$, there is a Boolean function of n variables that can be computed by COMMON in t steps using n processors and memory cells, but cannot be computed by PRIORITY in $t-1$ steps using $n^{O(1)}$ processors or using $n^{O(1)}$ memory cells.

Thus, even one extra time step can be more useful than increasing the number of processors or shared memory cells by a polynomial factor or using a more powerful write resolution rule.

An early example of this approach was applied to the following problem [77].

SEARCH AN ORDERED LIST
Given $x_1, \ldots, x_n, y \in \{1, \ldots, r\}$ such that $x_1 \leq x_2 \leq \ldots \leq x_n$, determine that either $y < x_1$ or $x_n \leq y$ or find the index i such that $x_i \leq y < x_{i+1}$.

Using $(p+1)$-ary search, a CREW PRAM with p processors can solve this problem in $O(\log n/\log(p+1))$ steps. A nontrivial lower bound can be obtained on an EREW PRAM, even for the restricted version of the problem in which $x_1, \ldots, x_n, y \in \{0, 1\}$, by bounding the number of variables affecting processors and memory cells as a function of time. This lower bound remains valid even if concurrent writes are allowed.

EXERCISE 20.36 [77]
Prove that an EREW PRAM with p processors requires $\Omega(\log n - \log p)$ steps to SEARCH AN ORDERED LIST of length n.

EXERCISE 20.37 [77]
Prove that an EREW PRAM can SEARCH AN ORDERED LIST of length n in $O(\log n - \log p)$ steps using p processors, provided p copies of y are given as part of the input.

EXERCISE 20.38 [77]
Prove that an EREW PRAM can SEARCH AN ORDERED LIST of length n in $O(\sqrt{\log n})$ steps using n processors and n memory cells.

Even with an arbitrarily large number of processors, the upper bound in Exercise 20.38 cannot be improved. This is a consequence of the limited ability of the EREW PRAM to access the input variable y.

THEOREM 20.31 *[77]*
An EREW PRAM requires $\Omega(\sqrt{\log n})$ steps to SEARCH AN ORDERED list of length n, provided the domain size, r, is sufficiently large.

The idea of the proof is to show that any algorithm has a simple structure for a large subset of the inputs. Then a lower bound is obtained assuming this simple structure.

Two inputs $x, x' \in D^n$ are *order equivalent* if, for all $i, j \in \{1, \ldots, n\}$,

$$x_i < x_j \text{ if and only if } x'_i < x'_j.$$

A function $f : D^n \to R$ *depends only on the relative values of its variables* if $f(x) = f(x')$ for all order equivalent inputs $x, x' \in D^n$. The *address functions* of a PRAM algorithm are the functions of the input that describe where in shared memory each processor reads from and writes to at each step in the computation. A PRAM algorithm *depends only on the relative values of its variables* if its address functions depend only on the relative values of their variables.

LEMMA 20.1
Consider any EREW PRAM algorithm to SEARCH AN ORDERED LIST. If the domain size r is sufficiently large (as a function of the number of processors, memory cells, and time steps), then there is a large subset $S \subseteq \{1, \ldots, r\}$ such that the EREW PRAM algorithm depends only on the relative values of its variables when restricted to inputs $x_1, \ldots, x_n, y \in S$.

This lemma can be proved by applying the following result from Ramsey theory to each of the address functions.

THEOREM 20.32 *[44]*
Given a function $f : D^n \to R$, where $|D|$ is sufficiently large in terms of n, $|R|$, and s, there is a subset $S \subseteq D$, with $|S| \geq s$, such that $f|_{S^n}$ depends only on the relative values of its variables.

Finally, any EREW PRAM algorithm to SEARCH AN ORDERED LIST that depends only on the relative values of its variables can be shown to require $\Omega(\sqrt{\log n})$ steps, using arguments similar to those needed for Exercise 20.36 [77].

This result demonstrates that concurrent read can be more powerful than exclusive read, because a CREW PRAM with n processors can SEARCH AN ORDERED LIST of n numbers in $O(1)$ steps. Note, however, that the problem SEARCH AN ORDERED LIST does not have a complete domain.

Although the lower bounds in Exercise 20.36 and Theorem 20.31 also apply to any problem that has SEARCH AN ORDERED LIST as a special case, the natural extension of this problem to a complete domain, the problem of searching an unordered list, is also difficult for a CREW PRAM.

EXERCISE 20.39
Prove that a CREW PRAM requires $\Omega(\log n)$ steps to search an unordered list $x_1, \ldots, x_n$ for an element y.

It is possible to construct a problem with complete domain that has SEARCH AN ORDERED LIST as a special case and can be computed substantially more quickly by a CREW PRAM than by an EREW PRAM [40]. A *comparison tree* is a binary decision tree in which each internal node is labelled by a comparison between two input variables instead of by a single input variable. The two children of an internal node correspond to the $<$ and $\geq$ outcomes of the comparison. As in a decision tree, computation begins at the root and the output is the label of the leaf that is reached. Consider the following comparison tree with n internal nodes and depth $\lceil \log_2 n \rceil$, an example of which is illustrated in Figure 20.4. At the ith internal node (encountered in an inorder traversal), the input variables y and x_i are compared. The leaves are labelled sequentially from left to right with the numbers $0, 1, \ldots, n$. The problem is to determine the output of this comparison tree, given the input $x_1, \ldots, x_n, y \in \{1, \ldots, r\}$. When the input variables $x_1, \ldots, x_n$ are restricted to be in sorted order, this problem is equivalent to SEARCH AN ORDERED LIST. Thus an EREW PRAM requires $\Omega(\sqrt{\log n})$ steps to solve the comparison tree problem.

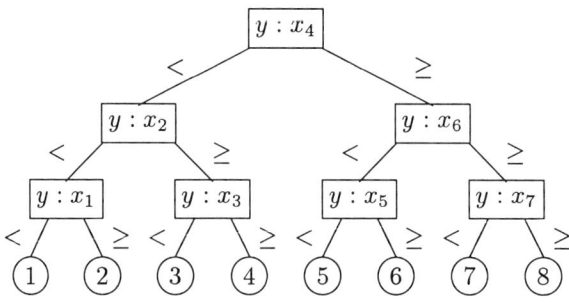

FIGURE 20.4
The comparison tree problem, for $n = 7$.

EXERCISE 20.40 [40]
Prove that on a CREW PRAM, the comparison tree problem can be solved in $O(\log \log n)$ steps, using n processors, and requires $\Omega(\log \log n)$ steps, regardless of the number of processors.

EXERCISE 20.41
Prove that an EREW PRAM with an infinite number of processors requires $\Omega(\sqrt{\log p}/\log \log p)$ steps to simulate a general step of a CREW PRAM with p processors, when the input domain is sufficiently large.

EXERCISE 20.42 [40]
Construct a function with domain $\{0,1\}^n$ and range $\{0,1\}$ that can be solved in $O(\log n \log \log p/\log p)$ steps on a CREW PRAM using p processors, but requires $\Omega(\log n - \log p)$ steps to be computed by an EREW PRAM with p processors.

Lower bounds for a number of other problems have been obtained using the Ramsey theory technique. The ELEMENT DISTINCTNESS problem was studied by Fich, Meyer auf der Heide, and Wigderson [31]. They showed that COMMON with n processors and an infinite amount of shared memory requires $\Omega(\log \log \log n)$ steps to solve this problem for a domain of size $2^{\Omega(n \log n)}$. Ragde, Steiger, Szemeredi, and Wigderson [69] improved the lower bound on time to $\Omega(\sqrt{\log n})$ and Boppana [10] improved it further to $\Omega(\log n/\log \log n)$, matching the upper bound. (See Theorem 20.7 and Exercise 20.12.) Both results require substantially larger domains. Recently, Edmonds [26] was able to obtain an $\Omega(\log n/\log \log n)$ lower bound using a domain with size only doubly exponential in n. Boppana [10] and Edmonds [26] also proved that PRIORITY with n processors requires $\Omega(\log n/\log \log n)$ steps to solve ELEMENT DISTINCTNESS on these domains, when the amount of shared memory does not increase as a function of the domain size r. On PRIORITY with n processors and infinite memory, Fich, Meyer auf der Heide, and Wigderson [60] showed that finding the MAXIMUM of n elements requires $\Omega(\log \log n)$ steps (matching Shiloach and Vishkin's upper bound [78]) and Meyer auf der Heide and Wigderson [60] showed that SORTING a list of length n requires $\Omega(\sqrt{\log n})$ steps. Schieber and Vishkin [81] used similar ideas to obtain a tight $\Omega(\log \log n)$ lower bound on the number of steps needed by PRIORITY with $n(\log n)^{0(1)}$ processors and infinite memory to merge two sorted lists of length n or find the nearest neighbour of each vertex in an n-vertex convex polygon.

A serious limitation of these lower bounds, and one that is inherent in the use of Ramsey theory, is that they are only applicable when the size of the problem domain is very large. For example, provided the domain size $r \in O(m)$, ARBITRARY can solve ELEMENT DISTINCTNESS in $O(1)$ steps using n processors and m shared memory cells. (See Exercise 20.12.) Berkman and Vishkin [14] showed that two sorted lists of length n containing numbers in the range $\{1, \ldots, r\}$ can be merged on a CREW PRAM in $O(\log \log \log r)$ steps using $n / \log \log \log r$ processors.

EXERCISE 20.43 [33, 27]
Prove that the MAXIMUM of n elements with values in $\{1, \ldots, n^{O(1)}\}$ can be found on COMMON in $O(1)$ steps using n processors and n memory cells.

At present, it is unknown whether an EREW PRAM can compute every Boolean function as quickly as a CREW PRAM can. However, the following Boolean function requires substantially more time to compute on an EROW PRAM than on a CROW PRAM [37].

BOOLEAN DECISION TREE EVALUATION
Given the values of 2^k Boolean variables $x_0, \ldots, x_{2^k-1}$ and (the binary encoding of) a complete decision tree of height h, in which each internal node is labelled by one of these 2^k variables and the leaves are alternately labelled 0 and 1, determine the label of the leaf that is reached. The size of the input is $n = 2^k + k(2^h - 1)$.

For example, when $k = 2$ and $h = 3$, the first four bits of the input 111000100111100110 represent the values $x_0 = 1$, $x_1 = 1$, $x_2 = 1$, and $x_3 = 0$. The remaining bits represent the indices of the labels of the interior nodes of the decision tree illustrated in Figure 20.5, when these nodes are arranged according to the inorder traversal of the tree.

THEOREM 20.33 [37]
For an appropriate choice of h and k, a CROW PRAM can solve the BOOLEAN DECISION TREE EVALUATION problem in $O(\log \log n)$ steps, but any (randomized) EROW PRAM that computes it requires $\Omega(\sqrt{\log n})$ (expected) steps.

From Theorem 20.19, it follows that a CROW PRAM can solve the BOOLEAN DECISION TREE EVALUATION problem in $O(\log k + \log h)$ steps. To prove the lower bound, it is sufficient to consider the behaviour of

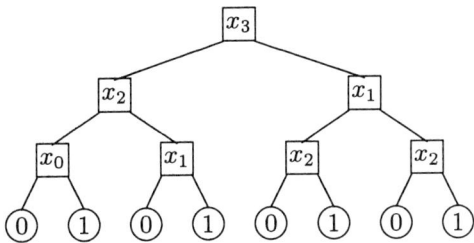

FIGURE 20.5
The Boolean decision tree of height 3 represented by the encoding 00100111100110.

deterministic EROW PRAM algorithms on a randomly chosen input from $\{0,1\}^n$ [90]. After each step of a EROW PRAM computation, a longer initial segment of the root to leaf path is revealed. It is very unlikely that any processor knows *both* the value of a variable that has not been revealed and (any bit of) the label of a node in the subtree rooted at the end of this initial segment. In particular, for sufficiently short computations, if a processor knows anything about which variable labels the parent of the leaf that is reached, then it is unlikely to know anything about the value of that variable. It is conjectured that an EREW PRAM requires $(\log n)^{\Omega(1)}$ steps to solve the BOOLEAN DECISION TREE EVALUATION problem, for appropriate h and k.

20.4.3 Small Memory

When proving lower bounds on a PRAM with a small amount of shared memory, it is particularly useful to focus attention on the sequence of contents of the m shared memory cells as the computation proceeds. This sequence is called the *history* of the computation. If a concurrent read PRAM has only one shared memory cell, we may assume that all processors read the shared memory cell at every step of the computation. Then a processor's knowledge can be expressed as a function of the history and any input values initially located in its local memory. A lower bound of $\Omega(t/m)$ steps on a PRAM using m shared memory cells can be obtained from a lower bound of t steps on a PRAM using one shared memory cell by applying the result of Exercise 20.2.

The following information theoretic lower bound is obtained by considering the partition of the set of all possible inputs into blocks that cause the shared memory to have exactly the same history up to a given point in time. Since two processors may not attempt to write different values in the same memory cell at the same time, it must be possible to infer some "mutual exclusion" information from the history. This mutual exclusion information takes time to set up and is not reusable, so the partition cannot become too fine too quickly.

THEOREM 20.34 [33]
To compute a surjective function $f : \{0,1\}^n \to R$, (randomized) COMMON with n processors and m shared memory cells requires $\Omega((\log|R|)/m)$ (expected) steps for some input.

EXERCISE 20.44 [33]
Prove that COMMON with p processors and m shared memory cells requires $\Omega(\log(p/m))$ steps to simulate one step of ARBITRARY or PRIORITY with p processors and m memory cells.

Adversary arguments are another way to prove lower bounds, especially for small amounts of shared memory. The idea is to define an adversary that fixes the history as the computation proceeds, by fixing the values of some of the input variables and, in the case of ARBITRARY, the outcomes of write conflicts. Then show that, unless the computation is sufficiently long, this information about the input is insufficient to determine the answer. The following lower bound is an example of this approach.

THEOREM 20.35 [33]
ARBITRARY with p processors and m shared memory cells requires $\Omega(\log(p/m))$ steps to simulate one step of PRIORITY with p processors and m memory cells.

PROOF
It suffices to prove the lower bound for the following restricted version of the LEFTMOST WRITERS problem. The p processors are divided into m groups of approximately equal size: group 1 consists of the first $\lfloor p/m \rfloor$ processors, group 2 consists of the next $\lfloor (p+1)/m \rfloor$ processors, ..., and group m consists of the last $\lfloor (p+m-1)/m \rfloor$ processors. Each processor in group i can have value 0 or i. The problem is to determine the processor of lowest index that has value i, for all $i \in \{1, \ldots, m\}$.

Consider any ARBITRARY algorithm that solves this problem. A processor whose value has not yet been fixed by the adversary is said to be *free*. The adversary fixes the values of certain processors after each step of the computation in a way that allows any free processor to be the leftmost writer in its group. This is accomplished by never fixing the value of a processor to anything but 0 if there is a free processor of lower index within the same group. As long as there is at least one free processor, the algorithm cannot have terminated.

Initially, the rightmost processor in each group has its value fixed to the index of the group. The values of the other processors are free. There are $\Omega((p/m)^m)$ possible answers. This is because any processor in group i can be the processor of lowest index with value i.

Once the history for the first t steps has been fixed, the action of each processor at step $t+1$ can be viewed as a function of its value. Based on these actions, the adversary fixes the contents of each of the m shared memory cells at step $t + 1$, as follows.

If possible, the adversary fixes the contents of a cell by choosing the value written at step $t+1$ by a processor whose value has already been fixed. If this is not possible, but there is a free processor that writes to the cell at step $t + 1$ when its value is 0, the adversary fixes the value of that processor to 0 and fixes the contents of the cell by choosing the value written by that processor.

The remaining unfixed cells are only written to by free processors when their values are nonzero. If possible, the adversary fixes the contents of such a cell by choosing the value written by a processor among the right half of the free processors in some group. The value of the processor is set to the index of its group to ensure that the write takes place. The value of every free processor of higher index within the same group is fixed to 0.

Note that each time the contents of a cell is fixed in one of these ways, the number of possible answers decreases by at most a factor of 2.

Finally, the left half of the free processors in each group have their values fixed to 0, ensuring that no processors write to the remaining unfixed cells. The adversary fixes the contents of those cells to be the same as at step t. This decreases the number of possible answers by a factor of 2^m.

At each step, the total number of possible answers decreases by a factor of at most 2^{2m}. Thus any algorithm must perform $\Omega(\log(p/m))$ steps to determine the answer in the worst case. ∎

EXERCISE 20.45 [29]
Prove that PRIORITY with an infinite number of processors and m shared memory cells requires $\Omega(n/m)$ steps to determine whether the input $x_1, \ldots, x_n$ contains two consecutive variables with value 1, assuming that the input is initially distributed among the processors' local memories.

EXERCISE 20.46
Prove that a CREW PRAM with an infinite number of processors and m shared memory cells requires $\Omega(n/m)$ steps to compute the OR of n input bits that are initially distributed among the processors' local memories.

20.4.4 Reductions

The most frequently used lower bound technique is to find a problem for a which a good lower bound is known and reduce it to the problem of interest. Specifically, let f be a problem with n input variables and let g be a problem with n' input variables. Suppose a PRAM can map each instance $(x_1, \ldots, x_n)$ of f to an instance $(y_1, \ldots, y_{n'})$ of g in t steps and can map the answer $g(y_1, \ldots, y_{n'})$ to the answer $f(x_1, \ldots, x_n)$ in t' steps. If the PRAM requires at least T steps to solve f, then it requires at least $T - t - t'$ steps to solve g. To get a nontrivial lower bound for g, it is important that the time $t + t'$ used to perform the reduction is significantly less than the lower bound T for f. For example, the problem of merging two sorted lists of length n can be reduced in constant time to the problem of triangulating a monotone polygon with $\Theta(n)$ vertices by an EREW PRAM using $O(n)$ processors [12]. Since the former problem requires $\Omega(\log \log n)$ steps on PRIORITY with $n(\log n)^{O(1)}$ processors (see Section 20.4.2), it follows that this lower bound also applies to the latter problem. More generally, if merging two sorted lists of length n can be reduced to a problem with $O(n)$ inputs on PRIORITY using $o(\log \log n)$ steps and $n(\log n)^{O(1)}$ processors, then that problem requires $\Omega(\log \log n)$ steps on PRIORITY with $n(\log n)^{O(1)}$ processors.

Sometimes it is possible to express each component of g's input as a function of at most one component of f's input and each component of f's answer as a function of at most one component of g's answer. Then f can be reduced to g by a CREW PRAM in constant time using one processor for each component of g's input and one processor for each component of f's answer. This type of reduction is called a *projection*. For example, a reduction can be used to reduce PARITY to LIST RANKING, a problem discussed in Chapter 3.

THEOREM 20.36
The PARITY of n bits can be reduced via a projection to LIST RANKING in a list of length $3n + 1$.

PROOF
Given an instance $(x_1, \ldots, x_n)$ of the PARITY problem, create an instance of LIST RANKING as follows. Arrange nodes $1, \ldots, 2n + 1$ in order starting at 1. If $x_i = 0$, then place node $2n + i + 1$ between nodes $n + i$ and $n + i + 1$ and if $x_i = 1$, then place it between nodes i and $i + 1$. Then the distance from node $n + 1$ to the end of the list is the number of input variables with value 1 plus twice the number of input variables with value 0. Hence, the least significant bit of the rank of node $n + 1$ is the PARITY of $x_1, \ldots, x_n$. ∎

From Theorem 20.20 and Corollary 20.5, it follows that LIST RANKING requires $\Omega(\log n)$ steps on a CREW PRAM using any number of processors and $\Omega(\log n/\log\log n)$ steps on PRIORITY using a polynomial number of processors.

EXERCISE 20.47 [35]
Prove that there is a projection mapping the PARITY of n bits to the MULTIPLICATION of two n bit numbers.

Many other examples of projections appear in [80] and [21].

An analogue of Turing reducibility is also useful for proving lower bounds. Let f be a problem with n inputs and let $\{g_i\}$ be a family of problems where g_i has i inputs. Suppose there is a PRAM that can solve f using t steps and p processors, given access to an oracle that solves any instance of g_i for $i \leq s(n)$. If the PRAM can solve g_i using $t'(i)$ steps and $p'(i)$ processors, then it can solve f using at most $t \cdot t'(s(n))$ steps and $p \cdot p'(s(n))$ processors. Furthermore, if no input to an oracle computation is a function of an output of another oracle computation, then the PRAM can solve f using only $t + t'(s(n))$ steps.

A special case of such a reduction is one that can be performed by a constant depth, polynomial size unbounded fan-in Boolean circuit with gates that compute any bit of the output of a problem in $\{g_i\}$. This reduction can also be performed by COMMON (and, hence, PRIORITY) in constant time using a polynomial number of processors. (See Theorem 20.10.) Constant depth, polynomial size reductions from determining the PARITY of n bits to ADDING, SORTING, and determining the MAJORITY of $O(n)$ bits and computing the TRANSITIVE CLOSURE of an $n+2$ node graph [35, 21] imply $\Omega(\log n/\log\log n)$ lower bounds for these problems on PRIORITY with $n^{0(1)}$

processors. Other examples of constant depth, polynomial size reductions appear in [35, 21, 1].

EXERCISE 20.48 [21]
Prove that SORTING n n-bit integers is reducible to determining (the bits of) the binary representation of the sum of n bits and vice versa.

Acknowledgements

I am grateful to Jeff Edmonds, David Neto, Naomi Nishimura, Prabhakar Ragde, Jeannine St. Jacques, Paul Beame, and two anonymous referees for carefully reading various drafts of this chapter and making valuable comments. Preparation of this chapter was supported in part by the Natural Sciences and Engineering Research Council of Canada under grant A9176, the Information Technology Research Centre of Ontario, and the Defense Advanced Research Projects Agency under grant N00014-91-J-1698.

Bibliography

[1] A. Aggarwal, B. Chazelle, L. Guibas, C. O'Dunlaing, and C. Yap. Parallel computational geometry. *Algorithmica*, 3, pages 293–327, 1988.

[2] H. Alt, T. Hagerup, K. Mehlhorn, and F. Preparata. Deterministic simulation of idealized parallel computers on more realistic ones. *SIAM Journal on Computing*, 16(5), pages 808–835, 1987.

[3] A. Aho, J. Hopcroft, and J. Ullman. *The Design and Analysis of Computer Algorithms*. Addison-Wesley, 1974.

[4] P. Beame. *Lower Bounds in Parallel Machine Computation*. PhD thesis, University of Toronto, 1987. Department of Computer Science Tech. Report 198/87.

[5] P. Beame. Limits on the power of concurrent-write parallel machines. *Information and Computation*, 76(1), pages 13–28, 1988.

[6] S. Bellantoni. Parallel RAMs with bounded memory wordsize. Master's thesis, University of Toronto, 1988.

[7] S. Bellantoni. Parallel RAMs with bounded memory wordsize. *Information and Computation*, 91(2), pages 259–273, 1991.

[8] P. Beame and J. Hastad. Optimal bounds for decision problems on the CRCW PRAM. *Journal of the ACM*, 36(3), pages 643–670, 1989.

[9] M. Blum and R. Impagliazzo. Generic oracles and oracle classes. In *Proc. 28th Annual IEEE Symposium on Foundations of Computer Science*, pages 118–126, 1987.

[10] R. Boppana. Optimal separations between concurrent-write parallel machines. In *Proc. 21st Annual ACM Symposium on Theory of Computing*, pages 320–326, 1989.

[11] A. Borodin. On relating time and space to size and depth. *SIAM Journal on Computing*, 6(4), pages 733–744, 1977.

[12] O. Berkman, B. Schieber, and U. Vishkin. Some doubly logarithmic optimal parallel algorithms based on finding all nearest smaller values. To appear in *Journal of Algorithms*.

[13] S. Bublitz, U. Schürfeld, B. Voigt, and I. Wegener. Properties of complexity measures for PRAMs and WRAMs. *Theoretical Computer Science*, 48, pages 53–73, 1986.

[14] O. Berkman and U. Vishkin. Recursive *-tree parallel data structure. In *Proc. 30th Annual IEEE Symposium on Foundations of Computer Science*, pages 196–202, 1989. To appear in *SIAM Journal on Computing*.

[15] B. Chlebus, K. Diks, T. Hagerup, and T. Radzik. Efficient simulations between concurrent-read concurrent-write PRAM models. In *Proc. 13th Symposium on Mathematical Foundations of Computer Science, Lecture Notes in Computer Science*, volume 324, pages 231–239, 1988.

[16] S. Cook, C. Dwork, and R. Reischuk. Upper and lower bounds for parallel random access machines without simultaneous writes. *SIAM Journal on Computing*, 15, pages 87–97, 1986.

[17] S. Chaudhuri. Tight bounds for the chaining problem. In *Proc. 3rd Annual ACM Symposium on Parallel Algorithms and Architectures*, pages 62–70, 1991.

[18] S. Cook. Towards a complexity theory of synchronous parallel computation. *Enseign. Math.*, 27(2), pages 99–124, 1981.

[19] S. Cook. A taxonomy of problems with fast parallel algorithms. *Information and Control*, 64, pages 2–22, 1984.

[20] S. Cook and R. Reckow. Time bounded random access machines. *Journal of Computer and System Sciences*, 7(4), pages 354–375, 1973.

[21] A. Chandra, L. Stockmeyer, and U. Vishkin. Constant depth reducibility. *SIAM Journal on Computing*, 13, pages 423–439, 1984.

[22] P. Dymond, F. Fich, N. Nishimura, P. Ragde, and W.L. Ruzzo. Pointers versus arithmetic in PRAMs. Manuscript, 1992.

[23] M. Dietzfelbinger, M. Kutyłowski, and R. Reischuk. Exact time bounds for computing Boolean functions on PRAMs without simultaneous writes. In *Proc. 2nd Annual ACM Symposium on Parallel Algorithms and Architectures*, pages 125–137, 1990. To appear in *Journal of Computer and System Sciences*.

[24] P. Dymond and W.L. Ruzzo. Parallel RAMs with owned global memory and deterministic context-free language recognition. In *Proc. 13th International Colloquium on Automata, Languages, and Programming*, pages 95–104, 1986.

[25] D. Eckstein. Simultaneous memory access. Technical Report TR-79-6, Iowa State University, 1979.

[26] J.A. Edmonds. Lower bounds with smaller domain size on concurrent write parallel machines. In *Proc. 6th Annual Conference on Structures in Complexity Theory*, pages 322–333, 1991.

[27] D. Eppstein and Z. Galil. Parallel algorithmic techniques for combinatorial computing. *Ann. Rev. Comput. Sci.*, 3, pages 233–283, 1988.

[28] F. Fich, R. Impagliazzo, B. Kapron, V. King, and M. Kutyłowski. Limits on the power of parallel random access machines with weak forms of write conflict resolution. Manuscript, 1992.

[29] F. Fich, M. Li, P. Ragde, and Y. Yesha. Lower bounds for parallel random access machines with read only memory. *Information and Computation*, 83(2), pages 234–244, 1989.

[30] F. Fich, F. Meyer auf der Heide, P. Ragde, and A. Wigderson. One, two, three ... infinity: Lower bounds for parallel communication. In *Proc. 17th Annual ACM Symposium on Theory of Computing*, pages 48–58, 1985.

[31] F. Fich, F. Meyer auf der Heide, and A. Wigderson. Lower bounds for parallel random access machines with unbounded shared memory. In *Advances in Computing Research 4: Parallel and Distributed Computing*, F. Preparata, editor, pages 1–15. JAI Press Inc., Greenwich, Conn., 1987.

[32] F. Fich and V. Ramachandran. Lower bounds for parallel computation on linked structures. In *Proc. 2nd ACM Symposium on Parallel Algorithms and Architectures*, 1990.

[33] F. Fich, P. Ragde, and A. Wigderson. Relations between concurrent-write models of parallel computation. *SIAM Journal on Computing*, 17, pages 606–627, 1988.

[34] F. Fich, P. Ragde, and A. Wigderson. Simulations among concurrent-write PRAMs. *Algorithmica*, 3, pages 43–51, 1988.

[35] M. Furst, J.B. Saxe, and M. Sipser. Parity, circuits, and the polynomial-time hierarchy. *Math. Sys. Th.*, 17, pages 13–27, 1984.

[36] S. Fortune and J. Wyllie. Parallelism in random access machines. In *Proc. 10th Annual ACM Symposium on Theory of Computing*, pages 114–118, 1978.

[37] F. Fich and A. Wigderson. Towards understanding exclusive read. *SIAM Journal on Computing*, 19(4), pages 718–727, 1990.

[38] A. Gottlieb, R. Grishman, K. McAuliffe, C. Kruskal, L. Rudolph, and M. Snir. The NYU ultracomputer–designing an MIMD parallel machine. *IEEE Trans. Comput.*, 32, pages 175–189, 1983.

[39] J. Gil, Y. Matias, and U. Vishkin. Towards a theory of nearly constant time parallel algorithms. In *Proc. 32nd Annual IEEE Symposium on Foundations of Computer Science*, pages 698–710, 1991.

[40] E. Gafni, J. Naor, and P. Ragde. On separating the EREW and CROW models. *Theoretical Computer Science*, 68(3), pages 343–346, 1989.

[41] L. Goldschlager. A unified approach to models of synchronous parallel machines. *Journal of the ACM*, 29, pages 1073–1086, 1982.

[42] M. Goodrich. Intersecting line segments in parallel with an output-sensitive number of processors. In *Proc. ACM Symposium on Parallel Algorithms and Architectures*, pages 127–137, 1989.

[43] V. Grolmusz and P. Ragde. Incomparability in parallel computation. *Discrete Applied Mathematics*, 29(1), pages 63–78, 1990.

[44] R.L. Graham, B.L. Rothschild, and J.H. Spencer. *Ramsey Theory*. John Wiley, New York, 1980.

[45] T. Hagerup. Fast and optimal simulations between CRCW PRAMs. In *Proc. 9th Symposium on Theoretical Aspects of Computer Science*, pages 45–56, 1992.

[46] J. Hastad. *Computational Limitations for Small Depth Circuits*. MIT Press, Cambridge, Mass., 1987.

[47] J. Hastad. Almost optimal lower bounds for small depth circuits. In *Advances in Computing Research 5: Randomness and Computation*, S. Micali, editor, pages 143–170. JAI Press, Greenwich, Conn., 1989.

[48] J. Hartmanis and L. Hemachandra. One-way functions, robustness, and the non-isomorphism of NP-complete sets. In *Proc. 2nd Annual Conference on Structures in Complexity Theory*, pages 160–174, 1987.

[49] J. Hoover, M. Klawe, and N. Pippenger. Bounding fan-out in logical networks. *Journal of the ACM*, 31, pages 13–18, 1984.

[50] S. Hornick and F. Preparata. Deterministic P-RAM simulation with constant redundancy. In *Proc. ACM Symposium on Parallel Algorithms and Architectures*, pages 103–109, 1989.

[51] T. Hagerup and T. Radzik. Every robust CRCW PRAM can efficiently simulate a PRIORITY PRAM. In *Proc. 2nd Annual ACM Symposium on Parallel Algorithms and Architectures*, pages 117–124, 1990.

[52] R. Karp and V. Ramachandran. A survey of parallel algorithms for shared-memory machines. In *Handbook of Theoretical Computer Science*, volume A, J. van Leeuwen, editor, pages 869–941. MIT Press, 1990.

[53] C. Kruskal, L. Rudolph, and M. Snir. A complexity theory of efficient parallel algorithms. *Theoretical Computer Science*, 71(1), pages 95–132, 1990.

[54] C. Kruskal. Algorithms for replace-add based paracomputers. In *Proc. International Conference on Parallel Processing*, pages 219–223, 1982.

[55] L. Kučera. Parallel computation and conflicts in memory access. *Information Processing Letters*, 14(2), pages 93–96, 1982.

[56] G. Lev, N. Pippenger, and L.G. Valiant. A fast parallel algorithm for routing in permutation networks. *IEEE Trans. Comput.*, C-30, pages 93–100, 1981.

[57] M. Li and Y. Yesha. New lower bounds for parallel computation. *Journal of the ACM*, 36(3), pages 671–680, 1989.

[58] F. Meyer auf der Heide and R. Reischuk. On the limits to speed up parallel machines by large hardware and unbounded communication. In *Proc. 25th Annual IEEE Symposium on Foundations of Computer Science*, pages 56–64, 1984.

[59] K. Mehlhorn and U. Vishkin. Randomized and deterministic simulation of prams by parallel machines with restricted granularity of parallel memories. *Acta Informatica*, 21, pages 339–374, 1984.

[60] F. Meyer auf der Heide and A. Wigderson. The complexity of parallel sorting. *SIAM Journal on Computing*, 16(1), pages 100–107, 1987.

[61] N. Nisan. CREW PRAMS and decision trees. In *Proc. 21st Annual ACM Symposium on Theory of Computing*, pages 327–335, 1989.

[62] D. Nassimi and S. Sahni. Data broadcasting in SIMD computers. *IEEE Trans. Comput.*, C-30, pages 101–107, 1981.

[63] N. Nisan and M. Szegedy. On the degree of Boolean functions as real polynomials. In *Proc. 24th Annual ACM Symposium on Theory of Computing*, 1992.

[64] I. Parberry and P. Yan. Improved upper and lower time bounds for parallel random access machines without simultaneous writes. *SIAM Journal on Computing*, 20, pages 88–99, 1991.

[65] P. Ragde. Manuscript.

[66] P. Ragde. Processor-time tradeoffs in PRAM simulations. *Journal of Computer and System Sciences*, 44(1), pages 103–113, 1992.

[67] A. Ranade. How to emulate shared memory. In *Proc. 28th Annual IEEE Symposium on Foundations of Computer Science*, pages 185–194, 1987.

[68] R. Reischuk. Simultaneous writes of PRAMs do not help to compute simple arithmetic functions. *Journal of the ACM*, 34, pages 163–178, 1987.

[69] P. Ragde, W. Steiger, E. Szemeredi, and A. Wigderson. The parallel complexity of element distinctness is $\omega((\log n)^{1/2})$. *SIAM Journal of Discrete Mathematics*, 1, pages 399–410, 1988.

[70] D. Rubinstein. Personal communication.

[71] W.L. Ruzzo. Personal communication, cited in [8, 4].

[72] J. T. Schwartz. Ultracomputers. *ACM Trans. Programming Lang. Systems*, 2, pages 484–521, 1980.

[73] H. Simon. A tight $\omega(\log \log n)$ bound on the time for parallel RAM's to compute nondegenerate Boolean functions. *Information and Control*, 55, pages 102–107, 1982.

[74] Burton Smith. The Tera computer system. In *Proc. ACM International Conference on Supercomputing*, pages 1–7, 1990.

[75] R. Smolensky. Algebraic methods in the theory of lower bounds for Boolean circuit complexity. In *Proc. 19th Annual ACM Symposium on Theory of Computing*, pages 77–82, 1987.

[76] M. Snir. Personal communication.

[77] M. Snir. On parallel searching. *SIAM Journal on Computing*, 14(3), pages 688–708, 1985.

[78] Y. Shiloach and U. Vishkin. Finding the maximum, merging, and sorting in a parallel computation model. *Journal of Algorithms*, 2, pages 88–102, 1981.

[79] L. Stockmeyer and U. Vishkin. Simulation of parallel random access machines by circuits. *SIAM Journal on Computing*, 13, pages 404–422, 1984.

[80] S. Skyum and L. Valiant. A complexity theory based on Boolean algebra. *Journal of the ACM*, 32, pages 484–502, 1985.

[81] B. Schieber and U. Vishkin. Finding all nearest neighbors for convex polygons in parallel: A new lower bound technique and a matching algorithm. *Discrete Applied Mathematics*, 29(1), pages 97–112, 1990.

[82] G. Szegedy. *Algebraic Methods in Lower Bounds for Computational Models with Limited Communication.* PhD thesis, University of Chicago, 1989.

[83] G. Tardos. Query complexity, or why is it difficult to separate $NP^A \cap co\text{-}NP^A$ from P^A by a random oracle A? *Combinatorica*, 9(4), pages 385–392, 1989.

[84] J. Trahan, M.C. Loui, and V. Ramachandran. Multiplication, division, and shift instruction in parallel random access machines. In *Proc. 22nd Conf. Inf. Sci. Syst.*, pages 126–130, 1988.

[85] J. Trahan. *Instruction Sets for Parallel Random Access Machines.* PhD thesis, University of Illinois, Urbana-Champaign, 1988.

[86] E. Upfal and A. Wigderson. How to share memory in a distributed system. *Journal of the ACM*, 34, pages 116–127, 1987.

[87] U. Vishkin. Implementation of simultaneous memory access in models that forbid it. *Journal of Algorithms*, 4, pages 45–50, 1983.

[88] U. Vishkin and A. Wigderson. Trade-offs between depth and width in parallel computation. *SIAM Journal on Computing*, 14, pages 303–314, 1985.

[89] I. Wegener and L. Zádori. A note on the relations between critical and sensitive complexity. Technical Report 256, Universitat Dortmund, 1988.

[90] A. Yao. Probabilistic computations: Toward a unified measure of complexity. In *Proc. 18th Annual IEEE Symposium on Foundations of Computer Science*, pages 222–227, 1977.

21

Polynomial Completeness and Parallel Computation

Raymond Greenlaw

Department of Computer Science
University of New Hampshire
Durham, New Hampshire 03824
greenlaw@cs.unh.edu

21.1
Introduction

Parallel computers represent some of the most powerful computing devices available today. As expected, many computational problems can be solved much faster on a parallel computer than on a single processor machine. If a sequential algorithm runs in time T on a sequential computer then by applying p processors to the problem one might hope to achieve a speed up by a full factor of p. If this were possible then the running time for the problem would be reduced to T/p. To achieve this running time, the problem would have to be totally parallelizable. Although such dramatic speed ups do not always seem to be possible, the algorithms presented in the previous chapters indicate that many problems benefit significantly from using parallelism.

The question arises as to whether all problems can be solved "substantially" faster on a parallel machine. By substantially, it is meant achieving an exponential speed up using only a polynomial number of processors. For example, this would mean converting a sequential algorithm that runs in polynomial time to a poly-log, $\log^k n$ for a constant k, parallel algorithm that uses a polynomial number of processors. The wide variety of algorithms presented up to this point may seem to suggest that all problems can be solved substantially faster by parallel machines.

Despite all of the available parallel techniques and their wide applications, it seems that not all problems can be solved substantially faster by parallel machines. In fact, many practical and important problems do not appear to have solutions that are computable efficiently by parallel machines. A theory has been developed called *polynomial completeness theory* that is useful for proving certain problems do not adapt well to parallelism. This chapter describes the important aspects of this theory, presents the techniques employed by the theory, discusses some problems that are highly sequential in nature and describes a framework for analyzing the parallel complexity of an algorithm.

There are several major motivations for studying P-completeness theory. Since it is difficult to prove lower bounds in complexity theory, it is useful to classify problems as being at least of comparable difficulty to other problems. The classic example of such a completeness theory is the theory of *NP*-completeness. If a problem is *NP*-complete then it is very likely that the problem is intractable, i.e., there is no deterministic polynomial time algorithm solving the problem. Similarly, if a problem is P-complete then it is very likely that there is no poly-log time parallel algorithm solving the problem while using only a polynomial number of processors. Thus, showing a problem is P-complete provides strong evidence that the problem is highly sequential. The reductions used in P-completeness theory are similar in flavor to those of NP-completeness theory.

In terms of practical benefits, P-completeness results have applications in designing parallel algorithms. As an example suppose in designing a parallel algorithm a procedure is needed to compute level numbers in a graph. A natural algorithm for computing level numbers in a graph is a breadth first search. One natural implementation of a breadth first search might use a queue as an underlying data structure, whereas, a second natural implementation might use a stack. It turns out that the queue based algorithm can be implemented very efficiently in parallel although the stack based algorithm is highly sequential. This indicates the design of the subroutine should incorporate a queue and not a stack.

In general, the goals of P-completeness theory are to advance the understanding of parallel computation—to figure out why certain problems do not adapt well to parallelism. By classifying these problems certain commonalities among their properties can be identified as being the cause for their highly sequential nature. Such constructs can then be avoided by programmers. The process of studying P-complete problems can also lead to the development of new algorithmic approaches that may lead to fast parallel algorithms. Finally, if a researcher knows a problem is P-complete then they realize the problem is unlikely to have a poly-log time parallel solution using only a polynomial number of processors. Therefore, effort can be focused on finding an alternative solution that does not involve a P-complete problem or on developing a fast parallel approximation to the problem, instead of devoting large amounts of time to finding a poly-log parallel algorithm that very likely does not exist.

21.2
The Complexity Classes P and NC

In this chapter the primary interest is in categorizing the problems in P as being either in the complexity class NC or being P-complete. NC represents the group of problems that can be solved very fast in parallel using a polynomial number of processors. P represents the class of problems that are tractable for sequential machines. The class of P-complete problems is comprised of problems having a highly sequential nature. Some background material is presented, and then the classes NC and P are defined. The class of P-complete problems is defined in the next section.

An *alphabet* is a finite set of symbols. A *language* over an alphabet is a set of strings consisting only of symbols from the alphabet. A *decision problem* Π is defined to consist of a set of *instances* D_Π and a set $Y_\Pi \subseteq D_\Pi$ of *yes instances*. A decision problem either has a *yes* or *no* answer.

An example of an alphabet is $\{0, 1\}$. An example of a language over this alphabet is

$$L = \{0, 00, 000, \ldots\} \cup \{1, 11, 111, \ldots\}.$$

L is made up of strings containing either all 0's or all 1's. An example of a decision problem is the *clique problem*.

DEFINITION 21.1
Clique Problem:
Instance: An undirected graph G and an integer k.
Problem: Is there a **clique**, *a subset of completely connected vertices of G, that has at least k vertices?*

Figure 21.1 depicts a *yes* instance of the clique problem for the case where $k = 3$. That is, the graph shown in Figure 21.1 has a clique of size 3. Namely, the clique consisting of vertices b, c, and d. The example also serves as a *no* instance for the case where $k = 4$, since there is no clique of size 4 in the graph.

A decision problem is a useful concept that can be used to provide valuable information about algorithms solving specific problems. Frequently, decision problems are based on specific algorithms. There may be several natural decision problems for a given algorithm. As an example, consider

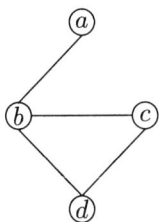

FIGURE 21.1
A *yes* instance of the clique problem where $k = 3$.

the *maximum flow problem*—a natural decision problem defined for general networks. First, the concepts of networks and flow functions are defined.

DEFINITION 21.2
A **network** *is comprised of the following three items:*

1. *A finite directed graph $G = (V, E)$.*
2. *A vertex s called the* **source** *and a vertex t called the* **sink**.
3. *A capacity function $c : E \to \mathcal{Z}^+$. For $e \in E$, $c(e)$ is called the* **capacity** *of edge e.*

DEFINITION 21.3
A function $f : E \to \mathcal{R}$ is called a **flow function** *(or a* **feasible flow function***) if the following two conditions are met:*

1. *For each $e \in E$, $0 \le f(e) \le c(e)$.*
2. *Let in(e) (out(e)) denote edges going into (out of) e. For every $v \in V - \{s, t\}$*

$$0 = \sum_{e \in in(v)} f(e) - \sum_{e \in out(v)} f(e). \qquad (21.1)$$

The **total flow** *F of f is defined by*

$$F = \sum_{e \in in(t)} f(e) - \sum_{e \in out(t)} f(e). \qquad (21.2)$$

The **maximum flow** *is the maximum value over all flow functions of the total flow.*

The first condition in the definition of the flow function ensures that each arc can only transport its capacity; the second condition enforces the *Law of Conservation of Flow*. An example of a flow network is shown in Figure 21.2. The arcs are labeled with their capacities. You can think of inserting fluid at the source s and having it flow downward towards the sink t. The maximum flow problem is defined below.

DEFINITION 21.4
Maximum Flow Problem (MF):
Instance: A directed graph G with non-negative integer edge labels.
Problem: Is the i-th bit of the value of the maximum flow from source s to sink t in G a 1?

906 Chapter 21. Polynomial Completeness and Parallel Computation

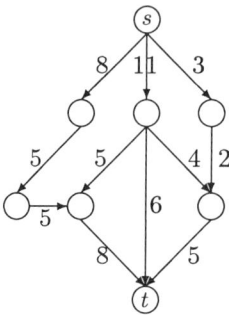

FIGURE 21.2
A sample flow network with maximum flow 18.

Notice, any algorithm designed to compute a maximum flow must also compute the solution to MF, or it at least can easily be converted to an algorithm that solves the decision problem stated in MF. Simply write down the value of the flow computed and check the i-th bit against 1. A general decision problem of the nature of MF seems to say something about any algorithm designed to compute a maximum flow. This is partly what motivates interest in decision problems. An example of another natural decision problem for maximum flow would be to check if the value of the maximum flow was even.

The maximum flow for the network shown in Figure 21.2 is 18. Represented in binary this is 10010. An instance of MF for this graph asking whether or not the third bit is 0 would be a no instance of MF. An instance using the alternative decision problem for MF and asking whether the flow was even would be a yes instance.

The classes P and NC are defined below in terms of language recognition questions. Shortly, language recognition questions are related to decision problems.

DEFINITION 21.5
The class P is the set of all languages (subsets of $\{0,1\}^*$) that can be recognized by deterministic Turing machines that run in polynomial time.

The class P remains unchanged for all of the standard models of sequential computation. For example, P could be defined in terms of random access machines. Running times of algorithms for recognizing the same language will vary over different models but only by a polynomial factor. To emphasize our focus on parallel computation it is convenient to view the languages in P as being solvable in polynomial time using a polynomial number of processors.

LEMMA 21.1
Let $P_\|$ be the set of all languages that can be recognized by a PRAM that uses $n^{O(1)}$ processors and $n^{O(1)}$ time. Then $P = P_\|$.

The proof of the lemma is left as an exercise. We have left the particular version of the PRAM in Theorem 21.1 unspecified. The theorem holds for any of the standard versions of the PRAM such as the EREW, CREW or CRCW. This alternative view of P is useful in the context of parallel computation.

NC was originally defined to represent those languages recognizable quickly in parallel using a feasible amount of hardware. A polynomial number of processors is generally considered feasible, whereas, an exponential number of processors is infeasible.

DEFINITION 21.6
The class NC is the set of all languages that can be recognized by a PRAM using $n^{O(1)}$ processors and $O(\log^k n)$ time, where k is a constant.

The class NC remains the same when defined using other standard models of parallel computation assuming the same resource bounds are imposed on hardware and time. This is in part what makes NC an appealing complexity class; it is robust with respect to many different models of computation. For example, the class has a similar definition using the Boolean circuit model described in Section 21.4. NC is an important theoretical class with emphasis on maximum parallelism. Languages in NC that have small exponents on the resource bounds are thought to have fast, feasible, parallel solutions. Lemma 21.1 makes it clear that NC is a subset of P. Whether P is a subset of NC is one of the major open questions in complexity theory.

Both P and NC were defined in terms of language recognition questions in order to be as general as possible. Usually, we will be interested in studying decision problems—problems with yes or no solutions. A decision problem can easily be encoded by a string and reformulated as a language recognition question. Such a string is in the language if and only if it is a yes instance of the decision problem. Frequently, the word *problem* is used as a shorthand for decision problem. We do not express all problems as language recognition questions in order to simplify the discussion. This allows us to avoid specifying specific encodings.

In general, there are many reasonable possible encodings for a problem. The main requirement for a reasonable encoding is that it is "compact." For example, an integer N is represented using $O(\log |N|)$ bits and not $O(N)$ bits. Reasonable encodings can be translated into one another and vice versa using a log space Turing machine. For example, in representing a graph one could use either edge lists or adjacency lists. The conversion of either of these forms to the other can be done in a straightforward way by a Turing machine that uses only log space. Since the specification of the particular codings used is not critical, we are somewhat informal and neglect to give the exact details for all the languages described. Languages in P are also referred to as problems when it is clear what the corresponding language recognition question is. Thus, the terms language and problem will be used interchangeably.

There is a group of problems in P that appear to be more difficult to solve than all other problems in P but which curiously are of the same relative difficulty. These problems are called *complete* problems for P. They are called complete because if a method for finding a solution to any one of them existed, then a polynomial method for solving any other problem in P could be derived from it. The cost incurred in solving the problem would be the expense of converting one problem to the other problem plus the cost of solving the new instance. The group of all complete problems in P is collectively called the class of *P-complete* problems. The next section expands on the ideas presented in this paragraph, and formal definitions of these concepts are given.

21.3
Reducibility and *P*-completeness

Polynomial completeness theory involves identifying the hardest problems in P—the P-complete problems. To identify the computationally difficult languages in P a method of relating one language to another is required. A type of reduction called a *log space reduction* is defined that is useful for doing this.

DEFINITION 21.7
Let L be a language over alphabet Σ and L' be a language over alphabet Σ'. A language L is **log space reducible** *to a language L' if there exists a function $f : \Sigma \to \Sigma'$ such that $x \in L$ if and only if $f(x) \in L'$, and f can be computed by a deterministic Turing machine that uses only log space.*

The definition can easily be modified to incorporate decision problems. For example, if A and B are decision problems then $f : D_A \to D_B$, and

$x \in D_A$ if and only if $f(x) \in D_B$. Intuitively, the definition means problem A is no harder to solve than problem B. Additionally, an answer to an instance of problem A can be computed by solving the equivalent version of problem B that the log space Turing machine created. Log space reducibility for P is analogous to polynomial time reducibility for NP. In fact, most of the reductions in NP-completeness proofs are actually log space reductions. There are other useful notions of reducibility for relating problems in P to one another although we will not pursue those further. We can now present the definition for a P-complete language.

DEFINITION 21.8
A language L is **P-complete under log space reducibility** if

1. $L \in P$,
2. For all other languages $L' \in P$, L' is log space reducible to L.

Figure 21.3 shows the structure of P with the hardest problems appearing at the top of the figure. Although the class P is drawn as three separate segments, it is possible that P equals NC. This open question is similar to the P versus NP question. A breakthrough in complexity theory is required to prove the result one way or the other. Most researchers currently believe that P does not equal NC. The following theorem constitutes additional support to the school of thought that P is different from NC.

THEOREM 21.1
Let L be a P-complete language. If L is in NC then NC equals P.

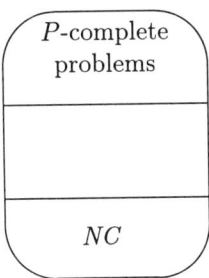

FIGURE 21.3
The structure of P with the hardest problems appearing at the top of the figure.

PROOF

Clearly, $NC \subseteq P$ based on Lemma 21.1. We need to show $P \subseteq NC$ under the assumptions given in the theorem. Let L' be any language in P. Since L is P-complete there is a log space reduction f between L' and L. It is known that a log space reduction can be performed in NC. Thus, f can be computed in NC. Let x be a string in L'. Using the NC algorithm for recognizing L and the reduction f, we can determine if $x \in L'$ in NC by checking if $f(x) \in L$. Note, since f is a log space reduction, $f(x)$ can only be polynomially longer than x. This shows L' is in NC. Since L' was an arbitrary language in P, this proves $P \subseteq NC$. ∎

There are many problems that have resisted attempts at being parallelized efficiently. In fact, many of these problems have been proven P-complete. In light of Theorem 21.1, it seems unlikely that NC equals P, since this result would imply that all of the P-complete problems do indeed have fast parallel solutions. Researchers would like to believe that if all of these problems had fast parallel solutions then someone would have discovered one by now. It is in this sense that the P-complete problems are called highly sequential or inherently sequential problems. That is, they do not appear to have NC algorithms.

To prove a problem A is P-complete one must first prove that the problem is in P. This is done by presenting an algorithm for A and showing that the algorithm has polynomial running time. In a P-completeness proof this is usually the easy step although there are exceptions. Linear programming is one notable example. The second step is to provide a log space reduction showing all other problems in P reduce to A. The first P-completeness proof, proving the *path systems problem* was P-complete, involved a direct simulation of an arbitrary polynomial time Turing machine. Fortunately, many problems have been proven P-complete and it suffices to reduce one of these to A since log space reductions are transitive. This helps avoid generic reductions. The final step in a P-completeness proof is to prove that the reduction is correct and requires only log space. Frequently, the proof of correctness for the reduction is omitted when it is clear how the reduction works. The reader who is familiar with NP-completeness theory will recognize these three steps. Figure 21.4 outlines these steps.

A wide variety of problems have been proven P-complete from areas such as circuit complexity, graph theory, combinatorial optimization, networking, graph searching, logic, formal language theory, and algebra. Having a considerable number of P-complete problems available greatly simplifies finding reductions to show additional problems are P-complete. This is because one

> **Steps in a P-completeness proof.**
>
> 1. Provide a polynomial time algorithm demonstrating the problem is in P.
> 2. Provide a log space reduction from an existing P-complete problem.
> 3. Prove the reduction is correct and requires only log space.

FIGURE 21.4
Outline of the steps in a P-completeness proof.

no longer needs to provide a direct simulation of a Turing machine but only a reduction from one of the existing P-complete problems. In later sections techniques are developed that allow for selecting the appropriate P-complete problem to reduce to and for finding the appropriate log space reduction. Creating a basis of complete problems can help to simplify the complexity of P-completeness reductions. This is because with a large group of problems to relate to it is more likely that a problem of a similar flavor can be found. In the next section a reduction that involves a direct simulation of a Turing machine by a family of Boolean circuits is given. This proof illustrates a generic P-completeness proof.

21.4 The Circuit Value Problem Is P-complete

In this section the most fundamental P-complete problem, the *circuit value problem*, is described. One motivation for researching this problem is that computers are currently built using circuits. Thus, studying methods for evaluating circuits is an extremely important task. An additional motivation is that circuits are mathematically easy to work with. They do not have instructions in the same sense as a PRAM and so seem simpler conceptually. The "program" for a circuit is the circuit itself. There are equivalences between circuits and many other models of parallel computation. For example, equivalences exist for the PRAM model when the appropriate resources are compared. Thus, results about circuits can easily be compared with those for other models. Definitions for the Boolean circuit model are presented below.

Let $B_k = \{f \mid f : \{0,1\}^k \to \{0,1\}\}$ denote the set of all k-ary Boolean functions. Examples of such functions are "AND," "OR," and "NOT." For example, if f represents the AND function on two variables then $f(0,0) = f(0,1) = f(1,0) = 0$, and $f(1,1) = 1$.

DEFINITION 21.9
A **circuit** α is a labeled finite directed acyclic graph in which each node v has a type $\tau(v) \in \{x_1, \ldots, x_g\} \cup \{y_1, \ldots, y_h\} \cup B_0 \cup B_1 \cup B_2$. A node v with $\tau(v) = x_i$ must have indegree 0 and is called an **input** node. A node v with $\tau(v) = y_i$ must have indegree 1 and outdegree 0, and is called an **output** node. For each $z \in \{x_1, \ldots, x_g\} \cup \{y_1, \ldots, y_h\}$ there is a unique vertex v with label z. A node v with $\tau(v) \in B_i$ is called a **gate** and must have indegree i. There is exactly one edge into v for each argument of the function $\tau(v)$. When the input variables x_i are assigned values from $\{0,1\}$, every node v assumes the unique value from $\{0,1\}$ given by applying $\tau(v)$ to the values on the edges into v.

Gates of type B_0 have indegree 0. They are often called constant inputs since they always have a value of either 0 or 1. A circuit α computes a function $f : \{0,1\}^g \to \{0,1\}^h$ in the following way: The inputs to α are propagated to the gates occurring at the first level of α and these gates are evaluated. When the value of a gate has been computed, the gate's value is propagated along its output edges. In general, gates compute their outputs when all of their inputs are available. Eventually, the output nodes of the circuit are evaluated. This propagation of values through the levels of a circuit seems to be a highly sequential process. In fact, this computation process can be classified as inherently sequential as shown later in this section.

Figure 21.5 depicts an example of a simple circuit. The circuit has two AND gates and an OR gate. The AND gates are numbered 5 and 7, and the OR gate is numbered 6. Suppose $x_1 = x_2 = x_3 = 1$ and $x_4 = 0$. With these inputs gate 5 evaluates to 1, gate 6 evaluates to 1, and gate 7 evaluates to 1. Thus, for the specified input the circuit's outputs y_1 and y_2 are both 1.

The resource bounds of interest for a circuit are *size* and *depth*.

DEFINITION 21.10
The **size** of a circuit α, denoted $c(\alpha)$, is the number of nodes in α. The **depth** of a circuit α, denoted $d(\alpha)$, is the length of (number of edges in) the longest path in α from an input node to an output node.

21.4. The Circuit Value Problem Is P-complete

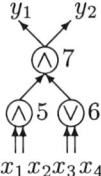

FIGURE 21.5
Example of a Boolean circuit.

In Figure 21.5 the sample circuit has size 9 and depth 3. The circuits of interest in this chapter are those having polynomial size. Each circuit α can be viewed as an abstract parallel machine that is described by a string. A convenient way of coding this string precisely is presented below.

DEFINITION 21.11
The **standard encoding** *$\bar{\alpha}$ of a circuit α is a string in $\{0,1\}^*$ grouped into 4-tuples (v, g, l, r), one tuple for each node of α. Node number v is a g-gate, where $g \in \{x_i, y_i\} \cup B_0 \cup B_1 \cup B_2$. The left (right) input to v, if any, is numbered l (r); the nodes of α are uniquely (but arbitrarily) numbered in the range $1, \ldots, c(\alpha)^{O(1)}$. Each tuple can be encoded in binary as a string of $O(\log c(\alpha))$ bits. The order of the tuples is arbitrary with the exception that the tuples corresponding to the input x_i's must be in order at the beginning of $\bar{\alpha}$, and the tuples corresponding to the output y_i's must be in order at the end of $\bar{\alpha}$. The length of $\bar{\alpha}$ is $O(c(\alpha) \log c(\alpha))$.*

Using the standard encoding, the circuit presented in Figure 21.5 could be encoded as follows: (1,in,-,-), (2,in,-,-), (3,in,-,-), (4,in,-,-), (6,∨,3,4), (5,∧,1,2), (7,∧,5,6), (8,out,-,-), (9,out,-,-), assuming the numbering of gates given in the figure, and that y_1 is numbered 8 and y_2 is numbered 9.

An individual circuit is a computational object computing a function from strings of a fixed length to strings of a fixed length. In Figure 21.5 inputs must be of length four and outputs are of length two. In order to speak of general functions over strings of varying lengths one can introduce the notion of a circuit family. This is just a collection of circuits with one circuit for each input size.

DEFINITION 21.12
A **circuit family** *$\langle \alpha_n \rangle$ is a collection of circuits. The n-th circuit α_n computes a function $f^n : \{0,1\}^n \to \{0,1\}^m$.*

Our focus will be on circuits as language acceptors so in general the circuits considered have only one output. Thus, the output will be either 0 or 1 and m will have a value of 1 in the definition of circuit family. In defining a circuit family a problem arises since it may be possible that circuits of one input size are structurally unrelated to circuits of other input sizes. Such *non-uniform* circuit families are known to be very powerful, i.e., they can compute non-recursive functions. We would like to be able to study a more restricted class of circuits. To do this a notion called *uniformity* is introduced.

DEFINITION 21.13
A family $\langle \alpha_n \rangle$ of circuits is **log space uniform** *if the transformation $1^n \to \bar{\alpha}_n$ can be computed in $O(\log c(\alpha_n))$ space on a deterministic Turing machine.*

Uniformity conditions reflect a practical consideration since non-uniform circuit families can be very difficult to construct. An important theoretical decision problem based on circuits is defined below.

DEFINITION 21.14
Circuit Value Problem (CVP):
Instance: An encoding $\bar{\alpha}$ of a Boolean circuit α and values for its inputs.
Problem: Does α evaluate to 1 on its input?

The method of encoding a circuit in an instance of CVP is traditionally done by using the standard encoding, although any other reasonable encoding suffices. With the inputs $x_1 = x_2 = x_3 = 1$ and $x_4 = 0$, the instance of CVP shown in Figure 21.5 represents a yes instance.

In general, the gates of α can be any Boolean function. It is convenient to focus on a small subset of these functions that can simulate all other functions. Traditionally, researchers have concentrated on NOT, OR, and AND gates. Thus, when CVP is referred to in general, we usually have in mind that the instance is comprised of these gate types, unless otherwise specified.

An interesting computational question to ask is whether there is a correspondence between the sequential computation of a Turing machine and the parallel computation of a circuit. The circuit value problem can be used to establish a relationship between polynomial time bounded Turing machines and polynomial size circuits. In fact, it turns out that CVP is P-complete.

To show CVP is P-complete requires showing CVP is in P, and that each language L in P is log space reducible to CVP. It is quite easy to see that given $\bar{\alpha}$ and the values of α inputs, one can compute the value of each gate in α in a polynomial (in the size of α) number of steps. The *topological circuit evaluation algorithm* described below evaluates the gates of a circuit in polynomial time. The algorithm is called a topological algorithm because it evaluates the gates level by level starting at the inputs and proceeding towards the outputs. Note, we assume the gates are topologically ordered. This in not necessary since they could be topologically ordered easily in polynomial time. It is easy to see the algorithm runs in polynomial time, so Lemma 21.2 is established.

ALGORITHM 21.1
Topological Circuit Evaluation
Input: A circuit C coded using the standard encoding with gates numbered in topological order $g_1, \ldots, g_n$.
Output: The value of gate g_n.

begin

    if the input is not a proper instance of CVP **then** HALT;

    (g_{iL} is the left input to gate i and g_{iR} is the right input to gate i.)
    for $i := 1$ **to** n **do** $g_i := g_{iL}$ $type(g_i)$ g_{iR};

    print g_n

end.

LEMMA 21.2
The circuit value problem is in P.

The more difficult step in proving CVP P-complete is showing there is a log space reduction between every language in P and CVP. This is proved by simulating an arbitrary Turing machine with a family of circuits. A special *normal form* for Turing machines is used in order to simplify the proof. For each language L in P, there is a 1-tape Turing machine M_L that on input $x = x_1, \ldots, x_n$ halts in time bound $t(n) = n^{O(1)}$ with output y equal to 1 if and only if x is in L. The lemma proved below shows how to simulate such a machine by a family of circuits.

LEMMA 21.3
Let M be an arbitrary 1-tape deterministic Turing machine with time bound $t(n) = n^{O(1)}$ operating on input $x_1, \ldots, x_n$. There exists a log space uniform circuit family $\langle \beta_n \rangle$ such that the nth circuit β_n simulates M on inputs of length n.

PROOF

Assume M has its input in adjacent positions on the input tape with x_1 at the initial position. Without loss of generality, assume M never moves left of the initial position of its input head and halts with its head back at the initial position with the output y in the initial position. Consider an input to M of length n. Since M runs in time $t(n)$, it can use at most $t(n)$ tape squares. A circuit β_n is constructed that simulates M on the first $t(n)$ tape squares.

The circuit simulating M consists of a $t(n) \times t(n)$ array of identical subcircuits connected in a regular way as follows. The subcircuits are numbered (i,j), where $i,j \in \{1,\ldots,t(n)\}$. Subcircuit (i,j) simulates tape square j at time step i. Associated with each tape square are two quantities—contents and state. The subcircuits keep track of these values. Detailed descriptions of the subcircuits are not given; however, it is easy to construct the subcircuits from NOT, OR, and AND gates. The contents indicates the value of the tape square. The state of a tape square specifies whether the head of M is currently positioned over the tape square. If the head is positioned over the tape square, then the state also specifies the current internal state of M.

The inputs to each subcircuit are listed below.

- v_{in}—the value of the associated tape square at the beginning of the time step.
- v_{-1}, v_0, v_1—the states of the associated tape squares above to the left, directly above, and above to the right at the beginning of the time step.

There are two outputs generated by each subcircuit.

- v_{out}—the value of the tape square at the end of the time step.
- s_{out}—a constant number of bits specifying the state of the tape square at the end of this time step. This value may get fanned out to several subcircuits.

Figure 21.6 depicts the (i,j)-th subcircuit. The double lines indicate the transmission of a constant number of bits. The four inputs and two outputs to each circuit are shown in the figure. The outputs may be fanned out to several other subcircuits. Each subcircuit (i,j) performs as described below based on the transition function of M.

21.4. The Circuit Value Problem Is P-complete

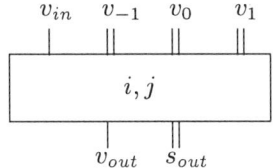

FIGURE 21.6
The (i,j)-th subcircuit.

1. If the head position is currently over the j-th tape square, then the j-th subcircuit computes the next state of M. If movement of the head is required the current state and control are passed to either the left or right neighbor subcircuit just below it—to subcircuit numbered $(i+1, j-1)$ or $(i+1, j+1)$. If there is no head movement the state information is passed directly to subcircuit j's successor—subcircuit $(i+1, j)$. In either of these cases subcircuit j passes its current value to subcircuit $(i+1, j)$.
2. If the head position is not over the j-th cell, then the subcircuit (i,j) passes its current value to subcircuit $(i+1, j)$.

The inputs to the first row of subcircuits are x_1, q_0 to the first subcircuit, $x_2, \ldots, x_n$ for the next $n-1$ subcircuits, and 0 for the remaining ones. A section of the overall subcircuit is shown in Figure 21.7. The wires for v_{in} and v_{out} consist of single bits, whereas the wires for the other quantities carry a constant number of bits. The circuit simulates the behavior of M for $t(n)$ steps with subcircuits $(t(n), 1), \ldots, (t(n), t(n))$ containing the final values of the tape squares used by M possibly padded with 0's.

At this point it should be clear how the circuit family simulates M. We omit the proof of correctness of the reduction. The final step in the proof is to establish that $\bar{\beta}_n$ is computable in log space. It is easy to see that the following items can be computed by a log space Turing machine.

1. The value of $t(n)$. Note, since $t(n) = n^{O(1)}$ there is a constant k such that $t(n) \leq n^k$ for sufficiently large n. The value of k can be hard-wired into the finite control of the log space Turing machine.

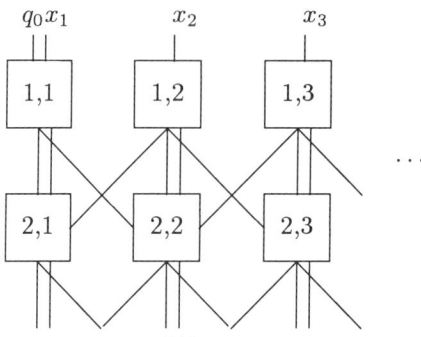

FIGURE 21.7
A section of the i, j array used to simulate a Turing machine. The v_{-1} value for cell (2,2) comes from cell (1,1); the v_0 value from cell (1,2); and the v_1 value from cell (1,3).

   2. The index pairs for the subcircuits and the connection pattern between subcircuits.
   3. The description of each subcircuit based on the transition function of M.

This shows $\langle \beta_n \rangle$ is a log space uniform circuit family. The details for the proof of correctness and of log space reducibility are left as an exercise for the reader. This completes the proof of Lemma 21.3. ∎

Note, the overall circuit constructed in Lemma 21.3 is comprised of NOT, OR, and AND gates.

THEOREM 21.2
The circuit value problem (CVP) is P-complete under log space reducibility.

PROOF
Lemmas 21.2 and 21.3 combined prove the result. ∎

Theorem 21.2 indicates CVP is P-complete. This means CVP is very likely to be inherently sequential, unless NC equals P. What is the intuition as to why CVP is so sequential? The values in a circuit have to be propagated layer by layer so that gates at depth d in the circuit require time d before they can be evaluated. If a separate processor is assigned to each gate on a given level, then the outputs from a given level can be computed in constant time; however, parallel computation of general circuits seems to require time proportional to the depth of the circuit. Unfortunately, there appears to be no technique for reducing the depth of a general circuit.

The circuit value problem is called the fundamental P-complete problem because it is used frequently in P-completeness proofs, was one of the first P-complete problems and has been one of the most widely studied P-complete problems. Its flexibility adapts well for showing other problems are P-complete. CVP plays the same role in P-completeness theory that *satisfiability* does in NP-completeness theory. Just as many versions of satisfiability are NP-complete, there are many variations on the circuit value problem that are also P-complete. Several of these, including the monotone circuit value problem, the planar circuit value problem, and the NAND circuit value problem, are described in the next section. These results represent a basis of P-complete problems useful for proving other problems are P-complete.

21.5
Variations of the Circuit Value Problem

A wide variety of instances of CVP are described in this section. Since the problems are all restricted versions of CVP, they are all in P. Reductions are given proving the problems are P-complete. The group of P-complete instances of CVP presented consists of those most frequently used in proving other problems are P-complete. It is enlightening to compare the various instances of CVP along with how additional restrictions impact the reductions proving they are P-complete.

The first variant of CVP described is the *monotone* version. A circuit is *monotone* if it consists of only AND and OR gates (technically, if it has only monotone gates). There are no NOT gates allowed in the circuit. In addition, to the usual inputs to an instance of CVP we also provide their complements.

DEFINITION 21.15
Monotone Circuit Value Problem (MCVP):
Instance: An encoding of a monotone Boolean circuit α, plus inputs $x_1, \ldots, x_n$, and their complements $\bar{x}_1, \ldots, \bar{x}_n$.
Problem: Does α on input $x_1, \ldots, x_n, \bar{x}_1, \ldots, \bar{x}_n$ output 1?

THEOREM 21.3
The monotone circuit value problem (MCVP) is P-complete under log space reducibility.

PROOF

An instance of general CVP is reduced to an instance of MCVP. Suppose an instance of CVP $\bar{\alpha}$ with inputs $x_1, \ldots, x_n$ is given. For convenience this instance is referred to as α. An instance of MCVP β is contructed from α such that β is true on its input if and only if α is true on its input. The idea behind the reduction is to replace each gate v_i of α by a corresponding pair of gates $u_i, \bar{u}_i$ in β. By using a pair of gates in β to represent each gate in α, we can keep track of the value of the original gate and its negation. To maintain two values for each gate in the original circuit, double rail logic is constructed as follows:

1. For input x_k in α connect $u_k \leftarrow x_k$ and $\bar{u}_k \leftarrow \bar{x}_k$ in β,
2. for an AND gate $v_k \leftarrow v_i \wedge v_j$ of the original circuit connect $u_k \leftarrow u_i \wedge u_j$ and $\bar{u}_k \leftarrow \bar{u}_i \vee \bar{u}_j$ in β,
3. for an OR gate $v_k \leftarrow v_i \vee v_j$ of the original circuit connect $u_k \leftarrow u_i \vee u_j$ and $\bar{u}_k \leftarrow \bar{u}_i \wedge \bar{u}_j$ in β and
4. for a NOT gate $v_k \leftarrow \neg v_i$ in α connect $u_k \leftarrow \bar{u}_i \wedge \bar{u}_i$ and $\bar{u}_k \leftarrow u_i \wedge u_i$ in β.

The construction presented above is shown in Figure 21.8. It shows how the value of each gate in α and its negation can be maintained in β. The output gate in α is true if and only if the corresponding output gate in β is true. A formal proof that the reduction is correct could be derived by an easy induction showing that gates on level $k+1$ of the new circuit properly simulate gates on level k of α (an extra level was added in β for inputs).

It is easy to see that the reduction involves only local connections with easy modifications to α. Thus, it can be performed in log space. Since CVP was P-complete and MCVP is log space reducible to it, this completes the proof of Theorem 21.3. ∎

Theorem 21.3 proves that all NOT gates can be eliminated from an instance of CVP and pushed up to the inputs. The next couple of theorems presented show that certain other restricted versions of MCVP are P-complete. Each reduction presented illustrates an important proof technique of P-completeness theory. The impact of the various restrictions on the reductions should be compared. This helps build an intuition as to how certain restrictions affect the complexity of a problem.

21.5. Variations of the Circuit Value Problem

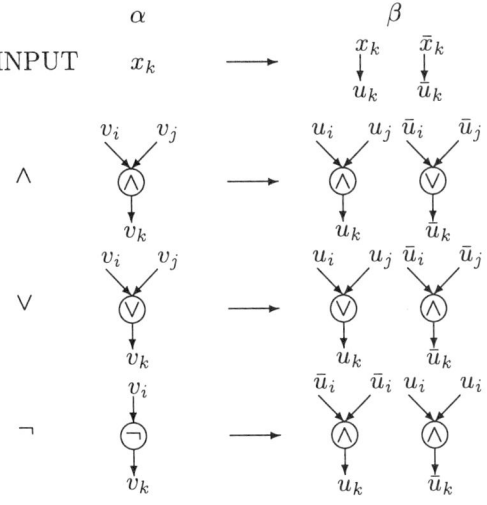

FIGURE 21.8
The double rail logic constructed to reduce an instance of CVP to an instance of MCVP.

The next version of CVP we present has the *alternating* property. A monotone circuit is *alternating* if on any path from an input to the output gate or to a gate whose outputs are unconnected, the gates alternate between OR and AND gates. Recall, we are thinking of our circuits as having only one output. The edges in the path must all be directed towards the output gate. Additionally, as a convention the inputs are all fed into OR gates and the output gate is required to be an OR gate.

DEFINITION 21.16
Monotone, Alternating CVP (MACVP):
Instance: An encoding of a monotone, alternating Boolean circuit α, plus inputs $x_1, \ldots, x_n$, and their complements $\bar{x}_1, \ldots, \bar{x}_n$.
Problem: Does α on input $x_1, \ldots, x_n, \bar{x}_1, \ldots, \bar{x}_n$ output 1?

The proof of Theorem 21.4 shows how to make a general instance of MCVP alternating.

THEOREM 21.4
The monotone, alternating circuit value problem (MACVP) is P-complete under log space reducibility.

PROOF

The idea is to begin with an instance of MCVP and convert it to a circuit having the alternating property while maintaining monotonicity. Since the steps used in the reduction can all be computed in log space and since MCVP is P-complete, it will follow that MACVP is also P-complete. The first step in the reduction is to replace all inputs with OR gates in the original circuit. Figure 21.9 illustrates this process pictorially.

The second step is to obtain the alternating property for the "interior" of the circuit. It is convenient to assume the circuit instance being modified does not have any parallel edges. If it did then they could be eliminated by doubling the size of the circuit. A second copy of the circuit could be made and parallel edges could be broken by exchanging copies of the parallel edges between the two instances.

To obtain the alternating property we could attempt to identify runs of AND and OR gates, and patch them up. This type of procedure seems to involve some sort of path tracing and involves examining the circuit in a global manner over these paths. Instead of doing this, a local translation of the circuit is performed that is log space computable. Each AND gate is examined locally and identity OR gates are "inserted" on each of the AND gates output wires that are inputs to other AND gates. A similar process is performed for OR gates.

Examples of the transformation are shown in Figure 21.10. The new subcircuits introduce the alternation property. Suppose the constant inputs and OR gates that replaced inputs are numbered using $1, \ldots, 3n$. For a gate labeled i in the original graph, it can be numbered $ni + 3n$ and the gates it spawns can be numbered $ni + 3n + 1, \ldots, ni + 4n - 1$. In cases where fewer than the maximum possible number of gates are spawned some numbers remain unused.

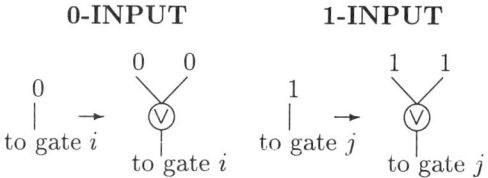

FIGURE 21.9
Method of replacing inputs with OR gates in step 1 of Theorem 21.4.

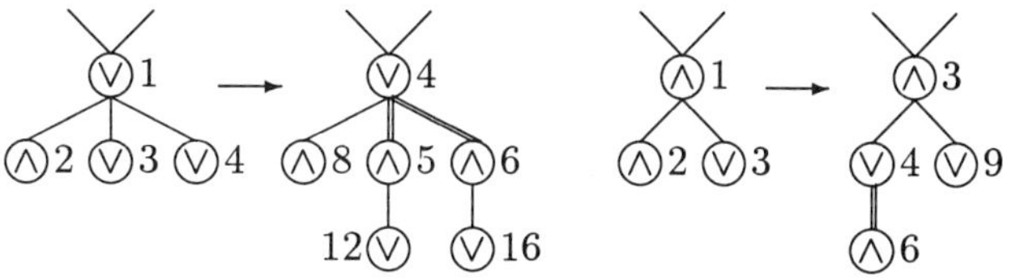

21.5. Variations of the Circuit Value Problem

FIGURE 21.10
Examples of the subcircuits replacing AND and OR gates to introduce the alternation property in the second step of Theorem 21.4.

The final step in the reduction is to add an OR gate after any AND gate of out-degree zero. This makes the last level of the circuit consist only of OR gates. Since gate type can be determined in log space, the overall reduction can be computed in log space. The proof of correctness of the reduction is left as an exercise for the reader. ∎

It is natural to ask if restricting the fan-out of gates to be small affects the complexity of MACVP. That is, if we allow gates to have only two outputs, is MACVP still P-complete? It is known that both MCVP and MACVP remain P-complete even with a fan-out restriction on the gates. In the latter case the fan-in and fan-out can be restricted to being exactly two for all gates. We present the definition of this problem below.

DEFINITION 21.17
Monotone, Alternating, Fan-in 2, Fan-out 2 CVP (MACVP2):
Instance: An encoding of a monotone, alternating Boolean circuit α, plus inputs $x_1, \ldots, x_n$, and their complements $\bar{x}_1, \ldots, \bar{x}_n$. The circuit is restricted so gates have fan-out and fan-in of exactly two.
Problem: Does α on input $x_1, \ldots, x_n, \bar{x}_1, \ldots, \bar{x}_n$ output 1?

THEOREM 21.5
The monotone, alternating, fan-in 2, fan-out 2 circuit value problem (MACVP2) is P-complete under log space reducibility.

PROOF
The technique used in the proof of the theorem is a general one that is often used in circuit complexity. We take an instance of MACVP and convert it so that the fan-in and fan-out restrictions stated in the theorem are met. Since MACVP is P-complete and since the modifications

made to it can be performed by a log space Turing machine, this is enough to prove the theorem.

The first step in the proof is to increase the fan-in of each node to two. If any node has fan-in one then add an edge parallel to its original input. This may increase the fan-out of some vertices to more than two; however, this is patched up in the next step.

Suppose there is a node k having fan-out greater than two. A fan-out two tree, as shown in Figure 21.11, can be added in so that the fan-out two requirement is met. Each node in the tree is labeled with AND or OR in such a way as to preserve the alternation property. AND (OR) gates can be made to behave like identity gates by having the constant input 1 (0) connected to them. In this way the tree reduces the fan-out of all large fan-out nodes to a group of nodes all having fan-out two.

The third step in the reduction is to increase the fan-out of nodes having only fan-out one. In the circuit constructed thus far, all AND gates have fan-in exactly two, no input node is connected to an AND gate, and all OR gates have fan-out at most two. It follows that the total number of OR gates with fan-out one must be even. By pairing the OR nodes with fan-out one together and introducing a new AND gate for them to connect to, the fan-out of all such OR nodes can be increased to two. A similar procedure can be used to increase the fan-out of all AND nodes to two.

Since it is easy to count the number of outputs of each gate in log space, we can determine the gates that need to be replaced by fan-out trees. The fan-out trees can then be inserted where necessary using log space. The other steps in the reduction can also be performed in log space. It should be clear that the new instance of MACVP2 is true if and only if the original instance of MACVP is true. This completes the proof of the theorem. ∎

There is one final restriction we place on MACVP2 and the problem remains P-complete. This restriction is the *synchronous* condition. A circuit is *synchronous* if each level of gates in the circuit receive inputs only from gates on the preceding level. This condition combined with the restrictions already imposed on the problem make for a highly regular version of CVP. In addition to presenting another interesting P-completeness technique via the proof that the synchronous version of MACVP2 is P-complete, we shall demonstrate the utility of this version of CVP by proving other problems are

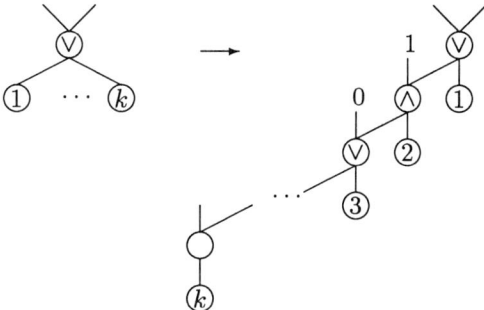

FIGURE 21.11
Fan-out tree used in the proof of Theorem 21.5.

P-complete by reductions from MACVP2. The reductions are much simpler than the original ones.

DEFINITION 21.18
Monotone, Alternating, Synchronous Fan-in 2, Fan-out 2 CVP (MASCVP2):
Instance: An encoding of a monotone, alternating, synchronous Boolean circuit α with the stated fan-in and fan-out restrictions, plus inputs $x_1, \ldots, x_n$, and their complements $\bar{x}_1, \ldots, \bar{x}_n$.
Problem: Does α on input $x_1, \ldots, x_n, \bar{x}_1, \ldots, \bar{x}_n$ output 1?

The proof that MASCVP2 is P-complete involves modifying an instance of MACVP2 and making it synchronous, while maintaining the previously imposed restrictions on it.

THEOREM 21.6
The monotone, alternating, synchronous, fan-in 2, fan-out 2 circuit value problem (MASCVP2) is P-complete under log space reducibility.

PROOF
We begin with an instance α of MACVP2 and show it can be made synchronous. The reduction used can be performed in log space on a Turing machine. Since MACVP2 is P-complete, it will follow that MASCVP2 is also P-complete.

Rather than attempting to transform various subcircuits that are not synchronous, which involves path tracing, a global transformation is applied to the circuit. Let n be the number of gates in α and let m be the number of inputs. We make $n/2$ copies of α. Each of these copies can be viewed as consisting of only two levels. The first level consisting of inputs to the circuit and AND gates is called the *AND level*. The second level consisting only of OR gates is called the *OR level*. Thus, in odd (even) numbered levels all gates are OR (AND). We describe below how to connect the $n/2$ copies to make a new circuit that is synchronous.

The edges between the AND and OR levels are preserved within each copy. For all c, $1 \leq c \leq n/2$, if OR gate i is connected to AND gate j in the original circuit then connect gate i in the $c-1$-st copy to gate j in the c-th copy.

Input and output nodes must be handled specially. In order to deliver an input node to the k-th copy, a chain of $k-1$ alternating identity nodes can be constructed out of OR and AND gates. An output node from the l-th copy can be handled similarly. A chain of alternating identity gates of length $n/2 - l$ can be constructed to deliver the output value synchronously to the last level.

Note, the monotonicity and alternating restrictions are preserved by the reduction. It is easy to see that the steps in the reduction can be performed in log space so this completes the proof of Theorem 21.6. ∎

The proof techniques applied in Theorems 21.4 and 21.6 indicate that it is sometimes useful to apply a global transformation to a problem rather than attempt to patch up the parts that do not meet the requirements. In some sense more modifications than are absolutely necessary are made. This is an important proof technique that has other applications in complexity theory.

MASCVP2 represents an extremely restricted version of CVP. Below we consider an instance of CVP consisting only of NAND gates and use MASCVP2 to prove it is P-complete.

DEFINITION 21.19
NAND Circuit Value Problem (NAND CVP):
Instance: An encoding of a Boolean circuit α constructed only of NAND gates plus inputs $x_1, \ldots, x_n$.
Problem: Does α on inputs $x_1, \ldots, x_n$ output 1?

THEOREM 21.7
The NAND circuit value problem (NAND CVP) is P-complete under log space reducibility.

21.5. Variations of the Circuit Value Problem

PROOF

The proof involves a reduction from MASCVP2 to NAND CVP. We modify an instance of MASCVP2 by replacing AND gates with a gadget consisting only of NAND gates and similarly for OR gates. To generate the complements of the inputs, we can feed each original input into a NAND gate that has as its other input the constant 1. Such a device simulates a NOT gate. Since the reduction involves only a local replacement of gates by gadgets, it is clear the reduction can be performed in log space. Additionally, the proof of correctness of the reduction essentially involves showing that the gadgets actually simulate AND and OR gates correctly. This is easy to check.

We replace x AND y in an instance of MASCVP2 by $(x$ NAND $y)$ NAND $(x$ NAND $y)$. x OR y is replaced by $\bar{x}$ NAND $\bar{y}$. Figure 21.12 depicts the gadgets with $\bar{\wedge}$ representing NAND. ∎

Any *complete basis* of gates suffices in order to prove an instance of CVP P-complete. A *complete basis* is a set of Boolean functions that can simulate all other Boolean functions. For example, NOR forms a complete basis. NOR CVP is defined analogously to NAND CVP and is also a useful problem for proving other problems are P-complete. NOR and NAND CVP are useful because they have only one gate type. This can often prove very beneficial in P-completeness proofs because reductions involving them become more uniform. Only one gate type needs to be simulated. We now turn our focus to a different type of restriction on an instance of CVP—planarity.

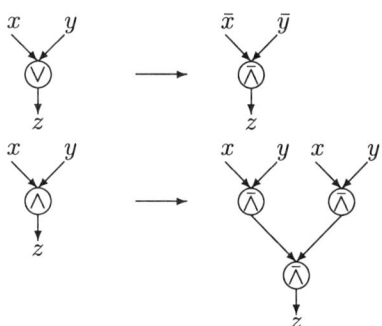

FIGURE 21.12
Gadgets used to replace OR and AND gates in converting an instance of MASCVP2 to NAND CVP.

DEFINITION 21.20
Planar Circuit Value Problem (PCVP):
Instance: An encoding of a planar Boolean circuit α plus inputs $x_1, \ldots, x_n$.
Problem: Does α on input $x_1, \ldots, x_n$ output 1?

THEOREM 21.8
The planar circuit value problem (PCVP) is P-complete under log space reducibility.

PROOF
The original reduction was from CVP. Without loss of generality, we can assume the gates are numbered in topological order since CVP remains P-complete with this requirement. The idea will be to take an instance of CVP and convert it to a planar circuit. The obvious approach to the reduction is to try to construct a planar cross-over circuit from equivalence subcircuits. Such a cross-over circuit is shown in Figure 21.13. The NAND gates are comprised of NOT and AND gates. Notice, the circuit switches the order of x and y, and is planar.

The straightforward approach of replacing every wire cross-over in the original circuit by a planar cross-over circuit does not seem to work

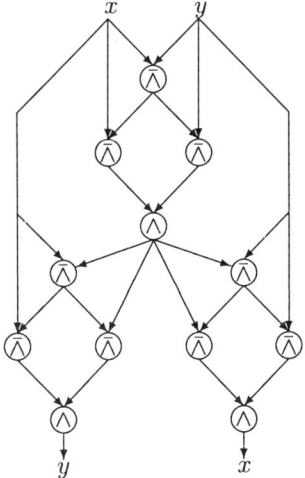

FIGURE 21.13
Planar cross-over circuit comprised of NAND and AND gates.

because it is difficult to find the required set of cross-overs in log space. Fortunately, the cross-over idea will work if the original circuit is laid out on a grid so that column i in the grid corresponds to gate i in the circuit. The cross-over circuit can then be used to route signals horizontally. For example, to evaluate AND gate $i = \text{AND}(i_1, i_2)$ we can do the following: move i_1 directly to the left of i_2, insert an AND gate in the grid at that point, evaluate the AND gate, immediately move i_1 back to where it began from, and move i to the far right of the grid. Other gate types are handled similarly. A specific example is presented in Figure 21.14. There the construction is given for $6 = \text{AND}(1,4)$. The rectangles represent the cross-over gadget shown in Figure 21.13.

The size of the new circuit constructed is $O(n^3)$. This is because for each gate we need a component of size $O(n^2)$ in order to shift things into the correct places. It is clear the new instance of CVP is also in P. The procedure described in the proof is very uniform. Therefore, the reduction can be performed in log space. It should be clear that the reduction generates a planar circuit and computes the same output as the original circuit. ∎

The proof of Theorem 21.8 illustrates again the point that applying a global transformation rather than a local patching up is a fundamental idea. The cross-over gadget in Theorem 21.8 has a fan-out of four on some nodes. Is it possible to derive a cross-over circuit with fan-out only 2 or 3 if NOT gates

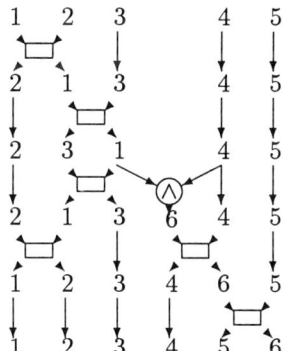

FIGURE 21.14
A specific example illustrating the construction that proves the planar circuit value problem is P-complete. Gate $6 = \text{AND}(1,4)$.

930 Chapter 21. Polynomial Completeness and Parallel Computation

are only allowed to have fan-out 1? We leave this question as an exercise. Interestingly, the monotone version of PCVP can be solved very quickly in parallel, i.e., it is in NC^2. The planar and monotone restrictions together seem to reduce the complexity of CVP. An an exercise, the reader is asked to find other versions of CVP that can be solved quickly in parallel.

The final version of CVP we describe involves *arithmetic circuits*. Thus far, we have examined only Boolean circuits. The inputs to an arithmetic circuit come from any *finite* ring consisting of c elements. Each input is coded using $O(\log c)$ bits. 0 denotes the additive identity and 1 denotes the multiplicative identity of the ring. $-a$ denotes the inverse of element a with respect to addition. A sample circuit is shown in Figure 21.15 with + and × denoting the usual addition and multiplication respectively.

DEFINITION 21.21
Arithmetic Circuit Value Problem (ACVP):
Instance: An encoding of an arithmetic circuit α with inputs $x_1, \ldots, x_n$.
Problem: Does α on input $x_1, \ldots, x_n$ output the multiplicative identity?

THEOREM 21.9
The arithmetic circuit value problem (ACVP) is P-complete under log space reducibility.

PROOF
The proof involves a direct reduction from an instance of CVP comprised of AND, NOT, and OR gates. Let $R = (S, +, \times, 0, 1)$ denote a ring, where S is finite. Let $-a$ denote the inverse of a with respect to $+$. The idea in the reduction is to represent TRUE and FALSE in the circuit respectively by the × and + identities of the ring. By replacing each gate

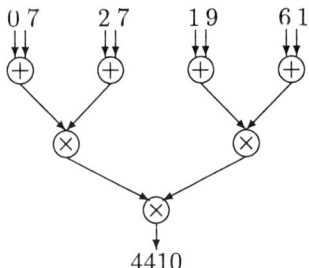

FIGURE 21.15
Example of an arithmetic circuit.

type by a combination of the × and + operations, the original circuit can then be simulated.

More formally, let TRUE inputs be replaced by the multiplicative identity and let FALSE inputs be replaced by the additive identity. A NOT gate with input x in the original circuit is replaced by $1 + (-x)$ in the instance of ACVP. An AND gate with inputs x_1 and x_2 in the instance of CVP is replaced by $x_1 \times x_2$ in the instance of ACVP. An OR gate with inputs x_1 and x_2 is replaced by the expression $(x_1 + x_2) + -(x_1 \times x_2)$ in the instance of ACVP.

It is easy to see that the output of the arithmetic circuit is 1 if and only if the original instance of CVP is TRUE. The reduction is easily computable in log space. ∎

ACVP is not necessarily in P for infinite rings like Z or Q, since the output of the circuit is not necessarily polynomial in length. The problem is *P-hard* in any ring since the proof of Theorem 21.9 only makes use of the ring identities and the ring operations. *P-hard* means the problem satisfies the completeness criterion but is not necessarily in P.

21.5.1 Summary of CVP

The ideas, proof techniques and intuition behind *P*-completeness results relative to the fundamental *P*-complete problem were described in this section. Table 21.1 presents a summary of these results. The wide range of instances of CVP proven *P*-complete provides a useful group of problems for showing other problems are *P*-complete. In the next section we present some applications to other important problems.

TABLE 21.1
A summary of the versions of the circuit value problem proved *P*-complete in this section.

monotone circuit value problem
monotone, alternating circuit value problem
monotone, alternating, fan-in and fan-out 2 circuit value problem
monotone, alternating, synchronous, fan-in and fan-out 2 circuit value problem
NAND circuit value problem
planar circuit value problem
arithmetic circuit value problem

21.6
P-completeness Techniques and Problems

The best way to learn the techniques of P-completeness theory is to examine a variety of P-completeness proofs. In review, there are three basic steps to a P-completeness proof for a specific problem. The first step is to show the problem is in P by presenting a polynomial time algorithm for the problem, the second is to present a reduction showing all other problems in P can be transformed to it, and the third is to prove the reduction is correct and that it can be performed in log space. Having a group of problems to work with in looking for new reductions greatly enhances the chance of success in showing another problem is P-complete. It is more likely that a problem similar to the one you are attempting to show complete can be found. This can often simplify the reduction. Sometimes if the "proper" problem is chosen to reduce to, the P-completeness result follows relatively easily. Thus, this selection process requires a thorough understanding of P-completeness techniques and of the known P-complete problems.

In this section several additional important P-complete problems are presented. This will help widen the base of P-complete problems available for showing others are P-complete. The first problem examined is related to breadth first search. There are several versions of graph searching that are P-complete and this problem illustrates some of the techniques involved in these proofs. The second result presented is for network flow. The reduction given proves that the maximum flow problem is P-complete. This proof illustrates the general techniques used in showing other versions of flow problems are P-complete as well. The third result described is related to linear programming. The linear inequalities problem is proved P-complete and then using linear inequalities it is easy to show a decision problem based on linear programming is P-complete.

How are P-completeness results derived and what are candidate P-complete problems? Generally a researcher is trying to find a fast parallel algorithm for a problem. If these attempts are unsuccessful and the problem seems to be highly sequential, then an attempt can be made to show the problem is P-complete. If the problem is proven P-complete then this indicates that the problem probably does not have a fast parallel solution. The best way to develop an intuition as to whether or not a particular problem is P-complete is to examine similar problems that are P-complete. We describe problems that are useful for building up our intuition about the nature of P-complete problems.

21.6. P-completeness Techniques and Problems

Several instances of graph search problems are known to be P-complete. The version presented here is based on a breadth first search algorithm that uses a stack as the underlying data structure. The algorithm is given below.

ALGORITHM 21.2
Stack Breadth First Search
Input: A graph $G = (V, E)$ represented by fixed ordered adjacency lists and a node s.
Output: Starting from s all nodes visited are labeled with the level in which they occur and are numbered consecutively in the order in which they were visited within their level.
Comment: The procedure next(v) returns the next unlabeled vertex on v's adjacency list. The procedure unlabeled(v) returns the total number of unvisited vertices adjacent to v. The arrays *level* and *order* are such that level[v] contains v's level number and order[v] contains v's number within a given level.

begin

  level[s] := 0; order[s] := 1; inlevel := 1; push s on stack S;
  while stack $S \neq$ null **do**

    v := pop S;
    T := unlabeled(v);
    for $i := 1$ **to** T **do**

      u := next(v);
      order[u] := inlevel;
      inlevel:= inlevel +1;
      level[u] := level[v] +1;
      push u on stack S_1;

    endfor

  endwhile
  if stack $S_1 \neq$ null **then**

    switch stacks(S, S_1);
    inlevel := 1;
    goto while loop;

  endif

end.

Based on the algorithm shown in Figure 21.2, the following natural decision problem can be defined.

DEFINITION 21.22
Stack Breadth First Search Problem (SBFS):
Instance: A graph G with a numbering on the vertices and three designated vertices u, v and s.
Problem: Determine if vertex u is visited before vertex v in the stack breadth first search of G originating from s and induced by the vertex numbering.

This type of decision problem is often used in examining the complexity of graph search algorithms.

THEOREM 21.10
The stack breadth first search problem (SBFS) is P-complete under log space reducibility.

PROOF
The SBFS algorithm is easily seen to run in polynomial time. To prove the theorem we provide a log space reduction from MASCVP2 to SBFS. Consider an input α to MASCVP2 having M gates. Further, we assume α has only one output and that the output gate is numbered M. Suppose α has inputs $x_1, \ldots, x_n$ and is coded by the standard encoding.

The idea behind the reduction is to construct from α a graph G that can be used to determine the value of the output gate of α. By sorting the input values to α, the search conducted by the SBFS algorithm on G partitions gates evaluating to TRUE at each level from those evaluating to FALSE. An extra column of nodes added to G forms the boundary of the partition. α is simulated by the search order on G in the following way: on OR levels outputs that were TRUE on the preceding AND level are used first in the search, and thus, we can pick out all TRUE OR gates first; on AND levels output values that were FALSE on the preceding OR level are used first in the search, and thus, we can pick out all FALSE AND gates first. Based on the order in which vertices are visited within their level relative to the extra column vertex appearing in that level, the value of any gate in α can be computed.

The graph G constructed has the same basic structure as α. Figure 21.16 shows the graph and illustrates the extra column of vertices. For each gate including inputs, there is a corresponding vertex in G. The input

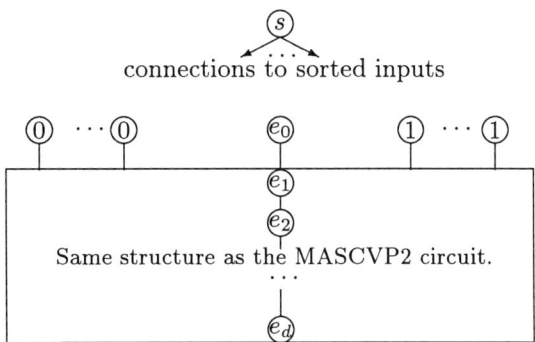

FIGURE 21.16
Graph G used in reducing MASCVP2 to SBFS in Theorem 21.10.

gates are sorted. Namely, vertex numbers are assigned in consecutive order to the FALSE inputs, the shallowest extra vertex, and then the original TRUE inputs. The remainder of the nodes are assigned numbers arbitrarily. Let $e_0, \ldots, e_d$ denote the numbers of the extra column starting from the inputs and continuing to the outputs. For each arc (u, v) in α there is a corresponding edge in G connecting the vertices that corresponded to gates u and v of the circuit. A special vertex s is added from which the search is to originate; s is connected to e_0 and all vertices corresponding to inputs. The following lemma establishes the correctness of the reduction.

LEMMA 21.4
Vertex e_l for l odd (even) corresponding to an OR (AND) level in α is visited in G by the SBFS algorithm before vertex v in level l if and only if the gate corresponding to vertex v evaluates to FALSE (TRUE) in α.

PROOF
Let F_i (T_i) denote the decreasing sequence of nodes corresponding to gates evaluating to FALSE (TRUE) on level i. The lemma is proved by induction.

For the base case we show the stack contents is $\langle T_0, e_0, F_0 \rangle$ after visiting the inputs. From s vertices are visited and pushed on the stack based on their vertex numbers. It is easy to see that the manner in which the inputs were sorted yields this ordering on the stack.

For the induction hypothesis suppose that if the stack contents is $\langle F_i, e_i, T_i \rangle$ ($\langle T_i, e_i, F_i \rangle$) after level i has been visited, then the next level is an AND (OR) level and the new stack contents after visiting level $i + 1$ is $\langle T_{i+1}, e_{i+1}, F_{i+1} \rangle$ ($\langle F_{i+1}, e_{i+1}, T_{i+1} \rangle$).

Suppose level $i + 1$ is an AND level. By the induction hypothesis the stack contents after visiting level $i + 1$ is $\langle T_{i+1}, e_{i+1}, F_{i+1} \rangle$. The search proceeds from level $i + 1$ by visiting OR gates on level $i + 2$. The gates visited from the T_{i+1}'s receive a TRUE input so they evaluate to TRUE. After all such OR gates are visited, e_{i+2} is visited from e_{i+1}. The remaining OR gates are visited from the F_{i+1}'s. All of these receive FALSE inputs only and so evaluate to FALSE. Thus, the stack contents at the end of visiting level $i + 2$ is $\langle F_{i+2}, e_{i+2}, T_{i+2} \rangle$.

The case in which level $i+1$ is assumed to be an OR level is similar. This completes the induction. The statement of the lemma follows directly from the induction. ∎

Since the lemma shows the reduction is correct and because it is easy to see that the reduction is a log space reduction, the proof of Theorem 21.10 is complete. ∎

The next problem we describe is related to network flows. The maximum flow problem was defined in Section 21.2. The decision problem we use here is whether the value of the maximum flow is odd. The theorem below shows this problem is P-complete.

THEOREM 21.11
The maximum flow problem (MF) is P-complete under log space reducibility.

PROOF
It is well-known that the maximum flow problem is in P. The reduction proving the problem complete is from the monotone CVP with fan-out restricted to 2 (MCVP2). This version of CVP is P-complete. The idea is to construct a flow network based on an instance $\alpha = (\alpha_n, \ldots, \alpha_0)$ of MCVP2 in which determining whether the value of the flow is odd allows us to determine whether the output of α is TRUE. We assume the nodes

21.6. P-completeness Techniques and Problems

of α are topologically ordered with the output gate first and the inputs last, that all inputs are connected to gates, and that the output gate is an OR gate. A description of how to build the required flow network G is presented below.

Gates α_i and connections e_{ij} of α are associated with nodes α_i' and edges e_{ij}' of the flow network G. G has additional nodes s and t. For each AND (OR) gate α_i, there is an *overflow* edge (α_i', t) $((s, \alpha_i'))$ in G. For each input gate α_i, there is an additional edge in G, (s, α_i'). There is also an edge (α_0', t). The capacity of the edges are specified below.

1. If α_i is an input then edge (s, α_i') has capacity 0 if α_i is FALSE, and capacity 2^i if α_i is TRUE.
2. If $\alpha_i = \text{AND}(j, k)$ then edge (α_j', α_i') has capacity 2^j, (α_k', α_i') has capacity 2^k, and (α_i', t) has capacity $2^j + 2^k - d2^i$, where d is the fan-out of α_i.
3. If $\alpha_i = \text{OR}(j, k)$ then edge (α_j', α_i') has capacity 2^j, (α_k', α_i') has capacity 2^k, and (α_i', s) has capacity $2^j + 2^k - d2^i$, where d is the fan-out of α_i.
4. (α_0', t) has capacity 1.

Figure 21.17 illustrates the construction described above for the circuit shown in Figure 21.18 assuming the output AND gate is switched to an OR gate. Notice, the gates are numbered in reverse topological order. The capacities are written on the lefthand side of the edges.

We define a flow function f_s that allows G to simulate α. f_s is defined as follows for all i, $0 \leq i \leq n$,

1. If α_i is an input then the flow in (s, α_i') is equal to its capacity.
2. For all (i, j), $0 \leq i, j \leq n$, the flow in edge (α_i', α_j') is 2^i if α_i evaluates to TRUE and is 0 is α_i evaluates to FALSE.
3. If $\alpha_i = \text{AND}(j, k)$ then the flow on edge (α_i', t) is equal to its capacity if both of α_i's inputs are TRUE. Otherwise, the flow is equal to the sum of the flows on edges (α_j', α_i') and (α_k', α_i').
4. If $\alpha_i = \text{OR}(j, k)$ then the flow on edge (α_i', s) is equal to the sum of the flows assigned to edges (α_j', α_i') and (α_k', α_i') minus $d2^i$ if either of α_i's inputs are TRUE. Otherwise, the flow is 0.

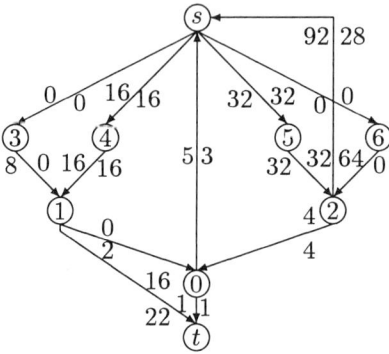

FIGURE 21.17
An example illustrating the construction presented in Theorem 21.11.

LEMMA 21.5
f_s is a feasible flow and achieves the maximum possible flow in G.

PROOF
Exercise. ∎

Note that all edge capacities into t are even except the one from α'_0 to t. From this observation in conjunction with the previous lemma, it is not hard to see that the maximum flow for G is odd if and only if α outputs TRUE. It is also easy to see that the reduction is a log space reduction. This completes the proof of Theorem 21.11. ∎

In Figure 21.17 the actual flow assigned by f_s is given on the right of each edge. The maximum flow is 17 which f_s achieves. This number is odd and note the circuit does evaluate to TRUE.

The reduction in Theorem 21.11 produces exponential edge capacities in G. This is because the flow into an input gate may need to be distributed in the worst case up to the depth of the circuit number of times. When the edge capacities are "small," MF is known to be in the class random NC. MF has been widely studied and there are several other versions of it that are also P-complete. Several different restricted versions have fast parallel algorithms. For example, if the flow network is restricted to being planar then there is a fast parallel algorithm for the problem. The next problem studied is related to linear programming.

21.6. P-completeness Techniques and Problems

DEFINITION 21.23

Linear Inequalities (LI):

Instance: An integer $n \times d$ matrix A and an integer $n \times 1$ vector b.
Problem: Determine if there exists a rational $d \times 1$ vector $x > 0$ such that $Ax \leq b$.

The definition of LI does not require that we find such an x. The problem is only to determine whether such an x exists.

THEOREM 21.12
The linear inequalities problem (LI) is P-complete under log space reducibility.

PROOF
It is known that LI is in P. The reduction proving completeness that we present involves showing CVP is log space reducible to LI. The instance of CVP used involves NOT, OR and AND gates. The idea in the reduction is to introduce equations that simulate these Boolean functions. The four parts of the reduction are enumerated below.

1. If input x_i is TRUE (FALSE) then it is represented by the equation $x_i = 1$ ($x_i = 0$).
2. A NOT gate with input u and output w, computing $w \leftarrow \neg u$ is represented by the inequalities $w = 1 - u$ and $0 \leq w \leq 1$.
3. An AND gate with inputs u, v computing $w \leftarrow u \wedge v$ is represented by the inequalities $0 \leq w \leq 1$, $w \leq u$, $w \leq v$, and $u + v - 1 \leq w$.
4. An OR gate with inputs u, v computing $w \leftarrow u \vee v$ is represented by the inequalities $0 \leq w \leq 1$, $u \leq w$, $v \leq w$, and $w \leq u + v$.

To determine the output z of the circuit, inequalities are added to force $z = 1$ ($1 \leq z, z \leq 1$). The system has a solution if and only if the output of α is TRUE. It should be clear that the reduction can be performed in log space. ∎

An example of the reduction given in Theorem 21.12 is shown in Figure 21.18. The corresponding inequalities constructed are shown below and the resulting system is shown in Figure 21.19. For the circuits inputs we get the following inequalities:

$$x_1 \leq 0, \; x_1 \geq 0; \; x_2 \leq 1, \; x_2 \geq 1; \; x_3 \leq 1, \; x_3 \geq 1; \; x_4 \leq 0, \; x_4 \geq 0.$$

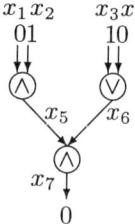

FIGURE 21.18
Example used to illustrate the reduction between the circuit value problem and linear inequalities.

For the gates of the circuits we get the following inequalities:

$$0 \leq x_5,\ x_5 \leq 1,\ x_5 \leq x_2,\ x_5 \leq x_1,\ x_1 + x_2 - 1 \leq x_5$$
$$0 \leq x_6,\ x_6 \leq 1,\ x_3 \leq x_6,\ x_4 \leq x_6,\ x_6 \leq x_3 + x_4$$
$$0 \leq x_7,\ x_7 \leq 1,\ x_7 \leq x_5,\ x_7 \leq x_6,\ x_5 + x_6 - 1 \leq x_7,\ x_7 \geq 1.$$

The inequalities do not have a solution because they imply that $x_5 = 0$, $x_7 = 1$, and $x_7 \leq x_5$. It is easy to see that the instance of CVP evaluates to FALSE. Since there was no solution to the system of inequalities, the circuit should evaluate to FALSE and this checks.

The linear equalities (LE) problem is defined analogously to LI except "$\leq$" is replaced by "$=$" in the definition. This problem is also P-complete. LE is log space reducible to LI since $Ax \leq b$ and $-Ax \leq -b$ if and only if $Ax = b$. Thus, LE is in P. For completeness an instance of LI can be reduced to LE as follows: for each inequality in LI there is a corresponding equality in LE with an additional variable that is used to make the inequality into an equality. If LE is restricted so the coefficients of A and b are either $-1, 0,$ or 1 then LE remains P-complete. Below we define the linear programming problem and sketch the proof that it is P-complete.

DEFINITION 21.24

Linear Programming (LP):

Instance: An integer $n \times d$ matrix A, an integer $n \times 1$ vector b, and an integer $1 \times d$ vector c.

Problem: Find a rational $d \times 1$ vector x such that $Ax \leq b$ and cx is maximized.

As is standard, we have stated LP as a *search problem*.

$$\begin{pmatrix} -1 & 0 & 0 & 0 & 0 & 0 & 0 \\ 1 & 0 & 0 & 0 & 0 & 0 & 0 \\ 0 & 1 & 0 & 0 & 0 & 0 & 0 \\ 0 & -1 & 0 & 0 & 0 & 0 & 0 \\ 0 & 0 & 1 & 0 & 0 & 0 & 0 \\ 0 & 0 & -1 & 0 & 0 & 0 & 0 \\ 0 & 0 & 0 & 1 & 0 & 0 & 0 \\ 0 & 0 & 0 & -1 & 0 & 0 & 0 \\ 0 & 0 & 0 & 0 & 1 & 0 & 0 \\ 0 & 0 & 0 & 0 & -1 & 0 & 0 \\ 0 & -1 & 0 & 0 & 1 & 0 & 0 \\ -1 & 0 & 0 & 0 & 1 & 0 & 0 \\ 1 & 1 & 0 & 0 & -1 & 0 & 0 \\ & & & \vdots & & & \end{pmatrix} \begin{pmatrix} x_1 \\ x_2 \\ x_3 \\ x_4 \\ x_5 \\ x_6 \\ x_7 \end{pmatrix} \leq \begin{pmatrix} 0 \\ 0 \\ 1 \\ 1 \\ 1 \\ 1 \\ 0 \\ 0 \\ 0 \\ 1 \\ 0 \\ 0 \\ 1 \\ \vdots \end{pmatrix}$$

FIGURE 21.19
The system of inequalities constructed as in the proof of Theorem 21.12 based on the circuit shown in Figure 21.18.

THEOREM 21.13
The linear programming problem (LP) is P-complete under log space reducibility.

PROOF
The linear programming problem is known to be in P. The idea of the reduction is to reduce LI to LP by picking any cost vector c and checking whether the resulting linear program is feasible. ∎

21.7
Greedy Algorithms

We consider the selection of basketball teams at a neighborhood playground in order to illustrate the greedy method. Usually the two "best" players are "designated" captains. All other players line up and the captains alternate choosing one player at a time. Typically, they choose the best player who has not been picked yet. The players are picked using a greedy strategy. Usually, the captains do not try to field a well-balanced basketball team by choosing two guards, two forwards and a center. Instead, they simply choose the best player remaining. This system of selection by choosing the best, most obvious or most convenient remaining choice is called the *greedy method*. An algorithm using this type of selection process is called a *greedy algorithm*. Often greedy algorithms lead to efficient sequential solutions to problems. Additionally, they are easy to implement. Unfortunately, greedy algorithms do not always seem to lead to solutions that are efficiently computable in parallel.

A greedy algorithm that constructs a *maximal clique* is presented in order to further illustrate some of the important aspects of greedy algorithms. A *clique* is a subset of vertices of a graph that are fully connected. The subset is *maximal* if no other vertex can be added to the subset while maintaining the clique property. An algorithm that computes a maximal clique is given below. The algorithm is a greedy algorithm because it cycles through the vertices in the order in which they are numbered picking the most obvious vertex next. The procedure always adds the lowest numbered vertex that has yet to be tried, if possible.

ALGORITHM 21.3
Greedy Maximal Clique
Input: Undirected graph $G = (V, E)$ with the vertices ordered $v_1, \ldots, v_{|V|}$.
Output: A maximal clique of G.

begin

$C := \emptyset$;

for $i := 1$ **to** $|V|$ **do**
if v_i is connected to all vertices in C **then** $C := C \cup \{v_i\}$

endfor

print C

end.

Many greedy algorithms compute lexicographically least solutions. The *lexicographically least solution* is the one that would appear first in an alphabetized list of all possible solutions. Our maximal clique algorithm computes the lexicographically first maximal clique. We can define a natural decision problem based on the algorithm.

DEFINITION 21.25
Lexicographically first maximal clique problem (LFMC):
Instance: An undirected graph G with an ordering on the vertices and a designated vertex v.
Problem: Is vertex v in the lexicographically first maximal clique of G?

The decision problem posed in LFMC is a standard question arising in many decision problems that are based on graph algorithms. An example is presented in Figure 21.20 to make the concepts just described more concrete. The lexicographically first maximal clique computed by our algorithm with the graph shown in Figure 21.20 as input is $\{v_1, v_2\}$. The lexicographically next maximal clique is $\{v_2, v_3, v_4\}$, which is the largest possible. The following theorem shows LFMC is P-complete. This suggests that finding the clique the greedy maximal clique algorithm computes is a highly sequential procedure.

THEOREM 21.14
The lexicographically first maximal clique problem (LFMC) is P-complete under log space reducibility.

PROOF
The reduction is from MCVP. Without loss of generality, assume the instance α of MCVP has its gates numbered in topological order with inputs numbered first and outputs last. A graph G is constructed to be input to LFMC. The graph G is constructed so each gate v of α is associated with nodes v' and v'' of G in such a way that the lexicographically first maximal clique C of G includes v' (v'') if and only if the value of v in α is 1 (0). G is described below.

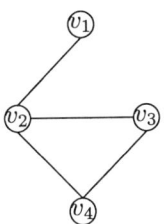

FIGURE 21.20
An example used to illustrate the lexicographically first maximal clique problem.

1. If an input x is 1 (0) then x' precedes x'' (x'' precedes x'), and x' (x'') is adjacent to all input nodes except x'' (x').
2. If v is an AND gate with inputs u, w then v' precedes v'', v' is adjacent to all preceding nodes except u'' and w'', and v'' is adjacent to all preceding nodes except v'.
3. If v is an OR gate with inputs u, w then v'' precedes v', v'' is adjacent to all preceding nodes except u' and w', and v' is adjacent to all preceding nodes except v''. ∎

There are many other decision problems based on greedy algorithms that are P-complete. Several other examples are lexicographically first depth first search, lexicographically first $\Delta + 1$ vertex coloring, and first fit decreasing bin packing. These results indicate that in general greedy algorithms can not be parallelized efficiently. Although many greedy algorithms appear to be highly sequential, there are some greedy algorithms that can be parallelized efficiently. One example is the greedy algorithm for computing a minimum spanning tree.

21.8
Inherently Sequential Algorithms

Thus far our focus in this chapter has been on the fundamental concepts of P-completeness theory. The theory is used to classify decision problems as inherently sequential. For many decision problems, there is an underlying algorithm that does not seem to be parallelizable. There are several questions about algorithmic complexity we cannot answer through our focus on decision problems. Is it possible that there are two natural algorithms with the same sequential complexity that solve the same problem yet have different parallel complexities? Is it possible there is one algorithm solving a problem A, which has a natural decision problem based on it that is P-complete, while another algorithm solving A is highly parallel? It turns out that the answers to questions like these are yes. In order to prove this, a model whose focus is directly on algorithms is required. In this section the description of such a model is given.

The model needs to be powerful enough to classify algorithms with respect to their parallel complexities. The intermediate computation values algorithms compute are enough to distinguish their parallel complexities. The model therefore must be capable of capturing the intermediate values an algorithm computes. It is convenient to base the description of the model

21.8. Inherently Sequential Algorithms

on the random access machine (RAM), with which we assume the reader is familiar with. The idea behind the model is to assign a value to each step of an algorithm. Examining the sequence of values an algorithm generates on its input can then lead to determining its parallel complexity. A formal description of the model is given below.

A *RAM algorithm* is a RAM program whose statements are numbered consecutively. Each RAM algorithm statement is given an associated value. The value associated with each statement is the valuation of the left hand side of the statement. The instruction is assumed to have already been executed. Table 21.2 shows the value associated with each RAM instruction. In the table $c(i)$ denotes the contents of register i. Note, register 0 of the RAM serves as the accumulator and all computation is performed in this register. $v(a)$ denotes the value of operand a as specified below.

$$\begin{aligned} v(=i) &= i \\ v(i) &= c(i) \\ v(*i) &= c(c(i)). \end{aligned}$$

Note, for STORE *i and READ *i there are two values associated with the instruction. For indirect addressing instructions, the second value determines which cell was read or written. Corresponding to each step of a RAM algorithm executed, there is an associated value generated. Each statement generates as many values as the number of times it is executed.

A function is associated with a particular RAM algorithm by considering the sequence of statements executed by the RAM algorithm and the values computed by those statements.

DEFINITION 21.26
Let N denote the natural numbers. Given any RAM algorithm A over an alphabet Σ with statements numbered sequentially starting from one, the **RAM flow function** $f_A : \Sigma^* \times N \to (N \times N)^*$ *corresponding to A is defined to be*

$$f_A(x_1 \cdots x_n, t) = v_1, v_2, \ldots, v_t$$

where the v_i's represent the statement numbers, value(s) pairs associated with the execution of t statements of A on input $x_1, \ldots, x_n$.

Let $n = |x|$. For all algorithms that halt on all inputs, define $f_A(x)$ to be $f_A(x, T(n))$, where $T(n)$ denotes the maximum running time of algorithm A on inputs of length n. The length of the flow function for algorithm A,

TABLE 21.2
Each RAM instruction and its associated value(s).

Instruction	Associated Value
LOAD a	v(c(0))
STORE i	v(c(i))
STORE *i	v(c(i)); v(c(c(i)))
ADD a	v(c(0))
SUB a	v(c(0))
MULT a	v(c(0))
DIV a	v(c(0))
READ i	v(c(i))
READ *i	v(c(i)); v(c(c(i)))
WRITE a	v(a)
JUMP b	v(b)
JGTZ b	v(b)
JZERO b	v(b)
HALT	—

denoted $|f_A|$, is related to the running time of A. In particular, if the log cost model is assumed then $|f_A|$ is just twice the running time of A. In the definition statement numbers of A and the values generated by each statement are represented by integers. Any reasonable encoding scheme suffices. The flow function plays a major role in relating algorithms and their parallel complexities as the following definition shows.

DEFINITION 21.27

Let A be a RAM algorithm with flow function f_A. A **RAM algorithm A is inherently sequential** if the language $L_A = \{\langle x, i, j \rangle \mid $ the i-th bit of $f_A(x) = j\}$. associated with its flow function is P-complete under log space reducibility.

The definition states that the "underlying flow function" of an algorithm must be at least as difficult to compute as a P-complete language in order for the algorithm to be inherently sequential. The theorem below shows the definition is such that any polynomial time algorithm computing the solution to a P-complete problem meets the requirements of the definition. This shows the model captures the right notion of highly sequential.

THEOREM 21.15
If polynomial time RAM algorithm A accepts a P-complete language L, then A is inherently sequential.

PROOF
Exercise. ∎

Theorem 21.15 shows the model has a desired fundamental property. That is, any algorithm solving a P-complete problem is inherently sequential. P-complete problems are highly sequential in nature, so algorithms solving them should be highly sequential. Several results illustrating how the model can be used to classify algorithms are given below.

The first example is an algorithm for biconnectivity. From a result proving a depth first search problem is P-complete and Theorem 21.15, it follows that the greedy depth first search algorithm is inherently sequential. Thus, an algorithm for determining the biconnected components of a graph using the greedy depth first search algorithm is inherently sequential. However, there is a parallel algorithm for computing the biconnected components of a graph that runs in time $O(\log n)$ time using n^2 processors on a CRCW PRAM. This algorithm computes the same output as the one using the greedy depth first search as a subroutine but it is in NC. In this case, the individual steps of the algorithms need to be considered before their parallel complexities can be compared.

There are some search problems for which one algorithm may find a solution to the problem quickly in parallel, whereas, another might be inherently sequential. Usually, the solutions computed to the problem are different. Such search problems are interesting since alternative parallel algorithms can sometimes be found for the problems that are in NC. It is often the case that only one solution to a problem is required and it is usually not important which solution is found. Thus, finding an alternative parallel algorithm for such a problem is very useful.

An example of such a search problem involves maximal paths. The problem of computing a lexicographic minimum maximal path is P-complete. It follows from Theorem 21.15 and this result that the greedy algorithm for computing such a path is inherently sequential. Using an alternative parallel approach yields an algorithm for computing a maximal path that is in random NC. This path is not the lexicographic minimum maximal path; however, it is a maximal path.

There are many other examples with a similar flavor to these. Earlier we described one related to breadth first search. A stack based breadth first search was inherently sequential, whereas a queue based breadth first search was highly parallel. A second example is related to Gaussian elimination. If one uses Gaussian elimination with partial pivoting to solve a system of equations then the process will be highly sequential since this problem is P-complete. Using another approach, though, a solution can be found quickly in parallel. The point is the intermediate values an algorithm computes and not just its final answer are important in determining complexity.

These results focusing on algorithms are important because they suggest inherently sequential algorithms will not be amenable to automatic parallelization by compilers even though the problems they solve have highly parallel solutions. This is because compilers are not sophisticated enough to figure out the alternative parallel approach. In addition, proving an algorithm is inherently sequential is a first step towards showing that a problem is not likely to be parallelizable or at least towards showing a different algorithmic approach needs to be found for the problem.

21.9 Summary

This chapter focused on identifying problems whose solutions are highly sequential—problems for which there appears to be no fast parallel algorithm. The intuition behind identifying such problems and the techniques involved in proving P-completeness results were described. Many variations of the circuit value problem were proven P-complete as were several other fundamental problems.

The conclusions we can draw from polynomial completeness theory is that not all problems seem to have fast parallel algorithms. There are important, practical problems that cannot be solved efficiently in parallel. Additionally, there are sequential algorithms that can not be parallelized effectively. These types of algorithms can be identified using the techniques presented and avoided by programmers. Although parallelism allows for fast solutions to many problems, one should not expect to achieve significant speedups for all problems.

21.10 Exercises

21.1   Cast the clique problem and the maximum flow problem in terms of language recognition questions.

21.2   Prove Lemma 21.1 using the CREW PRAM model.

21.3 Is log space reducibility an equivalence relation? That is, is log space reducibility reflexive, symmetric, and transitive? Carefully prove or disprove each property.

21.4 Extend the definition of a P-complete language to functions. That is, define what it means for a polynomial time computable function to be complete for the set of polynomial time computable functions.

21.5 Write down the standard encodings for the circuits shown in Figures 21.11 and 21.18.

21.6 Prove that the reduction given in Theorem 21.4 is correct.

21.7 Find a restricted version of CVP that can be solved quickly in parallel.

21.8 Define a variant of CVP as follows:
Given: An encoding of a (min, +) circuit α and inputs $x_1, \ldots, x_n$. The values in the circuit come from any finite ordered ring. min denotes the dyadic minimum operator and + denotes addition in the ring. 0 denotes the additive identity and 1 denotes the element succeeding 0 in the ring.
Problem: Does α on input $x_1, \ldots, x_n$ output 1?
Prove this version of CVP is P-complete or provide an NC algorithm solving the problem.

21.9 Is it possible to have a planar cross-over circuit of fan-out 2 or 3?

21.10 What is the smallest planar cross-over circuit you can construct from Boolean gates?

21.11 Prove the flow function f_s defined in Theorem 21.11 is feasible.

21.12 Prove the flow function f_s defined in Theorem 21.11 achieves the maximum possible flow in the flow network G.

21.13 Prove Theorem 21.15—if polynomial time RAM algorithm A accepts a P-complete language L, then A is inherently sequential.

21.14 Prove the following fact: Let A and B be RAM algorithms. Let f_A and f_B be their corresponding flow functions. If f_A is log space reducible to f_B and algorithm A is inherently sequential then algorithm B is also inherently sequential.

21.15 Prove the following fact: Let A be an inherently sequential RAM algorithm. Let f_A be the corresponding RAM flow function for A. If there exists a family $\langle \alpha_n \rangle$ of NC circuits computing f_A then $P = NC$.

21.16 Let $G = (V, E)$ be a graph. Let $\Delta = \max \{$ degree of $v \mid v \in V \}$. G can always be colored by $\Delta + 1$ colors. Prove the lexicographically first $\Delta + 1$ coloring is P-complete by a reduction from a variant of the circuit value problem. Can you construct a graph using only 3 colors? How about a graph that uses more than 3 colors but is smaller?

Bibliography

[1] A.V. Aho, J.E. Hopcroft, and J.D. Ullman. *The Design and Analysis of Computer Algorithms*. Addison-Wesley, Reading, Mass., 1974. One of the standard textbooks about computer algorithms. Our description of the RAM model is from here.

[2] R. Anderson and E. Mayr. Parallelism and the maximal path problem. *Information Processing Letters*, 24(2):121–126, 1987. This paper proves the lexmin maximal path problem is *P*-complete.

[3] R. Anderson, E. Mayr, and M. Warmuth. Parallel approximation algorithms for bin packing. *Information and Computation*, 82(3):262–277, 1989.

[4] A. Borodin. On relating time and space to size and depth. *SIAM Journal on Computing*, 6(4):733–744, 1977. A fundamental paper relating Turing machine space complexity to circuit depth complexity.

[5] S. A. Cook. Reduction of the circuit value problem to feasible linear programming. Private communication cited in Greenlaw, Hoover and Ruzzo, 1982.

[6] S.A. Cook. An observation on time-storage trade off. *Journal of Computer and System Sciences*, 9(3):308–316, 1974. Contains the first *P*-completeness result. The paper shows the path systems problem is *P*-complete.

[7] S.A. Cook. A taxonomy of problems with fast parallel algorithms. *Information and Control*, 64(1-3):2–22, 1985. This fundamental paper focuses primarily on *NC* and subclasses of *NC*. Several of the results we present are from this paper including the proof that LFMC is *P*-complete and that the greedy algorithm for minimum spanning tree is in *NC*.

[8] P. W. Dymond and S. A. Cook. Hardware complexity and parallel computation. In *21st Ann. Symp. on Foundations of Computer Science*, pages 360–372, Syracuse, New York, October 1980. This paper contains the result that monotone planar CVP is in *NC*.

[9] T. Feather. The parallel complexity of some flow and matching problems. Master's thesis, Department of Computer Science, University of Toronto, 1984. Shows maximum flow in a network with edge capacities expressed in unary is in RNC^2.

[10] M.R. Garey and D.S. Johnson. *Computers and Intractability: A Guide to the Theory of NP-Completeness*. W.H. Freeman and Company, San Francisco, 1979. This book surveys the field of *NP*-completeness and lists many *NP*-complete problems.

[11] L. M. Goldschlager, R. A. Shaw, and J. Staples. The maximum flow problem is log space complete for *P*. *Theoretical Computer Science*, 21:105–111, 1982.

[12] L.M. Goldschlager. The monotone and planar circuit value problems are log space complete for *P*. *SIGACT News*, 9(2):25–29, 1977. Our discussions about MCVP and planar CVP are taken from here.

[13] R. Greenlaw. Ordered vertex removal and subgraph problems. *Journal of Computer and System Sciences*, 39(3):323–342, 1989. The paper contains numerous *P*-completeness results for graph related problems.

[14] R. Greenlaw. A model classifying algorithms as inherently sequential with applications to graph searching. *Information and Computation*, to appear.

[15] R. Greenlaw. The parallel complexity of approximation algorithms for the acyclic subgraph

problem. Technical report 90-61, University of New Hampshire, 1990.

[16] R. Greenlaw, H.J.Hoover, and W.L. Ruzzo. A compendium of problems complete for *P* part I: Theory and open problems. Technical Report, University of Alberta 90-36 and University of Washington, 1991. Contains an advanced discussion of *P*-completeness theory as well as a substantial list of open problems.

[17] R. Greenlaw, H.J.Hoover, and W.L. Ruzzo. A compendium of problems complete for *P* part II: *P*-complete problems. Technical Report, University of Alberta 90-36 and University of Washington, 1991. This paper contains a list of over one hundred *P*-complete problems and sketches many of the reductions.

[18] D. B. Johnson and S. M. Venkatesan. Parallel algorithms for minimum cuts and maximum flows in planar networks. In 23rd *Ann. Symp. on Foundations of Computer Science*, pages 244–254, Chicago, Illinois, 1982. IEEE Computer Society. Contains a result showing flows in planar networks can be computed in *NC*.

[19] R. M. Karp, E. Upfal, and A. Wigderson. Constructing a perfect matching is in random *NC*. In 17th *Ann. Symp. on Theory of Computing*, pages 22–37, Providence, Rhode Island, 1985. Association for Computing Machinery. Contains an RNC^2 procedure for finding the maximum flow when capacities are expressed in unary.

[20] L. G. Khachian. A polynomial time algorithm for linear programming. *Doklady Akad. Nauk SSSR*, 244(5):1093–1096, 1979. Translated in Soviet Math. Doklady, 20, pp. 191–194. Contains the first algorithm for linear programming that is polynomial time in the worst case.

[21] R.E. Ladner. The circuit value problem is log space complete for *P*. *SIGACT News*, 7(1):18–20, January 1975. CVP is proved *P*-complete by a generic reduction from a Turing machine to a family of circuits.

[22] M. Luby. A simple parallel algorithm for the maximal independent set problem. *SIAM Journal on Computing*, 15(4):1036–1053, 1986.

[23] N. Pippenger. On simultaneous resource bounds. In *20th Ann. Symp. on Foundations of Computer Science*, pages 307–311, San Juan, Puerto Rico, 1979. IEEE Computer Society. The original paper defining the class *NC*.

[24] J. Reif. Depth-first search is inherently sequential. *Information Processing Letters*, 20(5):229–234, 1985. The first paper proving a problem based on a greedy search algorithm is *P*-complete.

[25] W. L. Ruzzo. On uniform circuit complexity. *Journal of Computer and System Sciences*, 22(3):365–383, June 1981. The fundamental paper on circuit uniformity.

[26] R.E. Tarjan. Depth first search and linear graph algorithms. *SIAM Journal of Computing*, 1(2):146–160, 1972. The fundamental paper about sequential depth first search with several applications including finding biconnected components.

[27] R.E. Tarjan and U. Vishkin. An efficient parallel biconnectivity algorithm. *SIAM Journal of Computing*, 14(4):862–874, 1985. This paper introduces several important algorithmic techniques including the Euler tour technique on trees. They give an *NC* algorithm for finding biconnected components.

[28] S. Vavasis. Gaussian elimination with pivoting is *P*-complete. *SIAM Journal on Discrete Mathematics*, 2(3):413–423, 1989. Contains an interesting reduction for a highly algebraic problem.

[29] H. Venkateswaran. Private communication cited in Greenlaw, Hoover and Ruzzo, 1983. Venkat proved that ACVP is *P*-complete as well as the problem given in Exercise 2.8.

Part IX

Asynchronous Parallel Computation

22

Asynchronous PRAM Algorithms

Phillip B. Gibbons

AT&T Bell Laboratories
600 Mountain Avenue
Murray Hill, NJ 07974
gibbons@research.att.com

22.1
Introduction

Most of the parallel algorithms presented in this book are designed for and analyzed using the PRAM model of computation. The PRAM model is relatively simple to use: most of the low-level details of parallel machines, e.g., interprocessor communication, memory management, and synchronization, are hidden in the model. For this reason, algorithm design is simplified and PRAM algorithms can be readily adapted to run on a variety of parallel machines.

Nevertheless, it is natural to consider modifying the PRAM model to make it more closely model real machines. In this chapter, we make a few simple adjustments to the PRAM model, making it more realistic, and then study the effect these adjustments have on algorithm design and analysis. In particular, in the new model, processors will no longer execute their instructions in lock-step with one another and accessing the shared global memory will no longer be a unit-time operation.

These two modifications to the PRAM model are motivated by the realities of existing MIMD parallel computers. Existing MIMD machines are asynchronous, i.e., the processors are not constrained to operate in lock-step. Each processor can proceed through its program at its own speed, constrained by the progress of other processors only at explicit synchronization points. However, in the PRAM model, all processors execute in lock-step with one another. Thus in order to safely execute a PRAM program on an asynchronous machine, there must be a synchronization point among all the processors after *each* PRAM instruction. Synchronizing after each instruction is inherently inefficient since the ability of the machine to run asynchronously is not fully exploited and there is a (potentially large) overhead in performing the synchronization. Therefore our first modification to the PRAM model will be to permit the processors to run asynchronously and then charge for any needed synchronization.

Second, in existing MIMD machines, operations involving interprocessor communication take considerably longer to complete than local operations. The time to read or write a global memory location can be over a hundred times slower than a local operation, due in part to the increased distances that data must travel and the overheads incurred when communicating between various components of the machine (e.g., processors, memory system, network). Therefore our second modification will be to charge more for a global memory access than for a local operation. To keep the model simple, we will use a single parameter d to quantify the **communication delay** to memory. This parameter is intended to capture the ratio of the median time for a global memory access to the median time for a local operation.

Both of these new features can have a large impact on algorithm design. Many examples will be given in this chapter after the formal model is defined. For now, we present two examples: the first demonstrates the impact of explicitly charging for synchronization; the second demonstrates the impact of considering the communication delay to memory.

Consider the problem of determining which node in a linked list is the first node in the list. The PRAM algorithm for this problem is simple. Assume that there is one processor per node.

Step 1 Each processor i initializes global memory location M_i to 0.

Step 2 If node i has a successor j, processor i writes 1 to global memory location M_j.

Step 3 Node i is the head of the list if and only if $M_i = 0$ after Step 2.

This algorithm runs in $O(1)$ time on an EREW PRAM with n processors. However, suppose the processors run asynchronously and are subject to varying delays. In the absence of any synchronization points, each processor i cannot distinguish the case where there is no writer to M_i from the case where the writer exists but has been delayed. In particular, the processor assigned to the node at the head of the linked list must synchronize with all its fellow processors before it can safely conclude that it is indeed the head of the list. Thus, in a model that charges, say, $\log p$ for synchronizing p processors, the running time is $O(\log n)$ when n processors are used.[1]

Assigning a non-unit cost to a global memory access, our second modification, can also impact the design and analysis of parallel algorithms. Consider the problem of broadcasting a single value in global memory to all the processors. The usual EREW PRAM algorithm for this problem consists of fanning out copies of the value in a binary tree fashion. The algorithm is depicted pictorially in Figure 22.1(a). Initially, at the root of the tree, one

---

[1] Unless stated otherwise, all logarithms are base two logarithms.

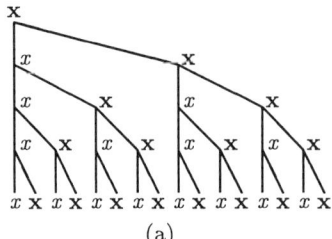
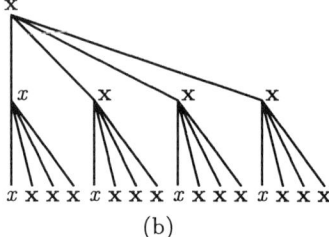

FIGURE 22.1
Accounting for communication delay when broadcasting a value. (a) The value **x** is broadcast to 16 processors in a binary tree fashion. (b) The value **x** is broadcast to 16 processors in a d-ary tree fashion, where $d = 4$.

processor has the value, **x**, to be broadcast. At each level of the tree in turn, each active processor reads a copy of the value and makes a new copy in the global memory. In the figure, each new copy of the value written this round is indicated by an **x**. After $O(\log n)$ such rounds, we have n copies of **x**. This algorithm runs in $O(d \log n)$ time when we account for the communication delay to memory, since each level takes $O(d)$ time.

Now consider the algorithm for broadcasting depicted in Figure 22.1(b), in which each active processor at a level writes $d - 1$ copies of **x**. These $d - 1$ writes to global memory can be overlapped ("pipelined"): an active processor may issue the write requests in a round one after another, without waiting for the earlier requests to reach the global memory. If d writes can be completed in $O(d)$ time by pipelining, then using a d-ary tree to fan-out the copies, as in Figure 22.1(b), leads to a faster algorithm. In particular, since only $O(\log_d n)$ rounds are needed to make n copies of **x**, the algorithm runs in $O(d \log_d n)$ time. This is an improvement over the binary tree algorithm of Figure 22.1(a) by a multiplicative factor of $\log d$.

In the next section (Section 22.2), we formally define the Asynchronous PRAM model, a variant of the PRAM that incorporates both asynchrony and communication delay. Section 22.3 contains Asynchronous PRAM algorithms for the all-prefix-sums operation and other important primitives. Section 22.4 presents algorithms, techniques, and lower bounds for DAG-based computations. Section 22.5 discusses an important variation on the Asynchronous PRAM model. Finally, Section 22.6 summarizes this chapter.

22.2
The Asynchronous PRAM Model

The Asynchronous PRAM model of computation consists of a collection of p sequential processors, each with its own private local memory, communicating with one another through a shared global memory. Each processor has its own local program. Unlike the PRAM, the processors of an Asynchronous PRAM run asynchronously, i.e., each processor executes its instructions independently of the timing of the other processors. Any desired timing dependencies between processors must be explicitly incorporated into the programs of the processors. There is no global clock.

A processor can issue up to one instruction per tick of its local clock. An instruction completes after some unbounded, but finite, number of ticks.

22.2.1 Instruction Types in the Model

There are four types of instructions.

Global read. Read the contents of a global memory location into a local memory location.

Local operation. Perform any RAM operation where the operands are in local memory and the result is stored in local memory.

Global write. Write the contents of a local memory cell into a global memory cell.

Synchronization step. A synchronization step among a set, S, of processors is a logical point in a computation where each processor in S waits for all the processors in S to arrive before continuing in its local program.

For simplicity, we will first consider a model in which a synchronization step involves all processors executing in a program, an operation denoted a synchronization *barrier*. Later, in Section 22.5, we will relax this assumption.

A computation in this model is a series of global phases in which the processors run asynchronously, separated by synchronization barriers (see Figure 22.2). The final instruction in each local program is a synchronization barrier, informing the processors that the program has completed.

Processors can read and write to the global memory asynchronously, but no two processors may access the same location in a phase. Thus in Figure 22.2, for example, the read of A by processor 2 is separated from

	processor 1	processor 2	$\cdots$	processor p
phase 1	read x_1 read x_2 * write to A	read x_3 * write to B write to C		read x_n * * write to D
phase 2	read B * write to B	read A * write to D		read C *
phase 3	* read D *	write to C		read B read A * write to B

FIGURE 22.2
An Asynchronous PRAM computation. There are three phases, terminated by synchronization barriers. Each column contains the instructions executed by a processor. A horizontal line represents a synchronization barrier. An asterisk (*) represents a local operation, i.e., an operation other than a global read, global write, or barrier. In phase 1, for example, processor 1 reads global memory locations x_1 and x_2, performs some local operation, and then writes to global memory location A. Note that no two processors access the same location in the same phase.

the write of A by processor 1 by a barrier, as is shown. Since accesses to a location by different processors are always separated by a synchronization barrier, the varying delays for instruction completion do not affect the program computation.

22.2.2 Computation Costs in the Model

The Asynchronous PRAM defines the following cost measures.

22.2. The Asynchronous PRAM Model

Asynchronous PRAM cost measures	
local operation	1
global read or write	d
k global reads or writes	$d + k - 1$
synchronization barrier	B

As discussed in Section 22.1, d is a parameter quantifying the communication delay. It represents the median time to read a word from the global memory, or to write a word to the global memory and receive an acknowledgement that the write has completed. The parameter d increases with the number of processors in the machine.

An Asynchronous PRAM processor can pipeline its instructions, issuing instructions $i+1$, $i+2$, and so forth, of its local program before instruction i has completed. The pipelining by a processor of its instructions in a phase is limited only by the dependencies (if any) between the instructions. Interdependencies between instructions in a local program arise if, for example, the value returned by a global read dictates the location to be read by a subsequent global read (as in the case of traversing a linked list). The cost in the model to complete a sequence of k global read or write instructions with no interdependencies, issued one after another by a processor, is $d + k - 1$.

The parameter $B = B(p)$, the time to synchronize all the processors, is a nondecreasing function of the number of processors, p, used by the program. In the Asynchronous PRAM model, the parameters are assumed to obey only the following constraints: $2 \leq d \leq B \leq p$. However, a reasonable assumption for modeling most machines is that $B(p) \in O(d \log p)$ or $B(p) \in O(d \log p / \log d)$ (see Exercises).

The running time for a program is computed on a global phase-by-phase basis. The time cost for a phase is the maximum over all processors of the cost of the instructions executed by a processor during the phase. The running time for a program is the sum of the time costs for each phase plus B times the number of synchronization barriers. For notational simplicity, we will use a single parameter B to denote the cost for a barrier, regardless of the number of processors. Note, however, that since $B = B(p)$ is typically a strictly increasing function of the number of processors, p, using more processors in a program results in a larger value for B.

Compared to the PRAM, the Asynchronous PRAM more closely models parallel machines, while preserving much of the simplicity of the PRAM. Lock-step execution of the processors is not assumed; a program must be correct regardless of the varying delays the processors may encounter. On the other hand, for the purpose of measuring the cost of a computation, the model assumes that the processors can execute similar instructions in roughly the same amount of time. Thus the cost for a local operation is 1 and the cost for a global memory access is d, independent of which processor issued the instruction. This is a reasonable first approximation to the behavior of parallel machines.

This concludes the formal description of the Asynchronous PRAM. As with the PRAM model, there are many variants of Asynchronous PRAM models. In the interest of clarity, we leave discussions of other variants to the end of this chapter.

22.3
Primitive Operations Revisited

In this section, we revisit some of the problems discussed in earlier chapters (e.g., all-prefix-sums, multiprefix) and present algorithms for these problems that are suitable for the Asynchronous PRAM. Previous chapters have demonstrated that devising fast algorithms for these primitives can lead to fast algorithms for many other problems.

First we observe that any algorithm for the EREW PRAM can be adapted to run on the Asynchronous PRAM as follows: insert a synchronization barrier after each read step and after each write step of the PRAM program. This forces the Asynchronous PRAM to execute in lock-step, ensuring that (a) no two processors access the same location in a phase, and (b) the program is correct regardless of the delays that may occur. The cost to simulate a single PRAM instruction (involving a read step, a compute step, and a write step) on the Asynchronous PRAM is d for the read step, B for the barrier after the read step, 1 for the compute step, d for the write step, and B for the barrier after the write step. Thus a single PRAM instruction can be simulated in $2B + 2d + 1$ time, which is $O(B)$ time, since $d \leq B$. It follows that an EREW PRAM algorithm running in time t using p processors can be run on an Asynchronous PRAM in $O(Bt)$ time using p processors.

We can simulate an EREW PRAM within the same time bounds using fewer Asynchronous PRAM processors as follows. The idea is to balance the time spent synchronizing with the time spent accessing memory and computing.

LEMMA 22.1
An EREW PRAM algorithm running in time t using p processors can be simulated by an Asynchronous PRAM running in time $O(Bt)$ with p/B processors.

PROOF
Let the PRAM processors be numbered 0 to $p-1$ and consider a single PRAM instruction. Each Asynchronous PRAM processor simulates the read step of B PRAM processors, numbered Bi to $B(i+1)-1$, synchronizes, simulates the compute step and the write step of these same B PRAM processors, and synchronizes again. The cost to simulate a single PRAM instruction on the Asynchronous PRAM is $d+B-1$ for the read step, B for the barrier, B for the compute step, $d+B-1$ for the write step, and B for the second barrier. Thus a single PRAM instruction can be simulated in $5B + 2d - 2$ time, which is $O(B)$ time. ∎

As we shall see, for many problems, algorithms exist that achieve better results than those obtained by applying this lemma to the best known EREW PRAM algorithms.

22.3.1 All Prefix Sums

Our first primitive is the all-prefix-sums operation, introduced in Chapter 1.

DEFINITION 22.1
*The **all-prefix-sums operation** takes a binary associative operator $\oplus$ with identity i, and an ordered set $[a_0, a_1, \ldots, a_{n-1}]$ of n elements, and returns the ordered set $[i, a_0, (a_0 \oplus a_1), \ldots, (a_0 \oplus a_1 \oplus \cdots \oplus a_{n-2})]$.*

(Note that the output $[a_0, (a_0 \oplus a_1), \ldots, (a_0 \oplus a_1 \oplus \cdots \oplus a_{n-1})]$ can be readily obtained from this operation as well.)

There is a simple linear time sequential algorithm and an $O(\log n)$ time, $n/\log n$ processor EREW PRAM algorithm for this problem (see Chapter 1). The PRAM algorithm involves computing partial results in a binary tree fashion. Applying Lemma 22.1 yields an $O(B \log n)$ time Asynchronous PRAM algorithm with $n/(B \log n)$ processors.

This can be improved to $O(B \log n / \log B)$ time by using a B-ary tree, instead of a binary tree. For simplicity, we will describe only the first half of the all-prefix-sums computation, in which the summation of the n input numbers is computed (the second half is left as an exercise). At each level of the tree, each active processor reads $B-1$ values, computes their sum, writes the result, and then synchronizes.

ALGORITHM 22.1
Computing the sum of n numbers.
Input: n numbers stored in global memory.
Output: Processor 0 holds the sum of these numbers.
Comment: Each processor i, $0 \leq i < n$, executes the following program.

>read input i from global memory;
>$cum\_sum :=$ value of input i;
>$level := 0$;
>
>**while** $level < \lceil \log_B n \rceil$
>
>>**if** left-most among your siblings at the current level of the B-ary tree **then**
>>
>>>read from global memory the values of siblings $2, 3, \ldots, B$;
>>>$cum\_sum :=$ the sum of the values of all B siblings;
>>>write the value of cum_sum to global memory;
>>
>>**endif;**
>>
>>**barrier;**
>>
>>$level := level + 1$;
>
>**endwhile;**

Note that the global reads can be pipelined. The cost for each level of the tree can be calculated according to the following table.

Time Complexity for Algorithm 22.1	
while loop test	1
left-most sibling test	1
read sibling values	$d + B - 2$
sum sibling values	$B - 1$
write sum to memory	d
synchronization barrier	B
increment level counter	1
time per level	$3B + 2d$

There are $\lceil \log n/ \log B\rceil$ levels, so Algorithm 22.1 runs in $O(B \log n/ \log B)$ time with n processors.

We can achieve an (optimal) linear processor-time product as follows. Let $T = B \log n/ \log B$. Have each of n/T Asynchronous PRAM processors initially read in T inputs, sum them, and then synchronize. Since the global reads can be pipelined, this initial phase requires $(d+T-1)+(T-1)$ time, i.e., $O(T)$ time since $d \leq B \leq T$. Next sum the n/T partial sums using a B-ary tree, as in Algorithm 22.1. In this way, n/T Asynchronous PRAM processors suffice to achieve $O(T)$ time. Moreover, if B is a strictly increasing function of the number of processors, using fewer processors results in faster barriers and hence a faster algorithm.

22.3.2 Brent's Scheduling Principle

Before presenting more new algorithms, we present results for simulating an Asynchronous PRAM with p_0 processors on an Asynchronous PRAM with $p < p_0$ processors. In each case, for comparison, we present the corresponding PRAM lemma followed by the Asynchronous PRAM version. We will refer to the processors of the (Asynchronous) PRAM being simulated as *virtual processors*, and the processors of the simulating (Asynchronous) PRAM as *machine processors*.

We begin with the following simple lemma for the PRAM.

LEMMA 22.2
A PRAM program using p_0 processors and running in time t can be simulated by a PRAM using $p < p_0$ processors in $O((p_0/p)t)$ time.

PROOF
For each PRAM instruction, each machine processor simulates p_0/p virtual processors in $O(p_0/p)$ time. ∎

In the corresponding lemma for the Asynchronous PRAM, we simulate a phase at a time.

LEMMA 22.3
An Asynchronous PRAM program using p_0 processors and running in $t + B(p_0)s$ time, where s is the number of synchronization barriers, can be simulated by an Asynchronous PRAM using $p < p_0$ processors in $O((p_0/p)t + B(p)s)$ time.

PROOF
For each phase, each machine processor simulates the instructions in the phase for p_0/p virtual processors and then synchronizes. ∎

In the remainder of Section 22.3.2, we prove a more general result, corresponding to Brent's scheduling principle for the PRAM.

LEMMA 22.4 *Brent's scheduling principle for the PRAM*
Let the work in a PRAM algorithm be the sum over all processors of the number of instructions performed by a processor. A PRAM program using p_0 processors, a total of w work, and t time can be simulated by a PRAM using $p < p_0$ processors in $w/p + t$ time.

PROOF
Let w_i be the number of processors performing an instruction in step i of the program, $1 \leq i \leq t$. By dividing the work for step i evenly among the processors, each machine processor simulates either $\lceil w_i/p \rceil$ or $\lfloor w_i/p \rfloor$ virtual processors for step i. Therefore,

$$\text{total time} \leq \sum_{i=1}^{t} \lceil w_i/p \rceil \leq \sum_{i=1}^{t}((w_i/p) + 1) = (\sum_{i=1}^{t} w_i)/p + t = w/p + t.$$

∎

Thus a technique to minimize the time of a simulation in which the number of processors is being reduced is to try to balance the work evenly among the machine's processors. Note that this lemma does not account for scheduling overheads such as the cost of determining the assignment of work to the processors at each step.

Lemma 22.4 applies to models in which the processors execute in lockstep and all instructions take unit time. In the context of the Asynchronous PRAM, then, we need a revised definition of the work an algorithm, a revised statement of the theorem, and a new proof.

DEFINITION 22.2
*The **work** of an Asynchronous PRAM algorithm is the sum over all processors of the cost of the instructions performed by a processor in the algorithm, not counting synchronization barriers.*

THEOREM 22.1 *Scheduling principle for the Asynchronous PRAM*
An Asynchronous PRAM program using p_0 processors, s synchronization steps, a total of w work, and $t + B(p_0)s$ time can be simulated by an Asynchronous PRAM using $p < p_0$ processors in $w/p + t + B(p)s$ time.

As in Lemma 22.3, the proof of this theorem involves simulating the virtual processors on a phase-by-phase basis. We begin with the following lemma.

22.3. Primitive Operations Revisited

LEMMA 22.5
Let w_α (t_α) be the work (time) for phase α of an Asynchronous PRAM program using p_0 processors. Then the phase can be simulated by an Asynchronous PRAM using $p < p_0$ processors in $w_\alpha/p + t_\alpha$ time.

PROOF
Let I_j be the instructions performed in phase α by the virtual processor with the jth largest cost in the phase. Thus $\text{cost}(I_1) \geq \text{cost}(I_2) \geq \cdots \geq \text{cost}(I_{p_0})$. Each sequence of instructions I_j in phase α can be viewed as a series of unit-time steps (see Figure 22.3). Let v_k be the number of virtual processors with at least k steps in phase α. Thus $w_\alpha = \sum_{k=1}^{t_\alpha} v_k$.

Assign machine processor i to perform the instructions in I_i, I_{i+p}, I_{i+2p}, and so forth. Processor 1 has the most work and it performs $\lceil v_k/p \rceil$ unit-

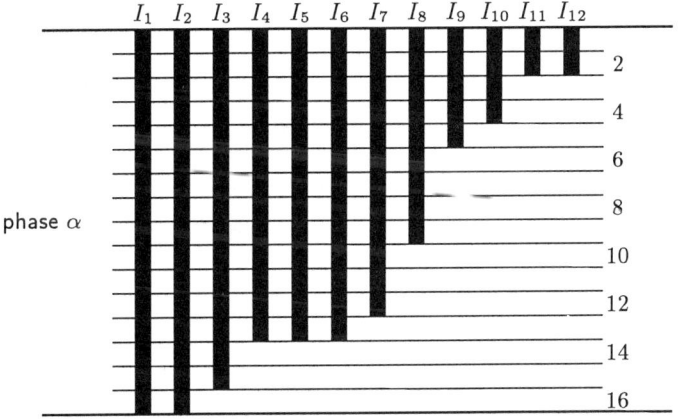

FIGURE 22.3
An example distribution of the work performed by each virtual processor in phase α, as discussed in the proof of Lemma 22.5. In this figure, $p_0 = 12$ virtual processors are active. The time, t_α, for the phase is 16 time steps; the work, w_α, for the phase is $16 + 16 + 15 + \cdots + 2 = 120$. Given $p = 4$ machine processors, processor 1 performs the instructions in I_1, then those in I_5, and finally those in I_9, for a total of 34 time steps of work. Thus the time for phase α with 4 processors is 34, which is less than $w_\alpha/p + t_\alpha = 120/4 + 16 = 46$, in accordance with Lemma 22.5.

time steps that are the kth step for some virtual processor. Hence, the time for phase α on the Asynchronous PRAM with p processors is

$$\text{time} \leq \sum_{k=1}^{t_\alpha} \lceil v_k/p \rceil \leq \sum_{k=1}^{t_\alpha} (v_k/p + 1) \leq w_\alpha/p + t_\alpha.$$

∎

Now we can prove Theorem 22.1.

PROOF
For each phase, each machine processor simulates the instructions in the phase for $\lceil p_0/p \rceil$ or $\lfloor p_0/p \rfloor$ virtual processors and then synchronizes. Let w_α (t_α) be the work (time) for phase α. Then $w = \sum_{\alpha=1}^{s} w_\alpha$ and $t = \sum_{\alpha=1}^{s} t_\alpha$. Let T_{ap} be the total time taken by the simulating Asynchronous PRAM. By Lemma 22.5,

$$T_{ap} \leq \sum_{\alpha=1}^{s} (w_\alpha/p + t_\alpha + B(p)) \leq w/p + t + B(p)s.$$

∎

Note that, as in Brent's scheduling principle for the PRAM model (Lemma 22.4), Theorem 22.1 does not account for scheduling overheads.

22.3.3 Multiprefix

In this section, we revisit the multiprefix operation introduced in Chapter 2.

DEFINITION 22.3
The **multiprefix operation** *takes a binary associative operator $\oplus$ with identity i, an ordered set $[a_0, a_1, \ldots, a_{n-1}]$ of n elements, and an ordered set $[l_0, l_1, \ldots, l_{n-1}]$ of n labels. It returns the ordered set $[b_0, b_1, \ldots, b_{n-1}]$ such that, for $0 \leq j \leq n-1$, $b_j = (i \oplus a_{z_0} \oplus \cdots \oplus a_{z_k})$, where $z_0 < \cdots < z_k$ and for $0 \leq x < j$,*

$$l_x = l_j \text{ if and only if there exists } y, 0 \leq y \leq k, \text{ such that } x = z_y.$$

The multiprefix operation computes all-prefix-sums on a per-label basis. There is a simple linear time sequential algorithm and an $O(\log n)$ time, n processor EREW PRAM algorithm for the multiprefix problem (see Chapter 2). Applying Lemma 22.1 yields an $O(B \log n)$ time Asynchronous PRAM algorithm with n/B processors.

22.3. Primitive Operations Revisited

The all-prefix-sums algorithm of Section 22.3.1 can be used to compute the prefixes for each possible label. Suppose the labels are integers in the range of 1 to L. Let $T = B \log n / \log B$. By assigning n/T processors per tree, one tree per label, a multiprefix problem of n data items can be solved by applying the all-prefix-sums algorithm to each tree in parallel. This approach uses $O(T)$ time, nL/T processors, and $O(nT + \lceil n/T \rceil L)$ memory cells (see Exercises).

The number of processors (and the number of memory cells) in the above approach increases proportionally with the number of labels. We next describe an algorithm that runs in $O(T)$ time using n processors, regardless of the number of labels. To simplify the description of our algorithm, we will consider the case where the associative operation, $\oplus$, is addition.

Consider a forest of L trees, one for each possible label. Each tree is an (implicit) complete B-ary tree on n leaves. In the first half of the algorithm (the "summation" pass), the per-label sums are computed by progressing level-by-level up each tree from its leaves to its root. However, unlike Algorithm 22.1, we have only x_j processors working on the tree for label j, where x_j is the number of data items with label j. Thus a more greedy approach is used: processor i starts at leaf i of the tree for its label and greedily works its way towards the root of the tree, stopping only if it encounters a smaller-numbered processor with the same label (see Figure 22.4).

ALGORITHM 22.2
Computing the multiprefix of n numbers ("summation" pass only).
Input: n number/label pairs stored in global memory.
Output: The per-label summation of the n numbers.
Comment: The "partial sums" pass of the full algorithm to compute a multiprefix is omitted here, and left as an exercise. Each processor i executes the following program.

    read from global memory the inputs a_i and l_i;
    cum_sum := value of input a_i;
    $active$:= TRUE;
    $level$:= 0;

    while $level < \lceil \log_B n \rceil$

        (Consider a $B \times B$ matrix for each node of the next level in the forest.)
        if $active$ = TRUE **then**

write the value of cum_sum in a *column* of the matrix for your parent in the tree for label l_i;
(Columns such that no processors have the appropriate label will remain all zero.)

endif;

barrier;

if *active* = TRUE **then**

read from global memory the B values in your *row* in the matrix for your parent in the tree for label l_i;
$cum\_sum :=$ the sum of the B values in your row;

if not left-most processor with your label among your $B - 1$ siblings at the current level of the tree **then**
  $active :=$ FALSE;
endif;

endif;

$level := level + 1$;

endwhile;

For simplicity, assume that all input values are positive. For each node visited by an active processor, i, the processor reads B values in a row: a value is nonzero if and only if there exists an active processor with the same label as i among its B siblings at this level of the tree. In this way, an active processor can determine whether or not it is the left-most sibling at this level of the tree, and if so, continue to the next level of the tree. If input values less than or equal to zero are permitted, flags can be used to distinguish a value that is a cum_sum for some active processor from a value that is not.

The "partial sums" pass of Algorithm 22.2, in which the partial sums are computed in a pass from the root of each tree back to its leaves, is left as an exercise.

Algorithm 22.2 runs in $O(B \log n / \log B)$ time using n processors. The number of global memory cells needed can be calculated as follows. There are L B-ary trees with n leaves each. B^2 cells are used for the matrix associated with each non-leaf node in the tree, for a total of $O(LnB)$ memory cells. Further refinements reduce the memory requirements to $O(Ln)$ cells and include

22.3. Primitive Operations Revisited

	0	1	2	3	4	5	6	7	8	9	10	11	12	13	14	15
label	3	7	3	3	4	7	1	3	1	6	9	9	4	3	3	9
value	2	–	1	3	–	–	–	2	–	–	–	–	–	1	1	–
sum	0	–	2	3	–	–	–	6	–	–	–	–	–	8	9	–

(a)

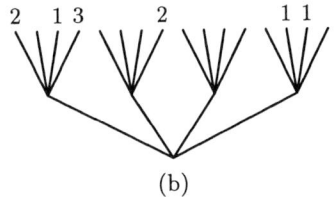

(b)

p_0	p_2	p_3	
2	0	1	3
2	0	1	3
2	0	1	3
2	0	1	3

			p_7
0	0	0	2
0	0	0	2
0	0	0	2
0	0	0	2

0	0	0	0
0	0	0	0
0	0	0	0
0	0	0	0

	p_{13}	p_{14}	
0	1	1	0
0	1	1	0
0	1	1	0
0	1	1	0

p_0	p_7		p_{13}
6	2	0	2
6	2	0	2
6	2	0	2
6	2	0	2

(c)

FIGURE 22.4
A multiprefix computation. (a) A multiprefix problem on $n = 16$ inputs, and its desired output. Only the values and prefix sums relevant to label 3 are shown. (b) Algorithm 22.2 uses implicit B-ary trees for each label. Shown here is the tree for label 3, where $B = 4$. The six processors with label 3 start at the appropriate six leaves of the tree for label 3 and work together to compute its prefix sums. (c) The sum for label 3 is computed by the summation pass of the algorithm. The intermediate results at two levels of the tree are shown. As a final step in this pass, processor p_0 sums $6 + 2 + 0 + 2$ to get the total sum, 10, of the values with label 3.

initialization costs, while maintaining the same time and processor bounds for any integer value of L (see Exercises).

22.4 DAG-Based Computations

In this section, we study problems whose PRAM algorithms can be viewed as a family of directed graphs, and show how these algorithms can be restructured to achieve improved time complexity on the Asynchronous PRAM.

DEFINITION 22.4

*A **computation graph** is a directed graph G such that (a) the nodes of G are labeled with either input values, unary operators, or binary operators, and (b) there is a directed edge in G from one node to another if and only if the former node computes an operand of the operation at the latter node.*

Thus input nodes have indegree 0, unary nodes have indegree 1, and binary nodes have indegree 2.

A computation graph for the summation problem on n inputs is a complete binary tree of n leaves. Other examples will be given later in this section.

We begin in Section 22.4.1 with an example problem, computing the Fast Fourier Transform. In Section 22.4.2 we show how to generalize the algorithm used for the Fast Fourier Transform to an algorithm for any operation that can be expressed as a family of directed graphs. This generalized algorithm may use more processors than necessary. In Section 22.4.3, we present an algorithm for merging two sorted lists which uses fewer processors than the generalized algorithm. Finally, in Section 22.4.4, we prove a lower bound on the time to solve a large class of dag-based problems in the Asynchronous PRAM model.

22.4.1 Fast Fourier Transform

The Fourier Transform and its inverse are functions used to convert between two different mathematical representations for a function. For example, for a given waveform, f, the Fourier transform and its inverse are used to convert between f expressed as a function of time and f expressed as a function of frequency (i.e., a sum of sinusoids of different frequencies). The utility of the transform is that often a computation to be performed will be far simpler using one representation than the other. The Fast Fourier Transform

(FFT) is an $O(n \log n)$ time sequential algorithm for computing an n-point discrete approximation to the Fourier transform. The FFT is used extensively in scientific applications.

The details of the FFT algorithm are not important for the discussion in this chapter, except for the fact that the fastest PRAM implementation of the FFT algorithm has a computation graph that is a butterfly graph of n rows and $\log n + 1$ columns, where n is the number of inputs (see Figure 22.5). A formal description of the butterfly graph is as follows. Assume that n is a power of two. The rows of the butterfly are numbered 0 to $n-1$; the columns are numbered 0 to $\log n$. For $0 \leq i < \log n$, there is a directed arc from the node in row r, column i

to the node in row r, column $i+1$, and

to the node in row $r' \bmod (n/2^i)$, column $i+1$, where $r' = r + n/2^{i+1}$.

The n input nodes comprise column 0; the n output nodes comprise column $\log n$.

By simulating one column of the butterfly at a time, an EREW PRAM can compute the Fast Fourier Transform in $O(\log n)$ time with n processors. Applying Lemma 22.1 yields an $O(B \log n)$ time Asynchronous PRAM algorithm with n/B processors.

We can improve the time complexity by using the following algorithm instead. The butterfly computation graph has the following properties. The value of any node in column $i \geq 1$ of the butterfly is a function of the values of its two neighbors in column $i-1$. Alternatively, the value of any node in column $i \geq 2$ can be computed from the values of four nodes in column $i-2$. In general, the value of any node, v, in column $i \geq k$ can be computed from the values of 2^k nodes in column $i-k$, by simulating the binary tree with these nodes as leaves and node v as the root. A processor can compute the value of a column i node from the values of B column $(i - \log B)$ nodes in $O(B)$ time by reading in these B values and simulating the binary tree.

This leads to the following algorithm for n processors. We partition the columns into $\log n / \log B$ stages of $\log B$ consecutive columns each. By the structure of the butterfly, the value of each node in the last column of its stage can be computed by a binary tree whose B leaves are in the last column of the previous stage (see Figure 22.5).

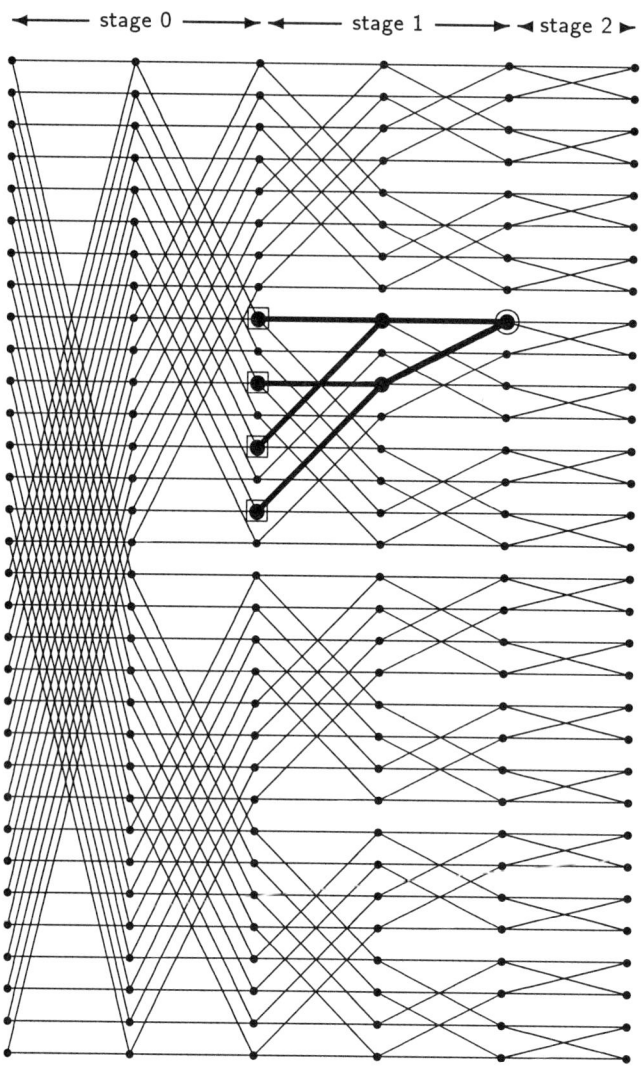

FIGURE 22.5
A butterfly graph with 32 rows. In Algorithm FFT, the graph is divided into stages of $\log B$ columns. Shown here are the stages when $B = 4$. The circled node is in the last column of stage 1. The binary tree for computing the value of the circled node is shown in bold, with its leaf nodes enclosed in squares.

22.4. DAG-Based Computations

ALGORITHM 22.3
Computing the FFT of n numbers.
Input: n numbers stored in global memory.
Output: The n values resulting from a Fast Fourier Transform on the inputs.
Comment: Each processor i executes the following program.

$stage := 0$;

while $stage < \lceil \log_B n \rceil$

(Processor i computes the value for the row i node in the last column of this stage.)
read from global memory the values of the B leaves for the row i node;
compute the value at the node by simulating the binary tree for the node;
write B copies of the value to the global memory;

barrier;

$stage := stage + 1$;

endwhile;

Note that B copies are written to the global memory of each value computed. This is done to avoid concurrent access to a location in a phase, since B processors read the value during the next stage. Each phase (i.e., each iteration of the while loop) takes $O(B)$ time, so Algorithm 22.3 runs in $O(B \log n / \log B)$ time using n processors.

This algorithm can be improved to use only $n \log B / B$ processors with the same time complexity (see Exercises), for an optimal $O(n \log n)$ processor-time product.

22.4.2 DAG-Based Algorithms

The FFT algorithm for the Asynchronous PRAM is an example of a general technique for improving the time complexity of DAG-based computations. This technique applies to any problem whose PRAM algorithm can be viewed as a family of computation graphs with the following properties.

1. There is one computation graph for each input size.
2. The computation graph is acyclic (a DAG).
3. Each node has (indegree and) outdegree at most two.

Problems of this type include all problems in the class *NC*, defined in Chapter 21.

DEFINITION 22.5
The **level** *of a node in a computation graph is the number of arcs in the longest path from an input to the node. The* **depth** *of a computation graph is the maximum level of any node in the graph. The* **width** *of a computation graph is the maximum number of nodes at any one level.*

Consider a family of computation graphs satisfying the above properties, and let G_n be the graph for input size n. Let $D(n) \geq \log n$ be the depth of G_n and $W(n)$ be its width. In the worst case, a sequential algorithm runs in $D(n)\,W(n)$ time and an EREW PRAM algorithm runs in $D(n)$ time with $W(n)$ processors. Applying Lemma 22.1 yields an $O(B\,D(n))$ time, $W(n)/B$ processor Asynchronous PRAM algorithm.

We can improve this time bound as follows. As in Algorithm 22.3, G_n is partitioned into stages of $\log B$ consecutive levels each. The value of each node, v, in a stage can be computed from the values of at most B nodes from earlier stages, by simulating the subgraph of the DAG that has these nodes as leaves and v as its root (see Figure 22.6). No more than B leaf nodes are needed since there are $\log B$ levels in a stage and the indegree of each internal node is at most two.

This leads to the following algorithm.

ALGORITHM 22.4
A general technique for evaluating a DAG of depth D.
Input: A DAG of depth D stored in global memory.
Output: The outputs of the DAG.
Comment: Each processor i executes the following program.

> $stage := 0;$
>
> **while** $stage < \lceil D/\log B \rceil$
>
>> (Processor i computes the value for the ith node in this stage.)
>> read from global memory the values of up to B leaves for your node;
>> compute the value at the node using local operations that mimic the DAG;
>> write B copies of the value to the global memory;

barrier;

stage := *stage* + 1;

endwhile;

The subgraph for a node has depth at most $\log B$. Since the outdegree of a node is at most two, each node, v, is a leaf node for at most B nodes.

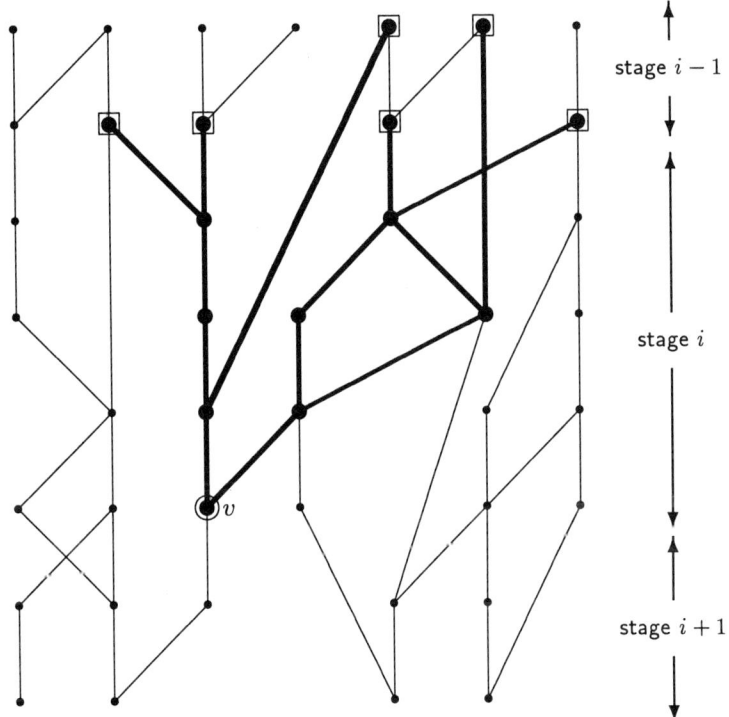

FIGURE 22.6
Part of an example computation graph. The arcs of the graph are directed downward (not shown). In Algorithm DAG, the graph is divided into stages of $\log B$ levels. Shown here is stage i when $B = 16$. The subgraph for computing the value of the circled node, v, is shown in bold. In Algorithm DAG, the processor for this node reads from global memory the values of the six leaves for the node (nodes enclosed in squares), and then computes the value using local operations that mimic the subgraph.

Thus in Algorithm 22.4, as in Algorithm 22.3, a processor writes B copies of a value it computes in order to ensure that each subsequent reader of the value has access to a unique copy. This avoids concurrent access to a location in a phase.

Since there are $\log B$ levels in each stage, there are at most $W(n)\log B$ nodes in a stage. The values for all the nodes in a stage can be computed in $O(B)$ time on the Asynchronous PRAM. Thus Algorithm 22.4 runs in $O(B\ D(n)/\log B)$ time with $W(n)\log B$ processors. This represents an improvement by a factor of $\log B$ over the running time resulting from simply synchronizing at each level of the computation graph. On the other hand, we require a factor of $\log B$ more processors.

Often, the processor bounds resulting from the general technique described above can be improved. For example, as we have seen, there are algorithms for FFT and all-prefix-sums that use fewer processors. In the next section another example is presented in which the number of processors required is less than $W(n)\log B$.

22.4.3 Bitonic Merge

DEFINITION 22.6
A sequence of elements over a totally ordered set is **bitonic** *if it is the cyclic shift of a monotonically increasing subsequence followed by a monotonically decreasing subsequence. The* **bitonic merge problem** *is to sort a bitonic sequence.*

For example, the sequence

$$\{12, 16, 23, 36, 45, 48, 42, 34, 28, 25, 20, 9, 8, 4, 6, 10\}$$

is a bitonic sequence.

An algorithm for the bitonic merge problem can be used to merge two sorted lists, or as a subroutine for a bitonic sorting algorithm. There is an $O(n)$ time sequential algorithm and an $O(\log n)$ time, n processor EREW PRAM algorithm for the bitonic merge problem. The computation graph for the bitonic merge problem on n inputs is a butterfly graph of $\log n$ depth and n width (i.e., n rows and $\log n + 1$ columns).

Applying Lemma 22.1 yields an $O(B\log n)$ time, n/B processor Asynchronous PRAM algorithm. The technique of the previous section yields an $O(B\log n/\log B)$ time algorithm using $n\log B$ processors. In what follows, we show how to reduce the processor bound to only n/B processors, while maintaining the same time bound.

22.4. DAG-Based Computations

We partition the graph into $\log n / \log B$ stages, each consisting of $\log B$ columns. For each stage j, $0 \leq j < \log n / \log B$, the rows of the butterfly graph can be partitioned into sets $S_{j,1}, S_{j,2}, \ldots, S_{j,n/B}$ of size B with the following property: for each stage j, the outputs of the rows in $S_{j,i}$ are the result of sorting the inputs to these same rows. Each stage will have a different partitioning of the rows. The rows in a set will be evenly spaced, but not adjacent (except for the final stage). All comparisons in a stage are between elements of rows that are in the same set (see Figure 22.7).

This leads to the following Asynchronous PRAM algorithm.

ALGORITHM 22.5
Bitonic merge.
Input: Two bitonic sequences of length $n/2$ stored in global memory.
Output: A bitonic sequence of length n resulting from merging the two input sequences.
Comment: Each processor i executes the following program.

$j := 0;$

while $j < \lceil \log_B n \rceil$

(Processor i sorts the B elements corresponding to the "rows" for set $S_{j,i}$.)
read from global memory the B elements in $S_{j,i}$;
sort these B elements using a sequential sorting algorithm;

barrier;

write back the B elements sorted order;

barrier;

$j := j + 1;$

endwhile;

If the B elements in a set were in arbitrary order, the sorting step above would require $\Omega(B \log B)$ time, since general comparison-based sorting of m elements requires $\Omega(m \log m)$ comparisons in the worst case. However, the sorting step can in fact be done in $O(B)$ time since the B elements in a set form a bitonic sequence. Thus Algorithm 22.5 runs in $O(B \log n / \log B)$ time on an Asynchronous PRAM with n/B processors. The proof of the correctness of this algorithm is left as an exercise.

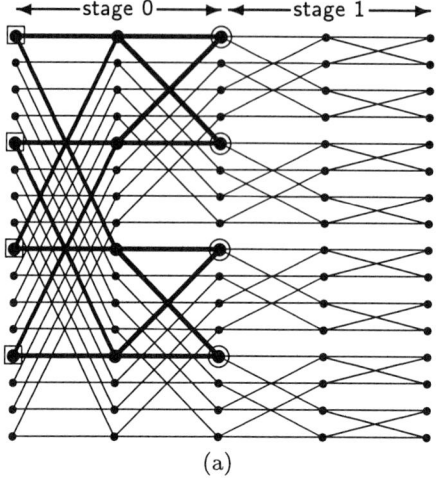

12	12	8	6	4
16	16	4	4	6
23	20	6	8	8
36	9	9	9	9
45	8	**12**	12	10
48	4	16	10	12
42	6	20	20	16
34	10	10	16	20
28	28	**28**	23	23
25	25	25	25	25
20	23	23	28	28
9	36	34	34	34
8	45	**45**	42	36
4	48	48	36	42
6	42	42	45	45
10	34	36	48	48

(b)

FIGURE 22.7
A bitonic merge computation. (a) The computation graph for a bitonic merge problem with 16 inputs. The graph is divided into stages of $\log B$ columns, where $B = 4$. The subgraph corresponding to the rows in the set $S_{0,4}$ is shown in bold. (b) The progression of intermediate values in the bitonic merge computation. The first column is the inputs and the last is the sorted outputs.

This is an example of a general paradigm for saving processors in the Asynchronous PRAM by (a) using a known parallel algorithm for the problem to structure the computation, while (b) performing the individual steps by reading B values, running a sequential algorithm on these values, and then writing the results.

22.4.4 Lower Bounds

Most of the Asynchronous PRAM algorithms presented thus far have run in $O(B \log n / \log B)$ time. The natural question is whether this time bound can be improved. We conclude our discussion of DAG-based computations with an indication that $\Theta(B \log n / \log B)$ time is the best that can be achieved for many of these problems. In particular, the following theorem states that $\Omega(B \log n / \log B)$ time is required to sum n numbers on an Asynchronous PRAM, matching the upper bound of Section 22.3.1.

THEOREM 22.2
Given n numbers, stored one per global memory location, and the following four types of instructions: $L := G$, $L := L + L$, $G := L$, and "barrier", where L is a local cell and G is a global cell, then the sum of n numbers on an Asynchronous PRAM with this instruction set requires $\Omega(B \log n / \log B)$ time, regardless of the number of processors.

PROOF
Fix the number of processors p. We will show that the running time for computing the sum is $\Omega(B(p) \log n / \log B(p))$. Let the fastest algorithm have s phases, of time $t_1, t_2, \ldots, t_s$. In a phase of time k, each processor can at best compute the sum of k numbers, thus the number of partial sums reduces by a factor of k at best. So in order to produce the sum, we must have $n/(t_1 t_2 \cdots t_s) \leq 1$. It can be readily proved (see Exercises) that the time $t = \sum_{i=1}^{s} t_i$ is minimized when $t_i = n^{1/s}$ for all i. Thus $t \geq s n^{1/s}$.

The running time, τ, to produce the sum is $t + Bs$. Let $\alpha = s \log B / \log n$. If $\alpha \geq 1$, then $s \geq \log n / \log B$, so

$$\tau \geq Bs \geq B \log n / \log B.$$

If $\alpha < 1$, then

$$\begin{aligned} \tau > t \geq s n^{1/s} &= (\alpha \log n / \log B) n^{\log B / (\alpha \log n)} \\ &= \alpha B^{1/\alpha} \log n / \log B \\ &> (B/2) \log n / \log B. \end{aligned}$$

In either case, the theorem is proved. ∎

This argument can be applied to any n input, m output function f, where (a) at least one of the outputs of f depends on all the inputs, and (b) the basic step permitted is combining two partial results to get one.

22.5
Subset Synchronization

The only synchronization step in the Asynchronous PRAM, as described thus far, is a synchronization barrier among all the processors in the program. Because of the global nature of the synchronization, this model is relatively simple to use since the processors reach a well-defined state at each barrier. For many problems, however, this global synchronization is overly constraining. Often each processor needs to synchronize with only a small subset of the other processors in order to enforce the desired ordering of read and write accesses. In this section, we describe a variant of the Asynchronous PRAM model which permits sets of processors to synchronize independently and in parallel. The processors in a set wait for only those processors in the same set. We will refer to this less restrictive model as an Asynchronous PRAM *with subset synchronization*.

22.5.1 Definition of the Model

In the Asynchronous PRAM with subset synchronization, the local program for a processor consists of a series of phases in which the processor runs independently, separated by synchronization steps involving at least one other processor. A synchronization step among a set, S, of processors is a logical point in a computation where each processor in S waits for all the processors in S to arrive before continuing in its local program. The set S may not be known at the beginning of the phase, e.g., it may be data dependent.

Processors can read and write to the global memory asynchronously, but no two processors may access the same memory location unless there is a synchronization step involving both processors between the two accesses. For example, if processor 1 writes to global memory location A and processor 2 reads from A, then there must be a synchronization step among processors 1 and 2 between these two accesses of A.

The computation costs in the Asynchronous PRAM with subset synchronization are the same as before (Section 22.2.2), with the following generalizations. The cost of a synchronization step among a set of processors is $B(x)$, a nondecreasing function of x, where x is the number of processors in the set. The parameters are assumed to obey only the following constraints: $2 \leq d \leq B(x) \leq p$, where p is the number of processors used by the program and $2 \leq x \leq p$. However, a reasonable assumption is that $B(x) \in \Theta(d\lceil \log x / \log d \rceil)$ (see Exercises). For example, $B(d) \in \Theta(d)$. The completion time for the synchronization step is $B(x)$ plus the time that the *last* processor in the set reached the step.

Formally, the *completion time* for a program is defined inductively as follows. Initially, all processors begin their local programs at time zero. Inductively, consider a phase, i, that is followed by a synchronization step among a set, S, of processors. The completion time for phase i for a processor in S (not counting the synchronization step) is defined to be the completion time for the processor's prior synchronization step plus the cost for its instructions in phase i. The completion time for the synchronization step is the maximum completion time for phase i over all processors in S plus the cost, $B(|S|)$, of the synchronization step itself. In this way, the completion time for a local program is defined. The *running time* for an algorithm is defined to be the maximum, over all processors j, of the completion time for processor j's local program.

22.5.2 A Program with Subset Synchronization

Although the cost measure for the Asynchronous PRAM with subset synchronization is more complicated, many algorithms are fairly easy to analyze. Consider, for example, the following algorithm for computing the sum of n numbers.

ALGORITHM 22.6
Computing the sum of n numbers.
Input: n numbers stored in global memory.
Output: Processor 0 holds the sum of these numbers.
Comment: Unlike Algorithm 22.1, this algorithm is tailored for an Asynchronous PRAM with subset synchronization, using a d-ary tree instead of a B-ary tree. Each processor i, $0 \leq i < n$, executes the following program.

    read input i from global memory;
    $cum\_sum :=$ value of input i;
    $level := 0$;

    while $level < \lceil \log_d n \rceil$

if left-most among your siblings at the current level of the d-ary tree **then**

> read from global memory the values of siblings $2, 3, \ldots, d$;
> $cum\_sum :=$ the sum of the values of all d siblings;
> write the value of cum_sum to global memory;

endif;

synchronize with your $d-1$ siblings at the next level of the tree;

$level := level + 1;$

endwhile;

All processors that remain active perform the same amount of work. The cost for each level of the tree is as follows.

Time Complexity for Algorithm 22.6

while loop test	1
left-most sibling test	1
read sibling values	$d + d - 2$
sum sibling values	$d - 1$
write sum to memory	d
synchronization	$B(d)$
increment level counter	1
time per level	$B(d) + 4d$

There are $\lceil \log n / \log d \rceil$ levels, so Algorithm 22.6 runs in $O(B(d) \log n / \log d)$ time with n processors. Fewer processors can be used if initially each processor sums $O(B(d) \log n / \log d)$ inputs prior to its first synchronization step. With this modification, $n \log d / B(d) \log n$ processors suffice, so the processor-time product is $O(n)$.

When $B(d) \in O(d)$ as argued above, the running time is $O(d \log n / \log d)$ with $n \log d / d \log n$ processors.

22.6
Summary

The Asynchronous PRAM model of computation is a variant of the PRAM that more closely models MIMD machines of today and the near future. Lock-step execution of the processors is not assumed; instead, any desired ordering among global reads or writes by different processors must be enforced using explicit synchronization instructions. Moreover, the model accounts for the fact that global reads or writes take considerably longer than local operations by charging more for the former than the latter.

Any EREW PRAM algorithm can be simulated on the Asynchronous PRAM by explicitly synchronizing all the processors after each read step and after each write step of the PRAM computation. However, typically, faster Asynchronous PRAM algorithms can be achieved by restructuring the computation so that the processors need to synchronize less frequently. Examples include all-prefix-sums, multiprefix, list ranking (see Exercises), Fast Fourier Transform, and bitonic merge, as well as all problems solved using these important primitives (given throughout this book). Techniques developed in this chapter can be used to reduce both the time and processor complexity of parallel algorithms developed for the PRAM, resulting in algorithms more suited for real parallel machines.

22.7
Exercises

22.1 Write an Asynchronous PRAM algorithm to compute the OR of n bits. What are the running time and number of processors required? How does this compare with other PRAM algorithms for the same problem?

22.2 (a) Describe a technique for synchronizing p processors on an asynchronous parallel machine in $O(\log p)$ rounds, where in each round up to $p/2$ disjoint pairs of processors can synchronize. (b) If a pair of processors can synchronize using the global memory in $O(d)$ time, the approach in part (a) results in an $O(d \log p)$ barrier implementation. Show how to improve this to $O(d \log p / \log d)$ on an asynchronous parallel machine in which k pairwise synchronizations can be completed in $O(d + k)$ time by pipelining.

22.3 (a) Give an $O(B \log n / \log B)$ time, n processor Asynchronous PRAM algorithm for computing the all-prefix-sums of n inputs. The "summation" pass of such an algorithm is given as Algorithm 22.1. Complete the algorithm by describing the missing "partial sums" pass. (Refer to Chapter 1 as needed.) (b) Describe how to reduce the processor-time product to $O(n)$ while maintaining the same time bound.

22.4 Consider the multiprefix operation described in Section 22.3.3. Let n be the number of input elements and the labels be integers between 1 and L, inclusive.

Chapter 22. Asynchronous PRAM Algorithms

1. Show that the straight-forward approach of applying Algorithm 22.1 to each of L trees in parallel uses $O(nT + \lceil n/T \rceil L)$ memory cells.
2. Give an $O(B \log n / \log B)$ time, n processor Asynchronous PRAM algorithm for computing the multiprefix. The "summation" pass of such an algorithm is given as Algorithm 22.2. Complete the algorithm by describing the missing "partial sums" pass.
3. Refine Algorithm 22.2 to handle uninitialized memory cells. The algorithm in Section 22.3.3 assumes that all memory cells contain a zero prior to the first write to the cell. Suppose instead that a cell can contain any value prior to the first time it is written, and refine the algorithm to work regardless of the initial values in the cells. (A technique is needed for distinguishing true values from arbitrary initial values.) The resulting algorithm should run in $O(B \log n / \log B)$ time, using n processors and $O(n(B+L))$ memory cells.
4. Compare the processor and memory requirements of the algorithms in parts (a) and (c). Show how to reduce the memory requirements to $O(nL)$ by combining the two algorithms.

22.5 Devise an algorithm for the FFT problem that runs in $O(B \log n / \log B)$ time using only $n \log B / B$ processors. Hint: Use stages with slightly fewer columns and have each processor compute the value for more than one node per stage.

22.6 Prove the correctness of Algorithm 22.5.

1. Give a precise description of the elements in the sets $S_{j,i}$.
2. Show that the elements in a set ordered by increasing row number form a bitonic sequence.
3. Assume n is a power of B. Prove that an initial bitonic sequence is sorted after $2 \log_B n$ phases.

22.7 (a) Devise an Asynchronous PRAM algorithm for multiplying two $n \times n$ matrices that runs in $O(B \log n / \log B)$ time with an $O(n^3)$ processor-time product. (b) Use this matrix multiplication algorithm to devise an Asynchronous PRAM algorithm for the dynamic parallel evaluation of DAGs (defined in Chapter 18).

22.8 Complete the proof of Theorem 22.2 of Section 22.4.4. Show that when
$$n/(t_1 t_2 \cdots t_s) \leq 1$$
then the sum $t = \sum_{i=1}^{s} t_i$ is minimized when $t_i = n^{1/s}$ for all i.

22.9 Show that the list ranking problem (defined in Chapter 2) can be solved in $O(B \log n / \log(B/d))$ time on an Asynchronous PRAM with n processors. Be careful to avoid concurrent access to a location in a phase. Hint: In $O(B)$ time, a processor can traverse B/d links of a linked list.

22.10 The Asynchronous PRAM model forbids two processors from accessing the same global memory location in the same phase. A possible variant on the model is to permit multiple processors to access a location in the same phase as long as *all* such processors are reading from it. This variant is called the *CREW Asynchronous PRAM*. Similarly, the *CRCW Asynchronous PRAM* permits multiple processors to access a location in the same phase as long as either (i) all such processors are reading from the location or (ii) all such processors are writing to it. Consider the following problem.

> DEFINITION 22.7
> The **set search problem** *is to determine if an element x is among a set of n elements stored in an array A of size n. The output is 1 if x is in A, otherwise the output is 0.*

1. Show that the set search problem can be done in $O(n/p + B \log n / \log B)$ time on an Asynchronous PRAM with p processors.
2. Show that the problem can be done in $O(B)$ time on a CRCW Asynchronous PRAM.
3. Show that the problem can be done in $O(B)$ time on a CREW Asynchronous PRAM if the set elements are unique.
4. How do these results compare with the time complexity needed for this problem on the EREW PRAM, CREW PRAM, and CRCW PRAM models?

22.11 The careful reader may have observed that the communication delay d does not appear in the complexity bounds for any Asynchronous PRAM algorithm presented in this chapter (outside of Section 22.5). Recall that a sequence of global reads with no interdependencies can be pipelined by a processor. Denote such a sequence of reads as an *oblivious sequence*. An *intraphase oblivious* algorithm is one in which, in each phase, each processor first issues an oblivious sequence of global reads, then issues a sequence of local computes, and finally issues a sequence of global writes.

1. Show that the asymptotic running time of an intraphase oblivious algorithm is the same for all values of d (such that $2 \leq d \leq B$).
2. Is this claim true for intraphase oblivious algorithms in the Asynchronous PRAM model with subset synchronization?

990   Chapter 22.   Asynchronous PRAM Algorithms

22.12 Consider the Asynchronous PRAM model with subset synchronization, and assume that $B(x) \in \Theta(d\lceil \log x/\log d\rceil)$. Show that the FFT of n numbers can be computed in $O(d \log n/ \log d)$ time using n processors in this model.

Notes and References

Examples of existing MIMD machines are the Intel iPSC (see, e.g., Seitz [32]), the IBM RP3 (Pfister *et al.* [28]), and the BBN Butterfly (see, e.g., LeBlanc, Scott, and Brown [19]). The importance of considering the communication delay of such machines is discussed by Athas and Seitz [3]. The Asynchronous PRAM model described in this chapter was introduced by Gibbons [12, 13]. The all-prefix-sums operation (Ladner and Fischer [17], Blelloch [5]), Brent's scheduling principle [6], the multiprefix operation (Ranade [30]), the Fast Fourier Transform (Brigham [7], Ullman [33]), and the bitonic merge problem (Batcher [4]) have each been studied extensively by parallel algorithm designers.

There are a number of papers on topics related to this chapter, including work on other asynchronous PRAM models, direct simulations of PRAMs on networks of processors, other models for studying communication delay, and other cost measures for asynchronous computation. In the remainder of this section, we highlight some of this work. A word of caution, however: many of the results described below are not practical, due to the overheads involved in the techniques.

Other Asynchronous PRAM Models

Kruskal, Rudolph, and Snir [16] studied an asynchronous model for parallel computers based on an accounting scheme for asynchronous computation due to Lynch and Fischer [22]. In this scheme, each processor completes an instruction in one *local* clock cycle, where clock speeds are not known in advance and may vary during the computation. Time is measured using the slowest local clock. In one time-unit or *round*, each processor executes at least one instruction (the slowest processor executes one, faster processors execute more). Kruskal, Rudolph, and Snir show that an asynchronous PRAM with this accounting scheme can simulate a CRCW PRAM of p processors using $O(\log p)$ rounds per CRCW PRAM instruction.

Cole and Zajicek [8] introduced the *APRAM model*, an asynchronous variant of the CRCW PRAM also based on the Lynch and Fischer accounting scheme. They present APRAM algorithms for important problems such as the all-prefix-sums operation and computing the connected components of an undirected graph. In a follow-on paper, Cole and Zajicek [9] introduced the *Variable Speed APRAM model* for studying the cost of constraining the processors to lock-step execution. In their *unbounded delays* version of the model, at each clock tick, each processor is delayed (idle) with probability p and successfully completes an instruction with probability $1 - p$, where p is a parameter of the model. In their *bounded delays* version of the model, each processor takes k clock ticks to complete its current instruction with probability p and 1 tick with probability $1 - p$, where p and k are parameters of the model. They present results showing that certain *adaptive* algorithms, in which faster processors can dynamically perform a larger share of the work than slower processors, outperform non-adaptive algorithms for the all-prefix-sums operation and other problems. Each of their APRAM models permits multiple processors to read (or write) a single location in one step, which may be unrealistic, and none specifically account for communication delay (thus, for example, memory accesses and local operations are equally likely to complete in one cycle).

Another approach to accounting for the varying delays encountered by processors in MIMD machines is due to Nishimura [26]. Many parallel programs are written assuming a *sequential consistency* memory model. In the *sequential consistency model* (Lamport [18]) each processor completes its instructions one at a time in program order and at most one processor completes an instruction at any given time. (In practice, a MIMD machine supporting sequential consistency will enforce an ordering between a pair of instructions only where necessary to preserve the illusion of sequential consistency, thereby permitting many instructions to complete in parallel (see, e.g., Gibbons, Merritt, and Gharachorloo [14]).

The programmer, however, has no control over *which* processor completes an instruction at a particular time. Nishimura studied a model in which each processor has an equal probability of completing an instruction at any given time. The running time for a program is defined to be the expectation of the maximum over all processors of the number of instructions completed by a processor. Nishimura presents results comparing various adaptive strategies for important primitives such as all-prefix-sums and list ranking (defined in Chapter 2), as well as results comparing the model to various (synchronous) PRAM models. The model does not specifically account for communication delay.

None of the models discussed thus far are suitable for machines in which processor failures are likely. Martel, Park, and Subramonian [23] introduced an asynchronous CRCW PRAM model in which processors can experience arbitrary unbounded delays in executing instructions. Whereas in the unbounded delays model of Cole and Zajicek long delays were very unlikely, in the Martel, Park, and Subramonian model, long delays or even processor failures may be frequent. The complexity measure for the model is the expected amount of *work* done, where *work* is defined to be the sum over all processors of the number of instructions completed by a processor. The model does not specifically account for communication delay. They show how important primitives such as all-prefix-sums and list ranking can be solved efficiently on this model using the following strategy: place the set of tasks to be done in global memory; at each step, each idle processor selects its next task at random and (attempts to) complete it. Tasks are selected until all tasks have been completed; the same task may be selected by multiple processors. In this way, processors that are delayed, or even fail, do not unduly slow down the computation: the faster processors will simply perform more tasks. In a follow-on paper, Martel, Subramonian, and Park [24] extended their earlier results to yield a technique for simulating a p-processor CRCW PRAM on their asynchronous model using only $O(p)$ expected work per PRAM instruction.

Simulating PRAMs Directly

Considerable research effort has focused on simulating the PRAM on realistic (asynchronous) networks of processors. In these simulations, the communication delay is calculated by modeling the transfer of messages through the interconnection network, including any contention for network links. A long series of papers, highlighted by the early work in this area by Valiant and Brebner [36] culminated with the result by Ranade [29] that a p-processor CRCW PRAM can be simulated on a butterfly network of p processors using, with very high probability, only $O(\log p)$ time per PRAM instruction. This is optimal, for a single memory request may need to pass through $\log p$ nodes to reach its destination.

Valiant [35] introduced an asynchronous model similar to the Asynchronous PRAM called the *Bulk-Synchronous Parallel Computer (BSPC)*. Valiant shows that given sufficient *parallel slackness*, i.e., given that there are sufficiently many virtual PRAM processors for each machine processor, an algorithm running in time t on either the BSPC or the PRAM can be simulated on a realistic network in time $O(t)$, with very high probability (the time t must be greater than the communication delay and greater than the time for a barrier synchronization).

The most efficient *deterministic* simulation of a PRAM on a network of processors is for the *multi-butterfly* network, a generalization of the butterfly network. The PRAM can be simulated on the multi-butterfly using $O(\log p)$ time per instruction (Upfal [34], Leighton and Maggs [20]).

Sorting algorithms for networks of processors (e.g., Nassimi and Sahni [25], Reif and Valiant [31], Cypher and Sanz [10]) can also be used to efficiently simulate the memory requests of a single PRAM instruction on networks such as the butterfly network.

Communication Delay

The effect on algorithm design of accounting for communication delay has been studied by Aggarwal, Chandra, and Snir [1, 2] and Papadimitriou and Yannakakis [27] in the context of synchronous models. Aggarwal, Chandra, and Snir introduced two synchronous models that are based on the PRAM, the *LPRAM* and the *BPRAM*. The LPRAM does not permit the pipelining of memory requests, while the BPRAM permits blocks of *consecutive* memory locations to be accessed in a pipelined fashion. In the model of Papadimitriou and Yannakakis, arbitrary pipelining of memory requests is permitted. These papers present algorithms for important primitives such as all-prefix-sums, FFT, and sorting. A more detailed model was studied by Gannon and Van Rosendale [11] in the context of numerical algorithms (e.g., algorithms for solving systems of equations). Variants of their model account for communication delay, pipelining rates, network bandwidth, and network topology.

Other Cost Measures

Greenberg, Lubachevsky, and Odlyzko [15] introduced a model for asynchronous parallel machines suitable for analyzing algorithms in which the inputs arrive staggered in time. In this context, the important metric is the average *response time* for a processor, namely the time from when its input is first ready until it completes its participation in the computation. They study the problem of finding the maximum of a set of inputs that arrive staggered in time. Algorithms are presented that achieve provably optimal response times. Their model permits multiple processors to read (or write) a single location in one step and does not specifically account for communication delay.

In the area of iterative numerical algorithms, models for asynchronous parallel machines have been defined in which the metric of interest is the *convergence rate* of the iterative process. For example, Lubachevsky and Mitra [21] introduced a model for iterative numerical algorithms in which arbitrary delays occur, but these delays are uniformly bounded by some finite value. Because of the delays, processors typically receive data computed several iterations earlier. They present an iterative algorithm for computing the fixed point of an important class of matrices that achieves provably fast convergence despite these communication delays.

Bibliography

[1] Aggarwal, A., and Chandra, A.K. Communication complexity of PRAMs. In T. Lepistö and A. Salomaa, editors, *Automata, Languages and Programming, Lecture Notes in Computer Science, Vol. 317*, pages 1–18. Springer-Verlag, Berlin, July 1988. Proceedings 15th ICALP.

[2] Aggarwal, A., Chandra, A.K., and Snir, M. On communication latency in PRAM computations. In *Proceedings 1st ACM Symposium on Parallel Algorithms and Architectures*, pages 11–21, Santa Fe, New Mexico, June 1989.

[3] Athas, W.C., and Seitz, C.L. Multicomputers: Message-passing concurrent computers. *IEEE Computer*, 21(8):9–24, 1988.

[4] Batcher, K.E. Sorting networks and their applications. In *Proceedings AFIPS Spring Joint Summer Computer Conference*, pages 307–314, 1968.

[5] Blelloch, G. Scans as primitive parallel operations. In *Proceedings 1987 International Conference on Parallel Processing*, pages 355–362, August 1987.

[6] Brent, R.P. The parallel evaluation of general arithmetic expressions. *Journal of the ACM*, 21(2):201–208, 1974.

[7] Brigham, E.O. *The Fast Fourier Transform*. Prentice-Hall, Englewood Cliffs, New Jersey, 1974.

[8] Cole, R., and Zajicek, O. The APRAM: Incorporating asynchrony into the PRAM model. In *Proceedings 1st ACM Symposium on Parallel Algorithms and Architectures*, pages 169–178, Santa Fe, New Mexico, June 1989.

[9] Cole, R., and Zajicek, O. The expected advantage of asynchrony. In *Proceedings 2nd ACM Symposium on Parallel Algorithms and Architectures*, pages 85–94, Crete, Greece, July 1990.

[10] Cypher, R., and Sanz, J.L.C. Cubesort: An optimal sorting algorithm for feasible parallel computers. In J.H. Reif, editor, *VLSI Algorithms and Architectures, Lecture Notes in Computer Science, Vol. 319*, pages 456–464. Springer-Verlag, Berlin, June 1988. Proceedings 3rd AWOC.

[11] Gannon, D.B., and Van Rosendale, J. On the impact of communication complexity on the design of parallel numerical algorithms. *IEEE Transactions on Computers*, C-33(12):1180–1194, 1984.

[12] Gibbons, P.B. *The Asynchronous PRAM: A semi-synchronous model for shared memory MIMD machines*. PhD thesis, Computer Science Division, University of California at Berkeley, Berkeley, California, December 1989. Technical Report TR-89-062, International Computer Science Institute, Berkeley, California.

[13] Gibbons, P.B. A more practical PRAM model. In *Proceedings 1st ACM Symposium on Parallel Algorithms and Architectures*, pages 158–168, Santa Fe, New Mexico, June 1989.

[14] Gibbons, P.B., Merritt, M., and Gharachorloo, K. Proving sequential consistency of high-performance shared memories. In *Proceedings 3rd ACM Symposium on Parallel Algorithms and Architectures*, Hilton Head, South Carolina, July 1991.

[15] Greenberg, A.G., Lubachevsky, B.D., and Odlyzko, A.M. Simple, efficient, asynchronous parallel algorithms for maximization. *ACM Transactions on Programming Languages and Systems*, 10(2):313–337, 1988.

[16] Kruskal, C.P., Rudolph, L., and Snir, M. A complexity theory of efficient parallel algorithms. Technical Report RC 13572, IBM T.J. Watson Research Center, Yorktown Heights, New York, March 1988.

[17] Ladner, R.E., and Fischer, M.J. Parallel prefix computation. *Journal of the ACM*, 27(4):831–838, 1980.

[18] Lamport, L. How to make a multiprocessor computer that correctly executes multiprocess programs. *IEEE Transactions on Computers*, C-28(9):241–248, 1979.

[19] LeBlanc, T.J., Scott, M.L., and Brown, C.M. Large-scale parallel programming: Experience with the BBN Butterfly parallel processor. In *Proceedings 1st ACM Symposium on Parallel Programming*, pages 161–172,

New Haven, Connecticut, July 1988. *SIGPLAN Notices*, 23(9), September 1988.

[20] Leighton, T., and Maggs, B. Expanders might be practical: Fast algorithms for routing around faults on multibutterflies. In *Proceedings 30th IEEE Symposium on Foundations of Computer Science*, pages 384–389, Research Triangle, North Carolina, October 1989.

[21] Lubachevsky, B.D., and Mitra, D. A chaotic asynchronous algorithm for computing the fixed point of a nonnegative matrix of unit spectral radius. *Journal of the ACM*, 33(1):130–150, 1986.

[22] Lynch, N.A., and Fischer, M.J. On describing the behavior and implementation of distributed systems. *Theoretical Computer Science*, 13(1):17–43, 1981.

[23] Martel, C., Park, A., and Subramonian, R. Optimal asynchronous algorithms for shared memory parallel computers. Technical Report CSE-89-8, Division of Computer Science, University of California, Davis, California, July 1989.

[24] Martel, C., Subramonian, R., and Park, A. Asynchronous PRAMs are (almost) as good as synchronous PRAMs. In *Proceedings 31st IEEE Symposium on Foundations of Computer Science*, pages 590–599, St. Louis, Missouri, October 1990.

[25] Nassimi, D., and Sahni, S. Data broadcasting in SIMD computers. *IEEE Transactions on Computers*, C-30(2):101–106, 1981.

[26] Nishimura, N. Asynchronous shared memory parallel computation. In *Proceedings 2nd ACM Symposium on Parallel Algorithms and Architectures*, pages 76–84, Crete, Greece, July 1990.

[27] Papadimitriou, C.H., and Yannakakis, M. Towards an architecture-independent analysis of parallel algorithms. In *Proceedings 20th ACM Symposium on Theory of Computing*, pages 510–513, Chicago, Illinois, May 1988.

[28] Pfister, G.F., *et al.* The IBM Research parallel processor prototype (RP3): Introduction and architecture. In *Proceedings 1985 International Conference on Parallel Processing*, pages 764–771, August 1985.

[29] Ranade, A.G. How to emulate shared memory. In *Proceedings 28th IEEE Symposium on Foundations of Computer Science*, pages 185–194, Los Angeles, California, October 1987.

[30] Ranade, A.G. *Fluent parallel computation*. PhD thesis, Department of Computer Science, Yale University, New Haven, Connecticut, May 1989.

[31] Reif, J.H., and Valiant, L.G. A logarithmic time sort for linear size networks. *Journal of the ACM*, 34(1):60–76, 1987.

[32] Seitz, C.L. The cosmic cube. *Communications of the ACM*, 28(1):22–33, 1985.

[33] Ullman, J.D. *Computational Aspects of VLSI*. Computer Science Press, Rockville, Maryland, 1984.

[34] Upfal, E. An $O(\log n)$ deterministic packet routing scheme. In *Proceedings 21st ACM Symposium on Theory of Computing*, pages 241–250, Seattle, Washington, May 1989.

[35] Valiant, L.G. A bridging model for parallel computation. *Communications of the ACM*, 33(8):103–111, 1990.

[36] Valiant, L.G., and Brebner, G.J. Universal schemes for parallel communication. In *Proceedings 13th ACM Symposium on Theory of Computing*, pages 263–277, Milwaukee, Wisconsin, May 1981.

Index

AC, 844, 846, 847, 849–852, 855, 858, 860–866, 868, 870, 882, 886, 889, 891, 892. *See also* circuit, arithmetic
ACVP. *See* arithmetic circuit value problem
addressing function, 873
admissible (arcs), 818, 819, 822, 825, 826, 828, 829, 831, 835
affects, 59, 561, 870–873, 923
affine, 681, 682, 684, 686, 687, 712–716
  space, 684, 687, 713
algebraic circuit compression, 734–736
algebraic circuit model, 724, 725, 729, 734
algebraic geometry
  real, 711
algebraic set, 684, 702, 714
algorithm
  branch-and bound, 54
  chess-playing, 54

divide-and-conquer, 36, 54, 146, 191, 351, 367, 520, 523, 528
dynamic-programming, 36
line-drawing, 55, 59
line-of-sight, 45
optimal deterministic, 38, 145, 446
sequential, 3, 14–16, 36, 38, 40, 50, 62, 63, 81, 116–118, 146, 163, 164, 179, 260, 276, 283, 336, 337, 342, 343, 346, 353, 357, 373, 377, 379, 385, 414, 416, 421, 425, 437, 445, 447, 462, 490, 498, 528, 582, 610, 612, 633, 638, 649, 661, 681, 682, 726, 797, 814, 822, 825, 828, 846, 902, 944, 948, 965, 970, 975, 978, 980, 983
all-prefix-sums, 36–38, 58, 59, 181, 960, 964, 970, 971, 980, 987, 990, 992, 993. *See also* prefix sums
allocating processors, 54
ancestor
  least common, 20, 171, 278, 280, 282

1000   Index

lowest common, 8, 21, 163, 173, 260, 261
maximum value of, 162, 163
approximate median, 445
approximate rank, 445
APRAM, 990, 991
ARBITRARY, 847, 848, 854, 858–860, 864, 886, 888
arithmetic Boolean circuits. *See* Boolean circuit, arithmetic
arithmetic circuit. *See* circuit, arithmetic
arithmetic circuit value problem (ACVP), 930, 931
assignment problem, 423
asynchrony, 960
atom, 28, 766, 767, 774, 831–836
  ground, 766, 767, 769, 774
attachment, 172, 173, 177, 179, 187
augmenting path, 820

BASIS, 599
basis, 9, 26, 559, 596, 599–601, 683, 686, 687, 689, 692, 703, 704, 716, 717, 742, 743, 752, 776, 911, 919, 927
Berkowitz's algorithm, 689, 697–699
binary associative operator, 36, 38, 39, 48, 965, 970
binary tree, 21, 39, 163, 165, 191, 192, 260–264, 268, 270, 272, 474, 492, 520, 523, 524, 578–580, 641, 730, 750, 828, 850, 856, 959, 960, 965, 974–977
binary representation, 90, 261, 265–267, 269, 547, 570, 892
bit reduction, 548, 552
bitonic merge, 491, 980–982, 987, 990
blocking flow, 814–817, 830
body, 117, 528, 680, 760–762
Boolean circuit, 15, 17, 31, 414, 540, 602, 606, 607, 609, 611, 733, 745, 746, 860–864, 866, 891, 907, 911–914, 919, 921, 923, 925, 926, 928, 930
  arithmetic, 602–610, 733

bounded fan-in, 540
  gates, 576, 577, 592, 603, 604, 607–609, 861, 866, 880, 891, 912–914, 916, 918, 919, 921–924, 926–930, 934–937, 939, 940, 943, 949
  input nodes, 11, 540, 750, 861, 944, 974, 975
  output nodes, 540, 740, 750, 751, 861, 912, 926, 975
  size, 12, 540, 607–609, 861–863, 866, 912–914, 922, 929
  unbounded fan-in, 607, 860–864, 866, 891
BOOLEAN DECISION TREE EVALUATION, 886, 887
bounded degree tree, 177–180
breadth first search, 903, 932–934, 947
breadth-first search trees, 361
Brent's scheduling principle, 88, 100, 104, 106, 225, 227, 623, 645, 967, 968, 970, 990
bridge, 171–177, 179–181, 187, 255, 260, 292–297, 299, 301–303, 305–308, 310–316, 318, 320–324, 334, 335
  m-, 172–176, 179–181
  edge, 173–175, 180
  leaf, 173, 174, 176, 179–181, 187
  top, 173–177, 179, 180, 187
Bulk-Synchronous Parallel Computer (BSPC), 992

C-equivalent, 172
candidate lists, 303–305, 307, 309, 313, 325, 332–334
cardinality of a set, 626
Cayley–Hamilton theorem, 590, 644
certificate, 875–877, 880
chain, 10, 14, 19, 64, 118, 119, 124, 131–137, 139, 141, 142, 145, 148, 150, 151, 153, 158–160, 163, 164, 175, 181, 183–187, 191, 200, 260, 264, 268, 270, 499, 500, 509–511, 654, 926

matrix, 732–734, 735
chain program, 770, 771, 773–776, 778
characteristic polynomial, 26, 574, 583, 586, 588–590, 593, 595, 598, 609, 610, 682, 708–710, 712, 715
  generalized, 682, 712, 715
CHARPOLY, 592, 593, 595
Chernoff bounds, 417, 419, 424, 430, 439
Chinese remainder algorithm, 611
Chistov's algorithm, 583, 689, 697–699, 710
chordal graphs, 22, 23, 342, 346–349, 351, 358
circuit, 11, 12, 25, 58, 59, 221, 252, 654, 699, 706, 910–925, 928–931, 935, 937–941, 948, 949
  algebraic, 28, 724–754
  arithmetic, 15, 575–580, 587, 588, 706, 930, 931. See also Boolean circuit, arithmetic
  Boolean. See Boolean circuit
  cross-over, 928, 929, 949
  depth of, 456
  size, 414, 540, 587, 912–914, 977, 978
  sorting, 454
  topological, 914, 915
  See also Euler circuit; odd-even sorting circuit
circuit family, 14, 540–542, 544, 547, 562, 564, 567, 569, 579, 607, 610, 913–915, 917, 918
circuit value problem (CVP), 911, 914, 915, 918, 919, 921, 923–931, 940, 948, 949
circuits with division, 746, 747, 750
clique, 22, 23, 342–344, 346–350, 352, 353, 356–362, 366–371, 373, 376, 388–393, 395–400, 403, 904, 942, 943, 948
clique cover, 349, 350, 352, 353, 362, 364
clique management, 368, 369, 388
clique tree, 351, 352, 403

clique-separator, 356, 357
coloring, 129–131, 135–137, 326, 342, 343, 349–353, 366–368, 370–373, 388, 944, 949
  k-, 130
combinatorics, 350
COMMON, 847, 848, 853–855, 857–864, 866, 882, 885, 886, 888, 891
common tangent, 502, 503, 505–508, 527
communication delay, 959, 960, 963, 989–994
comparison tree, 414, 884, 885
complete basis, 927. See also basis
complete binary tree, 260–264, 268, 270, 474, 492, 520, 524, 974. See also binary tree
complexity class, 574, 594, 598, 602, 606–610, 862, 864, 903, 907
composition, 9, 155, 161, 162, 191, 427
compress, 13, 20, 21, 147–155, 157, 163–165, 169, 170, 181–183, 186, 187, 192, 237, 250, 464, 522, 526, 734, 737, 738, 740, 741, 744–746, 748, 749, 752, 769
compression, 13, 20, 183, 734, 737, 738, 740, 741, 744–746, 748, 749, 752, 769
computation tree, 724, 725, 729
computational model, 2, 4, 540, 574, 726, 860–868
  comparison model, 490, 491, 501
  parallel model, 11–15, 216, 414
  sequential model, 8–11, 906
computer algebra, 576, 611
concurrent-write, 57, 216
configuration graph of a Turing machine, 862, 863
connected component, 8, 18, 19, 62–80, 89, 112, 198–204, 206, 228, 229, 233, 243, 260, 277, 278, 289, 291, 297, 317, 322, 323, 323–336, 357, 369, 387–389, 393, 400–403, 786, 787, 798, 991

connectivity, 17–19, 21–23, 62, 198, 199, 211–213, 216, 228–232, 242, 251–254, 279, 291, 324, 335, 336, 338, 389, 393, 643, 654, 655
consistency condition, 367, 371
context-free grammar (CFG), 771
contraction/contracting, 233. *See also* list contraction; tree contraction; stream contraction; graph contraction
convex hull, 4, 6, 25, 415, 436, 440, 442, 443, 447, 448, 498–501, 508, 509, 513, 526–529
Cramer's rule, 583, 585, 594, 694, 696
critical, 36, 74, 175, 482, 799, 870, 908
  m-, 171–180
critical complexity, 870
cross-edge, 355, 356, 358, 365
Csanky's algorithm, 583, 589, 689
CVP. *See* circuit value problem

datalog, 762–770, 774, 778
decision problem, 602, 784, 793, 904–908, 914, 932, 934, 936, 943, 944
decision tree, 861, 867, 868, 876–878, 880, 884, 886, 887
  complexity, 867, 868, 875–878, 880
  nondeterministic complexity, 875
  root of, 867, 868
degree
  of algebraic circuit, 725–735, 738–741, 743–745, 747–750, 752–754
  of a Boolean function, 880
  of a circuit, 28
  of a graph, 7
  of a monomial, 683, 687, 703, 716
  of a polynomial, 580, 582, 584, 585, 610–612, 681, 683, 686, 688–690, 693, 698–701, 703, 705–708, 710–712, 715, 717, 735, 790–792, 802, 803, 808
  of a supervertex, 199, 232, 233

  of a vertex, 198, 205, 235, 277, 293, 785–787, 802
  See also bounded degree tree; unbounded degree tree
depends on coordinate, 873
depth-first search, 63, 105, 108–111, 163, 276, 283, 284, 288, 296, 336, 337, 354, 356, 832
depth-first search tree, 283–285, 288, 337, 354, 356
DET_F, 594, 595, 598–600, 602, 609
DETERMINANT, 591, 592, 594, 595, 608
determinant, 26, 29, 574, 582–585, 593, 594, 602, 609, 610, 682, 688, 689, 692, 695, 696, 708, 710, 712, 725–731, 733, 745–748, 751–754, 788, 791, 796, 804, 805
deterministic coin tossing, 19, 62, 88–90, 92, 103, 104, 112, 116, 119, 129, 131, 221, 222, 227, 252, 253
deterministic list ranking, 128, 129, 131, 132, 145
digital differential analyzer, 57
directed acyclic graph (DAG), 6, 11, 20, 28, 292, 354, 540, 576, 578, 579, 602, 724, 729, 731, 734, 861, 912, 960, 974, 977–979, 983, 988
Discrete Fourier Transforms (DFTs), 549
discriminant, 718
divide and conquer
  parallel, 432
divide-and-conquer, 36, 54, 146, 191, 351, 367, 520, 523, 528
  cascading, 520, 523, 528
division with remainder, 574, 611, 694, 696
division-free circuits, 726
DLOG, 862–864
doubling, 64, 248, 483, 787, 922. *See also* shortcutting
down-sweep, 41–43
dual transform, 448
dynamic programming, 649, 728, 729, 734

ear graphs, 303, 304, 313, 314, 324.
  See also open ear decomposition
easy-case algorithm, 203, 205, 211
edge
  extrovert, 208
  introvert, 208
edge colouring, 787, 788
effective homogeneous
  Nullstellensatz, 703, 704
efficient parallel algorithm, 15, 27,
  36, 105, 146, 260, 276, 280, 288,
  296, 326, 334, 336, 342, 358, 498,
  500, 571, 589, 590, 622, 681, 682,
  717
efficient parallel reciprocal circuit,
  563
element distinctness, 858, 859, 885,
  886
elimination ordering, 347–349,
  351–353, 366, 373, 379, 624, 627,
  628. *See also* perfect elimination
  ordering
elimination theory, 680, 681, 702, 705
elimination tree, 354, 356–358,
  361–365, 369, 403
EOL (end-of-list), 38
EQ, 600, 601
equivalent, 48, 80, 153, 162, 174, 179,
  278, 359, 386, 387, 417, 500, 509,
  527, 547, 553, 570, 581, 593, 602,
  608, 636, 641, 652, 686, 689, 691,
  711, 724, 725, 764, 883, 884, 909
error, 412, 413, 543, 545, 555, 563,
  566, 569, 570, 624, 701, 791, 799
  1-sided, 413
  2-sided, 413, 443
estimation lemma, 429
Euclidean, 513, 611, 682, 684, 689,
  690, 693, 752
  algorithm, 684, 689, 690
  ring, 684
  scheme, 611, 682, 689, 690, 693
Euler circuit, 243, 786, 787
Euler Tour, 19–22, 62, 82, 105–111,
  113, 165, 177–180, 244, 253, 255,
  261, 265, 269, 271, 272, 282

technique, 19, 20, 62, 82, 105, 106,
  109, 113, 253, 261, 265, 269, 271,
  272, 282
exponentiation, 574, 610, 611
expression evaluation, 146, 147, 152,
  154, 155, 157, 160, 161
expression tree model, 724, 725
extended Euclidean scheme, 611,
  682, 689, 690, 693
extended Euclidean scheme (EES),
  689, 693
extensional database, 761

fan-out, 6, 724, 729, 740, 750, 751,
  861–863, 923–925, 929–931, 936,
  937, 949, 960
Fast Fourier Transform, 549, 569,
  733, 974, 975, 977, 987, 990
Fibonacci numbers, 50, 58
Fibonacci search, 436
fill-in, 624, 626, 627, 632, 633, 639,
  640, 643, 664
filtering scheme, 440
finite element method, 626
fixed-depth search, 54
flow, 29–31, 163, 191, 378, 476, 650,
  661, 726, 808, 814–820, 825, 828,
  830–833, 835, 836, 905, 906, 932,
  936–938, 945, 948, 949. *See also*
  blocking flow; maximum flow
flow function, 829, 905, 937, 945,
  946, 949
free vertex, 182, 183, 185, 186
function composition, 155, 161, 162,
  191

Gaussian elimination, 22, 347, 348,
  583, 625, 664, 726, 746–748, 948
generalized nested dissection
  algorithm, 633, 658, 659
geometric series, 212, 581, 585
geometry
  algebraic, 606, 612, 680, 683, 711
  real, 682, 715
graded ring, 683, 685–687, 717

graph
  associated, 624, 627
  computation, 974, 975, 977–980, 982
  elimination, 627, 640, 664
  interval graphs, 23, 342, 343, 345, 346, 348, 350, 351
  path algebra computations in, 649
  planar, 27, 146, 253, 632, 637, 650, 651, 661
  separatable associated, 627
  two-dimensional finite element, 632
graph algorithm, 4, 5, 10, 18, 20, 21, 28, 254, 943
graph connectivity, 17–19, 21–23, 228, 254, 335, 336, 654
graph contraction, 146–192
graph-theoretic definitions, 291
greatest common divisor (gcd), 27, 611, 612, 681, 684, 689, 690, 697–702, 718
greedy, 162, 349, 351, 362, 365, 366, 403, 941–943, 947, 971
  algorithms, 941, 942, 944
  maximal independent set, 362

half-planes, 435–443, 508, 509
Hauptidealsatz, 703
head, 89, 116, 124–126, 130, 132, 133, 137, 153, 183–186, 236, 237, 244–246, 249, 250, 271, 443, 444, 760, 761, 765, 766, 768–771, 862, 916, 917, 959
hierarchy, 541, 607
high order convergence, 560
Hilbert Basis Theorem, 622, 683
hits, 198, 209
homogeneous, 682, 685–687, 702–714, 717, 718, 733
  equations, 710
  ideal, 686, 687
  Nullstellensatz, 687, 703, 704, 711
  polynomial, 685, 687, 702–710, 712, 713, 717, 718
homogenization, 713

hooking, 19, 62, 64, 65, 68, 69, 72–75, 78, 79, 229–231, 242, 245
hyperplane, 448, 631
  at infinity, 687, 713, 714

ideal, 15, 120, 683–687, 697, 703, 711, 716, 865
ideal PRAM, 848, 858, 862, 865, 869, 883, 885, 888
improper solutions, 713, 714
INDEPENDENCE, 600, 602
independent set, 91, 162, 343, 349–353, 362–366, 373, 403, 632, 807
induction hypothesis, 43, 175, 201, 320, 328, 545, 551, 557–559, 578, 647, 742, 743, 876, 877, 936
inherently sequential, 36, 37, 498, 583, 910, 912, 918, 944, 946–949
inhomogeneous, 682, 712–714, 716
  polynomials, 712–714, 716
inorder numbering, 261, 262, 272
integer
  division, 25, 540, 547, 569
  reciprocals, 543, 561, 566
  powering, 546, 560
intensional database, 761
intersection graph, 351
inversion, 579, 581, 582, 586, 589, 590, 592, 611, 622, 623, 627, 635, 638, 640, 644, 649, 658, 797
isolated cell, 131, 133, 139, 140
isolated chain, 181–186
isolating lemma, 785, 794, 795, 797, 808, 809
iterated refinement, 377, 378, 381, 385, 402
iterated sum, 579
ITMATPROD, 593, 595, 598, 599, 609
ITPOLYPROD, 592

jump over, 118, 120–125, 153–155, 158, 159, 164, 181, 183

kingdom, 130, 131
Kolmogorov complexity, 866

language, 28, 31, 38, 58, 413, 591,
    654, 734, 753, 760, 771–776, 846,
    904, 906–910, 914, 915, 946–949
Las Vegas algorithm, 413, 798
leaf attachment, 173, 177, 179, 187
leftmost prisoner problem, 850,
    856–860
leftmost writers problem, 888
lexicographic breadth-first search,
    379, 381, 403
lexicographically first maximal clique
    problem, 943
lexicographically highest, 379
lexicographically least solution, 942
line-drawing, 54–57, 59
    routine, 54, 56
linear
    inequalities, 31, 932, 939, 940
    programming, 31, 815, 910, 932,
        938, 940, 941
    linear algebra, 26, 27, 346, 574, 575,
        583, 598, 602, 612, 634, 652, 653,
        682, 683, 694, 703, 784, 785, 788
    linear datalog program, 767, 768
    linear equations, 574, 582, 583, 595,
        600–602, 610, 612, 637, 712
    linear recurrences, 49, 589
linked-list, 37, 38, 59
list ranking,
    algorithm, 109, 112, 117–119, 178,
        891
    optimal, 20, 62, 261, 265, 269
    optimal deterministic, 21, 128–145
    optimal randomized, 21, 25, 123,
        126, 445, 447
    parallel algorithm, 15, 17, 18, 216,
        226, 227
    problem, 14, 19, 80–86, 108, 116,
        117
    randomized, 119–128, 252, 253
    Wyllie's algorithm, 116–119, 121,
        128, 129, 132, 145, 164, 183, 243,
        244, 252
    See also rank
log space reducible, 908, 909, 914,
    920, 939, 940, 949

log space uniform, 914, 915, 918
logical rule, 28, 760, 764
logical inference model, 760
lower bounds, 30, 417, 439, 574, 579,
    846, 848, 859–865, 867, 869, 874,
    878, 881, 884, 885, 887, 888, 891,
    902, 960, 983
lower envelope, 523, 526

MACVP. *See* monotone alternating
    circuit value problem
Markov's inequality, 420, 438
MASCVP, 925–927, 934, 935
MASCVP2, 925–927, 934, 935
matching, 17, 18, 27, 29, 378, 403,
    784, 787–791, 793–800, 802, 803,
    805–808, 815, 846, 858, 860, 885,
    983
    exact, 808
    lexicographically first, 793, 807,
        943, 944, 949
    maximal, 784, 785, 787, 807
    maximum, 403, 784, 789, 791–794,
        798, 799, 807, 808
    perfect, 29, 784, 785, 787–793,
        795–797, 799, 800, 802, 803,
        805–808
    vertex-weighted, 797
    weighted, 807
MATINV, 592, 594, 595
MATPOWERS, 592–594
MATPROD, 592, 609
matrix. *See* multiplication, matrix
matrix rank, 26, 574, 595–598, 605,
    609, 612, 704–706, 716, 717, 792
matroid intersection, 808
maximal cliques, 22, 342–345, 348,
    349, 351, 358, 360, 361
MAXIMUM, 870, 885, 886
maximum clique, 343, 350
maximum flow, 30, 31, 378, 650, 661,
    814–818, 820, 836, 905, 932, 936,
    938, 948
maximum flow problem, 814, 815,
    817, 905, 932, 936, 948

maximum independent set, 162, 349–353, 362–364, 366, 373, 403
maximum-weight clique, 342, 349, 351–353, 361
MAXMINOR, 599, 601
MCVP, 919–923, 936, 943
memoization, 729
merge, 25, 54, 367, 370, 371, 415, 421, 425, 454, 456–464, 470, 472–474, 476–478, 480, 481, 483, 485, 490–492, 501, 519–523, 525, 526, 885, 886, 980–982, 987, 990
merge sort, 24, 25, 54, 415, 454, 456, 457, 463, 501, 519, 520
MIMD, 11, 13, 958, 986, 990, 991
min-max inequality, 350
minimum coloring, 137, 342, 343, 349, 351–353, 366, 368, 373, 388
minimum degree ordering, 628, 664
modified factorization, 626
monotone alternating circuit value problem (MACVP), 921–925, 931
monotone circuit value problem (MCVP), 919, 931
Monte Carlo algorithm, 413, 748, 749, 784, 798, 799
Mulmuley's algorithm, 706, 716
multiplication, 540–543, 546, 548, 555, 558, 567–570, 579–582, 635, 639, 649, 653–658, 661, 664, 683, 725, 727–732, 734–739, 741, 743–746, 748, 815, 863, 864, 891, 930, 988
  iterated matrix of, 580
  matrix, 580, 587, 588, 622–624, 637, 638, 646–648, 649, 655, 664, 728, 732, 736, 738, 739, 748, 815, 988
multiplicity, 596, 711
multiprecision numbers, 37
multiprefix, 964, 970, 971, 973, 987, 988, 990

NAND circuit value problem, 919, 926, 927, 931
NC (Nick's Class), 14, 15, 26, 28–31, 574, 606, 607, 611, 613, 622, 650, 702, 784, 785, 805, 806, 814, 862–865, 867, 903, 904, 906, 907, 909, 910, 918, 930, 938, 947, 949, 978
NC^k, 574, 606, 607
NC_F^k, 574, 606, 607
network, 11–13, 24, 25, 29–31, 58, 178, 422, 445–447, 529, 633, 649, 650, 654, 661, 662, 814–817, 825, 830, 831, 865, 905, 906, 910, 932, 936–938, 949, 958, 990, 992, 993
Newton approximation, 541, 542–546, 560, 561
Newton identities, 590
Newton Iteration, 25, 26, 540, 570, 612
NLOG, 862–864
node-induced subgraph, 342, 347, 350, 394
NOR CVP, 927
normal forms, 574, 610, 611
  Smith, 610
NULLSPACE, 600, 601
Nullstellensatz, 680, 681, 684, 685, 687, 703, 704, 711, 716
  effective, 703, 704
  homogeneous, 687, 703, 704, 711
  strong form, 685
  weak form, 684

odd-even sorting circuit, 454, 456, 457–462, 488–490
one-dimensional array, 37
open ear decomposition, 22, 260, 276, 279, 280, 285–292, 294–300, 303–307, 309, 310, 313–316, 324, 334–338
optimal coloring, 343, 370
optimal matrix chain multiplication, 728
optimal tree contraction (binary trees), 163
optimal tree contraction (bounded degree trees), 147, 170
optimal tree contraction (unbounded degree trees), 147

Index **1007**

OR, 853, 854, 856, 861, 868–870, 872, 877–879, 890

P-completeness, 31, 611, 902–904, 908–911, 914, 915, 918–920, 922–929, 931–933, 936, 938, 940, 943, 944, 946–949
P-uniform families, 606, 611
parallel optimization algorithms, 351
parallel prefix, 11, 14–16, 18, 21, 32, 152, 216–218, 225, 235, 237, 251, 252, 370–372, 379, 394–396, 398, 400, 404, 522, 526, 528, 814, 825, 826, 830
parallel prefix computation, 15, 370, 371, 379, 394–396, 398, 400, 404, 522, 526, 814, 825, 826, 830
parallel slackness, 494, 992
parallel solution, 13, 18, 20, 25, 27, 59, 582, 610, 622, 625, 903, 907, 910, 932, 948
parallelization, 414, 610, 636, 724, 948
parity, 30, 325–332, 607, 846, 864–867, 870, 881, 890, 891
parsimony condition, 371, 372
partial derivatives, 750–752
partial fraction decomposition, 611
path compression, 769
path systems, 27, 910
perfect elimination ordering (peo), 342, 347–349, 351–353, 366, 373, 379. *See also* elimination ordering
perfect graph, 350
permutation group, 574, 613
phantom leaf, 189, 190
pipelining, 446, 492, 651, 960, 963, 987, 993
pivot, 52–54, 58, 660, 948
planar circuit value problem, 919, 928, 929, 931
point
 maxima, 499, 522
 visibility, 44, 442, 522–524, 528
point dominance, 520, 522

pointer jumping, 13, 19, 64, 117, 118, 120, 127, 147, 153, 155, 158, 164, 181, 183, 242, 243, 334, 868. *See also* shortcutting
polling, 25, 438–440, 443, 445
polling lemma, 440
polygon, 58, 296, 309, 310, 447, 499, 501, 505, 509–513, 515, 516, 519, 526–528, 885, 890
 kernel of, 511, 512
 star-shaped, 511, 527
polynomial
 characteristic, 26, 574, 583, 586, 588–590, 593, 595, 598, 609, 610, 682, 708–710, 712, 715
 completeness, 31, 902, 908, 948. *See also* P-completeness
 division, 694, 696, 728
 equations, 27, 611, 680, 705, 706, 711, 714, 716
 factorization of, 611
 fringe property, 28, 766–769, 771, 773–775, 777, 778
 inversion, 581, 582, 586, 592
 iterated product, 580, 582, 586
 remainder sequence (PRS), 27, 681, 689, 690
 root of, 697–702, 713, 715, 717, 718, 725
polynomial stack theorem, 771
polynomial time, 14, 606, 706, 710, 716, 768, 864, 902, 906, 907, 909–911, 914, 915, 932, 934, 946, 947, 949
POLYPROD, 591, 592
postorder, 20, 105, 110, 111, 272
power reduction, 555
powering, 540, 546–550, 552, 555, 556, 560, 563, 570
PRAM
 abstract, 865–867
 arbitrary-write, 76, 78
 asynchronous, 31, 32, 961–964, 974, 984–987, 990–993
 CRCW, 64, 129, 147, 148, 151, 188, 198, 201, 217, 224–227, 251, 253,

305, 329, 336, 353, 359, 366, 415,
417, 418, 422, 430, 432, 444, 445,
447, 491, 494, 527, 769, 860, 862,
865, 947, 989, 990, 992
CREW, 39, 46, 260, 351, 359, 361,
362, 415, 425, 426, 442, 444, 454,
456, 487, 491, 501, 503, 504, 508,
511–516, 519, 522, 526, 847, 848,
853, 854, 869–875, 877–879,
881–886, 890, 891, 948, 989
CROW, 861, 867–870, 877, 878,
881, 886
efficient algorithm, 2, 86, 191, 253,
301, 325, 342, 373, 388, 443, 700,
846
EREW, 39, 43, 45, 46, 49, 50, 57,
81, 82, 85, 90, 95, 98, 103, 106,
109, 118, 122, 126, 127, 129, 147,
164, 165, 167, 169–171, 176, 181,
183, 187, 252, 260, 261, 265, 362,
415, 444, 492, 493, 528, 623, 650,
847–850, 852, 857, 863, 869, 870,
873, 878, 882–887, 890, 959, 964,
965, 970, 975, 978, 980, 987, 989
EROW, 854, 861–865, 886, 887
model 37, 59, 218, 446, 462, 463,
609, 844–848, 911, 912, 944–948,
958–964
optimal algorithm, 43, 58, 62, 80,
93, 94, 109, 112, 223, 231, 233,
236, 241, 243, 335, 422, 431, 436,
447, 542, 574, 680, 846
priority-write, 80
randomized, 25, 447
word size, 844, 845, 861, 863, 866,
867
prefix computation, 2, 9, 15, 23, 370,
371, 379, 394–396, 398, 400, 404,
422, 522, 526, 814, 825, 826, 830
prefix sum, 18, 21, 36, 40, 62, 80, 86,
96, 98–100, 105, 116, 124, 129,
147, 178, 180, 217–221, 224–226,
235, 237, 245, 246, 249–253, 372,
424, 428, 434, 965, 973
preflow, 816–821, 831
preorder, 20, 41, 105, 110, 116,
265–268, 272, 278, 280, 282–284,
289, 291
preorder numbering, 265
Preparata's algorithm, 425, 431
preprocessing, 146, 260, 264, 265,
267, 269, 270, 272, 326, 327, 329
prescan, 38–40, 42–44, 46, 57
+-, 44, 47, 55
left-to-right, 43
primitive element, 715
PRIORITY, 30, 80, 847, 849, 850,
852–855, 857–860, 864, 881, 882,
885, 888, 890, 891
projection, 162, 191, 702, 890, 891
Projective Dimension Theorem, 703
projective space, 681, 683, 686, 687,
702, 712, 713
proof tree, 28, 766–771, 774, 775, 777
pruning terminal branches, 352, 363
push, 30, 158, 772, 773, 814, 815,
817–822, 824, 825, 828, 829, 834,
835, 920, 933, 935
pushdown automaton (PDA), 772

quantifier elimination, 716
query, 229, 260, 261, 264, 270, 764,
765, 769
quicksort, 23, 37, 51–54, 58, 59, 412,
444

radix-sort, 37, 44–46, 57, 59, 421,
426, 427
rake, 20, 21, 146–153, 156–158, 160,
163, 164, 169, 182, 185–187,
190–192, 735–737, 739, 740, 742,
746
RAM, 9, 10, 13, 14, 16, 90, 112, 116,
423, 738, 744, 844, 846, 945–947,
949, 961
algorithm, 10, 946, 947, 949
flow function, 945, 949
Ramsey theory, 883, 885
Random-Mate algorithm, 199, 213
random mating, 18–21, 122–124, 211
random sampling, 17, 18, 23, 25, 26,
28, 29, 412, 415, 421, 429, 434,
438, 443, 444, 446, 447

random sum, 445
randomization, 119, 120, 129, 145,
  147, 198, 252, 412–414, 416, 445,
  623, 632, 784, 785
randomized algorithms, 16, 17, 19,
  112, 120, 198, 252, 412, 414, 416,
  448, 612
randomized parallel connectivity, 21,
  198
randomness, 89, 697, 699, 700
rank
  in an array, 458, 502, 827, 830
  in a linked list, 83–88, 99–104, 249
  in graph matching, 804, 805
  in a metalinear program, 768
  of a key, 416, 425, 432, 433, 441
  See also matrix rank
real closed field, 611, 682
reciprocal function, 542
recurrence equations, 47, 58
recurrences
  first-order, 48
  higher-order, 50
recursive s(n)-factorization, 633, 636
recursive $s(n)$-factorization, 634,
  635, 640, 646
reduce operation, 39, 40, 43, 44
reducible, 541, 594, 608, 611, 892,
  908, 909, 914, 920, 939, 940, 949
reductions, 21, 31, 253, 541, 542,
  547, 555–557, 567, 569, 570, 574,
  591, 593, 599, 602, 890–892, 903,
  909, 911, 919, 920, 925, 927, 932
  dense-to-easy, 205, 206
  sparse-to-dense, 211
  Reischuk's algorithm, 431, 448
relabel, 280, 281, 285, 289, 817–826,
  828, 830
resampling, 434, 438
resolvent, 702
restricted instruction set PRAM,
  863–865
resultant, 18, 27, 28, 426, 427,
  680–683, 685, 688, 694, 697–700,
  702, 704–708, 711–714, 718
  -u, 682, 710–713

  multivariate, 682, 685, 694, 702,
    704, 718
  polynomial, 689, 707, 709
  system, 697, 699, 702, 705–707
  univariate, 683, 688, 702, 718
resultant matrix, 688
reversing a singly linked list, 878
robotics, 498, 681, 711
root (of a tree), 8, 19, 21, 39, 41,
  64–66, 69, 72, 73, 105, 106,
  108–111, 199, 201, 202, 206, 207,
  228–230, 260, 261, 265–269, 280,
  286, 287, 289, 316, 318–320, 322,
  323, 357, 399, 400, 433, 434, 631,
  634, 641, 850
  and its role in tree contraction, 131,
    146–148, 150, 153, 154, 157, 158,
    160, 161, 163–167, 169–171,
    173–179, 181, 184, 187–189, 191,
  extrovert, 206
ruler, 130, 132–135, 137–139, 141,
  142
ruling set, 19, 20, 100, 129–132, 135,
  145
  k-, 129, 130

$s(n)$-separator family, 630, 631, 633,
  635, 641, 642, 644, 645
$s(n)$-separator tree, 630, 632, 634,
  635, 641
SAC_F^k, 574, 607
sampling lemmas, 415
scan
  algorithm, 48, 49
  copy, 53, 55
  segmented, 51, 52
search an ordered list, 882–884
segment flags, 53, 55
selection, 10, 18, 23, 24, 87, 91, 106,
  235, 236, 241, 244, 249, 379,
  415–419, 421, 444, 447, 490, 491,
  494, 574, 603, 649, 833, 932, 941,
  942
  extremal, 416, 417, 445
selection gate, 603

semiorder, 373, 374, 376–379, 381–388, 390–392, 394–401, 403
sensitivity, 870–877, 880, 881
  block, 870, 874–877, 880, 881
sequential optimization algorithms, 349. *See also* algorithm, sequential
set search, 989
shared memory model, 414, 415
shortcutting, 13, 19, 20, 64, 68, 73–76, 78, 82, 83, 85, 376
shunt, 20, 21, 163–166, 168–170, 191, 192, 734, 735, 737–739, 742, 746
simple chain program, 776, 778
simplicial node, 347, 348, 373
SINGULAR, 600, 602
solid modeling, 681
SOLVABILITY, 599
sorted list, 94, 884–886, 890
sorting
  algorithms, 4, 6, 10, 17, 18, 421–423, 445–448, 462–488, 493, 739, 814
  circuits, 15, 454–457, 461–463, 488–490, 493
  general, 444, 446
  in constructing 2-ruling set, 94–98, 112, 223–226
  in rake operation, 188
  in ear decomposition, 280, 282, 285, 289, 291
  in finding candidate lists, 333
  in finding peo, 379
  in solving geometric problems, 520, 525
  integer, 112, 421–432, 444, 446, 447
  of an array, 82, 463
  of bitonic sequence, 980–982
  of inputs, 934–936
  of leaves, 189
  using random sampling, 18, 23, 25, 415–421, 445
  using deterministic techniques, 24, 25
sorts
  bucket sort 335, 336, 432–435
  flashsort, 445, 446

GENERAL_SORT, 425, 426, 429, 431, 432
quicksort, 37, 52, 54
See also odd-even sorting circuit; merge sort; radix-sort
sparse linear systems, 27, 622, 624, 633
splicing, 119–121, 125, 126, 128, 132–135, 137, 142, 160, 363, 364–366,
split-radix-sort, 46. *See also* radix-sort
square root, 569, 626
stack breadth first search, 933, 934
standard embedding, 687, 712
standard encoding, 913–915, 934, 949
star, 64–66, 68, 72–76, 78, 79, 199–204, 208, 228–230, 293, 295, 296, 300–304, 309, 311–313, 321–325, 329–334, 337, 393, 777
straight-line program, 575, 724, 731, 734
stratification, 389
stream contraction, 666
strong orientation, 255, 260
subexpression evaluation, 158, 169, 171
subject, 132–135, 137–139, 141, 143, 145
sublist compaction, 87, 98–105, 226, 227
subresultant matrix, 692, 693
subset synchronization, 984, 985, 989, 990
suffix sum, 80, 82–87, 100–103, 110–112, 226, 227
supervertex, 64, 66, 199–201, 205, 206, 209, 228–237, 239, 240, 242–248, 250, 251
Sylvester matrix, 688, 689, 694, 698, 699, 704
symmetric functions, 725, 729
symmetric polynomials, 725
synchronization barrier, 961–964, 966–968, 984

synchronous, 11, 13, 31, 189, 414, 844, 924–926, 931, 991, 993

table look-up, 862
test gate, 603
Toeplitz matrices, 591
topological circuit evaluation algorithm, 914, 915
total degree, 28, 686, 700, 713
trace, 57, 58, 514, 586, 590, 831–833
TRANSITIVE CLOSURE, 891
traversal, 31, 41, 105, 179, 261, 265, 272, 833, 884, 886
tree contraction, 18, 20–22, 28, 116, 146–192, 253, 337, 734, 863
  asynchronous, 11, 31, 32, 59, 187–189, 192, 958, 959, 961, 967, 984, 987, 990, 993
  basic, 151, 160, 192
  m-, 170, 171, 175–178, 180, 181, 187, 192
  optimal (unbounded degree trees), 147, 152, 174, 180, 181, 187, 188, 192
tree traversal, 261, 265, 272
triangular matrix, 26, 583, 589, 625, 627, 663, 726
triconnected components, 254, 276, 291, 295–297, 305, 307, 309, 312, 334–338

triconnectivity, 21, 22, 276, 291, 296, 297, 313, 325, 335, 336
tridiagonal linear system, 37, 664
Turing machine, 606, 607, 764, 860, 862–864, 906, 908–911, 914, 915, 917, 918, 924, 925
  alternating, 607, 764
Tutte matrix, 784, 788, 789, 791–793, 795, 802, 808
two-dimensional image, 37
2-edge connected graph, 280, 281
2-ruling set, 62, 91, 93–95, 98–105, 112, 129, 221, 223, 224, 226, 227

unbounded degree tree, 180, 181, 187–190, 192. *See also* tree contraction, optimal
uniformity, 14, 377, 382, 383, 386, 606, 607, 610, 914
univariate resultant, 683, 688, 702, 718

visibility, 522–524, 528

weighted tree, 189, 190
Wyllie's algorithm. *See* list ranking, Wyllie's algorithm

-Brault **DATE DUE**